Beginning and Intermediate Algebra
Northeastern State University Edition

Fourth Edition

R. David Gustafson
Rock Valley College

Peter D. Frisk
Rock Valley College

THOMSON
BROOKS/COLE

Australia · Canada · Mexico · Singapore · Spain · United Kingdom · United States

THOMSON

BROOKS/COLE

Beginning and Intermediate Algebra
Gustafson/Frisk

Executive Editors:
Michele Baird, Maureen Staudt &
Michael Stranz

Project Development Manager:
Linda de Stefano

Marketing Coordinators:
Lindsay Annett and Sara Mercurio

Production/Manufacturing Supervisor:
Donna M. Brown

Pre-Media Services Supervisor:
Dan Plofchan

Rights and Permissions Specialists:
Kalina Hintz and Bahman Naraghi

Cover Image
Getty Images*

Printed in the
United States of America
1 2 3 4 5 6 7 8 08 07 06 05

For more information, please contact Thomson Custom Solutions, 5191 Natorp Boulevard, Mason, OH 45040. Or you can visit our Internet site at www.thomsoncustom.com

The Adaptable Courseware Program consists of products and additions to existing Brooks/Cole products that are produced from camera-ready copy. Peer review, class testing, and accuracy are primarily the responsibility of the author(s).

Beginning and Intermediate Algebra / Gustafson/Frisk – Fourth Edition

ISBN 0-495-26648-5

International Divisions List

Asia (Including India):
Thomson Learning
(a division of Thomson Asia Pte Ltd)
5 Shenton Way #01-01
UIC Building
Singapore 068808
Tel: (65) 6410-1200
Fax: (65) 6410-1208

Australia/New Zealand:
Thomson Learning Australia
102 Dodds Street
Southbank, Victoria 3006
Australia

Latin America:
Thomson Learning
Seneca 53
Colonia Polano
11560 Mexico, D.F., Mexico
Tel (525) 281-2906
Fax (525) 281-2656

Canada:
Thomson Nelson
1120 Birchmount Road
Toronto, Ontario
Canada M1K 5G4
Tel (416) 752-9100
Fax (416) 752-8102

UK/Europe/Middle East/Africa:
Thomson Learning
High Holborn House
50-51 Bedford Row
London, WC1R 4L$
United Kingdom
Tel 44 (020) 7067-2500
Fax 44 (020) 7067-2600

Spain (Includes Portugal):
Thomson Paraninfo
Calle Magallanes 25
28015 Madrid
España
Tel 34 (0)91 446-3350
Fax 34 (0)91 445-6218

Custom Contents

Preface

To the Instructor

Beginning and Intermediate Algebra, Fourth Edition, combines the topics of beginning and intermediate algebra. This type of book has many advantages:

1. By combining topics, much of the overlap and redundancy of the material can be eliminated. For this reason, the instructor will have time to teach for mastery of the material.
2. For many students, the purchase of a single book will save money.
3. A combined approach in one book will enable some colleges to cut back on the number of hours needed for mathematics remediation.

However, there are three concerns inherent in a combined approach:

1. The first half of the book must include enough beginning algebra to ensure that students who complete the first half of the book and then transfer to another college will have the necessary prerequisites to enroll in an intermediate algebra course.
2. The beginning algebra material should not get too difficult too fast.
3. Intermediate algebra students beginning in the second half of the book must get some review of basic topics so that they can compete with students continuing on from the first course.

Unlike many other texts, this book uses an *integrated approach,* which addresses each of the previous three concerns by

- including a full course in beginning algebra in the first six chapters,
- delaying the presentation of intermediate algebra topics until Chapter 7 or later,
- providing a quick review of basic topics for those who begin in the second half of the book.

Organization

The first six chapters present all of the topics usually associated with a first course in algebra, except for a detailed discussion of manipulating radical expressions and the quadratic formula. These topics can be omitted because they will be carefully introduced and taught in any intermediate algebra course. Harder topics, such as absolute value inequalities and synthetic division, are left until Chapter 7.

Chapter 7 is the entry-level chapter for students enrolling in intermediate algebra. As such, it quickly reviews the topics taught in the first six chapters and extends these topics to the intermediate algebra level.

Chapter 3 discusses graphs of linear equations and systems of two equations in two variables. Systems of three equations in three variables and the methods for solving them are left until Chapter 13.

Chapters 8 through 14 are written at the intermediate algebra level, and include a quick review of important topics as needed. For example, Chapter 8 begins with a review of the rectangular coordinate system and graphing linear equations, a topic first taught in Chapter 3. It then moves on to the topics of writing equations of lines, nonlinear functions, and variation. As another example, Chapter 13 begins with a review of solving simple systems of equations, a topic first taught in Chapter 3. It then moves on to solving more difficult systems by matrices and determinants.

Goals of the Book

In addition to using a truly integrated approach, our goal was to write a book that

1. is enjoyable to read,
2. is easy to understand,
3. is relevant,
4. integrates the topic of function throughout the book, and
5. will develop the necessary skills for success in future academic courses or on the job.

Although the material has been extensively revised, this Fourth Edition retains the basic philosophy of the highly successful previous editions. The revisions include several improvements in line with the NCTM and AMATYC standards and the current trends in mathematics reform. For example, more emphasis has been placed on graphing and problem solving.

General Changes for the Fourth Edition

The overall effects of the changes made to the Fourth Edition are as follows:

- *To make the text more inviting to students,* we have incorporated a new design that is more open and easier to read. All calculator art has been rendered by an artist to be more visually appealing and to allow the labeling of graphs. Each section now includes a few sentences of introduction. We continue to use color not just as a design feature, but to highlight terms that instructors would point to in a classroom discussion.
- *To make the chapter openers more effective,* they have been redesigned and their relevant applications problems have been placed in the exercise sets.

- *To give students real-life applications of the material in the chapters,* each chapter opener contains an InfoTrac® College Edition project.
- *To increase the emphasis on problem solving through realistic applications,* we have increased the number of applications problems. The text now includes an index of applications.
- *To improve the flow of ideas and for clarity,* we have fine-tuned the presentation of many topics.
- *We have increased the emphasis on learning mathematics through graphing.* Although graphing calculators are used often, their use is not required. All of the topics are fully discussed in traditional ways. Of course, we recommend that instructors use the graphing calculator material.

Specific Changes in the Fourth Edition

Chapter One covers the real-number system, arithmetic fractions, order of operations, arithmetic of signed numbers, and an introduction to algebraic expressions. In Section 1.4, new rules are given for adding real numbers and the section has been reorganized. New rules for multiplying and dividing real numbers are given in Section 1.5.

Chapter Two covers basic work on solving equations and inequalities. Section 2.1 has been reorganized and the multiplication property of equality now comes before the division property of equality. In Section 2.7, material is now included to help students translate from words to inequality statements.

Chapter Three covers the coordinate system, graphing linear equations, and solving basic systems of equations and inequalities. In Exercise 3.1, some graphing problems have been included where the graph passes through the origin.

Chapter Four covers the rules for exponents, scientific notation, and the arithmetic of polynomials. This chapter contains only minor changes.

Chapter Five covers factoring. To allow students to focus completely on factoring techniques when first encountering them, the work on solving quadratic equations has been moved into a separate section toward the end of the chapter. To maintain the number of sections, the work on the greatest common factor and factoring by grouping have been combined into one section. Factoring the sum and difference of two cubes has been included in this chapter.

Chapter Six covers ratios, proportions, and the arithmetic of rational expressions. Material has been added listing the steps to use when solving equations containing fractions.

Chapter Seven remains the entry level chapter for intermediate algebra. It reviews much of basic algebra and extends the discussion to include equations and inequalities that contain absolute values and synthetic division. Since the work on factoring the sum and difference of two cubes has been included in Chapter 5, this material has been incorporated into the section on reviewing factoring.

Chapter Eight reviews the rectangular coordinate system and graphing linear equations. It then discusses slope of a nonvertical line and writing equations of lines. After reviewing functions, it considers graphs of nonlinear functions and covers variation.

Chapter Nine covers radicals, rational exponents, and complex numbers. Additional problems have been added to Exercises 9.1 and 9.2. The topic of complex numbers has been moved to this chapter as Section 9.7. In this section, the material on powers of *i* has been moved later in the section. A new project has been added.

Chapter Ten covers quadratic functions, inequalities, and algebra of functions. This chapter has been revised extensively.

The work on completing the square and the quadratic formula is now covered in two sections. Section 10.1 reviews solving quadratics by factoring and covers completing the square. The material has been expanded to cover quadratic equations with imaginary and complex roots. A graphical explanation of completing the square is now included. The exercise set has been greatly revised and includes new applications.

Section 10.2 covers solving quadratic equations using the quadratic formula. The material has been expanded to cover quadratic equations with complex roots. Again, the exercise set has been greatly revised.

To improve the transition from the work on the quadratic formula to the work on the discriminant, the material on the discriminant and equations that can be written in quadratic form has been moved to Section 10.3.

A better introduction to the construction of sign graphs has been included in Section 10.5.

Section 10.7 has been reorganized to improve the flow of topics.

Chapter Eleven covers exponential and logarithmic functions. This chapter contains only minor changes.

Chapter Twelve covers the conic sections and more graphing. This chapter contains only minor changes.

Chapter Thirteen begins by reviewing the methods used to solve systems of two equations, each with two variables. It then extends the discussion to include solutions of systems of three equations, each with three variables, and covers how to solve systems using matrices and determinants. Finally, it discusses how to solve systems involving second-degree equations and inequalities. This chapter has only minor changes.

Chapter Fourteen covers the binomial theorem, sequences, and permutations and combinations. This chapter contains only minor changes.

Calculators

The use of calculators is assumed throughout the text. We believe that students should learn calculator skills in the mathematics classroom. They will then be prepared to use calculators in science and business classes and for nonacademic purposes. The directions within each exercise set indicate which exercises require the use of a calculator.

Since many beginning algebra students now have graphing calculators, keystrokes are given for both scientific and graphing calculators.

Ancillaries for the Instructor

Annotated Instructor's Edition (0-534-46378-9)
This special version of the complete student text contains a Resource Integration Guide and all answers are printed next to their respective exercises.

Test Bank (0-534-46382-7)
The *Test Bank* includes 8 tests per chapter, as well as 3 final exams. The tests contain a combination of multiple-choice, free-response, true/false, and fill-in-the-blank questions.

Complete Solutions Manual (0-534-46381-9)
The *Complete Solutions Manual* provides worked-out solutions to all of the problems in the text.

Text-Specific Videotapes (0-534-46383-5)
The text-specific videotape set, available at no charge to qualified adopters of the text, features 10- to 20-minute problem-solving lessons that cover each section of every chapter.

BCA/iLrn 3.5 *Instructor Version* (0-534-46384-3)
BCA/iLrn Instructor Version is made up of two components: *BCA/iLrn Testing* and *BCA/iLrn Tutorial. BCA/iLrn Testing* is a revolutionary, Internet-ready, text-specific testing suite that allows you to customize exams and track student progress in an accessible, browser-based format. *BCA/iLrn* offers full algorithmic generation of problems and free-response mathematics. *BCA/iLrn Tutorial* is a text-specific, interactive tutorial that is delivered via the Web (at http://bca.brookscole.com) and is offered in both student and instructor versions. The tracking program built into the instructor version enables you to monitor student progress carefully. The complete integration of the testing, tutorial, and course management components simplifies your routine tasks. Results flow automatically to your gradebook and you can easily communicate with individuals, sections, and entire courses.

BCA/iLrn has been updated with a diagnostic tool to assess and place students in any course and provide a personalized study plan. Other updates include:

- The new rich text form (RTF) conversion tool that allows you to open tests in most word processors for further formatting and customization,
- Several new problem types that provide greater variety and more diverse challenges in your tests, and
- A redesigned *Tutorial* interface that effectively engages students and helps them learn math concepts faster.

TLE Labs (0-534-46448-3)
Think of *TLE Labs* as electronic labs that upgrade any *BCA/iLrn Tutorial. TLE Labs* introduce and explore key concepts. *BCA/iLrn Tutorials* reinforce those concepts with unlimited practice. Exploring these core concepts interactively at their own pace prepares students for the work of the traditional course, allows them to perform better in the course overall, and prepares them for the next level of mathematics.

- *TLE Labs* are now available with all © 2005 developmental mathematics titles.
- *TLE Labs* include 15 lessons adapted from the full version of *The Learning Equation.* The lessons correlate to the key concepts of this book and the course in general.
- In addition to the printed textbook, students receive everything they need to succeed in the course: an online version of the text, access to the *TLE* lessons, text-specific interactive tutorials, and access to *vMentor* (Web-based, real-time, live tutors), all in the same single, unified environment.

Ask your Thomson•Brooks/Cole representative about *TLE.*

WebTutor™ ToolBox on WebCT and *Blackboard*:
0-534-27488-9 (WebCT), 0-534-27489-7 (Blackboard)
Preloaded with content and available free via PIN code when packaged with this text, *WebTutor ToolBox for WebCT* and *Blackboard* pairs the content of this text's rich Book Companion Web Site with all the sophisticated course management functionality of a WebCT and Blackboard product. You can assign materials (including online quizzes) and have the results flow automatically to your gradebook. *ToolBox* is ready to use as soon as you log on. You also can customize its preloaded content by uploading images and other resources, adding Web links, or creating your own practice materials. Students have access only to student resources on the Web site. Instructors can enter a PIN code for access to password-protected instructor resources.

MyCourse 2.1
Ask us about our new free online course builder. Whether you want only the easy-to-use tools to build it or the content to furnish it, Brooks/Cole offers you a simple solution for a custom course Web site that allows you to assign, track, and report on student progress; load your syllabus; and more. Highlights of Version 2.1 include new features that let you import your class roster and grade book, see assignments as students will see them using the Student Preview Mode, create individual performance grade books, and copy content from an existing course to a new course. Contact your Thomson•Brooks/Cole representative for details or visit http://mycourse.thomson learning.com.

Ancillaries for the Student

Student Solutions Manual (0-534-46380-0)
The *Student Solutions Manual* provides worked out solutions to the odd-numbered problems in the text.

Brooks/Cole Mathematics Web site, http://mathematics.brookscole.com
When you adopt a Thomson•Brooks/Cole mathematics text, you and your students will have access to a variety of teaching and learning resources. This Web site features everything from book-specific resources to newsgroups. It's a great way to make teaching and learning an interactive and intriguing experience.

InfoTrac® College Edition
Incorporate an array of current applications or the latest research into your lectures with four months of free access to InfoTrac College Edition. In addition to robust tutorial services, your students also receive anytime, anywhere access to InfoTrac College Edition. This online library offers the full text of articles from almost 5,000 scholarly and popular publications, updated daily and going back as far as 22 years. Both adopters and their students receive unlimited access for four months.

BCA/iLrn Tutorial Student Version (0-534-46385-1)
Free access to this text-specific, interactive, Web-based tutorial system is included with the text. *BCA/iLrn Tutorial Student Version* is browser-based, making it an intuitive mathematical guide, even for students with little technological proficiency. *BCA/iLrn Tutorial* allows students to work with real math notation in real time, providing instant analysis and feedback. The entire textbook is available in PDF format through *BCA/iLrn Tutorial,* as are section-specific video tutorials, unlimited practice problems, and additional student resources, such as a glossary, Web links, and more. *BCA/iLrn Tutorial* is also offered on a dual-platform CD-ROM. And, when students get stuck on a particular problem or concept, they need only log on to *vMentor,* ac-

cessed through *BCA/iLrn Tutorial,* where they can talk (using their own computer microphones) to *vMentor* tutors who will skillfully guide them through the problem using the interactive whiteboard for illustration.

A personalized study plan also can be generated by the course-specific diagnostic built into *BCA/iLrn Tutorial.* With a personalized study plan, students can focus their time where they need it the most, creating a positive learning environment and paving a pathway to success in their mathematics course.

vMentor
Packaged free with every text and accessed seamlessly through *BCA/iLrn Tutorial, vMentor* provides tutorial help that can substantially improve student performance, increase test scores, and enhance technical aptitude. Students will have access, via the Web, to highly qualified tutors. When students get stuck on a particular problem or concept, they need only log on to *vMentor,* where they can talk (using their own computer microphones) to *vMentor* tutors who will skillfully guide them through the problem using the interactive whiteboard for illustration. Brooks/Cole also offers *vClass,* an online virtual classroom environment that is customizable and easy to use. *vClass* keeps students engaged with full two-way audio, instant messaging, and an interactive whiteboard—all in one intuitive, graphical interface. For information about obtaining a *vClass* site license, please contact your Thomson•Brooks/Cole representative.

Interactive Video Skillbuilder CD-ROM (0-534-46379-7)
The *Interactive Video Skillbuilder* CD-ROM contains more than eight hours of video instruction. The problems worked during each video lesson are shown next to the viewing screen so that students can try working them before watching the solution. To help students evaluate their progress, each section contains a 10-question Web quiz, the results of which can be emailed to the instructor, and each chapter contains a chapter test, with the answer to each problem on each test. Also included is *MathCue* tutorial and testing software, which is keyed to the text and includes these components:

- *MathCue Skill Builder* presents problems to solve, evaluates answers, and tutors students by displaying complete solutions with step-by-step explanations.
- *MathCue Quiz* allows students to generate large numbers of quiz problems keyed to problem types from each section of the book.
- *MathCue Chapter Test* also provides large numbers of problems keyed to problem types from each chapter.
- *MathCue Solution Finder* allows students to enter their own basic problems and receive step-by-step help as if they were working with a tutor.
- Score reports for any *MathCue* session can be printed and handed in for credit or extra-credit, or sent by email via *MathCue's* secure email score system.

To the Student

Congratulations! You now own a state-of-the-art textbook that has been written especially for you. We have tried to write a book that you can read and understand. The text includes carefully written narrative and an extensive number of worked examples with Self Checks.

To get the most out of this course, you must read and study the textbook properly. We recommend that you work the examples on paper first, and then work the

Self Checks. Only after you thoroughly understand the concepts taught in the examples should you attempt to work the exercises. A *Student Solutions Manual* is available, which contains the worked out solutions to the odd-numbered exercises.

Since the material presented in *Beginning and Intermediate Algebra: An Integrated Approach, Fourth Edition,* will be of value to you in later years, we suggest that you keep this text. It will be a good source of reference in the future and will keep at your fingertips the material that you have learned here.

We wish you well.

Hints on Studying Algebra

The phrase "Practice makes perfect" is not quite true. It is "*Perfect* practice that makes perfect." For this reason, it is important that you learn how to study algebra to get the most out of this course.

Although we all learn differently, here are some hints on studying algebra that most students find useful.

Plan a Strategy for Success

To get where you want to be, you need a goal and a plan. Your goal should be to pass this course with a grade of A or B. To earn one of these grades, you must have a plan to achieve it. A good plan involves several points:

- Getting ready for class,
- Attending class,
- Doing homework,
- Arranging for special help when you need it, and
- Having a strategy for taking tests.

Getting Ready for Class

To get the most out of every class period, you will need to prepare for class. One of the best things you can do is to preview the material in the text that your instructor will be discussing in class. Perhaps you will not understand all of what you read, but you will be better able to understand your instructor when he or she discusses the material in class.

Do your work every day. If you get behind, you will become frustrated and discouraged. Make a promise that you will always prepare for class, and then keep that promise.

Attending Class

The classroom experience is your opportunity to learn from your instructor. Make the most of it by attending every class. Sit near the front of the room where you can easily see and hear. Remember that it is your responsibility to follow the discussion, even though that takes concentration and hard work.

Pay attention to your instructor, and jot down the important things that he or she says. However, do not spend so much time taking notes that you fail to concentrate on what your instructor is explaining. Listening and understanding the big picture is much better than just copying solutions to problems.

Don't be afraid to ask questions when your instructor asks for them. Asking questions will make you an active participant in the class. This will help you pay attention and keep you alert and involved.

Doing Homework

It requires practice to excel at tennis, master a musical instrument, or learn a foreign language. In the same way, it requires practice to learn mathematics. Since practice in mathematics is homework, homework is your opportunity to practice your skills and experiment with ideas.

It is important for you to pick a definite time to study and do homework. Set a formal schedule and stick to it. Try to study in a place that is comfortable and quiet. If you can, do some homework shortly after class, or at least before you forget what was discussed in class. This quick follow-up will help you remember the skills and concepts your instructor taught that day.

Each formal study session should include three parts:

1. Begin every study session with a review period. Look over previous chapters and see if you can do a few problems from previous sections. Keeping old skills alive will greatly reduce the amount of time you will need to prepare for tests.

2. After reviewing, read the assigned material. Resist the temptation of diving into the problems without reading and understanding the examples. Instead, work the examples and Self Checks with pencil and paper. Only after you completely understand the underlying principles behind them should you try to work the exercises.

Once you begin to work the exercises, check your answers with the printed answers in the back of the text. If one of your answers differs from the printed answer, see if the two can be reconciled. Sometimes, answers have more than one form. If you decide that your answer is incorrect, compare your work to the example in the text that most closely resembles the exercise, and try to find your mistake. If you cannot find an error, consult the *Student Solutions Manual.* If nothing works, mark the problem and ask about it in your next class meeting.

3. After completing the written assignment, preview the next section. This preview will be helpful when you hear that material discussed during the next class period.

You probably already know the general rule of thumb for college homework: two hours of practice for every hour you spend in class. If mathematics is hard for you, plan on spending even more time on homework.

To make doing homework more enjoyable, study with one or more friends. The interaction will clarify ideas and help you remember them. If you must study alone, a good study technique is to explain the material to yourself out loud.

Arranging for Special Help

Take advantage of any special help that is available from your instructor. Often, your instructor can clear up difficulties in a short period of time.

Find out whether your college has a free tutoring program. Peer tutors can often be of great help.

Taking Tests

Students often get nervous before taking a test because they are afraid that they will do poorly. To build confidence in your ability to take tests, rework many of the problems in the exercise sets, work the exercises in the Chapter Summaries, and take the Chapter Tests. Check all answers with the answers printed at the back of the text.

Then guess what the instructor will ask, build your own tests, and work them. Once you know your instructor, you will be surprised at how good you can get at picking test questions. With this preparation, you will have some idea of what will be on the test, and you will have more confidence in your ability to do well.

When you take a test, work slowly and deliberately. Scan the test and work the easy problems first. Tackle the hardest problems last.

Acknowledgments

We are grateful to the following people, who reviewed the manuscript at various stages. They all had valuable suggestions that have been incorporated into the text.

Cynthia Broughtou
Arizona Western College

David Byrd
Enterprise State Junior College

Michael F. Cullinan
Glendale Community College

Karen Driskell
Calhoun Community College

Harold Farmer
Wallace Community College-Hanceville

Mark Foster
Santa Monica College

Dorothy K. Holtgrefe
Seminole Community College

Mike Judy
Fullerton College

Janet Mazzarella
Southwestern College

Donald J. McCarthy
Glendale Community College

Andrew P. McKintosh
Glendale Community College

Christian R. Miller
Glendale Community College

Feridoon Moinian
Cameron University

Daniel F. Mussa
Southern Illinois University

Joanne Peeples
El Paso Community College

Mary Ann Petruska
Pensacola Junior College

Janet Ritchie
SUNY-Old Westbury

April D. Strom
Glendale Community College

Hattie White
St. Phillip's College

Victoria Wacek
Missouri Western State College

George J. Witt
Glendale Community College

We are grateful to the staff at Brooks/Cole, especially our editor Jennifer Huber. We also thank Bob Pirtle, Rachael Sturgeon, Leah Thomson, Rebecca Subity, Hal Humphrey, and Vernon Boes.

We are indebted to Ellen Brownstein, our production service, Diane Koenig, who read the entire manuscript and worked every problem, and Mike Welden, who prepared the *Student Solutions Manual.* Finally, we thank Lori Heckleman for her fine artwork and Graphic World for their excellent typesetting.

R. David Gustafson
Peter D. Frisk

Math Tutorial Quick Start Guide

BCA/iLrn Can Help You Succeed in Math

BCA/iLrn is an online program that facilitates math learning by providing resources and practice to help you succeed in your math course. Your instructor chose to use *BCA/iLrn* because it provides online opportunities for learning (Explanations found by clicking **Read Book**), practice (Exercises), and evaluating (Quizzes). It also gives you a way to keep track of your own progress and manage your assignments.

The mathematical notation in *BCA/iLrn* is the same as that you see in your textbooks, in class, and when using other math tools like a graphing calculator. *BCA/iLrn* can also help you run calculations, plot graphs, enter expressions, and grasp difficult concepts. You will encounter various problem types as you work through *BCA/iLrn,* all of which are designed to strengthen your skills and engage you in learning in different ways.

Logging in to BCA/iLrn

Registering with the PIN Code on the BCA/iLrn Card

Situation: Your instructor has not given you a PIN code for an online course, but you have a textbook with a BCA/iLrn product PIN code.

Initial Login

1. Go to **http://iLrn.com.**
2. In the menu at the left, click on **Student Tutorial.**
3. Make sure that the name of your school appears in the "School" field. If your school name does not appear, follow steps a–d below. If your school is listed, go to step 4.
   - **a.** Click on **Find Your School.**
   - **b.** In the "State" field, select your state from the drop-down menu.
   - **c.** In the "Name of school" field, type the first few letters of your school's name; then click on **Search.** The school list will appear at the right.
   - **d.** Click on your school. The "First Time Users" screen will open.
4. In the "PIN Code" field, type the BCA/iLrn PIN code supplied on your BCA/iLrn card.
5. In the "ISBN" field, type the ISBN of your book (from the textbook's back cover), for example, 0-534-46332-0.
6. Click on **Register.**
7. Enter the appropriate information. Fields marked with a red asterisk must be filled in.
8. Click on **Register** and **Begin BCA/iLrn.** The "My Assignments" page will open.

You will be asked to select a user name and password. Save your user name and password in a safe place. You will need them to log in the next time you use BCA/iLrn. Only your user name and password will allow you to reenter BCA/iLrn.

Subsequent Login

1. Go to **http://iLrn.com.**
2. Click on **Login.**
3. Make sure the name of your school appears in the "School" field. If not, then follow steps 3a–d under "Initial Login" to identify your school.
4. Type your user name and password (see boxed information above); then click on **Login.** The "My Assignments" page will open.

Navigating through BCA/iLrn

To navigate between chapters and sections, use the drop-down menu below the top navigation bar. This will give you access to the study activities available for each section.

The view of a tutorial in BCA/iLrn looks like this.

Math Toolbar

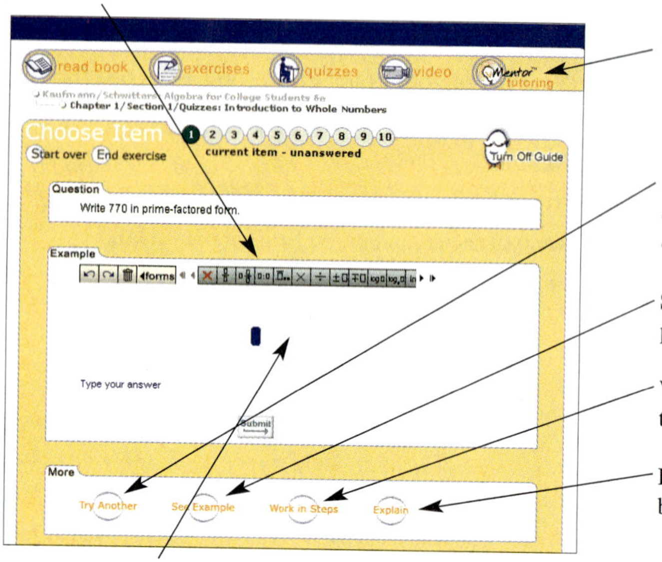

vMentor: Live online tutoring is only a click away. Tutors can take screen shots of your book and lead you through a problem with voice-over and visual aids.

Try Another: Click here to have BCA/iLrn create a new question or a new set of problems.

See Examples: Preworked examples provide you with additional help.

Work in Steps: BCA/iLrn can guide you through a problem step-by-step.

Explain: Additional explanation from your book can help you with a problem.

Type your answer here.

Online Tutoring with *vMentor*

Access to BCA/iLrn also means access to online tutors and support through vMentor, which provides live homework help and tutorials. To access vMentor while you are working in the Exercises or "Tutorial" areas in BCA/iLrn, click on the **vMentor Tutoring** button at the top right of the navigation bar above the problem or exercise.

Next, click on the **vMentor** button; you will be taken to a Web page that lists the steps for entering a vMentor classroom. If you are a first-time user of vMentor, you might need to download Java software before entering the class for the first time. You can either take an Orientation Session or log in to a vClass from the links at the bottom of the opening screen.

All vMentor Tutoring is done through a vClass, an Internet-based virtual classroom that features two-way audio, a shared whiteboard, chat, messaging, and experienced tutors.

You can access vMentor Sunday through Thursday, as follows:

5 p.m. to 9 p.m. Pacific Time

6 p.m. to 10 p.m. Mountain Time

7 p.m. to 11 p.m. Central Time

8 p.m. to midnight Eastern Time

If you need additional help using vMentor, you can access the Participant Guide at this Website: **http://www.elluminate.com/support/guide.pdf.**

Index of Applications

Examples that are applications are shown with boldface page numbers.
Exercises that are applications are shown with lightface page numbers.

Recreation, Travel, and Hobbies

Science and Engineering

Sports

Statistics

1 Real Numbers and Their Basic Properties

© Keren Su/CORBIS

InfoTrac Project

Do a subject guide search on "Siberia" and find the newspaper article, "Cold wave breathes fire in Siberia." Write a summary of the article and find the difference between the record low in Western Siberia and the lowest temperature the houses have been built to withstand.

The formula $C = \frac{5}{9}(F - 32)$ is used to convert degree measurement in Fahrenheit to degree measurement in Celsius. Using this formula, convert the following Fahrenheit temperatures to degrees Celsius: 212°, 32°, 98.6°, −10°, −20°, −30°, −40°, −50°, and −60°. Did you discover anything unusual? Write a statement about your findings.

The formula to convert Celsius degrees to Fahrenheit degrees is $F = \frac{9}{5}C + 32$. Use this formula to convert the following temperatures in degrees Celsius to degrees Fahrenheit: 125°, 46°, −20°, and −55°.

Complete this project after studying Section 1.6.

Mathematics in Data Processing

In electronic data processing, the process of sorting records into sequential order is a common task. One sorting technique, called a **selection sort,** requires C comparisons to sort N records, where C and N are related by the formula

$$C = \frac{N(N-1)}{2}$$

Exercise Set 1.6
Problem 95

How many comparisons are necessary to sort 10,000 records?

In Chapter 1, we discuss the various types of numbers that will be used throughout this course. We also review the basic arithmetic of fractions, explain how to add, subtract, multiply, and divide real numbers, introduce algebraic expressions, and summarize the properties of real numbers.

1.1 Real Numbers and Their Graphs

- Sets of Numbers
- Equality, Inequality Symbols, and Variables
- The Number Line
- Graphing Subsets of the Real Numbers
- Absolute Value of a Number

Getting Ready

1. Give an example of a number that is used for counting.
2. Give an example of a number that is used when dividing a pizza.
3. Give an example of a number that is used for measuring temperatures that are below zero.
4. What other types of numbers can you think of?

Sets of Numbers

A **set** is a collection of objects. For example, the set

$\{1, 2, 3, 4, 5\}$ Read as "the set with elements 1, 2, 3, 4, and 5."

contains the numbers 1, 2, 3, 4, and 5. The members, or **elements,** in a set are listed within braces { }.

Two basic sets of numbers are the **natural numbers** (often called the **positive integers**) and the **whole numbers.**

The Set of Natural Numbers (Positive Integers) $\{1, 2, 3, 4, 5, 6, 7, 8, 9, 10, \ldots\}$

The Set of Whole Numbers $\{0, 1, 2, 3, 4, 5, 6, 7, 8, 9, 10, \ldots\}$

The three dots in the previous definitions, called **ellipses,** indicate that each list of numbers continues on forever.

We can use whole numbers to describe many real-life situations. For example, some cars might get 30 miles per gallon (mpg) of gas, and some students might pay $1,750 in tuition.

Numbers that show a loss or a downward direction are called **negative numbers,** and they are denoted with a $-$ sign. For example, a debt of $1,500 can be denoted as $-\$1,500$, and a temperature of 20° below zero can be denoted as $-20°$.

The negatives of the natural numbers and the whole numbers together form the set of integers.

The Set of Integers

$\{\ldots, -5, -4, -3, -2, -1, 0, 1, 2, 3, 4, 5, \ldots\}$

Because the set of natural numbers and the set of whole numbers are included within the set of integers, these sets are called **subsets** of the set of integers.

Integers cannot describe every real-life situation. For example, a student might study $3\frac{1}{2}$ hours, or a television set might cost $217.37. To describe these situations, we need fractions, more formally called **rational numbers.**

The Set of Rational Numbers

{All numbers that can be written as a fraction with an integer in its numerator and a nonzero integer in its denominator.}

Some examples of rational numbers are

$$\frac{3}{2}, \quad \frac{17}{12}, \quad -\frac{43}{8}, \quad 0.25, \quad \text{and} \quad -0.66666\ldots$$

The decimals 0.25 and $-0.66666\ldots$ are rational numbers, because 0.25 can be written as the fraction $\frac{1}{4}$, and $-0.66666\ldots$ can be written as the fraction $-\frac{2}{3}$.

Since every integer can be written as a fraction with a denominator of 1, every integer is also a rational number. Since every integer is a rational number, the set of integers is a subset of the rational numbers.

Comment Because division by 0 is undefined, expressions such as $\frac{6}{0}$ and $\frac{8}{0}$ do not represent any number.

Since π and $\sqrt{2}$ cannot be written as fractions with an integer numerator and a nonzero integer denominator, they are not rational numbers. They are called **irrational numbers.** We can find their decimal approximations using a calculator. For example,

$\pi \approx 3.141592654$ — On a scientific calculator, press π. On a graphing calculator, press π ENTER. Read $\approx$ as "is approximately equal to."

$\sqrt{2} \approx 1.414213562$ — On a scientific calculator, press 2 $\sqrt{\ }$. On a graphing calculator, press $\sqrt{\ }$ 2 ENTER.

If we combine the rational and the irrational numbers, we have the set of real numbers.

The Set of Real Numbers {All numbers that are either a rational number or an irrational number.}

Figure 1-1 shows how the various sets of numbers are interrelated.

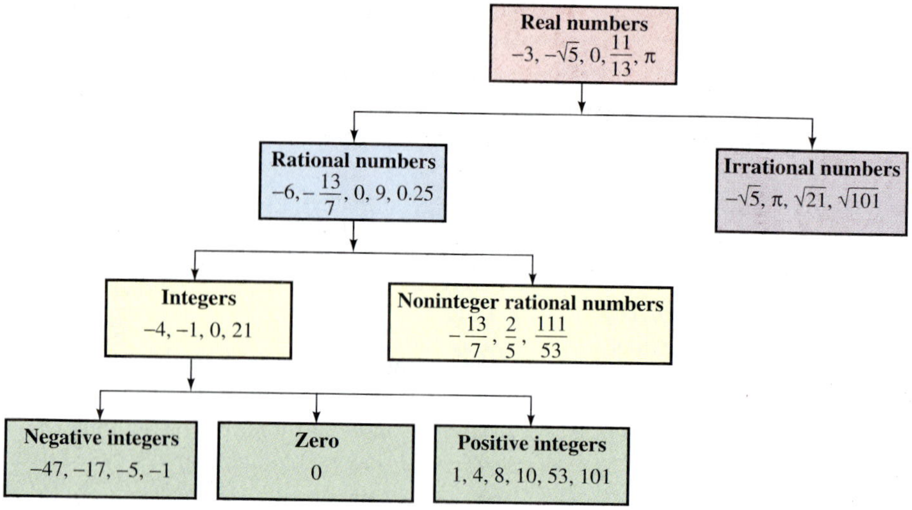

Figure 1-1

EXAMPLE 1 Which numbers in the set $\left\{-3, 0, \frac{1}{2}, 1.25, \sqrt{3}, 5\right\}$ are **a.** natural numbers, **b.** whole numbers, **c.** negative integers, **d.** rational numbers, **e.** irrational numbers, and **f.** real numbers?

Solution

a. The only natural number is 5.

b. The whole numbers are 0 and 5.

c. The only negative integer is -3.

d. The rational numbers are -3, 0, $\frac{1}{2}$, 1.25, and 5. (1.25 is rational, because 1.25 can be written in the form $\frac{5}{4}$.)

e. The only irrational number is $\sqrt{3}$.

f. All of the numbers are real numbers.

Self Check Which numbers in the set $\left\{-2, 0, 1.5, \sqrt{5}, 7\right\}$ are **a.** positive integers and **b.** rational numbers?

A natural number greater than 1 that can be divided evenly only by 1 and itself is called a **prime number.**

The set of prime numbers:

$\{2, 3, 5, 7, 11, 13, 17, 19, 23, 29, \ldots\}$

A nonprime natural number greater than 1 is called a **composite number.**

The set of composite numbers:

$\{4, 6, 8, 9, 10, 12, 14, 15, 16, 18, 20, 21, 22, \ldots\}$

An integer that can be divided evenly by 2 is called an **even integer.** An integer that cannot be divided evenly by 2 is called an **odd integer.**

PERSPECTIVE

Algebra is an extension of arithmetic. In algebra, the operations of addition, subtraction, multiplication, and division are performed on numbers and letters, with the understanding that the letters represent numbers.

The origins of algebra are found in a papyrus written before 1600 B.C. by an Egyptian priest named Ahmes. This papyrus contains 84 algebra problems and their solutions.

Further development of algebra occurred in the ninth century in the Middle East. In A.D. 830, an Arabian mathematician named al-Khowarazmi wrote a book called *Ihm al-jabr wa'l muqabalah.* This title was shortened to *al-Jabr.* We now know the subject as *algebra.* The French mathematician François Vieta (1540–1603) later simplified algebra by developing the symbolic notation that we use today.

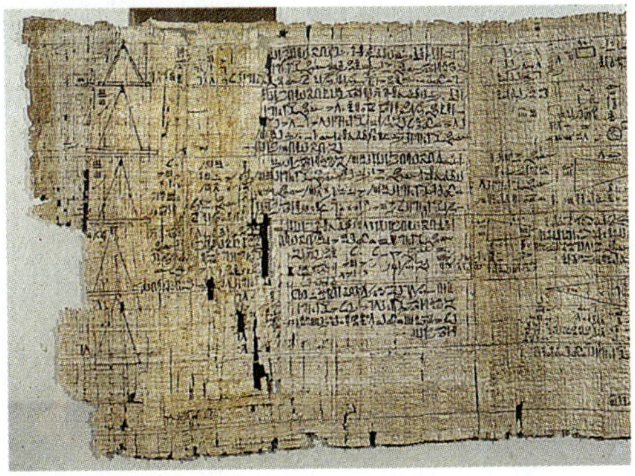

The Ahmes Papyrus © Copyright the British Museum

The set of even integers:

$\{\ldots, -10, -8, -6, -4, -2, 0, 2, 4, 6, 8, 10, \ldots\}$

The set of odd integers:

$\{\ldots, -9, -7, -5, -3, -1, 1, 3, 5, 7, 9, \ldots\}$

EXAMPLE 2 Which numbers in the set $\{-3, -2, 0, 1, 2, 3, 4, 5, 9\}$ are **a.** prime numbers, **b.** composite numbers, **c.** even integers, and **d.** odd integers?

Solution **a.** The prime numbers are 2, 3, and 5.

b. The composite numbers are 4 and 9.

c. The even integers are -2, 0, 2, and 4.

d. The odd integers are -3, 1, 3, 5, and 9.

Self Check Which numbers in the set $\{-5, 0, 1, 2, 4, 5\}$ are **a.** prime numbers and **b.** even integers?

Equality, Inequality Symbols, and Variables

To show that two expressions represent the same number, we use the = sign. Since 4 + 5 and 9 represent the same number, we can write

$4 + 5 = 9$ Read as "the sum of 4 and 5 is equal to 9."

Likewise, we can write

$5 - 3 = 2$ Read as "the difference between 5 and 3 equals 2," or "5 minus 3 equals 2."

$4 \cdot 5 = 20$ Read as "the product of 4 and 5 equals 20," or "4 times 5 equals 20."

and

$30 \div 6 = 5$ Read as "the quotient obtained when 30 is divided by 6 is 5," or "30 divided by 6 equals 5."

We can use **inequality symbols** to show that expressions are not equal.

Symbol	Read as	Symbol	Read as
$\approx$	"is approximately equal to"	$\neq$	"is not equal to"
$<$	"is less than"	$>$	"is greater than"
$\leq$	"is less than or equal to"	$\geq$	"is greater than or equal to"

EXAMPLE 3 **Inequality symbols**

a. $\pi \approx 3.14$ Read as "pi is approximately equal to 3.14."

b. $6 \neq 9$ Read as "6 is not equal to 9."

c. $8 < 10$ Read as "8 is less than 10."

d. $12 > 1$ Read as "12 is greater than 1."

e. $5 \leq 5$ Read as "5 is less than or equal to 5." (Since $5 = 5$, this is a true statement.)

f. $9 \geq 7$ Read as "9 is greater than or equal to 7." (Since $9 > 7$, this is a true statement.)

Self Check Tell whether each statement is true or false: **a.** $12 \neq 12$, **b.** $7 \geq 7$, and **c.** $125 < 137$.

Inequality statements can be written so that the inequality symbol points in the opposite direction. For example,

$$5 < 7 \qquad \text{and} \qquad 7 > 5$$

both indicate that 5 is a smaller number than 7. Likewise,

$$12 \geq 3 \qquad \text{and} \qquad 3 \leq 12$$

both indicate that 12 is greater than or equal to 3.

In algebra, we use letters, called **variables,** to represent real numbers. For example,

- If x represents 4, then $x = 4$.
- If y represents any number greater than 3, then $y > 3$.
- If z represents any number less than or equal to -4, then $z \leq -4$.

Comment In algebra, we usually do not use the times sign ($\times$) to indicate multiplication. It might be mistaken for the variable x.

The Number Line

We can use the **number line** shown in Figure 1-2 to represent sets of numbers. The number line continues forever to the left and to the right. Numbers to the left of 0 (the **origin**) are negative, and numbers to the right of 0 are positive.

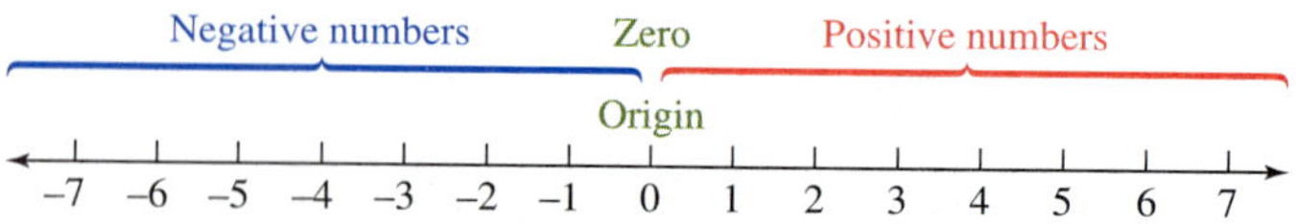

Figure 1-2

Comment The number 0 is neither positive nor negative.

The number that corresponds to a point on the number line is called the **coordinate** of that point. For example, the coordinate of the origin is 0.

Many points on the number line do not have integer coordinates. For example, the point midway between 0 and 1 has the coordinate $\frac{1}{2}$, and the point midway between -3 and -2 has the coordinate $-\frac{5}{2}$ (see Figure 1-3).

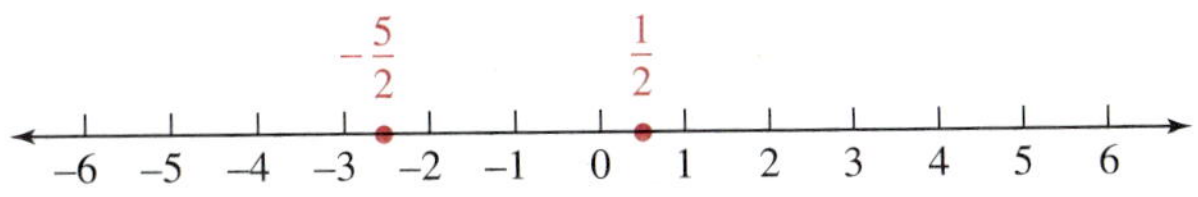

Figure 1-3

Numbers represented by points that lie on opposite sides of the origin and at equal distances from the origin are called **negatives** (or **opposites**) of each other. For example, 5 and -5 are negatives (or opposites). We need parentheses to express the opposite of a negative number. For example, $-(-5)$ represents the opposite of -5, which we know to be 5. Thus,

$$-(-5) = 5$$

This suggests the following rule.

Double Negative Rule

If x represents a real number, then

$$-(-x) = x$$

If one point lies to the *right* of a second point on a number line, its coordinate is the *greater.* Since the point with coordinate 1 lies to the right of the point with coordinate -2 (see Figure 1-4 (a)), it follows that $1 > -2$.

If one point lies to the *left* of another, its coordinate is the *smaller* (see Figure 1-4(b)). The point with coordinate -6 lies to the left of the point with coordinate -3, so it follows that $-6 < -3$.

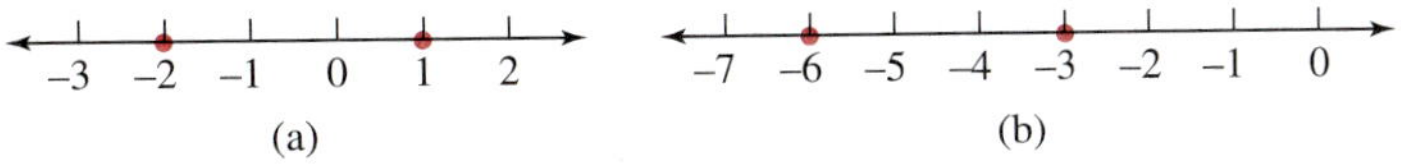

Figure 1-4

Graphing Subsets of the Real Numbers

Figure 1-5 shows the graph of the natural numbers from 2 to 8. The points on the line are called **graphs** of their corresponding coordinates.

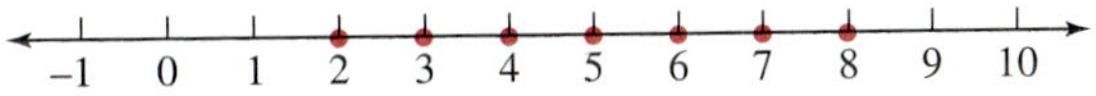

Figure 1-5

EXAMPLE 4 Graph the set of integers between -3 and 3.

Solution The integers between -3 and 3 are -2, -1, 0, 1, and 2. The graph is shown in Figure 1-6.

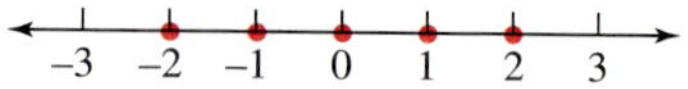

Figure 1-6

Self Check Graph the set of integers between -4 and 0.

Graphs of many sets of real numbers are **intervals** on the number line. For example, two graphs of all real numbers x such that $x > -2$ are shown in Figure 1-7. The parenthesis or the open circle at -2 shows that this point is not included in the graph. The arrow pointing to the right shows that all numbers to the right of -2 are included.

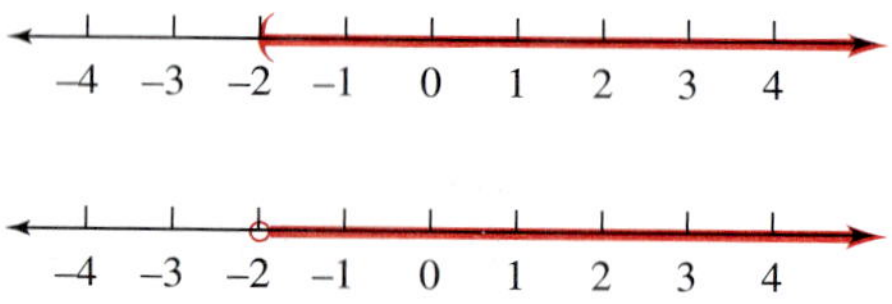

Figure 1-7

Figure 1-8 shows two graphs of the set of real numbers x between -2 and 4. This is the graph of all real numbers x such that $x > -2$ and $x < 4$. The parentheses or open circles at -2 and 4 show that these points are not included in the graph. However, all the numbers between -2 and 4 are included.

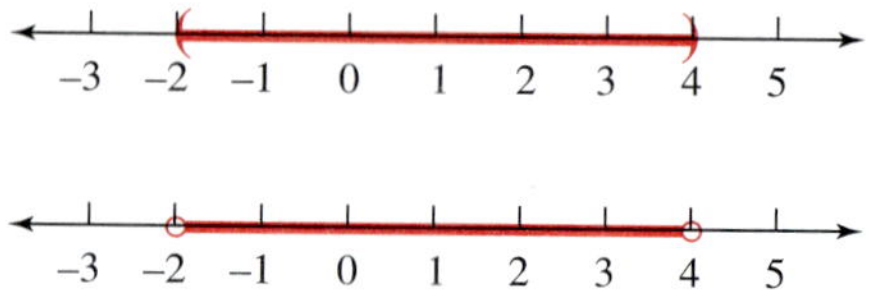

Figure 1-8

EXAMPLE 5 Graph all real numbers x such that $x < -3$ or $x > 1$.

Solution The graph of all real numbers less than -3 includes all points on the number line that are to the left of -3. The graph of all real numbers greater than 1 includes all points that are to the right of 1. The two graphs are shown in Figure 1-9.

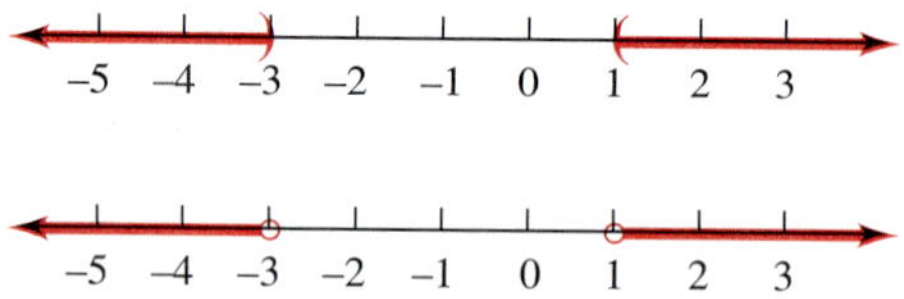

Figure 1-9

Self Check Graph all real numbers x such that $x < -1$ or $x > 0$. Use parentheses.

EXAMPLE 6 Graph the set of all real numbers from -5 to -1.

Solution The set of all real numbers from -5 to -1 includes -5 and -1 and all the numbers in between. In the graphs shown in Figure 1-10, the brackets or the solid circles at -5 and -1 show that these points are included.

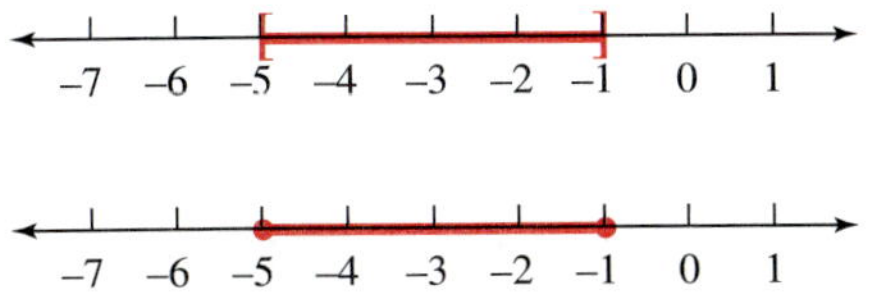

Figure 1-10

Self Check Graph the set of real numbers from -2 to 1. Use brackets.

Absolute Value of a Number

On a number line, the distance between a number x and 0 is called the **absolute value** of x. For example, the distance between 5 and 0 is 5 units (see Figure 1-11). Thus, the absolute value of 5 is 5:

$|5| = 5$ Read as "The absolute value of 5 is 5."

Since the distance between -6 and 0 is 6,

$|-6| = 6$ Read as "The absolute value of -6 is 6."

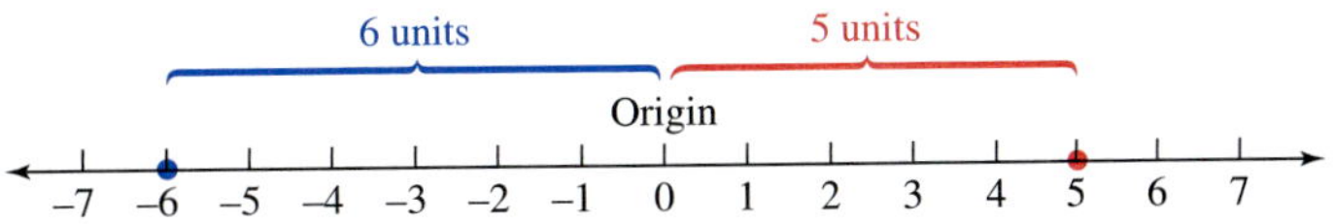

Figure 1-11

Because the absolute value of a real number represents that number's distance from 0 on the number line, the absolute value of every real number x is either positive or 0. In symbols, we say

$|x| \geq 0$ **for every real number x**

EXAMPLE 7 Evaluate: **a.** $|6|$, **b.** $|-3|$, **c.** $|0|$, and **d.** $-|2 + 3|$.

Solution **a.** $|6| = 6$, because 6 is six units from 0.

b. $|-3| = 3$, because -3 is three units from 0.

c. $|0| = 0$, because 0 is zero units from 0.

d. $-|2 + 3| = -|5| = -5$

Self Check Evaluate: **a.** $|8|$, **b.** $|-8|$, and **c.** $-|-8|$.

Self Check Answers

1. a. 7, **b.** $-2, 0, 1.5, 7$ **2. a.** 2, 5, **b.** 0, 2, 4 **3. a.** false, **b.** true, **c.** true

4. 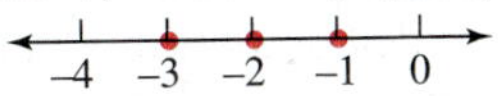

5. 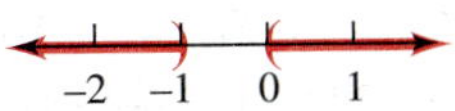

6. 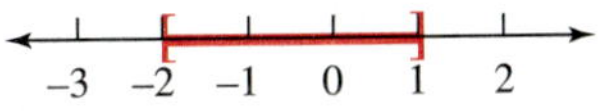

7. a. 8, **b.** 8, **c.** -8

Orals *Describe each set of numbers.*

1. natural numbers
2. whole numbers
3. integers
4. rational numbers
5. real numbers
6. prime numbers
7. composite numbers
8. even integers
9. odd integers
10. irrational numbers

Find each value.

11. $-|15|$
12. $|-25|$

1.1 EXERCISES

VOCABULARY AND CONCEPTS *Fill in the blanks.*

1. A ___ is a collection of objects.

2. The numbers 1, 2, 3, 4, 5, . . . form the set of _______ numbers.

3. The set of ______ numbers is the set $\{0, 1, 2, 3, 4, 5, \ldots\}$.

4. The set of ________ is the set $\{\ldots, -3, -2, -1, 0, 1, 2, 3, \ldots\}$.

5. Since every whole number is also an integer, the set of whole numbers is called a ______ of the set of integers.

6. $\sqrt{2}$ is an example of an ________ number.

7. If a natural number is greater than 1 and can be divided exactly only by 1 and itself, it is called a ______ number.

8. A composite number is a ______ number that is greater than 1 and is not ______.

9. The symbol $\neq$ means __________.

10. The symbol ___ means "is less than."

11. The symbol $\geq$ means ______________.

12. The opposite of -7 is ___.

13. The figure (number line: −3 −2 −1 0 1 2 3) is called a ______ line.

14. The distance between 8 and 0 on a number line is called the ___________ of 8.

PRACTICE *Which numbers in the set* $\{-3, -\frac{1}{2}, -1, 0, 1, 2, \frac{5}{3}, \sqrt{7}, 3.25, 6, 9\}$ *are*

15. Natural numbers
16. Whole numbers
17. Positive integers
18. Negative integers
19. Integers
20. Rational numbers
21. Real numbers
22. Irrational numbers
23. Odd integers
24. Even integers
25. Composite numbers
26. Prime numbers

Simplify each expression. Then classify the result as a natural number, an even integer, an odd integer, a prime number, a composite number, and/or a whole number.

27. $4 + 5$
28. $7 - 2$
29. $15 - 15$
30. $0 + 7$
31. $3 \cdot 8$
32. $8 \cdot 9$
33. $24 \div 8$
34. $3 \div 3$

Place one of the symbols =, <, or > in each box to make a true statement.

35. 5 □ 3 + 2
36. 9 □ 7
37. 25 □ 32
38. 2 + 3 □ 17
39. 5 + 7 □ 10
40. 3 + 3 □ 9 − 3
41. 3 + 9 □ 20 − 8
42. 19 − 3 □ 8 + 6
43. 4 · 2 □ 2 · 4
44. 7 · 9 □ 9 · 6

45. $8 \div 2 \;\square\; 4 + 2$

46. $0 \div 7 \;\square\; 1$

47. $3 + 2 + 5 \;\square\; 5 + 2 + 3$

48. $8 + 5 + 2 \;\square\; 5 + 2 + 8$

Write each sentence as a mathematical expression.

49. Seven is greater than three.

50. Five is less than thirty-two.

51. Eight is less than or equal to eight.

52. Twenty-five is not equal to twenty-three.

53. The result of adding three and four is equal to seven.

54. Thirty-seven is greater than the result of multiplying three and four.

55. $\sqrt{2}$ is approximately equal to 1.41.

56. x is greater than or equal to 12.

Write each inequality as an equivalent inequality in which the inequality symbol points in the opposite direction.

57. $3 \le 7$

58. $5 > 2$

59. $6 > 0$

60. $34 \le 40$

61. $3 + 8 > 8$

62. $8 - 3 < 8$

63. $6 - 2 < 10 - 4$

64. $8 \cdot 2 \ge 8 \cdot 1$

65. $2 \cdot 3 < 3 \cdot 4$

66. $8 \div 2 \ge 9 \div 3$

67. $\frac{12}{4} < \frac{24}{6}$

68. $\frac{2}{3} \le \frac{3}{4}$

Graph each pair of numbers on a number line. In each pair, indicate which number is the greater and which number lies farther to the right.

69. 3, 6

70. 4, 7

71. 11, 6

72. 12, 10

73. 0, 2

74. 4, 10

75. 8, 0

76. 20, 30

Graph each set of numbers on the number line.

77. The natural numbers between 2 and 8

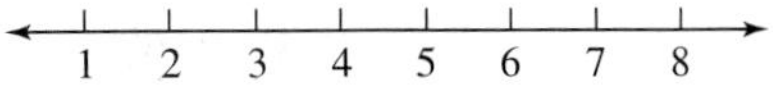

78. The prime numbers from 10 to 20

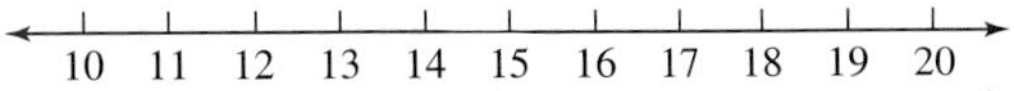

79. The even integers greater than 10 but less than 20

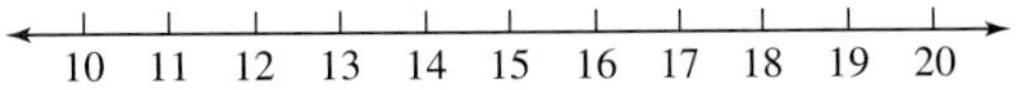

80. The even integers that are also prime numbers

81. The numbers that are whole numbers but not natural numbers

82. The prime numbers between 5 and 15

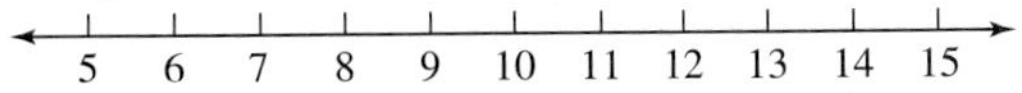

83. The natural numbers between 15 and 25 that are exactly divisible by 6

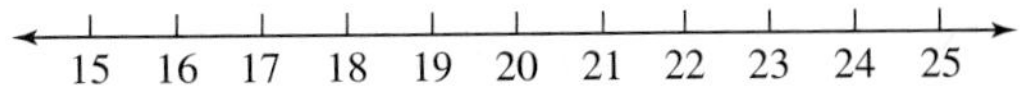

84. The odd integers between -5 and 5 that are exactly divisible by 3

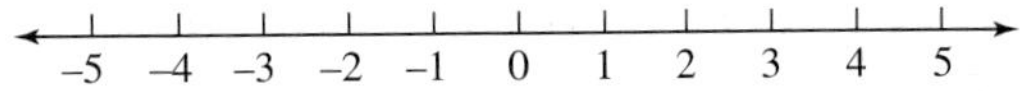

85. The real numbers between 1 and 5

86. The real numbers greater than or equal to 8

87. The real numbers greater than or equal to 3 or less than or equal to -3

88. The real numbers greater than -2 and less than 3

Find each absolute value.

89. $|36|$

90. $|-30|$

91. $|0|$

92. $|120|$

93. $|-230|$

94. $|18 - 12|$

95. $|12 - 4|$

96. $|100 - 100|$

WRITING

97. Explain why there is no greatest natural number.

98. Explain why 2 is the only even prime number.

99. Explain how to determine the absolute value of a number.

100. Explain why zero is an even integer.

SOMETHING TO THINK ABOUT *Consider the following sets: the integers, natural numbers, even and odd integers, positive and negative numbers, prime and composite numbers, and rational numbers.*

101. Find a number that fits in as many of these categories as possible.

102. Find a number that fits in as few of these categories as possible.

1.2 Fractions

- Fractions
- Simplifying Fractions
- Multiplying Fractions
- Dividing Fractions
- Adding Fractions
- Subtracting Fractions
- Mixed Numbers
- Decimals
- Rounding Decimals
- Applications

Getting Ready

1. Add: $\begin{array}{r} 132 \\ 45 \\ \underline{73} \end{array}$

2. Subtract: $\begin{array}{r} 321 \\ \underline{173} \end{array}$

3. Multiply: $\begin{array}{r} 437 \\ \underline{38} \end{array}$

4. Divide: $37\overline{)3{,}885}$

In this section, we will review arithmetic fractions. This will help us prepare for algebraic fractions, which we will encounter later in the book.

Fractions

In the **fractions**

$$\frac{1}{2}, \frac{3}{5}, \frac{2}{17}, \text{ and } \frac{37}{7}$$

the number above the bar is called the **numerator,** and the number below the bar is called the **denominator.**

We often use fractions to indicate parts of a whole. In Figure 1-12(a), a rectangle has been divided into 5 equal parts, and 3 of the parts are shaded. The fraction $\frac{3}{5}$ indicates how much of the figure is shaded. In Figure 1-12(b), $\frac{5}{7}$ of the rectangle is shaded. In either example, the denominator of the fraction shows the total number of equal parts into which the whole is divided, and the numerator shows how many of these equal parts are being considered.

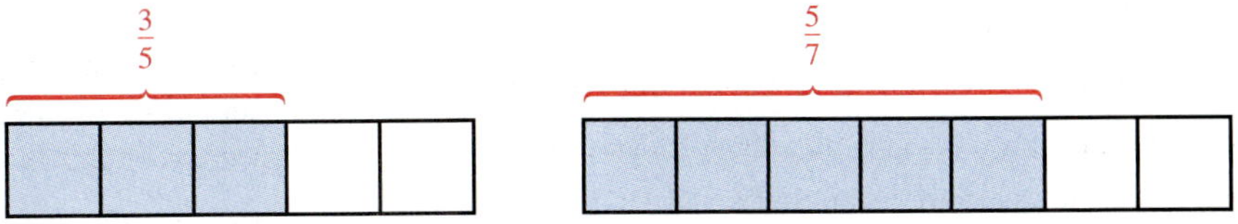

(a) (b)

Figure 1-12

We can also use fractions to indicate division. For example, the fraction $\frac{8}{2}$ indicates that 8 is to be divided by 2:

$$\frac{8}{2} = 8 \div 2 = 4$$

Comment Note that $\frac{8}{2} = 4$, because $4 \cdot 2 = 8$, and that $\frac{0}{7} = 0$, because $0 \cdot 7 = 0$. However, $\frac{6}{0}$ is undefined, because no number multiplied by 0 gives 6. Since every number multiplied by 0 gives 0, $\frac{0}{0}$ is indeterminate. Remember that the denominator of a fraction cannot be 0.

Simplifying Fractions

A fraction is in **lowest terms** when no integer other than 1 will divide both its numerator and its denominator exactly. The fraction $\frac{6}{11}$ is in lowest terms, because only 1 divides both 6 and 11 exactly. The fraction $\frac{6}{8}$ is not in lowest terms, because 2 divides both 6 and 8 exactly.

We can **simplify** a fraction that is not in lowest terms by dividing its numerator and its denominator by the same number. For example, to simplify $\frac{6}{8}$, we divide the numerator and the denominator by 2.

$$\frac{6}{8} = \frac{6 \div \mathbf{2}}{8 \div \mathbf{2}} = \frac{3}{4}$$

From Figure 1-13, we see that $\frac{6}{8}$ and $\frac{3}{4}$ are equal fractions, because each one represents the same part of the rectangle.

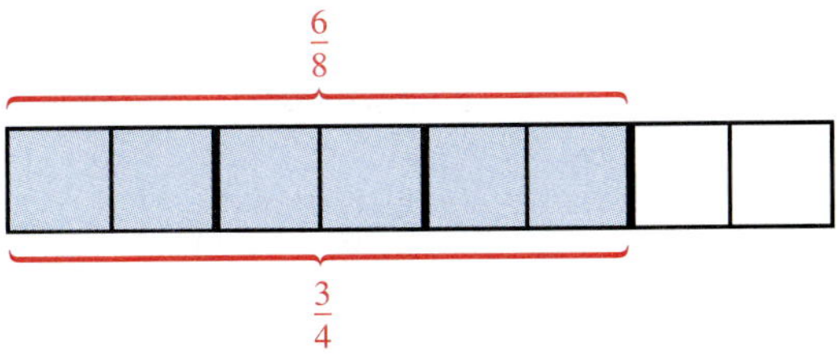

Figure 1-13

When a composite number has been written as the product of other natural numbers, we say that it has been **factored.** For example, 15 can be written as the product of 5 and 3.

$$15 = 5 \cdot 3$$

The numbers 5 and 3 are called **factors** of 15. When a composite number is written as the product of prime numbers, we say that it is written in **prime-factored form.**

EXAMPLE 1 Write 210 in prime-factored form.

Solution We can write 210 as the product of 21 and 10 and proceed as follows:

$$210 = \mathbf{21} \cdot \mathbf{10}$$

$$210 = \mathbf{3} \cdot \mathbf{7} \cdot \mathbf{2} \cdot \mathbf{5} \quad \text{Factor 21 as } 3 \cdot 7 \text{ and factor 10 as } 2 \cdot 5.$$

Since 210 is now written as the product of prime numbers, its prime-factored form is $210 = 2 \cdot 3 \cdot 5 \cdot 7$.

Self Check Write 70 in prime-factored form.

To simplify a fraction, we factor its numerator and denominator and divide out all factors that are common to the numerator and denominator. For example,

$$\frac{6}{8} = \frac{3 \cdot 2}{4 \cdot 2} = \frac{3 \cdot \overset{1}{\cancel{2}}}{4 \cdot \underset{1}{\cancel{2}}} = \frac{3}{4} \quad \text{and} \quad \frac{15}{18} = \frac{5 \cdot 3}{6 \cdot 3} = \frac{5 \cdot \overset{1}{\cancel{3}}}{6 \cdot \underset{1}{\cancel{3}}} = \frac{5}{6}$$

Comment Remember that a fraction is in lowest terms only when its numerator and denominator have no common factors.

EXAMPLE 2 Simplify: **a.** $\frac{6}{30}$ and **b.** $\frac{33}{40}$, if possible.

Solution **a.** To simplify $\frac{6}{30}$, we factor the numerator and denominator and divide out the common factor of 6.

$$\frac{6}{30} = \frac{6 \cdot 1}{6 \cdot 5} = \frac{\overset{1}{\cancel{6}} \cdot 1}{\underset{1}{\cancel{6}} \cdot 5} = \frac{1}{5}$$

b. To simplify $\frac{33}{40}$, we factor the numerator and denominator and hope to divide out any common factors.

$$\frac{33}{40} = \frac{3 \cdot 11}{2 \cdot 2 \cdot 2 \cdot 5}$$

Since the numerator and denominator have no common factors, $\frac{33}{40}$ is in lowest terms.

Self Check Simplify: $\frac{14}{35}$.

The preceding examples illustrate the **fundamental property of fractions.**

The Fundamental Property of Fractions

If a, b, and x are real numbers,

$$\frac{a \cdot x}{b \cdot x} = \frac{a}{b} \quad (b \neq 0 \text{ and } x \neq 0)$$

Multiplying Fractions

Multiplying Fractions

To multiply fractions, we multiply their numerators and multiply their denominators. In symbols, if a, b, c, and d are real numbers,

$$\frac{a}{b} \cdot \frac{c}{d} = \frac{a \cdot c}{b \cdot d} \quad (b \neq 0 \text{ and } d \neq 0)$$

For example,

$$\frac{4}{7} \cdot \frac{2}{3} = \frac{4 \cdot 2}{7 \cdot 3} = \frac{8}{21} \qquad \frac{4}{5} \cdot \frac{13}{9} = \frac{4 \cdot 13}{5 \cdot 9} = \frac{52}{45}$$

To justify the rule for multiplying fractions, we consider the square in Figure 1-14. Because the length of each side of the square is 1 unit and the area is the product of the lengths of two sides, the area is 1 square unit.

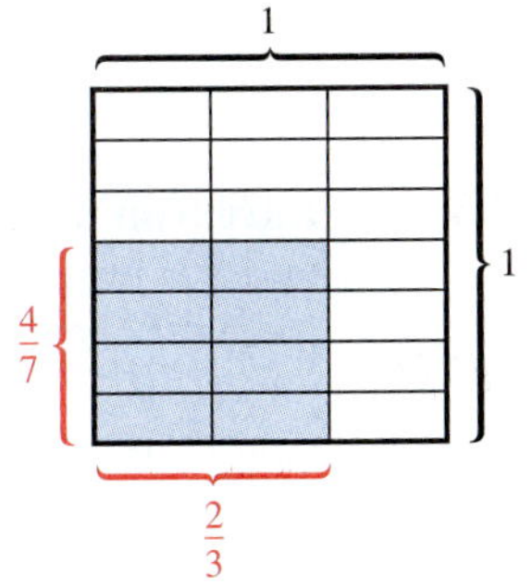

Figure 1-14

If this square is divided into 3 equal parts vertically and 7 equal parts horizontally, it is divided into 21 equals parts, and each represents $\frac{1}{21}$ of the total area. The area of the shaded rectangle in the square is $\frac{8}{21}$, because it contains 8 of the 21 parts. The width, w, of the shaded rectangle is $\frac{4}{7}$; its length, l, is $\frac{2}{3}$; and its area, A, is the product of l and w:

$$A = l \cdot w$$

$$\frac{8}{21} = \frac{2}{3} \cdot \frac{4}{7}$$

This suggests that we can find the product of

$$\frac{4}{7} \text{ and } \frac{2}{3}$$

by multiplying their numerators and multiplying their denominators.

Fractions whose numerators are less than their denominators, such as $\frac{8}{21}$, are called **proper fractions.** Fractions whose numerators are greater than or equal to their denominators, such as $\frac{52}{45}$, are called **improper fractions.**

EXAMPLE 3 Perform each multiplication.

a. $\frac{3}{7} \cdot \frac{13}{5} = \frac{3 \cdot 13}{7 \cdot 5}$ Multiply the numerators and multiply the denominators.

$= \frac{39}{35}$

b. $5 \cdot \frac{3}{15} = \frac{5}{1} \cdot \frac{3}{15}$ Write 5 as the improper fraction $\frac{5}{1}$.

$= \frac{5 \cdot 3}{1 \cdot 15}$ Multiply the numerators and multiply the denominators.

$= \frac{5 \cdot 3}{1 \cdot 5 \cdot 3}$ To attempt to simplify the fraction, factor the denominator.

$= \frac{\overset{1}{\cancel{5}} \cdot \overset{1}{\cancel{3}}}{1 \cdot \underset{1}{\cancel{5}} \cdot \underset{1}{\cancel{3}}}$ Divide out the common factors of 3 and 5.

$= 1$ $\frac{1 \cdot 1}{1 \cdot 1 \cdot 1} = 1.$

Self Check Multiply: $\frac{5}{9} \cdot \frac{7}{10}$.

EXAMPLE 4 **European travel** Out of 36 students in a history class, three-fourths have signed up for a trip to Europe. If there are 28 places available on the flight, will there be room for one more student?

Solution We first find three-fourths of 36.

$\frac{3}{4} \cdot 36 = \frac{3}{4} \cdot \frac{36}{1}$ Write 36 as $\frac{36}{1}$.

$= \frac{3 \cdot 36}{4 \cdot 1}$ Multiply the numerators and multiply the denominators.

$= \frac{3 \cdot 4 \cdot 9}{4 \cdot 1}$ To simplify the fraction, factor the numerator.

$= \frac{3 \cdot \overset{1}{\cancel{4}} \cdot 9}{\underset{1}{\cancel{4}} \cdot 1}$ Divide out the common factor of 4.

$= \frac{27}{1}$

$= 27$

Twenty-seven students plan to go on the trip. Since there is room for 28 passengers, there is room for one more.

Self Check If five-sixths of the 36 students have signed up, will there be room for one more?

Dividing Fractions

One number is called the **reciprocal** of another if their product is 1. For example, $\frac{3}{5}$ is the reciprocal of $\frac{5}{3}$, because

$$\frac{3}{5} \cdot \frac{5}{3} = \frac{15}{15} = 1$$

Dividing Fractions

To divide two fractions, we multiply the first fraction by the reciprocal of the second fraction. In symbols, if a, b, c, and d are real numbers,

$$\frac{a}{b} \div \frac{c}{d} = \frac{a}{b} \cdot \frac{d}{c} = \frac{a \cdot d}{b \cdot c} \quad (b \neq 0, c \neq 0, \text{ and } d \neq 0)$$

EXAMPLE 5 Perform each division.

a. $\frac{3}{5} \div \frac{6}{5} = \frac{3}{5} \cdot \frac{5}{6}$ Multiply $\frac{3}{5}$ by the reciprocal of $\frac{6}{5}$.

$= \frac{3 \cdot 5}{5 \cdot 6}$ Multiply the numerators and multiply the denominators.

$= \frac{3 \cdot 5}{5 \cdot 2 \cdot 3}$ Factor the denominator.

$= \frac{\overset{1}{\cancel{3}} \cdot \overset{1}{\cancel{5}}}{\underset{1}{\cancel{5}} \cdot 2 \cdot \underset{1}{\cancel{3}}}$ Divide out the common factors of 3 and 5.

$= \frac{1}{2}$

b. $\frac{15}{7} \div 10 = \frac{15}{7} \div \frac{10}{1}$ Write 10 as the improper fraction $\frac{10}{1}$.

$= \frac{15}{7} \cdot \frac{1}{10}$ Multiply $\frac{15}{7}$ by the reciprocal of $\frac{10}{1}$.

$= \frac{15 \cdot 1}{7 \cdot 10}$ Multiply the numerators and multiply the denominators.

$= \frac{3 \cdot \overset{1}{\cancel{5}}}{7 \cdot 2 \cdot \underset{1}{\cancel{5}}}$ Factor the numerator and the denominator, and divide out the common factor of 5.

$= \frac{3}{14}$

Self Check Divide: $\frac{13}{6} \div \frac{26}{8}$.

Adding Fractions

Adding Fractions with the Same Denominator

To add fractions with the same denominator, we add the numerators and keep the common denominator. In symbols, if a, b, and d are real numbers,

$$\frac{a}{d} + \frac{b}{d} = \frac{a + b}{d} \quad (d \neq 0)$$

For example,

$$\frac{3}{7} + \frac{2}{7} = \frac{3 + 2}{7} \quad \text{Add the numerators and keep the common denominator.}$$

$$= \frac{5}{7}$$

Figure 1-15 illustrates why $\frac{3}{7} + \frac{2}{7} = \frac{5}{7}$.

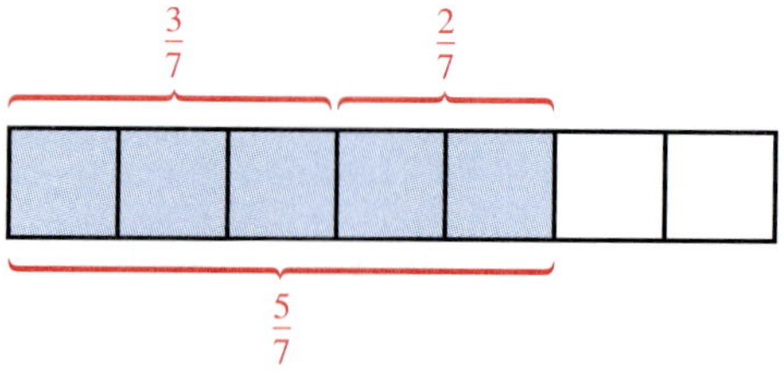

Figure 1-15

To add fractions with unlike denominators, we write the fractions so that they have the same denominator. For example, we can multiply both the numerator and denominator of $\frac{1}{3}$ by 5 to obtain an equivalent fraction with a denominator of 15:

$$\frac{1}{3} = \frac{1 \cdot 5}{3 \cdot 5} = \frac{5}{15}$$

To write $\frac{1}{5}$ as an equivalent fraction with a denominator of 15, we multiply the numerator and the denominator by 3:

$$\frac{1}{5} = \frac{1 \cdot 3}{5 \cdot 3} = \frac{3}{15}$$

Since 15 is the smallest number that can be used as a common denominator for $\frac{1}{3}$ and $\frac{1}{5}$, it is called the **least** or **lowest common denominator** (the **LCD**).

To add the fractions $\frac{1}{3}$ and $\frac{1}{5}$, we rewrite each fraction as an equivalent fraction having a denominator of 15, and then we add the results:

$$\frac{1}{3} + \frac{1}{5} = \frac{1 \cdot 5}{3 \cdot 5} + \frac{1 \cdot 3}{5 \cdot 3}$$

$$= \frac{5}{15} + \frac{3}{15}$$

$$= \frac{5 + 3}{15}$$

$$= \frac{8}{15}$$

EXAMPLE 6 Add: $\frac{3}{10} + \frac{5}{28}$.

Solution To find the LCD, we find the prime factorization of each denominator and use each prime factor the greatest number of times it appears in either factorization:

$$\left.\begin{aligned} 10 &= 2 \cdot 5 \\ 28 &= 2 \cdot 2 \cdot 7 \end{aligned}\right\} \text{LCD} = 2 \cdot 2 \cdot 5 \cdot 7 = 140$$

Since 140 is the smallest number that 10 and 28 divide exactly, we write both fractions as fractions with denominators of 140.

$$\begin{aligned} \frac{3}{10} + \frac{5}{28} &= \frac{3 \cdot 14}{10 \cdot 14} + \frac{5 \cdot 5}{28 \cdot 5} && \text{Write each fraction as a fraction with a denominator of 140.} \\ &= \frac{42}{140} + \frac{25}{140} \\ &= \frac{42 + 25}{140} && \text{Add the numerators and keep the denominator.} \\ &= \frac{67}{140} \end{aligned}$$

Since 67 is a prime number, it has no common factor with 140. Thus, $\frac{67}{140}$ is in lowest terms.

Self Check Add: $\frac{3}{8} + \frac{5}{12}$.

Subtracting Fractions

Subtracting Fractions with the Same Denominator

To subtract fractions with the same denominator, we subtract their numerators and keep their common denominator. In symbols, if a, b, and d are real numbers,

$$\frac{a}{d} - \frac{b}{d} = \frac{a - b}{d} \quad (d \neq 0)$$

For example,

$$\frac{7}{9} - \frac{2}{9} = \frac{7 - 2}{9} = \frac{5}{9}$$

To subtract fractions with unlike denominators, we write them as equivalent fractions with a common denominator. For example, to subtract $\frac{2}{5}$ from $\frac{3}{4}$, we write $\frac{3}{4} - \frac{2}{5}$, find the LCD of 20, and proceed as follows:

$$\begin{aligned} \frac{3}{4} - \frac{2}{5} &= \frac{3 \cdot 5}{4 \cdot 5} - \frac{2 \cdot 4}{5 \cdot 4} \\ &= \frac{15}{20} - \frac{8}{20} \end{aligned}$$

$$= \frac{15 - 8}{20}$$

$$= \frac{7}{20}$$

EXAMPLE 7 Subtract 5 from $\frac{23}{3}$.

Solution

$$\frac{23}{3} - 5 = \frac{23}{3} - \frac{5}{1} \quad \text{Write 5 as the improper fraction } \tfrac{5}{1}.$$

$$= \frac{23}{3} - \frac{5 \cdot 3}{1 \cdot 3} \quad \text{Write } \tfrac{5}{1} \text{ as a fraction with a denominator of 3.}$$

$$= \frac{23}{3} - \frac{15}{3}$$

$$= \frac{23 - 15}{3} \quad \text{Subtract the numerators and keep the denominator.}$$

$$= \frac{8}{3}$$

Self Check Subtract: $\frac{5}{6} - \frac{3}{4}$.

Mixed Numbers

The **mixed number** $3\frac{1}{2}$ represents the sum of 3 and $\frac{1}{2}$. We can write $3\frac{1}{2}$ as an improper fraction as follows:

$$3\frac{1}{2} = 3 + \frac{1}{2}$$

$$= \frac{6}{2} + \frac{1}{2} \quad 3 = \tfrac{6}{2}.$$

$$= \frac{6 + 1}{2} \quad \text{Add the numerators and keep the denominator.}$$

$$= \frac{7}{2}$$

To write the fraction $\frac{19}{5}$ as a mixed number, we divide 19 by 5 to get 3, with a remainder of 4.

$$\frac{19}{5} = 3 + \frac{4}{5} = 3\frac{4}{5}$$

EXAMPLE 8 Add: $2\frac{1}{4} + 1\frac{1}{3}$.

Solution We first change each mixed number to an improper fraction.

$$2\frac{1}{4} = 2 + \frac{1}{4} \qquad 1\frac{1}{3} = 1 + \frac{1}{3}$$

$$= \frac{8}{4} + \frac{1}{4} \qquad = \frac{3}{3} + \frac{1}{3}$$

$$= \frac{9}{4} \qquad = \frac{4}{3}$$

Then we add the fractions.

$$2\frac{1}{4} + 1\frac{1}{3} = \frac{9}{4} + \frac{4}{3}$$

$$= \frac{9 \cdot 3}{4 \cdot 3} + \frac{4 \cdot 4}{3 \cdot 4} \quad \text{Change each fraction into a fraction with the LCD of 12.}$$

$$= \frac{27}{12} + \frac{16}{12}$$

$$= \frac{43}{12}$$

Finally, we change $\frac{43}{12}$ to a mixed number.

$$\frac{43}{12} = 3 + \frac{7}{12} = 3\frac{7}{12}$$

Self Check Add: $5\frac{1}{7} + 4\frac{2}{3}$.

EXAMPLE 9 **Fencing land** The three sides of a triangular lot measure $33\frac{1}{4}$, $57\frac{3}{4}$, and $72\frac{1}{2}$ meters. How much fencing will be needed to enclose the area?

Solution We can find the sum of the lengths (called the **perimeter**) by adding the whole-number parts and the fractional parts of the dimensions separately:

$$33\frac{1}{4} + 57\frac{3}{4} + 72\frac{1}{2} = 33 + 57 + 72 + \frac{1}{4} + \frac{3}{4} + \frac{1}{2}$$

$$= 162 + \frac{1}{4} + \frac{3}{4} + \frac{2}{4}$$

Change $\frac{1}{2}$ to $\frac{2}{4}$ to obtain a common denominator.

$$= 162 + \frac{6}{4}$$

Add the fractions by adding the numerators and keeping the common denominator.

$$= 162 + \frac{3}{2}$$

$$\frac{6}{4} = \frac{2 \cdot 3}{2 \cdot 2} = \frac{\overset{1}{\cancel{2}} \cdot 3}{\underset{1}{\cancel{2}} \cdot 2} = \frac{3}{2}.$$

$$= 162 + 1\frac{1}{2}$$

Change $\frac{3}{2}$ to a mixed number.

$$= 163\frac{1}{2}$$

To enclose the area, $163\frac{1}{2}$ meters of fencing will be needed.

Self Check A rectangular plot is $85\frac{1}{2}$ feet wide and $140\frac{2}{3}$ feet deep. Find its perimeter.

Decimals

Rational numbers can always be changed to decimal form. For example, to write $\frac{1}{4}$ and $\frac{5}{22}$ as decimals, we use long division:

$$
\begin{array}{r}
0.25 \\
4\overline{)1.00} \\
\underline{8} \\
20 \\
\underline{20}
\end{array}
\qquad\qquad
\begin{array}{r}
0.22727\ldots \\
22\overline{)5.00000} \\
\underline{4\,4} \\
60 \\
\underline{44} \\
160 \\
\underline{154} \\
60 \\
\underline{44} \\
160
\end{array}
$$

The decimal 0.25 is called a **terminating decimal.** The decimal 0.2272727. . . (often written as $0.2\overline{27}$) is called a **repeating decimal,** because it repeats the block of digits 27. Every rational number can be changed into either a terminating or a repeating decimal.

Terminating decimals	***Repeating decimals***
$\frac{1}{2} = 0.5$	$\frac{1}{3} = 0.33333. . .$ or $0.\overline{3}$
$\frac{3}{4} = 0.75$	$\frac{1}{6} = 0.16666. . .$ or $0.1\overline{6}$
$\frac{5}{8} = 0.625$	$\frac{5}{22} = 0.2272727. . .$ or $0.2\overline{27}$

The decimal 0.5 has one **decimal place,** because it has one digit to the right of the decimal point. The decimal 0.75 has two decimal places, and 0.625 has three.

To *add* or *subtract* decimal fractions, we align their decimal points and then add or subtract.

$$
\begin{array}{r}
25.568 \\
\underline{+2.74} \\
28.308
\end{array}
\qquad\qquad
\begin{array}{r}
25.568 \\
\underline{-2.74} \\
22.828
\end{array}
$$

To perform the previous operations with a calculator, we enter these numbers and press these keys:

25.568 + 2.74 = and 25.568 − 2.74 = On a scientific calculator.

25.568 + 2.74 ENTER and 25.568 − 2.74 ENTER On a graphing calculator.

To *multiply* decimals, we multiply the numbers and place the decimal point so that the number of decimal places in the answer is equal to the sum of the decimal places in the factors.

$$
\begin{array}{r l}
3.453 & \text{Here there are three decimal places.} \\
\underline{\times\ \ 9.25} & \text{Here there are two decimal places.} \\
17265 & \\
6906 & \\
\underline{31\,077} & \\
31.94025 & \text{The product has } 3 + 2 = 5 \text{ decimal places.}
\end{array}
$$

To perform this multiplication with a calculator, we enter these numbers and press these keys:

3.453 × 9.25 =	On a scientific calculator.
3.453 × 9.25 ENTER	On a graphing calculator.

To *divide* decimals, we move the decimal point in the divisor to the right to make the divisor a whole number. We then move the decimal point in the dividend the same number of places to the right.

$1.23\overline{)30.258}$ Move the decimal point in both the divisor and the dividend two places to the right.

We align the decimal point in the quotient with the repositioned decimal point in the dividend and use long division.

$$\begin{array}{r} 24.6 \\ 123\overline{)3025.8} \\ \underline{246} \\ 565 \\ \underline{492} \\ 73\;8 \\ \underline{73\;8} \end{array}$$

To perform the previous division with a calculator, we enter these numbers and press these keys:

30.258 ÷ 1.23 =	On a scientific calculator.
30.258 ÷ 1.23 ENTER	On a graphing calculator.

Rounding Decimals

We often **round** long decimals to a specific number of decimal places. For example, the decimal 25.36124 rounded to one place (or to the nearest tenth) is 25.4. Rounded to two places (or to the nearest one-hundredth), the decimal is 25.36. To round decimals, we use the following rules.

Rounding Decimals

1. Determine to how many decimal places you wish to round.
2. Look at the first digit to the right of that decimal place.
3. If that digit is 4 or less, drop it and all digits that follow. If it is 5 or greater, add 1 to the digit in the position to which you wish to round, and drop all digits that follow.

Applications

A **percent** is the numerator of a fraction with a denominator of 100. For example, $6\frac{1}{4}$ percent, written $6\frac{1}{4}\%$, is the fraction $\frac{6.25}{100}$, or the decimal 0.0625. In problems involving percent, the word *of* usually indicates multiplication. For example, $6\frac{1}{4}\%$ of 8,500 is the product 0.0625(8,500).

EXAMPLE 10 **Auto loans** Juan signs a one-year note to borrow \$8,500 to buy a car. If the rate of interest is $6\frac{1}{4}\%$, how much interest will he pay?

Solution For the privilege of using the bank's money for one year, Juan must pay $6\frac{1}{4}\%$ of \$8,500. We calculate the interest, i, as follows:

$$i = 6\tfrac{1}{4}\% \text{ of } 8{,}500$$
$$= 0.0625 \cdot 8{,}500 \quad \text{The word } \textit{of} \text{ means } \textit{times}.$$
$$= 531.25$$

Juan will pay \$531.25 interest.

Self Check How much interest will Juan pay if the rate is 9%?

Self Check Answers

1. $2 \cdot 5 \cdot 7$ **2.** $\frac{2}{5}$ **3.** $\frac{7}{18}$ **4.** no **5.** $\frac{2}{3}$ **6.** $\frac{19}{24}$ **7.** $\frac{1}{12}$ **8.** $9\frac{17}{21}$ **9.** $452\frac{1}{3}$ ft **10.** \$765

Orals *Simplify each fraction.*

1. $\frac{3}{6}$ **2.** $\frac{5}{10}$ **3.** $\frac{10}{20}$ **4.** $\frac{25}{75}$

Perform each operation.

5. $\frac{5}{6} \cdot \frac{1}{2}$ **6.** $\frac{3}{4} \cdot \frac{3}{5}$ **7.** $\frac{2}{3} \div \frac{3}{2}$ **8.** $\frac{3}{5} \div \frac{5}{2}$

9. $\frac{4}{9} + \frac{7}{9}$ **10.** $\frac{6}{7} - \frac{3}{7}$ **11.** $\frac{2}{3} - \frac{1}{2}$ **12.** $\frac{3}{4} + \frac{1}{2}$

13. $2.5 + 0.36$ **14.** $3.45 - 2.21$

15. $0.2 \cdot 2.5$ **16.** $0.3 \cdot 13$

Round each decimal to two decimal places.

17. 3.244993 **18.** 3.24521

1.2 EXERCISES

REVIEW *Decide whether the following statements are true or false.*

1. 6 is an integer.

2. $\frac{1}{2}$ is a natural number.

3. 21 is a prime number.

4. No prime number is an even number.

5. $8 > -2$

6. $-3 < -2$

7. $9 \le |-9|$

8. $|-11| \ge 10$

Place an appropriate symbol in each box to make the statement true.

9. $3 + 7 \ \square\ 10$

10. $\frac{3}{7} \ \square\ \frac{2}{7} = \frac{1}{7}$

11. $|-2| \ \square\ 2$

12. $4 + 8 \ \square\ 11$

VOCABULARY AND CONCEPTS *Fill in the blanks.*

13. The number above the bar in a fraction is called the ________.

14. The number below the bar in a fraction is called the ________.

15. To ______ a fraction, we divide its numerator and denominator by the same number.

16. To write a number in prime-factored form, we write it as the product of ______ numbers.

17. If the numerator of a fraction is less than the denominator, the fraction is called a ______ fraction.

18. If the numerator of a fraction is greater than the denominator, the fraction is called an ______ fraction.

19. If the product of two numbers is __, the numbers are called reciprocals.

20. $\frac{ax}{bx} = \_$

21. To multiply two fractions, ______ the numerators and multiply the denominators.

22. To divide two fractions, multiply the first fraction by the ______ of the second fraction.

23. To add fractions with a common denominator, add the ______ and keep the common ______.

24. To subtract fractions with a common ______, subtract the numerators and keep the common denominator.

25. $75\frac{2}{3}$ means 75 ___ $\frac{2}{3}$.

26. 0.75 is an example of a ______ decimal.

27. $5.3\overline{27}$ is an example of a ______ decimal.

28. A ______ is the numerator of a fraction whose denominator is 100.

PRACTICE *Write each fraction in lowest terms. If the fraction is already in lowest terms, so indicate.*

29. $\frac{6}{12}$

30. $\frac{3}{9}$

31. $\frac{15}{20}$

32. $\frac{22}{77}$

33. $\frac{24}{18}$

34. $\frac{35}{14}$

35. $\frac{72}{64}$

36. $\frac{26}{21}$

Perform each operation. Simplify each result when possible.

37. $\frac{1}{2} \cdot \frac{3}{5}$

38. $\frac{3}{4} \cdot \frac{5}{7}$

39. $\frac{4}{3} \cdot \frac{6}{5}$

40. $\frac{7}{8} \cdot \frac{6}{15}$

41. $\frac{5}{12} \cdot \frac{18}{5}$

42. $\frac{5}{4} \cdot \frac{12}{10}$

43. $\frac{17}{34} \cdot \frac{3}{6}$

44. $\frac{21}{14} \cdot \frac{3}{6}$

45. $12 \cdot \frac{5}{6}$

46. $9 \cdot \frac{7}{12}$

47. $\frac{10}{21} \cdot 14$

48. $\frac{5}{24} \cdot 16$

49. $\frac{3}{5} \div \frac{2}{3}$

50. $\frac{4}{5} \div \frac{3}{7}$

51. $\frac{3}{4} \div \frac{6}{5}$

52. $\frac{3}{8} \div \frac{15}{28}$

53. $\frac{2}{13} \div \frac{8}{13}$

54. $\frac{4}{7} \div \frac{20}{21}$

55. $\frac{21}{35} \div \frac{3}{14}$

56. $\frac{23}{25} \div \frac{46}{5}$

57. $6 \div \frac{3}{14}$

58. $23 \div \frac{46}{5}$

59. $\frac{42}{30} \div 7$

60. $\frac{34}{8} \div 17$

61. $\frac{3}{5} + \frac{3}{5}$

62. $\frac{4}{7} - \frac{2}{7}$

63. $\frac{4}{13} - \frac{3}{13}$

64. $\frac{2}{11} + \frac{9}{11}$

65. $\frac{1}{6} + \frac{1}{24}$

66. $\frac{17}{25} - \frac{2}{5}$

67. $\frac{3}{5} + \frac{2}{3}$

68. $\frac{4}{3} + \frac{7}{2}$

69. $\frac{9}{4} - \frac{5}{6}$

70. $\frac{2}{15} + \frac{7}{9}$

71. $\frac{7}{10} - \frac{1}{14}$

72. $\frac{7}{25} + \frac{3}{10}$

73. $3 - \frac{3}{4}$

74. $5 + \frac{21}{5}$

75. $\frac{17}{3} + 4$

76. $\frac{13}{9} - 1$

77. $4\frac{3}{5} + \frac{3}{5}$

78. $2\frac{1}{8} + \frac{3}{8}$

79. $3\frac{1}{3} - 1\frac{2}{3}$

80. $5\frac{1}{7} - 3\frac{2}{7}$

81. $3\frac{3}{4} - 2\frac{1}{2}$

82. $15\frac{5}{6} + 11\frac{5}{8}$

83. $8\frac{2}{9} - 7\frac{2}{3}$

84. $3\frac{4}{5} - 3\frac{1}{10}$

85. $23.45 + 135.2$

86. $345.213 - 27.35$

87. $67.235 - 22.45$

88. $12.17 + 3.457$

89. $3.4 \cdot 13.2$

90. $4.21 \cdot 2.73$

91. $0.23\overline{)1.0465}$

92. $4.7\overline{)10.857}$

Use a calculator to perform each operation. Round each of the answers to two decimal places.

93. $323.24 + 27.2543$

94. $843.45213 - 712.765$

95. $55.77443 - 0.568245$

96. $0.62317 + 1.3316$

97. $25.25 \cdot 132.179$

98. $234.874 \cdot 242.46473$

99. $0.456\overline{)4.5694323}$

100. $43.225\overline{)32.465748}$

APPLICATIONS

101. Spring plowing A farmer has plowed $12\frac{1}{3}$ acres of a $43\frac{1}{2}$-acre field. How much more needs to be plowed?

102. Perimeter of a garden The four sides of a garden measure $7\frac{2}{3}$ feet, $15\frac{1}{4}$ feet, $19\frac{1}{2}$ feet and $10\frac{3}{4}$ feet. Find the length of the fence needed to enclose the garden.

103. Making clothes A designer needs $3\frac{1}{4}$ yards of material for each dress he makes. How much material will he need to make 14 dresses?

The following circle graph shows the various sources of retirement income for a typical retired person in the year 2001.

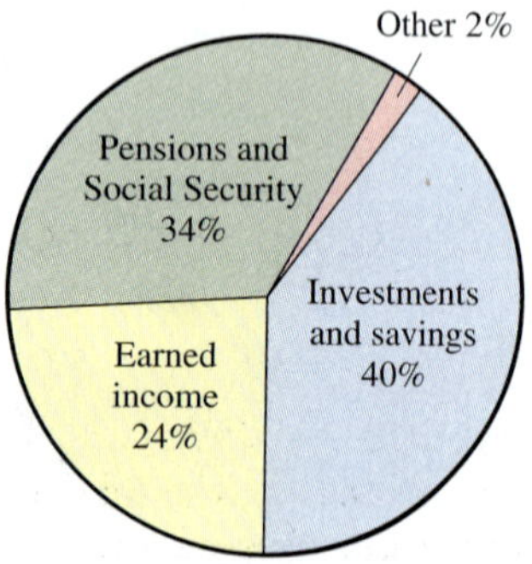

104. Retirement income If a retiree has \$36,000 of income, how much is expected to come from pensions and social security?

105. Retirement income If a retiree has \$42,500 of income, how much is expected to come from earned income?

106. Minority population 26.5% of the 12,419,000 citizens of Illinois are nonwhite. How many are nonwhite?

107. Quality control In the manufacture of active-matrix color LCD computer displays, many units must be rejected as defective. If 23% of a production run of 17,500 units is defective, how many units are acceptable?

108. Freeze-drying Almost all of the water must be removed when food is preserved by freeze-drying. Find the weight of the water removed from 750 pounds of a food that is 36% water.

109. Planning for growth This year, sales at Positronics Corporation totaled \$18.7 million. If the projection of 12% annual growth is true, what will be next year's sales?

110. Speed skating In tryouts for the Olympics, a speed skater had times of 44.47, 43.24, 42.77, and 42.05 seconds. Find the average time. (*Hint:* Add the numbers and divide by 4.)

111. Cost of gasoline Otis drove his car 15,675.2 miles last year, averaging 25.5 miles per gallon of gasoline. If the average cost of gasoline was \$1.67 per gallon, find the fuel cost to drive the car.

112. Paying taxes A woman earns \$48,712.32 in taxable income. She must pay 15% tax on the first \$23,000 and 28% on the rest. In addition, she must pay a Social Security tax of 15.4% on the total amount. How much tax will she need to pay?

113. Sealing asphalt A rectangular parking lot is 253.5 feet long and 178.5 feet wide. A 55-gallon drum of asphalt sealer covers 4,000 square feet and costs \$97.50. Find the cost to seal the parking lot. (Sealer can only be purchased in full drums.)

114. Inventory costs Each television a retailer buys costs \$3.25 per day for warehouse storage. What does it cost to store 37 television sets for three weeks?

115. Manufacturing profits A manufacturer of computer memory boards has a profit of \$37.50 on each standard-capacity memory board, and \$57.35 on each high-capacity board. The sales department has orders for 2,530 standard boards and 1,670 high-capacity boards. Which order will produce the greater profit?

116. Dairy production A Holstein cow will produce 7,600 pounds of milk each year, with a $3\frac{1}{2}\%$ butterfat content. Each year, a Guernsey cow will produce about 6,500 pounds of milk that is 5% butterfat. Which cow produces more butterfat?

117. Feeding dairy cows Each year, a typical dairy cow will eat 12,000 pounds of food that is 57% silage. To feed 30 cows, how much silage will a farmer use in a year?

118. Comparing bids Two contractors bid on a home remodeling project. The first bids \$9,350 for the entire job. The second contractor will work for \$27.50 per hour, plus \$4,500 for materials. He estimates that the job will take 150 hours. Which contractor has the lower bid?

119. Choosing a furnace A high-efficiency home heating system can be installed for \$4,170, with an average monthly heating bill of \$57.50. A regular furnace can be installed for \$1,730, but monthly heating bills average \$107.75. After three years, which system has cost more altogether?

120. Choosing a furnace Refer to Exercise 119. Decide which furnace system will have cost more after five years.

WRITING

121. Describe how you would find the common denominator of two fractions.

122. Explain how to convert an improper fraction into a mixed number.

123. Explain how to convert a mixed number into an improper fraction.

124. Explain how you would decide which of two decimal fractions is the larger.

SOMETHING TO THINK ABOUT

125. In what situations would it be better to leave an answer in the form of an improper fraction?

126. When would it be better to change an improper-fraction answer into a mixed number?

127. Can the product of two proper fractions be larger than either of the fractions?

128. How does the product of one proper and one improper fraction compare with the two factors?

1.3 Exponents and Order of Operations

- Exponents
- Order of Operations
- Geometry

Getting Ready *Perform the operations.*

1. $2 \cdot 2$ **2.** $3 \cdot 3$ **3.** $3 \cdot 3 \cdot 3$ **4.** $2 \cdot 2 \cdot 2$

5. $\frac{1}{2} \cdot \frac{1}{2}$ **6.** $\frac{1}{3} \cdot \frac{1}{3} \cdot \frac{1}{3}$ **7.** $\frac{2}{5} \cdot \frac{2}{5} \cdot \frac{2}{5}$ **8.** $\frac{3}{10} \cdot \frac{3}{10} \cdot \frac{3}{10}$

In algebra, we will encounter many expressions that contain exponents, a shortcut method of showing repeated multiplication. In this section, we will introduce exponential notation and discuss the rules for the order of operations.

Exponents

To show how many times a number is to be used as a factor in a product, we use *exponents.* In the expression 2^3, 2 is called the base and 3 is called the **exponent.**

$$\text{Base} \rightarrow 2^3 \leftarrow \text{Exponent}$$

The exponent of 3 indicates that the base of 2 is to be used as a factor three times:

$$2^3 = \underbrace{2 \cdot 2 \cdot 2}_{\text{3 factors of 2}} = 8$$

Comment Note that $2^3 = 8$. This is not the same as $2 \cdot 3 = 6$.

In the expression x^5 (called an **exponential expression** or a **power of *x***), x is the base and 5 is the exponent. The exponent of 5 indicates that a base of x is to be used as a factor five times.

$$x^5 = \underbrace{x \cdot x \cdot x \cdot x \cdot x}_{\text{5 factors of } x}$$

In expressions such as x or y, the exponent is understood to be 1:

$$x = x^1 \qquad \text{and} \qquad y = y^1$$

In general, we have the following definition.

Natural-Number Exponents

If n is a natural number, then

$$x^n = \underbrace{x \cdot x \cdot x \cdot \cdots \cdot x}_{n \text{ factors of } x}$$

EXAMPLE 1 Write each expression without exponents.

a. $4^2 = 4 \cdot 4 = 16$ — Read 4^2 as "4 squared" or as "4 to the second power."

b. $5^3 = 5 \cdot 5 \cdot 5 = 125$ — Read 5^3 as "5 cubed" or as "5 to the third power."

c. $6^4 = 6 \cdot 6 \cdot 6 \cdot 6 = 1{,}296$ — Read 6^4 as "6 to the fourth power."

d. $\left(\frac{2}{3}\right)^5 = \frac{2}{3} \cdot \frac{2}{3} \cdot \frac{2}{3} \cdot \frac{2}{3} \cdot \frac{2}{3} = \frac{32}{243}$ — Read $\left(\frac{2}{3}\right)^5$ as "$\frac{2}{3}$ to the fifth power."

Self Check Evaluate: **a.** 7^2 and **b.** $\left(\frac{3}{4}\right)^3$.

We can find powers using a calculator. For example, to find 2.35^4, we enter these numbers and press these keys:

2.35 y^x 4 = — On a scientific calculator.

2.35 ^ 4 ENTER — On a graphing calculator.

Either way, the display will read 30.49800625. Some scientific calculators have an x^y key rather than a y^x key.

In the next example, the base of an exponential expression is a variable.

EXAMPLE 2 Write each expression without exponents.

a. $y^6 = y \cdot y \cdot y \cdot y \cdot y \cdot y$ Read y^6 as "y to the sixth power."

b. $x^3 = x \cdot x \cdot x$ Read x^3 as "x cubed" or as "x to the third power."

c. $z^2 = z \cdot z$ Read z^2 as "z squared" or as "z to the second power."

d. $a^1 = a$ Read a^1 as "a to the first power."

Self Check Write each expression without exponents: **a.** a^3 and **b.** b^4.

Order of Operations

Suppose you are asked to contact a friend if you see a Rolex watch for sale while traveling in Switzerland. After locating the watch, you send the following message to your friend.

E-Mail

ROLEX WATCH $6,200. SHOULD I BUY IT FOR YOU?

The next day, you receive this response.

E-Mail

NO PRICE TOO HIGH! REPEAT...NO! PRICE TOO HIGH.

The first statement says to buy the watch at any price. The second says not to buy it, because it is too expensive. The placement of the exclamation point makes these statements read differently, resulting in different interpretations.

When reading a mathematical statement, the same kind of confusion is possible. To illustrate, we consider the expression $2 + 3 \cdot 4$, which contains the operations of addition and multiplication. We can calculate this expression in two different ways. We can perform the addition first and then perform the multiplication. Or we can perform the multiplication first and then perform the addition. However, we will get different results.

Method 1: Add first		***Method 2: Multiply first***	
$2 + 3 \cdot 4 = 5 \cdot 4$	Add 2 and 3.	$2 + 3 \cdot 4 = 2 + 12$	Multiply 3 and 4.
$= 20$	Multiply 5 and 4.	$= 14$	Add 2 and 12.

Different results

To eliminate the possibility of getting different answers, we will agree to perform multiplications before additions. The correct calculation of $2 + 3 \cdot 4$ is

$$\begin{aligned} 2 + 3 \cdot 4 &= 2 + 12 && \text{Do the multiplication first.} \\ &= 14 \end{aligned}$$

To indicate that additions should be done before multiplications, we use **grouping symbols** such as parentheses ( ), brackets [ ], or braces { }. In the expression $(2 + 3)4$, the parentheses indicate that the addition is to be done first:

$$\begin{aligned} (2 + 3)4 &= 5 \cdot 4 && \text{Do the addition within the parentheses first.} \\ &= 20 \end{aligned}$$

To guarantee that calculations will have one correct result, we will always perform calculations in the following order.

Rules for the Order of Operations

Use the following steps to perform all calculations within each pair of grouping symbols, working from the innermost pair to the outermost pair.

1. Find the values of any exponential expressions.
2. Perform all multiplications and divisions, working from left to right.
3. Perform all additions and subtractions, working from left to right.

When all grouping symbols have been removed, repeat the rules above to finish the calculation.

In a fraction, simplify the numerator and the denominator separately. Then simplify the fraction, whenever possible.

Comment Note that $4(2)^3 \neq (4 \cdot 2)^3$:

$$4(2)^3 = 4 \cdot 2 \cdot 2 \cdot 2 = 4(8) = 32 \quad \text{and} \quad (4 \cdot 2)^3 = 8^3 = 8 \cdot 8 \cdot 8 = 512$$

Likewise, $4x^3 \neq (4x)^3$ because

$$4x^3 = 4xxx \quad \text{and} \quad (4x)^3 = (4x)(4x)(4x) = 64x^3$$

EXAMPLE 3 Evaluate: $5^3 + 2(8 - 3 \cdot 2)$.

Solution We perform the work within the parentheses first and then simplify.

$$\begin{aligned} 5^3 + 2(8 - 3 \cdot 2) &= 5^3 + 2(8 - 6) && \text{Do the multiplication within the parentheses.} \\ &= 5^3 + 2(2) && \text{Do the subtraction within the parentheses.} \\ &= 125 + 2(2) && \text{Find the value of the exponential expression.} \\ &= 125 + 4 && \text{Do the multiplication.} \\ &= 129 && \text{Do the addition.} \end{aligned}$$

Self Check Evaluate: $5 + 4 \cdot 3^2$.

EXAMPLE 4 Evaluate: $\dfrac{3(3 + 2) + 5}{17 - 3(4)}$.

Solution We simplify the numerator and denominator separately and then simplify the fraction.

$$\frac{3(3 + 2) + 5}{17 - 3(4)} = \frac{3(5) + 5}{17 - 3(4)}$$ Do the addition within the parentheses.

$$= \frac{15 + 5}{17 - 12}$$ Do the multiplications.

$$= \frac{20}{5}$$ Do the addition and the subtraction.

$$= 4$$ Do the division.

Self Check Evaluate: $\frac{4 + 2(5 - 3)}{2 + 3(2)}$.

EXAMPLE 5 Evaluate: $\frac{3(4^2) - 2(3)}{2(4 + 3)}$.

Solution

$$\frac{3(4^2) - 2(3)}{2(4 + 3)} = \frac{3(16) - 2(3)}{2(7)}$$ Find the value of 4^2 in the numerator and do the addition in the denominator.

$$= \frac{48 - 6}{14}$$ Do the multiplications.

$$= \frac{42}{14}$$ Do the subtraction.

$$= 3$$ Do the division.

Self Check Evaluate: $\frac{2^2 + 6(5)}{2(2 + 5) + 3}$.

Geometry

To find perimeters and areas of geometric figures, substituting numbers for variables is often required. The **perimeter** of a geometric figure is the distance around it, and the **area** of a figure is the amount of surface that it encloses. The perimeter of a circle is called its **circumference.**

EXAMPLE 6 **Circles** Find **a.** the circumference and **b.** the area of the circle shown in Figure 1-16.

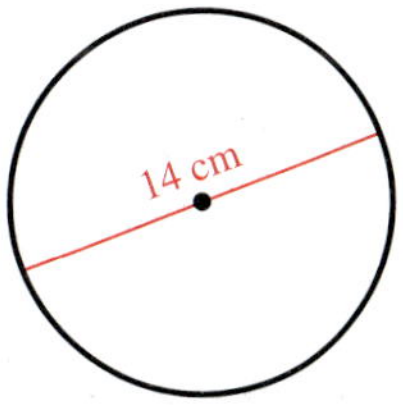

Figure 1-16

Solution **a.** The formula for the circumference of a circle is

$$C = \pi D$$

where C is the circumference, π is approximately $\frac{22}{7}$, and D is the **diameter**—a line segment that passes through the center of the circle and joins two points on the circle. We can approximate the circumference by substituting $\frac{22}{7}$ for π and 14 for D in the formula and simplifying.

$$C = \pi D$$

$$C \approx \frac{22}{7} \cdot 14 \quad \text{Read} \approx \text{as "is approximately equal to."}$$

$$C \approx \frac{22 \cdot \overset{2}{\cancel{14}}}{\underset{1}{\cancel{7}} \cdot 1} \quad \text{Multiply the fractions and simplify.}$$

$$C \approx 44$$

The circumference is approximately 44 centimeters. To use a calculator, we enter these numbers and press these keys:

π × 14 = On a scientific calculator.

π × 14 ENTER On a graphing calculator.

Either way, the display will read 43.98229715. . . . The result is not 44, because a calculator uses a better approximation for π than $\frac{22}{7}$.

Comment A segment drawn from the center of a circle to a point on the circle is called a **radius.** Since the diameter D of a circle is twice as long as its radius r, we have $D = 2r$. If we substitute $2r$ for D in the formula $C = \pi D$, we obtain an alternate formula for the circumference of a circle: $C = 2\pi r$.

b. The formula for the area of a circle is

$$A = \pi r^2$$

where A is the area, $\pi \approx \frac{22}{7}$, and r is the **radius** of the circle. We can approximate the area by substituting $\frac{22}{7}$ for π and 7 for r in the formula and simplifying.

$$A = \pi r^2$$

$$A \approx \frac{22}{7} \cdot 7^2$$

$$A \approx \frac{22}{7} \cdot \frac{49}{1} \quad \text{Evaluate the exponential expression.}$$

$$A \approx \frac{22 \cdot \overset{7}{\cancel{49}}}{\underset{1}{\cancel{7}} \cdot 1} \quad \text{Multiply the fractions and simplify.}$$

$$A \approx 154$$

The area is approximately 154 square centimeters. To use a calculator, we enter these numbers and press these keys:

π $\times$ 7 x^2 = On a scientific calculator.

π $\times$ 7 x^2 ENTER On a graphing calculator.

The display will read 153.93804.

Self Check Find **a.** the circumference and **b.** the area of a circle with a diameter of 28 meters. (Use $\frac{22}{7}$ as an estimate for π.) Check your results with a calculator.

Table 1-1 shows the formulas for the perimeter and area of several geometric figures.

Figure	Name	Perimeter	Area
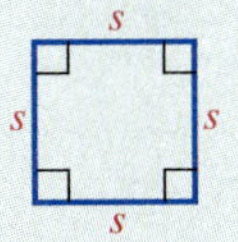	Square	$P = 4s$	$A = s^2$
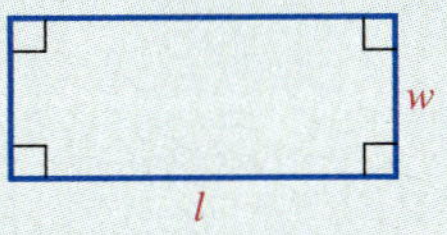	Rectangle	$P = 2l + 2w$	$A = lw$
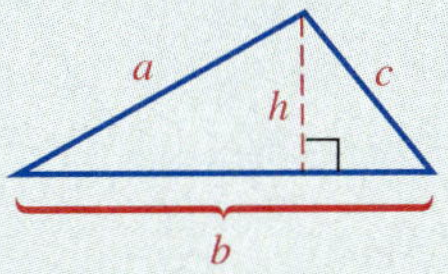	Triangle	$P = a + b + c$	$A = \frac{1}{2}bh$
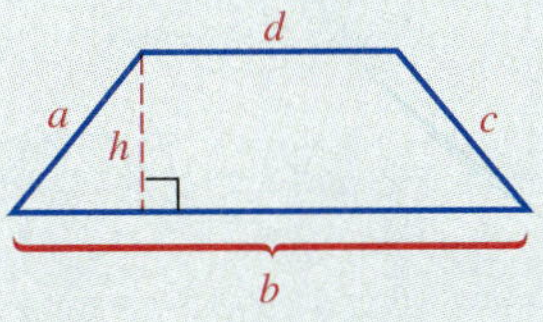	Trapezoid	$P = a + b + c + d$	$A = \frac{1}{2}h(b + d)$
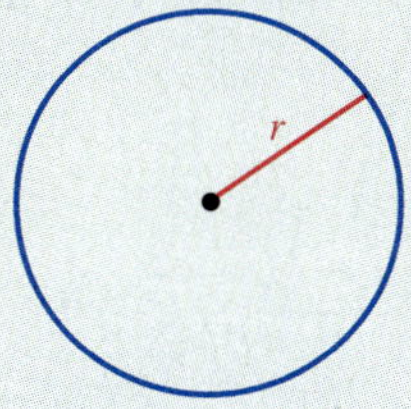	Circle	$C = \pi D = 2\pi r$	$A = \pi r^2$

Table 1-1

The **volume** of a three-dimensional geometric solid is the amount of space it encloses. Table 1-2 shows the formulas for the volume of several solids.

Figure	Name	Volume
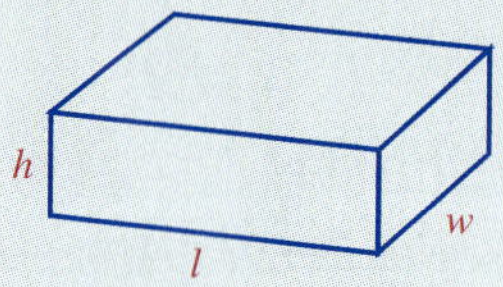	Rectangular solid	$V = lwh$
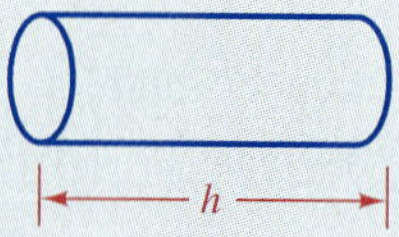	Cylinder	$V = Bh$, where B is the area of the base
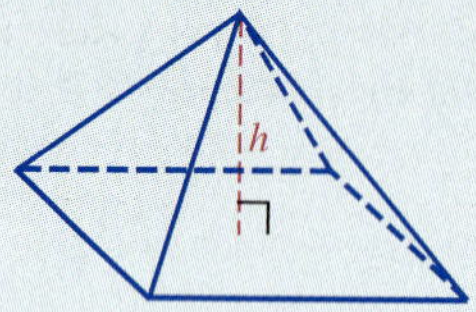	Pyramid	$V = \frac{1}{3}Bh$, where B is the area of the base
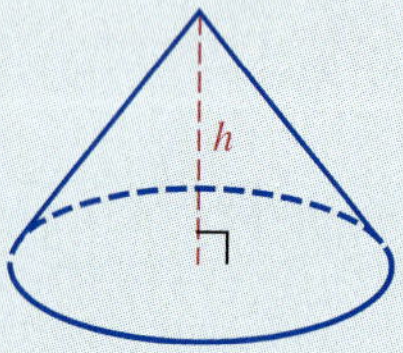	Cone	$V = \frac{1}{3}Bh$, where B is the area of the base
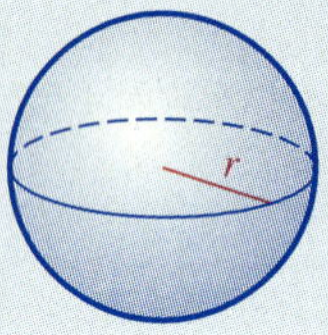	Sphere	$V = \frac{4}{3}\pi r^3$

Table 1-2

EXAMPLE 7 **Winter driving** Find the number of cubic feet of road salt in the conical pile shown in Figure 1-17. Round the answer to two decimal places.

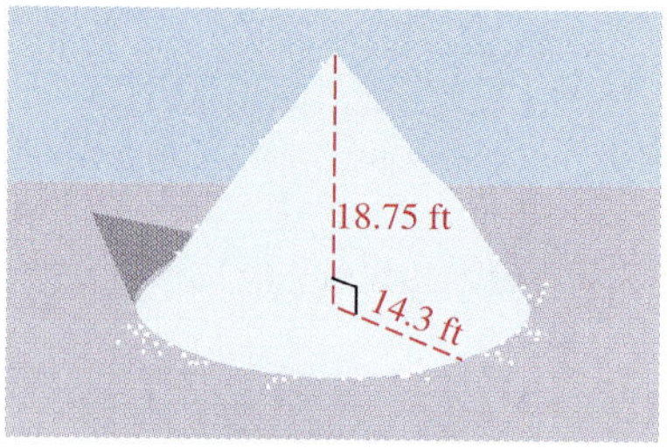

Figure 1-17

Solution We can find the area of the circular base by substituting $\frac{22}{7}$ for π and 14.3 for the radius.

$$A = \pi r^2$$

$$\approx \frac{22}{7}(14.3)^2$$

$$\approx 642.6828571 \quad \text{Use a calculator.}$$

We then substitute 642.6828571 for B and 18.75 for h in the formula for the volume of a cone.

$$V = \frac{1}{3}Bh$$
$$\approx \frac{1}{3}(642.6828571)(18.75)$$
$$\approx 4{,}016.767857 \qquad \text{Use a calculator.}$$

To two decimal places, there are 4,016.77 cubic feet of salt in the pile.

Self Check To the nearest hundredth, find the number of cubic feet of water that can be contained in a spherical tank that has a radius of 9 feet. (Use $\pi \approx \frac{22}{7}$.)

Self Check Answers

1. a. 49, **b.** $\frac{27}{64}$ **2. a.** $a \cdot a \cdot a$, **b.** $b \cdot b \cdot b \cdot b$ **3.** 41 **4.** 1 **5.** 2 **6. a.** 88 m, **b.** 616 m^2
7. 3,054.86 ft^3

Orals *Find the value of each expression.*

1. 2^5 **2.** 3^4 **3.** 4^3 **4.** 5^3

Simplify each expression.

5. $3(2)^3$ **6.** $(3 \cdot 2)^2$ **7.** $3 + 2 \cdot 4$
8. $10 - 3^2$ **9.** $4 + 2^2 \cdot 3$ **10.** $2 \cdot 3 + 2 \cdot 3^2$

1.3 EXERCISES

REVIEW

1. On the number line, graph the prime numbers between 10 and 20.

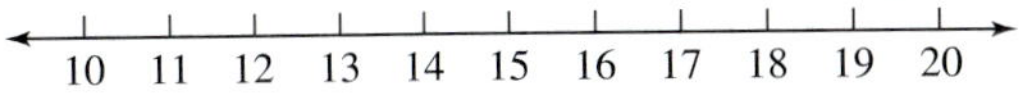

2. Write the inequality $7 \le 12$ as an inequality using the symbol $\ge$.

3. Classify the number 17 as a prime number or a composite number.

4. Evaluate: $\frac{3}{5} - \frac{1}{2}$

VOCABULARY AND CONCEPTS *Fill in the blanks.*

5. An ________ indicates how many times a base is to be used as a factor in a product.

6. In the expression x^5, x is called the _____ and 5 is called an ________.

7. In the expression $3 + 4 \cdot 5$, the ____________ should be done first.

8. Parentheses, brackets, and braces are called ________ symbols.

Write the appropriate formula to find each quantity.

9. The perimeter of a square ______

10. The area of a square ______

11. The perimeter of a rectangle __________

12. The area of a rectangle ______

13. The perimeter of a triangle ____________

14. The area of a triangle _______

15. The perimeter of a trapezoid _______________

16. The area of a trapezoid ____________

17. The circumference of a circle _________________

18. The area of a circle _______

19. The volume of a rectangular solid _______

20. The volume of a cylinder ______

21. The volume of a pyramid ______

22. The volume of a cone ______

23. The volume of a sphere ______

24. In Exercises 20-22, B is the ____ of the base.

PRACTICE *Find the value of each expression.*

25. 4^2 **26.** 5^2

27. 6^2 **28.** 7^3

29. $\left(\frac{1}{10}\right)^4$ **30.** $\left(\frac{1}{2}\right)^6$

Use a calculator to find each power.

31. 7.9^3 **32.** 0.45^4

33. 25.3^2 **34.** 7.567^3

Write each expression without using exponents.

35. x^2 **36.** y^3

37. $3z^4$ **38.** $5t^2$

39. $(5t)^2$ **40.** $(3z)^4$

41. $5(2x)^3$ **42.** $7(3t)^2$

Find the value of each expression.

43. $4(3^2)$ **44.** $4(2^3)$

45. $(5 \cdot 2)^3$ **46.** $(2 \cdot 2)^4$

47. $2(3^2)$ **48.** $3(2^3)$

49. $(3 \cdot 2)^3$ **50.** $(2 \cdot 3)^2$

Simplify each expression.

51. $3 \cdot 5 - 4$ **52.** $6 + 4 \cdot 3$

53. $3(5 - 4)$ **54.** $4(6 + 5)$

55. $3 + 5^2$ **56.** $4^2 - 2^2$

57. $(3 + 5)^2$ **58.** $(5 - 2)^3$

59. $2 + 3 \cdot 5 - 4$ **60.** $12 + 2 \cdot 3 + 2$

61. $64 \div (3 + 1)$ **62.** $16 \div (5 + 3)$

63. $(7 + 9) \div (2 \cdot 4)$ **64.** $(7 + 9) \div 2 \cdot 4$

65. $(5 + 7) \div 3 \cdot 4$ **66.** $(5 + 7) \div (3 \cdot 4)$

67. $24 \div 4 \cdot 3 + 3$ **68.** $36 \div 9 \cdot 4 - 2$

69. $3^2 + 2(1 + 4) - 2$

70. $4 \cdot 3 + 2(5 - 2) - 2^3$

71. $5^2 - (7 - 3)^2$ **72.** $3^3 + (3 - 1)^3$

73. $(2 \cdot 3 - 4)^3$ **74.** $(3 \cdot 5 - 2 \cdot 6)^2$

75. $\frac{3}{5} \cdot \frac{10}{3} + \frac{1}{2} \cdot 12$ **76.** $\frac{15}{4}\left(1 + \frac{3}{5}\right)$

77. $\left[\frac{1}{3} - \left(\frac{1}{2}\right)^2\right]^2$ **78.** $\left[\left(\frac{2}{3}\right)^2 - \frac{1}{3}\right]^2$

Use a calculator to simplify each fraction.

79. $\frac{(3 + 5)^2 + 2}{2(8 - 5)}$ **80.** $\frac{25 - (2 \cdot 3 - 1)}{2 \cdot 9 - 8}$

81. $\frac{(5 - 3)^2 + 2}{4^2 - (8 + 2)}$ **82.** $\frac{(4^2 - 2) + 7}{5(2 + 4) - 3^2}$

83. $\frac{2[4 + 2(3 - 1)]}{3[3(2 \cdot 3 - 4)]}$ **84.** $\frac{3[9 - 2(7 - 3)]}{(8 - 5)(9 - 7)}$

85. $\frac{3 \cdot 7 - 5(3 \cdot 4 - 11)}{4(3 + 2) - 3^2 + 5}$ **86.** $\frac{2 \cdot 5^2 - 2^2 + 3}{2(5 - 2)^2 - 11}$

Insert parentheses in the expression $3 \cdot 8 + 5 \cdot 3$ *to make its value equal to the given number.*

87. 39 **88.** 117

89. 87 **90.** 69

Find the perimeter of each figure.

91.

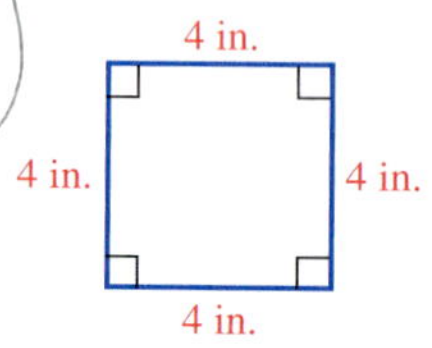

92.

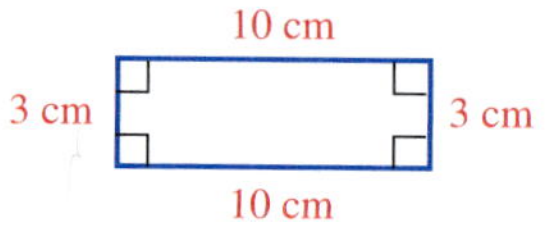

93.

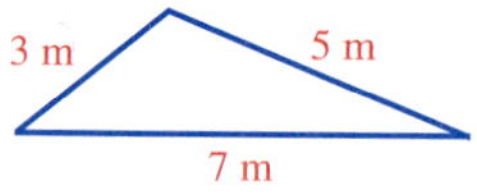

94.

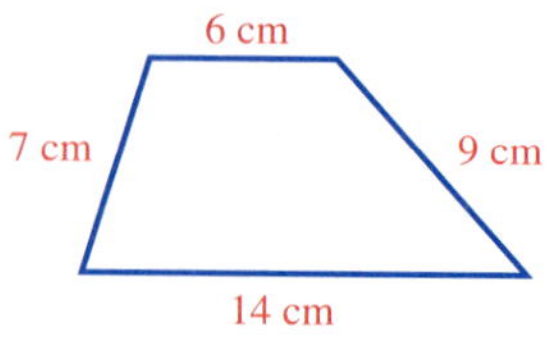

Find the area of each figure.

95.

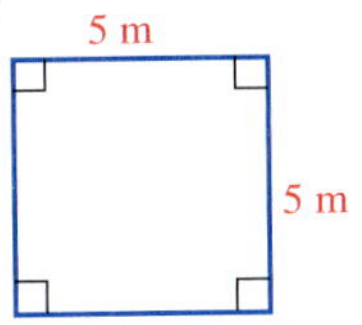

96.

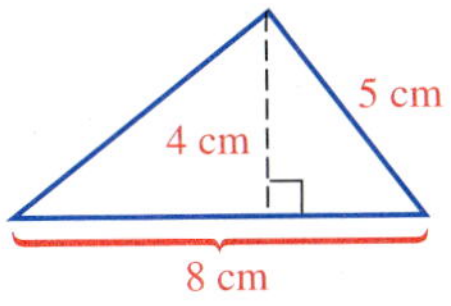

97.

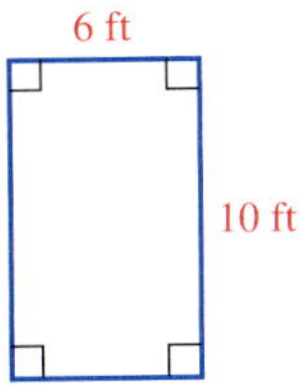

98.

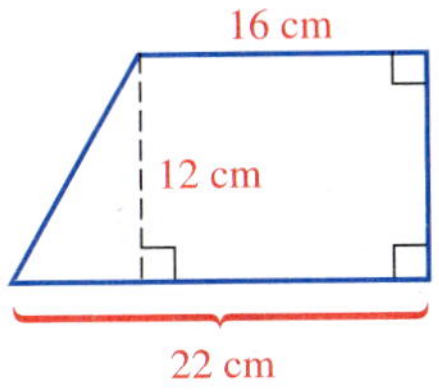

Find the circumference of each circle. Use $\pi \approx \frac{22}{7}$.

99.

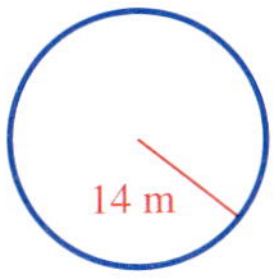

100.

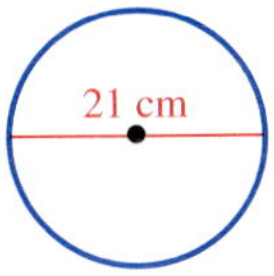

Find the area of each circle. Use $\pi \approx \frac{22}{7}$.

101.

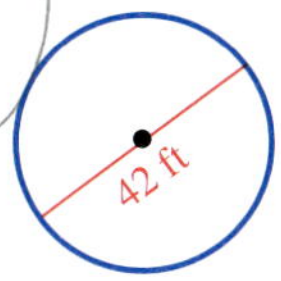

102.

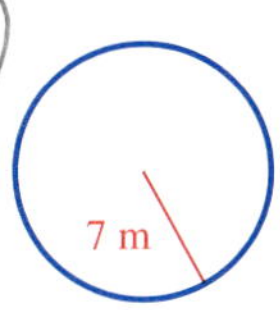

Find the volume of each solid. Use $\pi \approx \frac{22}{7}$.

103.

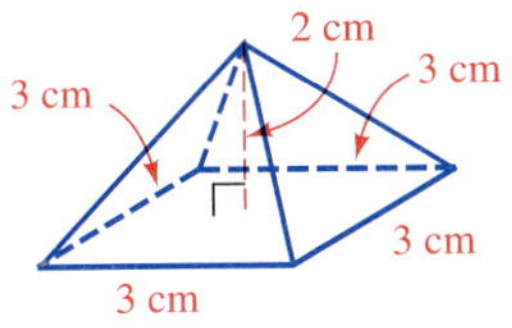

104.

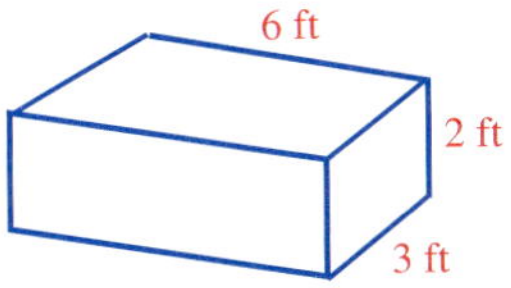

105.

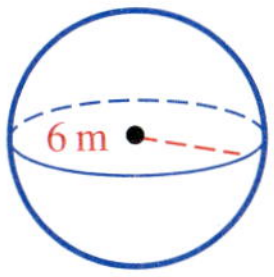

106.

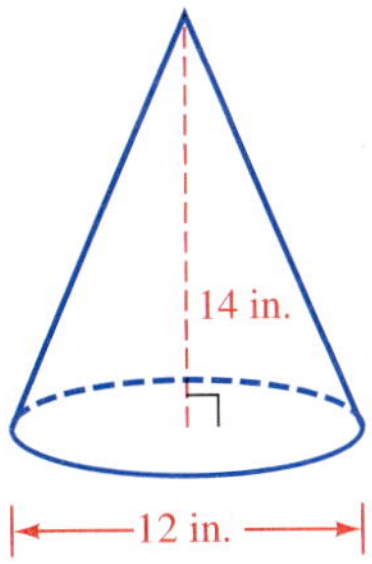

107.

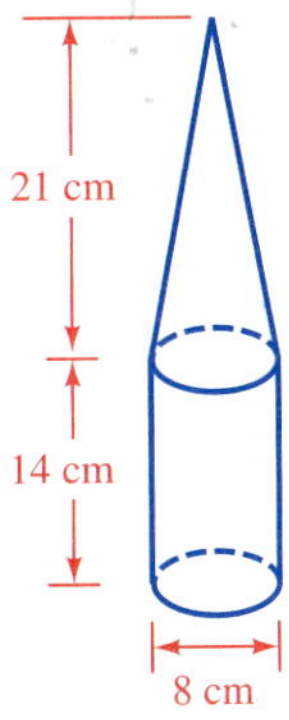

108.

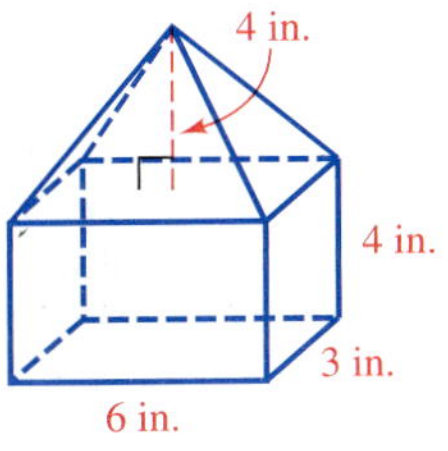

APPLICATIONS

109. Buying fencing How many meters of fencing are needed to enclose the square field shown in the illustration?

110. Installing carpet What will it cost to carpet the area shown in the illustration with carpet that costs \$29.79 per square yard? (One square yard is 9 square feet.)

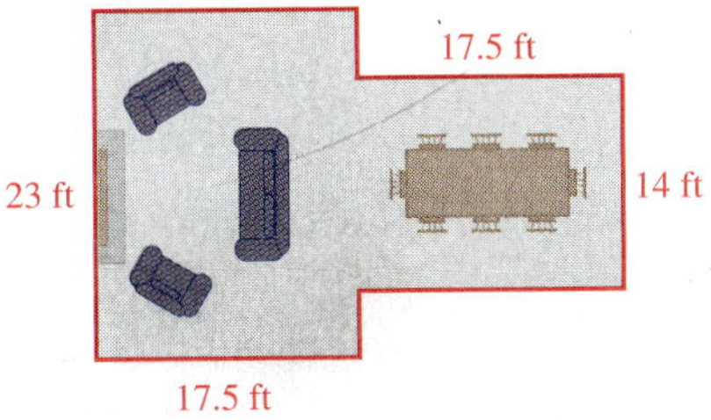

Use a calculator. For π, use the π key. Round to two decimal places.

111. Volume of a tank Find the number of cubic feet of water in a spherical tank with a radius of 21.35 feet.

112. Storing solvents A hazardous solvent fills a rectangular tank with dimensions of 12 inches by 9.5 inches by 7.3 inches. For disposal, it must be transferred to a cylindrical canister 7.5 inches in diameter and 18 inches high. How much solvent will be left over?

113. Volume of a classroom Thirty students are in a classroom with dimensions of 40 feet by 40 feet by 9 feet. How many cubic feet of air are there for each student?

114. Wallpapering One roll of wallpaper covers about 33 square feet. At $27.50 per roll, how much would it cost to paper two walls 8.5 feet high and 17.3 feet long? (*Hint:* Wallpaper can only be purchased in full rolls.)

115. Focal length The focal length f of a double-convex thin lens is given by the formula

$$f = \frac{rs}{(r + s)(n - 1)}$$

If $r = 8$, $s = 12$, and $n = 1.6$, find f.

116. Resistance The total resistance R of two resistors in parallel is given by the formula

$$R = \frac{rs}{r + s}$$

If $r = 170$ and $s = 255$, find R.

WRITING

117. Explain why the symbols $3x$ and x^3 have different meanings.

118. Students often say that x^n means "x multiplied by itself n times." Explain why this is not correct.

SOMETHING TO THINK ABOUT

119. If x were greater than 1, would raising x to higher and higher powers produce bigger numbers or smaller numbers?

120. What would happen in Exercise 119 if x were a positive number that was less than 1?

1.4 Adding and Subtracting Real Numbers

- Adding Real Numbers with Like Signs
- Adding Real Numbers with Unlike Signs
- Subtracting Real Numbers
- Using a Calculator to Add and Subtract Real Numbers

Getting Ready Perform each operation.

1. $14.32 + 3.2$

2. $5.54 - 2.6$

3. $4.2 - (3 - 0.8)$

4. $(5.42 - 4.22) - 0.2$

5. $(437 - 198) - 143$

6. $437 - (198 - 143)$

In this section, we will discuss how to add and subtract real numbers. To develop the rules for adding real numbers, we will use the number line.

Adding Real Numbers with Like Signs

Since the positive direction on the number line is to the right, positive numbers can be represented by arrows pointing to the right. Negative numbers can be represented by arrows pointing to the left.

To add +2 and +3, we can represent +2 with an arrow the length of 2, pointing to the right. We can represent +3 with an arrow of length 3, also pointing to the right. To add the numbers, we place the arrows end to end, as in Figure 1-18. Since the endpoint of the second arrow is the point with coordinate +5, we have

$$(+2) + (+3) = +5$$

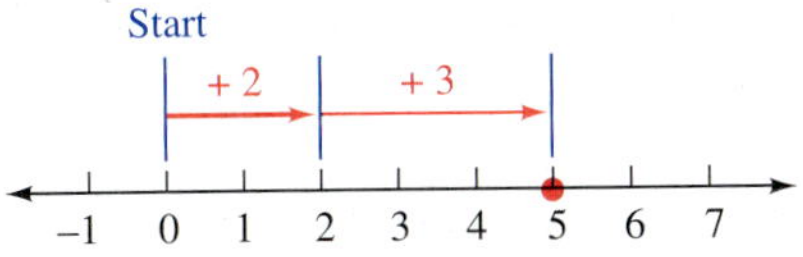

Figure 1-18

As a check, we can think of this problem in terms of money. If you had \$2 and earned \$3 more, you would have a total of \$5.

The addition

$$(-2) + (-3)$$

can be represented by the arrows shown in Figure 1-19. Since the endpoint of the final arrow is the point with coordinate −5, we have

$$(-2) + (-3) = -5$$

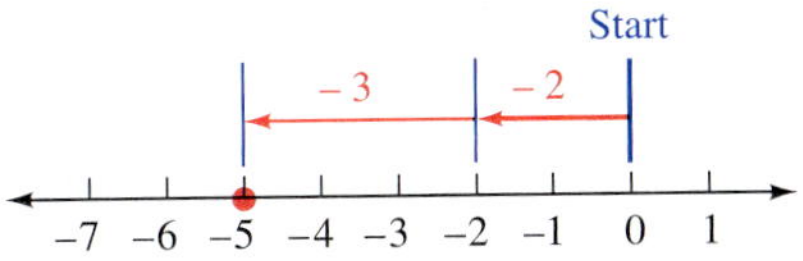

Figure 1-19

As a check, we can think of this problem in terms of money. If you lost \$2 and then lost \$3 more, you would have lost a total of \$5.

Because two real numbers with the same sign can be represented by arrows pointing in the same direction, we have the following rule.

Adding Real Numbers with Like Signs

1. *To add two positive numbers,* add their absolute values and make the answer positive.
2. *To add two negative numbers,* add their absolute values and make the answer negative.

EXAMPLE 1 **Adding real numbers**

a. $(+4) + (+6) = +(4 + 6)$
$\qquad = 10$

b. $(-4) + (-6) = -(4 + 6)$
$\qquad = -10$

c. $+5 + (+10) = +(5 + 10)$
$\qquad = 15$

d. $-\frac{1}{2} + \left(-\frac{3}{2}\right) = -\left(\frac{1}{2} + \frac{3}{2}\right)$
$\qquad = -\frac{4}{2}$
$\qquad = -2$

Self Check Add: **a.** $(+0.5) + (+1.2)$ and **b.** $(-3.7) + (-2.3)$.

Comment We do not need to write a + sign in front of a positive number.

$$+4 = 4 \quad \text{and} \quad +5 = 5$$

However, we must always write a − sign in front of a negative number.

Adding Real Numbers with Unlike Signs

Real numbers with unlike signs can be represented by arrows on a number line pointing in opposite directions. For example, the addition

$$(-6) + (+2)$$

can be represented by the arrows shown in Figure 1-20. Since the endpoint of the final arrow is the point with coordinate −4, we have

$$(-6) + (+2) = -4$$

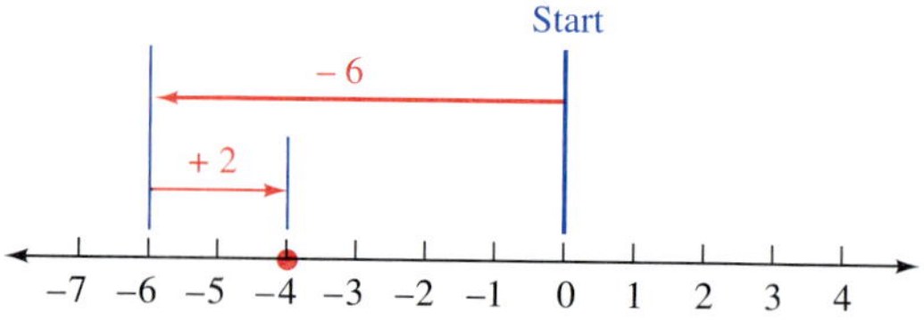

Figure 1-20

As a check, we can think of this problem in terms of money. If you lost \$6 and then earned \$2, you would still have a loss of \$4.

The addition

$$(+7) + (-4)$$

can be represented by the arrows shown in Figure 1-21. Since the endpoint of the final arrow is the point with coordinate +3, we have

$$(+7) + (-4) = +3$$

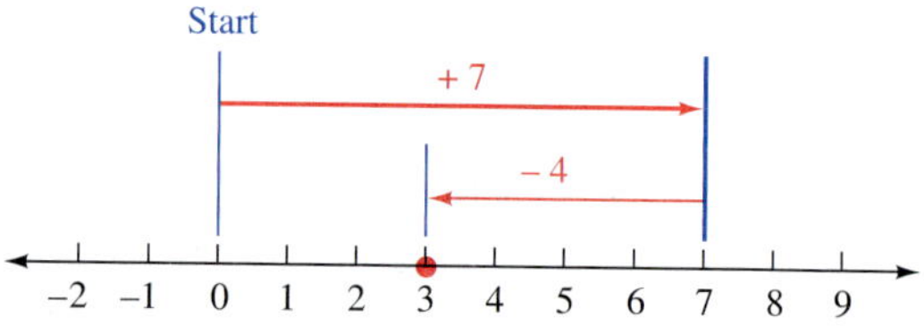

Figure 1-21

As a check, you can think of this problem in terms of money. If you had \$7 and then lost \$4, you would still have a gain of \$3.

Because two real numbers with unlike signs can be represented by arrows pointing in opposite directions, we have the following rule.

Adding Real Numbers with Unlike Signs

To add a positive and a negative number, subtract the smaller absolute value from the larger.

1. If the positive number has the larger absolute value, the answer is positive.
2. If the negative number has the larger absolute value, the answer is negative.

EXAMPLE 2 **Adding real numbers**

a. $(+6) + (-5) = +(6 - 5)$
$= 1$

b. $(-2) + (+3) = +(3 - 2)$
$= 1$

c. $+6 + (-9) = -(9 - 6)$
$= -3$

d. $-\frac{2}{3} + \left(+\frac{1}{2}\right) = -\left(\frac{2}{3} - \frac{1}{2}\right)$
$= -\left(\frac{4}{6} - \frac{3}{6}\right)$
$= -\frac{1}{6}$

Self Check Add: **a.** $(+3.5) + (-2.6)$ and **b.** $(-7.2) + (+4.7)$.

When adding three or more real numbers, we use the rules for the order of operations.

EXAMPLE 3 **Working with grouping symbols**

a. $[(+3) + (-7)] + (-4) = [-4] + (-4)$ Do the work within the brackets first.
$= -8$

b. $-3 + [(-2) + (-8)] = -3 + [-10]$ Do the work within the brackets first.
$= -13$

c. $2.75 + [8.57 + (-4.8)] = 2.75 + 3.77$ Do the work within the brackets first.
$= 6.52$

Self Check Add: $-2 + [(+5.2) + (-12.7)]$.

Sometimes numbers are added vertically, as shown in the next example.

EXAMPLE 4 **Adding numbers in a vertical format**

a. $\begin{array}{r} +5 \\ \underline{+2} \\ +7 \end{array}$ **b.** $\begin{array}{r} +5 \\ \underline{-2} \\ +3 \end{array}$ **c.** $\begin{array}{r} -5 \\ \underline{+2} \\ -3 \end{array}$ **d.** $\begin{array}{r} -5 \\ \underline{-2} \\ -7 \end{array}$

Self Check Add: **a.** $\begin{array}{r} +3.2 \\ \underline{-5.4} \end{array}$ and **b.** $\begin{array}{r} -13.5 \\ \underline{-4.3} \end{array}$

Words such as *found, gain, credit, up, increase, forward, rises, in the future,* and *to the right* indicate a positive direction. Words such as *lost, loss, debit, down, backward, falls, in the past,* and *to the left* indicate a negative direction.

EXAMPLE 5 **Account balance** The treasurer of a math club opens a checking account by depositing \$350 in the bank. The bank debits the account \$9 for check printing, and the treasurer writes a check for \$22. Find the balance after these transactions.

Solution The deposit can be represented by +350. The debit of \$9 can be represented by −9, and the check written for \$22 can be represented by −22. The balance in the account after these transactions is the sum of 350, −9, and −22.

$$\begin{aligned} 350 + (-9) + (-22) &= 341 + (-22) \quad \text{Work from left to right.} \\ &= 319 \end{aligned}$$

The balance is \$319.

Self Check Find the balance if another deposit of \$17 is made.

Subtracting Real Numbers

In arithmetic, subtraction is a take-away process. For example,

$$7 - 4 = 3$$

can be thought of as taking 4 objects away from 7 objects, leaving 3 objects.

For algebra, a better approach treats the subtraction problem

$$7 - 4$$

as the equivalent addition problem:

$$7 + (-4)$$

In either case, the answer is 3.

$$7 - 4 = 3 \quad \text{and} \quad 7 + (-4) = 3$$

Thus, to subtract 4 from 7, we can add the negative (or opposite) of 4 to 7. In general, to subtract one real number from another, we add the negative (or opposite) of the number that is being subtracted.

Subtracting Real Numbers

If a and b are two real numbers, then

$$a - b = a + (-b)$$

EXAMPLE 6 Evaluate: **a.** $12 - 4$, **b.** $-13 - 5$, and **c.** $-14 - (-6)$.

Solution

a. $$\begin{aligned} 12 - 4 &= 12 + (-4) \quad \text{To subtract 4, add the opposite of 4.} \\ &= 8 \end{aligned}$$

b. $$\begin{aligned} -13 - 5 &= -13 + (-5) \quad \text{To subtract 5, add the opposite of 5.} \\ &= -18 \end{aligned}$$

c. $$\begin{aligned} -14 - (-6) &= -14 + [-(-6)] \quad \text{To subtract } -6\text{, add the opposite of } -6. \\ &= -14 + 6 \quad \text{The opposite of } -6 \text{ is } 6. \\ &= -8 \end{aligned}$$

Self Check Evaluate: **a.** $-12.7 - 8.9$ and **b.** $15.7 - (-11.3)$.

To use a vertical format for subtracting real numbers, we add the opposite of the number that is to be subtracted by changing the sign of the lower number (called the **subtrahend**) and proceeding as in addition.

EXAMPLE 7 Perform each subtraction by doing an equivalent addition.

a. The subtraction $\begin{array}{r} 5 \\ -\underline{-4} \end{array}$ becomes the addition $\begin{array}{r} 5 \\ +\underline{+4} \\ 9 \end{array}$

b. The subtraction $\begin{array}{r} -8 \\ -\underline{+3} \end{array}$ becomes the addition $\begin{array}{r} -8 \\ +\underline{-3} \\ -11 \end{array}$

Self Check Do the subtraction: $\begin{array}{r} 5.8 \\ -\underline{-4.6} \end{array}$

When dealing with three or more real numbers, we use the rules for the order of operations.

EXAMPLE 8 Simplify: **a.** $3 - [4 + (-6)]$ and **b.** $[-5 + (-3)] - [-2 - (+5)]$.

Solution **a.**

$$\begin{aligned} 3 - [4 + (-6)] &= 3 - (-2) && \text{Do the addition within the brackets first.} \\ &= 3 + [-(-2)] && \text{To subtract } -2\text{, add the opposite of } -2.\ -(-2) = 2. \\ &= 3 + 2 \\ &= 5 \end{aligned}$$

b.

$$\begin{aligned} &[-5 + (-3)] - [-2 - (+5)] \\ &= [-5 + (-3)] - [-2 + (-5)] && \text{To subtract } +5\text{, add the opposite of } 5. \\ &= -8 - (-7) && \text{Do the work within the brackets.} \\ &= -8 + [-(-7)] && \text{To subtract } -7\text{, add the opposite of } -7. \\ &= -8 + 7 && -(-7) = 7. \\ &= -1 \end{aligned}$$

Self Check Simplify: $[7.2 - (-3)] - [3.2 + (-1.7)]$.

EXAMPLE 9 Evaluate: **a.** $\dfrac{-3 - (-5)}{7 + (-5)}$ and **b.** $\dfrac{6 + (-5)}{-3 - (-5)} - \dfrac{-3 - 4}{7 + (-5)}$.

Solution **a.**

$$\begin{aligned} \frac{-3 - (-5)}{7 + (-5)} &= \frac{-3 + [-(-5)]}{2} && \text{To subtract } -5\text{, add the opposite of } -5.\ 7 + (-5) = 2. \\ &= \frac{-3 + 5}{2} && -(-5) = 5. \\ &= \frac{2}{2} \\ &= 1 \end{aligned}$$

b.

$$\begin{aligned} \frac{6 + (-5)}{-3 - (-5)} - \frac{-3 - 4}{7 + (-5)} &= \frac{1}{-3 + 5} - \frac{-3 + (-4)}{2} && 6 + (-5) = 1;\ -(-5) = +5;\ -3 - 4 = -3 + (-4);\ 7 + (-5) = 2. \\ &= \frac{1}{2} - \frac{-7}{2} && -3 + (-4) = -7;\ -3 + 5 = 2. \end{aligned}$$

$$= \frac{1 - (-7)}{2}$$

$$= \frac{1 + [-(-7)]}{2} \qquad \text{To subtract } -7\text{, add the opposite of } -7.$$

$$= \frac{1 + 7}{2} \qquad -(-7) = 7.$$

$$= \frac{8}{2}$$

$$= 4$$

Self Check Evaluate: $\dfrac{7 - (-3)}{-5 - (-3) + 3}$.

EXAMPLE 10 **Temperature change** At noon, the temperature was 7° above zero. At midnight, the temperature was 4° below zero. Find the difference between these two temperatures.

Solution A temperature of 7° above zero can be represented as +7. A temperature of 4° below zero can be represented as −4. To find the difference between these temperatures, we can set up a subtraction problem and simplify.

$$\begin{aligned} 7 - (-4) &= 7 + [-(-4)] && \text{To subtract } -4\text{, add the opposite of } -4. \\ &= 7 + 4 && -(-4) = 4. \\ &= 11 \end{aligned}$$

The difference between the temperatures is 11°. Figure 1-22 shows this difference.

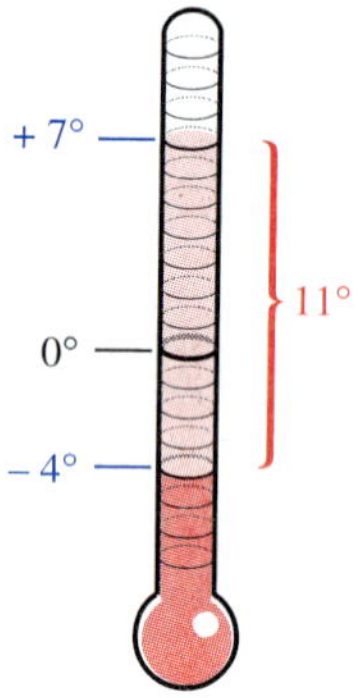

Figure 1-22

Self Check Find the difference between temperatures of 32° and −10°.

Using a Calculator to Add and Subtract Real Numbers

A calculator can add positive and negative numbers.

- You do not have to do anything special to enter positive numbers. When you press 5, for example, a positive 5 is entered.
- To enter −5 into a calculator with a [+/−] key, called the *plus–minus* or *change-of-sign* key, you must press the [+/−] key. To enter −5 into a calculator with a [(−)] key, you must press the [(−)] key.

To evaluate −345.678 + (−527.339), we enter these numbers and press these keys:

345.678 [+/−] [+] 527.339 [+/−] [=] — On a calculator with a [+/−] key.

[(−)] 345.678 [+] [(−)] 527.339 [ENTER] — On a graphing calculator.

The display will read −873.017.

Self Check Answers

1. a. 1.7, **b.** −6 **2. a.** 0.9, **b.** −2.5 **3.** −9.5 **4. a.** −2.2, **b.** −17.8 **5.** $336 **6. a.** −21.6, **b.** 27 **7.** 10.4 **8.** 8.7 **9.** 10 **10.** 42°

Orals *Find each value.*

1. $2 + 3$
2. $2 + (-5)$
3. $-4 + 7$
4. $-5 + (-6)$
5. $6 - 2$
6. $-8 - 4$
7. $-5 - (-7)$
8. $12 - (-4)$
9. $-5 + (3 - 4)$
10. $(-5 + 3) - 4$

1.4 EXERCISES

REVIEW *Simplify each expression.*

1. $5 + 3(7 - 2)$
2. $(5 + 3)(7 - 2)$
3. $5 + 3(7) - 2$
4. $(5 + 3)7 - 2$

VOCABULARY AND CONCEPTS *Fill in the blanks.*

5. Positive and negative numbers can be represented by ______ on the number line.

6. To find the sum of two real numbers with like signs, ____ their absolute values and _____ their common sign.

7. To find the sum of two real numbers with unlike signs, ________ their absolute values and use the sign of the number with the _______ absolute value.

8. $a - b =$ ________

9. To subtract a number, we ____ its ________.

10. The subtraction

$$\begin{array}{r} 35 \\ -45 \\ \hline \end{array}$$

is equivalent to the subtraction 35 _____.

Find each sum.

11. $4 + 8$
12. $(-4) + (-2)$
13. $(-3) + (-7)$
14. $(+4) + 11$
15. $6 + (-4)$
16. $5 + (-3)$
17. $9 + (-11)$
18. $10 + (-13)$
19. $(-0.4) + 0.9$
20. $(-1.2) + (-5.3)$
21. $\frac{1}{5} + \left(+\frac{1}{7}\right)$
22. $\frac{2}{3} + \left(-\frac{1}{4}\right)$
23. $\begin{array}{r} 5 \\ +\,-4 \\ \hline \end{array}$
24. $\begin{array}{r} -20 \\ +\,-17 \\ \hline \end{array}$
25. $\begin{array}{r} -1.3 \\ +\;\;3.5 \\ \hline \end{array}$
26. $\begin{array}{r} 1.3 \\ +\,-2.5 \\ \hline \end{array}$

Evaluate each expression.

27. $5 + [4 + (-2)]$
28. $-6 + [(-3) + 8]$
29. $-2 + (-4 + 5)$
30. $5 + [-4 + (-6)]$
31. $(-3 + 5) + 2$
32. $-7 + [-3 + (-7)]$
33. $-15 + (-4 + 12)$
34. $-27 + [-12 + (-13)]$
35. $[-4 + (-3)] + [2 + (-2)]$
36. $[3 + (-1)] + [-2 + (-3)]$
37. $-4 + (-3 + 2) + (-3)$
38. $5 + [2 + (-5)] + (-2)$
39. $-|8 + (-4)| + 7$
40. $\left|\frac{3}{5} + \left(-\frac{4}{5}\right)\right|$
41. $-5.2 + |-2.5 + (-4)|$
42. $6.8 + |8.6 + (-1.1)|$

Find each difference.

43. $8 - 4$
44. $-8 - 4$
45. $8 - (-4)$
46. $-9 - (-5)$
47. $0 - (-5)$
48. $0 - 75$
49. $\frac{5}{3} - \frac{7}{6}$
50. $-\frac{5}{9} - \frac{5}{3}$
51. $-3\frac{1}{2} - 5\frac{1}{4}$
52. $2\frac{1}{2} - \left(-3\frac{1}{2}\right)$
53. $-6.7 - (-2.5)$
54. $25.3 - 17.5$

55. $\begin{array}{r} 8 \\ -\underline{4} \end{array}$

56. $\begin{array}{r} 8 \\ -\underline{-3} \end{array}$

57. $\begin{array}{r} -10 \\ -\underline{-\ 3} \end{array}$

58. $\begin{array}{r} -13 \\ -\underline{\ \ 5} \end{array}$

Evaluate each quantity.

59. $+3 - [(-4) - 3]$

60. $-5 - [4 - (-2)]$

61. $(5 - 3) + (3 - 5)$

62. $(3 - 5) - [5 - (-3)]$

63. $5 - [4 + (-2) - 5]$

64. $3 - [-(-2) + 5]$

65. $\dfrac{5 - (-4)}{3 - (-6)}$

66. $\dfrac{2 + (-3)}{-3 - (-4)}$

67. $\dfrac{-4 - 2}{-[2 + (-3)]}$

68. $\dfrac{5}{-4 - (-6)} - \dfrac{-4}{8 + (-6)}$

69. $\left(\dfrac{5}{2} - 3\right) - \left(\dfrac{3}{2} - 5\right)$

70. $\left(\dfrac{7}{3} - \dfrac{5}{6}\right) - \left[\dfrac{5}{6} - \left(-\dfrac{7}{3}\right)\right]$

71. $(5.2 - 2.5) - (5.25 - 5)$

72. $\left(3\dfrac{1}{2} - 2\dfrac{1}{2}\right) - \left[5\dfrac{1}{3} - \left(-5\dfrac{2}{3}\right)\right]$

Use a calculator to evaluate each quantity. Round the answers to one decimal place.

73. $(2.34)^3 - 3.47 + (0.72)^2$

74. $3.47 - (0.72)^2 - (2.34)^2$

75. $(2.34)^2 - (3.47)^2 - (0.72)^2$

76. $(0.72)^3 - (2.34)^2 + (3.47)^3$

APPLICATIONS *Use signed numbers to help solve each problem.*

77. **College tuition** A student owes \$575 in tuition. If she is awarded a scholarship that will pay \$400 of the bill, what will she still owe?

78. **Dieting** Scott weighed 212 pounds but lost 24 pounds during a diet. What does Scott weigh now?

79. **Temperatures** The temperature rose 13 degrees in 1 hour and then dropped 4 degrees in the next hour. What signed number represents the net change in temperature?

80. **Mountain climbing** A team of mountaineers climbed 2,347 feet one day and then came down 597 feet to a good spot to make camp. What signed number represents their net change in altitude?

81. **Temperatures** The temperature fell from zero to 14° below one night. By 5:00 P.M. the next day, the temperature had risen 10 degrees. What was the temperature at 5.00 P.M.?

82. **History** In 1897, Joseph Thompson discovered the electron. Fifty-four years later, the first fission reactor was built. Nineteen years before the reactor was erected, James Chadwick discovered the neutron. In what year was the neutron discovered?

83. **History** The Greek mathematician Euclid was alive in 300 B.C. The English mathematician Sir Isaac Newton was alive in A.D. 1700. How many years apart did they live?

84. **Banking** Abdul deposited \$212 in a new checking account, wrote a check for \$173, and deposited another \$312. Find the balance in his account.

85. **Military science** An army retreated 2,300 meters. After regrouping, it moved forward 1,750 meters. The next day it gained another 1,875 meters. What was the army's net gain?

86. **Football** A football player gained and lost the yardage shown in the illustration on six consecutive plays. How many total yards were gained or lost on the six plays?

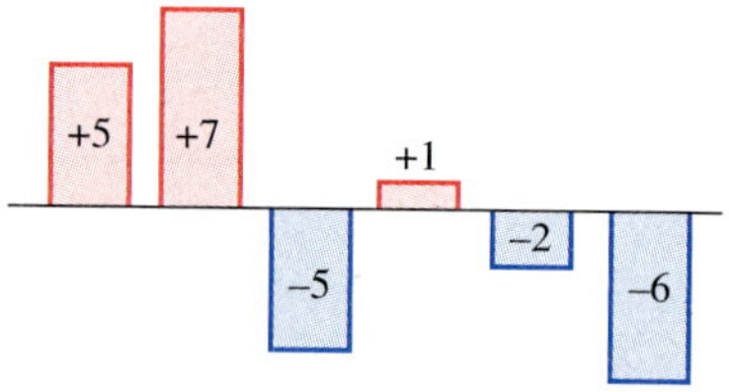

Gains and Losses

87. **Aviation** A pilot flying at 32,000 feet is instructed to descend to 28,000 feet. How many feet must he descend?

88. **Stock market** Tuesday's high and low prices for Transitronics stock were $37\frac{1}{8}$ and $31\frac{5}{8}$. Find the range of prices for this stock.

89. **Temperatures** Find the difference between a temperature of 32° above zero and a temperature of 27° above zero.

90. Temperatures Find the difference between a temperature of 3° below zero and a temperature of 21° below zero.

91. Stock market At the opening bell on Monday, the Dow Jones Industrial Average was 12,153. At the close, the Dow was down 23 points, but news of a half-point drop in interest rates on Tuesday sent the market up 57 points. What was the Dow average after the market closed on Tuesday?

92. Stock market On a Monday morning, the Dow Jones Industrial Average opened at 11,917. For the week, the Dow rose 29 points on Monday and 12 points on Wednesday. However, it fell 53 points on Tuesday and 27 points on both Thursday and Friday. Where did the Dow close on Friday?

93. Stock splits A man owned 500 shares of Transitronics Corporation before the company declared a two-for-one stock split. After the split, he sold 300 shares. How many shares does the man now own?

94. Small business Maria earned $2,532 in a part-time business. However, $633 of the earnings went for taxes. Find Maria's net earnings.

Use a calculator to help solve each problem.

95. Balancing the books On January 1, Sally had $437.45 in the bank. During the month, she had deposits of $25.17, $37.93, and $45.26, and she had withdrawals of $17.13, $83.44, and $22.58. How much was in her account at the end of the month?

96. Small business The owner of a small business has a gross income of $97,345.32. However, he paid $37,675.66 in expenses plus $7,537.45 in taxes, $3,723.41 in health care premiums, and $5,767.99 in pension payments. Find his profit.

97. Closing a real estate transaction A woman sold her house for $115,000. Her fees at closing were $78 for preparing a deed, $446 for title work, $216 for revenue stamps, and a sales commission of $7,612.32. In addition, there was a deduction of $23,445.11 to pay off her old mortgage. As part of the deal, the buyer agreed to pay half of the title work. How much money did the woman receive after closing?

98. Winning the lottery Mike won $500,000 in a state lottery. He will get $\frac{1}{20}$ of the sum each year for the next 20 years. After he receives his first installment, he plans to pay off a car loan of $7,645.12 and give his son $10,000 for college. By paying off the car loan, he will receive a rebate of 2% of the loan. If he must pay income tax of 28% on his first installment, how much will he have left to spend?

WRITING

99. Explain why the sum of two negative numbers is always negative, and the sum of two positive numbers is always positive.

100. Explain why the sum of a negative number and a positive number could be either negative or positive.

SOMETHING TO THINK ABOUT

101. Think of two numbers. First, add the absolute values of the two numbers, and write your answer. Second, add the two numbers, take the absolute value of that sum, and write that answer. Do the two answers agree? Can you find two numbers that produce different answers? When do you get answers that agree, and when don't you?

102. "Think of a very small number," requests the teacher. "One one-millionth," answers Charles. "Negative one million," responds Mia. Explain why either answer might be considered correct.

1.5 Multiplying and Dividing Real Numbers

- Multiplying Real Numbers
- Dividing Real Numbers
- Using a Calculator to Multiply and Divide Real Numbers

Getting Ready *Find each product or quotient.*

1. 8×7 **2.** 9×6 **3.** 8×9 **4.** 7×9

5. $\frac{81}{9}$ **6.** $\frac{48}{8}$ **7.** $\frac{64}{8}$ **8.** $\frac{56}{7}$

In this section, we will develop the rules for multiplying and dividing real numbers. We will see that the rules for multiplication and division are very similar.

Multiplying Real Numbers

Because the times sign, $\times$, looks like the letter x, it is seldom used in algebra. Instead, we will use a dot, parentheses, or no symbol at all to denote multiplication. Each of the following expressions indicates the **product** obtained when two real numbers x and y are multiplied.

$$x \cdot y \qquad (x)(y) \qquad x(y) \qquad (x)y \qquad xy$$

To develop rules for multiplying real numbers, we rely on the definition of multiplication. The expression $5 \cdot 4$ indicates that 4 is to be used as a term in a sum five times.

$$5(4) = 4 + 4 + 4 + 4 + 4 = 20$$ Read 5(4) as "5 times 4."

Likewise, the expression $5(-4)$ indicates that -4 is to be used as a term in a sum five times.

$$5(-4) = (-4) + (-4) + (-4) + (-4) + (-4) = -20$$ Read $5(-4)$ as "5 times negative 4."

If multiplying by a positive number indicates repeated addition, it is reasonable that multiplication by a negative number indicates repeated subtraction. The expression $(-5)4$, for example, means that 4 is to be used as a term in a repeated subtraction five times.

$$\begin{aligned}(-5)4 &= -(4) - (4) - (4) - (4) - (4)\\ &= (-4) + (-4) + (-4) + (-4) + (-4)\\ &= -20\end{aligned}$$

Likewise, the expression $(-5)(-4)$ indicates that -4 is to be used as a term in a repeated subtraction five times.

$$\begin{aligned}(-5)(-4) &= -(-4) - (-4) - (-4) - (-4) - (-4)\\ &= -(-4) + [-(-4)] + [-(-4)] + [-(-4)] + [-(-4)]\\ &= 4 + 4 + 4 + 4 + 4\\ &= 20\end{aligned}$$

The expression $0(-2)$ indicates that -2 is to be used zero times as a term in a repeated addition. Thus,

$$0(-2) = 0$$

Finally, the expression $(-3)(1) = -3$ suggests that the product of any number and 1 is the number itself.

The previous results suggest the following rules.

Rules for Multiplying Signed Numbers

To multiply two real numbers, multiply their absolute values.

1. *If the numbers are positive,* the product is positive.
2. *If the numbers are negative,* the product is positive.
3. *If one number is positive and the other is negative,* the product is negative.
4. Any number multiplied by 0 is 0: $a \cdot 0 = 0 \cdot a = 0$
5. Any number multiplied by 1 is the number itself: $a \cdot 1 = 1 \cdot a = a$

EXAMPLE 1 Find each product: **a.** $4(-7)$, **b.** $(-5)(-4)$, **c.** $(-7)(6)$, **d.** $8(6)$, **e.** $(-3)^2$, **f.** $(-3)^3$, **g.** $(-3)(5)(-4)$, and **h.** $(-4)(-2)(-3)$.

Solution

a. $4(-7) = -(4 \cdot 7)$
$= -28$

b. $(-5)(-4) = +(5 \cdot 4)$
$= +20$

c. $(-7)(6) = -(7 \cdot 6)$
$= -42$

d. $8(6) = +(8 \cdot 6)$
$= +48$

e. $(-3)^2 = (-3)(-3)$
$= 9$

f. $(-3)^3 = (-3)(-3)(-3)$
$= 9(-3)$
$= -27$

g. $(-3)(5)(-4) = (-15)(-4)$
$= +60$

h. $(-4)(-2)(-3) = 8(-3)$
$= -24$

Self Check Find each product: **a.** $-7(5)$, **b.** $-12(-7)$, **c.** $(-5)^2$, and **d.** $-2(-4)(-9)$.

EXAMPLE 2 Evaluate: **a.** $2 + (-3)(4)$, **b.** $-3(2 - 4)$, **c.** $(-2)^2 - 3^2$, and **d.** $-(-2)^2 + 4$.

Solution

a. $2 + (-3)(4) = 2 + (-12)$
$= -10$

b. $-3(2 - 4) = -3[2 + (-4)]$
$= -3(-2)$
$= 6$

c. $(-2)^2 - 3^2 = 4 - 9$
$= -5$

d. $-(-2)^2 + 4 = -4 + 4$
$= 0$

Self Check Evaluate: **a.** $-4 - (-3)(5)$ and **b.** $(-3.2)^2 - 2(-5)^3$.

EXAMPLE 3 Find each product: **a.** $\left(-\frac{2}{3}\right)\left(-\frac{6}{5}\right)$ and **b.** $\left(\frac{3}{10}\right)\left(-\frac{5}{9}\right)$.

Solution a. $\left(-\frac{2}{3}\right)\left(-\frac{6}{5}\right) = +\left(\frac{2}{3} \cdot \frac{6}{5}\right)$

$= +\frac{2 \cdot 6}{3 \cdot 5}$

$= +\frac{12}{15}$

$= +\frac{4}{5}$

b. $\left(\frac{3}{10}\right)\left(-\frac{5}{9}\right) = -\left(\frac{3}{10} \cdot \frac{5}{9}\right)$

$= -\frac{3 \cdot 5}{10 \cdot 9}$

$= -\frac{15}{90}$

$= -\frac{1}{6}$

Self Check Evaluate: **a.** $\frac{3}{5}\left(-\frac{10}{9}\right)$ and **b.** $-\left(\frac{15}{8}\right)\left(-\frac{16}{5}\right)$.

EXAMPLE 4 **Temperature change** If the temperature is dropping 4° each hour, how much warmer was it 3 hours ago?

Solution A temperature drop of 4° per hour can be represented by $-4°$ per hour. "Three hours ago" can be represented by -3. The temperature 3 hours ago is the product of -3 and -4.

$$(-3)(-4) = +12$$

The temperature was 12° warmer 3 hours ago.

Self Check How much colder will it be after 5 hours?

Dividing Real Numbers

We know that 8 divided by 4 is 2 and 18 divided by 6 is 3.

$$\frac{8}{4} = 2, \text{ because } 2 \cdot 4 = 8 \qquad \frac{18}{6} = 3, \text{ because } 3 \cdot 6 = 18$$

These examples suggest that the following rule

$$\frac{a}{b} = c \quad \text{if and only if} \quad c \cdot b = a$$

is true for the division of any real number a by any nonzero real number b. For example,

$$\frac{+10}{+2} = +5, \text{ because } (+5)(+2) = +10$$

$$\frac{-10}{-2} = +5, \text{ because } (+5)(-2) = -10$$

$$\frac{+10}{-2} = -5, \text{ because } (-5)(-2) = +10$$

$$\frac{-10}{+2} = -5, \text{ because } (-5)(+2) = -10$$

Furthermore,

$\frac{-10}{0}$ is undefined, because no number multiplied by 0 gives -10.

$\frac{0}{0}$ is indeterminate, because every number multiplied by 0 gives 0.

$\frac{0}{-10} = 0$, because $0(-10) = 0$.

These examples suggest the rules for dividing real numbers.

Rules for Dividing Signed Numbers

To divide two real numbers, find the quotient of their absolute values.

1. *If the numbers are positive,* the quotient is positive.
2. *If the numbers are negative,* the quotient is positive.
3. *If one number is positive and the other is negative,* the quotient is negative.
4. $\frac{a}{0}$ is undefined; $\frac{0}{0}$ is indeterminate.
5. If $a \neq 0$, then $\frac{0}{a} = 0$.

EXAMPLE 5 Find each quotient: **a.** $\frac{36}{18}$, **b.** $\frac{-44}{11}$, **c.** $\frac{27}{-9}$, and **d.** $\frac{-64}{-8}$.

Solution

a. $\frac{36}{18} = +\frac{36}{18} = 2$ — The quotient of two numbers with like signs is the quotient of their absolute values.

b. $\frac{-44}{11} = -\frac{44}{11} = -4$ — The quotient of two numbers with unlike signs is the negative of the quotient of their absolute values.

c. $\frac{27}{-9} = -\frac{27}{9} = -3$ — The quotient of two numbers with unlike signs is the negative of the quotient of their absolute values.

d. $\frac{-64}{-8} = +\frac{64}{8} = 8$ — The quotient of two numbers with like signs is the quotient of their absolute values.

Self Check Find each quotient: **a.** $\frac{-72.6}{12.1}$ and **b.** $\frac{-24.51}{-4.3}$.

EXAMPLE 6 Evaluate: **a.** $\frac{16(-4)}{-(-64)}$ and **b.** $\frac{(-4)^3(16)}{-64}$.

Solution

a. $\frac{16(-4)}{-(-64)} = \frac{-64}{+64}$

$= -1$

b. $\frac{(-4)^3(16)}{-64} = \frac{(-64)(16)}{(-64)}$

$= 16$

Self Check Evaluate: $\frac{-64 + 16}{-(-4)^2}$.

When dealing with three or more real numbers, we use the rules for the order of operations.

EXAMPLE 7 Evaluate: **a.** $\dfrac{(-50)(10)(-5)}{-50 - 5(-5)}$ and **b.** $\dfrac{3(-50)(10) + 2(10)(-5)}{2(-50 + 10)}$.

Solution **a.**

$$\begin{aligned}\frac{(-50)(10)(-5)}{-50 - 5(-5)} &= \frac{(-500)(-5)}{-50 + 25}\\ &= \frac{2{,}500}{-25}\\ &= -100\end{aligned}$$

b.

$$\begin{aligned}\frac{3(-50)(10) + 2(10)(-5)}{2(-50 + 10)} &= \frac{-150(10) + (20)(-5)}{2(-40)}\\ &= \frac{-1{,}500 - 100}{-80}\\ &= \frac{-1{,}600}{-80}\\ &= 20\end{aligned}$$

Self Check Evaluate: $\dfrac{2(-50)(10) - 3(-5) - 5}{3[10 - (-5)]}$.

EXAMPLE 8 **Stock reports** In its annual report, a corporation reports its performance on a per-share basis. When a company with 35 million shares loses \$2.3 million, what will be the per-share loss?

Solution A loss of \$2.3 million can be represented by −2,300,000. Because there are 35 million shares, the per-share loss can be represented by the quotient $\frac{-2{,}300{,}000}{35{,}000{,}000}$.

$$\frac{-2{,}300{,}000}{35{,}000{,}000} \approx -0.065714286 \qquad \text{Use a calculator.}$$

The company lost about 6.6¢ per share.

Self Check If the company earns \$1.5 million in the following year, find its per-share gain for that year.

Remember these facts about dividing real numbers.

Division

1. $\dfrac{a}{0}$ is undefined. **2.** If $a \neq 0$, then $\dfrac{0}{a} = 0$.

3. $\dfrac{a}{1} = a$. **4.** If $a \neq 0$, then $\dfrac{a}{a} = 1$.

Using a Calculator to Multiply and Divide Real Numbers

A calculator can be used to multiply and divide positive and negative numbers. To evaluate $(-345.678)(-527.339)$, we enter these numbers and press these keys:

345.678 +/− × 527.339 +/− = On a calculator with a +/− key.

(−) 345.678 × (−) 527.339 ENTER On a graphing calculator.

The display will read 182289.4908.

To evaluate $\frac{-345.678}{-527.339}$, we enter these numbers and press these keys:

345.678 +/− ÷ 527.339 +/− = On a calculator with a +/− key.

(−) 345.678 ÷ (−) 527.339 ENTER On a graphing calculator.

The display will read 0.655513815.

Self Check Answers

1. a. -35, **b.** 84, **c.** 25, **d.** -72 **2. a.** 11, **b.** 260.24 **3. a.** $-\frac{2}{3}$, **b.** 6 **4.** 20° colder **5. a.** -6, **b.** 5.7 **6.** 3 **7.** -22 **8.** about 4.3¢

Orals *Find each product or quotient.*

1. $1(-3)$ **2.** $-2(-5)$ **3.** $-3(-6)$ **4.** $4(-6)$

5. $-2(3)(-4)$ **6.** $-2(-3)(-4)$ **7.** $\frac{-12}{6}$

8. $\frac{-10}{-5}$ **9.** $\frac{3(6)}{-2}$ **10.** $\frac{(-2)(-3)}{-6}$

1.5 EXERCISES

REVIEW

1. A concrete block weighs $37\frac{1}{2}$ pounds. How much will 30 of these blocks weigh?
2. If one brick weighs 1.3 pounds, how much will 500 bricks weigh?
3. Evaluate: $5^3 - 8(3)^2$.
4. Put <, =, or > in the box to make a true statement: $-2(-3 + 4)$ ☐ $-3[3 - (-4)]$

VOCABULARY AND CONCEPTS *Fill in the blanks.*

5. The product of two positive numbers is ______.
6. The product of a ______ number and a negative number is negative.
7. The product of two negative numbers is ______.
8. The quotient of a ______ number and a positive number is negative.
9. The quotient of two negative numbers is ______.
10. Any number multiplied by __ is 0.
11. $a \cdot 1 =$ __
12. The symbol $\frac{a}{0}$ is ______.
13. If $a \neq 0$, $\frac{0}{a} =$ __.
14. If $a \neq 0$, $\frac{a}{a} =$ __.

PRACTICE *Perform the operations.*

15. $(+6)(+8)$
16. $(-9)(-7)$
17. $(-8)(-7)$
18. $(9)(-6)$
19. $(+12)(-12)$
20. $(-9)(12)$
21. $\left(\frac{1}{2}\right)(-32)$
22. $\left(-\frac{3}{4}\right)(12)$
23. $\left(-\frac{3}{4}\right)\left(-\frac{8}{3}\right)$
24. $\left(-\frac{2}{5}\right)\left(\frac{15}{2}\right)$
25. $(-3)\left(-\frac{1}{3}\right)$
26. $(5)\left(-\frac{2}{5}\right)$
27. $(3)(-4)(-6)$
28. $(-1)(-3)(-6)$
29. $(-2)(3)(4)$
30. $(5)(0)(-3)$
31. $(2)(-5)(-6)(-7)$
32. $(-3)(-5)(-5)(-2)$
33. $(-2)(-2)(-2)(-3)(-4)$
34. $(-5)(4)(3)(-2)(-1)$
35. $(-2)^2$
36. $(-1)^3$
37. $(-4)^3$
38. $(-6)^2$
39. $(-7)^2$
40. $(-2)^3$
41. $-(-3)^2$
42. $-(-1)(-3)^2$
43. $(-1)(2^3)$
44. $[2(-3)]^2$
45. $2 + (-1)(-3)$
46. $-3 - (-1)(2)$
47. $(-1 + 2)(-3)$
48. $2[-1 - (-3)]$
49. $[-1 - (-3)]\,[-1 + (-3)]$
50. $[2 + (-3)]\,[-1 - (-3)]$
51. $-1(2) + 2(-3)$
52. $-3(-1) - (-3)(2)$
53. $-1(2)(-3) + 6$
54. $(-1)^2(2) + (-3)$
55. $(-1)^2[2 - (-3)]$
56. $2^2[-1 - (-3)]$
57. $(-1)^3(-2)^2 + (-3)^2$
58. $(-2)^3[3 - (-5)]$
59. $\frac{80}{-20}$
60. $\frac{-66}{33}$
61. $\frac{-110}{-55}$
62. $\frac{200}{40}$
63. $\frac{-160}{40}$
64. $\frac{-250}{-25}$
65. $\frac{320}{-16}$
66. $\frac{180}{-36}$
67. $\frac{8 - 12}{-2}$
68. $\frac{16 - 2}{2 - 9}$
69. $\frac{20 - 25}{7 - 12}$
70. $\frac{2(15)^2 - 2}{-2^3 + 1}$
71. $\frac{3(4)}{-2}$
72. $\frac{4(5)}{-2}$
73. $\frac{5(-18)}{3}$
74. $\frac{-18}{-2(3)}$
75. $\frac{4 + (-18)}{-2}$
76. $\frac{-2(3)(4)}{3 - 1}$
77. $\frac{-2(5)(4)}{-3 + 1}$
78. $\frac{-2 + 3 - (-18)}{4(-5) + 1}$
79. $\frac{1}{2} - \frac{2}{3}$
80. $-\frac{2}{3} - \frac{3}{4}$
81. $\frac{1}{2} - \frac{2}{3} - \frac{3}{4}$
82. $-\frac{2}{3} + \frac{1}{2} + \frac{3}{4}$
83. $\left(\frac{1}{2} - \frac{2}{3}\right)\left(\frac{1}{2} + \frac{2}{3}\right)$
84. $\left(\frac{1}{2} + \frac{3}{4}\right)\left(\frac{1}{2} - \frac{3}{4}\right)$

Use a calculator to evaluate each expression.

85. $\frac{(-6)^3 + 4(-3)^3}{(4 - 6)^2}$
86. $\frac{4^2 - 2(4)(-3) + (-3)^2}{4 - (-6) - 3}$
87. $\frac{4(-6)^2(-3) + 4^2(-6)}{2(-6) - 2(-3)}$
88. $\frac{[4^2 - 2(-6)](-3)^2}{-4(-3)}$

APPLICATIONS *Use signed numbers to help solve each problem.*

89. Temperature change If the temperature is increasing 2 degrees each hour for 3 hours, what product of signed numbers represents the temperature change?

90. Temperature change If the temperature is decreasing 2 degrees each hour for 3 hours, what product of signed numbers represents the temperature change?

91. Gambling In Las Vegas, Robert lost \$30 per hour playing the slot machines for 15 hours. What product of signed numbers represents the change in his financial condition?

92. Draining a pool A pool is emptying at the rate of 12 gallons per minute. What product of signed numbers would represent how much more water was in the pool 2 hours ago?

93. Filling a pool Water from a pipe is filling a pool at the rate of 23 gallons per minute. What product of signed numbers represents the amount of water in the pool 2 hours ago?

94. Mowing lawns Rafael worked all day mowing lawns and was paid \$8 per hour. If he had \$94 at the end of an 8-hour day, how much did he have before he started working?

95. Temperature Suppose that the temperature is dropping at the rate of 3 degrees each hour. If the temperature has dropped 18 degrees, what signed number expresses how many hours the temperature has been falling?

96. Dieting A man lost 37.5 pounds. If he lost 2.5 pounds each week, how long has he been dieting?

Use a calculator and signed numbers to help solve each problem.

97. Stock market Over a 7-day period, the Dow Jones Industrial Average had gains of 26, 35, and 17 points. In that period, there were also losses of 25, 31, 12, and 24 points. Find the average daily performance over the 7-day period.

98. Astronomy Light travels at the rate of 186,000 miles per second. How long will it take light to travel from the sun to Venus? (*Hint:* The distance from the sun to Venus is 67,000,000 miles.)

99. Saving for school A student has saved \$15,000 to attend graduate school. If she estimates that her expenses will be \$613.50 a month while in school, does she have enough to complete an 18-month master's degree program?

100. Earnings per share Over a five-year period, a corporation reported profits of \$18 million, \$21 million, and \$33 million. It also reported losses of \$5 million and \$71 million. Find the average gain (or loss) each year.

WRITING

101. Explain how you would decide whether the product of several numbers is positive or negative.

102. Describe two situations in which negative numbers are useful.

SOMETHING TO THINK ABOUT

103. If the quotient of two numbers is undefined, what would their product be?

104. If the product of five numbers is negative, how many of the factors could be negative?

105. If x^5 is a negative number, can you decide whether x is negative too?

106. If x^6 is a positive number, can you decide whether x is positive too?

1.6 Algebraic Expressions

- Algebraic Expressions
- Evaluating Algebraic Expressions
- Algebraic Terms

Getting Ready *Identify each of the following as a sum, difference, product, or quotient.*

1. $x + 3$

2. $57x$

3. $\frac{x}{9}$

4. $19 - y$

5. $\frac{x-7}{3}$

6. $x - \frac{7}{3}$

7. $5(x + 2)$

8. $5x + 10$

A fundamental concept in the study of algebra is the algebraic expression. They convey mathematical operations and are the building blocks of many equations, the main topic of the next chapter.

Algebraic Expressions

Variables and numbers can be combined with the operations of arithmetic to produce **algebraic expressions.** For example, if x and y are variables, the algebraic expression $x + y$ represents the **sum** of x and y, and the algebraic expression $x - y$ represents their **difference.**

There are many other ways to express addition or subtraction with algebraic expressions, as shown in Tables 1-3 and 1-4.

The phrase	translates into the algebraic expression
the *sum* of t and 12	$t + 12$
5 *plus s*	$5 + s$
7 *added to a*	$a + 7$
10 *more than q*	$q + 10$
12 *greater than m*	$m + 12$
l increased by m	$l + m$
exceeds p by 50	$p + 50$

Table 1-3

The phrase	translates into the algebraic expression
the *difference* of 50 and r	$50 - r$
1,000 *minus q*	$1{,}000 - q$
15 *less than w*	$w - 15$
t decreased by q	$t - q$
12 *reduced by m*	$12 - m$
l subtracted from 250	$250 - l$
2,000 *less p*	$2{,}000 - p$

Table 1-4

EXAMPLE 1 Let x represent a certain number. Write an expression that represents **a.** the number that is 5 more than x and **b.** the number 12 decreased by x.

Solution **a.** The number "5 more than x" is the number found by adding 5 to x. It is represented by $x + 5$.

b. The number "12 decreased by x" is the number found by subtracting x from 12. It is represented by $12 - x$.

Self Check Let y represent a certain number. Write an expression that represents y increased by 25.

EXAMPLE 2 **Income taxes** Bob worked x hours preparing his income tax return. He worked 3 hours less than that on his son's return. Write an expression that represents **a.** the number of hours he spent preparing his son's return and **b.** the total number of hours he worked.

Solution **a.** Because he worked x hours on his own return and 3 hours less on his son's return, he worked $(x - 3)$ hours on his son's return.

b. Because he worked x hours on his own return and $(x - 3)$ hours on his son's return, the total time he spent on taxes was $[x + (x - 3)]$ hours.

Self Check Javier deposited $\$d$ in a bank account. Later, he withdrew \$500. Write an expression that represents the difference of d and 500.

There are several ways to indicate the product of two numbers with algebraic expressions, as shown in Table 1-5.

The phrase	translates into the algebraic expression
the *product* of 100 and a	$100a$
25 *times* B	$25B$
twice x	$2x$
$\frac{1}{2}$ *of* z	$\frac{1}{2}z$
12 *multiplied by* m	$12m$

Table 1-5

EXAMPLE 3 Let x represent a certain number. Denote a number that is **a.** twice as large as x, **b.** 5 more than 3 times x, and **c.** 4 less than $\frac{1}{2}$ of x.

Solution

a. The number "twice as large as x" is found by multiplying x by 2. It is represented by $2x$.

b. The number "5 more than 3 times x" is found by adding 5 to the product of 3 and x. It is represented by $3x + 5$.

c. The number "4 less than $\frac{1}{2}$ of x" is found by subtracting 4 from the product of $\frac{1}{2}$ and x. It is represented by $\frac{1}{2}x - 4$.

Self Check Find the product of 40 and t.

EXAMPLE 4 **Stock valuation** Jim owns x shares of Transitronic stock, valued at \$29 a share; y shares of Positone stock, valued at \$32 a share; and 300 shares of Baby Bell, valued at \$42 a share.

a. How many shares of stock does he own?

b. What is the value of his stock?

Solution

a. Because there are x shares of Transitronic, y shares of Positone, and 300 shares of Baby Bell, his total number of shares is $x + y + 300$.

b. The value of x shares of Transitronic is $\$29x$, the value of y shares of Positone is $\$32y$, and the value of 300 shares of Baby Bell is \$42(300). The total value of the stock is $\$(29x + 32y + 12{,}600)$.

Self Check If water softener salt costs $\$p$ per bag, find the cost of 25 bags.

There are also several ways to indicate the quotient of two numbers with algebraic expressions, as shown in Table 1-6.

The phrase	translates into the algebraic expression
the *quotient* of 470 and A	$\frac{470}{A}$
B *divided by* C	$\frac{B}{C}$
the *ratio* of h *to* 5	$\frac{h}{5}$
x *split into* 5 equal parts	$\frac{x}{5}$

Table 1-6

EXAMPLE 5 Let x and y represent two numbers. Write an algebraic expression that represents the sum obtained when 3 times the first number is added to the quotient obtained when the second number is divided by 6.

Solution Three times the first number x is denoted as $3x$. The quotient obtained when the second number y is divided by 6 is the fraction $\frac{y}{6}$. Their sum is expressed as $3x + \frac{y}{6}$.

Self Check If the cost c of a meal is split equally among 4 people, what is each person's share?

EXAMPLE 6 **Cutting rope** A 5-foot section is cut from the end of a rope that is l feet long. If the remaining rope is divided into three equal pieces, find an expression for the length of each of the equal pieces.

Solution After a 5-foot section is cut from one end of l feet of rope, the rope that remains is $(l - 5)$ feet long. When that remaining rope is cut into 3 equal pieces, each piece will be $\frac{l-5}{3}$ feet long. See Figure 1-23.

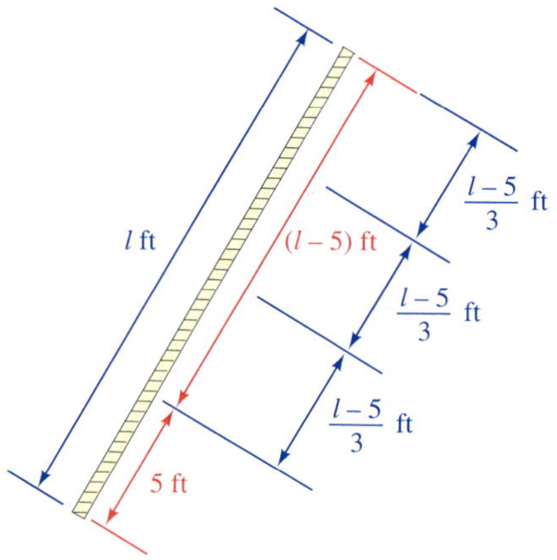

Figure 1-23

Self Check If a 7-foot section is cut from a rope that is l feet long and the remaining rope is divided into two equal pieces, find an expression for the length of each piece.

Evaluating Algebraic Expressions

Since variables represent numbers, algebraic expressions also represent numbers. We can evaluate algebraic expressions when we know the values of the variables.

EXAMPLE 7 If $x = 8$ and $y = 10$, evaluate **a.** $x + y$, **b.** $y - x$, **c.** $3xy$, and **d.** $\frac{5x}{y - 5}$.

Solution We substitute 8 for x and 10 for y in each expression and simplify.

a. $x + y = 8 + 10$
$= 18$

b. $y - x = 10 - 8$
$= 2$

c. $3xy = (3)(8)(10)$
$= (24)(10)$ — Do the multiplications from left to right.
$= 240$

d. $\frac{5x}{y - 5} = \frac{5 \cdot 8}{10 - 5}$
$= \frac{40}{5}$ — Simplify the numerator and the denominator separately.
$= 8$ — Simplify the fraction.

Comment After substituting numbers for variables in a product, it is often necessary to insert a dot or parentheses to show multiplication. Otherwise (3)(8)(10) might be mistaken for 3,810, and $5 \cdot 8$ might be mistaken for 58.

Self Check If $a = -2$ and $b = 5$, evaluate $\frac{6b + 2}{a + 2b}$.

EXAMPLE 8 If $x = -4$, $y = 8$, and $z = -6$, evaluate **a.** $\frac{7x^2y}{2(y - z)}$ and **b.** $\frac{3xz^2}{y(x + z)}$.

Solution We substitute -4 for x, 8 for y, and -6 for z in each expression and simplify.

a. $\frac{7x^2y}{2(y - z)} = \frac{7(-4)^2(8)}{2[8 - (-6)]}$
$= \frac{7(16)(8)}{2(14)}$ — $(-4)^2 = 16$; $8 - (-6) = 14$.
$= 32$ — Simplify.

b. $\frac{3xz^2}{y(x + z)} = \frac{3(-4)(-6)^2}{8[-4 + (-6)]}$
$= \frac{3(-4)(36)}{8(-10)}$ — $(-6)^2 = 36$; $-4 + (-6) = -10$.
$= \frac{3\overset{1}{\cancel{(-2)}}\overset{1}{\cancel{(2)}}\overset{1}{\cancel{(4)}}(9)}{\underset{1}{\cancel{2}}\underset{1}{\cancel{(4)}}\underset{1}{\cancel{(-2)}}(5)}$ — Factor the numerator and denominator and divide out all common factors.
$= \frac{27}{5}$ — $3(9) = 27$; $1(5) = 5$.

Self Check If $a = -3$, $b = -2$, and $c = -5$, evaluate $\frac{b(a + c^2)}{abc}$.

Algebraic Terms

Numbers without variables, such as 7, 21, and 23, are called **constants.** Expressions such as 37, xyz, and $32t$, which are constants, variables, or products of constants and variables, are called **algebraic terms.**

- The expression $3x + 5y$ contains two terms. The first term is $3x$, and the second term is $5y$.
- The expression $xy + (-7)$ contains two terms. The first term is xy, and the second term is -7.
- The expression $3 + x + 2y$ contains three terms. The first term is 3, the second term is x, and the third term is $2y$.

Numbers and variables that are part of a product are called **factors.** For example,

- The product $7x$ has two factors, which are 7 and x.
- The product $-3xy$ has three factors, which are -3, x, and y.
- The product $\frac{1}{2}abc$ has four factors, which are $\frac{1}{2}$, a, b, and c.

The number factor of a product is called its **numerical coefficient.** The numerical coefficient (or just the *coefficient*) of $7x$ is 7. The coefficient of $-3xy$ is -3, and the coefficient of $\frac{1}{2}abc$ is $\frac{1}{2}$. The coefficient of terms such as x, ab, and rst is understood to be 1.

$$x = 1x, \qquad ab = 1ab, \qquad \text{and} \qquad rst = 1rst$$

EXAMPLE 9

a. The expression $5x + y$ has two terms. The coefficient of its first term is 5. The coefficient of its second term is 1.

b. The expression $-17wxyz$ has one term, which contains the five factors -17, w, x, y, and z. Its coefficient is -17.

c. The expression 37 has one term, the constant 37. Its coefficient is 37.

d. The expression $3x^2 - 2x$ has two terms. The coefficient of the first term is 3. Since $3x^2 - 2x$ can be written as $3x^2 + (-2x)$, the coefficient of the second term is -2.

Self Check How many terms does the expression $3x^2 - 2x + 7$ have? Find the sum of the coefficients.

Self Check Answers

1. $y + 25$ **2.** $d - 500$ **3.** $40t$ **4.** $\$25p$ **5.** $\frac{c}{4}$ **6.** $\frac{l - 7}{2}$ ft **7.** 4 **8.** $\frac{22}{15}$ **9.** 3, 8

Orals *If $x = -2$ and $y = 3$, find the value of each expression.*

1. $x + y$ **2.** $7x$ **3.** $7x + y$ **4.** $7(x + y)$

5. $4x^2$ **6.** $(4x)^2$ **7.** $-3x^2$ **8.** $(-3x)^2$

1.6 EXERCISES

REVIEW *Evaluate each expression.*

1. $0.14 \cdot 3{,}800$

2. $\frac{3}{5}$ of 4,765

3. $\frac{-4 + (7 - 9)}{(-9 - 7) + 4}$

4. $\frac{5}{4}\left(1 - \frac{3}{5}\right)$

VOCABULARY AND CONCEPTS *Fill in the blanks.*

5. The answer to an addition problem is called a ____.

6. The answer to a ________ problem is called a difference.

7. The answer to a ________ problem is called a product.

8. The answer to a division problem is called a ______.

9. An ______ expression is a combination of variables, numbers, and the operation symbols for addition, subtraction, multiplication, or division.

10. To ______ an algebraic expression, we substitute values for the variables and simplify.

11. Letters that stand for numbers are called ______.

12. Terms that have no variables are called ______.

PRACTICE *Let x, y, and z represent three real numbers. Write an algebraic expression to denote each quantity.*

13. The sum of x and y

14. The product of x and y

15. The product of x and twice y

16. The sum of twice x and twice y

17. The difference obtained when x is subtracted from y

18. The difference obtained when twice x is subtracted from y

19. The quotient obtained when y is divided by x

20. The quotient obtained when the sum of x and y is divided by z

21. The sum obtained when the quotient of x divided by y is added to z

22. y decreased by x

23. z less the product of x and y

24. z less than the product of x and y

25. The product of 3, x, and y

26. The quotient obtained when the product of 3 and z is divided by the product of 4 and x

27. The quotient obtained when the sum of x and y is divided by the sum of y and z

28. The quotient obtained when the product of x and y is divided by the sum of x and z

29. The sum of the product xy and the quotient obtained when y is divided by z

30. The number obtained when x decreased by 4 is divided by the product of 3 and y

Write each algebraic expression as an English phrase.

31. $x + 3$

32. $y - 2$

33. $\frac{x}{y}$

34. xz

35. $2xy$

36. $\frac{x + y}{2}$

37. $\frac{5}{x + y}$

38. $\frac{3x}{y + z}$

39. $\frac{3 + x}{y}$

40. $3 + \frac{x}{y}$

41. $xy(x + y)$

42. $(x + y + z)(xyz)$

Let $x = 8$, $y = 4$, and $z = 2$. Write each phrase as an algebraic expression, and evaluate it.

43. The sum of x and z

44. The product of x, y, and z

45. z less than y

46. The quotient obtained when y is divided by z

47. 3 less than the product of y and z

48. 7 less than the sum of x and y

49. The quotient obtained when the product of x and y is divided by z

50. The quotient obtained when 10 greater than x is divided by z

If $x = -2$, $y = 5$, and $z = -3$, evaluate each expression.

51. $x + y$

52. $x - z$

53. xyz

54. x^2z

55. $\dfrac{yz}{x}$

56. $\dfrac{xy - 2}{z}$

57. $\dfrac{3(x + z)}{y}$

58. $\dfrac{x + y + z}{y^2}$

59. $\dfrac{3(x^2 + z^2) + 4}{y(x - z)}$

60. $\dfrac{x(y^2 - 2z) - 1}{z(y - x^2)}$

Give the number of terms in each algebraic expression and also give the numerical coefficient of the first term.

61. $6d$

62. $-4c + 3d$

63. $-xy - 4t + 35$

64. xy

65. $3ab + bc - cd - ef$

66. $-2xyz + cde - 14$

67. $-4xyz + 7xy - z$

68. $5uvw - 4uv + 8uw$

69. $3x + 4y + 2z + 2$

70. $7abc - 9ab + 2bc + a - 1$

Consider the algebraic expression $29xyz + 23xy + 19x$.

71. What are the factors of the third term?

72. What are the factors of the second term?

73. What are the factors of the first term?

74. What factor is common to all three terms?

Consider the algebraic expression $3xyz + 5xy + 17xz$.

75. What are the factors of the first term?

76. What are the factors of the second term?

77. What are the factors of the third term?

78. What factor is common to all three terms?

Consider the algebraic expression $5xy + yt + 8xyt$.

79. Find the numerical coefficients of each term.

80. What factor is common to all three terms?

81. What factors are common to the first and third terms?

82. What factors are common to the second and third terms?

Consider the algebraic expression $3xy + y + 25xyz$.

83. Find the numerical coefficient of each term and find their product.

84. Find the numerical coefficient of each term and find their sum.

85. What factors are common to the first and third terms?

86. What factor is common to all three terms?

APPLICATIONS

87. Course loads A man enrolls in college for c hours of credit, and his sister enrolls for 4 more hours than her brother. Write an expression that represents the number of hours the sister is taking.

88. Antique cars An antique Ford has 25,000 more miles on its odometer than a newer car. If the newer car has traveled m miles, find an expression that represents the mileage on the Ford.

89. T-bills Write an expression that represents the value of t T-bills, each worth \$9,987.

90. Real estate Write an expression that represents the value of a vacant lots if each lot is worth \$35,000.

91. Cutting rope A rope x feet long is cut into 5 equal pieces. Find an expression for the length of each piece.

92. Plumbing A plumber cuts a pipe that is 12 feet long into x equal pieces. Find an expression for the length of each piece.

93. Comparing assets A girl had d dollars, and her brother had \$5 more than three times that amount. How much did the brother have?

94. Comparing investments Wendy has x shares of stock. Her sister has 2 fewer shares than twice Wendy's shares. How many shares does her sister have?

95. Sorting records In electronic data processing, the process of sorting records into sequential order is a common task. One sorting technique, called a **selection sort,** requires C comparisons to sort N records, where C and N are related by the formula

$$C = \frac{N(N - 1)}{2}$$

How many comparisons are necessary to sort 10,000 records?

96. Sorting records How many comparisons are necessary to sort 50,000 records? See Exercise 95.

100. In words, xy is "the product of x and y." However, $\frac{x}{y}$ is "the quotient obtained when x is divided by y." Explain why the extra words are needed.

WRITING

97. Distinguish between the meanings of these two phrases: "3 less than x" and "3 is less than x."

98. Distinguish between *factor* and *term*.

99. What is the purpose of using variables? Why aren't ordinary numbers enough?

SOMETHING TO THINK ABOUT

101. If the value of x were doubled, what would happen to the value of $37x$?

102. If the values of both x and y were doubled, what would happen to the value of $5xy^2$?

1.7 Properties of Real Numbers

- The Closure Properties
- The Commutative Properties
- The Associative Properties
- The Distributive Property
- The Identity Elements
- Inverses for Addition and Multiplication

Getting Ready *Perform the operations.*

1. $3 + (5 + 9)$

2. $(3 + 5) + 9$

3. $23.7 + 14.9$

4. $14.9 + 23.7$

5. $7(5 + 3)$

6. $7 \cdot 5 + 7 \cdot 3$

7. $125.3 + (-125.3)$

8. $125.3\left(\frac{1}{125.3}\right)$

9. $777 + 0$

10. $777 \cdot 1$

Basic to understanding algebra is knowing the properties that govern the operations of addition, subtraction, multiplication, and division of real numbers. These properties enable us to write expressions in equivalent forms, often making our work easier.

The Closure Properties

The **closure properties** guarantee that the sum, difference, product, or quotient (except for division by zero) of any two real numbers is also a real number.

Closure Properties

If a and b are real numbers, then

$a + b$ is a real number. $a - b$ is a real number.

ab is a real number. $\frac{a}{b}$ is a real number ($b \neq 0$).

EXAMPLE 1 Let $x = 8$ and $y = -4$. Find the real-number answers to show that **a.** $x + y$, **b.** $x - y$, **c.** xy, and **d.** $\frac{x}{y}$ all represent real numbers.

Solution We substitute 8 for x and -4 for y in each expression and simplify.

a. $x + y = 8 + (-4)$
$= 4$

b. $x - y = 8 - (-4)$
$= 8 + 4$
$= 12$

c. $xy = 8(-4)$
$= -32$

d. $\frac{x}{y} = \frac{8}{-4}$
$= -2$

Self Check Assume that $a = -6$ and $b = 3$. Find the real-number answers to show that **a.** $a - b$ and **b.** $\frac{a}{b}$ are real numbers.

The Commutative Properties

The **commutative properties** (from the word *commute,* which means to go back and forth) guarantee that addition or multiplication of two real numbers can be done in either order.

Commutative Properties

If a and b are real numbers, then

$a + b = b + a$ commutative property of addition
$ab = ba$ commutative property of multiplication

EXAMPLE 2 Let $x = -3$ and $y = 7$. Show that **a.** $x + y = y + x$ and **b.** $xy = yx$.

Solution **a.** We can show that the sum $x + y$ is the same as the sum $y + x$ by substituting -3 for x and 7 for y in each expression and simplifying.

$x + y = -3 + 7 = 4$ and $y + x = 7 + (-3) = 4$

b. We can show that the product xy is the same as the product yx by substituting -3 for x and 7 for y in each expression and simplifying.

$xy = -3(7) = -21$ and $yx = 7(-3) = -21$

Self Check Let $a = 6$ and $b = -5$. Show that **a.** $a + b = b + a$ and **b.** $ab = ba$.

The Associative Properties

The **associative properties** guarantee that three real numbers can be regrouped in an addition or multiplication.

Associative Properties

If a, b, and c are real numbers, then

$(a + b) + c = a + (b + c)$ associative property of addition
$(ab)c = a(bc)$ associative property of multiplication

Because of the associative property of addition, we can group (or *associate*) the numbers in a sum in any way that we wish. For example,

$$(3 + 4) + 5 = 7 + 5 \qquad 3 + (4 + 5) = 3 + 9$$
$$= 12 \qquad\qquad = 12$$

The answer is 12 regardless of how we group the three numbers.

The associative property of multiplication permits us to group (or *associate*) the numbers in a product in any way that we wish. For example,

$$(3 \cdot 4) \cdot 7 = 12 \cdot 7 \qquad 3 \cdot (4 \cdot 7) = 3 \cdot 28$$
$$= 84 \qquad\qquad = 84$$

The answer is 84 regardless of how we group the three numbers.

The Distributive Property

The **distributive property** shows how to multiply the sum of two numbers by a third number. Because of this property, we can often add first and then multiply, or multiply first and then add.

For example, $2(3 + 7)$ can be calculated in two different ways. We can add and then multiply, or we can multiply each number within the parentheses by 2 and then add.

$$2(3 + 7) = 2(10) \qquad 2(3 + 7) = 2 \cdot 3 + 2 \cdot 7$$
$$= 20 \qquad\qquad = 6 + 14$$
$$= 20$$

Either way, the result is 20.

In general, we have the following property.

Distributive Property

If a, b, and c are real numbers, then

$$a(b + c) = ab + ac$$

Because multiplication is commutative, the distributive property can also be written in the form

$$(b + c)a = ba + ca$$

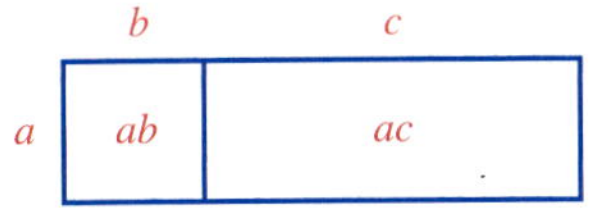

Figure 1-24

We can interpret the distributive property geometrically. Since the area of the largest rectangle in Figure 1-24 is the product of its width a and its length $b + c$, its area is $a(b + c)$. The areas of the two smaller rectangles are ab and ac. Since the area of the largest rectangle is equal to the sum of the areas of the smaller rectangles, we have $a(b + c) = ab + ac$.

EXAMPLE 3 Evaluate each expression in two different ways: **a.** $3(5 + 9)$ and **b.** $-2(-7 + 3)$.

Solution **a.** $3(5 + 9) = 3(14)$
$= 42$

$3(5 + 9) = 3 \cdot 5 + 3 \cdot 9$
$= 15 + 27$
$= 42$

b. $-2(-7 + 3) = -2(-4)$
$= 8$

$-2(-7 + 3) = -2(-7) + (-2)3$
$= 14 + (-6)$
$= 8$

Self Check Evaluate $-5(-7 + 20)$ in two different ways.

The distributive property can be extended to three or more terms. For example, if a, b, c, and d are real numbers, then

$$a(b + c + d) = ab + ac + ad$$

EXAMPLE 4 Write $3.2(x + y + 2.7)$ without using parentheses.

Solution

$$3.2(x + y + 2.7) = 3.2x + 3.2y + (3.2)(2.7) \quad \text{Distribute the multiplication by 3.2.}$$
$$= 3.2x + 3.2y + 8.64$$

Self Check Write $-6.3(a + 2b + 3.7)$ without using parentheses.

The Identity Elements

The numbers 0 and 1 play special roles in arithmetic. The number 0 is the only number that can be added to another number (say, a) and give an answer that is the same number a:

$$0 + a = a + 0 = a$$

The number 1 is the only number that can be multiplied by another number (say, a) and give an answer that is the same number a:

$$1 \cdot a = a \cdot 1 = a$$

Because adding 0 to a number or multiplying a number by 1 leaves that number the same (identical), the numbers 0 and 1 are called **identity elements.**

Identity Elements

0 is the **identity element for addition.**

1 is the **identity element for multiplication.**

Inverses for Addition and Multiplication

If the sum of two numbers is 0, the numbers are called **negatives,** or **additive inverses,** of each other. Since $3 + (-3) = 0$, the numbers 3 and -3 are negatives or additive inverses of each other. In general, because

$$a + (-a) = 0$$

the numbers represented by a and $-a$ are negatives or additive inverses of each other.

If the product of two numbers is 1, the numbers are called **reciprocals,** or **multiplicative inverses,** of each other. Since $7\left(\frac{1}{7}\right) = 1$, the numbers 7 and $\frac{1}{7}$ are reciprocals. Since $(-0.25)(-4) = 1$, the numbers -0.25 and -4 are reciprocals. In general, because

$$a\left(\frac{1}{a}\right) = 1 \qquad \text{provided } a \neq 0$$

the numbers represented by a and $\frac{1}{a}$ are reciprocals (or multiplicative inverses) of each other.

Additive and Multiplicative Inverses

Because $a + (-a) = 0$, the numbers a and $-a$ are called **negatives** or **additive inverses.**

Because $a\left(\frac{1}{a}\right) = 1$ $(a \neq 0)$, the numbers a and $\frac{1}{a}$ are called **reciprocals** or **multiplicative inverses.**

EXAMPLE 5 The property in the right column justifies the statement in the left column.

$3 + 4$ is a real number	closure property of addition
$\frac{8}{3}$ is a real number	closure property of division
$3 + 4 = 4 + 3$	commutative property of addition
$-3 + (2 + 7) = (-3 + 2) + 7$	associative property of addition
$(5)(-4) = (-4)(5)$	commutative property of multiplication
$(ab)c = a(bc)$	associative property of multiplication
$3(a + 2) = 3a + 3 \cdot 2$	distributive property
$3 + 0 = 3$	additive identity property
$3(1) = 3$	multiplicative identity property
$2 + (-2) = 0$	additive inverse property
$\left(\frac{2}{3}\right)\left(\frac{3}{2}\right) = 1$	multiplicative inverse property

Self Check Which property justifies each statement? **a.** $a + 7 = 7 + a$ **b.** $3(y + 2) = 3y + 3 \cdot 2$ **c.** $3 \cdot (2 \cdot p) = (3 \cdot 2) \cdot p$

The properties of the real numbers are summarized as follows.

Properties of Real Numbers

For all real numbers a, b, and c,

Closure properties	$a + b$ is a real number. $a - b$ is a real number.	$a \cdot b$ is a real number. $a \div b$ is a real number $(b \neq 0)$.
	Addition	***Multiplication***
Commutative properties	$a + b = b + a$	$a \cdot b = b \cdot a$
Associative properties	$(a + b) + c = a + (b + c)$	$(ab)c = a(bc)$
Identity properties	$a + 0 = a$	$a \cdot 1 = a$
Inverse properties	$a + (-a) = 0$	$a\left(\frac{1}{a}\right) = 1 \quad (a \neq 0)$
Distributive property	$a(b + c) = ab + ac$	

Self Check Answers

1. a. -9, **b.** -2 **2. a.** $a + b = 1$ and $b + a = 1$, **b.** $ab = -30$ and $ba = -30$ **3.** -65 **4.** $-6.3a - 12.6b - 23.31$ **5. a.** commutative property of addition, **b.** distributive property, **c.** associative property of multiplication

Orals *Give an example of each property.*

1. The associative property of multiplication
2. The additive identity property
3. The distributive property
4. The inverse property for multiplication

Provide an example to illustrate each statement.

5. Subtraction is not commutative.
6. Division is not associative.

1.7 EXERCISES

REVIEW

1. Write as a mathematical expression: The sum of x and the square of y is greater than or equal to z.
2. Write as an English phrase: $3(x + z)$.

Fill each box with an appropriate symbol.

3. For any number x, $|x| \geq$ ☐.
4. $x - y = x + ($ ☐ $)$

Fill in the blanks.

5. The product of two negative numbers is a ______ number.
6. The sum of two negative numbers is a ______ number.

VOCABULARY AND CONCEPTS *Fill in the blanks.*

7. If a and b are real numbers, $a + b$ is a ____ number.
8. If a and b are real numbers, $\frac{a}{b}$ is a real number, provided that ______.
9. $a + b = b +$ __
10. $a \cdot b =$ __ $\cdot\, a$
11. $(a + b) + c = a +$ ______
12. $(ab)c =$ __ $\cdot\, (bc)$
13. $a(b + c) = ab +$ __
14. $0 + a =$ __
15. $a \cdot 1 =$ __
16. 0 is the ______ element for ______.
17. 1 is the identity ______ for ______.
18. If $a + (-a) = 0$, then a and $-a$ are called ______ inverses.
19. If $a\left(\frac{1}{a}\right) = 1$, then __ and __ are called reciprocals.
20. $a(b + c + d) = ab +$ ______

PRACTICE *Assume that $x = 12$ and $y = -2$. Show that each expression represents a real number by finding the real-number answer.*

21. $x + y$
22. $y - x$
23. xy
24. $\frac{x}{y}$
25. x^2
26. y^2
27. $\frac{x}{y^2}$
28. $\frac{2x}{3y}$

Let $x = 5$ and $y = 7$. Show that the two expressions have the same value.

29. $x + y$; $y + x$
30. xy; yx
31. $3x + 2y$; $2y + 3x$
32. $3xy$; $3yx$
33. $x(x + y)$; $(x + y)x$
34. $xy + y^2$; $y^2 + xy$

Let $x = 2$, $y = -3$, and $z = 1$. Show that the two expressions have the same value.

35. $(x + y) + z; x + (y + z)$

36. $(xy)z; x(yz)$

37. $(xz)y; x(yz)$

38. $(x + y) + z; y + (x + z)$

39. $x^2(yz^2); (x^2y)z^2$

40. $x(y^2z^3); (xy^2)z^3$

Use the distributive property to write each expression without parentheses. Simplify each result, if possible.

41. $3(x + y)$

42. $4(a + b)$

43. $x(x + 3)$

44. $y(y + z)$

45. $-x(a + b)$

46. $a(x + y)$

47. $4(x^2 + x)$

48. $-2(a^2 + 3)$

49. $-5(t + 2)$

50. $2x(a - x)$

51. $-2a(x + a)$

52. $-p(p - q)$

Give the additive and the multiplicative inverse of each number, if possible.

53. 2

54. 3

55. $\frac{1}{3}$

56. $-\frac{1}{2}$

57. 0

58. -2

59. $-\frac{5}{2}$

60. 0.5

61. -0.2

62. 0.75

63. $\frac{4}{3}$

64. -1.25

State which property of real numbers justifies each statement.

65. $3 + x = x + 3$

66. $(3 + x) + y = 3 + (x + y)$

67. $xy = yx$

68. $(3)(2) = (2)(3)$

69. $-2(x + 3) = -2x + (-2)(3)$

70. $x(y + z) = (y + z)x$

71. $(x + y) + z = z + (x + y)$

72. $3(x + y) = 3x + 3y$

73. $5 \cdot 1 = 5$

74. $x + 0 = x$

75. $3 + (-3) = 0$

76. $9 \cdot \frac{1}{9} = 1$

Use the given property to rewrite the expression in a different form.

77. $3(x + 2)$; distributive property

78. $x + y$; commutative property of addition

79. y^2x; commutative property of multiplication

80. $x + (y + z)$; associative property of addition

81. $(x + y)z$; commutative property of addition

82. $x(y + z)$; distributive property

83. $(xy)z$; associative property of multiplication

84. $1x$; multiplicative identity property

85. $0 + x$; additive identity property

86. $5 \cdot \frac{1}{5}$; multiplicative inverse property

WRITING

87. Explain why division is not commutative.

88. Describe two ways of calculating the value of $3(12 + 7)$.

SOMETHING TO THINK ABOUT

89. Suppose there were no other numbers than the odd integers.

- Would the closure property for addition still be true?
- Would the closure property for multiplication still be true?
- Would there still be an identity for addition?
- Would there still be an identity for multiplication?

90. Suppose there were no other numbers than the even integers. Answer the four parts of Exercise 89 again.

Projects

Project 1

The circumference of any circle and its diameter are related. When you divide the circumference by the diameter, the quotient is always the same number, **pi,** denoted by the Greek letter π.

- Carefully measure the circumference of several circles—a quarter, a dinner plate, a bicycle tire—whatever you can find that is round. Then calculate approximations of π by dividing each circle's circumference by its diameter.
- Use the π key on the calculator to obtain a more accurate value of π. How close were your approximations?

Project 2

a. The fraction $\frac{22}{7}$ is often used as an approximation of π. To how many decimal places is this approximation accurate?

b. Experiment with your calculator and try to do better. Find another fraction (with no more than three digits in either its numerator or its denominator) that is closer to π. Who in your class has done best?

Project 3

Write an essay answering this question.

When three professors attending a convention in Las Vegas registered at the hotel, they were told that the room rate was \$120. Each professor paid his \$40 share.

Later the desk clerk realized that the cost of the room should have been \$115. To fix the mistake, she sent a bellhop to the room to refund the \$5 overcharge. Realizing that \$5 could not be evenly divided among the three professors, the bellhop refunded only \$3 and kept the other \$2.

Since each professor received a \$1 refund, each paid \$39 for the room, and the bellhop kept \$2. This gives \$39 + \$39 + \$39 + \$2, or \$119. What happened to the other \$1?

Chapter Summary

CONCEPTS	REVIEW EXERCISES
1.1	**Real Numbers and Their Graphs**
Natural numbers: $\{1, 2, 3, 4, 5, \cdots\}$ **Whole numbers:** $\{0, 1, 2, 3, 4, 5, \cdots\}$ **Integers:** $\{\cdots, -3, -2, -1, 0, 1, 2, 3, \cdots\}$ **Rational numbers:** {Fractions with integer numerators and nonzero integer denominators} **Real numbers:** {Rational numbers or irrational numbers} **Prime numbers:** $\{2, 3, 5, 7, 11, 13, 17, \cdots\}$	Consider the set $\{0, 1, 2, 3, 4, 5\}$. **1.** Which numbers are natural numbers? **2.** Which numbers are prime numbers? **3.** Which numbers are odd natural numbers? **4.** Which numbers are composite numbers? Consider the set $\left\{-6, -\frac{2}{3}, 0, \sqrt{2}, 2.6, \pi, 5\right\}$. **5.** Which numbers are integers? **6.** Which numbers are rational numbers? **7.** Which numbers are prime numbers? **8.** Which numbers are real numbers? **9.** Which numbers are even integers?

Composite numbers:
$\{4, 6, 8, 9, 10, 12, 14, 15, \cdots\}$

Even integers:
$\{\cdots, -6, -4, -2, 0, 2, 4, 6, \cdots\}$

Odd integers:
$\{\cdots, -5, -3, -1, 1, 3, 5, \cdots\}$

10. Which numbers are odd integers?

11. Which numbers are not rational?

Place one of the symbols $=$, $<$, or $>$ in each box to make a true statement.

12. $-5 \;\square\; 12 - 12$

13. $\frac{24}{6} \;\square\; 5$

14. $13 - 13 \;\square\; 5 - \frac{25}{5}$

15. $\frac{21}{7} \;\square\; -33$

Double negative rule:
$-(-x) = x$

Simplify each expression.

16. $-(-8)$

17. $-(12 - 4)$

Sets of numbers can be graphed on the number line.

Draw a number line and graph each set of numbers.

18. The composite numbers from 14 to 20

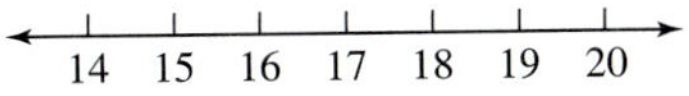

19. The whole numbers between 19 and 25

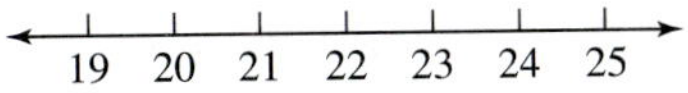

20. The real numbers less than or equal to -3 or greater than 2

21. The real numbers greater than -4 and less than 3

The **absolute value** of x, denoted as $|x|$, is the distance between x and 0 on the number line.
$|x| \geq 0$

Find each absolute value.

22. $|53 - 42|$

23. $|-31|$

1.2 Fractions

To simplify a fraction, factor the numerator and the denominator. Then divide out all common factors.

Simplify each fraction.

24. $\frac{45}{27}$

25. $\frac{121}{11}$

To multiply two fractions, multiply their numerators and multiply their denominators.

To divide two fractions, multiply the first by the reciprocal of the second.

To add (or subtract) two fractions with like denominators, add (or subtract) their numerators and keep their common denominator.

To add (or subtract) two fractions with unlike denominators, rewrite the fractions with the same denominator, add (or subtract) their numerators, and use the common denominator.

Before working with mixed numbers, convert them to improper fractions.

Perform each operation and simplify the answer, if possible.

26. $\frac{31}{15} \cdot \frac{10}{62}$

27. $\frac{25}{36} \cdot \frac{12}{15} \cdot \frac{3}{5}$

28. $\frac{18}{21} \div \frac{6}{7}$

29. $\frac{14}{24} \div \frac{7}{12} \div \frac{2}{5}$

30. $\frac{7}{12} + \frac{9}{12}$

31. $\frac{13}{24} - \frac{5}{24}$

32. $\frac{1}{3} + \frac{1}{7}$

33. $\frac{5}{7} + \frac{4}{9}$

34. $\frac{2}{3} - \frac{1}{7}$

35. $\frac{4}{5} - \frac{2}{3}$

36. $3\frac{2}{3} + 5\frac{1}{4}$

37. $7\frac{5}{12} - 4\frac{1}{2}$

Perform the operations.

38. $32.71 + 15.9$

39. $27.92 - 14.93$

40. $5.3 \cdot 3.5$

41. $21.83 \div 5.9$

Perform each operation and round to two decimal places.

42. $2.7(4.92 - 3.18)$

43. $\frac{3.3 + 2.5}{0.22}$

44. $\frac{12.5}{14.7 - 11.2}$

45. $(3 - 0.7)(3.63 - 2)$

46. Average study time Four students recorded the time they spent working on a take-home exam: 5.2, 4.7, 9.5, and 8 hours. Find the average time spent. (*Hint:* Add the numbers and divide by 4.)

47. Absenteeism During the height of the flu season, 15% of the 380 university faculty members were sick. How many were ill?

48. Packaging Four steel bands surround the shipping crate in the illustration. Find the total length of strapping needed.

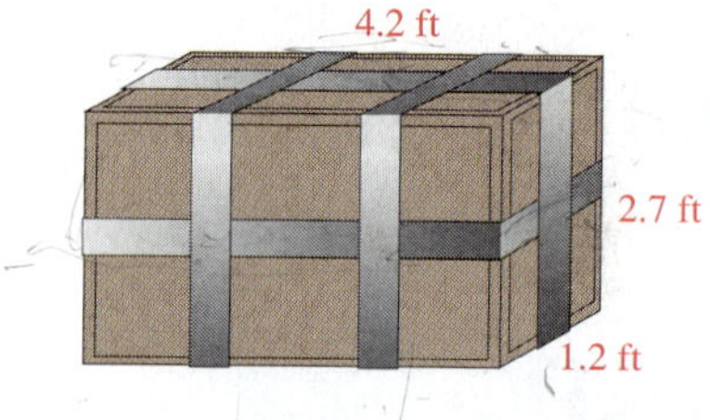

1.3 Exponents and Order of Operations

If n is a natural number, then

$$x^n = \underbrace{x \cdot x \cdot x \cdot x \cdot \cdots \cdot x}_{n \text{ factors of } x}$$

Order of operations
Within each pair of grouping symbols (working from the innermost pair to the outermost pair), perform the following operations:

1. Evaluate all exponential expressions.
2. Perform multiplications and divisions, working from left to right.
3. Perform additions and subtractions, working from left to right.

When the grouping symbols are gone, repeat the above rules to finish the calculation.

In a fraction, simplify the numerator and denominator separately. Then simplify the fraction, if possible.

Find the value of each expression.

49. 3^4

50. $\left(\frac{2}{3}\right)^2$

51. $(0.5)^2$

52. $5^2 + 2^3$

53. $3^2 + 4^2$

54. $(3 + 4)^2$

55. **Petroleum storage** Find the volume of the cylindrical storage tank in the illustration. Round to one decimal place.

Simplify each expression.

56. $5 + 3^3$

57. $7 \cdot 2 - 7$

58. $4 + (8 \div 4)$

59. $(4 + 8) \div 4$

60. $5^3 - \frac{81}{3}$

61. $(5 - 2)^2 + 5^2 + 2^2$

62. $\frac{4 \cdot 3 + 3^4}{31}$

63. $\frac{4}{3} \cdot \frac{9}{2} + \frac{1}{2} \cdot 18$

Evaluate each expression.

64. $8^2 - 6$

65. $(8 - 6)^2$

66. $\frac{6 + 8}{6 - 4}$

67. $\frac{6(8) - 12}{4 + 8}$

68. $2^2 + 2(3^2)$

69. $\frac{2^2 + 3}{2^3 - 1}$

1.4 Adding and Subtracting Real Numbers

To add two positive numbers, add their absolute values and make the answer positive.

To add two negative numbers, add their absolute values and make the answer negative.

Evaluate each expression.

70. $(+7) + (+8)$

71. $(-25) + (-32)$

72. $(-2.7) + (-3.8)$

73. $\frac{1}{3} + \frac{1}{6}$

To add a positive and a negative number, subtract the smaller absolute value from the larger.

1. If the positive number has the larger absolute value, the answer is positive.
2. If the negative number has the larger absolute value, the answer is negative.

If a and b are two real numbers, then

$a - b = a + (-b)$

74. $(+12) + (-24)$

75. $(-44) + (+60)$

76. $3.7 + (-2.5)$

77. $-5.6 + (+2.06)$

78. $15 - (-4)$

79. $-12 - (-13)$

80. $[-5 + (-5)] - (-5)$

81. $1 - [5 - (-3)]$

82. $\frac{5}{6} - \left(-\frac{2}{3}\right)$

83. $\frac{2}{3} - \left(\frac{1}{3} - \frac{2}{3}\right)$

84. $\left|\frac{3}{7} - \left(-\frac{4}{7}\right)\right|$

85. $\frac{3}{7} - \left|-\frac{4}{7}\right|$

1.5 Multiplying and Dividing Real Numbers

To multiply two real numbers, multiply their absolute values.

1. If the numbers are positive, the product is positive.
2. If the numbers are negative, the product is positive.
3. If one number is positive and the other is negative, the product is negative.
4. $a \cdot 0 = 0 \cdot a = 0$
5. $a \cdot 1 = 1 \cdot a = a$

To divide two real numbers, find the quotient of their absolute values.

1. If the numbers are positive, the quotient is positive.
2. If the numbers are negative, the quotient is positive.
3. If one number is positive and the other is negative, the quotient is negative.
4. $\frac{a}{0}$ is undefined; $\frac{0}{0}$ is indeterminate.
5. If $a \neq 0$, then $\frac{0}{a} = 0$.

Evaluate each expression.

86. $(+3)(+4)$

87. $(-5)(-12)$

88. $\left(-\frac{3}{14}\right)\left(-\frac{7}{6}\right)$

89. $(3.75)(0.37)$

90. $5(-7)$

91. $(-15)(7)$

92. $\left(-\frac{1}{2}\right)\left(\frac{4}{3}\right)$

93. $(-12.2)(3.7)$

94. $\frac{+25}{+5}$

95. $\frac{-14}{-2}$

96. $\frac{(-2)(-7)}{4}$

97. $\frac{-22.5}{-3.75}$

98. $\frac{-25}{5}$

99. $\frac{(-3)(-4)}{-6}$

100. $\left(\frac{-10}{2}\right)^2 - (-1)^3$

101. $\frac{[-3 + (-4)]^2}{10 + (-3)}$

102. $\left(\frac{-3 + (-3)}{3}\right)\left(\frac{-15}{5}\right)$

103. $\frac{-2 - (-8)}{5 + (-1)}$

1.6 Algebraic Expressions

Let x, y, and z represent three real numbers. Write an algebraic expression that represents each quantity.

104. The product of x and z

105. The sum of x and twice y

106. Twice the sum of x and y

107. x decreased by the product of y and z

Write each algebraic expression as an English phrase.

108. $3xy$

109. $5 - yz$

110. $yz - 5$

111. $\dfrac{x + y + z}{2xyz}$

Let $x = 2$, $y = -3$, and $z = -1$ and evaluate each expression.

112. $y + z$

113. $x + y$

114. $x + (y + z)$

115. $x - y$

116. $x - (y - z)$

117. $(x - y) - z$

Let $x = 2$, $y = -3$, and $z = -1$ and evaluate each expression.

118. xy

119. yz

120. $x(x + z)$

121. xyz

122. $y^2z + x$

123. $yz^3 + (xy)^2$

124. $\dfrac{xy}{z}$

125. $\dfrac{|xy|}{3z}$

126. How many terms does the expression $3x + 4y + 9$ have?

127. What is the numerical coefficient of the term $7xy$?

128. What is the numerical coefficient of the term xy?

129. Find the sum of the numerical coefficients in $2x^3 + 4x^2 + 3x$.

1.7 Properties of Real Numbers

The closure properties:

$x + y$ is a real number.

$x - y$ is a real number.

xy is a real number.

$\frac{x}{y}$ is a real number $(y \neq 0)$.

The commutative properties:

$x + y = y + x$

$xy = yx$

The associative properties:

$(x + y) + z = x + (y + z)$

$(xy)z = x(yz)$

The distributive property:

$x(y + z) = xy + xz$

The identity elements:

0 is the identity for addition.

1 is the identity for multiplication.

The additive and multiplicative inverse properties:

$x + (-x) = 0$

$x\left(\frac{1}{x}\right) = 1 \quad (x \neq 0)$

Tell which property of real numbers justifies each statement. Assume that all variables represent real numbers.

130. $x + y$ is a real number

131. $3 \cdot (4 \cdot 5) = (4 \cdot 5) \cdot 3$

132. $3 + (4 + 5) = (3 + 4) + 5$

133. $5(x + 2) = 5 \cdot x + 5 \cdot 2$

134. $a + x = x + a$

135. $3 \cdot (4 \cdot 5) = (3 \cdot 4) \cdot 5$

136. $3 + (x + 1) = (x + 1) + 3$

137. $x \cdot 1 = x$

138. $17 + (-17) = 0$

139. $x + 0 = x$

Chapter Test

1. List the prime numbers between 30 and 50.

2. What is the only even prime number?

3. Graph the composite numbers less than 10 on a number line.

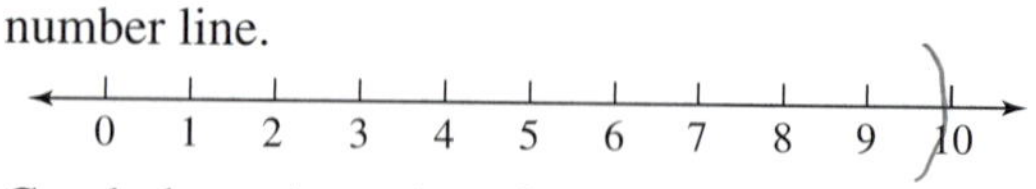

4. Graph the real numbers from 5 to 15 on a number line.

5. Evaluate: $-|23|$.

6. Evaluate: $-|7| + |-7|$.

Place one of the symbols =, <, or > in each box to make a true statement.

7. $3(4 - 2) \ \square \ -2(2 - 5)$

8. $1 + 4 \cdot 3 \ \square \ -2(-7)$

9. 25% of 136 $\square$ $\frac{1}{2}$ of 66

10. $-13.7 \ \square \ -|-13.7|$

Simplify each expression.

11. $\frac{26}{40}$

12. $\frac{7}{8} \cdot \frac{24}{21}$

13. $\frac{18}{35} \div \frac{9}{14}$

14. $\frac{24}{16} + 3$

15. $\dfrac{17-5}{36} - \dfrac{2(13-5)}{12}$

16. $\dfrac{|-7-(-6)|}{-7-|-6|}$

17. Find 17% of 457 and round the answer to one decimal place.

18. Find the area of a rectangle 12.8 feet wide and 23.56 feet long. Round the answer to two decimal places.

19. Find the area of the triangle in the illustration.

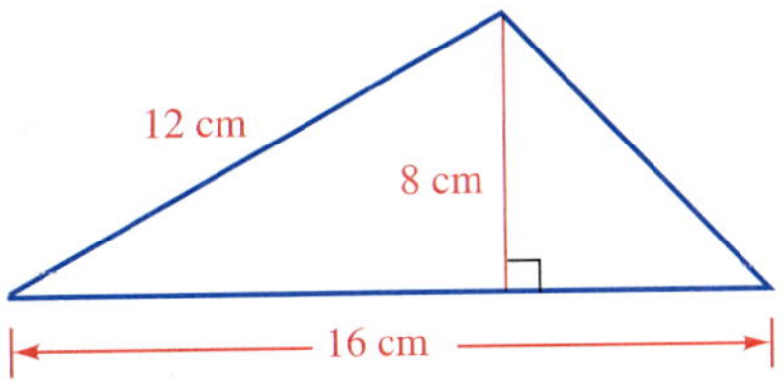

20. To the nearest cubic inch, find the volume of the solid in the illustration.

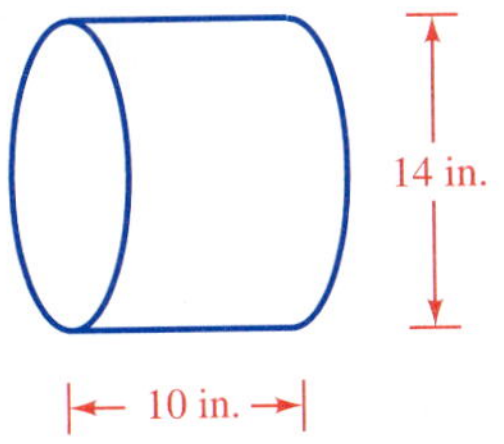

Let $x = -2$, $y = 3$, and $z = 4$. Evaluate each expression.

21. $xy + z$

22. $x(y + z)$

23. $\dfrac{z + 4y}{2x}$

24. $|x^3 - z|$

25. $x^3 + y^2 + z$

26. $|x| - 3|y| - 4|z|$

27. Let x and y represent two real numbers. Write an algebraic expression to denote the quotient obtained when the product of the two numbers is divided by their sum.

28. Let x and y represent two real numbers. Write an algebraic expression to denote the difference obtained when the sum of x and y is subtracted from the product of 5 and y.

29. A man lives 12 miles from work and 7 miles from the grocery store. If he made x round trips to work and y round trips to the store, write an expression to represent how many miles he drove?

30. A baseball costs \$$a$ and a glove costs \$$b$. Write an expression to represent how much it will cost a community center to buy 12 baseballs and 8 gloves.

31. What is the numerical coefficient of the term $3xy^2$?

32. How many terms are in the expression $3x^2y + 5xy^2 + x + 7$?

33. What is the identity element for addition?

34. What is the multiplicative inverse of $\frac{1}{5}$?

State which property of the real numbers justifies each statement.

35. $(xy)z = z(xy)$

36. $3(x + y) = 3x + 3y$

37. $2 + x = x + 2$

38. $7 \cdot \frac{1}{7} = 1$

2 Equations and Inequalities

InfoTrac Project

Do a keyword search on "cholesterol" and find an article titled "Cooking oil to fight fat and cholesterol." Write a summary of the article.

If a container of "functional oil" contains 1.8 liters of flaxseed oil, how many liters of "functional oil" are in the container? Write your equation and solve. Use this information to determine the quantity of each of the other types of oil that will be in the mixture.

Complete this project after studying Section 2.1.

© CORBIS

Mathematics in Retirement Planning

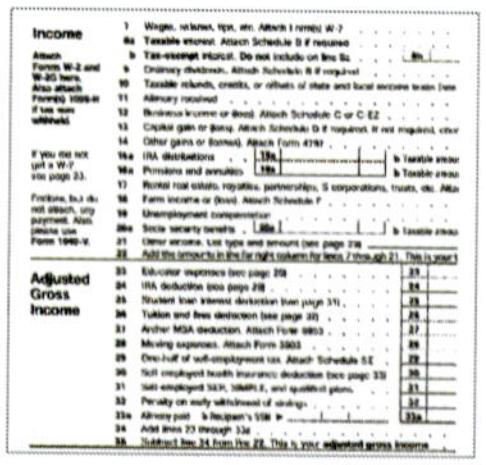

Employees of most corporations participate in retirement plans funded by contributions from both the company and the employee. People who are self-employed can also fund retirement plans. One such plan, called a **Simplified Employee Pension (SEP),** allows an annual contribution that does not exceed 15% of the income available after deductible expenses. This would seem to be an easy calculation—simply subtract deductible expenses from gross income and take 15% of what's left. However, the tax code is not so simple. The SEP contribution is considered a deductible expense, and the SEP

Exercise Set 2.6
Problems 53–54

contribution is 15% of what's left after subtracting deductible expenses and the amount of the SEP contribution. It would seem that to calculate the contribution, you must first know the contribution.

Calculate this contribution.

*In this chapter, we will learn how to solve basic equations and apply that knowledge to solving many types of applied problems. We then will consider special equations called **formulas** and conclude by solving inequalities.*

2.1 Solving Basic Equations

- Equations
- The Addition Property of Equality
- The Subtraction Property of Equality
- The Multiplication Property of Equality
- The Division Property of Equality
- Markdown and Markup
- Percent

Getting Ready *Fill in the blanks.*

1. $3 + \square = 0$ **2.** $(-7) + \square = 0$ **3.** $(-x) + \square = 0$

4. $\frac{1}{3} \cdot 3 = \square$ **5.** $5 \cdot \square = 1$ **6.** $\frac{-6}{-6} = \square$

7. $\frac{4(2)}{\square} = 2$ **8.** $5 \cdot \frac{4}{5} = \square$ **9.** $\frac{-5(3)}{-5} = \square$

To answer questions such as "How many?", "How far?", "How fast?", and "How heavy?", we will often use mathematical statements called *equations.* In this chapter, we will discuss this important idea.

Equations

An **equation** is a statement indicating that two quantities are equal. Some examples of equations are

$$x + 5 = 21, \quad 2x - 5 = 11, \quad \text{and} \quad 3x^2 - 4x + 5 = 0$$

The expression $3x + 2$ is not an equation, because it does not contain an = sign.

In the equation $x + 5 = 21$, the expression $x + 5$ is called the **left-hand side,** and 21 is called the **right-hand side.** The letter x is called the **variable** (or the **unknown**).

An equation can be true or false. The equation $16 + 5 = 21$ is true, but the equation $10 + 5 = 21$ is false. The equation $2x - 5 = 11$ might be true or false, depending on the value of x. If $x = 8$, the equation is true, because when we substitute 8 for x we get 11.

$$\begin{aligned} 2(8) - 5 &= 16 - 5 \\ &= 11 \end{aligned}$$

Any number that makes an equation true when substituted for its variable is said to *satisfy* the equation. All of the numbers that satisfy an equation are called its **solutions** or **roots.** Since 8 is the only number that satisfies the equation $2x - 5 = 11$, it is the only solution.

EXAMPLE 1 Is 6 a solution of $3x - 5 = 2x$?

Solution We substitute 6 for x and simplify.

$$\begin{aligned} 3x - 5 &= 2x \\ 3 \cdot 6 - 5 &\stackrel{?}{=} 2 \cdot 6 \quad \text{Substitute 6 for } x. \\ 18 - 5 &\stackrel{?}{=} 12 \\ 13 &= 12 \end{aligned}$$

Since $13 = 12$ is false, 6 is not a solution.

Self Check Is 1 a solution of $2x + 3 = 5$?

The Addition Property of Equality

To **solve an equation** means to find its solutions. To develop an understanding of how to solve equations, we refer to the scales shown in Figure 2-1. We can think of the scale shown in Figure 2-1(a) as representing the equation $x - 5 = 2$. The weight on the left-hand side of the scale is $(x - 5)$ grams, and the weight on the right-hand side is 2 grams. Because these weights are equal, the scale is in balance. To find x, we need to isolate it by adding 5 grams to the left-hand side of the scale. To keep the scale in balance, we must also add 5 grams to the right-hand side. After adding 5 grams to both sides of the scale, we can see from Figure 2-1(b) that x grams will be balanced by 7 grams. We say that we have solved the equation and that the solution is 7.

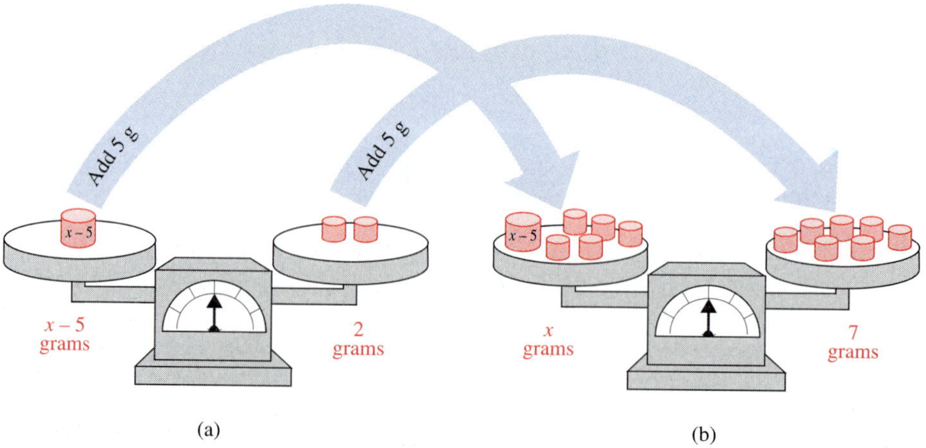

Figure 2-1

The example suggests the following property of equality: *If the same quantity is added to equal quantities, the results will be equal quantities.* We can express this property in symbols.

Addition Property of Equality

Suppose that a, b, and c are real numbers. Then

If $a = b$, then $a + c = b + c$.

When we use this property, the resulting equation will have the same solutions as the original one. We say that the equations are *equivalent.*

Equivalent Equations

Two equations are **equivalent equations** when they have the same solutions.

Using the scales shown in Figure 2-1, we found that $x - 5 = 2$ is equivalent to $x = 7$. In the next example, we use the addition property of equality to solve the equation $x - 5 = 2$ algebraically.

EXAMPLE 2 Solve: $x - 5 = 2$.

Solution To isolate x on one side of the = sign, we undo the subtraction of 5 by adding 5 to both sides of the equation.

$$x - 5 = 2$$
$$x - 5 + 5 = 2 + 5 \quad \text{Add 5 to both sides of the equation.}$$
$$x = 7 \quad -5 + 5 = 0 \text{ and } 2 + 5 = 7.$$

We check by substituting 7 for x in the original equation and simplifying.

$$x - 5 = 2$$
$$7 - 5 \stackrel{?}{=} 2 \quad \text{Substitute 7 for } x.$$
$$2 = 2$$

Since $2 = 2$, the solution checks.

Self Check Solve: $b - 21.8 = 13$.

PERSPECTIVE

To find answers to such questions as How many? How far? How fast? and How heavy?, we often make use of equations. This concept has a long history, and the techniques that we will study in this chapter have been developed over many centuries.

The mathematical notation that we use today is the result of thousands of years of development. The ancient Egyptians used a word for variables, best translated as *heap.* Others used the word *res,* which is Latin for *thing.* In the fifteenth century, the letters *p*: and *m*: were used for *plus* and *minus.* What we would now write as $2x + 3 = 5$ might have been written by those early mathematicians as

2 *res p*:3 *aequalis* 5.

The Subtraction Property of Equality

We can think of the scale shown in Figure 2-2(a) as representing the equation $x + 4 = 9$. The weight on the left-hand side of the scale is $(x + 4)$ grams, and the weight on the right-hand side is 9 grams. Because these weights are equal, the scale is in balance. To find x, we need to isolate it by removing 4 grams from the left-hand side. To keep the scale in balance, we must also remove 4 grams from the right-hand side. In Figure 2-2(b), we can see that x grams will be balanced by 5 grams. We have found that the solution is 5.

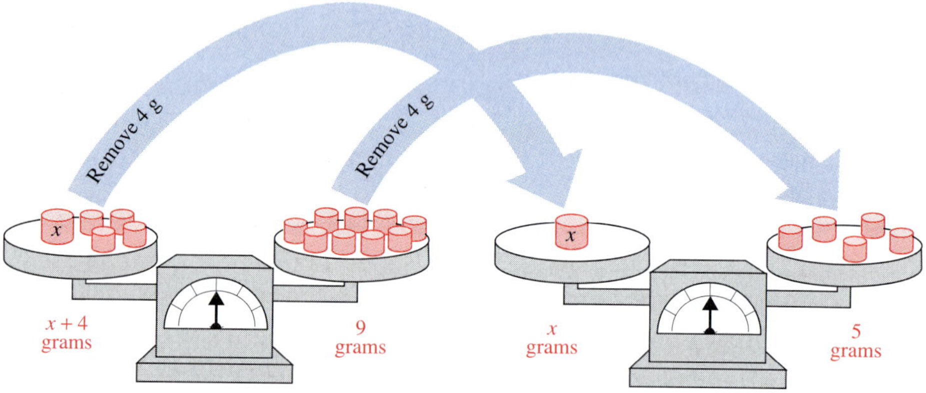

(a) (b)

Figure 2-2

The previous example suggests the following property of equality: *If the same quantity is subtracted from equal quantities, the results will be equal quantities.* We can express this property in symbols.

Subtraction Property of Equality

Suppose that a, b, and c are real numbers. Then

If $a = b$, then $a - c = b - c$.

When we use this property, the resulting equation will be equivalent to the original one.

In the next example, we use the subtraction property of equality to solve the equation $x + 4 = 9$ algebraically.

EXAMPLE 3 Solve: $x + 4 = 9$.

Solution To isolate x on one side of the = sign, we undo the addition of 4 by subtracting 4 from both sides of the equation.

$$x + 4 = 9$$

$$x + 4 - 4 = 9 - 4 \quad \text{Subtract 4 from both sides.}$$

$$x = 5 \quad 4 - 4 = 0 \text{ and } 9 - 4 = 5.$$

We can check by substituting 5 for x in the original equation and simplifying.

$$x + 4 = 9$$
$$5 + 4 \stackrel{?}{=} 9 \quad \text{Substitute 5 for } x.$$
$$9 = 9$$

The solution checks. Instead of subtracting 4 from both sides, we could just as well have added -4 to both sides.

Self Check Solve: $a + 17.5 = 12.2$.

The Multiplication Property of Equality

We can think of the scale shown in Figure 2-3(a) as representing the equation $\frac{x}{3} = 12$. The weight on the left-hand side of the scale is $\frac{x}{3}$ grams, and the weight on the right-hand side is 12 grams. Because these weights are equal, the scale is in balance. To find x, we can triple (or multiply by 3) the weight on each side. When we do this, the scale will remain in balance. From the scale shown in Figure 2-3(b), we can see that x grams will be balanced by 36 grams. Thus, $x = 36$.

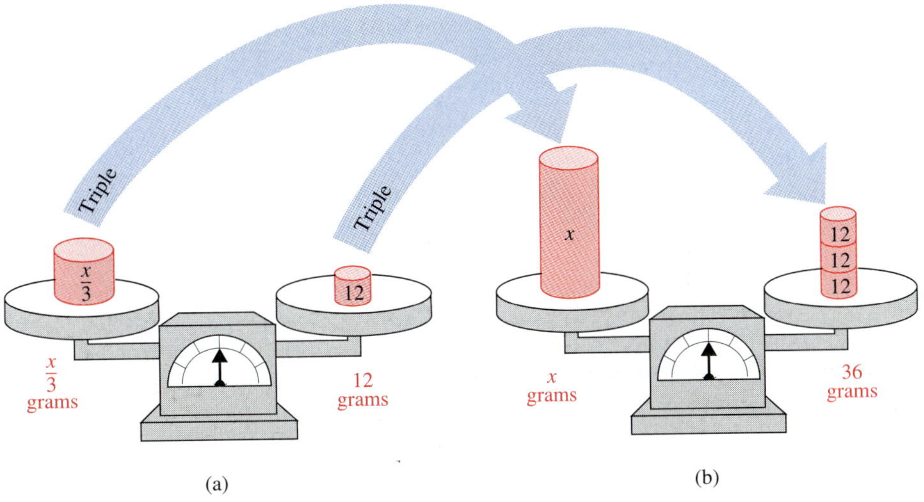

Figure 2-3

The previous example suggests the following property of equality: *If equal quantities are multiplied by the same nonzero quantity, the results will be equal quantities.* We can express this property in symbols.

Multiplication Property of Equality

Suppose that a, b, and c are real numbers and $c \neq 0$. Then

If $a = b$, then $ca = cb$.

When we use the multiplication property, the resulting equation will be equivalent to the original one.

To solve the equation $\frac{x}{3} = 12$ algebraically, we proceed as in Example 4.

EXAMPLE 4 Solve: $\frac{x}{3} = 12$.

Solution To find x, we undo the division by 3 by multiplying both sides of the equation by 3.

$$\frac{x}{3} = 12$$

$$3 \cdot \frac{x}{3} = 3 \cdot 12 \quad \text{Multiply both sides by 3.}$$

$$x = 36 \quad 3 \cdot \tfrac{x}{3} = x \text{ and } 3 \cdot 12 = 36.$$

Verify that the solution checks.

Self Check Solve: $\frac{x}{5} = -7$.

The Division Property of Equality

We will now consider how to solve the equation $2x = 6$. Since $2x$ means $2 \cdot x$, the equation can be written as $2 \cdot x = 6$. We can think of the scale shown in Figure 2-4(a) as representing this equation. The weight on the left-hand side of the scale is $2 \cdot x$ grams, and the weight on the right-hand side is 6 grams. Because these weights are equal, the scale is in balance. To find x, we remove half of the weight from each side. This is equivalent to dividing the weight on both sides by 2. When we do this, the scale will remain in balance. From the scale shown in Figure 2-4(b), we can see that x grams will be balanced by 3 grams. Thus, $x = 3$.

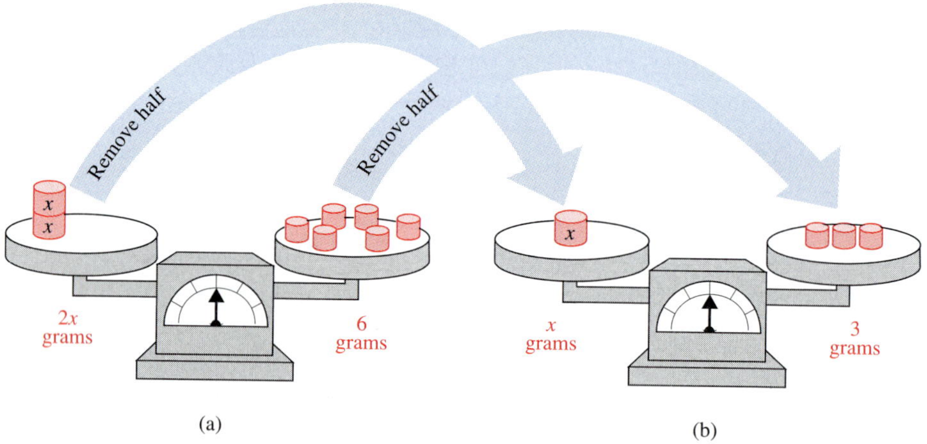

Figure 2-4

The previous example suggests the following property of equality: *If equal quantities are divided by the same nonzero quantity, the results will be equal quantities.* We can express this property in symbols.

Division Property of Equality

Suppose that a, b, and c are real numbers and $c \neq 0$. Then

If $a = b$, then $\frac{a}{c} = \frac{b}{c}$.

When we use the division property, the resulting equation will be equivalent to the original one.

To solve the equation $2x = 6$ algebraically, we proceed as in Example 5.

EXAMPLE 5 Solve: $2x = 6$.

Solution To isolate x on one side of the $=$ sign, we undo the multiplication by 2 by dividing both sides by 2.

$$2x = 6$$

$$\frac{2x}{2} = \frac{6}{2} \quad \text{Divide both sides by 2.}$$

$$x = 3 \quad \tfrac{2}{2} = 1 \text{ and } \tfrac{6}{2} = 3.$$

Verify that the solution is 3.

Self Check Solve: $-5x = 15$.

EXAMPLE 6 Solve: $3x = \frac{1}{5}$.

Solution To isolate x on the left-hand side of the equation, we could undo the multiplication by 3 by dividing both sides by 3. However, it is easier to isolate x by multiplying both sides by the reciprocal of 3, which is $\frac{1}{3}$.

$$3x = \frac{1}{5}$$

$$\frac{1}{3}(3x) = \frac{1}{3} \cdot \frac{1}{5} \quad \text{Multiply both sides by } \frac{1}{3}.$$

$$\left(\frac{1}{3} \cdot 3\right)x = \frac{1}{15} \quad \text{Use the associative property of multiplication.}$$

$$1x = \frac{1}{15} \quad \frac{1}{3} \cdot 3 = 1.$$

$$x = \frac{1}{15}$$

Verify that the solution is $\frac{1}{15}$.

Self Check Solve: $-5x = \frac{1}{3}$.

Markdown and Markup

When the price of merchandise is reduced, the amount of reduction is called the **markdown** or the **discount.** To find the sale price of an item, we subtract the markdown from the regular price.

EXAMPLE 7 **Buying a sofa** A sofa is on sale for \$650. If it has been marked down \$325, find its regular price.

Solution We can let r represent the regular price and substitute 650 for the sale price and 325 for the markdown in the following formula.

Sale price	equals	regular price	minus	markdown
650	=	r	−	325

We can use the addition property of equality to solve the equation.

$$650 = r - 325$$

$$650 + 325 = r - 325 + 325 \quad \text{Add 325 to both sides.}$$

$$975 = r \quad 650 + 325 = 975 \text{ and } -325 + 325 = 0.$$

The regular price is \$975.

Self Check Find the regular price of the sofa if the discount is \$275.

To make a profit, a merchant must sell an item for more than he paid for it. The retail price of the item is the sum of its wholesale cost and the **markup.**

EXAMPLE 8 **Buying a car** A car with a sticker price of \$17,500 has a markup of \$3,500. Find the invoice price (the wholesale price) to the dealer.

Solution We can let w represent the wholesale price and substitute 17,500 for the retail price and 3,500 for the markup in the following formula.

Retail price	equals	wholesale cost	plus	markup
17,500	=	w	+	3,500

We can use the subtraction property of equality to solve the equation.

$$17{,}500 = w + 3{,}500$$

$$17{,}500 - 3{,}500 = w + 3{,}500 - 3{,}500 \quad \text{Subtract 3,500 from both sides.}$$

$$14{,}000 = w \quad 17{,}500 - 3{,}500 = 14{,}000 \text{ and } 3{,}500 - 3{,}500 = 0.$$

The invoice price is \$14,000.

Self Check Find the invoice price of the car if the markup is \$6,700.

Percent

A percent is the numerator of a fraction with a denominator of 100. For example, $6\frac{1}{4}$ percent (written as $6\frac{1}{4}\%$) is the fraction $\frac{6.25}{100}$, or the decimal 0.0625. In problems involving percent, the word *of* usually means multiplication. For example $6\frac{1}{4}\%$ of 8,500 is the product of 0.0625 and 8,500.

$$6\tfrac{1}{4}\% \text{ of } 8{,}500 = 0.0625 \cdot 8{,}500$$

$$= 531.25$$

In the statement $6\frac{1}{4}\%$ of 8,500 = 531.25, the percent $6\frac{1}{4}\%$ is called a **rate,** 8,500 is called the **base,** and their product, 531.25, is called the **amount.** Every percent problem is based on the equation **rate · base = amount.**

Percent Formula

If r is the rate, b is the base, and a is the amount, then

$$rb = a$$

Percent problems involve questions such as the following.

- What is 30% of 1,000? — In this problem, we must find the amount.
- 45% of what number is 405? — In this problem, we must find the base.
- What percent of 400 is 60? — In this problem, we must find the rate.

When we use equations, these problems are easy to solve.

EXAMPLE 9 What is 30% of 1,000?

Solution In this problem, the rate r is 30% and the base is 1,000. We must find the amount.

Rate	·	base	=	amount
30%	of	1,000	is	the amount.

We can substitute these values into the percent formula and solve for a.

$$rb = a$$

$$30\% \cdot 1{,}000 = a \quad \text{Substitute 30\% for } r \text{ and 1,000 for } b.$$

$$0.30 \cdot 1{,}000 = a \quad \text{Change 30\% to the decimal 0.30.}$$

$$300 = a \quad \text{Multiply.}$$

Thus, 30% of 1,000 is 300.

Self Check Find 45% of 800.

EXAMPLE 10 45% of what number is 405?

Solution In this problem, the rate r is 45% and the amount a is 405. We must find the base.

Rate	·	base	=	amount
45%	of	what number	is	405?

We can substitute these values into the percent formula and solve for b.

$$rb = a$$

$$45\% \cdot b = 405 \quad \text{Substitute 45\% for } r \text{ and 405 for } a.$$

$$0.45 \cdot b = 405 \quad \text{Change 45\% to a decimal.}$$

$$\frac{0.45b}{0.45} = \frac{405}{0.45} \quad \text{To undo the multiplication by 0.45, divide both sides by 0.45.}$$

$$b = 900 \quad \frac{0.45}{0.45} = 1 \text{ and } \frac{405}{0.45} = 900.$$

Thus, 45% of 900 is 405.

Self Check 35% of what number is 306.25?

EXAMPLE 11 What percent of 400 is 60?

Solution In this problem, the base b is 400 and the amount a is 60. We must find the rate.

Rate	·	base	=	amount
What percent	of	400	is	60?

We can substitute these values in the percent formula and solve for r.

$$rb = a$$

$$r \cdot 400 = 60 \quad \text{Substitute 400 for } b \text{ and 60 for } a.$$

$$\frac{400r}{400} = \frac{60}{400} \quad \text{To undo the multiplication by 400, divide both sides by 400.}$$

$$r = 0.15 \quad \frac{400}{400} = 1 \text{ and } \frac{60}{400} = 0.15.$$

$$r = 15\% \quad \text{To change the decimal into a percent, we multiply by 100 and insert a \% sign.}$$

Thus, 15% of 400 is 60.

Self Check What percent of 600 is 150?

EXAMPLE 12 **Investing** At a stockholders meeting, 4.5 million shares were voted in favor of a proposal for a mandatory retirement age for the members of the board of directors. If this represented 75% of the number of shares outstanding, how many shares were outstanding?

Solution Let b represent the number of outstanding shares. Then 75% of b is 4.5 million. We can substitute 75% for r and 4.5 million for a in the percent formula and solve for b.

$$rb = a$$

$$75\% \cdot b = 4{,}500{,}000$$

$$0.75b = 4{,}500{,}000 \quad \text{Change 75\% to a decimal.}$$

$$\frac{0.75b}{0.75} = \frac{4{,}500{,}000}{0.75} \quad \text{To undo the multiplication of 0.75, divide both sides by 0.75.}$$

$$b = 6{,}000{,}000 \quad \frac{0.75}{0.75} = 1 \text{ and } \frac{4{,}500{,}000}{0.75} = 6{,}000{,}000.$$

There were 6 million shares outstanding.

Self Check If 60% of the shares outstanding were voted in favor of the proposal, how many shares were voted in favor?

EXAMPLE 13 **Quality control** After examining 240 sweaters, a quality-control inspector found 5 with defective stitching, 8 with mismatched designs, and 2 with incorrect labels. What percent were defective?

Solution Let r represent the percent that are defective. Then the base b is 240 and the amount a is the number of defective sweaters, which is $5 + 8 + 2 = 15$. We can find r by using the percent formula.

$$rb = a$$

$$r \cdot 240 = 15 \quad \text{Substitute 240 for } b \text{ and 15 for } a.$$

$$\frac{240r}{240} = \frac{15}{240} \quad \text{To undo the multiplication of 240, divide both sides by 240.}$$

$$r = 0.0625 \quad \frac{240}{240} = 1 \text{ and } \frac{15}{240} = 0.0625.$$

$$r = 6.25\% \quad \text{To change 0.0625 to a percent, multiply by 100 and insert a \% sign.}$$

The defect rate is $6\frac{1}{4}\%$.

Self Check If a second inspector found 3 sweaters with faded colors in addition to the defectives found by inspector 1, what percent were defective?

Self Check Answers

1. yes **2.** 34.8 **3.** -5.3 **4.** -35 **5.** -3 **6.** $-\frac{1}{15}$ **7.** \$925 **8.** \$10,800 **9.** 360 **10.** 875 **11.** 25% **12.** 3.6 million **13.** $7\frac{1}{2}\%$

Orals *Solve each equation.*

1. $x - 9 = 11$
2. $x - 3 = 13$
3. $w + 5 = 7$
4. $x + 32 = 36$
5. $3x = 3$
6. $5x = 5$
7. $-7x = 14$
8. $\frac{x}{5} = 2$
9. $\frac{x}{2} = -10$
10. Change 30% to a decimal.
11. Change 0.08 to a percent.

2.1 EXERCISES

REVIEW *Perform the operations. Simplify the result when possible.*

1. $\frac{4}{5} + \frac{2}{3}$
2. $\frac{5}{6} \cdot \frac{12}{25}$
3. $\frac{5}{9} \div \frac{3}{5}$
4. $\frac{15}{7} - \frac{10}{3}$
5. $2 + 3 \cdot 4$
6. $3 \cdot 4^2$
7. $3 + 4^3(-5)$
8. $\frac{5(-4) - 3(-2)}{10 - (-4)}$

VOCABULARY AND CONCEPTS *Fill in the blanks.*

9. An ________ is a statement that two quantities are equal.
10. A ________ or ______ of an equation is a number that satisfies the equation.
11. If two equations have the same solutions, they are called ________ equations.
12. To solve an equation, we isolate the ________ on one side of the equation.
13. If the same quantity is added to ______ quantities, the results will be equal quantities.
14. If the same quantity is subtracted from equal quantities, the results will be ______ quantities.
15. If equal quantities are divided by the same nonzero quantity, the results are ______ quantities.
16. If ______ quantities are multiplied by the same nonzero quantity, the results will be equal quantities.
17. Sale price = ________ − markdown
18. Retail price = wholesale cost + ________
19. A percent is the numerator of a fraction whose denominator is ______.
20. Rate · ______ = amount

Tell whether each statement is an equation.

21. $x = 2$
22. $y - 3$
23. $7x < 8$
24. $7 + x = 2$
25. $x + 7 = 0$
26. $3 - 3y > 2$
27. $1 + 1 = 3$
28. $5 = a + 2$

PRACTICE *Tell whether the given number is a solution of the equation.*

29. $x + 2 = 3; 1$
30. $x - 2 = 4; 6$
31. $a - 7 = 0; -7$
32. $x + 4 = 4; 0$
33. $\frac{y}{7} = 4; 28$
34. $\frac{c}{-5} = -2; -10$
35. $\frac{x}{5} = x; 0$
36. $\frac{x}{7} = 7x; 0$
37. $3k + 5 = 5k - 1; 3$
38. $2s - 1 = s + 7; 6$
39. $\frac{5 + x}{10} - x = \frac{1}{2}; 0$
40. $\frac{x - 5}{6} = 12 - x; 11$

Use a property of equality to solve each equation. ***Check all solutions.***

41. $x + 7 = 13$
42. $y + 3 = 7$
43. $y - 7 = 12$
44. $c - 11 = 22$
45. $1 = y - 5$
46. $0 = r + 10$
47. $p - 404 = 115$
48. $41 = 45 + q$
49. $-37 + z = 37$
50. $-43 + a = -43$
51. $-57 = b - 29$
52. $-93 = 67 + y$
53. $\frac{4}{3} = -\frac{2}{3} + x$
54. $z + \frac{5}{7} = -\frac{2}{7}$
55. $d + \frac{2}{3} = \frac{3}{2}$
56. $s + \frac{2}{3} = \frac{1}{5}$
57. $-\frac{3}{5} = x - \frac{2}{5}$
58. $b + 7 = \frac{20}{3}$
59. $r - \frac{1}{5} = \frac{3}{10}$
60. $t + \frac{4}{7} = \frac{11}{14}$
61. $\frac{x}{5} = 5$
62. $\frac{x}{15} = 3$
63. $\frac{x}{32} = -2$
64. $\frac{y}{16} = -5$
65. $\frac{b}{3} = 5$
66. $\frac{a}{5} = -3$
67. $-3 = \frac{s}{11}$
68. $\frac{s}{-12} = 4$
69. $6x = 18$
70. $25x = 625$
71. $-4x = 36$
72. $-16y = 64$
73. $4t = 108$
74. $-66 = -6t$
75. $11x = -121$
76. $-9y = -9$
77. $2x = \frac{1}{7}$
78. $4x = \frac{1}{3}$
79. $5x = \frac{5}{8}$
80. $6x = \frac{2}{3}$
81. $\frac{z}{7} = 14$
82. $-19x = -57$
83. $\frac{w}{7} = \frac{5}{7}$
84. $-17z = -51$
85. $\frac{s}{-3} = -\frac{5}{6}$
86. $1{,}228 = \frac{x}{0.25}$
87. $-32z = 64$
88. $15 = \frac{r}{-5}$
89. $18z = -9$
90. $-12z = 3$
91. $0.25x = 1{,}228$
92. $-0.2y = 51$
93. $\frac{b}{3} = \frac{1}{3}$
94. $\frac{a}{13} = \frac{1}{26}$
95. $-1.2w = -102$
96. $1.5a = -15$
97. $\frac{u}{5} = -\frac{3}{10}$
98. $\frac{t}{-7} = \frac{1}{2}$
99. $\frac{p}{0.2} = 12$
100. $\frac{t}{0.3} = -36$

Use the formula $rb = a$ to find each value.

101. What number is 40% of 200?
102. What number is 35% of 520?
103. What number is 50% of 38?
104. What number is 25% of 300?
105. 15% of what number is 48?
106. 26% of what number is 78?
107. 133 is 35% of what number?
108. 13.3 is 3.5% of what number?
109. 28% of what number is 42?
110. 44% of what number is 143?
111. What percent of 357.5 is 71.5?

112. What percent of 254 is 13.208?

113. 0.32 is what percent of 4?

114. 3.6 is what percent of 28.8?

115. 34 is what percent of 17?

116. 39 is what percent of 13?

APPLICATIONS *Use an equation to solve each problem.*

117. Buying a boat A boat is on sale for $7,995. Find its regular price if it has been marked down $1,350.

118. Buying a house A house that was priced at $105,000 has been discounted $7,500. Find the new asking price.

119. Buying clothes A sport jacket that sells for $175 has a markup of $85. Find the wholesale price.

120. Buying a vacuum cleaner A vacuum that sells for $97 has a markup of $37. Find the wholesale price.

121. Banking The amount A in an account is given by the formula

$$A = p + i$$

where p is the principal and i is the interest. How much interest was earned if an original deposit (the principal) of $4,750 has grown to be $5,010?

122. Depreciation The current value v of a car is given by the formula

$$v = c - d$$

where c is the original price and d is the depreciation. Find the original cost of a car that is worth $10,250 after depreciating $7,500.

123. Appreciation The value v of a house is given by the formula

$$v = p + a$$

where p is the original purchase price and a is the appreciation. Find the original purchase price of a house that is worth $150,000 and has appreciated $57,000.

124. Taxes The cost c of an item is given by the formula

$$c = p + t$$

where p is the price and t is the sales tax. Find the tax paid on an item that was priced at $37.10 and cost $39.32.

125. Buying carpet The cost c of carpet is given by the formula

$$c = p + t$$

where p is the price and t is cost of installation. How much did it cost to install $317 worth of carpet that cost $512?

126. Selling real estate The money m received from selling a house is given by the formula

$$m = s - c$$

where s is the selling price and c is the agent's commission. Find the selling price of a house if the seller received $217,000 and the agent received $13,020.

127. Buying real estate The cost of a condominium is $57,595 less than the cost of a house. If the house costs $202,744, find the cost of the condominium.

128. Buying paint After reading the ad below, a decorator bought one gallon of primer, one gallon of paint, and a brush. The total cost was $30.44. Find the cost of the brush.

129. Customer satisfaction One-third of the movie audience left the theater in disgust. If 78 angry patrons walked out, how many were there originally?

130. Stock split After a 3-for-2 stock split, each shareholder will own 1.5 times as many shares as before. If 555 shares are owned after the split, how many were owned before?

131. Off-campus housing One-seventh of the senior class is living in off-campus housing. If 217 students live off campus, how large is the senior class?

132. Union membership If 2,484 union members represent 90% of a factory's work force, how many workers are employed?

133. Shopper dissatisfaction Refer to the survey results shown in the table. What percent of those surveyed were not pleased?

Shopper Survey Results	
First-time shoppers	1,731
Major purchase today	539
Shopped within previous month	1,823
Satisfied with service	4,140
Seniors	2,387
Total surveyed	9,200

134. Charity overhead Out of $237,000 donated to a certain charity, $5,925 is used to pay for fund-raising expenses. What percent of the donations is overhead?

135. Selling price of a microwave oven The 5% sales tax on a microwave oven amounts to $13.50. What is the microwave's selling price?

136. Sales taxes Sales tax on a $12 compact disc is $0.72. At what rate is sales tax computed?

137. Hospital occupancy 18% of hospital patients stay for less than 1 day. If 1,008 patients in January stayed for less than 1 day, what total number of patients did the hospital treat in January?

138. Home prices The average price of homes in one neighborhood decreased 8% since last year, a drop of $7,800. What was the average price of a home last year?

WRITING

139. Explain what it means for a number to satisfy an equation.

140. How can you tell whether a number is the solution to an equation?

SOMETHING TO THINK ABOUT

141. The Ahmes papyrus mentioned on page 5 contains this statement: *A circle nine units in diameter has the same area as a square eight units on a side.* From this statement, determine the ancient Egyptians' approximation of π.

142. Calculate the Egyptians' **percent of error:** What percent of the actual value of π is the difference of the estimate obtained in Exercise 141 and the actual value of π?

2.2 Solving More Equations

- Solving More Complicated Equations
- Markup and Markdown

Getting Ready *Perform the operations.*

1. $7 + 3 \cdot 5$ **2.** $3(5 + 7)$ **3.** $\dfrac{3 + 7}{2}$ **4.** $3 + \dfrac{7}{2}$

5. $\dfrac{3(5 - 8)}{9}$ **6.** $3 \cdot \dfrac{5 - 8}{9}$ **7.** $\dfrac{3 \cdot 5 - 8}{9}$ **8.** $3 \cdot \dfrac{5}{9} - 8$

We have solved equations by using the addition, subtraction, multiplication, and division properties of equality. To solve more complicated equations, we need to use several of these properties in succession.

Solving More Complicated Equations

In the first example, we must use both the subtraction and division properties of equality.

EXAMPLE 1 Solve: $-12x + 5 = 17$.

Solution The left-hand side of the equation indicates that x is to be multiplied by -12 and then 5 is to be added to that product. To isolate x, we must undo these operations in the opposite order.

- To undo the addition of 5, we subtract 5 from both sides.
- To undo the multiplication by -12, we divide both sides by -12.

$$-12x + 5 = 17$$

$$-12x + 5 - 5 = 17 - 5 \quad \text{To undo the addition of 5, subtract 5 from both sides.}$$

$$-12x = 12 \quad 5 - 5 = 0 \text{ and } 17 - 5 = 12.$$

$$\frac{-12x}{-12} = \frac{12}{-12} \quad \text{To undo the multiplication by } -12\text{, divide both sides by } -12.$$

$$x = -1 \quad \frac{-12}{-12} = 1 \text{ and } \frac{12}{-12} = -1.$$

Check:

$$-12x + 5 = 17$$

$$-12(-1) + 5 \stackrel{?}{=} 17 \quad \text{Substitute } -1 \text{ for } x.$$

$$12 + 5 \stackrel{?}{=} 17 \quad \text{Simplify.}$$

$$17 = 17$$

Because $17 = 17$, the solution checks.

Self Check Solve: $2x + 3 = 15$.

EXAMPLE 2 Solve: $\frac{x}{3} - 7 = -3$.

Solution The left-hand side of the equation indicates that x is to be divided by 3 and then 7 is to be subtracted from that quotient. To isolate x, we must undo these operations in the opposite order.

- To undo the subtraction of 7, we add 7 to both sides.
- To undo the division by 3, we multiply both sides by 3.

$$\frac{x}{3} - 7 = -3$$

$$\frac{x}{3} - 7 + 7 = -3 + 7 \quad \text{To undo the subtraction of 7, add 7 to both sides.}$$

$$\frac{x}{3} = 4 \quad -7 + 7 = 0 \text{ and } -3 + 7 = 4.$$

$$3 \cdot \frac{x}{3} = 3 \cdot 4 \quad \text{To undo the division by 3, multiply both sides by 3.}$$

$$x = 12 \quad 3 \cdot \tfrac{1}{3} = 1 \text{ and } 3 \cdot 4 = 12.$$

Check: $\frac{x}{3} - 7 = -3$

$$\frac{12}{3} - 7 \stackrel{?}{=} -3 \quad \text{Substitute 12 for } x.$$

$$4 - 7 \stackrel{?}{=} -3 \quad \text{Simplify.}$$

$$-3 = -3$$

Since $-3 = -3$, the solution checks.

Self Check Solve: $\frac{x}{4} - 3 = 5$.

EXAMPLE 3 Solve: $\frac{x - 7}{3} = 9$.

Solution The left-hand side of the equation indicates that 7 is to be subtracted from x and that the difference is to be divided by 3. To isolate x, we must undo these operations in the opposite order.

- To undo the division by 3, we multiply both sides by 3.
- To undo the subtraction of 7, we add 7 to both sides.

$$\frac{x - 7}{3} = 9$$

$$3\left(\frac{x - 7}{3}\right) = 3(9) \quad \text{To undo the division by 3, multiply both sides by 3.}$$

$$x - 7 = 27 \quad 3 \cdot \tfrac{1}{3} = 1 \text{ and } 3(9) = 27.$$

$$x - 7 + 7 = 27 + 7 \quad \text{To undo the subtraction of 7, add 7 to both sides.}$$

$$x = 34 \quad -7 + 7 = 0 \text{ and } 27 + 7 = 34.$$

Verify that the solution checks.

Self Check Solve: $\frac{a - 3}{5} = -2$.

EXAMPLE 4 Solve: $\frac{3x}{4} + \frac{2}{3} = -7$.

Solution The left-hand side of the equation indicates that x is to be multiplied by 3, then $3x$ is to be divided by 4, and then $\frac{2}{3}$ is to be added to that result. To isolate x, we must undo these operations in the opposite order.

- To undo the addition of $\frac{2}{3}$, we subtract $\frac{2}{3}$ from both sides.
- To undo the division by 4, we multiply both sides by 4.
- To undo the multiplication by 3, we multiply both sides by $\frac{1}{3}$.

$$\frac{3x}{4} + \frac{2}{3} = -7$$

$$\frac{3x}{4} + \frac{2}{3} - \frac{2}{3} = -7 - \frac{2}{3}$$ To undo the addition of $\frac{2}{3}$, subtract $\frac{2}{3}$ from both sides.

$$\frac{3x}{4} = -\frac{23}{3}$$ $\frac{2}{3} - \frac{2}{3} = 0$ and $-7 - \frac{2}{3} = -\frac{23}{3}$.

$$4\left(\frac{3x}{4}\right) = 4\left(-\frac{23}{3}\right)$$ To undo the division by 4, multiply both sides by 4.

$$3x = -\frac{92}{3}$$ $4 \cdot \frac{3x}{4} = 3x$ and $4\left(-\frac{23}{3}\right) = -\frac{92}{3}$.

$$\frac{1}{3}(3x) = \frac{1}{3}\left(-\frac{92}{3}\right)$$ To undo the multiplication by 3, multiply both sides by $\frac{1}{3}$.

$$x = -\frac{92}{9}$$ $\frac{1}{3} \cdot 3x = x$ and $\frac{1}{3}\left(-\frac{92}{3}\right) = -\frac{92}{9}$.

Verify that the solution checks.

Self Check Solve: $\frac{2x}{3} - \frac{4}{5} = 3$.

EXAMPLE 5 **Advertising** A store manager hires a student to distribute advertising circulars door to door. The student will be paid \$24 a day plus 12¢ for every ad distributed. How many circulars must she distribute to earn \$42 in one day?

Solution We can let a represent the number of circulars that the student must distribute. Her earnings can be expressed in two ways: as \$24 plus the 12¢-apiece cost of distributing the circulars, and as \$42.

\$24	plus	*a* ads at \$0.12 each	is	\$42.
24	+	$0.12a$	=	42

12¢ = \$0.12.

We can solve this equation as follows:

$$24 + 0.12a = 42$$

$$24 - 24 + 0.12a = 42 - 24$$ To undo the addition of 24, subtract 24 from both sides.

$$0.12a = 18$$ $24 - 24 = 0$ and $42 - 24 = 18$.

$$\frac{0.12a}{0.12} = \frac{18}{0.12}$$ To undo the multiplication by 0.12, divide both sides by 0.12.

$$a = 150$$ $\frac{0.12}{0.12} = 1$ and $\frac{18}{0.12} = 150$.

The student must distribute 150 ads. Check the result.

Self Check How many circulars must the student deliver in one day to earn \$48?

Markup and Markdown

We have seen that the retail price of an item is the sum of the cost and the markup.

Retail price equals cost plus markup

Often, the markup is expressed as a **percent of cost.**

Markup equals percent of markup times cost

Suppose a store manager buys toasters for \$21 and sells them at a 17% markup. To find the retail price, the manager begins with his cost and adds 17% of that cost.

$$\begin{aligned}\text{Retail price} &= \text{cost} + \text{markup}\\ &= \text{cost} + \text{percent of markup} \cdot \text{cost}\\ &= 21 + 0.17 \cdot 21\\ &= 21 + 3.57\\ &= 24.57\end{aligned}$$

The retail price of a toaster is \$24.57.

EXAMPLE 6 **Antique cars** In 1956, a Chevrolet BelAir automobile sold for \$4,000. Today, it is worth about \$28,600. Find the **percent of increase.**

Solution We let p represent the percent of increase, expressed as a decimal.

Current price equals original price plus p(original price)

$$28{,}600 = 4{,}000 + p(4{,}000)$$

$$28{,}600 - \mathbf{4{,}000} = 4{,}000 - \mathbf{4{,}000} + 4{,}000p$$ To undo the addition of 4,000, subtract 4,000 from both sides.

$$24{,}600 = 4{,}000p$$ 28,600 − 4,000 = 24,600 and 4,000 − 4,000 = 0.

$$\frac{24{,}600}{\mathbf{4{,}000}} = \frac{4{,}000p}{\mathbf{4{,}000}}$$ To undo the multiplication by 4,000, divide both sides by 4,000.

$$6.15 = p$$ Simplify.

To convert 6.15 to a percent, we multiply by 100 and insert a % sign. Since the percent of increase is 615%, the car has appreciated 615%.

Self Check Find the percent of increase if the car sells for \$30,000.

We have seen that when the price of merchandise is reduced, the amount of reduction is the **markdown** (also called the **discount**).

Sale price equals regular price minus markdown

Usually, the markdown is expressed as a percent of the regular price.

Markdown equals percent of markdown times regular price

Suppose that a television set that regularly sells for \$570 has been marked down 25%. That means the customer will pay 25% less than the regular price. To find the sale price, we use the formula

Sale price	=	regular price	−	markdown		
	=	regular price	−	percent of markdown	·	regular price
	=	\$570	−	25%	of	\$570

$= \$570 - (0.25)(\$570)$ 25% = 0.25.

$= \$570 - \$142.50$

$= \$427.50$

The television set is selling for \$427.50.

EXAMPLE 7 **Buying a camera** A camera that was originally priced at \$452 is on sale for \$384.20. Find the percent of markdown.

Solution We let p represent the percent of markdown, expressed as a decimal, and substitute \$384.20 for the sale price and \$452 for the regular price.

Sale price	equals	regular price	minus	percent of markdown	times	regular price
384.20	=	452	−	p	·	452

$384.20 - 452 = 452 - 452 - p(452)$ To undo the addition of 452, subtract 452 from both sides.

$-67.80 = -p(452)$ $384.20 - 452 = -67.80$; $452 - 452 = 0$.

$\frac{-67.80}{-452} = \frac{-p(452)}{-452}$ To undo the multiplication by −452, divide both sides by −452.

$0.15 = p$ $\frac{-67.80}{-452} = 0.15$ and $\frac{-452}{-452} = 1$.

The camera is on sale at a 15% markdown.

Self Check If the camera is reduced another \$23, find the percent of discount.

Comment When a price increases from \$100 to \$125, the percent of increase is 25%. When the price *decreases* from \$125 to \$100, the percent of decrease is 20%. These different results occur because the percent of increase is a percent of the original (smaller) price, \$100. The percent of decrease is a percent of the original (larger) price, \$125.

Self Check Answers

1. 6 **2.** 32 **3.** −7 **4.** $\frac{57}{10}$ **5.** 200 **6.** 650% **7.** 20%

Orals *What would you do first when solving each equation?*

1. $5x - 7 = -12$

2. $15 = \frac{x}{5} + 3$

3. $\frac{x}{7} - 3 = 0$

4. $\frac{x - 3}{7} = -7$

5. $5w - 5 = 5$

6. $5w + 5 = 5$

7. $\frac{x-7}{3} = 5$

8. $\frac{3x-5}{2} + 2 = 0$

Solve each equation.

9. $7z - 7 = 14$

10. $\frac{t-1}{2} = 6$

2.2 EXERCISES

REVIEW *Refer to the formulas given in Section 1.3.*

1. Find the perimeter of a rectangle with sides measuring 8.5 and 16.5 cm.
2. Find the area of a rectangle with sides measuring 2.3 in. and 3.7 in.
3. Find the area of a trapezoid with a height of 8.5 in. and bases measuring 6.7 in. and 12.2 in.
4. Find the volume of a rectangular solid with dimensions of 8.2 cm by 7.6 cm by 10.2 cm.

VOCABULARY AND CONCEPTS *Fill in the blanks.*

5. Retail price = ____ + markup
6. Markup = percent of markup · ____
7. Markdown = _______ of markdown · regular price
8. Another word for markdown is _______.

PRACTICE *Solve each equation.* ***Check all solutions.***

9. $5x - 1 = 4$

10. $5x + 3 = 8$

11. $6x + 2 = -4$

12. $4x - 4 = 4$

13. $3x - 8 = 1$

14. $7x - 19 = 2$

15. $11x + 17 = -5$

16. $13x - 29 = -3$

17. $43t + 72 = 158$

18. $96t + 23 = -265$

19. $-47 - 21s = 58$

20. $-151 + 13s = -229$

21. $2y - \frac{5}{3} = \frac{4}{3}$

22. $9y + \frac{1}{2} = \frac{3}{2}$

23. $-0.4y - 12 = -20$

24. $-0.8y + 64 = -32$

25. $\frac{x}{3} - 3 = -2$

26. $\frac{x}{7} + 3 = 5$

27. $\frac{z}{9} + 5 = -1$

28. $\frac{y}{5} - 3 = 3$

29. $\frac{b}{3} + 5 = 2$

30. $\frac{a}{5} - 3 = -4$

31. $\frac{s}{11} + 9 = 6$

32. $\frac{r}{12} + 2 = 4$

33. $\frac{k}{5} - \frac{1}{2} = \frac{3}{2}$

34. $\frac{y}{5} - \frac{8}{7} = -\frac{1}{7}$

35. $\frac{w}{16} + \frac{5}{4} = 1$

36. $\frac{m}{7} - \frac{1}{14} = \frac{1}{14}$

37. $\frac{b+5}{3} = 11$

38. $\frac{2+a}{13} = 3$

39. $\frac{r+7}{3} = 4$

40. $\frac{t-2}{7} = -3$

41. $\frac{u-2}{5} = 1$

42. $\frac{v-7}{3} = -1$

43. $\frac{x-4}{4} = -3$

44. $\frac{3+y}{5} = -3$

45. $\frac{3x}{2} - 6 = 9$

46. $\frac{5x}{7} + 3 = 8$

47. $\frac{3y}{2} + 5 = 11$

48. $\frac{5z}{3} + 3 = -2$

49. $\frac{2x}{3} + \frac{1}{2} = 3$

50. $\frac{4x}{5} - \frac{1}{3} = 1$

51. $\frac{3x}{4} - \frac{2}{5} = 2$

52. $\frac{5x}{6} + \frac{3}{5} = 3$

53. $\frac{3x-12}{2} = 9$

54. $\frac{5x+10}{7} = 0$

55. $\frac{5k-8}{9} = 1$

56. $\frac{2x-1}{3} = -5$

57. $\frac{3z+2}{17} = 0$

58. $\frac{10t-4}{2} = 1$

59. $\frac{17k-28}{21} + \frac{4}{3} = 0$

60. $\frac{5a-2}{3} = \frac{1}{6}$

61. $-\frac{x}{3} - \frac{1}{2} = -\frac{5}{2}$

62. $\frac{17 - 7a}{8} = 2$

63. $\frac{9 - 5w}{15} = \frac{2}{5}$

64. $\frac{3t - 5}{5} + \frac{1}{2} = -\frac{19}{2}$

APPLICATIONS

65. **Integer problem** Six less than 3 times a certain number is 9. Find the number.

66. **Integer problem** If a certain number is increased by 7 and that result is divided by 2, the number 5 is obtained. Find the original number.

67. **Apartment rental** A student moves into a bigger apartment that rents for \$400 per month. That rent is \$100 less than twice what she had been paying. Find her former rent.

68. **Auto repair** A mechanic charged \$20 an hour to repair the water pump on a car, plus \$95 for parts. If the total bill was \$155, how many hours did the repair take?

69. **Boarding dogs** A sportsman boarded his dog at a kennel for a \$16 registration fee plus \$12 a day. If the stay cost \$100, how many days was the owner gone?

70. **Water billing** The city's water department charges \$7 per month, plus 42¢ for every 100 gallons of water used. Last month, one homeowner used 1,900 gallons and received a bill for \$17.98. Was the billing correct?

71. **Telephone charges** A call to Tucson from a pay phone in Chicago costs 85¢ for the first minute and 27¢ for each additional minute or portion of a minute. If a student has \$8.50 in change, how long can she talk?

72. **Monthly sales** A clerk's sales in February were \$2,000 less than 3 times her sales in January. If her February sales were \$7,000, by what amount did her sales increase?

73. **Ticket sales** A music group charges \$1,500 for each performance, plus 20% of the total ticket sales. After a concert, the group received \$2,980. How much money did the ticket sales raise?

74. **Getting an A** To receive a grade of A, the average of four 100-point exams must be 90 or better. If a student received scores of 88, 83, and 92 on the first three exams, what minimum score does he need on the fourth exam to earn an A?

75. **Getting an A** The grade in history class is based on the average of five 100-point exams. One student received scores of 85, 80, 95, and 78 on the first four exams. With an average of 90 needed, what chance does he have for an A?

76. **Excess inventory** From the portion of the following ad, determine the sale price of a shirt.

Clearance Sale
Save 40%

	Regularly	Sale
Sweaters	\$45.95	\$27.57
Shirts	\$37.50	\$

77. **Clearance sales** Sweaters already on sale for 20% off the regular price cost \$36 when purchased with a promotional coupon that allows an additional 10% discount. Find the original price. (*Hint:* When you save 20%, you are paying 80%.)

78. **Furniture sale** A \$1,250 sofa is marked down to \$900. Find the percent of markdown.

79. **Value of coupons** The percent discount offered by the following coupon depends on the amount purchased. Find the range of the percent discount.

Value coupon
Save \$15

on purchases of \$100 to \$250.

80. **Furniture pricing** A bedroom set selling for \$1,900 cost \$1,000 wholesale. Find the percent markup.

WRITING

81. In solving the equation $5x - 3 = 12$, explain why you would add 3 to both sides first, rather than dividing by 5 first.

82. To solve the equation $\frac{3x - 4}{7} = 2$, what operations would you perform, and in what order?

SOMETHING TO THINK ABOUT

83. Suppose you must solve the following equation but you can't quite read one number. It reads

$$\frac{7x + \#}{22} = \frac{1}{2}$$

If the solution of the equation is 1, what is the equation?

84. A store manager first increases his prices by 30% and then advertises as follows.

SALE!! 30% savings!!

What is the real percent discount to customers?

2.3 Simplifying Expressions to Solve Equations

- Like Terms
- Combining Like Terms
- Solving Equations
- Identities and Contradictions

Getting Ready *Use the distributive property to remove parentheses.*

1. $(3 + 4)x$ **2.** $(7 + 2)x$

3. $(8 - 3)w$ **4.** $(10 - 4)y$

Simplify each expression by performing the operations within the parentheses.

5. $(3 + 4)x$ **6.** $(7 + 2)x$

7. $(8 - 3)w$ **8.** $(10 - 4)y$

When algebraic expressions with the same variables occur in an equation, we can combine them. Then, we can solve the simplified equation.

Like Terms

Recall that a *term* is either a number or the product of numbers and variables. Some examples of terms are $7x$, $-3xy$, y^2, and 8. The number part of each term is called its **numerical coefficient** (or just the **coefficient**).

- The coefficient of $7x$ is 7.
- The coefficient of $-3xy$ is -3.
- The coefficient of y^2 is the understood factor of 1.
- The coefficient of 8 is 8.

Like Terms **Like terms,** or **similar terms,** are terms with exactly the same variables and exponents.

The terms $3x$ and $5x$ are **like terms,** as are $9x^2$ and $-3x^2$. The terms $4xy$ and $3x^2$ are **unlike terms,** because they have different variables. The terms $4x$ and $5x^2$ are unlike terms, because the variables have different exponents.

Combining Like Terms

The distributive property can be used to combine terms of algebraic expressions that contain sums or differences of like terms. For example, the terms in $3x + 5x$ and $9xy^2 - 11xy^2$ can be combined as follows:

$$3x + 5x = (3 + 5)x \qquad 9xy^2 - 11xy^2 = (9 - 11)xy^2$$
$$= 8x \qquad = -2xy^2$$

These examples suggest the following rule.

Combining Like Terms

> To combine like terms, add their coefficients and keep the same variables and exponents.

Comment If the terms of an expression are unlike terms, they cannot be combined. For example, since the terms in $9xy^2 - 11x^2y$ have variables with different exponents, they are unlike terms and cannot be combined.

EXAMPLE 1 Simplify: $3(x + 2) + 2(x - 8)$.

Solution

$3(x + 2) + 2(x - 8)$

$= 3x + 3 \cdot 2 + 2x - 2 \cdot 8$ — Use the distributive property to remove parentheses.

$= 3x + 6 + 2x - 16$ — $3 \cdot 2 = 6$ and $2 \cdot 8 = 16$.

$= 3x + 2x + 6 - 16$ — Use the commutative property of addition: $6 + 2x = 2x + 6$.

$= 5x - 10$ — Combine like terms.

Self Check Simplify: $-5(a + 3) + 2(a - 5)$.

EXAMPLE 2 Simplify: $3(x - 3) - 5(x + 4)$.

Solution

$3(x - 3) - 5(x + 4)$

$= 3(x - 3) + (-5)(x + 4)$ — $a - b = a + (-b)$.

$= 3x - 3 \cdot 3 + (-5)x + (-5)4$ — Use the distributive property to remove parentheses.

$= 3x - 9 + (-5x) + (-20)$ — $3 \cdot 3 = 9$ and $(-5)(4) = -20$.

$= -2x - 29$ — Combine like terms.

Self Check Simplify: $-3(b - 2) - 4(b - 4)$.

Solving Equations

To solve an equation, we must isolate the variable on one side. This is often a multistep process that may require combining like terms. As we solve equations, we will follow these steps.

Solving Equations

1. Clear the equation of fractions.
2. Use the distributive property to remove parentheses.
3. Combine like terms on each side of the equation, if necessary.
4. Undo the operations of addition and subtraction to get the variables on one side and the constants on the other.
5. Combine like terms and undo the operations of multiplication and division to isolate the variable.
6. Check the solution.

EXAMPLE 3 Solve: $3(x + 2) - 5x = 0$.

Solution

$$\begin{aligned}
3(x + 2) - 5x &= 0 \\
3x + 3 \cdot 2 - 5x &= 0 && \text{Use the distributive property to remove parentheses.} \\
3x - 5x + 6 &= 0 && \text{Rearrange terms and simplify.} \\
-2x + 6 &= 0 && \text{Combine like terms.} \\
-2x + 6 - 6 &= 0 - 6 && \text{Subtract 6 from both sides.} \\
-2x &= -6 && \text{Combine like terms.} \\
\frac{-2x}{-2} &= \frac{-6}{-2} && \text{Divide both sides by } -2. \\
x &= 3 && \text{Simplify.}
\end{aligned}$$

Check:

$$\begin{aligned}
3(x + 2) - 5x &= 0 \\
3(3 + 2) - 5 \cdot 3 &\stackrel{?}{=} 0 && \text{Substitute 3 for } x. \\
3 \cdot 5 - 5 \cdot 3 &\stackrel{?}{=} 0 \\
15 - 15 &\stackrel{?}{=} 0 \\
0 &= 0
\end{aligned}$$

Self Check Solve: $-2(y - 3) - 4y = 0$.

EXAMPLE 4 Solve: $3(x - 5) = 4(x + 9)$.

Solution

$$\begin{aligned}
3(x - 5) &= 4(x + 9) \\
3x - 15 &= 4x + 36 && \text{Remove parentheses.} \\
3x - 15 - 3x &= 4x + 36 - 3x && \text{Subtract } 3x \text{ from both sides.} \\
-15 &= x + 36 && \text{Combine like terms.} \\
-15 - 36 &= x + 36 - 36 && \text{Subtract 36 from both sides.} \\
-51 &= x && \text{Combine like terms.} \\
x &= -51
\end{aligned}$$

Check:

$$\begin{aligned}
3(x - 5) &= 4(x + 9) \\
3(-51 - 5) &\stackrel{?}{=} 4(-51 + 9) && \text{Substitute } -51 \text{ for } x. \\
3(-56) &\stackrel{?}{=} 4(-42) \\
-168 &= -168
\end{aligned}$$

Self Check Solve: $4(x + 3) = -3(x - 4)$.

EXAMPLE 5 Solve: $\frac{3x + 11}{5} = x + 3$.

Solution We first multiply both sides by 5 to clear the equation of fractions. When we multiply the right-hand side by 5, we must multiply the entire right-hand side by 5.

$$\frac{3x + 11}{5} = x + 3$$

$$\mathbf{5}\left(\frac{3x + 11}{5}\right) = \mathbf{5}(x + 3) \quad \text{[illegible]s by 5.}$$

$$3x + 11 = 5x + 15 \quad \text{[illegible]es.}$$

$$3x + 11 - \mathbf{11} = 5x + 15 - \mathbf{11} \quad \text{[illegible]oth sides.}$$

$$3x = 5x + 4 \quad \text{[illegible]s.}$$

$$3x - \mathbf{5x} = 5x + 4 - \mathbf{5x} \quad \text{Subtract } 5x \text{ from both sides.}$$

$$-2x = 4 \quad \text{Combine like terms.}$$

$$\frac{-2x}{\mathbf{-2}} = \frac{4}{\mathbf{-2}} \quad \text{Divide both sides by } -2.$$

$$x = -2 \quad \text{Simplify.}$$

Check:

$$\frac{3\mathbf{x} + 11}{5} = \mathbf{x} + 3$$

$$\frac{3(\mathbf{-2}) + 11}{5} \stackrel{?}{=} (\mathbf{-2}) + 3 \quad \text{Substitute } -2 \text{ for } x.$$

$$\frac{-6 + 11}{5} \stackrel{?}{=} 1 \quad \text{Simplify.}$$

$$\frac{5}{5} \stackrel{?}{=} 1$$

$$1 = 1$$

Self Check Solve: $\frac{2x - 5}{4} = x - 2$.

Comment Remember that when you multiply one side of an equation by a nonzero number, you must multiply the other side by the same number.

EXAMPLE 6 Solve: $0.2x + 0.4(50 - x) = 19$.

Solution Since $0.2 = \frac{2}{10}$ and $0.4 = \frac{4}{10}$, this equation contains fractions. To clear the fractions, we multiply both sides by 10.

$$0.2x + 0.4(50 - x) = 19$$

$$\mathbf{10[}0.2x + 0.4(50 - x)\mathbf{]} = \mathbf{10(}19\mathbf{)} \quad \text{Multiply both sides by 10.}$$

$$\mathbf{10[}0.2x\mathbf{]} + \mathbf{10[}0.4(50 - x)\mathbf{]} = \mathbf{10(}19\mathbf{)} \quad \text{Use the distributive property on the left-hand side.}$$

$$\begin{aligned} 2x + 4(50 - x) &= 190 && \text{Do the multiplications.} \\ 2x + 200 - 4x &= 190 && \text{Remove parentheses.} \\ -2x + 200 &= 190 && \text{Combine like terms.} \\ -2x &= -10 && \text{Subtract 200 from both sides.} \\ x &= 5 && \text{Divide both sides by } -2. \end{aligned}$$

Verify that the solution checks.

Self Check Solve: $0.3(20 - x) + 0.5x = 15$.

Identities and Contradictions

An equation that is true for all values of its variable is called an **identity.** For example, the equation $x + x = 2x$ is an identity because it is true for all values of x.

Because no number can equal a number that is 1 larger than itself, the equation $x = x + 1$ is not true for any number x. Such equations are called **contradictions.**

The equations in Examples 3–6 are called **conditional equations.** For these equations, some values of x are solutions, but other values of x are not.

EXAMPLE 7 Solve: $3(x + 8) + 5x = 2(12 + 4x)$.

Solution

$$\begin{aligned} 3(x + 8) + 5x &= 2(12 + 4x) \\ 3x + 24 + 5x &= 24 + 8x && \text{Remove parentheses.} \\ 8x + 24 &= 24 + 8x && \text{Combine like terms.} \\ 8x + 24 - 8x &= 24 + 8x - 8x && \text{Subtract } 8x \text{ from both sides.} \\ 24 &= 24 && \text{Combine like terms.} \end{aligned}$$

Since the result $24 = 24$ is true for every number x, every number x is a solution of the original equation. This equation is an identity.

Self Check Solve: $-2(x + 3) - 18x = 5(9 - 4x) - 51$.

EXAMPLE 8 Solve: $3(x + 7) - x = 2(x + 10)$.

Solution

$$\begin{aligned} 3(x + 7) - x &= 2(x + 10) \\ 3x + 21 - x &= 2x + 20 && \text{Remove parentheses.} \\ 2x + 21 &= 2x + 20 && \text{Combine like terms.} \\ 2x + 21 - 2x &= 2x + 20 - 2x && \text{Subtract } 2x \text{ from both sides.} \\ 21 &= 20 && \text{Combine like terms.} \end{aligned}$$

Since the result $21 = 20$ is false, the original equation has no solution. It is a contradiction.

Self Check Solve: $5(x - 2) - 2x = 3(x + 7)$.

Self Check Answers

1. $-3a - 25$ **2.** $-7b + 22$ **3.** 1 **4.** 0 **5.** $\frac{3}{2}$ **6.** 45 **7.** identity **8.** contradiction

Orals *Simplify by combining like terms.*

1. $3x + 5x$
2. $-2y + 3y$
3. $3x + 2x - 5x$
4. $3y + 2y - 7y$
5. $3(x + 2) - 3x + 6$
6. $3(x + 2) + 3x - 6$

Solve each equation, when possible.

7. $5x = 4x + 3$
8. $2(x - 1) = 2(x + 1)$
9. $3x = 2(x + 1)$
10. $x + 2(x + 1) = 3$

2.3 EXERCISES

REVIEW *Evaluate each expression when $x = -3$, $y = -5$, and $z = 0$.*

1. $x^2z(y^3 - z)$
2. $z - y^3$
3. $\dfrac{x - y^2}{2y - 1 + x}$
4. $\dfrac{2y + 1}{x} - x$

Perform the operations.

5. $\frac{6}{7} - \frac{5}{8}$
6. $\frac{6}{7} \cdot \frac{5}{8}$
7. $\frac{6}{7} \div \frac{5}{8}$
8. $\frac{6}{7} + \frac{5}{8}$

VOCABULARY AND CONCEPTS *Fill in the blanks.*

9. If terms have the same ________ with the same exponents, they are called ____ terms.

10. To combine like terms, ____ their numerical coefficients and _____ the same variables and exponents.

11. If an equation is true for all values of its variable, it is called an _______.

12. If an equation is true for some values of its variable, but not all, it is called a _________ equation.

PRACTICE *Simplify each expression, when possible.*

13. $3x + 17x$
14. $12y - 15y$
15. $8x^2 - 5x^2$
16. $17x^2 + 3x^2$
17. $9x + 3y$
18. $5x + 5y$
19. $3(x + 2) + 4x$
20. $9(y - 3) + 2y$
21. $5(z - 3) + 2z$
22. $4(y + 9) - 6y$
23. $12(x + 11) - 11$
24. $-3(3 + z) + 2z$
25. $8(y + 7) - 2(y - 3)$
26. $9(z + 2) + 5(3 - z)$
27. $2x + 4(y - x) + 3y$
28. $3y - 6(y + z) + y$
29. $(x + 2) - (x - y)$
30. $3z + 2(y - z) + y$
31. $2\left(4x + \frac{9}{2}\right) - 3\left(x + \frac{2}{3}\right)$
32. $7\left(3x - \frac{2}{7}\right) - 5\left(2x - \frac{3}{5}\right) + x$
33. $8x(x + 3) - 3x^2$
34. $2x + x(x + 3)$

Solve each equation. ***Check all solutions.***

35. $3x + 2 = 2x$
36. $5x + 7 = 4x$
37. $5x - 3 = 4x$
38. $4x + 3 = 5x$
39. $9y - 3 = 6y$
40. $8y + 4 = 4y$
41. $8y - 7 = y$
42. $9y - 8 = y$
43. $9 - 23w = 4w$
44. $y + 4 = -7y$
45. $22 - 3r = 8r$
46. $14 + 7s = s$
47. $3(a + 2) = 4a$
48. $4(a - 5) = 3a$
49. $5(b + 7) = 6b$
50. $8(b + 2) = 9b$
51. $2 + 3(x - 5) = 4(x - 1)$

52. $2 - (4x + 7) = 3 + 2(x + 2)$

53. $10x + 3(2 - x) = 5(x + 2) - 4$

54. $11x + 6(3 - x) = 3$

55. $3(a + 2) = 2(a - 7)$

56. $9(t - 1) = 6(t + 2) - t$

57. $9(x + 11) + 5(13 - x) = 0$

58. $3(x + 15) + 4(11 - x) = 0$

59. $\frac{3(t - 7)}{2} = t - 6$

60. $\frac{2(t + 9)}{3} = t - 8$

61. $\frac{5(2 - s)}{3} = s + 6$

62. $\frac{8(5 - s)}{5} = -2s$

63. $\frac{4(2x - 10)}{3} = 2(x - 4)$

64. $\frac{11(x - 12)}{2} = 9 - 2x$

65. $3.1(x - 2) = 1.3x + 2.8$

66. $0.6x - 0.8 = 0.8(2x - 1) - 0.7$

67. $2.7(y + 1) = 0.3(3y + 33)$

68. $1.5(5 - y) = 3y + 12$

69. $19.1x - 4(x + 0.3) = -46.5$

70. $18.6x + 7.2 = 1.5(48 - 2x)$

71. $14.3(x + 2) + 13.7(x - 3) = 15.5$

72. $1.25(x - 1) = 0.5(3x - 1) - 1$

Solve each equation. If it is an identity or a contradiction, so indicate.

73. $8x + 3(2 - x) = 5(x + 2) - 4$

74. $5(x + 2) = 5x - 2$

75. $2(s + 2) = 2(s + 1) + 3$

76. $21(b - 1) + 3 = 3(7b - 6)$

77. $\frac{2(t - 1)}{6} - 2 = \frac{t + 2}{6}$

78. $\frac{2(2r - 1)}{6} + 5 = \frac{3(r + 7)}{6}$

79. $2(3z + 4) = 2(3z - 2) + 13$

80. $x + 7 = \frac{2x + 6}{2} + 4$

81. $2(y - 3) - \frac{y}{2} = \frac{3}{2}(y - 4)$

82. $\frac{20 - a}{2} = \frac{3}{2}(a + 4)$

83. $\frac{3x + 14}{2} = x - 2 + \frac{x + 18}{2}$

84. $\frac{5(x + 3)}{3} - x = \frac{2(x + 8)}{3}$

WRITING

85. Explain why $3x^2y$ and $5x^2y$ are like terms.

86. Explain why $3x^2y$ and $3xy^2$ are unlike terms.

87. Discuss whether $7xxy^3$ and $5x^2yyy$ are like terms.

88. Discuss whether $\frac{3}{2}x$ and $\frac{3x}{2}$ are like terms.

SOMETHING TO THINK ABOUT

89. What number is equal to its own double?

90. What number is equal to one-half of itself?

2.4 Introduction to Problem Solving

- Problem Solving
- Number Problems
- Geometry Problems
- Investment Problems

Getting Ready

1. If one part of a pipe is x feet long and the other part is $(x + 2)$ feet long, find an expression that represents the total length of the pipe.

2. If one part of a board is x feet long and the other part is three times as long, find an expression that represents the length of board.

3. What is the formula for the perimeter of a rectangle?

4. Define a triangle.

In this section, we will discuss the solutions to several types of word problems. To do so, we will follow a five-step problem-solving technique.

Problem Solving

Algebra enables us to solve many problems. The key to successful problem solving is to thoroughly understand the problem and then devise a plan to solve it. The following list of steps provides a strategy to follow.

Problem Solving

1. **Analyze the problem** by reading it several times to understand the given facts. What information is given? What are you asked to find? Often a sketch, chart, or diagram will help you visualize the facts of the problem.
2. **Form an equation** by picking a variable to represent the quantity to be found. Then express all other unknown quantities in the problem as expressions involving that variable. Finally, write an equation expressing the quantity in two different ways.
3. **Solve the equation** found in Step 2.
4. **State the conclusion.**
5. **Check the result.**

In this section, we will use this five-step strategy to solve many problems.

Number Problems

EXAMPLE 1 A plumber wants to cut a 17-foot pipe into three parts. (See Figure 2-5.) If the longest part is to be 3 times as long as the shortest part, and the middle-sized part is to be 2 feet longer than the shortest part, how long should each part be?

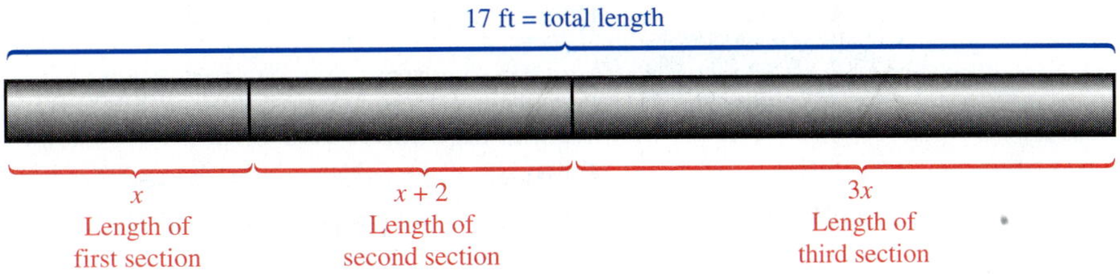

Figure 2-5

Analyze the problem The information is given in terms of the length of the shortest part. Therefore, we let a variable represent the length of the shortest part and express the other lengths in terms of that variable.

Form an equation Let x represent the length of the shortest part. Then $3x$ represents the length of the longest part, and $x + 2$ represents the length of the middle-sized part.

The sum of the lengths of these three parts is equal to the total length of the pipe.

The length of part 1	plus	the length of part 2	plus	the length of part 3	equals	the total length.
x	$+$	$x + 2$	$+$	$3x$	$=$	17

Solve the equation We can solve this equation as follows.

$$x + x + 2 + 3x = 17 \quad \text{The equation to solve.}$$
$$5x + 2 = 17 \quad \text{Combine like terms.}$$
$$5x = 15 \quad \text{Subtract 2 from both sides.}$$
$$x = 3 \quad \text{Divide both sides by 5.}$$

State the conclusion The shortest part is 3 feet long. Because the middle-sized part is 2 feet longer than the shortest, it is 5 feet long. Because the longest part is 3 times longer than the shortest, it is 9 feet long.

Check the result Because 3 feet, 5 feet, and 9 feet total 17 feet, the solution checks.

Geometry Problems

The geometric figure shown in Figure 2-6(a) is called an **angle.** Angles are measured in **degrees.** The angle shown in Figure 2-6(b) measures 45 degrees (denoted as 45°). If an angle measures 90°, as in Figure 2-6(c), it is called a **right angle.** If an angle measures 180°, as in Figure 2-6(d), it is called a **straight angle.**

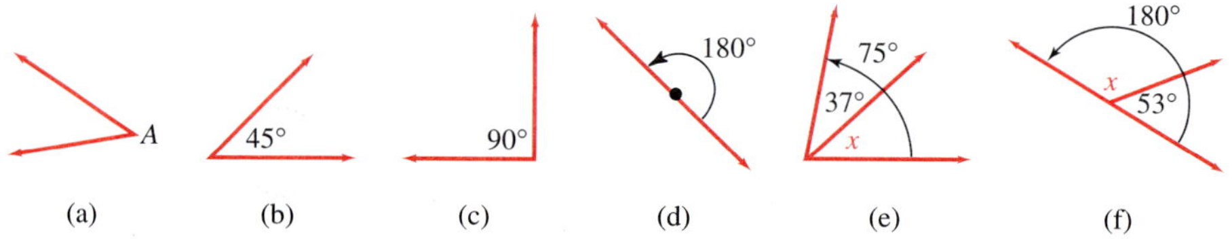

Figure 2-6

EXAMPLE 2 Refer to Figure 2-6(e) and find x.

Analyze the problem In Figure 2-6(e), we have two angles that are side by side. From the figure, we can see that the sum of their measures is 75°.

Form an equation Since the sum of x and 37° is equal to 75°, we can form the equation as follows.

The angle that measures x	plus	the angle that measures 37°	equals	the angle that measures 75°.
x	$+$	37	$=$	75

Solve the equation We can solve this equation as follows.

$$x + 37 = 75 \quad \text{The equation to solve.}$$
$$x + 37 - 37 = 75 - 37 \quad \text{Subtract 37 from both sides.}$$
$$x = 38 \quad 37 - 37 = 0 \text{ and } 75 - 37 = 38.$$

State the conclusion The value of x is 38°.

Check the result Since the sum of 38° and 37° is 75°, the solution checks. ∎

EXAMPLE 3 Refer to Figure 2-6(f) and find x.

Analyze the problem In Figure 2-6(f), we have two angles that are side by side, and the sum of their measures is 180°.

Form an equation Since the sum of x and 53° is equal to 180°, we can form the equation:

The angle that measures x	plus	the angle that measures 53°	equals	the angle that measures 180°.
x	$+$	53	$=$	180

Solve the equation We can solve this equation as follows.

$$x + 53 = 180 \qquad \text{The equation to solve.}$$
$$x + 53 - 53 = 180 - 53 \qquad \text{Subtract 53 from both sides.}$$
$$x = 127 \qquad 53 - 53 = 0 \text{ and } 180 - 53 = 127.$$

State the conclusion The value of x is 127°.

Check the result Since the sum of 127° and 53° is 180°, the solution checks. ∎

If the sum of two angles is 90°, the angles are called **complementary.** If the sum of two angles is 180°, the angles are called **supplementary.**

EXAMPLE 4 Find the complement of an angle measuring 30°.

Analyze the problem If two angles are complementary, either angle is called the **complement** of the other. To find the complement of a 30° angle, we must find an angle whose measure plus 30° equals 90°.

Form an equation We can let x represent the complement of 30°. Since the sum of two complementary angles is 90°, we can form the equation:

The angle that measures x	plus	the angle that measures 30°	equals	90°.
x	$+$	30	$=$	90

Solve the equation We can solve this equation as follows.

$$x + 30 = 90 \qquad \text{The equation to solve.}$$
$$x + 30 - 30 = 90 - 30 \qquad \text{Subtract 30 from both sides.}$$
$$x = 60 \qquad 30 - 30 = 0 \text{ and } 90 - 30 = 60.$$

State the result The complement of a 30° angle is a 60° angle.

Check the result Since the sum of 60° and 30° is 90°, the solution checks. ∎

EXAMPLE 5 Find the supplement of an angle measuring 50°.

Analyze the problem If two angles are supplementary, either angle is called the **supplement** of the other. To find the supplement of a 50° angle, we must find an angle whose measure plus 50° equals 180°.

Form an equation We can let x represent the supplement of 50°. Since the sum of two supplementary angles is 180°, we can form the equation:

The angle that measures x	plus	the angle that measures 50°	equals	180°.
x	$+$	50	$=$	180

Solve the equation We can solve this equation as follows.

$$x + 50 = 180 \quad \text{The equation to solve.}$$
$$x + 50 - 50 = 180 - 50 \quad \text{Subtract 50 from both sides.}$$
$$x = 130 \quad 50 - 50 = 0 \text{ and } 180 - 50 = 130.$$

State the conclusion The supplement of a 50° angle is a 130° angle.

Check the result Since the sum of 50° and 130° is 180°, the solution checks. ▮

EXAMPLE 6 The length of a rectangle is 4 meters longer than twice its width. If the perimeter of the rectangle is 26 meters, find its dimensions.

Analyze the problem To visualize the problem, we sketch the rectangle shown in Figure 2-7. Since the formula for the perimeter of a rectangle is $P = 2l + 2w$, the perimeter of the rectangle in the figure is $2(4 + 2w) + 2w$. We are also told that the perimeter is 26.

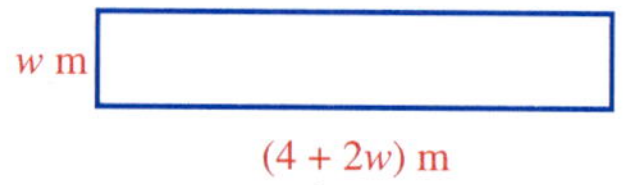

Figure 2-7

Form an equation Let w represent the width of the rectangle. Then $4 + 2w$ represents the length of the rectangle. We can form the equation as follows.

2	times	the length	plus	2	times	the width	equals	the perimeter.
2	$\cdot$	$(4 + 2w)$	$+$	2	$\cdot$	w	$=$	26

Solve the equation We can solve this equation as follows.

$$2(4 + 2w) + 2w = 26 \quad \text{The equation to solve.}$$
$$8 + 4w + 2w = 26 \quad \text{Remove parentheses.}$$
$$6w + 8 = 26 \quad \text{Combine like terms.}$$
$$6w = 18 \quad \text{Subtract 8 from both sides.}$$
$$w = 3 \quad \text{Divide both sides by 6.}$$

State the conclusion The width of the rectangle is 3 meters, and the length, $4 + 2w$, is 10 meters.

Check the result If the rectangle has a width of 3 meters and a length of 10 meters, the length is 4 meters longer than twice the width ($4 + 2 \cdot 3 = 10$), and the perimeter is $2 \cdot 10 + 2 \cdot 3 = 26$ meters. The solution checks. ■

EXAMPLE 7 The vertex angle of an isosceles triangle is 56°. Find the measure of each base angle.

Analyze the problem An **isosceles triangle** has two sides of equal length, which meet to form the **vertex angle.** See Figure 2-8. The angles opposite those sides, called **base angles,** are also equal. If we let x represent the measure of one base angle, the measure of the other base angle is also x. In any triangle, the sum of the measures of its three angles is 180°.

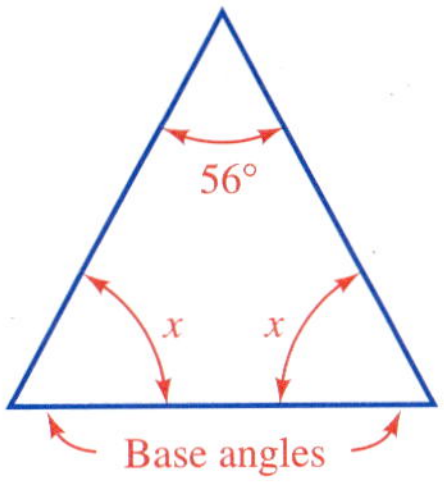

Figure 2-8

Form an equation Let x represent the measure of one base angle. Then x also represents the measure of the other base angle. We can form the equation:

One base angle	plus	the other base angle	plus	the vertex angle	equals	180°.
x	$+$	x	$+$	56	$=$	180

Solve the equation We can solve this equation as follows.

$$x + x + 56 = 180 \quad \text{The equation to solve.}$$
$$2x + 56 = 180 \quad \text{Combine like terms.}$$
$$2x = 124 \quad \text{Subtract 56 from both sides.}$$
$$x = 62 \quad \text{Divide both sides by 2.}$$

State the conclusion The measure of each base angle is 62°.

Check the result The measure of each base angle is 62°, and the vertex angle measures 56°. Since $62° + 62° + 56° = 180°$, the measures of these three angles total 180°. The solution checks. ■

Investment Problems

EXAMPLE 8 A teacher invests part of \$12,000 at 6% annual interest, and the rest at 9%. If the annual income from these investments was \$945, how much did the teacher invest at each rate?

Analyze the problem The interest i earned by an amount p invested at an annual rate r for t years is given by the formula $i = prt$. In this example, $t = 1$ year. Hence, if x dollars were invested at 6%, the interest earned would be $0.06x$ dollars. If x dollars were invested at 6%, the rest of the money, $(12{,}000 - x)$ dollars, would be invested at 9%. The interest earned on that money would be $0.09(12{,}000 - x)$ dollars. The total interest earned in dollars can be expressed in two ways: as 945 and as the sum $0.06x + 0.09(12{,}000 - x)$.

Form an equation Let x represent the amount of money invested at 6%. Then $12{,}000 - x$ represents the amount of money invested at 9%. We can form an equation as follows.

The interest earned at 6%	plus	the interest earned at 9%	equals	the total interest.
$0.06x$	$+$	$0.09(12{,}000 - x)$	$=$	945

Solve the equation We can solve this equation as follows.

$$0.06x + 0.09(12{,}000 - x) = 945 \quad \text{The equation to solve.}$$
$$6x + 9(12{,}000 - x) = 94{,}500 \quad \text{Multiply both sides by 100 to clear the equation of decimals.}$$
$$6x + 108{,}000 - 9x = 94{,}500 \quad \text{Remove parentheses.}$$
$$-3x + 108{,}000 = 94{,}500 \quad \text{Combine like terms.}$$
$$-3x = -13{,}500 \quad \text{Subtract 108,000 from both sides.}$$
$$x = 4{,}500 \quad \text{Divide both sides by } -3.$$

State the conclusion The teacher invested \$4,500 at 6% and \$12,000 − \$4,500 = \$7,500 at 9%.

Check the result The first investment earned 6% of \$4,500, or \$270. The second investment earned 9% of \$7,500, or \$675. Because the total return was \$270 + \$675, or \$945, the solutions check.

Orals

1. Find the complement of a 20° angle.
2. Find the supplement of a 70° angle.
3. Find the perimeter of a rectangle 4 feet wide and 6 feet long.
4. Find an expression that represents one year's interest on \$18,000, invested at an annual rate r.

2.4 EXERCISES

REVIEW *Refer to the formulas in Section 1.3.*

1. Find the volume of a pyramid that has a height of 6 centimeters and a square base, 10 centimeters on each side.
2. Find the volume of a cone with a height of 6 centimeters and a circular base with radius 6 centimeters. Use $\pi \approx \frac{22}{7}$.

Simplify each expression.

3. $3(x + 2) + 4(x - 3)$
4. $4(x - 2) - 3(x + 1)$
5. $\frac{1}{2}(x + 1) - \frac{1}{2}(x + 4)$
6. $\frac{3}{2}\left(x + \frac{2}{3}\right) + \frac{1}{2}(x + 8)$
7. The amount A on deposit in a bank account bearing simple interest is given by the formula

$$A = P + Prt$$

Find A when $P = \$1{,}200$, $r = 0.08$, and $t = 3$.

8. The distance s that a certain object falls in t seconds is given by the formula

$$s = 350 - 16t^2 + vt$$

Find s when $t = 4$ and $v = -3$.

VOCABULARY AND CONCEPTS *Fill in the blanks.*

9. The perimeter of a rectangle is given by the formula $P =$ ______.

10. An ______ triangle is a triangle with two sides of equal length.

11. The sides of equal length of an isosceles triangle meet to form the ______ angle.

12. The angles opposite the sides of equal length of an isosceles triangle are called ______ angles.

13. If the sum of the measures of two angles is 90°, the angles are called ______ angles.

14. If the sum of the measures of two angles is 180°, the angles are called ______ angles.

15. The sum of the measures of the angles of any triangle is ______.

16. The last step in the problem solving process is to ______ the result.

APPLICATIONS

17. Carpentry The 12-foot board in the illustration has been cut into two parts, one twice as long as the other. How long is each part?

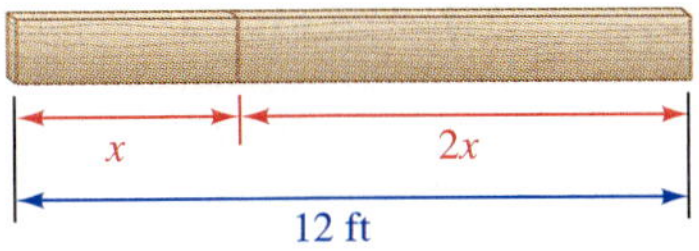

18. Plumbing A 20-foot pipe has been cut into two parts, one 3 times as long as the other. How long is each part?

19. Robotics If the robotic arm shown in the illustration will extend a total distance of 30 feet, how long is each section?

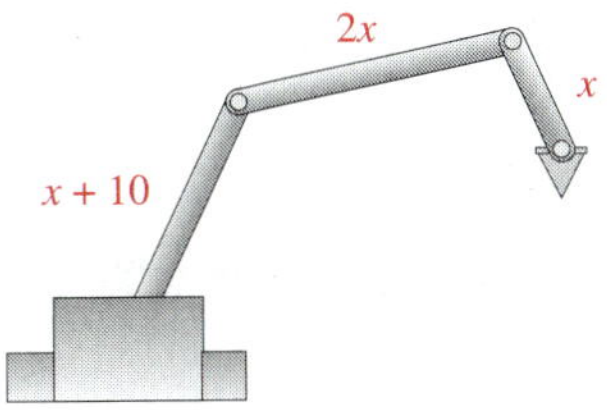

20. Statue of Liberty If the figure part of the Statue of Liberty is 3 feet shorter than the height of its pedestal base, find the height of the figure.

21. Window design The perimeter of the triangular window shown in the illustration is 24 feet. How long is each section?

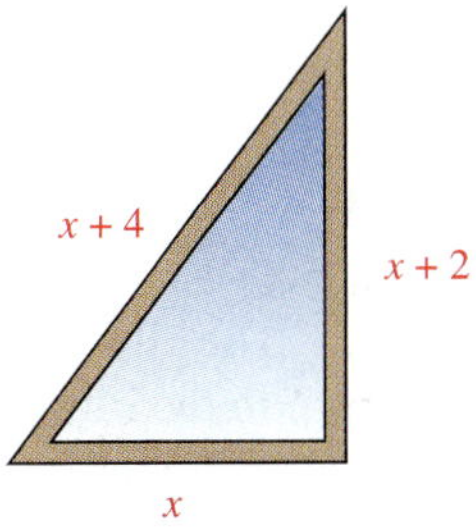

22. Football In 1967, Green Bay beat Kansas City by 25 points in the first Super Bowl. If a total of 45 points were scored, what was the final score of the game?

GEOMETRY *In Exercises 23–30, find x.*

23.

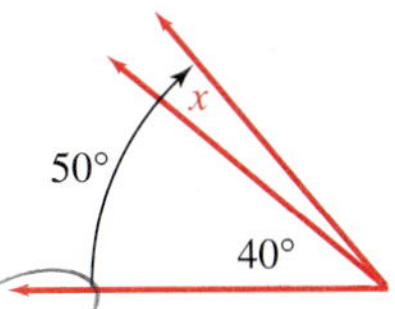

24.

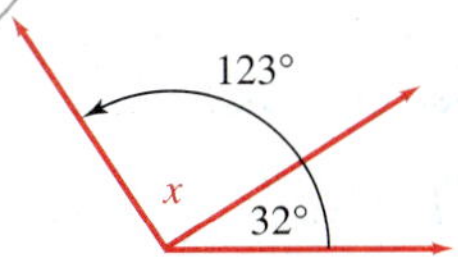

25.

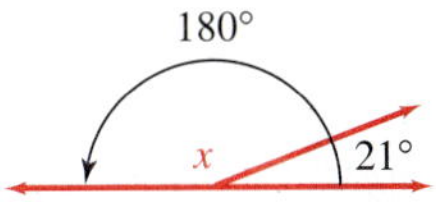

26.

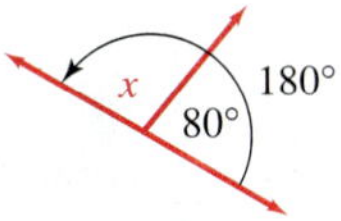

27.

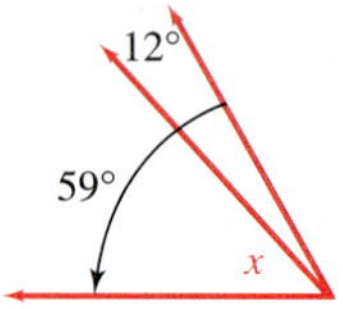

28.

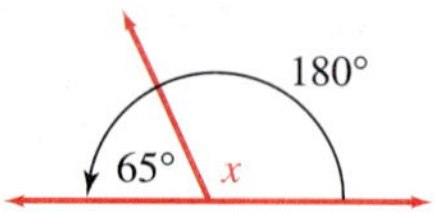

29.

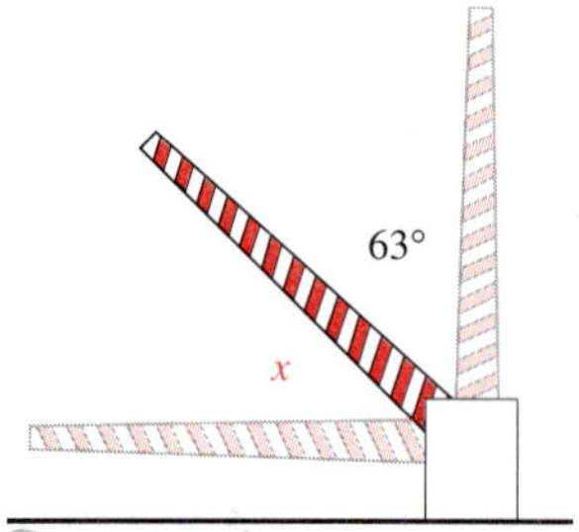

30.

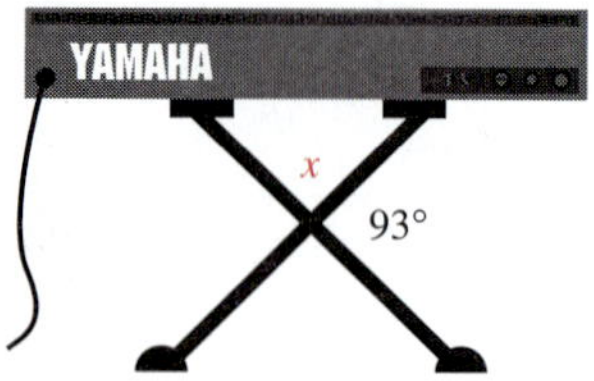

31. Find the complement of 37°.

32. Find the supplement of 37°.

33. Find the supplement of the complement of 40°.

34. Find the complement of the supplement of 140°.

35. Triangular bracing The outside perimeter of the triangular brace shown in the illustration is 57 feet. If all three sides are of equal length, find the length of each side.

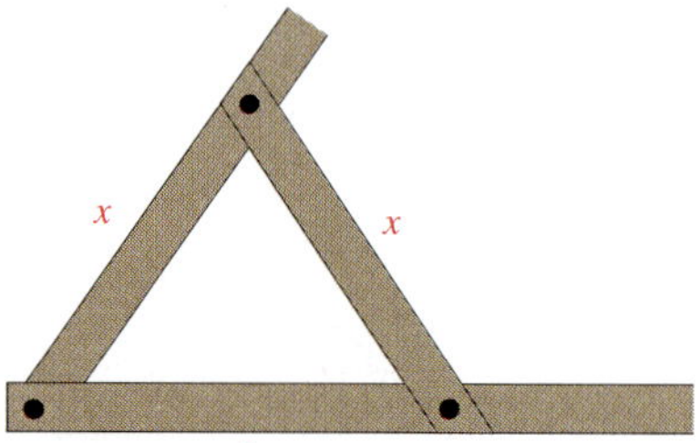

36. Circuit board The perimeter of the circuit board in the illustration is 90 centimeters. Find the dimensions of the board.

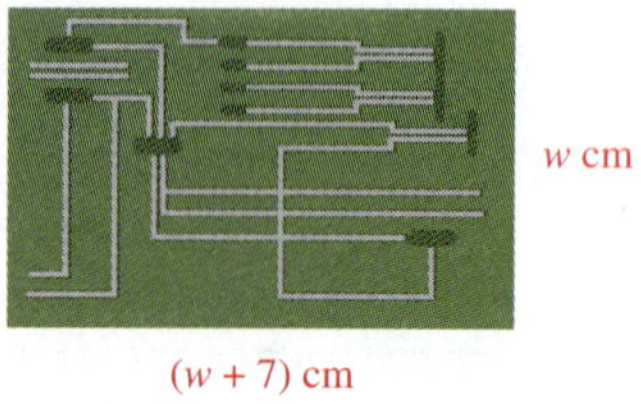

37. Swimming pool The width of a rectangular swimming pool is 11 meters less than the length, and the perimeter is 94 meters. Find its dimensions.

38. Wooden truss The truss in the illustration is in the form of an isosceles triangle. Each of the two equal sides is 4 feet less than the third side. If the perimeter is 25 feet, find the length of each side.

39. Framing a picture The length of a rectangular picture is 5 inches greater than twice the width. If the perimeter is 112 inches, find the dimensions of the frame.

40. Guy wires The two guy wires in the illustration form an isosceles triangle. One of the two equal angles of the triangle is 4 times the third angle (the vertex angle). Find the measure of the vertex angle.

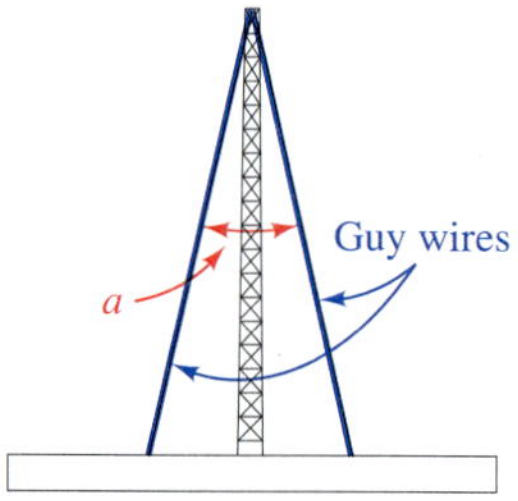

41. Equilateral triangles Find the measure of each angle of an equilateral triangle. (*Hint:* The three angles of an equilateral triangle are equal.)

42. Land area The perimeter of a square piece of land is twice the perimeter of an equilateral (equal-sided) triangular lot. If one side of the square is 60 meters, find the length of a side of the triangle.

43. Investment problem A student invested some money at an annual rate of 5%. If the annual income from the investment is \$300, how much did he invest?

44. Investment problem A student invested 90% of her savings in the stock market. If she invested \$4,050, what are her total savings?

45. Investment problem A broker invested $24,000 in two mutual funds, one earning 9% annual interest and the other earning 14%. After 1 year, his combined interest is $3,135. How much was invested at each rate?

46. Investment problem A rollover IRA of $18,750 was invested in two mutual funds, one earning 12% interest and the other earning 10%. After 1 year, the combined interest income is $2,117. How much was invested at each rate?

47. Investment problem One investment pays 8% and another pays 11%. If equal amounts are invested in each, the combined interest income for 1 year is $712.50. How much is invested at each rate?

48. Investment problem When equal amounts are invested in each of three accounts paying 7%, 8%, and 10.5%, one year's combined interest income is $1,249.50. How much is invested in each account?

49. Investment problem A college professor wants to supplement her retirement income with investment interest. If she invests $15,000 at 6% annual interest, how much more would she have to invest at 7% to achieve a goal of $1,250 in supplemental income?

50. Investment problem A teacher has a choice of two investment plans: an insured fund that has paid an average of 11% interest per year, or a riskier investment that has averaged a 13% return. If the same amount invested at the higher rate would generate an extra $150 per year, how much does the teacher have to invest?

51. Investment problem A financial counselor recommends investing twice as much in CDs (certificates of deposit) as in a bond fund. A client follows his advice and invests $21,000 in CDs paying 1% more interest than the fund. The CDs would generate $840 more interest than the fund. Find the two rates. (*Hint:* 1% = 0.01.)

52. Investment problem The amount of annual interest earned by $8,000 invested at a certain rate is $200 less than $12,000 would earn at a 1% lower rate. At what rate is the $8,000 invested?

WRITING

53. Write a paragraph describing the problem-solving process.

54. List as many types of angles as you can think of. Then define each type.

SOMETHING TO THINK ABOUT

55. If two lines intersect as in the illustration, angle 1 (denoted as $\angle 1$) and $\angle 2$, and $\angle 3$ and $\angle 4$, are called **vertical angles.** Let the measure of $\angle 1$ be various numbers and compute the values of the other three angles. What do you discover?

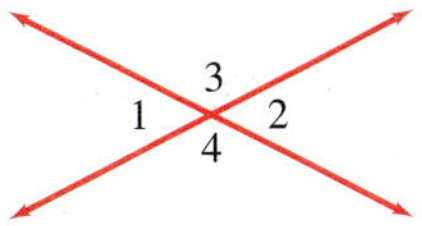

56. If two lines meet and form a right angle, the lines are said to be **perpendicular.** See the illustration. Find the measures of $\angle 1$, $\angle 2$, and $\angle 3$. What do you discover?

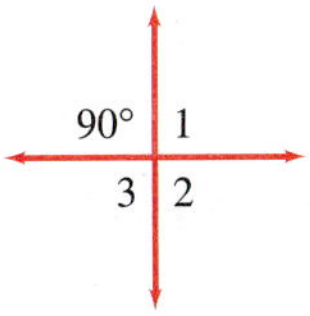

2.5 Motion and Mixture Problems

- Motion Problems
- Liquid Mixture Problems
- Dry Mixture Problems

Getting Ready

1. At 30 mph, how far would a bus go in 2 hours?

2. At 55 mph, how far would a car travel in 7 hours?

3. If 8 gallons of a mixture of water and alcohol is 70% alcohol, how much alcohol does the mixture contain?

4. At \$7 per pound, how many pounds of chocolate would be worth \$63?

In this section, we continue our discussion of problem solving by considering motion and mixture problems.

Motion Problems

EXAMPLE 1 Chicago and Green Bay are about 200 miles apart. If a car leaves Chicago traveling toward Green Bay at 55 mph at the same time as a truck leaves Green Bay bound for Chicago at 45 mph, how long will it take them to meet?

Analyze the problem Motion problems are based on the formula $d = rt$, where d is the distance traveled, r is the rate, and t is the time. We know that the two vehicles travel for the same amount of time—say, t hours. The faster car will travel $55t$ miles, and the slower truck will travel $45t$ miles. At the time they meet, the total distance traveled can be expressed in two ways: as the sum $55t + 45t$, and as 200 miles.

We can organize the information of this problem in a chart or a diagram, as shown in Figure 2-9.

	r	$\cdot\ t$	$= d$
Car	55	t	$55t$
Truck	45	t	$45t$

(a)

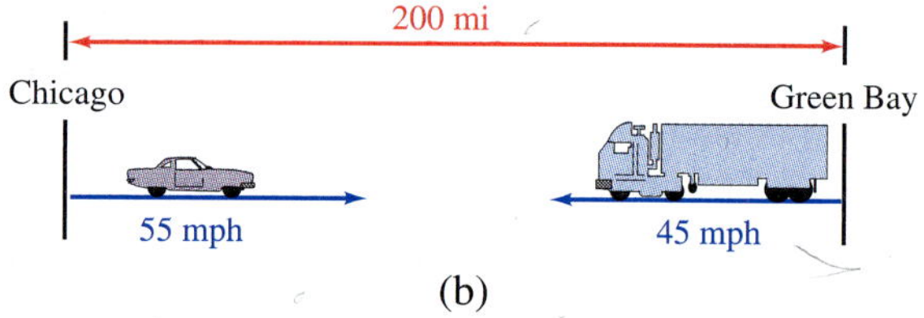

(b)

Figure 2-9

Form an equation Let t represent the time that the vehicles travel until they meet. Then $55t$ represents the distance traveled by the car, and $45t$ represents the distance traveled by the truck. After referring to Figure 2-9, we can form the equation:

The distance the car goes	plus	the distance the truck goes	equals	the total distance.
$55t$	$+$	$45t$	$=$	200

Solve the equation We can solve this equation as follows.

$$55t + 45t = 200 \quad \text{The equation to solve.}$$
$$100t = 200 \quad \text{Combine like terms.}$$
$$t = 2 \quad \text{Divide both sides by 100.}$$

State the conclusion The vehicles will meet in 2 hours.

Check the result In 2 hours, the car will travel $55 \cdot 2 = 110$ miles, while the truck will travel $45 \cdot 2 = 90$ miles. The total distance traveled will be $110 + 90 = 200$ miles. Since this is the total distance between Chicago and Green Bay, the solution checks. ▮

EXAMPLE 2 Two boats leave port, one heading east at 12 mph and one heading west at 10 mph. How long will it take before the boats are 33 miles apart?

Analyze the problem In this problem, the boats leave port at the same time and travel in opposite directions. We know that both boats travel for the same amount of time—say, t hours. The faster boat will travel $12t$ miles, and the slower boat will travel $10t$ miles. When they are 33 miles apart, the total distance traveled can be expressed in two ways: as the sum $12t + 10t$ and as 33 miles. We can organize the information of this problem in a chart or a diagram, as shown in Figure 2-10.

	r	$\cdot\ t$	$= d$
Faster boat	12	t	$12t$
Slower boat	10	t	$10t$

(a)

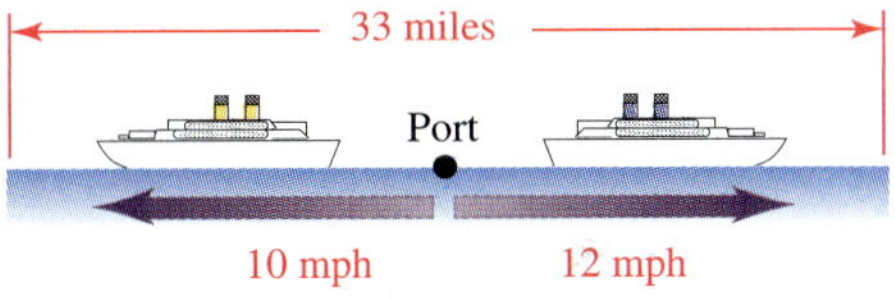

(b)

Figure 2-10

Form an equation Let t represent the time that each boat travels until they meet. Then $12t$ represents the distance traveled by the faster boat, and $10t$ represents the distance traveled by the slower boat. After referring to Figure 2-10, we can form the equation:

The distance the faster boat goes	plus	the distance the slower boat goes	equals	the total distance.
$12t$	$+$	$10t$	$=$	33

Solve the equation We can solve this equation as follows.

$$12t + 10t = 33 \quad \text{The equation to solve.}$$

$$22t = 33 \quad \text{Combine like terms.}$$

$$t = \frac{33}{22} \quad \text{Divide both sides by 22.}$$

$$t = \frac{3}{2} \quad \text{Simplify the fractions: } \frac{33}{22} = \frac{3 \cdot \overset{1}{\cancel{11}}}{2 \cdot \underset{1}{\cancel{11}}} = \frac{3}{2}.$$

State the conclusion The boats will be 33 miles apart in $\frac{3}{2}$ or 1.5 hours.

Check the result In 1.5 hours, the faster boat travels $12 \cdot 1.5 = 18$ miles, while the slower boat travels $10 \cdot 1.5 = 15$ miles. Since the total distance traveled is $18 + 15 = 33$ miles, the solution checks. ▮

EXAMPLE 3 A car leaves Beloit, heading east at 50 mph. One hour later, a second car leaves Beloit, heading east at 65 mph. How long will it take for the second car to overtake the first car?

Analyze the problem In this problem, the cars are going in the same direction, but the first car travels for one additional hour, because it got a 1-hour head start. If we let t represent the number of hours that the second car travels, then $t + 1$ will represent the number of hours that the first car travels. We know that car 1 travels at 50 mph and car 2 travels at 65 mph. When car 2 overtakes car 1, car 2 will have traveled $65t$ miles and car 1 will have traveled $50(t + 1)$ hours. The distance the cars travel can be expressed in two ways, as $50(t + 1)$ miles and as $65t$ miles.

We can organize the information in a chart or a diagram, as shown in Figure 2-11.

	r ·	t =	d
Car 1	50	$(t + 1)$	$50(t + 1)$
Car 2	65	t	$65t$

(a)

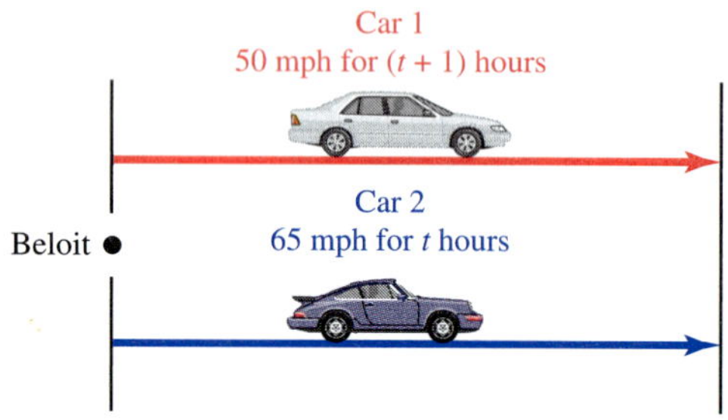

(b)

Figure 2-11

Form an equation Let t represent the time that car 2 travels. Then $t + 1$ represents the time that car 1 travels. Using the formula $r \cdot t = d$, we know that car 1 going 50 mph travels $50(t + 1)$ miles and that car 2 going 65 mph travels $65t$ miles. Since these distances are equal when car 2 overtakes car 1, we can form the equation:

The distance that car 1 goes	equals	the distance that car 2 goes.
$50(t + 1)$	=	$65t$

Solve the equation We can solve this equation as follows.

$$50(t + 1) = 65t \quad \text{The equation to solve.}$$

$$50t + 50 = 65t \quad \text{Use the distributive property to remove parentheses.}$$

$$50 = 15t \quad \text{Subtract } 50t \text{ from both sides.}$$

$$\frac{50}{15} = t \quad \text{Divide both sides by 15.}$$

$$t = \frac{10}{3} \quad \text{Simplify the fraction: } \frac{50}{15} = \frac{10 \cdot \overset{1}{\cancel{5}}}{3 \cdot \underset{1}{\cancel{5}}} = \frac{10}{3}.$$

State the conclusion Car 2 will overtake car 1 in $\frac{10}{3}$, or $3\frac{1}{3}$ hours.

Check the result In $3\frac{1}{3}$ hours, car 2 will have traveled $65\left(\frac{10}{3}\right)$, or $\frac{650}{3}$ miles. With a 1-hour head start, car 1 will have traveled $50\left(\frac{10}{3} + 1\right) = 50\left(\frac{13}{3}\right)$, or $\frac{650}{3}$ miles. Since these distances are equal, the solution checks. ▮

Liquid Mixture Problems

EXAMPLE 4 A chemist has one solution that is 50% sulfuric acid and another that is 20% sulfuric acid. How much of each should she use to make 12 liters of a solution that is 30% sulfuric acid?

Analyze the problem Liquid mixture problems are based on the percent formula $rb = a$, where b is the base, r is the rate, and a is the amount.

If we let x represent the number of liters of 50% solution to use, the amount of sulfuric acid in the solution will be $0.50x$ liters. Since there must be 12 liters of final mixture, $12 - x$ represents the number of liters of 20% solution to use. The amount of sulfuric acid in the 20% solution will be $0.20(12 - x)$ liters. The amount of sulfuric acid in the final mixture will be $0.30(12)$ liters. We can organize this information in a chart or a diagram, as shown in Figure 2-12.

	r ·	b =	a
50% solution	0.50	x	$0.50x$
20% solution	0.20	$12 - x$	$0.20(12 - x)$
30% solution	0.30	12	$0.30(12)$

(a)

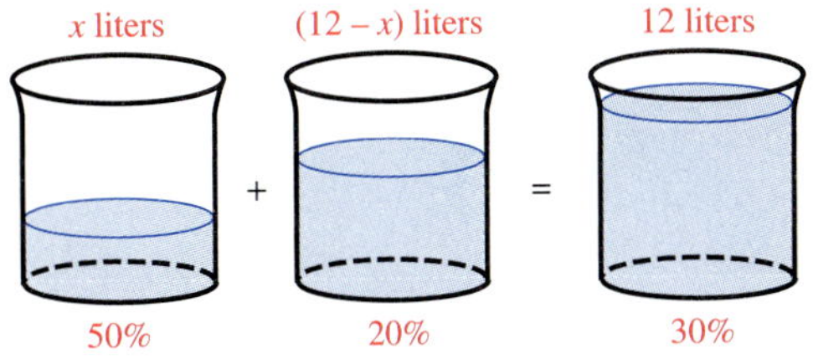

(b)

Figure 2-12

Form an equation Let x represent the number of liters of 50% solution that will be used. Then $12 - x$ represents the number of liters of 20% solution that will be used. Since the number of liters of sulfuric acid in the 50% solution plus the number of liters of sulfuric acid in the 20% solution will equal the number of liters of sulfuric acid in the mixture, we can form the equation:

The sulfuric acid in the 50% solution	plus	the sulfuric acid in the 20% solution	equals	the sulfuric acid in the final mixture.
50% of x	+	20% of $(12 - x)$	=	30% of 12

Solve the equation We can solve this equation as follows.

$$0.5x + 0.2(12 - x) = 0.3(12)$$ The equation to solve. $50\% = 0.5$, $20\% = 0.2$, and $30\% = 0.3$.

$$5x + 2(12 - x) = 3(12)$$ Multiply both sides by 10 to clear the equation of decimals.

$$5x + 24 - 2x = 36$$ Remove parentheses.

$$3x + 24 = 36$$ Combine like terms.

$$3x = 12$$ Subtract 24 from both sides.

$$x = 4$$ Divide both sides by 3.

State the conclusion The chemist must mix 4 liters of the 50% solution and $12 - 4 = 8$ liters of the 20% solution.

Check the result Check the result. ∎

Dry Mixture Problems

EXAMPLE 5 Fancy cashews are not selling at \$9 per pound, because they are too expensive. However, filberts are selling well at \$6 per pound. How many pounds of filberts should be combined with 50 pounds of cashews to obtain a mixture that can be sold at \$7 per pound?

Analyze the problem Dry mixture problems are based on the formula $v = pn$, where v is the value of the mixture, p is the price per pound, and n is the number of pounds. Suppose x pounds of filberts are used in the mixture. At \$6 per pound, they are worth $\$6x$. At \$9 per pound, the 50 pounds of cashews are worth $\$9 \cdot 50$, or \$450. The mixture will weigh $(50 + x)$ pounds, and at \$7 per pound, it will be worth $\$7(50 + x)$. The value of the ingredients, $\$(6x + 450)$, is equal to the value of the mixture, $\$7(50 + x)$. We can organize the information of the problem in a chart or a diagram, as shown in Figure 2-13.

	p ·	n =	v
Filberts	6	x	$6x$
Cashews	9	50	$9(50)$
Mixture	7	$50 + x$	$7(50 + x)$

(a)

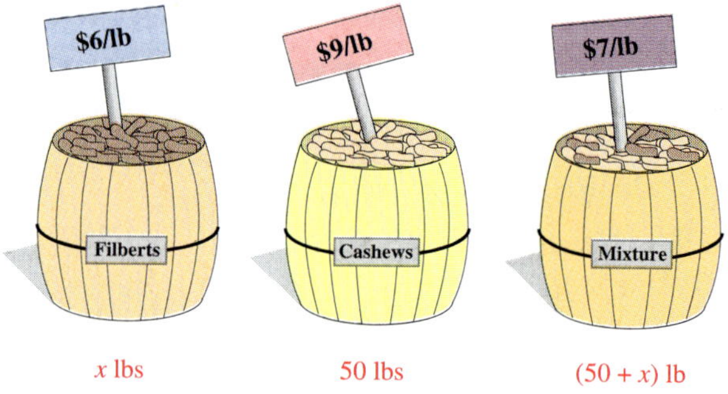

(b)

Figure 2-13

Form an equation Let x represent the number of pounds of filberts in the mixture. We can form the equation:

The value of the filberts	plus	the value of the cashews	equals	the value of the mixture.
$6x$	$+$	$9(50)$	$=$	$7(50 + x)$

Solve the equation We can solve this equation as follows.

$$6x + 9(50) = 7(50 + x) \quad \text{The equation to solve.}$$
$$6x + 450 = 350 + 7x \quad \text{Remove parentheses and simplify.}$$
$$100 = x \quad \text{Subtract } 6x \text{ and } 350 \text{ from both sides.}$$

State the conclusion The storekeeper should use 100 pounds of filberts in the mixture.

Check the result

The value of 100 pounds of filberts at \$6 per pound is	\$ 600
The value of 50 pounds of cashews at \$9 per pound is	\$ 450
The value of the mixture is	\$1,050

The value of 150 pounds of mixture at \$7 per pound is also \$1,050. ∎

Orals

1. How far will a car travel in h hours at a speed of 50 mph?
2. Two cars leave Midtown at the same time, one at 55 mph and the other at 65 mph. If they travel in the same direction, how far apart will they be in h hours?
3. How many ounces of alcohol are there in 12 ounces of a solution that is 40% alcohol?
4. Find the value of 7 pounds of coffee worth $\$d$ per pound.

2.5 EXERCISES

REVIEW *Simplify each expression.*

1. $3 + 4(-5)$
2. $\dfrac{-5(3) - 2(-2)}{6 - (-5)}$
3. $2^3 - 3^2$
4. $3^2 + 3(2) - (-5)$

Solve each equation.

5. $-2x + 3 = 9$
6. $\frac{1}{3}y - 4 = 2$
7. $\frac{2}{3}p + 1 = 5$
8. $2(z + 3) = 4(z - 1)$

VOCABULARY AND CONCEPTS *Fill in the blanks.*

9. Motion problems are based on the formula ______.
10. Liquid mixture problems are based on the formula ______.
11. Dry mixture problems are based on the formula ______.
12. The information in motion and mixture problems can be organized in the form of a _____ or a _______.

PRACTICE

13. **Travel time** Ashford and Bartlett are 315 miles apart. A car leaves Ashford bound for Barlett at 50 mph. At the same time, another car leaves Bartlett bound for Ashford at 55 mph. How long will it take them to meet?

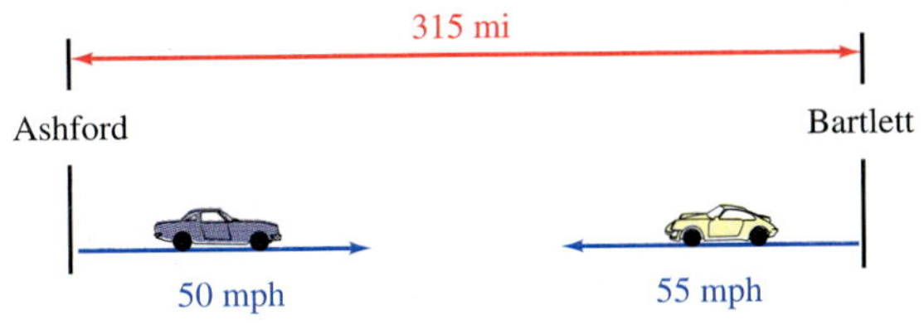

14. **Travel time** Granville and Preston are 535 miles apart. A car leaves Preston bound for Granville at 47 mph. At the same time, another car leaves Granville bound for Preston at 60 mph. How long will it take them to meet?
15. **Paving highway** Two crews working toward each other are 9.45 miles apart. One crew paves 1.5 miles of highway per day, and the other paves 1.2 miles per day. How long will it take them to meet?
16. **Biking** Two friends who live 33 miles apart ride bikes toward each other. One averages 12 mph, and the other averages 10 mph. How long will it take for them to meet?
17. **Travel time** Two cars leave Peoria at the same time, one heading east at 60 mph and the other west at 50 mph. How long will it take them to be 715 miles apart?

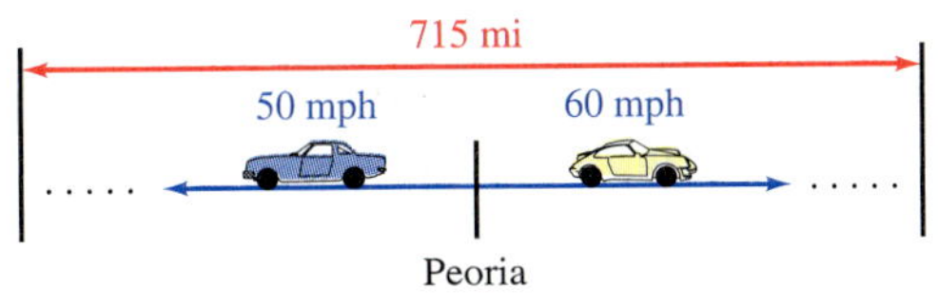

18. **Boating** Two boats leave port at the same time, one heading north at 35 knots (nautical miles per hour) and the other south at 47 knots. How long will it take them to be 738 nautical miles apart?
19. **Hiking** Two boys with two-way radios that have a range of 2 miles leave camp and walk in opposite directions. If one boy walks 3 mph and the other walks 4 mph, how long will it take before they lose radio contact?
20. **Biking** Two cyclists leave a park and ride in opposite directions, one averaging 9 mph and the other 6 mph. If they have two-way radios with a 5-mile range, for how many minutes will they remain in radio contact?

21. Complete the following table and compute how long it will take the car to overtake the bus if the bus had a 2-hour head start.

	r ·	t =	d
Car	60 mph	t	
Bus	50 mph	$t + 2$	

22. Hot pursuit Two crooks rob a bank and flee to the east at 66 mph. In 30 minutes, the police follow them in a helicopter, flying at 132 mph. How long will it take for the police to overtake the robbers?

23. Travel time Two cars start together and head east, one averaging 42 mph and the other averaging 53 mph. See the illustration. In how many hours will the cars be 82.5 miles apart?

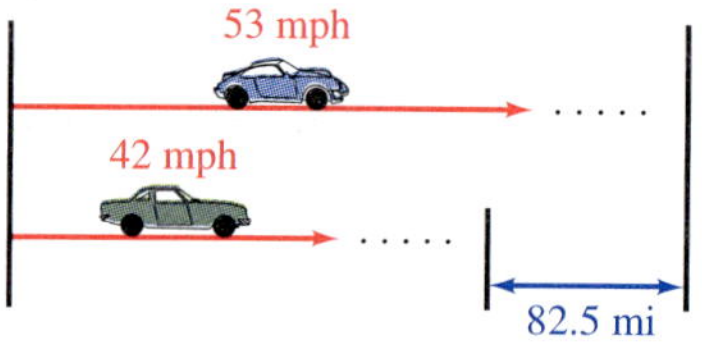

24. Aviation A plane leaves an airport and flies south at 180 mph. Later, a second plane leaves the same airport and flies south at 450 mph. If the second plane overtakes the first one in $1\frac{1}{2}$ hours, how much of a head start did the first plane have?

25. Speed of trains Two trains are 330 miles apart, and their speeds differ by 20 mph. They travel toward each other and meet in 3 hours. Find the speed of each train.

26. Speed of an airplane Two planes are 6,000 miles apart, and their speeds differ by 200 mph. They travel toward each other and meet in 5 hours. Find the speed of the slower plane.

27. Average speed An automobile averaged 40 mph for part of a trip and 50 mph for the remainder. If the 5-hour trip covered 210 miles, for how long did the car average 40 mph?

28. Vacation driving A family drove to the Grand Canyon, averaging 45 mph. They returned using the same route, averaging 60 mph. If they spent a total of 7 hours of driving time, how far is their home from the Grand Canyon?

29. Chemistry A solution contains 0.3 liters of sulfuric acid. If this represents 12% of the total amount, find the total amount.

30. Medicine A laboratory has a solution that contains 3 ounces of benzalkonium chloride. If this is 15% of the total solution, how many ounces of solution does the lab have?

31. Mixing fuels How many gallons of fuel costing \$1.15 per gallon must be mixed with 20 gallons of a fuel costing \$0.85 per gallon to obtain a mixture costing \$1 per gallon?

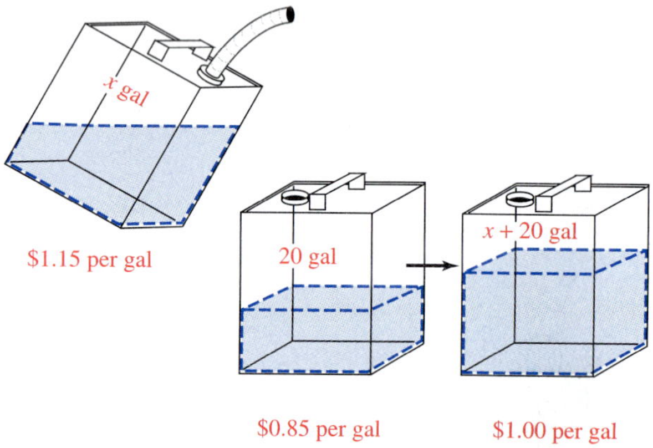

32. Mixing paint Paint costing \$19 per gallon is to be mixed with 5 gallons of paint thinner costing \$3 per gallon to make a paint that can be sold for \$14 per gallon. Refer to the table and compute how much paint will be produced.

	p ·	n =	v
Paint	\$19	x gal	\$19x
Thinner	\$3	5 gal	\$3(5)
Mixture	\$14	$(x + 5)$ gal	\$14$(x + 5)$

33. Brine solution How many gallons of a 3% salt solution must be mixed with 50 gallons of a 7% solution to obtain a 5% solution?

34. Making cottage cheese To make low-fat cottage cheese, milk containing 4% butterfat is mixed with 10 gallons of milk containing 1% butterfat to obtain a mixture containing 2% butterfat. How many gallons of the richer milk must be used?

35. Antiseptic solutions A nurse wishes to add water to 30 ounces of a 10% solution of benzalkonium chloride to dilute it to an 8% solution. How much water must she add?

36. Mixing photographic chemicals A photographer wishes to mix 2 liters of a 5% acetic acid solution with a 10% solution to get a 7% solution. How many liters of 10% solution must be added?

37. Mixing candy Lemon drops are to be mixed with jelly beans to make 100 pounds of mixture. Refer to the illustration and compute how many pounds of each candy should be used.

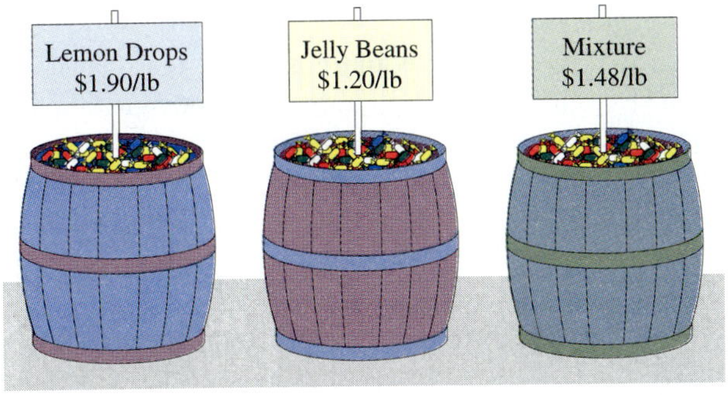

38. Blending gourmet tea One grade of tea, worth $3.20 per pound, is to be mixed with another grade worth $2 per pound to make 20 pounds that will sell for $2.72 per pound. How much of each grade of tea must be used?

39. Mixing nuts A bag of peanuts is worth 30¢ less than a bag of cashews. Equal amounts of peanuts and cashews are used to make 40 bags of a mixture that sells for $1.05 per bag. How much would a bag of cashews be worth?

40. Mixing candy Twenty pounds of lemon drops are to be mixed with cherry chews to make a mixture that will sell for $1.80 per pound. How much of the more expensive candy should be used? See the table.

	Price per pound
Peppermint patties	$1.35
Lemon drops	$1.70
Licorice lumps	$1.95
Cherry chews	$2.00

41. Coffee blends A store sells regular coffee for $4 a pound and a gourmet coffee for $7 a pound. To get rid of 40 pounds of the gourmet coffee, the shopkeeper plans to make a gourmet blend to sell for $5 a pound. How many pounds of regular coffee should be used?

42. Lawn seed blends A garden store sells Kentucky bluegrass seed for $6 per pound and ryegrass seed for $3 per pound. How much rye must be mixed with 100 pounds of bluegrass to obtain a blend that will sell for $5 per pound?

43. Mixing coffee A shopkeeper sells chocolate coffee beans for $7 per pound. A customer asks the shopkeeper to mix 2 pounds of chocolate beans with 5 pounds of hazelnut coffee beans. If the customer paid $6 per pound for the mixture, what is the price per pound of the hazelnut beans?

44. Trail mix Fifteen pounds of trail mix are made by mixing 2 pounds of raisins worth $3 per pound with peanuts worth $4 per pound and M&M's worth $5 per pound. How many pounds of peanuts must be used if the mixture is to be worth $4.20 per pound?

WRITING

45. Describe the steps you would use to analyze and solve a problem.

46. Create a mixture problem that could be solved by using the equation $4x + 6(12 - x) = 5(12)$.

47. Create a mixture problem of your own, and solve it.

48. In mixture problems, explain why it is important to distinguish between the quantity and the value of the materials being combined.

SOMETHING TO THINK ABOUT

49. Is it possible for the equation of a problem to have a solution, but for the problem to have no solution? For example, is it possible to find two consecutive even integers whose sum is 16?

50. Invent a motion problem that leads to an equation that has a solution, although the problem does not.

51. Consider the problem: How many gallons of a 10% and a 20% solution should be mixed to obtain a 30% solution? Without solving it, how do you know that the problem has no solution?

52. What happens if you try to solve Exercise 51?

2.6 Formulas

- Solving Formulas

Getting Ready *Fill in the blanks.*

1. $\frac{3x}{\square} = x$ 2. $\frac{-5y}{\square} = y$ 3. $\frac{rx}{\square} = x$ 4. $\frac{-ay}{\square} = y$

5. $\square \cdot \frac{x}{7} = x$ 6. $\square \cdot \frac{y}{12} = y$ 7. $\square \cdot \frac{x}{d} = x$ 8. $\square \cdot \frac{y}{s} = y$

Equations with several variables are called **literal equations.** Often these equations are **formulas** such as $A = lw$, the formula for finding the area of a rectangle. Suppose that we wish to find the lengths of several rectangles whose areas and widths are known. It would be tedious to substitute values for A and w into the formula and then repeatedly solve the formula for l. It would be easier to solve the formula $A = lw$ for l first and then substitute values for A and w and compute l directly.

Solving Formulas

To **solve a formula for a variable** means to isolate that variable on one side of the formula, with all other variables on the opposite side. We can isolate the variable by using the usual equation-solving techniques.

EXAMPLE 1 Solve $A = lw$ for l.

Solution To isolate l on the left-hand side, we undo the multiplication by w by dividing both sides of the equation by w.

$$A = lw$$

$$\frac{A}{w} = \frac{lw}{w} \quad \text{To undo the multiplication by } w\text{, divide both sides by } w.$$

$$\frac{A}{w} = l \quad \frac{w}{w} = 1.$$

$$l = \frac{A}{w}$$

Self Check Solve $A = lw$ for w.

EXAMPLE 2 Recall that the formula $A = \frac{1}{2}bh$ gives the area of a triangle with base b and height h. Solve the formula for b.

Solution

$$A = \frac{1}{2}bh$$

$$2A = 2 \cdot \frac{1}{2}bh \qquad \text{To eliminate the fraction, multiply both sides by 2.}$$

$$2A = bh \qquad 2 \cdot \tfrac{1}{2} = 1.$$

$$\frac{2A}{h} = \frac{bh}{h} \qquad \text{To undo the multiplication by } h\text{, divide both sides by } h.$$

$$\frac{2A}{h} = b \qquad \frac{h}{h} = 1.$$

$$b = \frac{2A}{h}$$

If the area A and the height h of a triangle are known, the base b is given by the formula $b = \frac{2A}{h}$.

Self Check Solve $A = \frac{1}{2}bh$ for h.

EXAMPLE 3 The formula $C = \frac{5}{9}(F - 32)$ is used to convert Fahrenheit temperature readings into their Celsius equivalents. Solve the formula for F.

Solution

$$C = \frac{5}{9}(F - 32)$$

$$\frac{9}{5}C = \frac{9}{5} \cdot \frac{5}{9}(F - 32) \qquad \text{To eliminate } \frac{5}{9}\text{, multiply both sides by } \frac{9}{5}.$$

$$\frac{9}{5}C = 1(F - 32) \qquad \frac{9}{5} \cdot \frac{5}{9} = \frac{9 \cdot 5}{5 \cdot 9} = 1.$$

$$\frac{9}{5}C = F - 32 \qquad \text{Remove parentheses.}$$

$$\frac{9}{5}C + 32 = F - 32 + 32 \qquad \text{To undo the subtraction of 32, add 32 to both sides.}$$

$$\frac{9}{5}C + 32 = F \qquad \text{Combine like terms.}$$

$$F = \frac{9}{5}C + 32$$

The formula $F = \frac{9}{5}C + 32$ is used to convert degrees Celsius to degrees Fahrenheit.

Self Check Solve $x = \frac{2}{3}(y + 5)$ for y.

EXAMPLE 4 Recall that the area A of the trapezoid shown in Figure 2-14 is given by the formula

$$A = \frac{1}{2}h(B + b)$$

where B and b are its bases and h is its height. Solve the formula for b.

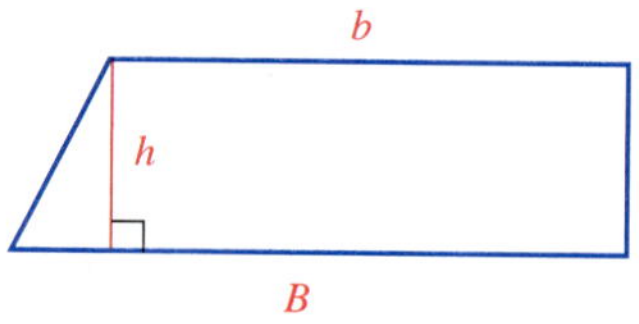

Figure 2-14

Solution **Method 1:** $A = \frac{1}{2}(B + b)h$

$$\begin{aligned}
2A &= 2 \cdot \frac{1}{2}(B + b)h && \text{Multiply both sides by 2.} \\
2A &= Bh + bh && \text{Simplify and remove parentheses.} \\
2A - Bh &= Bh + bh - Bh && \text{Subtract } Bh \text{ from both sides.} \\
2A - Bh &= bh && \text{Combine like terms.} \\
\frac{2A - Bh}{h} &= \frac{bh}{h} && \text{Divide both sides by } h. \\
\frac{2A - Bh}{h} &= b && \frac{h}{h} = 1.
\end{aligned}$$

Method 2: $A = \frac{1}{2}(B + b)h$

$$\begin{aligned}
2 \cdot A &= 2 \cdot \frac{1}{2}(B + b)h && \text{Multiply both sides by 2.} \\
2A &= (B + b)h && \text{Simplify.} \\
\frac{2A}{h} &= \frac{(B + b)h}{h} && \text{Divide both sides by } h. \\
\frac{2A}{h} &= B + b && \frac{h}{h} = 1. \\
\frac{2A}{h} - B &= B + b - B && \text{Subtract } B \text{ from both sides.} \\
\frac{2A}{h} - B &= b && \text{Combine like terms.}
\end{aligned}$$

Although they look different, the results of Methods 1 and 2 are equivalent.

Self Check Solve $A = \frac{1}{2}h(B + b)$ for B.

EXAMPLE 5 Solve the formula $P = 2l + 2w$ for l, and then find l when $P = 56$ and $w = 11$.

Solution We first solve the formula $P = 2l + 2w$ for l.

$$\begin{aligned}
P &= 2l + 2w \\
P - 2w &= 2l + 2w - 2w && \text{Subtract } 2w \text{ from both sides.} \\
P - 2w &= 2l && \text{Combine like terms.}
\end{aligned}$$

$$\frac{P - 2w}{\mathbf{2}} = \frac{2l}{\mathbf{2}}$$ Divide both sides by 2.

$$\frac{P - 2w}{2} = l$$ $\frac{2}{2} = 1.$

$$l = \frac{P - 2w}{2}$$

We then substitute 56 for P and 11 for w and simplify.

$$l = \frac{P - 2w}{2}$$

$$l = \frac{56 - 2(11)}{2}$$

$$= \frac{56 - 22}{2}$$

$$= \frac{34}{2}$$

$$= 17$$

Thus, $l = 17$.

Self Check Solve $P = 2l + 2w$ for w, and then find w when $P = 46$ and $l = 16$.

EXAMPLE 6 Recall that the volume V of the right-circular cone shown in Figure 2-15 is given by the formula

$$V = \frac{1}{3}Bh$$

where B is the area of its circular base and h is its height. Solve the formula for h, and find the height of a right-circular cone with a volume of 64 cubic centimeters and a base area of 16 square centimeters.

Solution We first solve the formula for h.

$$V = \frac{1}{3}Bh$$

$$\mathbf{3}V = \mathbf{3} \cdot \frac{1}{3}Bh$$ Multiply both sides by 3.

$$3V = Bh$$ $3 \cdot \frac{1}{3} = 1.$

$$\frac{3V}{\mathbf{B}} = \frac{Bh}{\mathbf{B}}$$ Divide both sides by B.

$$\frac{3V}{B} = h$$ $\frac{B}{B} = 1.$

$$h = \frac{3V}{B}$$

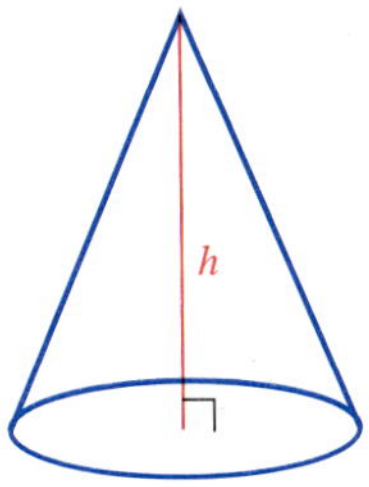

Figure 2-15

We then substitute 64 for V and 16 for B and simplify.

$$h = \frac{3V}{B}$$

$$h = \frac{3(64)}{16}$$

$$= 3(4)$$

$$= 12$$

The height of the cone is 12 centimeters.

Self Check Solve $V = \frac{1}{3}Bh$ for B, and find the area of the base when the volume is 42 cubic feet and the height is 6 feet.

Self Check Answers

1. $w = \frac{A}{l}$ **2.** $h = \frac{2A}{b}$ **3.** $y = \frac{3}{2}x - 5$ **4.** $B = \frac{2A - hb}{h}$ or $B = \frac{2A}{h} - b$ **5.** $w = \frac{P - 2l}{2}$, 7 **6.** $B = \frac{3V}{h}$, 21 square feet

Orals *Solve the equation $ab + c - d = 0$*

1. for a
2. for b
3. for c
4. for d

Solve the equation $a + b = \frac{c}{d}$

5. for a
6. for b
7. for c
8. for d

2.6 EXERCISES

REVIEW *Simplify each expression, if possible.*

1. $2x - 5y + 3x$

2. $2x^2y + 5x^2y^2$

3. $\frac{3}{5}(x + 5) - \frac{8}{5}(10 + x)$

4. $\frac{2}{11}(22x - y) + \frac{9}{11}y$

VOCABULARY AND CONCEPTS *Fill in the blanks.*

5. Equations that contain several variables are called ______ equations.

6. The equation $A = lw$ is an example of a ______.

7. To solve a formula for a variable means to ______ the variable on one side of the formula.

8. To solve the formula $d = rt$ for t, divide both sides of the formula by __.

9. To solve $A = p + i$ for p, ______ i from both sides.

10. To solve $t = \frac{d}{r}$ for d, ______ both sides by r.

PRACTICE *Solve each formula for the indicated variable.*

11. $E = IR$; for I

12. $i = prt$; for r

13. $V = lwh$; for w

14. $K = A + 32$; for A

15. $P = a + b + c$; for b

16. $P = 4s$; for s

17. $P = 2l + 2w$; for w

18. $d = rt$; for t

19. $A = P + Prt$; for t

20. $A = \frac{1}{2}(B + b)h$; for h

21. $C = 2\pi r$; for r

22. $I = \frac{E}{R}$; for R

23. $K = \frac{wv^2}{2g}$; for w

24. $V = \pi r^2 h$; for h

25. $P = I^2 R$; for R

26. $V = \frac{1}{3}\pi r^2 h$; for h

27. $K = \frac{wv^2}{2g}$; for g

28. $P = \frac{RT}{mV}$; for V

29. $F = \frac{GMm}{d^2}$; for M

30. $C = 1 - \frac{A}{a}$; for A

31. $F = \frac{GMm}{d^2}$; for d^2

32. $y = mx + b$; for x

33. $G = 2(r - 1)b$; for r

34. $F = f(1 - M)$; for M

Solve each formula for the indicated variable. Then substitute numbers to find the variable's value.

35. $d = rt$ Find t if $d = 135$ and $r = 45$.

36. $d = rt$ Find r if $d = 275$ and $t = 5$.

37. $i = prt$ Find t if $i = 12$, $p = 100$, and $r = 0.06$.

38. $i = prt$ Find r if $i = 120$, $p = 500$, and $t = 6$.

39. $P = a + b + c$ Find c if $P = 37$, $a = 15$, and $b = 19$.

40. $y = mx + b$ Find x if $y = 30$, $m = 3$, and $b = 0$.

41. $K = \frac{1}{2}h(a + b)$ Find h if $K = 48$, $a = 7$, and $b = 5$.

42. $\frac{x}{2} + y = z^2$ Find x if $y = 3$ and $z = 3$.

APPLICATIONS

43. **Ohm's law** The formula $E = IR$, called **Ohm's law,** is used in electronics. Solve for I, and then calculate the current I if the voltage E is 48 volts and the resistance R is 12 ohms. Current has units of *amperes.*

44. **Volume of a cone** The volume V of a cone is given by the formula $V = \frac{1}{3}\pi r^2 h$. Solve the formula for h, and then calculate the height h if V is 36π cubic inches and the radius r is 6 inches.

45. **Circumference of a circle** The circumference C of a circle is given by $C = 2\pi r$, where r is the radius of the circle. Solve the formula for r, and then calculate the radius of a circle with a circumference of 14.32 feet. Round to the nearest hundredth of a foot.

46. **Growth of money** At a simple interest rate r, an amount of money P grows to an amount A in t years according to the formula $A = P(1 + rt)$. Solve the formula for P. After $t = 3$ years, a girl has an amount $A = \$4,357$ on deposit. What amount P did she start with? Assume an interest rate of 6%.

47. **Power loss** The power P lost when an electric current I passes through a resistance R is given by the formula $P = I^2R$. Solve for R. If P is 2,700 watts and I is 14 amperes, calculate R to the nearest hundredth of an ohm.

48. **Geometry** The perimeter P of a rectangle with length l and width w is given by the formula $P = 2l + 2w$. Solve this formula for w. If the perimeter of a certain rectangle is 58.37 meters and its length is 17.23 meters, find its width. Round to two decimal places.

49. **Force of gravity** The masses of the two objects in the illustration are m and M. The force of gravitation F between the masses is given by

$$F = \frac{GmM}{d^2}$$

where G is a constant and d is the distance between them. Solve for m.

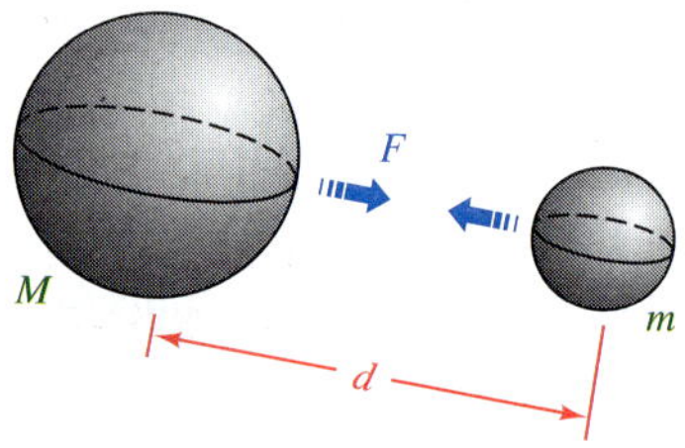

50. **Thermodynamics** In thermodynamics, the Gibbs free-energy equation is given by

$$G = U - TS + pV$$

Solve this equation for the pressure, p.

51. **Pulleys** The approximate length L of a belt joining two pulleys of radii r and R feet with centers D feet apart is given by the formula

$$L = 2D + 3.25(r + R)$$

Solve the formula for D. If a 25-foot belt joins pulleys with radii of 1 foot and 3 feet, how far apart are their centers?

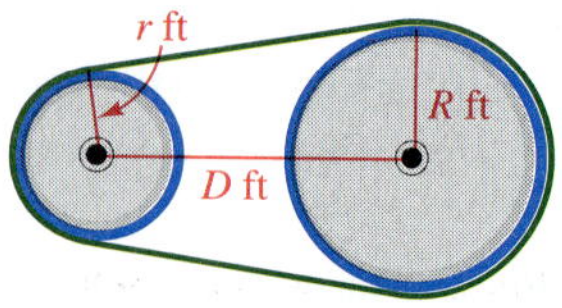

52. **Geometry** The measure a of an interior angle of a regular polygon with n sides is given by $a = 180°\left(1 - \frac{2}{n}\right)$. Solve the formula for n. How many sides does a regular polygon have if an interior angle is 108°? (*Hint:* Distribute first.)

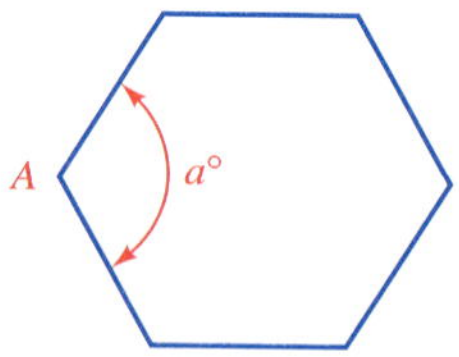

One common retirement plan for self-employed people is called a **Simplified Employee Pension Plan.** *It allows for a maximum annual contribution of 15% of taxable income (earned income minus deductible expenses). However, since the Internal Revenue Service considers the SEP contribution to be a deductible expense, the taxable income must be reduced by the amount of the contribution. Therefore, to calculate the maximum contribution C, we take 15% of what's left after we subtract the contribution C from the taxable income T.*

$$C = 0.15(T - C)$$

53. **Calculating a SEP contribution** Find the maximum allowable contribution to a SEP plan by solving the equation $C = 0.15(T - C)$ for C.

54. **Calculating a SEP contribution** Find the maximum allowable contribution to a SEP plan for a person who earns \$75,000 and has deductible expenses of \$27,540.

WRITING

55. The formula $P = 2l + 2w$ is also an equation, but an equation such as $2x + 3 = 5$ is not a formula. What equations do you think should be called formulas?

56. To solve the equation $s - A(s - 5) = r$ for the variable s, one student simply added $A(s - 5)$ to both sides to get $s = r + A(s - 5)$. Explain why this is not correct.

SOMETHING TO THINK ABOUT

57. The energy of an atomic bomb comes from the conversion of matter into energy, according to Einstein's formula $E = mc^2$. The constant c is the speed of light, about 300,000 meters per second. Find the energy in a mass m of 1 kilogram. Energy has units of **joules.**

58. When a car of mass m collides with a wall, the energy of the collision is given by the formula $E = \frac{1}{2}mv^2$. Compare the energy of two collisions: a car striking a wall at 30 mph, and at 60 mph.

2.7 Solving Inequalities

▮ Inequalities ▮ Compound Inequalities ▮ Applications

Getting Ready *Graph each set on the number line.*

1. All real numbers greater than -1.

2. All real number less than or equal to 5.

3. All real numbers between -2 and 4.

4. All real numbers less than -2 or greater than or equal to 4.

Many times, we will encounter mathematical statements indicating that two quantities are not necessarily equal. These statements are called *inequalities.*

Inequalities

Recall the meaning of the following symbols.

Inequality Symbols

$<$	means	"is less than"
$>$	means	"is greater than"
$\leq$	means	"is less than or equal to"
$\geq$	means	"is greater than or equal to"

An **inequality** is a statement that indicates that two quantities are not necessarily equal. A **solution of an inequality** is any number that makes the inequality true. The number 2 is a solution of the inequality

$$x \leq 3$$

because $2 \leq 3$.

This inequality has many more solutions, because any real number that is less than or equal to 3 will satisfy it. We can use a graph on the number line to represent the solutions of the inequality. The colored arrow in Figure 2-16 indicates all those points with coordinates that satisfy the inequality $x \leq 3$.

Figure 2-16

The bracket at the point with coordinate 3 indicates that the number 3 is a solution of the inequality $x \leq 3$.

The graph of the inequality $x > 1$ appears in Figure 2-17. The colored arrow indicates all those points whose coordinates satisfy the inequality. The parenthesis at the point with coordinate 1 indicates that 1 is not a solution of the inequality $x > 1$.

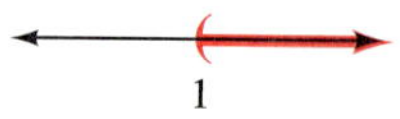

Figure 2-17

To solve more complicated inequalities, we need to use the addition, subtraction, multiplication, and division properties of inequalities. When we use any of these properties, the resulting inequality will have the same solutions as the original one.

Addition Property of Inequality

If a, b, and c are real numbers, and

If $a < b$, then $a + c < b + c$.

Similar statements can be made for the symbols $>$, $\leq$, and $\geq$.

The **addition property of inequality** can be stated this way: *If any quantity is added to both sides of an inequality, the resulting inequality has the same direction as the original inequality.*

Subtraction Property of Inequality

If a, b, and c are real numbers, and

If $a < b$, then $a - c < b - c$.

Similar statements can be made for the symbols $>$, $\leq$, and $\geq$.

The **subtraction property of inequality** can be stated this way: *If any quantity is subtracted from both sides of an inequality, the resulting inequality has the same direction as the original inequality.*

The subtraction property of inequality is included in the addition property: To *subtract* a number a from both sides of an inequality, we could instead *add* the *negative* of a to both sides.

EXAMPLE 1 Solve $2x + 5 > x - 4$ and graph the solution on a number line.

Solution To isolate the x on the left-hand side of the $>$ sign, we proceed as if we were solving equations.

$$2x + 5 > x - 4$$

$$2x + 5 - 5 > x - 4 - 5 \quad \text{Subtract 5 from both sides.}$$

$$2x > x - 9 \quad \text{Combine like terms.}$$

$$2x - x > x - 9 - x \quad \text{Subtract } x \text{ from both sides.}$$

$$x > -9 \quad \text{Combine like terms.}$$

The graph of the solution (see Figure 2-18) includes all points to the right of -9 but does not include -9 itself. For this reason, we use a parenthesis at -9.

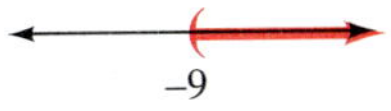

Figure 2-18

Self Check Graph the solution of $3x - 2 < x + 4$.

If both sides of the true inequality $2 < 5$ are multiplied by a *positive* number, such as 3, another true inequality results.

$$2 < 5$$
$$3 \cdot 2 < 3 \cdot 5 \quad \text{Multiply both sides by 3.}$$
$$6 < 15$$

The inequality $6 < 15$ is true. However, if both sides of $2 < 5$ are multiplied by a negative number, such as -3, the direction of the inequality symbol must be reversed to produce another true inequality.

$$2 < 5$$
$$-3 \cdot 2 > -3 \cdot 5 \quad \text{Multiply both sides by the negative number } -3 \text{ and reverse the direction of the inequality.}$$
$$-6 > -15$$

The inequality $-6 > -15$ is true, because -6 lies to the right of -15 on the number line. These examples suggest the following properties.

Multiplication Property of Inequality

If a, b, and c are real numbers, and

If $a < b$ and $c > 0$, then $ac < bc$.

If $a < b$ and $c < 0$, then $ac > bc$.

Similar statements can be made for the symbols $>$, $\leq$, and $\geq$.

The multiplication property of inequality can be stated this way:

If unequal quantities are multiplied by the same positive quantity, the results will be unequal and in the same direction.

If unequal quantities are multiplied by the same negative quantity, the results will be unequal but in the opposite direction.

There is a similar property for division.

Division Property of Inequality

If a, b, and c are real numbers, and

If $a < b$ and $c > 0$, then $\frac{a}{c} < \frac{b}{c}$.

If $a < b$ and $c < 0$, then $\frac{a}{c} > \frac{b}{c}$.

Similar statements can be made for the symbols $>$, $\leq$, and $\geq$.

The division property of inequality can be stated this way:

If unequal quantities are divided by the same positive quantity, the results will be unequal and in the same direction.

If unequal quantities are divided by the same negative quantity, the results will be unequal but in the opposite direction.

To *divide* both sides of an inequality by a nonzero number c, we could instead *multiply* both sides by $\frac{1}{c}$.

Comment Remember that if both sides of an inequality are multiplied by a *positive* number, the direction of the resulting inequality remains the same. However, if both sides of an inequality are multiplied by a *negative* number, the direction of the resulting inequality must be reversed.

Note that the procedures for solving inequalities are the same as for solving equations, except that we must reverse the inequality symbol whenever we multiply or divide by a negative number.

EXAMPLE 2 Solve $3x + 7 \le -5$ and graph the solution.

Solution

$$3x + 7 \le -5$$

$$3x + 7 - 7 \le -5 - 7 \quad \text{Subtract 7 from both sides.}$$

$$3x \le -12 \quad \text{Combine like terms.}$$

$$\frac{3x}{3} \le \frac{-12}{3} \quad \text{Divide both sides by 3.}$$

$$x \le -4$$

Figure 2-19

The solution consists of all real numbers that are less than or equal to -4. The bracket at -4 in the graph of Figure 2-19 indicates that -4 is one of the solutions.

Self Check Graph the solution of $2x - 5 \ge -3$.

EXAMPLE 3 Solve $5 - 3x \le 14$ and graph the solution.

Solution

$$5 - 3x \le 14$$

$$5 - 3x - 5 \le 14 - 5 \quad \text{Subtract 5 from both sides.}$$

$$-3x \le 9 \quad \text{Combine like terms.}$$

$$\frac{-3x}{-3} \ge \frac{9}{-3} \quad \text{Divide both sides by } -3 \text{ and reverse the direction of the } \le \text{ symbol.}$$

$$x \ge -3$$

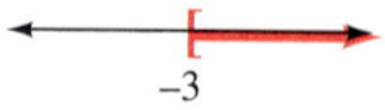

Figure 2-20

Since both sides of the inequality were divided by -3, the direction of the inequality was *reversed.* The graph of the solution appears in Figure 2-20. The bracket at -3 indicates that -3 is one of the solutions.

Self Check Graph the solution of $6 - 7x \ge -15$.

Compound Inequalities

Two inequalities can often be combined into a **double inequality** or **compound inequality** to indicate that numbers lie *between* two fixed values. For example, the inequality $2 < x < 5$ indicates that x is greater than 2 and that x is also less than 5. The solution of $2 < x < 5$ consists of all numbers that lie *between* 2 and 5. The graph of this set (called an **interval**) appears in Figure 2-21.

Figure 2-21

EXAMPLE 4 Solve $-4 < 2(x - 1) \leq 4$ and graph the solution.

Solution

$$-4 < 2(x - 1) \leq 4$$

$$-4 < 2x - 2 \leq 4 \quad \text{Remove parentheses.}$$

$$-2 < 2x \leq 6 \quad \text{Add 2 to all three parts.}$$

$$-1 < x \leq 3 \quad \text{Divide all three parts by 2.}$$

The graph of the solution appears in Figure 2-22.

Figure 2-22

Self Check Graph the solution of $0 \leq 4(x + 5) < 26$.

Applications

When solving application problems, there are certain words that help us translate a sentence into a mathematical inequality.

Words	Sentence	Inequality
at least	To earn a grade of A, you must score at least 90%.	$S \geq 90\%$
is less than	The perimeter is less than 30 feet.	$P < 30$ ft
is no less than	The perimeter is no less than 100 centimeters.	$P \geq 100$ cm
is more than	The area is more than 30 square inches.	$A > 30$ sq in.
exceeds	The car's speed exceeded the limit of 45 mph.	$S > 45$ mph
cannot exceed	The salary cannot exceed $50,000.	$S \leq \$50,000$
at most	The perimeter is at most 75 feet	$P \leq 75$
is between	The altitude was between 10,000 and 15,000 feet.	$10,000 < A < 15,000$

EXAMPLE 5 A student has scores of 72, 74, and 78 points on three mathematics examinations. How many points does he need on his last exam to earn a B, an average of at least 80 points?

Solution We can let x represent the score on the fourth (and last) exam. To find the average grade, we add the four scores and divide by 4. To earn a B, this average must be greater than or equal to 80 points.

The average of the four grades	$\geq$	80
$\frac{72 + 74 + 78 + x}{4}$	$\geq$	80

We can solve this inequality for x.

$$\frac{224 + x}{4} \geq 80 \quad 72 + 74 + 78 = 224.$$

$$224 + x \geq 320 \quad \text{Multiply both sides by 4.}$$

$$x \geq 96 \quad \text{Subtract 224 from both sides.}$$

To earn a B, the student must score at least 96 points.

EXAMPLE 6 If the perimeter of an equilateral triangle is less than 15 feet, how long could each side be?

Solution Recall that each side of an equilateral triangle is the same length and that the perimeter of a triangle is the sum of the lengths of its three sides. If we let x represent the length of one of the sides, then $x + x + x$ represents the perimeter. Since the perimeter is to be less than 15 feet, we have the following inequality:

$$x + x + x < 15$$

$$3x < 15 \quad \text{Combine like terms.}$$

$$x < 5 \quad \text{Divide both sides by 3.}$$

Each side of the triangle must be less than 5 feet long.

Self Check Answers

1. 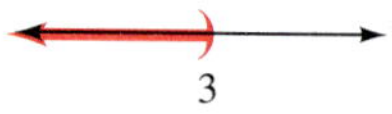

2. 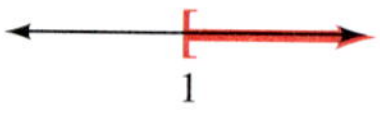

3. 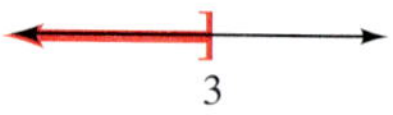

4. 

Orals *Solve each inequality.*

1. $2x < 4$
2. $x + 5 \geq 6$
3. $-3x \leq -6$
4. $-x > 2$
5. $2x - 5 < 7$
6. $5 - 2x < 7$

2.7 EXERCISES

REVIEW *Simplify each expression.*

1. $3x^2 - 2(y^2 - x^2)$
2. $5(xy + 2) - 3xy - 8$
3. $\frac{1}{3}(x + 6) - \frac{4}{3}(x - 9)$
4. $\frac{4}{5}x(y + 1) - \frac{9}{5}y(x - 1)$

VOCABULARY AND CONCEPTS *Fill in the blanks.*

5. The symbol $<$ means ________.
6. The symbol $>$ means ________.
7. The symbol ___ means "is greater than or equal to."
8. The symbol ___ means "is less than or equal to."
9. An ________ is a statement indicating that two quantities are not necessarily equal.
10. A ________ of an inequality is a number that makes the inequality true.

PRACTICE *Solve each inequality and graph the solution.*

11. $x + 2 > 5$
12. $x + 5 \geq 2$
13. $-x - 3 \leq 7$
14. $-x - 9 > 3$
15. $3 + x < 2$
16. $5 + x \geq 3$
17. $2x - 3 \leq 5$
18. $-3x - 5 < 4$
19. $-3x - 7 > -1$
20. $-5x + 7 \leq 12$
21. $-4x + 1 > 17$
22. $7x - 9 > 5$
23. $2x + 9 \leq x + 8$
24. $3x + 7 \leq 4x - 2$
25. $9x + 13 \geq 8x$
26. $7x - 16 < 6x$
27. $8x + 4 > 6x - 2$
28. $7x + 6 \geq 4x$
29. $5x + 7 < 2x + 1$
30. $7x + 2 > 4x - 1$
31. $7 - x \leq 3x - 1$
32. $2 - 3x \geq 6 + x$
33. $9 - 2x > 24 - 7x$
34. $13 - 17x < 34 - 10x$
35. $3(x - 8) < 5x + 6$
36. $9(x - 11) > 13 + 7x$
37. $8(5 - x) \leq 10(8 - x)$
38. $17(3 - x) \geq 3 - 13x$
39. $\frac{5}{2}(7x - 15) + x \geq \frac{13}{2}x - \frac{3}{2}$
40. $\frac{5}{3}(x + 1) \leq -x + \frac{2}{3}$
41. $\frac{3x - 3}{2} < 2x + 2$
42. $\frac{x + 7}{3} \geq x - 3$
43. $\frac{2(x + 5)}{3} \leq 3x - 6$
44. $\frac{3(x - 1)}{4} > x + 1$
45. $2 < x - 5 < 5$
46. $3 < x - 2 < 7$
47. $-5 < x + 4 \leq 7$
48. $-9 \leq x + 8 < 1$
49. $0 \leq x + 10 \leq 10$

50. $-8 < x - 8 < 8$

51. $4 < -2x < 10$

52. $-4 \leq -4x < 12$

53. $-3 \leq \frac{x}{2} \leq 5$

54. $-12 \leq \frac{x}{3} < 0$

55. $3 \leq 2x - 1 < 5$

56. $4 < 3x - 5 \leq 7$

57. $0 < 10 - 5x \leq 15$

58. $1 \leq -7x + 8 \leq 15$

59. $-6 < 3(x + 2) < 9$

60. $-18 \leq 9(x - 5) < 27$

61. $-4 < \frac{x - 2}{2} < 6$

62. $-1 \leq \frac{x + 1}{3} \leq 3$

APPLICATIONS *Express each solution as an inequality.*

63. **Calculating grades** A student has test scores of 68, 75, and 79 points. What must she score on the fourth exam to have an average score of at least 80 points?

64. **Calculating grades** A student has test scores of 70, 74, and 84 points. What must he score on the fourth exam to have an average score of at least 70 points?

65. **Fleet average** An automobile manufacturer produces three light trucks in equal quantities. One model has an economy rating of 17 miles per gallon, and the second model is rated for 19 mpg. If the manufacturer is required to have a fleet average of at least 21 mpg, what economy rating is required for the third model?

66. **Avoiding a service charge** When the average daily balance of a customer's checking account is less than \$500 in any week, the bank assesses a \$5 service charge. Bill's account balances for the week were as shown in the table. What must Friday's balance be to avoid the service charge?

Monday	\$540.00
Tuesday	\$435.50
Wednesday	\$345.30
Thursday	\$310.00

67. **Geometry** The perimeter of an equilateral triangle is at most 57 feet. What could be the length of a side? (*Hint:* All three sides of an equilateral triangle are equal.)

68. **Geometry** The perimeter of a square is no less than 68 centimeters. How long can a side be?

69. **Land elevations** The land elevations in Nevada fall between the 13,143-foot height of Boundary Peak and the Colorado River at 470 feet. To the nearest tenth, what is the range of these elevations in miles? (*Hint:* 1 mile is 5,280 feet.)

70. **Doing homework** A teacher requires that students do homework at least 2 hours a day. How many minutes should a student work each week?

71. **Plane altitudes** A pilot plans to fly at an altitude of between 17,500 and 21,700 feet. To the nearest tenth, what will be the range of altitudes in miles? (*Hint:* There are 5,280 feet in 1 mile.)

72. **Getting exercise** A certain exercise program recommends that your daily exercise period should exceed 15 minutes but should not exceed 30 minutes per day. Find the range of exercise time for one week.

73. Comparing temperatures To hold the temperature of a room between 19° and 22° Celsius, what Fahrenheit temperatures must be maintained? (*Hint:* Fahrenheit temperature (F) and Celsius temperature (C) are related by the formula $C = \frac{5}{9}(F - 32)$.)

74. Melting iron To melt iron, the temperature of a furnace must be at least 1,540°C but at most 1,650°C. What range of Fahrenheit temperatures must be maintained?

75. Phonograph records The radii of old phonograph records lie between 5.9 and 6.1 inches. What variation in circumference can occur? (*Hint:* The circumference of a circle is given by the formula $C = 2\pi r$, where r is the radius. Let $\pi = 3.14$.)

76. Pythons A large snake, the African Rock Python, can grow to a length of 25 feet. To the nearest hundredth, find the snake's range of lengths in meters. (*Hint:* There are about 3.281 feet in 1 meter.)

77. Comparing weights The normal weight of a 6 foot 2 inch man is between 150 and 190 pounds. To the nearest hundredth, what would such a person weigh in kilograms? (*Hint:* There are 2.2 pounds in 1 kilogram.)

78. Manufacturing The time required to assemble a television set at the factory is 2 hours. A stereo receiver requires only 1 hour. The labor force at the factory can supply at least 640 and at most 810 hours of assembly time per week. When the factory is producing 3 times as many television sets as stereos, how many stereos could be manufactured in 1 week?

79. Geometry A rectangle's length is 3 feet less than twice its width, and its perimeter is between 24 and 48 feet. What might be its width?

80. Geometry A rectangle's width is 8 feet less than 3 times its length, and its perimeter is between 8 and 16 feet. What might be its length?

WRITING

81. Explain why multiplying both sides of an inequality by a negative constant reverses the direction of the inequality.

82. Explain the use of parentheses and brackets in the graphing of the solution of an inequality.

SOMETHING TO THINK ABOUT

83. To solve the inequality $1 < \frac{1}{x}$, one student multiplies both sides by x to get $x < 1$. Why is this not correct?

84. Find the solution of $1 < \frac{1}{x}$. (*Hint:* Will any negative values of x work?)

Projects

Project 1

Build a scale similar to the one shown in Figure 2-1. Demonstrate to your class how you would use the scale to solve the following equations.

a. $x - 4 = 6$ **b.** $x + 3 = 2$ **c.** $2x = 6$

d. $\frac{x}{2} = 3$ **e.** $3x - 2 = 5$ **f.** $\frac{x}{3} + 1 = 2$

Project 2

Use a calculator to determine whether the following statements are true or false.

1. $7^5 = 5^7$ **2.** $2^3 + 7^3 = (2 + 7)^3$

3. $(-4)^4 = -4^4$ **4.** $\frac{10^3}{5^3} = 2^3$

5. $8^4 \cdot 9^4 = (8 \cdot 9)^4$ **6.** $2^3 \cdot 3^3 = 6^3$

7. $\frac{3^{10}}{3^2} = 3^5$ **8.** $[(1.2)^3]^2 = [(1.2)^2]^3$

9. $(7.2)^2 - (5.1)^2 = (7.2 - 5.1)^2$

Chapter Summary

CONCEPTS	REVIEW EXERCISES
2.1	**Solving Basic Equations**
An **equation** is a statement indicating that two quantities are equal.	Tell whether the given number is a solution of the equation. **1.** $3x + 7 = 1; -2$ **2.** $5 - 2x = 3; -1$ **3.** $2(x + 3) = x; -3$ **4.** $5(3 - x) = 2 - 4x; 13$ **5.** $3(x + 5) = 2(x - 3); -21$ **6.** $2(x - 7) = x + 14; 0$
Any real number can be added to (or subtracted from) both sides of an equation to form another equation with the same solutions as the original equation.	Solve each equation and check all solutions. **7.** $x - 7 = -6$ **8.** $y - 4 = 5$ **9.** $p + 4 = 20$ **10.** $x + \frac{3}{5} = \frac{3}{5}$ **11.** $y - \frac{7}{2} = \frac{1}{2}$ **12.** $z + \frac{5}{3} = -\frac{1}{3}$
Sales price = regular price − markdown Retail price = wholesale cost + markup	**13.** A necklace is on sale for \$69.95. If it has been marked down \$35.45, what is its regular price? **14.** A suit that has been marked up \$115.25 sells for \$212.95. Find its wholesale price.
Both sides of an equation can be multiplied (or divided) by any *nonzero* real number to form another equation with the same solutions as the original equation.	Solve each equation and check all solutions. **15.** $3x = 15$ **16.** $8r = -16$ **17.** $10z = 5$ **18.** $14s = 21$ **19.** $\frac{y}{3} = 6$ **20.** $\frac{w}{7} = -5$ **21.** $\frac{a}{-7} = \frac{1}{14}$ **22.** $\frac{t}{12} = \frac{1}{2}$
Amount = rate · base	Solve each problem. **23.** What number is 35% of 700? **24.** 72% of what number is 936? **25.** What percent of 2,300 is 851? **26.** 72 is what percent of 576?

2.2 Solving More Equations

Solve each equation and check all solutions.

27. $5y + 6 = 21$
28. $5y - 9 = 1$
29. $-12z + 4 = -8$
30. $17z + 3 = 20$
31. $13 - 13t = 0$
32. $10 + 7t = -4$
33. $23a - 43 = 3$
34. $84 - 21a = -63$
35. $3x + 7 = 1$
36. $7 - 9x = 16$
37. $\frac{b+3}{4} = 2$
38. $\frac{b-7}{2} = -2$
39. $\frac{x-8}{5} = 1$
40. $\frac{x+10}{2} = -1$
41. $\frac{2y-2}{4} = 2$
42. $\frac{3y+12}{11} = 3$
43. $\frac{x}{2} + 7 = 11$
44. $\frac{r}{3} - 3 = 7$
45. $\frac{a}{2} + \frac{9}{4} = 6$
46. $\frac{x}{8} - 2.3 = 3.2$
47. A compact disc player is on sale for \$240, a 25% savings from the regular price. Find the regular price.
48. A \$38 dictionary costs \$40.47, with sales tax. Find the tax rate.
49. A Turkish rug was purchased for \$560. If it is now worth \$1,100, find the percent of increase.
50. A clock on sale for \$215 was regularly priced at \$465. Find the percent of discount.

2.3 Simplifying Expressions to Solve Equations

Like terms can be combined by adding their numerical coefficients and using the same variables and exponents.

Simplify each expression, if possible.

51. $5x + 9x$
52. $7a + 12a$
53. $18b - 13b$
54. $21x - 23x$
55. $5y - 7y$
56. $19x - 19$
57. $7(x + 2) + 2(x - 7)$
58. $2(3 - x) + x - 6x$
59. $y^2 + 3(y^2 - 2)$
60. $2x^2 - 2(x^2 - 2)$

Solve each equation and check all solutions.

61. $2x - 19 = 2 - x$

62. $5b - 19 = 2b + 20$

63. $3x + 20 = 5 - 2x$

64. $0.9x + 10 = 0.7x + 1.8$

65. $10(t - 3) = 3(t + 11)$

66. $2(5x - 7) = 2(x - 35)$

67. $\dfrac{3u - 6}{5} = 3$

68. $\dfrac{5v - 35}{3} = -5$

69. $\dfrac{7x - 28}{4} = -21$

70. $\dfrac{27 + 9y}{5} = -27$

An **identity** is an equation that is true for all values of its variable.

A **contradiction** is an equation that is true for no values of its variable.

Classify each equation as an identity or a contradiction.

71. $2x - 5 = x - 5 + x$

72. $-3(a + 1) - a = -4a + 3$

73. $2(x - 1) + 4 = 4(1 + x) - (2x + 2)$

2.4 Introduction to Problem Solving

Equations are useful in solving applied problems.

74. A carpenter wants to cut an 8-foot board into two pieces so that one piece is 7 feet shorter than twice the longer piece. Where should he make the cut?

75. Find x.

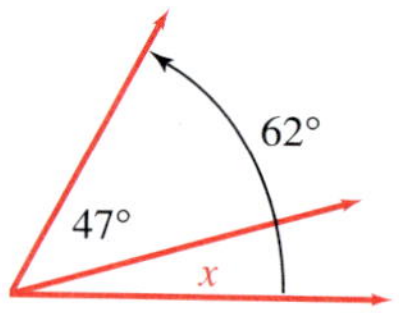

76. Find x.

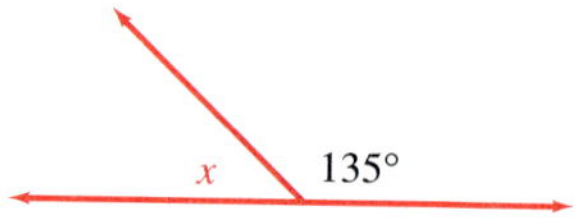

If the sum of the measures of two angles is 90°, the angles are complementary.

If the sum of the measures of two angles is 180°, the angles are supplementary.

77. Find the complement of an angle that measures 69°.

78. Find the supplement of an angle that measures 69°.

79. If the length of the rectangular painting in the illustration is 3 inches more than twice the width, how wide is the rectangle?

80. A woman has $27,000. Part is invested for 1 year in a certificate of deposit paying 7% interest, and the remaining amount in a cash management fund paying 9%. The total interest on the two investments is $2,110. How much does she invest at each rate?

2.5 Motion and Mixture Problems

81. A bicycle path is 5 miles long. A man walks from one end at the rate of 3 mph. At the same time, a friend bicycles from the other end, traveling at 12 mph. In how many minutes will they meet?

82. A container is partly filled with 12 liters of whole milk containing 4% butterfat. How much 1% milk must be added to get a mixture that is 2% butterfat?

83. A store manager mixes candy worth 90¢ per pound with gumdrops worth $1.50 per pound to make 20 pounds of a mixture worth $1.20 per pound. How many pounds of each kind of candy must he use?

84. An electric company charges $17.50 per month, plus 18¢ for every kilowatt-hour of energy used. One resident's bill was $43.96. How many kilowatt-hours were used that month?

85. A contractor charges $35 for the installation of rain gutters, plus $1.50 per foot. If one installation cost $162.50, how many feet of gutter were required?

2.6 Formulas

A literal equation or formula can often be solved for any of its variables.

Solve each equation for the indicated variable.

86. $E = IR$; for R

87. $i = prt$; for t

88. $P = I^2R$; for R

89. $d = rt$; for r

90. $V = lwh$; for h

91. $y = mx + b$; for m

92. $V = \pi r^2 h$; for h

93. $a = 2\pi rh$; for r

94. $F = \dfrac{GMm}{d^2}$; for G

95. $P = \dfrac{RT}{mV}$; for m

2.7 Solving Inequalities

Inequalities are solved by techniques similar to those used to solve equations, with this exception: *If both sides of an inequality are multiplied or divided by a negative number, the direction of the inequality must be reversed.*

The solution of an inequality can be graphed on the number line.

Graph the solution to each inequality.

96. $3x + 2 < 5$

97. $-5x - 8 > 7$

98. $5x - 3 \geq 2x + 9$

99. $7x + 1 \leq 8x - 5$

100. $5(3 - x) \leq 3(x - 3)$

101. $3(5 - x) \geq 2x$

102. $8 < x + 2 < 13$

103. $0 \leq 2 - 2x < 4$

104. By city ordinance, the perimeter of a rectangular swimming pool cannot exceed 68 feet. The width is 6 feet shorter than the length. What possible lengths will meet these conditions?

Chapter Test

Decide whether the given number is a solution of the equation.

1. $5x + 3 = -2; -1$

2. $3(x + 2) = 2x; -6$

3. $-3(2 - x) = 0; -2$

4. $3(x + 2) = 2x + 7; 1$

Solve each equation.

5. $x + 17 = -19$

6. $a - 15 = 32$

7. $12x = -144$

8. $\dfrac{x}{7} = -1$

9. $8x + 2 = -14$

10. $3 = 5 - 2x$

11. $\dfrac{2x - 5}{3} = 3$

12. $\dfrac{3x - 18}{2} = 6x$

13. $23 - 5(x + 10) = -12$ **14.** $\frac{7}{8}(x - 4) = 5x - \frac{7}{2}$

Simplify each expression.

15. $x + 5(x - 3)$

16. $3x - 5(2 - x)$

17. $-3x(x + 3) + 3x(x - 3)$

18. $-4x(2x - 5) - 7x(4x + 1)$

19. Find x.

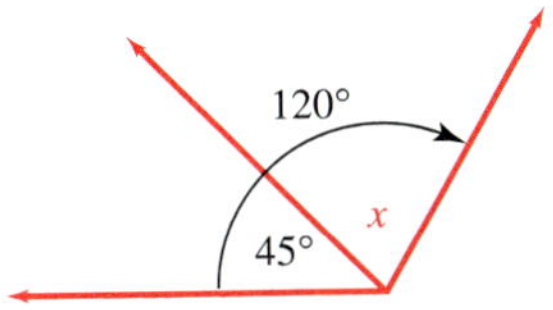

20. Find the supplement of a 105° angle.

21. A student invests part of \$10,000 at 6% annual interest and the rest at 5%. If the annual income from these investments is \$560, how much was invested at each rate?

22. A car leaves Rockford at the rate of 65 mph, bound for Madison. At the same time, a truck leaves Madison at the rate of 55 mph, bound for Rockford. If the cities are 72 miles apart, how long will it take for the car and the truck to meet?

23. How many liters of water must be added to 30 liters of a 10% brine solution to dilute it to an 8% solution?

Solve each equation for the variable indicated.

24. $d = rt$; for t

25. $P = 2l + 2w$; for l

26. $A = 2\pi rh$; for h

27. $A = P + Prt$; for r

Graph the solution of each inequality.

28. $8x - 20 \geq 4$

29. $x - 2(x + 7) > 14$

30. $-4 \leq 2(x + 1) < 10$

31. $-2 < 5(x - 1) \leq 10$

CUMULATIVE REVIEW EXERCISES

Classify each number as an integer, a rational number, an irrational number, a real number, a positive number, or a negative number. Each number may be in several classifications.

1. $\frac{27}{9}$

2. -0.25

Graph each set of numbers on the number line.

3. The natural numbers between 2 and 7

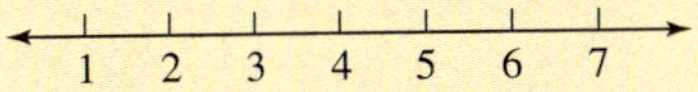

4. The real numbers between 2 and 7

Simplify each expression.

5. $\frac{|-3| - |3|}{|-3 - 3|}$

6. $\frac{5}{7} \cdot \frac{14}{3}$

7. $2\frac{3}{5} + 5\frac{1}{2}$

8. $35.7 - 0.05$

Let $x = -5$, $y = 3$, $z = 0$, and evaluate each expression.

9. $(3x - 2y)z$

10. $\frac{x - 3y + |z|}{2 - x}$

11. $x^2 - y^2 + z^2$

12. $\frac{x}{y} + \frac{y + 2}{3 - z}$

13. What is $7\frac{1}{2}\%$ of 330?

14. 1,688 is 32% of what number?

Consider the algebraic expression $3x^3 + 5x^2y + 37y$.

15. Find the coefficient of the second term.

16. List the factors of the third term.

Simplify each expression.

17. $3x - 5x + 2y$

18. $3(x - 7) + 2(8 - x)$

19. $2x^2y^3 - xy(xy^2)$

20. $x^2(3 - y) + x(xy + x)$

Solve each equation and check the result.

21. $3(x - 5) + 2 = 2x$

22. $\dfrac{x - 5}{3} - 5 = 7$

23. $\dfrac{2x - 1}{5} = \dfrac{1}{2}$

24. $2(a - 5) - (3a + 1) = 0$

25. Auto sales An auto dealer's promotional ad appears in the illustration. One car is selling for \$23,499. What was the dealer's invoice?

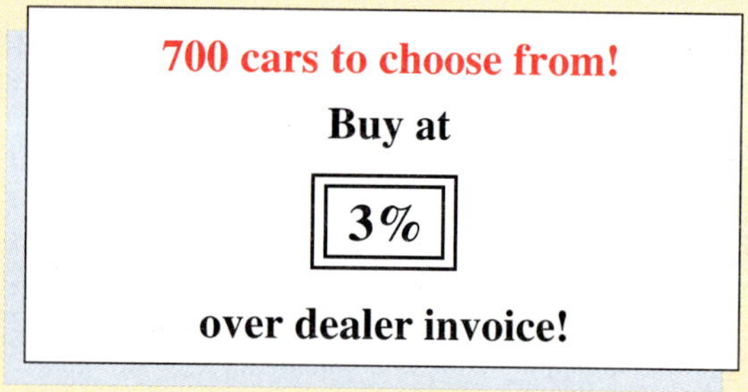

26. Furniture pricing A sofa and a \$300 chair are discounted 35%, and are priced at \$780 for both. Find the original price of the sofa.

27. Cost of a car The total cost of a new car, including an 8.5% sales tax, is \$13,725.25. Find the cost before tax.

28. Manufacturing concrete Concrete contains 3 times as much gravel as cement. How many pounds of cement are in 500 pounds of dry concrete mix?

29. Building construction A 35-foot beam, 1 foot wide and 2 inches thick, is cut into three sections. One section is 14 feet long. Of the remaining two sections, one is twice as long as the other. Will the shortest section span an 8-foot-wide doorway?

30. Installing solar heating One solar panel in the illustration is 3.4 feet wider than the other. Find the width of each panel.

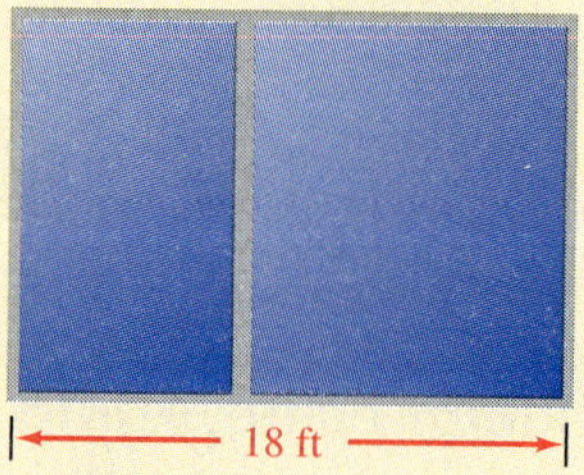

Solve each formula for the variable indicated.

31. $A = \frac{1}{2}h(b + B)$; for h

32. $y = mx + b$; for x

Evaluate each expression.

33. $4^2 - 5^2$

34. $(4 - 5)^2$

35. $5(4^3 - 2^3)$

36. $-2(5^4 - 7^3)$

Graph the solutions of each inequality.

37. $8(4 + x) > 10(6 + x)$

38. $-9 < 3(x + 2) \le 3$

3 Graphing and Solving Systems of Equations and Inequalities

© Wolfgang Kaehler/CORBIS

InfoTrac Project

Do a subject guide search on "break-even." Under "Break-Even Analysis," click on "see 12 subdivisions," and then click on "case studies." Find the article "Break-even analysis: Simon Harrison gives a revision reminder about a concept that seems straightforward, yet keeps causing problems in exams." Write a summary of the article and define "break-even."

The overhead cost of running a coffee shop is $4,000 a month (rent, utilities, etc.) and the cost to the business of each cup of coffee is $0.20 a cup. Write an equation where y represents the total cost of running the business and x represents the number of cups of coffee sold. If coffee sells for $1.50 a cup, write an equation showing the total revenue, where y is the revenue earned and x is the number of cups of coffee sold. Solve this system of equations to find the number of cups of coffee that must be sold to break-even. What will the total revenue be at the break-even point? Suppose the cost of making the coffee increases from $0.20 a cup to $0.35 a cup. How many cups of coffee must now be sold to break even? What will the total revenue be at the break-even point?

Complete this project after studying Section 3.6.

Mathematics in Electrical Engineering

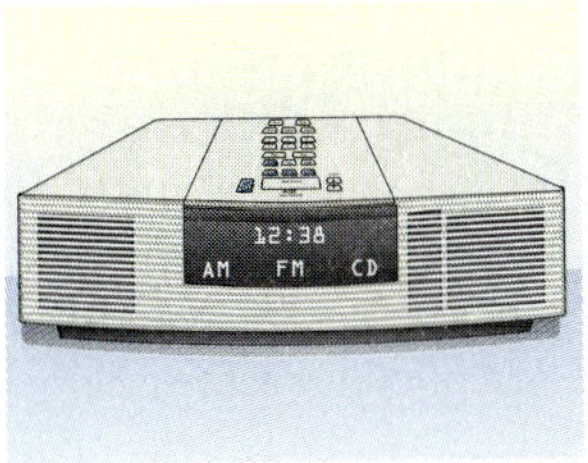

In a radio, an inductor and a capacitor are used in a resonant circuit to select a wanted radio station at a frequency f and reject all others. The inductance L and the capacitance C determine the inductive reactance X_L and the capacitive reactance X_C of that circuit, where

$$X_L = 2\pi f L \quad \text{and} \quad X_C = \frac{1}{2\pi f C}$$

Exercise Set 3.6 Problem 54

The radio station selected will be at the frequency f, where $X_L = X_C$. Write a formula for f^2 in terms of L and C.

In this chapter, we will discuss equations with two variables and see that the relationships set up by these equations can also be expressed in tables or graphs.

After discussing how to graph linear equations on a rectangular coordinate system, we will learn how to solve systems of two equations, each with two variables. After solving some application problems involving systems of equations, we will conclude the chapter by solving systems of linear inequalitites.

3.1 The Rectangular Coordinate System

- **The Rectangular Coordinate System**
- **Graphing Mathematical Relationships**
- **Reading Graphs**
- **Step Graphs**

Getting Ready *Graph each set of numbers on the number line.*

1. −2, 1, 3

2. All numbers greater than −2

3. All numbers less than or equal to 3

4. All numbers between −3 and 2

It is often said, "A picture is worth a thousand words." In this section, we will show how numerical relationships can be described using mathematical pictures called **graphs.** We will also show how we can obtain important information by reading graphs.

The Rectangular Coordinate System

When designing the Gateway Arch, shown in Figure 3-1(a), architects created a mathematical model of the arch called a **graph.** This graph, shown in Figure 3-1(b), is drawn on a grid called a **rectangular coordinate system.** This coordinate system is sometimes called a **Cartesian coordinate system** after the 17th-century French mathematician René Descartes.

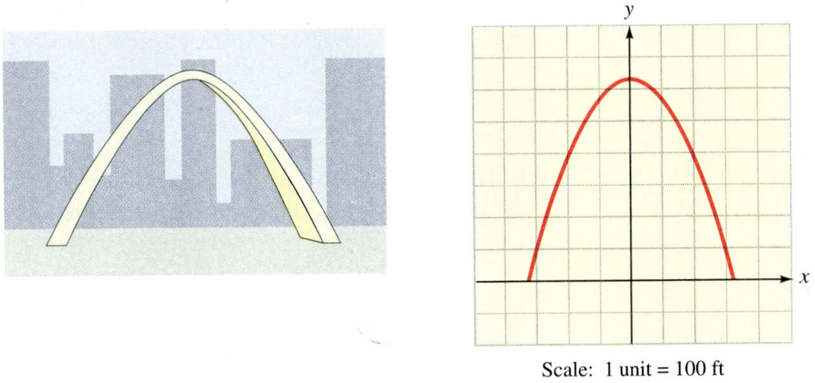

(a) (b)

Figure 3-1

A rectangular coordinate system (see Figure 3-2) is formed by two perpendicular number lines.

- The horizontal number line is called the ***x*-axis.**
- The vertical number line is called the ***y*-axis.**

The positive direction on the x-axis is to the right, and the positive direction on the y-axis is upward. The scale on each axis should fit the data. For example, the axes of the graph of the arch shown in Figure 3-1(b) are scaled in units of 100 feet. If no scale is indicated on the axes, we assume that the axes are scaled in units of 1.

The point where the axes cross is called the **origin.** This is the 0 point on each axis. The two axes form a **coordinate plane** and divide it into four regions called **quadrants,** which are numbered as shown in Figure 3-2.

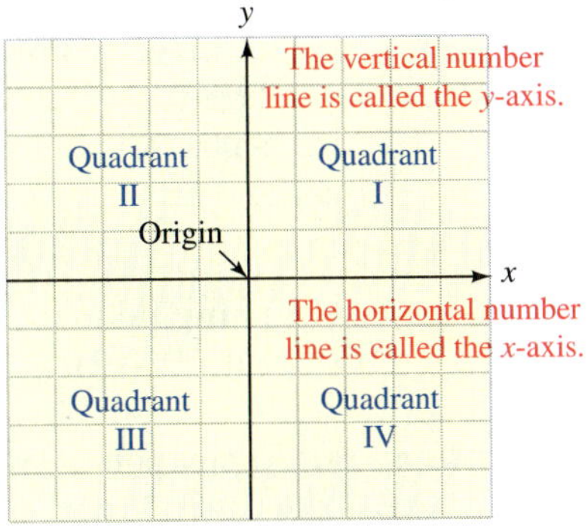

Figure 3-2

Each point in a coordinate plane can be identified by a pair of real numbers x and y, written as (x, y). The first number in the pair is the ***x*-coordinate,** and the second number is the ***y*-coordinate.** The numbers are called the **coordinates** of the point. Some examples of ordered pairs are $(3, -4)$, $\left(-1, -\frac{3}{2}\right)$, and $(0, 2.5)$.

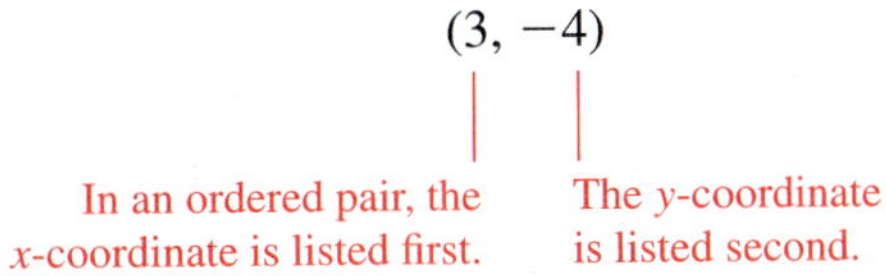

The process of locating a point in the coordinate plane is called **graphing** or **plotting** the point. In Figure 3-3(a), we show how to graph the point A with coordinates of $(3, -4)$. Since the x-coordinate is positive, we start at the origin and move 3 units to the right along the x-axis. Since the y-coordinate is negative, we then move down 4 units to locate point A. Point A is the **graph** of $(3, -4)$ and lies in quadrant IV.

To plot the point $B(-4, 3)$, we start at the origin, move 4 units to the left along the x-axis, and then move up 3 units to locate point B. Point B lies in quadrant II.

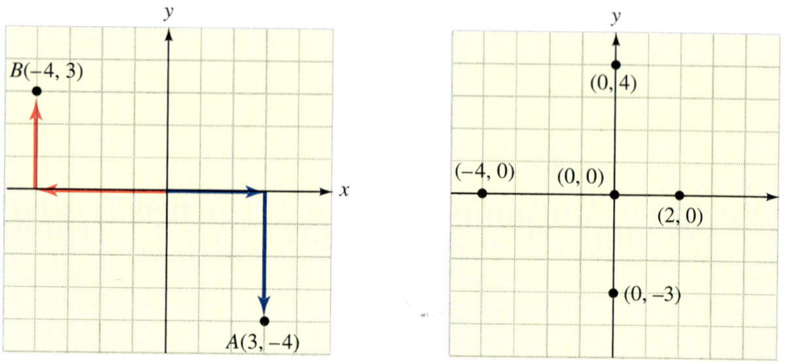

(a) (b)

Figure 3-3

Comment Note that point A with coordinates of $(3, -4)$ is not the same as point B with coordinates $(-4, 3)$. Since the order of the coordinates of a point is important, we call the pairs **ordered pairs.**

In Figure 3-3(b), we see that the points $(-4, 0)$, $(0, 0)$, and $(2, 0)$ lie on the x-axis. In fact, all points with a y-coordinate of 0 will lie on the x-axis.

From Figure 3-3(b), we also see that the points $(0, -3)$, $(0, 0)$, and $(0, 4)$ lie on the y-axis. All points with an x-coordinate of 0 lie on the y-axis. From the figure, we can also see that the coordinates of the origin are $(0, 0)$.

EXAMPLE 1 **Graphing points** Plot the points: **a.** $A(-2, 3)$, **b.** $B\left(-1, -\frac{3}{2}\right)$, **c.** $C(0, 2.5)$, and **d.** $D(4, 2)$.

Solution

a. To plot point A with coordinates $(-2, 3)$, we start at the origin, move 2 units to the *left* on the x-axis, and move 3 units *up*. Point A lies in quadrant II. (See Figure 3-4.)

b. To plot point B with coordinates of $\left(-1, -\frac{3}{2}\right)$, we start at the origin and move 1 unit to the *left* and $\frac{3}{2}$ $\left(\text{or } 1\frac{1}{2}\right)$ units *down*. Point B lies in quadrant III, as shown in Figure 3-4.

c. To graph point C with coordinates of $(0, 2.5)$, we start at the origin and move 0 units on the x-axis and 2.5 units *up*. Point C lies on the y-axis, as shown in Figure 3-4.

d. To graph point D with coordinates of $(4, 2)$, we start at the origin and move 4 units to the *right* and 2 units *up*. Point D lies in quadrant I, as shown in Figure 3-4.

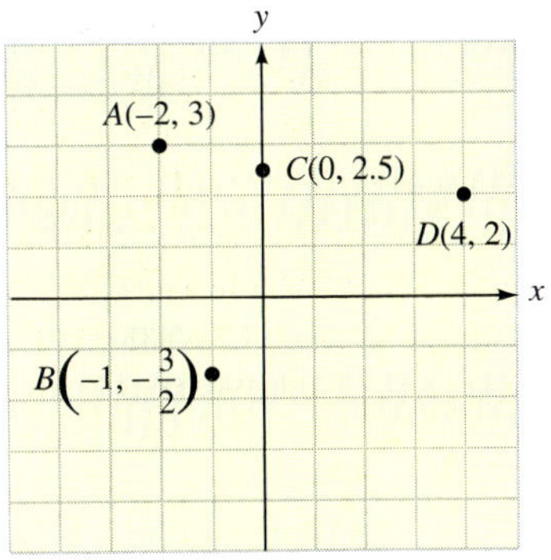

Figure 3-4

Self Check Plot the points: **a.** $E(2, -2)$, **b.** $F(-4, 0)$, **c.** $G\left(1.5, \frac{5}{2}\right)$, and **d.** $H(0, 5)$.

EXAMPLE 2 **Orbit of the Earth** The circle shown in Figure 3-5 is an approximate graph of the orbit of the Earth. The graph is made up of infinitely many points, each with its own x- and y-coordinates. Use the graph to find the coordinates of the Earth's position during the months of February, May, August, and December.

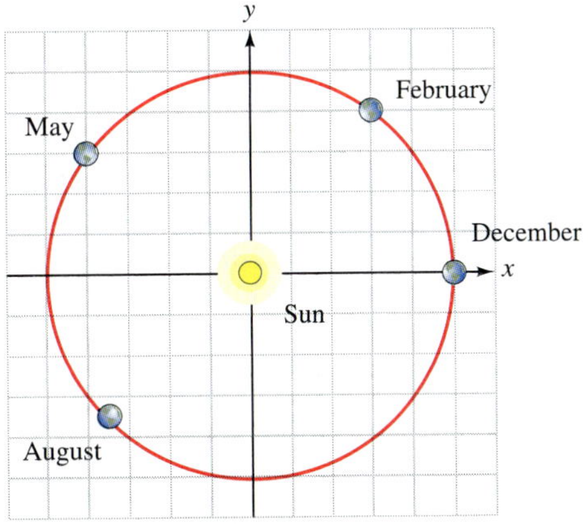

Figure 3-5

Solution To find the coordinates of each position, we start at the origin and move left or right along the x-axis to find the x-coordinate and then up or down to find the y-coordinate. See Table 3-1.

Month	Position of the Earth on graph	Coordinates
February	3 units to the *right,* then 4 units *up*	(3, 4)
May	4 units to the *left,* then 3 units *up*	(−4, 3)
August	3.5 units to the *left,* then 3.5 units *down*	(−3.5, −3.5)
December	5 units *right,* then no units *up* or *down*	(5, 0)

Table 3-1

PERSPECTIVE

As a child, René Descartes was frail and often sick. To improve his health, eight-year-old René was sent to a Jesuit school. The headmaster encouraged him to sleep in the morning as long as he wished. As a young man, Descartes spent several years as a soldier and world traveler, but his interests included mathematics and philosophy, as well as science, literature, writing, and taking it easy. The habit of sleeping late continued throughout his life. He claimed that his most productive thinking occurred when he was lying in bed. According to one story, Descartes first thought of analytic geometry as he watched a fly walking on his bedroom ceiling.

Descartes might have lived longer if he had stayed in bed. In 1649, Queen Christina of Sweden decided that she needed a tutor in philosophy, and she requested the services of Descartes. Tutoring would not have been difficult, except that the queen scheduled her lessons before dawn in her library with her windows open. The cold Stockholm mornings were too much for a man who was used to sleeping past noon. Within a few months, Descartes developed a fever and died, probably of pneumonia.

Graphing Mathematical Relationships

Every day, we deal with quantities that are related.

- The distance that we travel depends on how fast we are going.
- Our weight depends on how much we eat.
- The amount of water in a tub depends on how long the water has been running.

We can often use graphs to visualize relationships between two quantities. For example, suppose that we know the number of gallons of water that are in a tub at several time intervals after the water has been turned on. We can list that information in a **table of values** (see Figure 3-6).

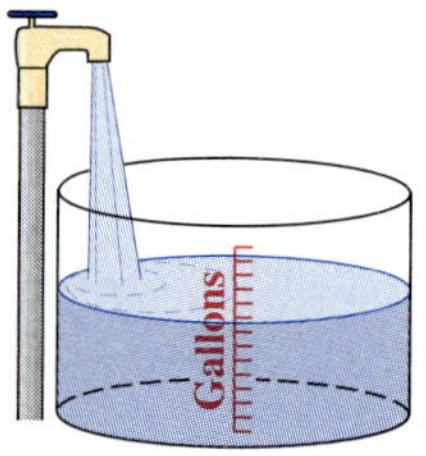

At various times, the amount of water in the tub was measured and recorded in the table of values.

Time (minutes)	Water in tub (gallons)	
0	0	→ (0, 0)
1	8	→ (1, 8)
3	24	→ (3, 24)
4	32	→ (4, 32)

x-coordinate — y-coordinate — The data in the table can be expressed as ordered pairs (x, y).

Figure 3-6

The information in the table can be used to construct a graph that shows the relationship between the amount of water in the tub and the time the water has been running. Since the amount of water in the tub depends on the time, we will associate *time* with the x-axis and the *amount of water* with the y-axis.

To construct the graph in Figure 3-7, we plot the four ordered pairs and draw a line through the resulting data points.

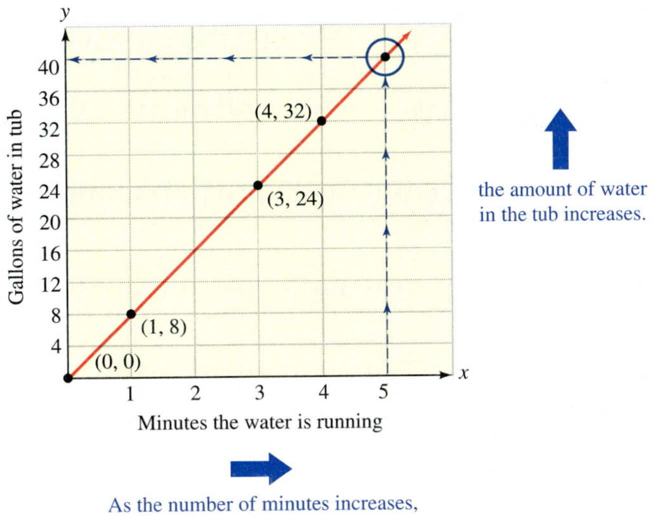

Figure 3-7

From the graph, we can see that the amount of water in the tub increases as the water is allowed to run. We can also use the graph to make observations about the amount of water in the tub at other times. For example, the dashed line on the graph shows that in 5 minutes, the tub will contain 40 gallons of water.

Reading Graphs

In the next example, we show that valuable information can be obtained from a graph.

EXAMPLE 3 **Reading a graph** The graph in Figure 3-8 shows the number of people in an audience before, during, and after the taping of a television show. On the x-axis, 0 represents the time when taping began. Use the graph to answer the following questions, and record each result in a table of values.

a. How many people were in the audience when taping began?

b. What was the size of the audience 10 minutes before taping began?

c. At what times were there exactly 100 people in the audience?

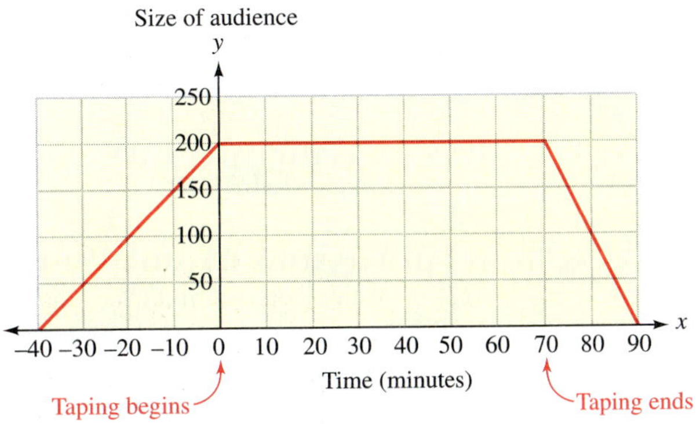

Figure 3-8

Solution **a.** The time when taping began is represented by 0 on the x-axis. Since the point on the graph directly above 0 has a y-coordinate of 200, the point (0, 200) is on the graph. The y-coordinate of this point indicates that 200 people were in the audience when the taping began.

Time	Audience
0	200

b. Ten minutes before taping began is represented by -10 on the x-axis. Since the point on the graph directly above -10 has a y-coordinate of 150, the point $(-10, 150)$ is on the graph. The y-coordinate of this point indicates that 150 people were in the audience 10 minutes before the taping began.

Time	Audience
-10	150

c. We can draw a horizontal line passing through 100 on the y-axis. Since this line intersects the graph twice, there were two times when 100 people were in the audience. One time was 20 minutes before taping began, and the other was 80 minutes after taping began. So the points $(-20, 100)$ and $(80, 100)$ are on the graph. The y-coordinates of these points indicate that there were 100 people in the audience 20 minutes before and 80 minutes after taping began.

Time	Audience
-20	100
80	100

Self Check Use the graph in Figure 3-8 to answer the following questions. **a.** At what times were there exactly 50 people in the audience? **b.** What was the size of the audience that watched the taping? **c.** How long did it take for the audience to leave the studio after taping ended?

Step Graphs

The graph in Figure 3-9 shows the cost of renting a trailer for different periods of time. For example, the cost of renting the trailer for 4 days is \$60, which is the y-coordinate of the point with coordinates of (4, 60). For renting the trailer for a period lasting over 4 and up to 5 days, the cost jumps to \$70. Since the jumps in cost form steps in the graph, we call the graph a **step graph.**

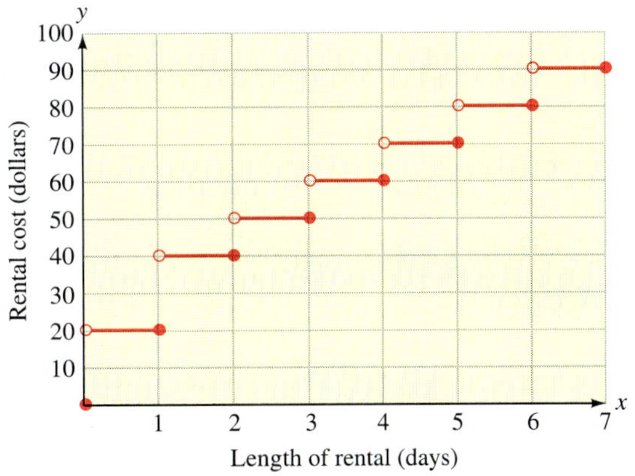

Figure 3-9

EXAMPLE 4 Use the information in Figure 3-9 to answer the following questions. Write the results in a table of values.

a. Find the cost of renting the trailer for 2 days.

b. Find the cost of renting the trailer for $5\frac{1}{2}$ days.

c. How long can you rent the trailer if you have $50?

d. Is the rental cost per day the same?

Solution **a.** We locate 2 days on the x-axis and move up to locate the point on the graph directly above the 2. Since the point has coordinates (2, 40), a two-day rental would cost $40. We enter this ordered pair in Table 3-2.

b. We locate $5\frac{1}{2}$ days on the x-axis and move straight up to locate the point on the graph with coordinates $\left(5\frac{1}{2}, 80\right)$, which indicates that a $5\frac{1}{2}$-day rental would cost $80. We enter this ordered pair in Table 3-2.

c. We draw a horizontal line through the point labeled 50 on the y-axis. Since this line intersects one step of the graph, we can look down to the x-axis to find the x-values that correspond to a y-value of 50. From the graph, we see that the trailer can be rented for more than 2 and up to 3 days for $50. We write (3, 50) in Table 3-2.

Length of rental (days)	Cost (dollars)
2	40
$5\frac{1}{2}$	80
3	50

Table 3-2

d. No, the cost per day is not the same. If we look at the y-coordinates, we see that for the first day, the rental fee is $20. For the second day, the cost jumps another $20. For the third day, and all subsequent days, the cost jumps only $10.

Self Check Answers

1.

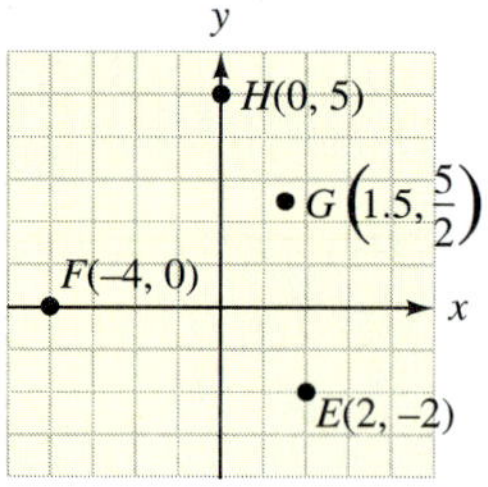

3. a. 30 min before and 85 min after taping began, **b.** 200, **c.** 20 min

Orals

1. Explain why the pair $(-2, 4)$ is called an ordered pair.

2. At what point do the coordinate axes intersect?

3. In which quadrant does the graph of $(3, -5)$ lie?

4. On which axis does the point $(0, 5)$ lie?

3.1 EXERCISES

REVIEW

1. Evaluate: $-3 - 3(-5)$.

2. Evaluate: $(-5)^2 + (-5)$.

3. What is the opposite of -8?

4. Simplify: $|-1 - 9|$.

5. Solve: $-4x + 7 = -21$.

6. Solve $P = 2l + 2w$ for w.

7. Evaluate $(x + 1)(x + y)^2$ for $x = -2$ and $y = -5$.

8. Simplify: $-6(x - 3) - 2(1 - x)$.

VOCABULARY AND CONCEPTS *Fill in the blanks.*

9. The pair of numbers $(-1, -5)$ is called an ________.

10. In the ordered pair $\left(-\frac{3}{2}, -5\right)$, -5 is called the __ coordinate.

11. The point with coordinates $(0, 0)$ is the ______.

12. The x- and y-axes divide the coordinate plane into four regions called ________.

13. The point with coordinates $(4, 2)$ can be graphed on a ________________ system.

14. The process of locating the position of a point on a coordinate plane is called ________________ the point.

Answer the question or fill in the blanks.

15. Do $(3, 2)$ and $(2, 3)$ represent the same point?

16. In the ordered pair $(4, 5)$, is 4 associated with the horizontal or the vertical axis?

17. To plot the point with coordinates $(-5, 4.5)$, we start at the ______, move 5 units to the ____, and then move 4.5 units ____.

18. To plot the point with coordinates $\left(6, -\frac{3}{2}\right)$, we start at the ______, move 6 units to the ______, and then move $\frac{3}{2}$ units ______.

19. In which quadrant do points with a negative x-coordinate and a positive y-coordinate lie?

20. In which quadrant do points with a positive x-coordinate and a negative y-coordinate lie?

21. Use the graph to complete the table.

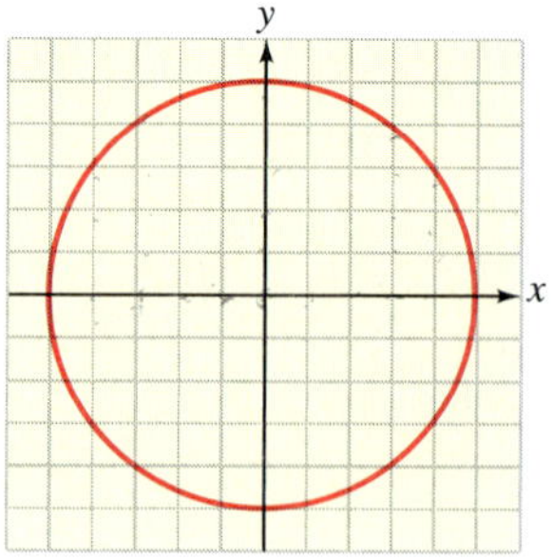

x	y
4	3
0	5, 5
−3	−4, 4
5, −5	0
−4	−3, 3
3	−4, 4

22. Use the graph to complete the table.

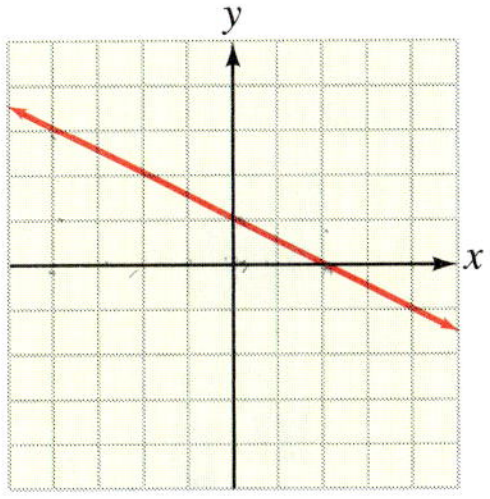

x	y
	0
	2
	−1
−4	
	1

The graph in the illustration gives the heart rate of a woman before, during, and after an aerobic workout. Use the graph to answer the following questions.

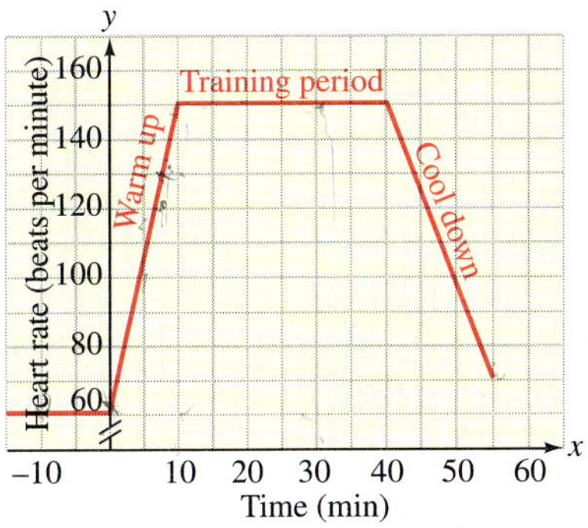

23. What information does the point (−10, 60) give us?

24. After beginning the workout, how long did it take the woman to reach her training-zone heart rate?

25. What was her heart rate one-half hour after beginning the workout?

26. For how long did she work out at the training-zone level?

27. At what times was her heart rate 100 beats per minute?

28. How long was her cool-down period?

29. What was the difference in her heart rate before the workout and after the cooldown period?

30. What was her approximate heart rate 8 minutes after beginning?

PRACTICE *Graph each point on the coordinate grid.*

31. $A(-3, 4)$, $B(4, 3.5)$, $C\left(-2, -\frac{5}{2}\right)$, $D(0, -4)$, $E\left(\frac{3}{2}, 0\right)$, $F(3, -4)$

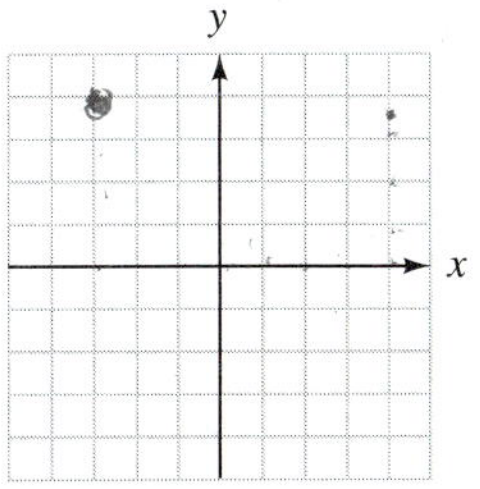

32. $G(4, 4)$, $H(0.5, -3)$, $I(-4, -4)$, $J(0, -1)$, $K(0, 0)$, $L(0, 3)$, $M(-2, 0)$

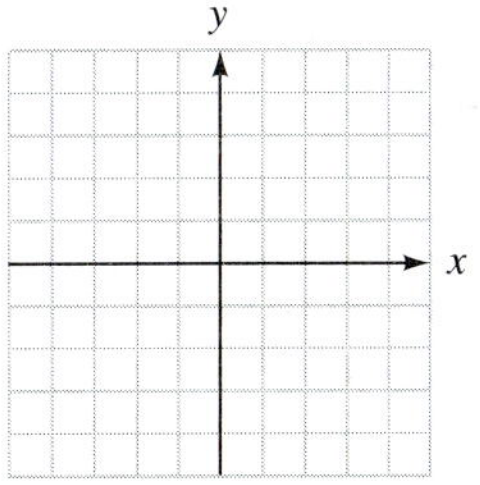

APPLICATIONS

33. Road maps Road maps usually have a coordinate system to help locate cities. Use the map in the illustration to locate Carbondale, Champaign, Chicago, Peoria, Rockford, Springfield, and St. Louis. Express each answer in the form (number, letter).

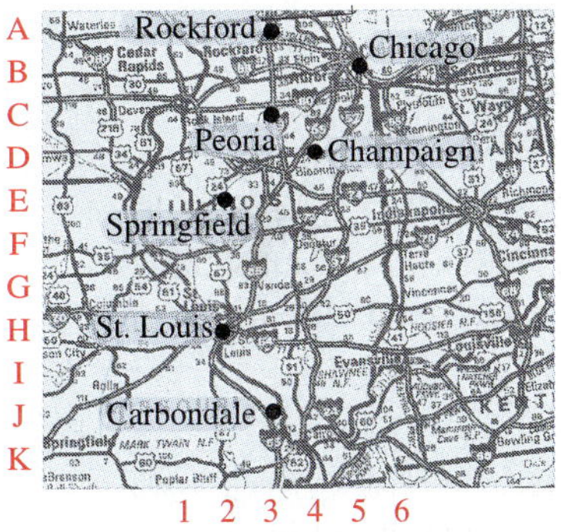

34. Battleship In the game Battleship, players use coordinates to drop depth charges from a battleship to hit a hidden submarine. What coordinates should be used to make three hits on the exposed submarine shown in the illustration? Express each answer in the form (letter, number).

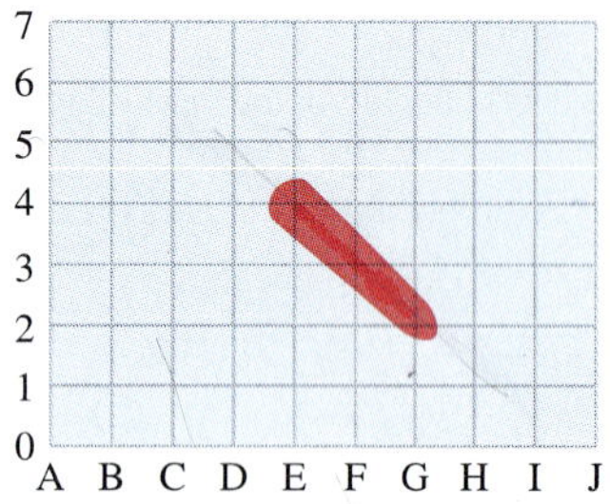

35. Water pressure The graphs in the illustration show the paths of two streams of water from the same hose held at two different angles.

a. At which angle does the stream of water shoot higher? How much higher?

b. At which angle does the stream of water shoot out farther? How much farther?

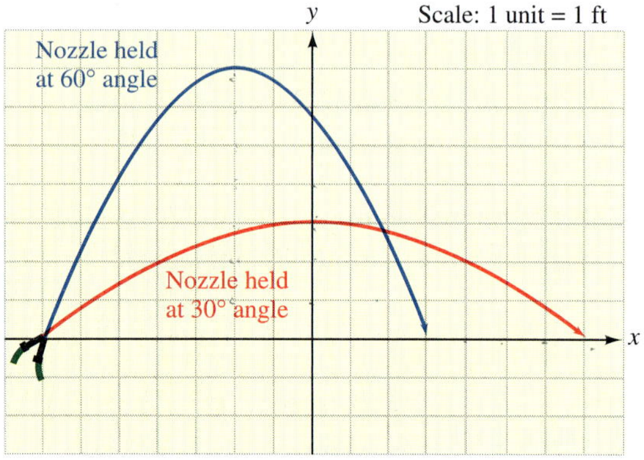

36. Golf swings To correct her swing, a golfer was videotaped and then had her image displayed on a computer monitor so that it could be analyzed by a golf pro. See the illustration. Give the coordinates of the points that are highlighted on the arc of her swing.

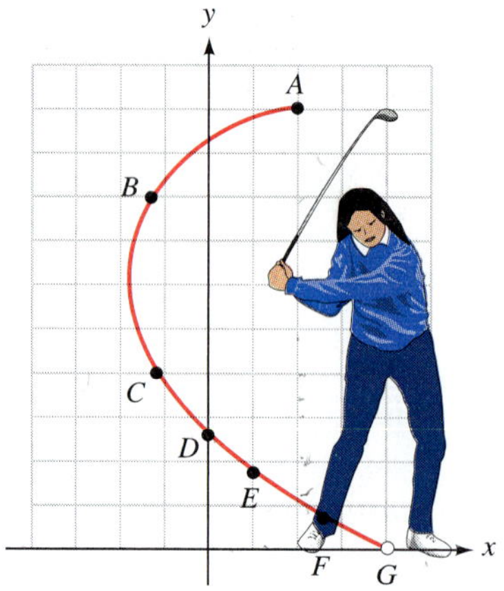

37. Video rental The charges for renting a movie are shown in the graph in the illustration.

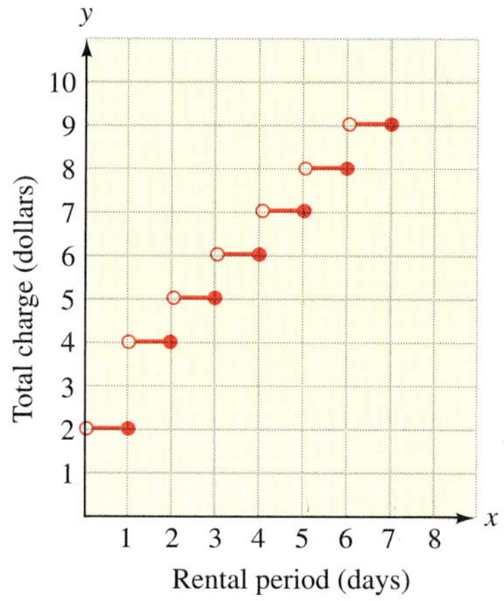

a. Find the charge for a 1-day rental.

b. Find the charge for a 2-day rental.

c. Find the charge if the tape is kept for 5 days.

d. Find the charge if the tape is kept for a week.

38. Postage rates The graph shown in the illustration gives the first-class postage rates for mailing parcels weighing up to 5 ounces.

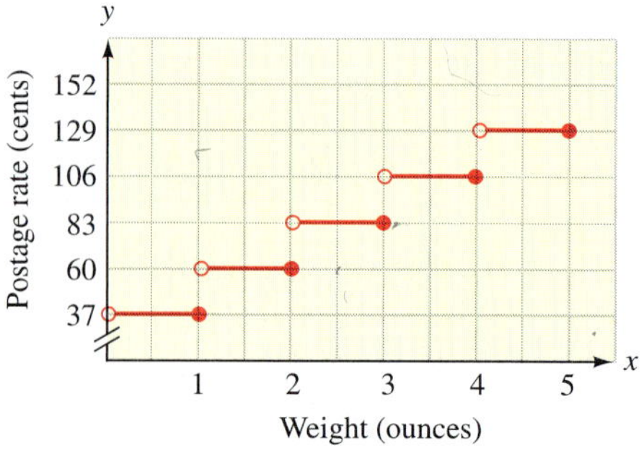

a. Find the cost of postage to mail each of the following letters first class: a 1-ounce letter, a 4-ounce letter, and a $2\frac{1}{2}$-ounce letter.

b. Find the difference in postage for a 3.75-ounce letter and a 4.75-ounce letter.

c. What is the heaviest letter that can be mailed first class for 60¢?

39. Gas mileage The table in the illustration gives the number of miles (y) that a truck can be driven on x gallons of gasoline. Plot the ordered pairs and draw a line connecting the points.

x	y
2	10
3	15
5	25

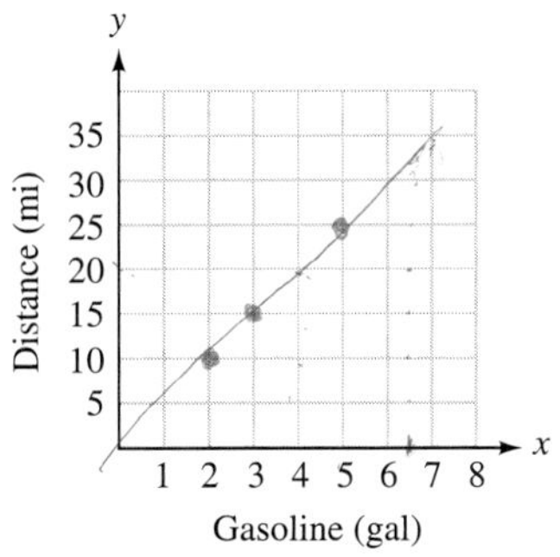

a. Estimate how far the truck can go on 7 gallons of gasoline.

b. How many gallons of gas are needed to travel a distance of 20 miles?

c. Estimate how far the truck can go on 6.5 gallons of gasoline.

40. Wages The table in the illustration gives the amount y (in dollars) that a student can earn by working x hours. Plot the ordered pairs and draw a line connecting the points.

x	y
3	18
6	36
7	42

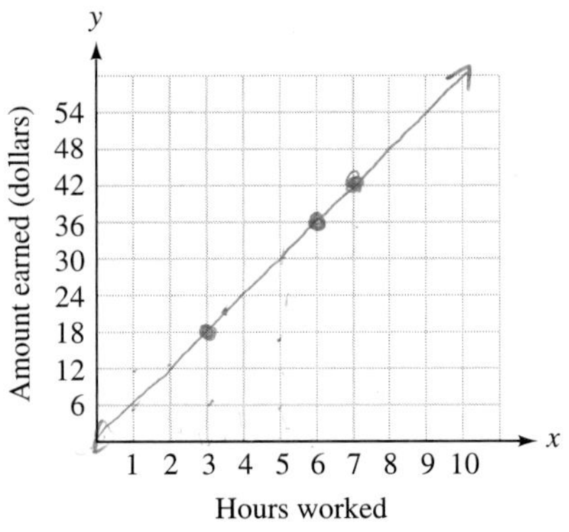

a. How much will the student earn in 5 hours?

b. How long would the student have to work to earn $12?

c. Estimate how much the student will earn in 3.5 hours.

41. Value of a car The table in the illustration shows the value y (in thousands of dollars) of a car that is x years old. Plot the ordered pairs and draw a line connecting the points.

a. What does the point (3, 7) on the graph tell you?

b. Estimate the value of the car when it is 7 years old.

c. After how many years will the car be worth $2,500?

x	y
3	7
4	5.5
5	4

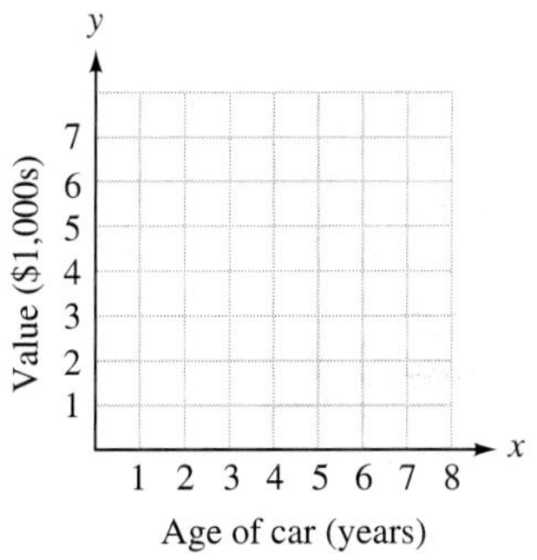

42. Depreciation As a piece of farm machinery gets older, it loses value. The table in the illustration shows the value y of a tractor that is x years old. Plot the ordered pairs and draw a line connecting them.

a. What does the point (0, 9) on the graph tell you?

b. Estimate the value of the tractor in 3 years.

c. When will the tractor's value fall below $30,000?

x	y
0	9
6	5
9	3

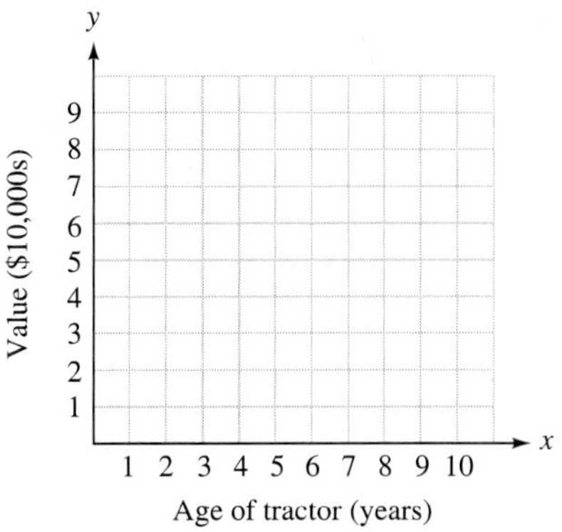

WRITING

43. Explain why the point with coordinates $(-3, 3)$ is not the same as the point with coordinates $(3, -3)$.

44. Explain what is meant when we say that the rectangular coordinate graph of the St. Louis Arch is made up of *infinitely many* points.

45. Explain how to plot the point with coordinates $(-2, 5)$.

46. Explain why the coordinates of the origin are $(0, 0)$.

SOMETHING TO THINK ABOUT

47. Could you have a coordinate system where the coordinate axes were not perpendicular? How would it be different?

48. René Descartes is famous for saying, "I think. Therefore I am." What do you think he meant by that?

3.2 Graphing Linear Equations

- Equations with Two Variables
- Constructing Tables of Values
- Graphing Equations
- The Intercept Method of Graphing a Line
- Graphing Horizontal and Vertical Lines
- An Application of Linear Equations

Getting Ready *In Problems 1–4, let* $y = 2x + 1$.

1. Find y when $x = 0$.

2. Find y when $x = 2$.

3. Find y when $x = -2$.

4. Find y when $x = \frac{1}{2}$.

5. Find five pairs of numbers with a sum of 8.

6. Find five pairs of numbers with a difference of 5.

In this section, we will discuss how to graph linear equations. We will then show how to make tables and graphs with a graphing calculator.

Equations with Two Variables

The equation $x + 2y = 5$ contains the two variables x and y. The solutions of such equations are ordered pairs of numbers. For example, the ordered pair $(1, 2)$ is a solution, because the equation is satisfied when $x = 1$ and $y = 2$.

$$x + 2y = 5$$

$$1 + 2(2) = 5 \quad \text{Substitute 1 for } x \text{ and 2 for } y.$$

$$1 + 4 = 5$$

$$5 = 5$$

EXAMPLE 1 Is the pair $(-2, 4)$ a solution of $y = 3x + 9$?

Solution We substitute -2 for x and 4 for y and see whether the resulting equation is true.

$y = 3x + 9$ The original equation.

$4 \stackrel{?}{=} 3(-2) + 9$ Substitute -2 for x and 4 for y.

$4 \stackrel{?}{=} -6 + 9$ Do the multiplication: $3(-2) = -6$.

$4 = 3$ Do the addition: $-6 + 9 = 3$.

Since the equation $4 = 3$ is false, the pair $(-2, 4)$ is not a solution.

Self Check Is $(-1, -5)$ a solution of $y = 5x$?

Constructing Tables of Values

To find solutions of equations in x and y, we can pick numbers at random, substitute them for x, and find the corresponding values of y. For example, to find some ordered pairs that satisfy $y = 5 - x$, we can let $x = 1$ (called the **input value**), substitute 1 for x, and solve for y (called the **output value**).

(1) $y = 5 - x$

x	y	(x, y)
1	4	(1, 4)

$y = 5 - x$ The original equation.

$y = 5 - 1$ Substitute the input value of 1 for x.

$y = 4$ The output is 4.

The ordered pair (1, 4) is a solution. As we find solutions, we will list them in a **table of values** like Table (1) at the left.

If $x = 2$, we have

(2) $y = 5 - x$

x	y	(x, y)
1	4	(1, 4)
2	3	(2, 3)

$y = 5 - x$ The original equation.

$y = 5 - 2$ Substitute the input value of 2 for x.

$y = 3$ The output is 3.

A second solution is (2, 3). We list it in Table (2) at the left.

If $x = 5$, we have

(3) $y = 5 - x$

x	y	(x, y)
1	4	(1, 4)
2	3	(2, 3)
5	0	(5, 0)

$y = 5 - x$ The original equation.

$y = 5 - 5$ Substitute the input value of 5 for x.

$y = 0$ The output is 0.

A third solution is (5, 0). We list it in Table (3) at the left.

If $x = -1$, we have

(4) $y = 5 - x$

x	y	(x, y)
1	4	(1, 4)
2	3	(2, 3)
5	0	(5, 0)
−1	6	(−1, 6)

$y = 5 - x$ The original equation.

$y = 5 - (-1)$ Substitute the input value of -1 for x.

$y = 6$ The output is 6.

A fourth solution is $(-1, 6)$. We list it in Table (4) at the left.

If $x = 6$, we have

(5) $y = 5 - x$

x	y	(x, y)
1	4	(1, 4)
2	3	(2, 3)
5	0	(5, 0)
−1	6	(−1, 6)
6	−1	(6, −1)

$y = 5 - x$ The original equation.

$y = 5 - 6$ Substitute the input value of 6 for x.

$y = -1$ The output is -1.

A fifth solution is $(6, -1)$. We list it in Table (5) at the left.

Since we can choose any real number for x, and since any choice of x will give a corresponding value of y, it is apparent that the equation $y = 5 - x$ has *infinitely many solutions*.

Graphing Equations

To graph the equation $y = 5 - x$, we plot the ordered pairs listed in the table on a rectangular coordinate system, as in Figure 3-10. From the figure, we can see that the five points lie on a line.

$y = 5 - x$

x	y	(x, y)
1	4	(1, 4)
2	3	(2, 3)
5	0	(5, 0)
−1	6	(−1, 6)
6	−1	(6, −1)

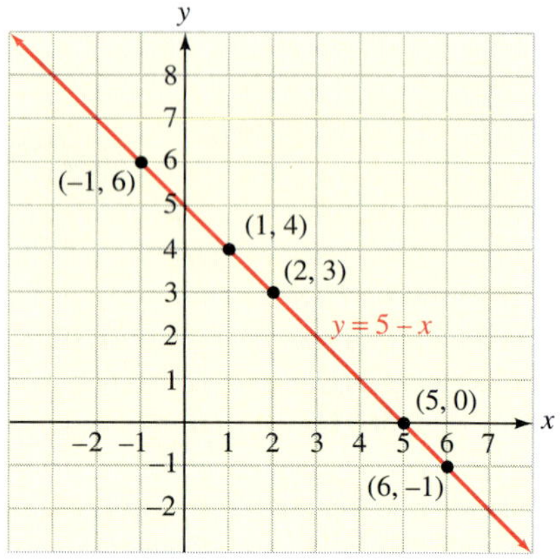

Figure 3-10

We draw a line through the points. The arrowheads on the line show that the graph continues forever in both directions. Since the graph of any solution of $y = 5 - x$ will lie on this line, the line is a picture of all of the solutions of the equation. The line is said to be the **graph** of the equation.

Any equation, such as $y = 5 - x$, whose graph is a line is called a **linear equation in two variables.** Any point on the line has coordinates that satisfy the equation, and the graph of any pair (x, y) that satisfies the equation is a point on the line.

Since we will usually choose a number for x first and then find the corresponding value of y, the value of y depends on x. For this reason, we call y the **dependent variable** and x the **independent variable.** The value of the independent variable is the input value, and the value of the dependent variable is the output value.

Although only two points are needed to graph a linear equation, we often plot a third point as a check. If the three points do not lie on a line, at least one of them is in error.

Graphing Linear Equations

1. Find two pairs (x, y) that satisfy the equation by picking arbitrary input values for x and solving for the corresponding output values of y. A third point provides a check.
2. Plot each resulting pair (x, y) on a rectangular coordinate system. If they do not lie on a line, check your calculations.
3. Draw the line passing through the points.

EXAMPLE 2 Graph: $y = 3x - 4$.

Solution We find three ordered pairs that satisfy the equation.

***If* $x = 1$**	***If* $x = 2$**	***If* $x = 3$**
$y = 3x - 4$	$y = 3x - 4$	$y = 3x - 4$
$y = 3(1) - 4$	$y = 3(2) - 4$	$y = 3(3) - 4$
$y = -1$	$y = 2$	$y = 5$

We enter the results in a table of values, plot the points, and draw a line through the points. The graph appears in Figure 3-11.

$y = 3x - 4$

x	y	(x, y)
1	−1	(1, −1)
2	2	(2, 2)
3	5	(3, 5)

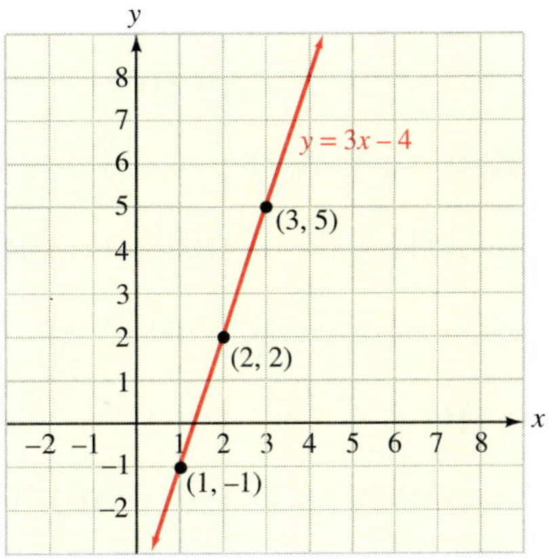

Figure 3-11

Self Check Graph: $y = 3x$.

EXAMPLE 3 Graph: $y - 4 = \frac{1}{2}(x - 8)$.

Solution We first solve for y and simplify.

$$y - 4 = \frac{1}{2}(x - 8)$$

$$y - 4 = \frac{1}{2}x - 4 \quad \text{Use the distributive property to remove parentheses.}$$

$$y = \frac{1}{2}x \quad \text{Add 4 to both sides.}$$

We now find three ordered pairs that satisfy the equation.

***If* $x = 0$**	***If* $x = 2$**	***If* $x = -4$**
$y = \frac{1}{2}x$	$y = \frac{1}{2}x$	$y = \frac{1}{2}x$
$y = \frac{1}{2}(0)$	$y = \frac{1}{2}(2)$	$y = \frac{1}{2}(-4)$
$y = 0$	$y = 1$	$y = -2$

We enter the results in a table of values, plot the points, and draw a line through the points. The graph appears in Figure 3-12.

$y - 4 = \frac{1}{2}(x - 8)$

x	y	(x, y)
0	0	(0, 0)
2	1	(2, 1)
−4	−2	(−4, −2)

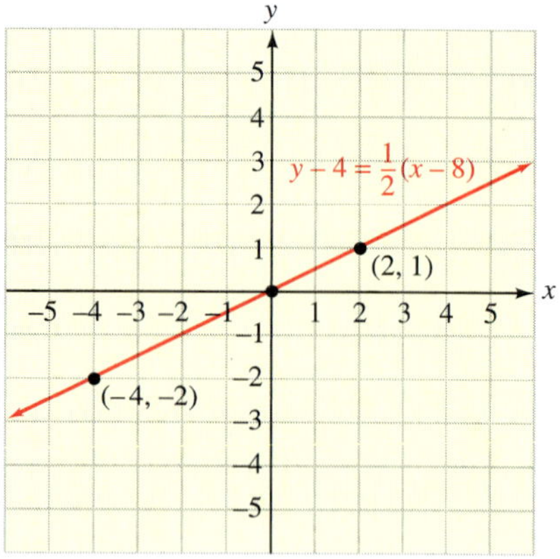

Figure 3-12

Self Check Graph: $y + 3 = \frac{1}{3}(x - 6)$.

The Intercept Method of Graphing a Line

The points where a line intersects the x- and y-axes are called **intercepts** of the line.

x- and y-Intercepts

The **x-intercept** of a line is a point $(a, 0)$ where the line intersects the x-axis. (See Figure 3-13.) To find a, substitute 0 for y in the equation of the line and solve for x.

A **y-intercept** of a line is a point $(0, b)$ where the line intersects the y-axis. To find b, substitute 0 for x in the equation of the line and solve for y.

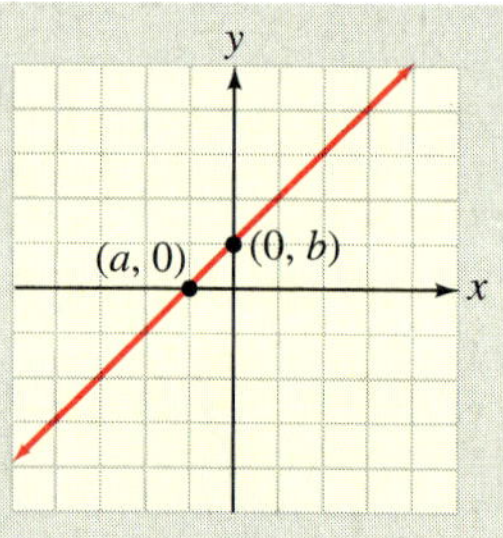

Figure 3-13

Plotting the x- and y-intercepts and drawing a line through them is called the **intercept method of graphing a line.** This method is useful for graphing equations written in **general form.**

General Form of the Equation of a Line

If A, B, and C are real numbers and A and B are not both 0, then the equation

$$Ax + By = C$$

is called the **general form** of the equation of a line.

Comment Whenever possible, we will write the general form $Ax + By = C$ so that A, B, and C are integers and $A \geq 0$. We will also make A, B, and C as small as possible. For example, the equation $6x + 12y = 24$ can be changed to $x + 2y = 4$ by dividing both sides by 6.

EXAMPLE 4 Graph: $3x + 2y = 6$.

Solution To find the y-intercept, we let $x = 0$ and solve for y.

$$
\begin{aligned}
3x + 2y &= 6 \\
3(0) + 2y &= 6 && \text{Substitute 0 for } x. \\
2y &= 6 && \text{Simplify.} \\
y &= 3 && \text{Divide both sides by 2.}
\end{aligned}
$$

The y-intercept is the pair $(0, 3)$. To find the x-intercept, we let $y = 0$ and solve for x.

$$
\begin{aligned}
3x + 2y &= 6 \\
3x + 2(0) &= 6 && \text{Substitute 0 for } y. \\
3x &= 6 && \text{Simplify.} \\
x &= 2 && \text{Divide both sides by 3.}
\end{aligned}
$$

The x-intercept is the pair $(2, 0)$. As a check, we plot one more point. If $x = 4$, then

$$
\begin{aligned}
3x + 2y &= 6 \\
3(4) + 2y &= 6 && \text{Substitute 4 for } x. \\
12 + 2y &= 6 && \text{Simplify.} \\
2y &= -6 && \text{Add } -12 \text{ to both sides.} \\
y &= -3 && \text{Divide both sides by 2.}
\end{aligned}
$$

The point $(4, -3)$ is on the graph. We plot these three points and join them with a line. The graph of $3x + 2y = 6$ is shown in Figure 3-14.

$3x + 2y = 6$

x	y	(x, y)
0	3	(0, 3)
2	0	(2, 0)
4	−3	(4, −3)

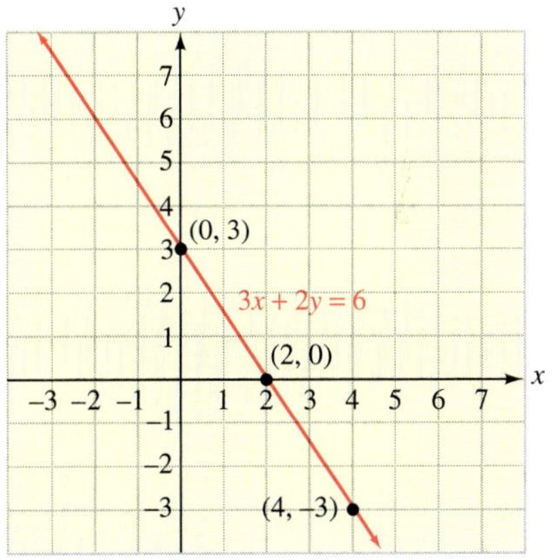

Figure 3-14

Self Check Graph: $4x + 3y = 6$.

Graphing Horizontal and Vertical Lines

Equations such as $y = 3$ and $x = -2$ are linear equations, because they can be written in the general form $Ax + By = C$.

$$
\begin{aligned}
y &= 3 && \text{is equivalent to} && 0x + 1y = 3 \\
x &= -2 && \text{is equivalent to} && 1x + 0y = -2
\end{aligned}
$$

Next, we discuss how to graph these types of linear equations.

EXAMPLE 5 Graph: **a.** $y = 3$ and **b.** $x = -2$.

Solution **a.** We can write the equation $y = 3$ in general form as $0x + y = 3$. Since the coefficient of x is 0, the numbers chosen for x have no effect on y. The value of y is always 3. For example, if we substitute -3 for x, we get

$$\begin{aligned} 0x + y &= 3 \\ 0(-3) + y &= 3 \\ 0 + y &= 3 \\ y &= 3 \end{aligned}$$

The table in Figure 3-15(a) gives several pairs that satisfy the equation $y = 3$. After plotting these pairs and joining them with a line, we see that the graph of $y = 3$ is a horizontal line that intersects the y-axis at 3. The y-intercept is (0, 3). There is no x-intercept.

b. We can write $x = -2$ in general form as $x + 0y = -2$. Since the coefficient of y is 0, the values of y have no effect on x. The number x is always -2. A table of values and the graph are shown in Figure 3-15(b). The graph of $x = -2$ is a vertical line that intersects the x-axis at -2. The x-intercept is $(-2, 0)$. There is no y-intercept.

$y = 3$

x	y	(x, y)
−3	3	(−3, 3)
0	3	(0, 3)
2	3	(2, 3)
4	3	(4, 3)

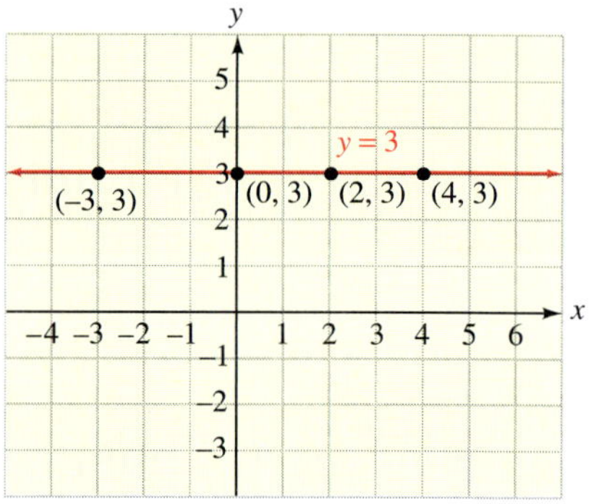

(a)

$x = -2$

x	y	(x, y)
−2	−2	(−2, −2)
−2	0	(−2, 0)
−2	2	(−2, 2)
−2	3	(−2, 3)

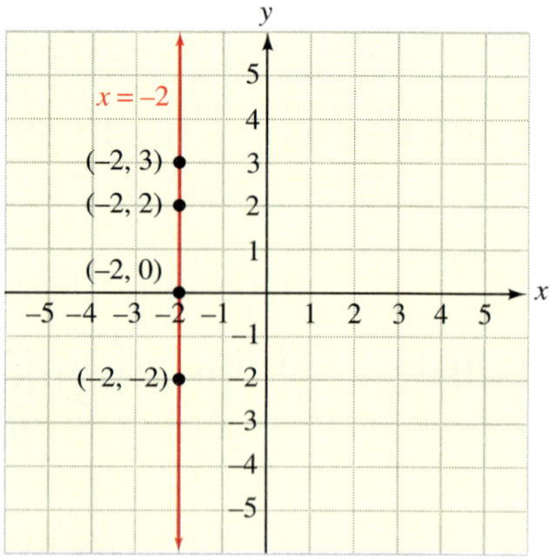

(b)

Figure 3-15

Self Check Identify the graph of each equation as a horizontal or a vertical line: **a.** $x = 5$, **b.** $y = -3$, and **c.** $x = 0$.

From the results of Example 5, we have the following facts.

Equations of Horizontal and Vertical Lines

The equation $y = b$ represents a horizontal line that intersects the y-axis at $(0, b)$. If $b = 0$, the line is the x-axis.

The equation $x = a$ represents a vertical line that intersects the x-axis at $(a, 0)$. If $a = 0$, the line is the y-axis.

An Application of Linear Equations

EXAMPLE 6 **Birthday parties** A restaurant offers a party package that includes food, drinks, cake, and party favors for a cost of \$25 plus \$3 per child. Write a linear equation that will give the cost for a party of any size. Then graph the equation.

Solution We can let c represent the cost of the party. Then c will be the sum of the basic charge of \$25 and the cost per child times the number of children attending. If n children attend, at \$3 per child, the total cost is $\$3n$.

The cost	equals	the basic \$25 charge	plus	\$3	times	the number of children.
c	$=$	25	$+$	3	$\cdot$	n

For the equation $c = 25 + 3n$, the independent variable (input) is n, the number of children. The dependent variable (output) is c, the cost of the party. We will find three points on the graph of the equation by choosing n-values of 0, 5, and 10 and finding the corresponding c-values. The results are recorded in the table.

***If* $n = 0$**

$c = 25 + 3(0)$

$c = 25$

***If* $n = 5$**

$c = 25 + 3(5)$

$c = 25 + 15$

$c = 40$

***If* $n = 10$**

$c = 25 + 3(10)$

$c = 25 + 30$

$c = 55$

$c = 25 + 3n$

n	c
0	25
5	40
10	55

Next, we graph the points in Figure 3-16 on the next page and draw a line through them. We don't draw an arrowhead on the left, because it doesn't make sense to have a *negative* number of children attend a party. We can use the graph to determine the cost of a party of any size. For example, to find the cost of a party with 8 children, we locate 8 on the horizontal axis and then move up to find a point on the graph directly above the 8. Since the coordinates of that point are (8, 49), the cost for 8 children would be \$49.

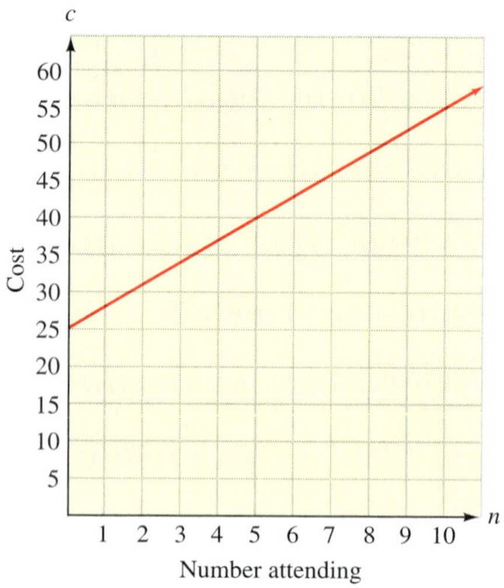

Figure 3-16

Accent on Technology

MAKING TABLES AND GRAPHS

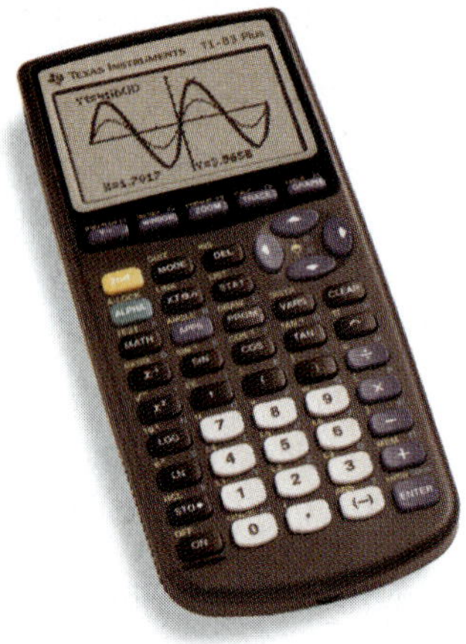

TI-83 Plus graphing calculator
Courtesy of Texas Instruments Incorporated

So far, we have graphed equations by making tables and plotting points. This method is often tedious and time-consuming. Fortunately, making tables and graphing equations is easier when we use a graphing calculator.

Although we will use calculators to make tables and graph equations, we will not show complete keystrokes for any specific brand. For these details, please consult your owner's manual.

All graphing calculators have a **viewing window** that is used to display tables and graphs. We will first discuss how to make tables and then discuss how to draw graphs.

Making Tables To construct a table of values for the equation $y = x^2$, simply press the Y = key, enter the expression x^2, and press the 2nd and TABLE keys to get a screen similar to Figure 3-17(a). You can use the up and down keys to scroll through the table to obtain a screen like Figure 3-17(b).

X	Y1	
0	0	
1	1	
2	4	
3	9	
4	16	
5	25	
6	36	

X = 0

(a)

X	Y1	
−5	25	
−4	16	
−3	9	
−2	4	
−1	1	
0	0	
1	1	

X = −5

(b)

Figure 3-17

Drawing Graphs To see the proper picture of a graph, we must often set the minimum and maximum values for the x- and y-coordinates. The standard window settings of

$$\text{Xmin} = -10 \qquad \text{Xmax} = 10 \qquad \text{Ymin} = -10 \qquad \text{Ymax} = 10$$

indicate that -10 is the minimum x- and y-coordinate to be used in the graph, and that 10 is the maximum x- and y-coordinate to be used. We will usually express window values in interval notation. In this notation, the standard settings are

$$X = [-10, 10] \qquad Y = [-10, 10]$$

To graph the equation $2x - 3y = 14$ with a calculator, we must first solve the equation for y.

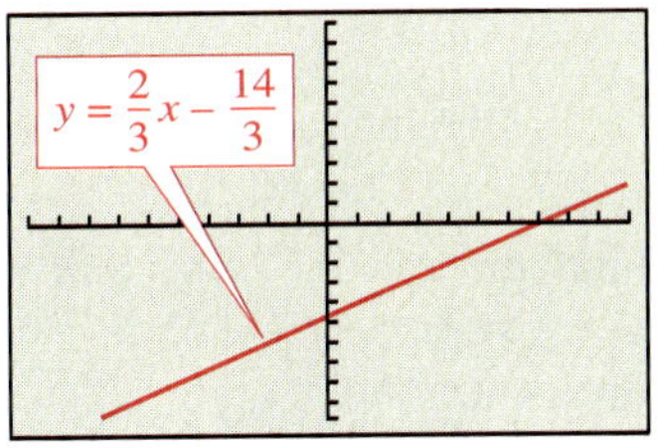

Figure 3-18

$$2x - 3y = 14$$

$$-3y = -2x + 14 \quad \text{Subtract } 2x \text{ from both sides.}$$

$$y = \frac{2}{3}x - \frac{14}{3} \quad \text{Divide both sides by } -3.$$

We now set the standard window values of $X = [-10, 10]$ and $Y = [-10, 10]$, press the Y = key and enter the equation as $(2/3)x - 14/3$, and press GRAPH to get the line shown in Figure 3-18.

Comment To graph an equation with a graphing calculator, the equation must be solved for y.

Using the Trace and Zoom Features With the trace feature, we can find the coordinates of any point on a graph. For example, to find the x-intercept of the line shown in Figure 3-18, we press the TRACE key and move the flashing cursor along the line until we approach the x-intercept, as shown in Figure 3-19(a). The x- and y-coordinates of the flashing cursor appear at the bottom of the screen.

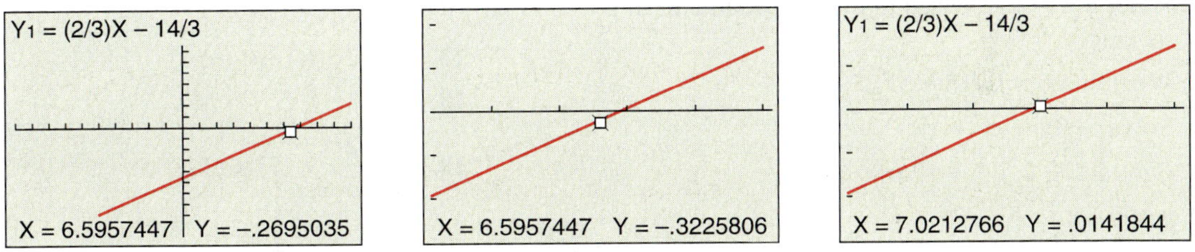

(a) (b) (c)

Figure 3-19

To get better results, we can press the ZOOM key to see a magnified picture of the line, as shown in Figure 3-19(b). We can trace again and move the cursor even closer to the x-intercept, as shown in Figure 3-19(c). Since the y-coordinate shown on the screen is close to 0, the x-coordinate shown on the screen is close to the x-value of the x-intercept. Repeated zooms will show that the x-intercept is $(7, 0)$.

Self Check Answers

1. yes

2.

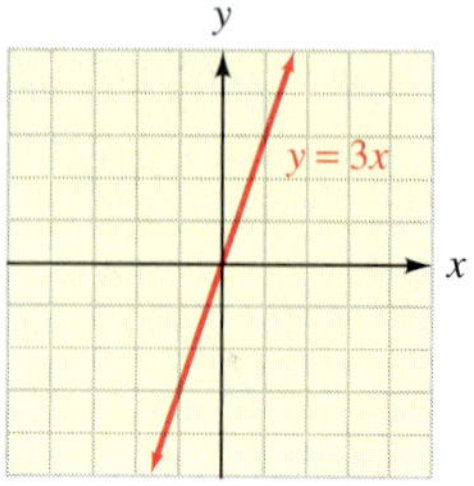

3.

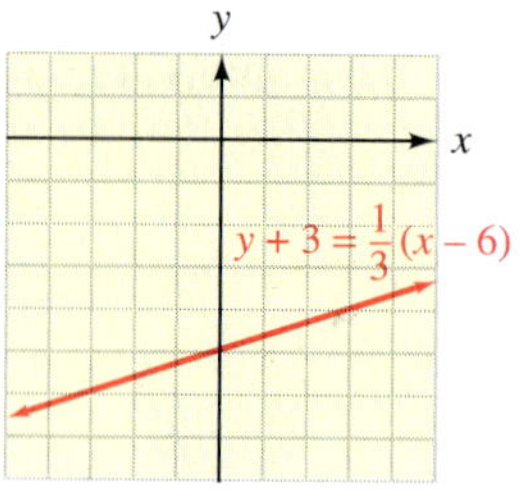

4.

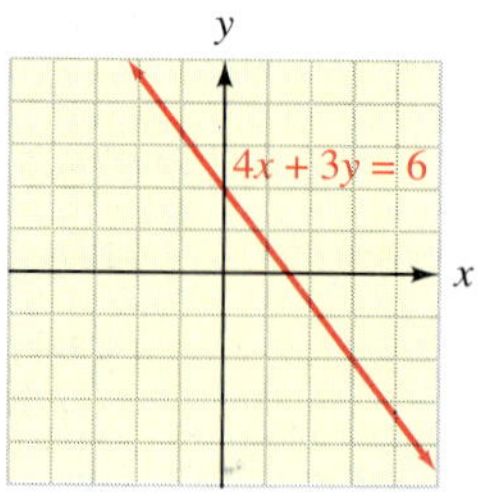

5. a. vertical **b.** horizontal **c.** vertical

Orals

1. How many points should be plotted to graph a line?

2. Define the intercepts of a line.

3. Find three pairs (x, y) that satisfy $x + y = 8$.

4. Find three pairs (x, y) that satisfy $x - y = 6$.

5. Which lines have no y-intercepts?

6. Which lines have no x-intercepts?

3.2 EXERCISES

REVIEW

1. Solve: $\frac{x}{8} = -12$.

2. Combine like terms: $3t - 4T + 5T - 6t$.

3. Is $\frac{x + 5}{6}$ an expression or an equation?

4. Write the formula used to find the perimeter of a rectangle.

5. What number is 0.5% of 250?

6. Solve: $-3x + 5 > 17$.

7. Subtract: $-2.5 - (-2.6)$.

8. Evaluate: $(-5)^3$.

VOCABULARY AND CONCEPTS *Fill in the blanks.*

9. The equation $y = x + 1$ is an equation in ____ variables.

10. An ordered pair is a _______ of an equation if the numbers in the ordered pair satisfy the equation.

11. In equations containing the variables x and y, x is called the __________ variable and y is called the ________ variable.

12. When constructing a _____ of values, the values of x are the _____ values and the values of y are the ______ values.

13. An equation whose graph is a line and whose variables are to the first power is called a _____ equation.

14. The equation $Ax + By = C$ is the ____________ form of the equation of a line.

15. The ________ of a line is the point $(0, b)$, where the line intersects the y-axis.

16. The ________ of a line is the point $(a, 0)$, where the line intersects the x-axis.

PRACTICE *Tell whether the ordered pair satisfies the equation.*

17. $x - 2y = -4$; $(4, 4)$

18. $y = 8x - 5$; $(4, 26)$

19. $y = \frac{2}{3}x + 5$; $(6, 12)$

20. $y = -\frac{1}{2}x - 2$; $(4, -4)$

Complete each table of values. Check your work with a graphing calculator.

21. $y = x - 3$

x	y
0	−3
1	
−2	
−4	

22. $y = x + 2$

x	$x + 2$
0	
−1	
−2	
3	

23. $y = -2x$

input	*output*
0	
1	
3	
−2	

24. $y = \dfrac{x}{2}$

x	$\frac{x}{2}$
0	
1	
−2	
−4	

Graph each equation. Check your work with a graphing calculator.

25. $y = 2x$

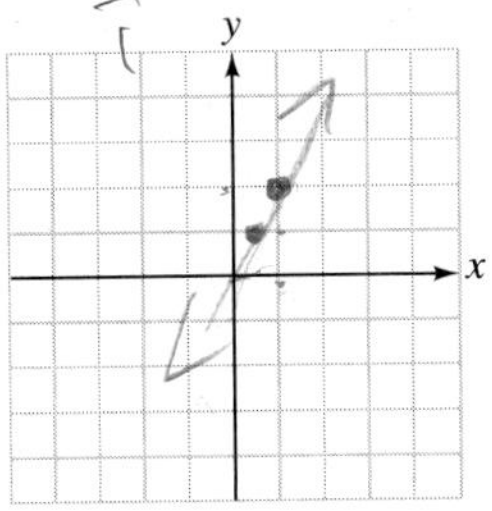

26. $y = -\dfrac{1}{2}x$

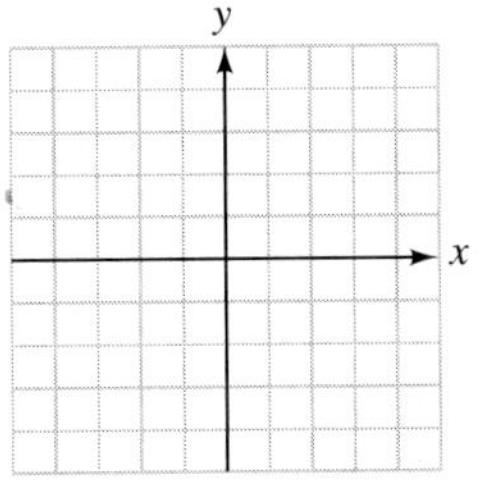

27. $y = 2x - 1$

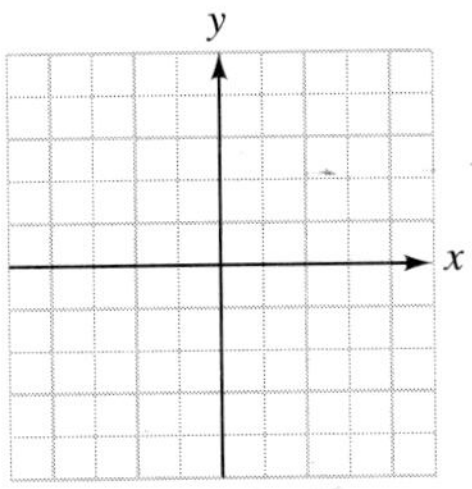

28. $y = 3x + 1$

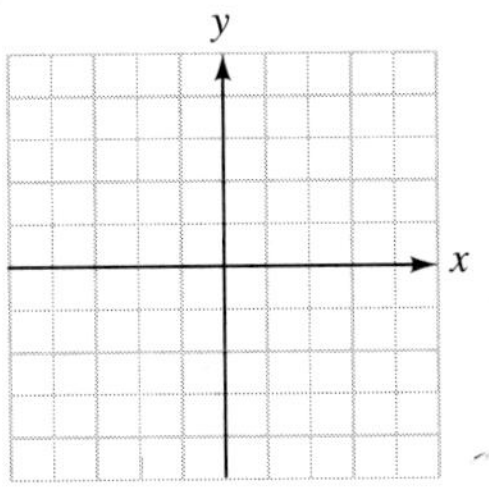

29. $y = \dfrac{x}{2} - 2$

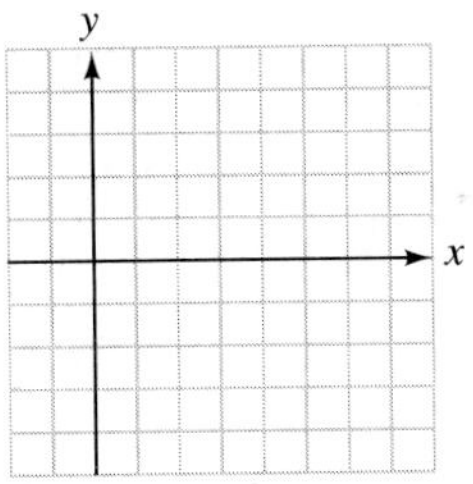

30. $y = \dfrac{x}{3} - 3$

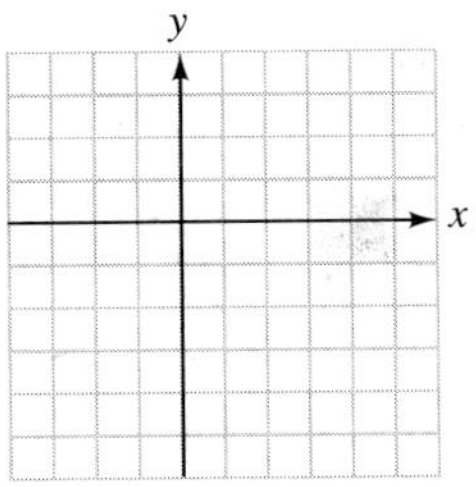

Write each equation in general form, if necessary, and graph each equation using the intercept method.

31. $x + y = 7$

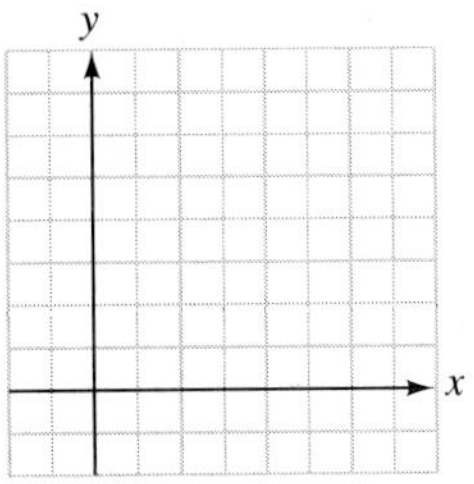

32. $x + y = -2$

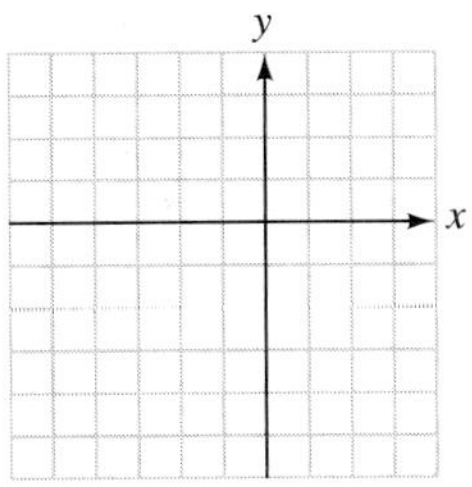

33. $x - y = 7$

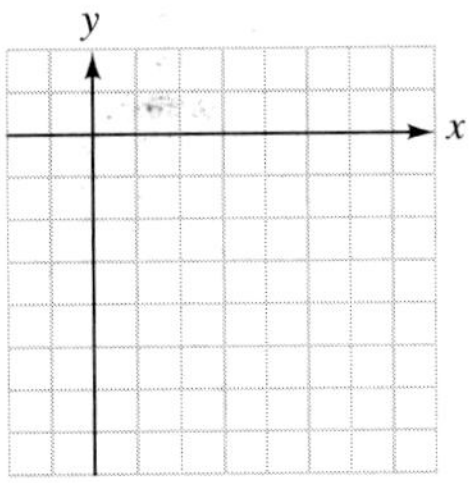

34. $x - y = -2$

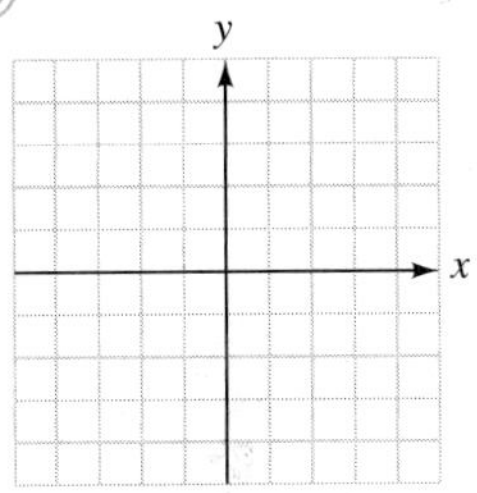

35. $y = -2x + 5$

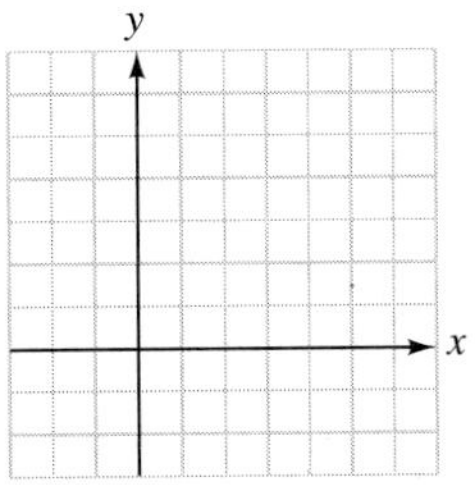

36. $y = -3x - 1$

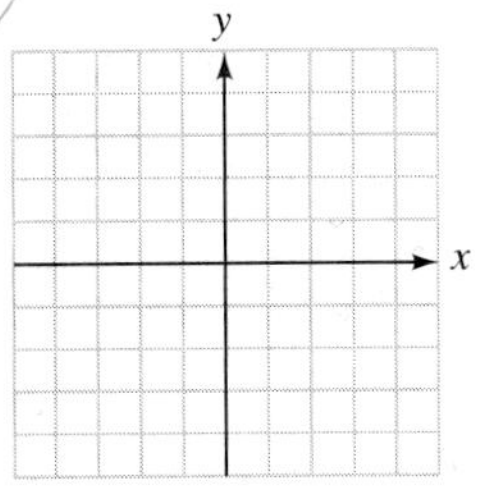

37. $2x + 3y = 12$

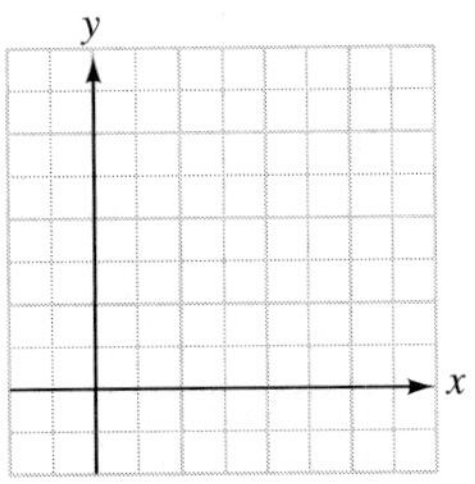

38. $3x - 2y = 6$

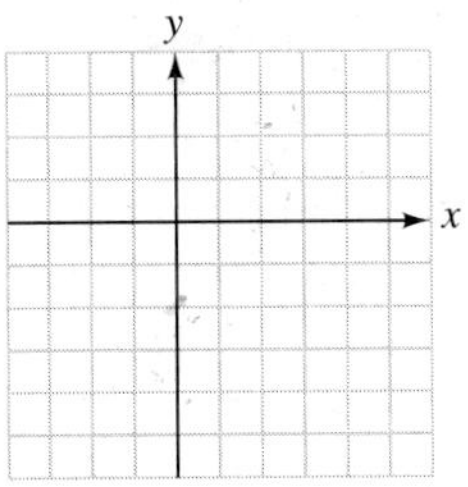

Graph each equation.

39. $y = -5$

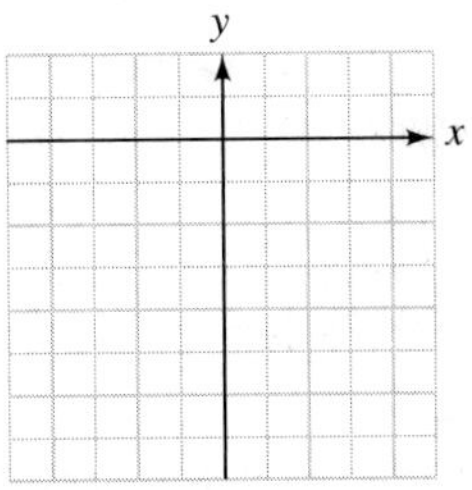

40. $x = 4$

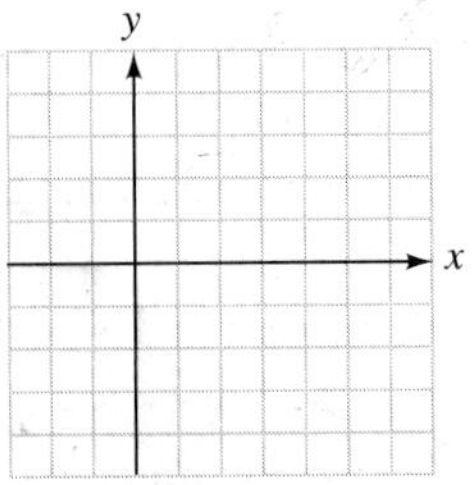

41. $x = 5$

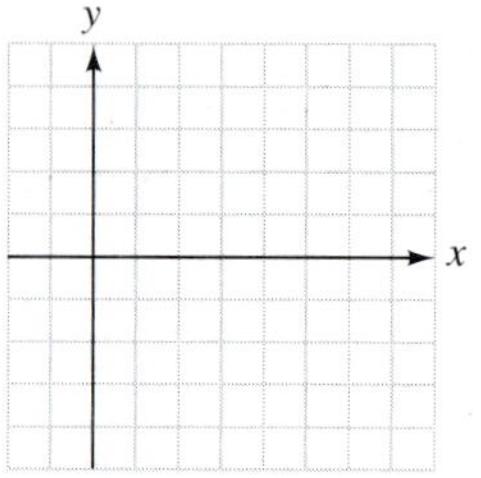

42. $y = 4$

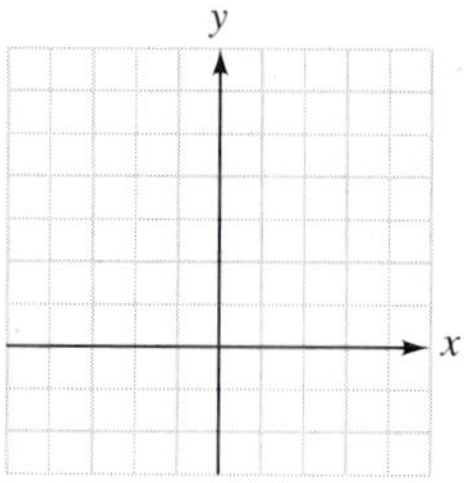

43. $y = 0$

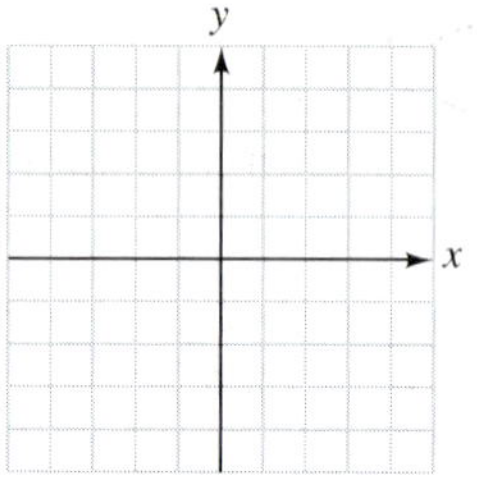

44. $x = 0$

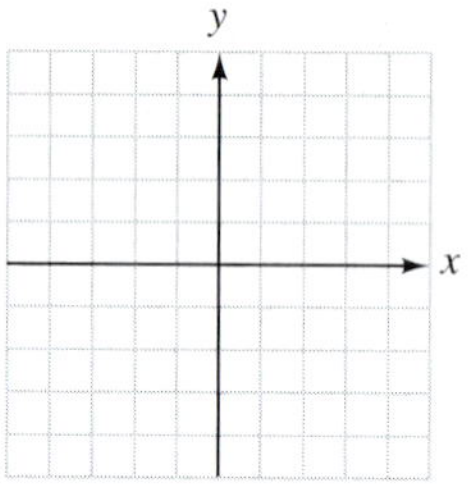

45. $2x = 5$

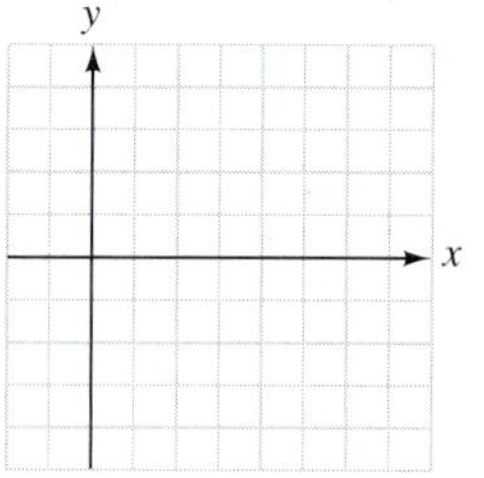

46. $3y = 7$

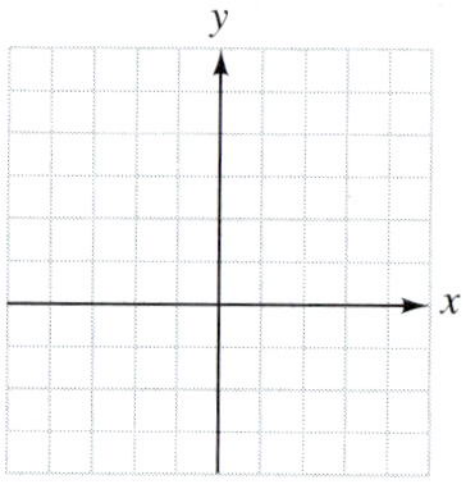

APPLICATIONS

47. Educational costs Each semester, a college charges a service fee of \$50 plus \$25 for each unit taken by a student.

a. Write a linear equation that gives the total enrollment cost c for a student taking u units.

b. Complete the table of values and graph the equation. See the illustration.

c. What does the y-intercept of the line tell you?

d. Use the graph to find the total cost for a student taking 18 units the first semester and 12 units the second semester.

u	c
4	
8	
14	

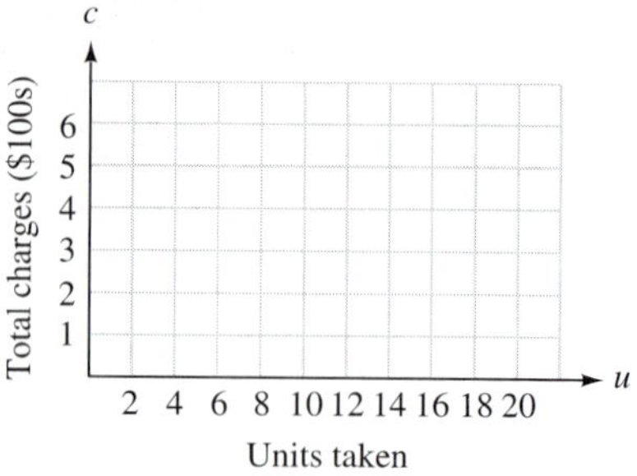

48. Group rates To promote the sale of tickets for a cruise to Alaska, a travel agency reduces the regular ticket price of \$3,000 by \$5 for each individual traveling in the group.

a. Write a linear equation that would find the ticket price T for the cruise if a group of p people travel together.

b. Complete the table of values and then graph the equation. See the illustration.

c. As the size of the group increases, what happens to the ticket price?

d. Use the graph to determine the cost of an individual ticket if a group of 25 will be traveling together.

p	T
10	
30	
60	

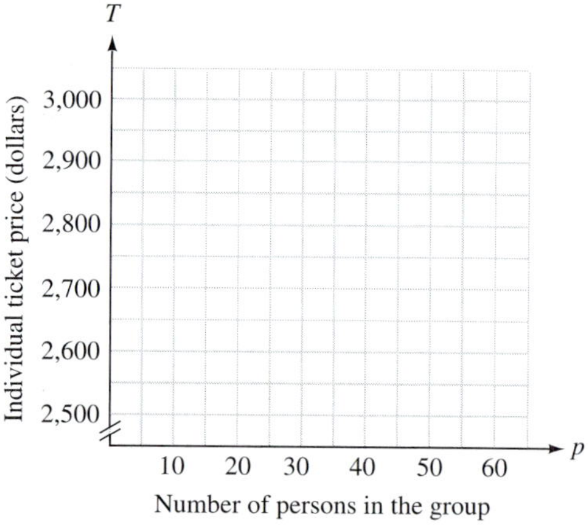

49. Physiology Physiologists have found that a woman's height h in inches can be approximated using the linear equation $h = 3.9r + 28.9$, where r represents the length of her radius bone in inches. See the illustration.

a. Complete the table of values (round to the nearest tenth), and then graph the equation.

b. Complete this sentence: From the graph, we see that the longer the radius bone, the . . .

c. From the graph, estimate the height of a girl whose radius bone is 7.5 inches long.

r	h
7	
8.5	
9	

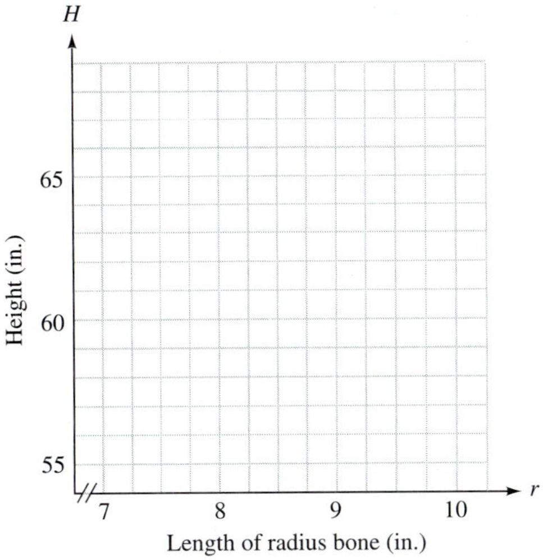

50. Research A psychology major found that the time t in seconds that it took a white rat to complete a maze was related to the number of trials n the rat had been given by the equation $t = 25 - 0.25n$. See the illustration.

a. Complete the table of values and then graph the equation.

b. Complete this sentence: From the graph, we see that the more trials the rat had, the . . .

c. From the graph, estimate the time it will take the rat to complete the maze on its 32nd trial.

n	t
4	
12	
16	

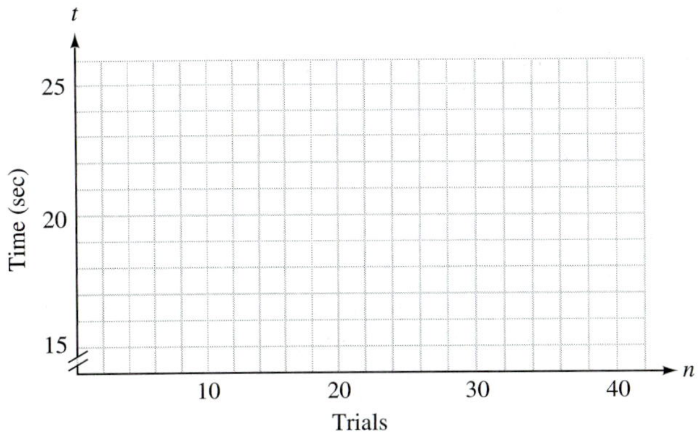

WRITING

51. From geometry, we know that two points determine a line. Explain why it is good practice when graphing linear equations to find and plot three points instead of just two.

52. Explain the process used to find the x- and y-intercepts of the graph of a line.

53. What is a table of values? Why is it often called a table of solutions?

54. When graphing an equation in two variables, how many solutions of the equation must be found?

55. Give examples of an equation in one variable and an equation in two variables. How do their solutions differ?

56. What does it mean when we say that an equation in two variables has infinitely many solutions?

SOMETHING TO THINK ABOUT *If points $P(a, b)$ and $Q(c, d)$ are two points on a rectangular coordinate system and point M is midway between them, then point M is called the **midpoint** of the line segment joining P and Q. (See the illustration.) To find the coordinates of the midpoint $M(x_M, y_M)$ of the segment PQ, we find the average of the x-coordinates and the average of the y-coordinates of P and Q.*

$$x_M = \frac{a + c}{2} \quad \text{and} \quad y_M = \frac{b + d}{2}$$

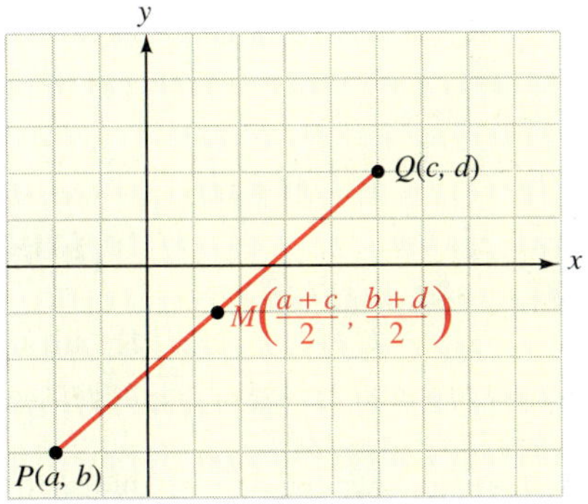

Find the coordinates of the midpoint of the line segment with the given endpoints.

57. $P(5, 3)$ and $Q(7, 9)$

58. $P(5, 6)$ and $Q(7, 10)$

59. $P(2, -7)$ and $Q(-3, 12)$

60. $P(-8, 12)$ and $Q(3, -9)$

61. $A(4, 6)$ and $B(10, 6)$

62. $A(8, -6)$ and the origin

3.3 Solving Systems of Equations by Graphing

- Systems of Equations
- The Graphing Method
- Inconsistent Systems
- Dependent Equations

Getting Ready *If* $y = x^2 - 3$, *find* y *when*

1. $x = 0$ **2.** $x = 1$ **3.** $x = -2$ **4.** $x = 3$

The lines graphed in Figure 3-20 approximate the per-person consumption of chicken and beef by Americans for the years 1995 to 2002. We can see that over this period, consumption of chicken increased, while that of beef decreased.

By graphing this *pair* of lines on the same coordinate system, it is apparent that Americans consumed equal amounts of chicken and beef in 1998—about 67 pounds each. In this section, we will work with pairs of linear equations whose graphs will often be intersecting lines.

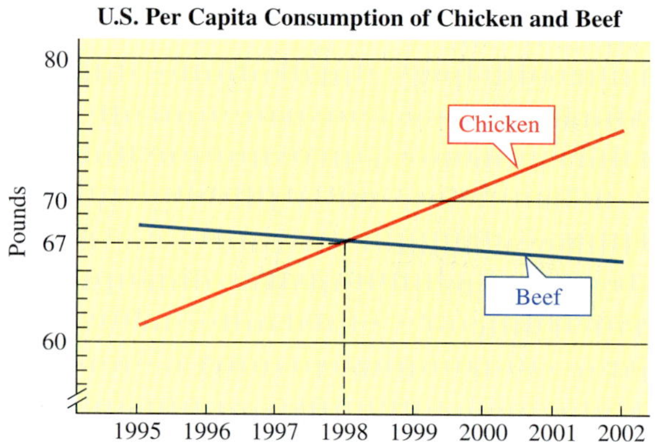

Figure 3-20

Systems of Equations

Recall that we have considered equations such as $x + y = 3$ that contain two variables. Because there are infinitely many pairs of numbers whose sum is 3, there are infinitely many pairs (x, y) that will satisfy this equation. Some of these pairs are listed in Table 3-3(a). Likewise, there are infinitely many pairs (x, y) that will satisfy the equation $3x - y = 1$. Some of these pairs are listed in Table 3-3(b).

$x + y = 3$

x	y
0	3
1	2
2	1
3	0

(a)

$3x - y = 1$

x	y
0	−1
1	2
2	5
3	8

(b)

Table 3-3

Although there are infinitely many pairs that satisfy each of these equations, only the pair (1, 2) satisfies both equations. The pair of equations

$$\begin{cases} x + y = 3 \\ 3x - y = 1 \end{cases}$$

is called a **system of equations.** Because the ordered pair (1, 2) satisfies both equations, it is called a **simultaneous solution** or just a **solution of the system of equations.** In this chapter, we will discuss three methods for finding the solution of a system of two equations, each with two variables. In this section, we consider the graphing method.

The Graphing Method

To use the method of graphing to solve the system

$$\begin{cases} x + y = 3 \\ 3x - y = 1 \end{cases}$$

we graph both equations on one set of coordinate axes using the intercept method. See Figure 3-21.

$x + y = 3$

x	y	(x, y)
0	3	(0, 3)
3	0	(3, 0)
2	1	(2, 1)

$3x - y = 1$

x	y	(x, y)
0	-1	$(0, -1)$
$\frac{1}{3}$	0	$\left(\frac{1}{3}, 0\right)$
2	5	(2, 5)

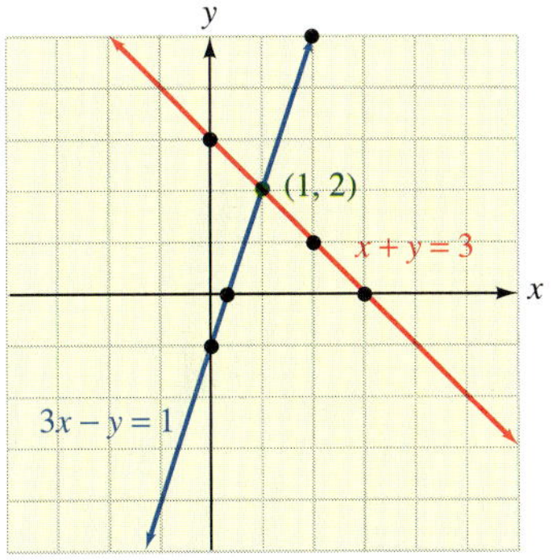

Figure 3-21

Although there are infinitely many pairs (x, y) that satisfy $x + y = 3$ and infinitely many pairs (x, y) that satisfy $3x - y = 1$, only the coordinates of the point where their graphs intersect satisfy both equations. The solution of the system is $x = 1$ and $y = 2$, or just (1, 2).

To check the solution, we substitute 1 for x and 2 for y in each equation and verify that the pair (1, 2) satisfies each equation.

First equation

$$x + y = 3$$
$$1 + 2 \stackrel{?}{=} 3$$
$$3 = 3$$

Second equation

$$3x - y = 1$$
$$3(1) - 2 \stackrel{?}{=} 1$$
$$3 - 2 \stackrel{?}{=} 1$$
$$1 = 1$$

PERSPECTIVE

To schedule a company's workers, managers must consider several factors to match a worker's ability to the demands of various jobs and to match company resources to the requirements of the job. To design bridges or office buildings, engineers must analyze the effects of thousands of forces to ensure that structures won't collapse. A telephone switching network decides which of thousands of possible routes is the most efficient and then rings the correct telephone in seconds. Each of these tasks requires solving systems of equations—not just two equations in two variables, but hundreds of equations in hundreds of variables. These tasks are common in every business, industry, educational institution, and government in the world. All would be much more difficult without a computer.

One of the earliest computers in use was the Mark I, which resulted from a collaboration between IBM and a Harvard mathematician, Howard Aiken. The Mark I was started in 1939 and finished in 1944. It was 8 feet tall, 2 feet thick, and over 50 feet long. It contained over 750,000 parts and performed 3 calculations per second.

Ironically, Aiken could not envision the importance of his invention. He advised the National Bureau of Standards that there was no point in building a better computer, because "there will never be enough work for more than one or two of these machines."

Mark I Relay Computer (1944)
Courtesy of IBM Corporation

When the graphs of two equations in a system are different lines, the equations are called **independent equations.** When a system of equations has a solution, the system is called a **consistent system.**

To solve a system of equations in two variables by graphing, we follow these steps.

The Graphing Method

1. Carefully graph each equation.
2. When possible, find the coordinates of the point where the graphs intersect.
3. Check the solution in the equations of the original system.

EXAMPLE 1 Use graphing to solve $\begin{cases} 2x + 3y = 2 \\ 3x = 2y + 16 \end{cases}$.

Solution Using the intercept method, we graph both equations on one set of coordinate axes, as shown in Figure 3-22.

$2x + 3y = 2$

x	y	(x, y)
0	$\frac{2}{3}$	$\left(0, \frac{2}{3}\right)$
1	0	$(1, 0)$
-2	2	$(-2, 2)$

$3x = 2y + 16$

x	y	(x, y)
0	-8	$(0, -8)$
$\frac{16}{3}$	0	$\left(\frac{16}{3}, 0\right)$
4	-2	$(4, -2)$

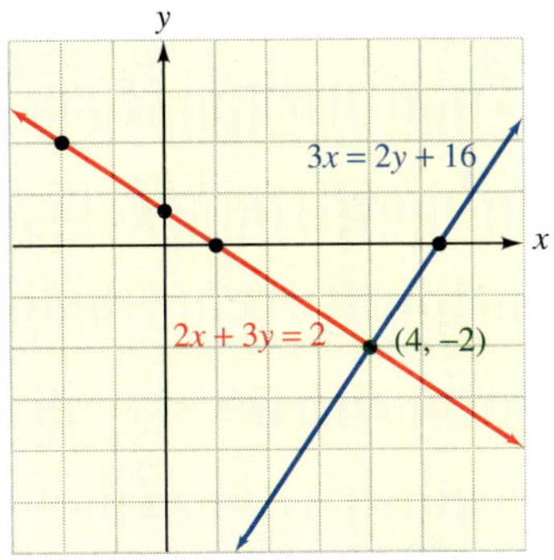

Figure 3-22

Although there are infinitely many pairs (x, y) that satisfy $2x + 3y = 2$ and infinitely many pairs (x, y) that satisfy $3x = 2y + 16$, only the coordinates of the point where the graphs intersect satisfy both equations. The solution is $x = 4$ and $y = -2$, or just $(4, -2)$.

To check, we substitute 4 for x and -2 for y in each equation and verify that the pair $(4, -2)$ satisfies each equation.

$$\begin{aligned} 2x + 3y &= 2 \\ 2(4) + 3(-2) &\stackrel{?}{=} 2 \\ 8 - 6 &\stackrel{?}{=} 2 \\ 2 &= 2 \end{aligned} \qquad \begin{aligned} 3x &= 2y + 16 \\ 3(4) &\stackrel{?}{=} 2(-2) + 16 \\ 12 &\stackrel{?}{=} -4 + 16 \\ 12 &= 12 \end{aligned}$$

The equations in this system are independent equations, and the system is a consistent system of equations.

Self Check Use graphing to solve $\begin{cases} 2x = y - 5 \\ x + y = -1 \end{cases}$.

Inconsistent Systems

Sometimes a system of equations will have no solution. These systems are called **inconsistent systems.**

EXAMPLE 2 Solve the system: $\begin{cases} 2x + y = -6 \\ 4x + 2y = 8 \end{cases}$.

Solution We graph both equations on one set of coordinate axes, as in Figure 3-23.

$2x + y = -6$

x	y	(x, y)
-3	0	$(-3, 0)$
0	-6	$(0, -6)$
-2	-2	$(-2, -2)$

$4x + 2y = 8$

x	y	(x, y)
2	0	$(2, 0)$
0	4	$(0, 4)$
1	2	$(1, 2)$

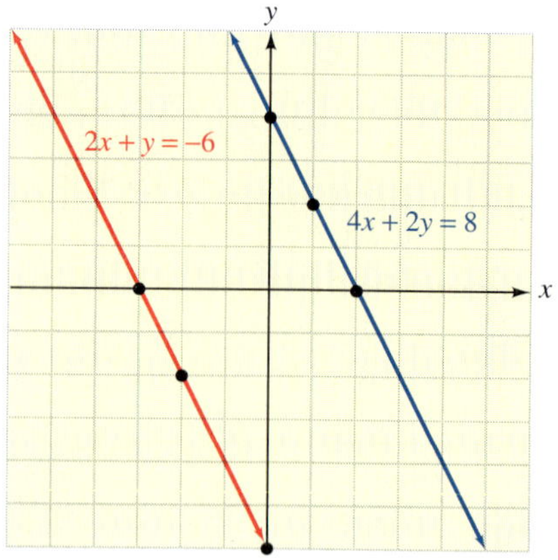

Figure 3-23

The lines in the figure are parallel. Because parallel lines do not intersect, the system has no solution, and the system is inconsistent. Since the graphs are different lines, the equations of the system are independent.

Self Check Solve the system: $\begin{cases} 2y = 3x \\ 3x - 2y = 6 \end{cases}$.

Dependent Equations

Sometimes a system will have infinitely many solutions. In this case, we say that the equations of the system are **dependent equations.**

EXAMPLE 3 Solve the system: $\begin{cases} y - 2x = 4 \\ 4x + 8 = 2y \end{cases}$.

Solution We graph each equation on one set of axes, as in Figure 3-24.

$y - 2x = 4$

x	y	(x, y)
0	4	$(0, 4)$
-2	0	$(-2, 0)$
1	6	$(1, 6)$

$4x + 8 = 2y$

x	y	(x, y)
0	4	$(0, 4)$
-2	0	$(-2, 0)$
-3	-2	$(-3, -2)$

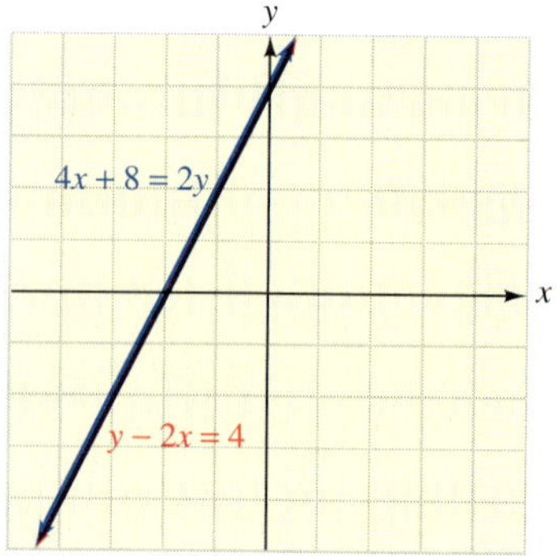

Figure 3-24

The lines in the figure are the same line. Since the lines intersect at infinitely many points, there are infinitely many solutions. Any pair (x, y) that satisfies one of the equations satisfies the other also.

From the graph, we can see that some solutions are $(0, 4)$, $(1, 6)$, and $(-1, 2)$, since each of these points lies on the one line that is the graph of both equations.

Self Check Solve the system: $\begin{cases} 6x - 2y = 4 \\ y + 2 = 3x \end{cases}$.

Table 3-4 summarizes the possibilities that can occur when two equations, each with two variables, are graphed.

Possible graph	If the	then
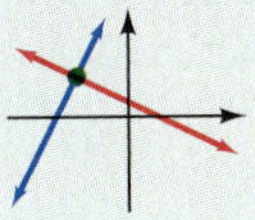	lines are different and intersect,	the equations are independent and the system is consistent. One solution exists.
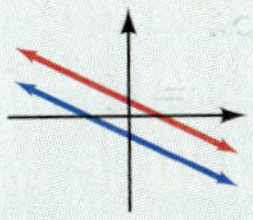	lines are different and parallel,	the equations are independent and the system is inconsistent. No solutions exist.
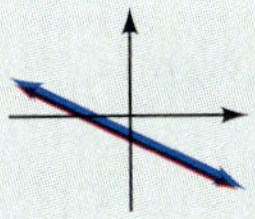	lines coincide (are the same line),	the equations are dependent and the system is consistent. Infinitely many solutions exist.

Table 3-4

EXAMPLE 4 Solve the system: $\begin{cases} \frac{2}{3}x - \frac{1}{2}y = 1 \\ \frac{1}{10}x + \frac{1}{15}y = 1 \end{cases}$.

Solution We can multiply both sides of the first equation by 6 to clear it of fractions.

$$\frac{2}{3}x - \frac{1}{2}y = 1$$

$$6\left(\frac{2}{3}x - \frac{1}{2}y\right) = 6(1)$$

(1) $$4x - 3y = 6$$

We then multiply both sides of the second equation by 30 to clear it of fractions.

$$\frac{1}{10}x + \frac{1}{15}y = 1$$

$$30\left(\frac{1}{10}x + \frac{1}{15}y\right) = 30(1)$$

(2) $$3x + 2y = 30$$

Equations 1 and 2 form the following equivalent system of equations, which has the same solutions as the original system.

$$\begin{cases} 4x - 3y = 6 \\ 3x + 2y = 30 \end{cases}$$

We can graph each equation of the previous system (see Figure 3-25) and find that their point of intersection has coordinates of (6, 6). The solution of the given system is $x = 6$ and $y = 6$, or just (6, 6).

To verify that (6, 6) satisfies each equation of the original system, we substitute 6 for x and 6 for y in each of the original equations and simplify.

$$\frac{2}{3}x - \frac{1}{2}y = 1$$
$$\frac{2}{3}(6) - \frac{1}{2}(6) \stackrel{?}{=} 1$$
$$4 - 3 \stackrel{?}{=} 1$$
$$1 = 1$$

$$\frac{1}{10}x + \frac{1}{15}y = 1$$
$$\frac{1}{10}(6) + \frac{1}{15}(6) \stackrel{?}{=} 1$$
$$\frac{3}{5} + \frac{2}{5} \stackrel{?}{=} 1$$
$$1 = 1$$

The equations in this system are independent, and the system is consistent.

$4x - 3y = 6$

x	y	(x, y)
0	−2	(0, −2)
3	2	(3, 2)
6	6	(6, 6)

$3x + 2y = 30$

x	y	(x, y)
10	0	(10, 0)
8	3	(8, 3)
6	6	(6, 6)

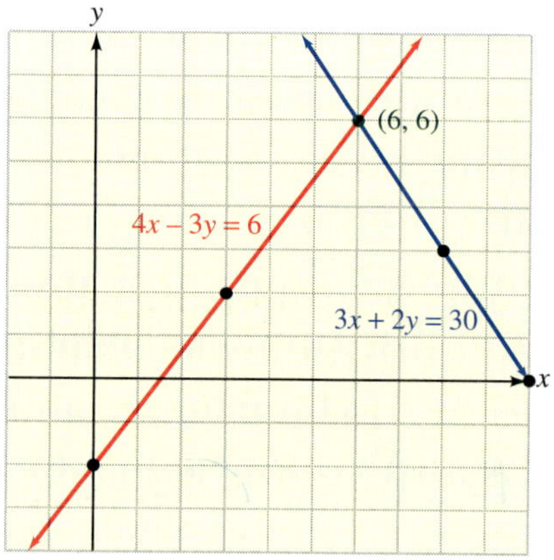

Figure 3-25

Self Check Solve the system: $\begin{cases} -\dfrac{x}{2} = \dfrac{y}{4} \\ \dfrac{1}{4}x - \dfrac{3}{8}y = -2 \end{cases}$.

Accent on Technology — SOLVING SYSTEMS OF EQUATIONS

We can use a graphing calculator to solve the system $\begin{cases} 2x + y = 12 \\ 2x - y = -2 \end{cases}$. However, before we can enter the equations into the calculator, we must solve them for y.

$$\begin{aligned} 2x + y &= 12 \\ y &= -2x + 12 \end{aligned} \qquad\qquad \begin{aligned} 2x - y &= -2 \\ -y &= -2x - 2 \\ y &= 2x + 2 \end{aligned}$$

We can now enter the resulting equations into a calculator and graph them. If we use standard window settings of $x = [-10, 10]$ and $y = [-10, 10]$, their graphs will look like Figure 3-26(a). We can trace to see that the coordinates of the intersection point are approximately

$$x = 2.5531915 \qquad \text{and} \qquad y = 6.893617$$

See Figure 3-26(b). For better results, we can zoom in on the intersection point and trace again to find that

$$x = 2.5 \qquad \text{and} \qquad y = 7$$

See Figure 3-26(c). Check the solution.

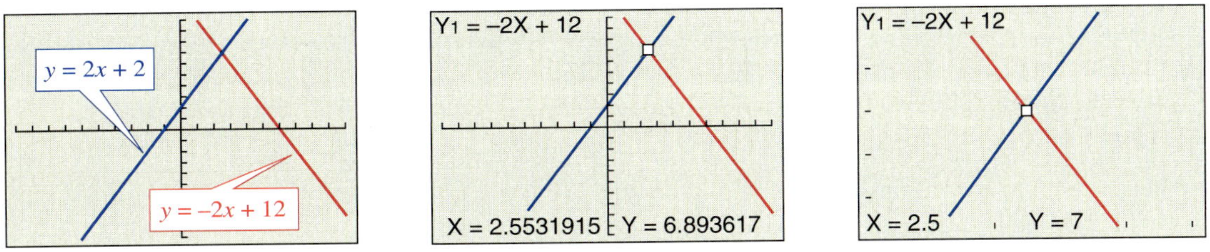

(a) (b) (c)

Figure 3-26

You can also find the intersection point by using the intersect command, found in the CALC menu.

Self Check Answers

1.

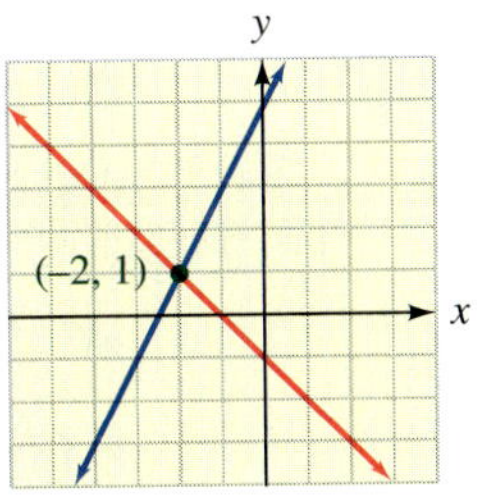

2. no solution

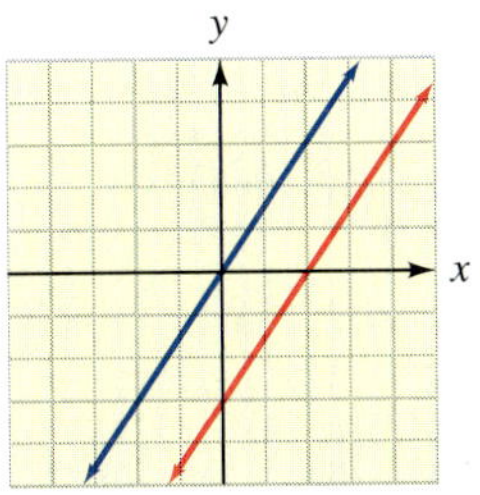

3. infinitely many solutions

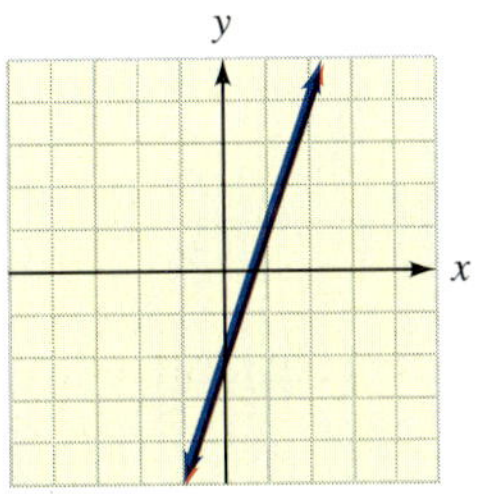

4.

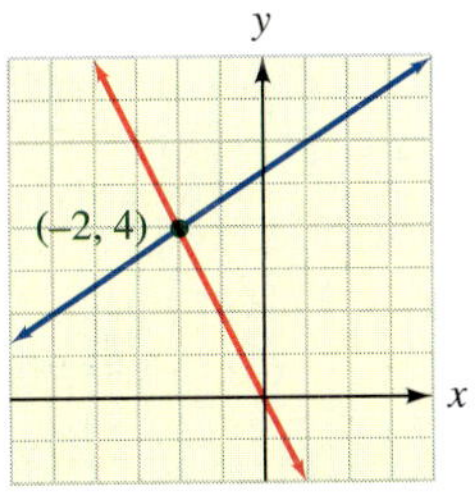

Orals *Tell whether the pair is a solution of the system.*

1. $(3, 2), \begin{cases} x + y = 5 \\ x - y = 1 \end{cases}$

2. $(1, 2), \begin{cases} x - y = -1 \\ x + y = 3 \end{cases}$

3. $(4, 1), \begin{cases} x + y = 5 \\ x - y = 2 \end{cases}$

4. $(5, 2), \begin{cases} x - y = 3 \\ x + y = 6 \end{cases}$

3.3 EXERCISES

REVIEW *Evaluate each expression. Assume that* $x = -3$.

1. $(-2)^4$

2. -2^4

3. $3x - x^2$

4. $\dfrac{-3 + 2x}{6x}$

VOCABULARY AND CONCEPTS *Fill in the blanks.*

5. The pair of equations $\begin{cases} x - y = -1 \\ 2x - y = 1 \end{cases}$ is called a ______ of equations.

6. Because the ordered pair (2, 3) satisfies both equations in Exercise 5, it is called a ______ of the system.

7. When the graphs of two equations in a system are different lines, the equations are called ______ equations.

8. When a system of equations has a solution, the system is called a ______.

9. Systems of equations that have no solution are called ______ systems.

10. When a system has infinitely many solutions, the equations of the system are said to be ______ equations.

Tell whether the ordered pair is a solution of the given system.

11. $(1, 1), \begin{cases} x + y = 2 \\ 2x - y = 1 \end{cases}$

12. $(1, 3), \begin{cases} 2x + y = 5 \\ 3x - y = 0 \end{cases}$

13. $(3, -2), \begin{cases} 2x + y = 4 \\ x + y = 1 \end{cases}$

14. $(-2, 4), \begin{cases} 2x + 2y = 4 \\ x + 3y = 10 \end{cases}$

15. $(4, 5), \begin{cases} 2x - 3y = -7 \\ 4x - 5y = 25 \end{cases}$

16. $(2, 3), \begin{cases} 3x - 2y = 0 \\ 5x - 3y = -1 \end{cases}$

17. $(-2, -3), \begin{cases} 4x + 5y = -23 \\ -3x + 2y = 0 \end{cases}$

18. $(-5, 1), \begin{cases} -2x + 7y = 17 \\ 3x - 4y = -19 \end{cases}$

19. $\left(\frac{1}{2}, 3\right), \begin{cases} 2x + y = 4 \\ 4x - 3y = 11 \end{cases}$

20. $\left(2, \frac{1}{3}\right), \begin{cases} x - 3y = 1 \\ -2x + 6y = -6 \end{cases}$

21. $\left(-\frac{2}{5}, \frac{1}{4}\right), \begin{cases} 5x - 4y = -6 \\ 8y = 10x + 12 \end{cases}$

22. $\left(-\frac{1}{3}, \frac{3}{4}\right), \begin{cases} 3x + 4y = 2 \\ 12y = 3(2 - 3x) \end{cases}$

Solve each system.

23. $\begin{cases} x + y = 2 \\ x - y = 0 \end{cases}$

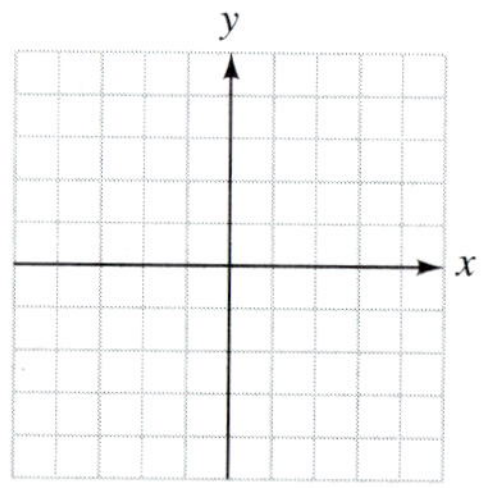

24. $\begin{cases} x + y = 4 \\ x - y = 0 \end{cases}$

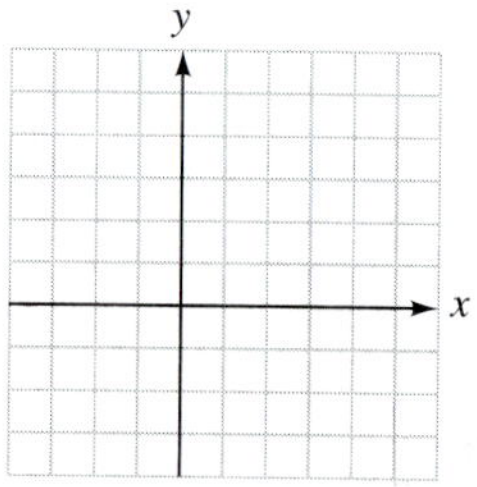

25. $\begin{cases} x + y = 2 \\ x - y = 4 \end{cases}$

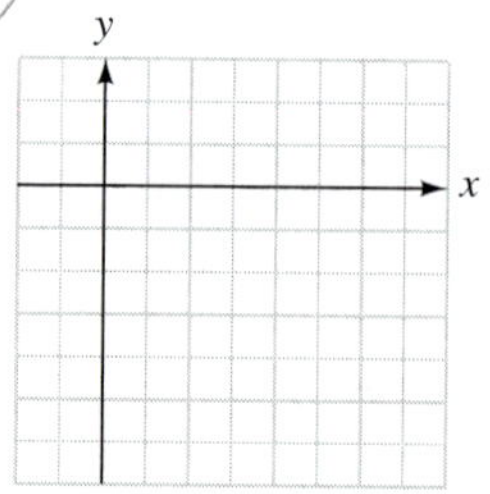

26. $\begin{cases} x + y = 1 \\ x - y = -5 \end{cases}$

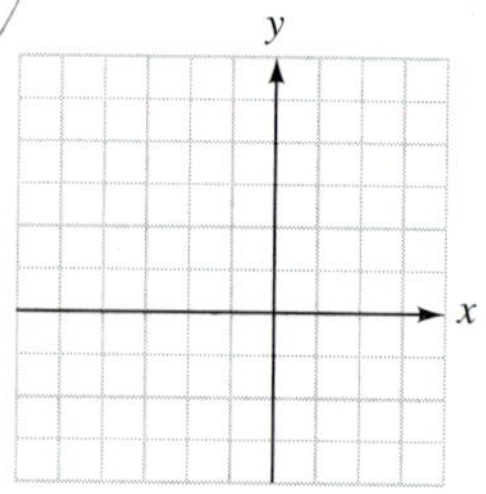

27. $\begin{cases} 3x + 2y = -8 \\ 2x - 3y = -1 \end{cases}$

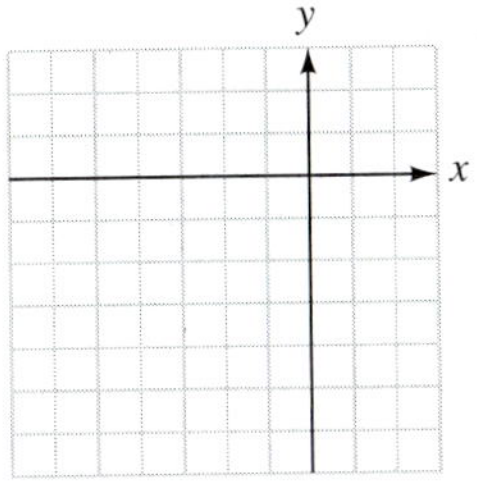

28. $\begin{cases} x + 4y = -2 \\ x + y = -5 \end{cases}$

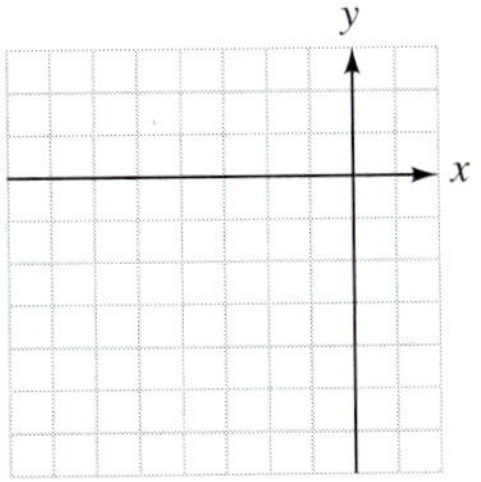

29. $\begin{cases} 4x - 2y = 8 \\ y = 2x - 4 \end{cases}$

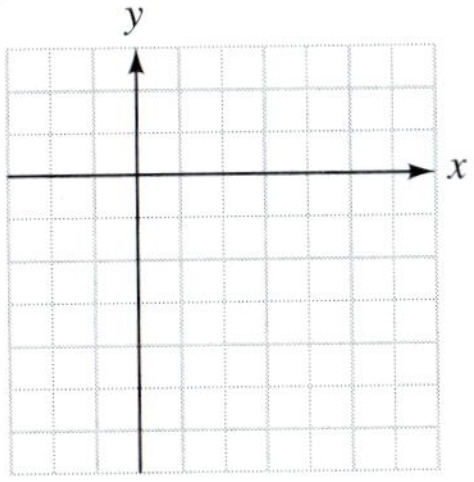

30. $\begin{cases} 3x - 6y = 18 \\ x = 2y + 3 \end{cases}$

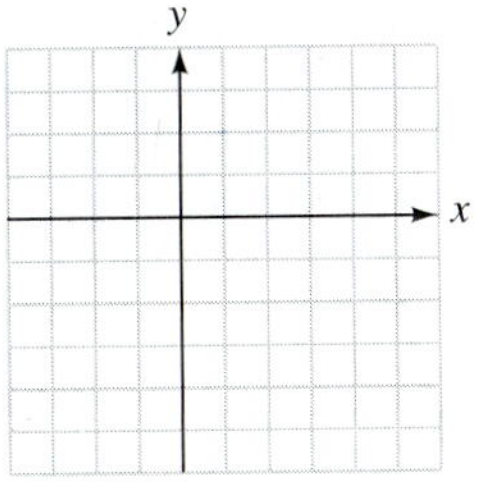

31. $\begin{cases} 2x - 3y = -18 \\ 3x + 2y = -1 \end{cases}$

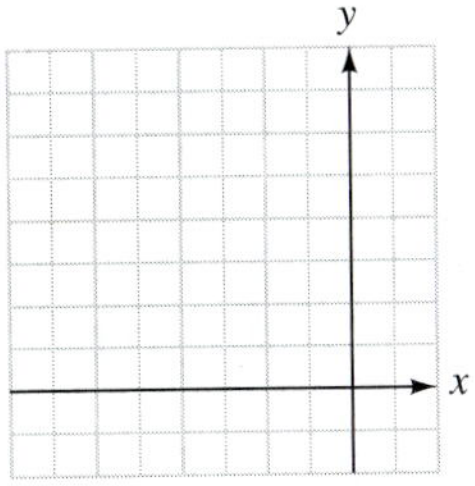

32. $\begin{cases} -x + 3y = -11 \\ 3x - y = 17 \end{cases}$

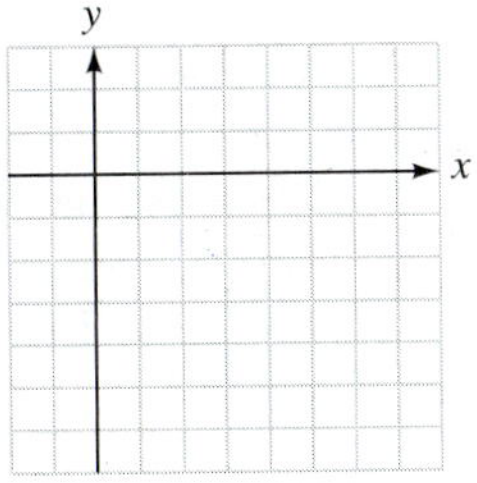

33. $\begin{cases} 4x = 3(4 - y) \\ 2y = 4(3 - x) \end{cases}$

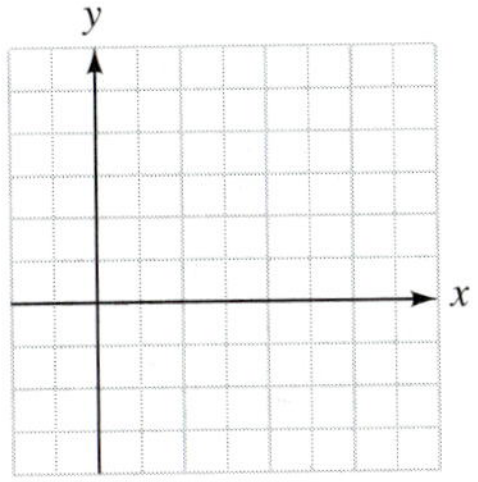

34. $\begin{cases} 2x = 3(2 - y) \\ 3y = 2(3 - x) \end{cases}$

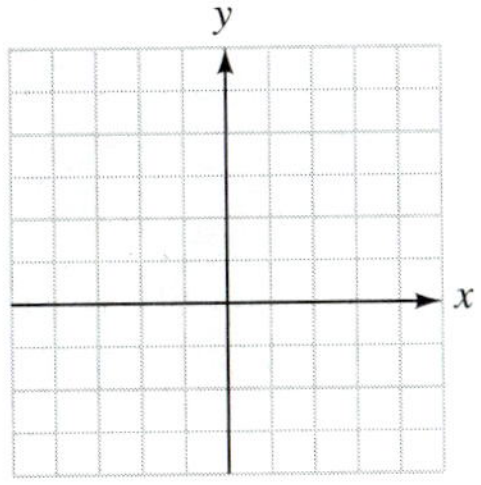

35. $\begin{cases} x + 2y = -4 \\ x - \frac{1}{2}y = 6 \end{cases}$

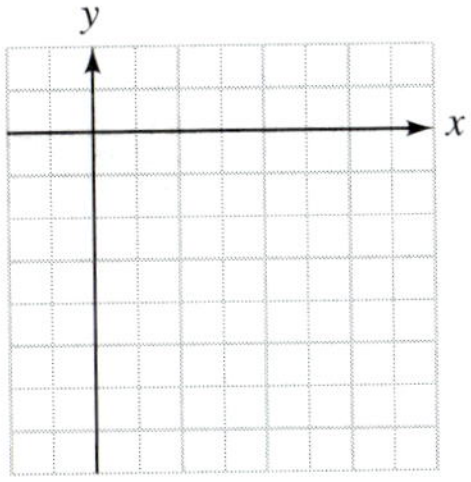

36. $\begin{cases} \frac{2}{3}x - y = -3 \\ 3x + y = 3 \end{cases}$

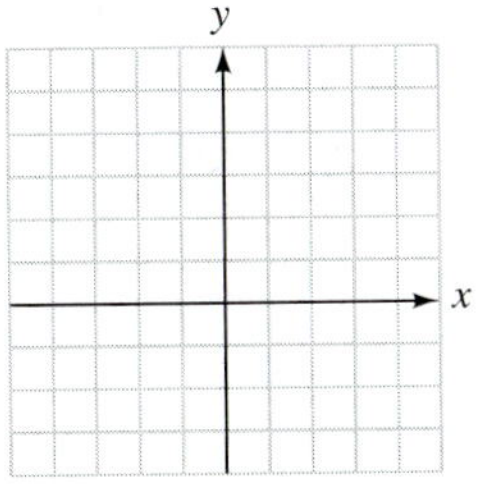

37. $\begin{cases} -\frac{3}{4}x + y = 3 \\ \frac{1}{4}x + y = -1 \end{cases}$

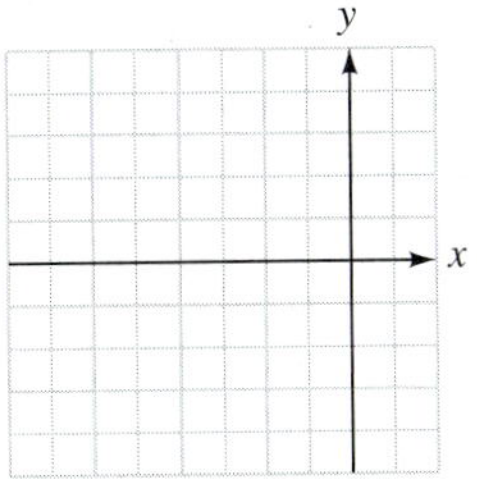

38. $\begin{cases} \frac{1}{3}x + y = 7 \\ \frac{2}{3}x - y = -4 \end{cases}$

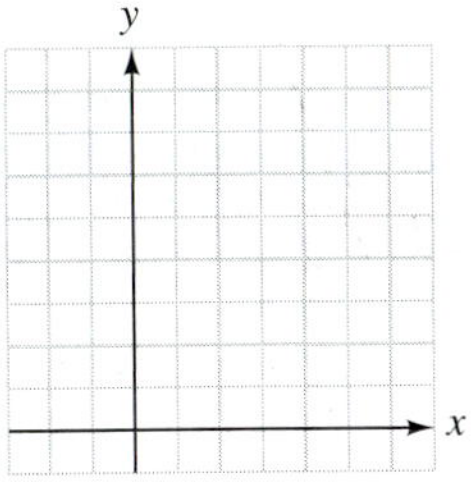

39. $\begin{cases} \frac{1}{2}x + \frac{1}{4}y = 0 \\ \frac{1}{4}x - \frac{3}{8}y = -2 \end{cases}$

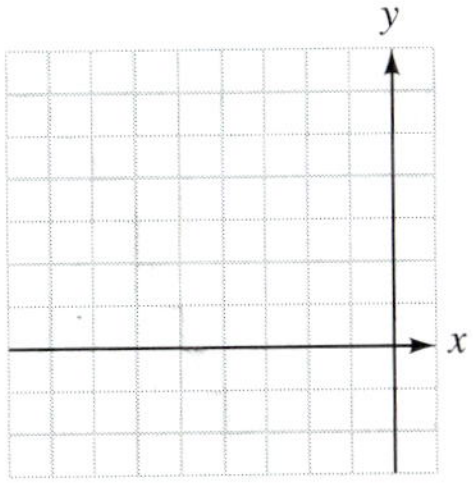

40. $\begin{cases} \frac{1}{2}x + \frac{2}{3}y = -5 \\ \frac{3}{2}x - y = 3 \end{cases}$

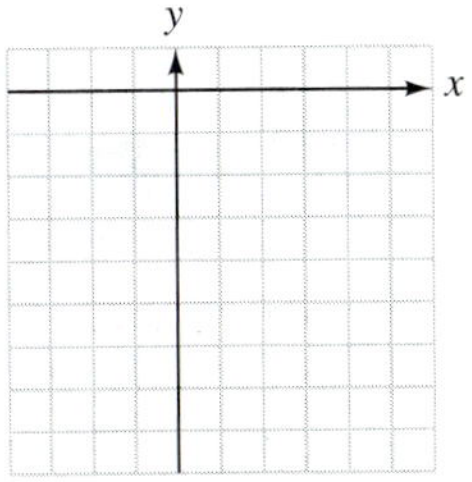

41. $\begin{cases} \frac{1}{3}x - \frac{1}{2}y = \frac{1}{6} \\ \frac{2}{5}x + \frac{1}{2}y = \frac{13}{10} \end{cases}$

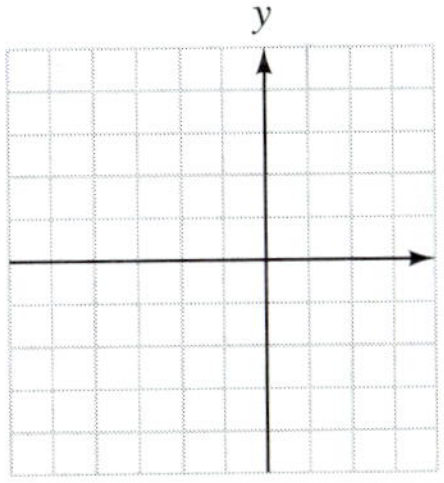

42. $\begin{cases} \frac{3}{4}x + \frac{2}{3}y = -\frac{19}{6} \\ y - x = -\frac{4}{3}x \end{cases}$

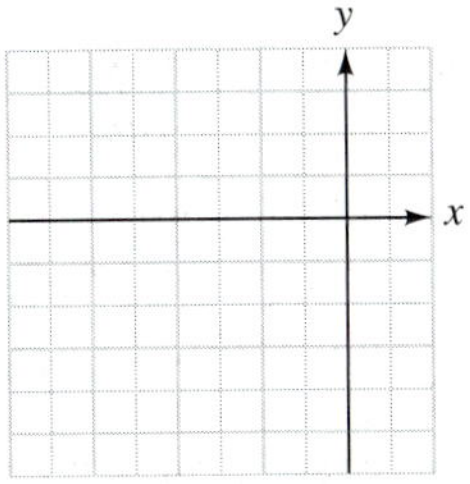

Use a graphing calculator to solve each system, if possible. If answers are not exact, round to the nearest hundredth.

43. $\begin{cases} y = 4 - x \\ y = 2 + x \end{cases}$

44. $\begin{cases} y = x - 2 \\ y = x + 2 \end{cases}$

45. $\begin{cases} 3x - 6y = 4 \\ 2x + y = 1 \end{cases}$

46. $\begin{cases} 4x + 9y = 4 \\ 6x + 3y = -1 \end{cases}$

APPLICATIONS

47. Transplants See the illustration.

a. What was the relationship between the number of donors and those awaiting a transplant in 1994?

b. In what year was the number of donors and the number waiting for a transplant the same? Estimate the number.

c. Explain the most recent trends.

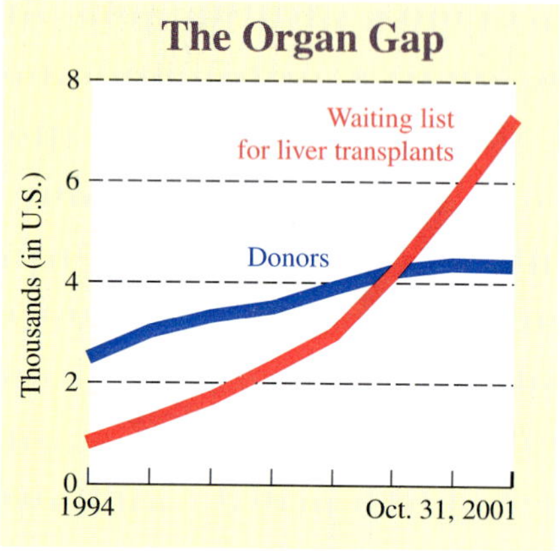

48. Daily tracking polls See the illustration.

a. Which candidate was ahead on October 28 and by how much?

b. On what day did the challenger pull even with the incumbent?

c. If the election was held November 4, whom did the poll predict as the winner and by how many percentage points?

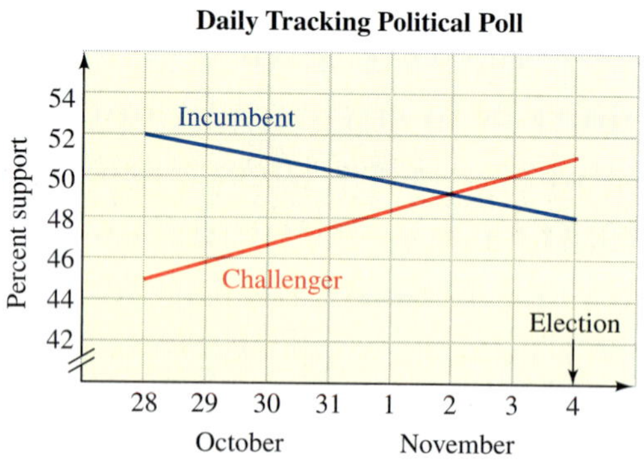

49. Latitude and longitude See the illustration.

a. Name three American cities that lie on the latitude line of 30° north.

b. Name three American cities that lie on the longitude line of 90° west.

c. What city lies on both lines?

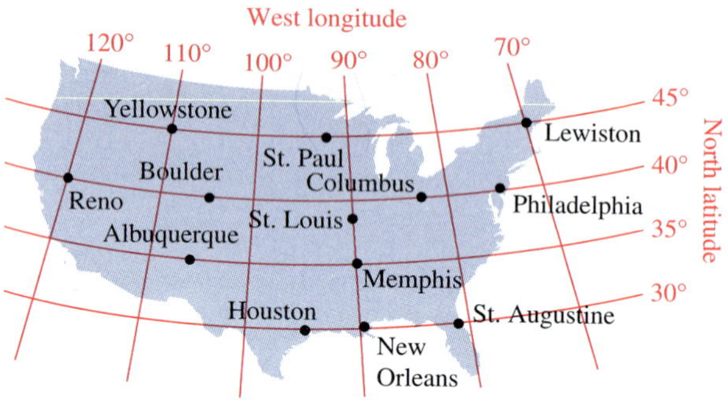

50. Economics The graph in the illustration illustrates the law of supply and demand.

a. Complete this sentence: "As the price of an item increases, the *supply* of the item ________."

b. Complete this sentence: "As the price of an item increases, the *demand* for the item ________."

c. For what price will the supply equal the demand? How many items will be supplied for this price?

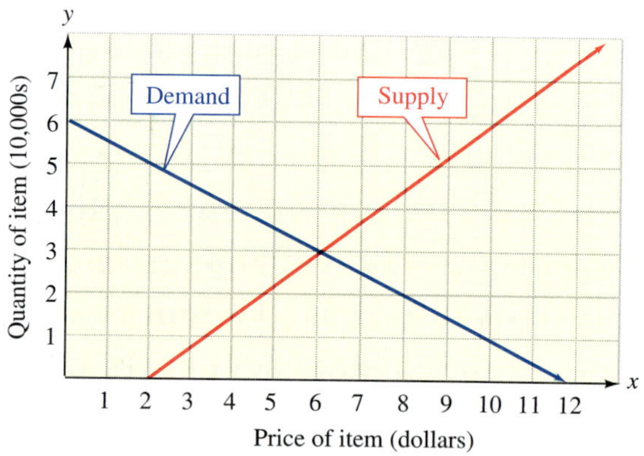

WRITING

51. Explain what we mean when we say "inconsistent system."

52. Explain what we mean when we say, "The equations of a system are dependent."

SOMETHING TO THINK ABOUT

53. Use a graphing calculator to solve the system

$$\begin{cases} 11x - 20y = 21 \\ -4x + 7y = 21 \end{cases}$$

What problems did you encounter?

54. Can the equations of an inconsistent system with two equations in two variables be dependent?

3.4 Solving Systems of Equations by Substitution

- The Substitution Method
- Inconsistent Systems
- Dependent Equations

Getting Ready *Remove parentheses.*

1. $2(3x + 2)$

2. $5(-5 - 2x)$

Substitute $x - 2$ for y and remove parentheses.

3. $2y$

4. $3(y - 2)$

The graphing method for solving systems of equations does not always provide exact solutions. For example, if the solution of a system is $x = \frac{11}{97}$ and $y = \frac{13}{97}$, it is unlikely that we could read this solution exactly from a graph. Fortunately, there are other methods that provide exact solutions. We now consider one of them, called the *substitution method.*

The Substitution Method

To solve the system

$$\begin{cases} y = 3x - 2 \\ 2x + y = 8 \end{cases}$$

by the substitution method, we note that $y = 3x - 2$. Because $y = 3x - 2$, we can substitute $3x - 2$ for y in the equation $2x + y = 8$ to get

$$2x + \mathbf{y} = 8$$
$$2x + (\mathbf{3x - 2}) = 8$$

The resulting equation has only one variable and can be solved for x.

$$2x + (3x - 2) = 8$$
$$2x + 3x - 2 = 8 \quad \text{Remove parentheses.}$$
$$5x - 2 = 8 \quad \text{Combine like terms.}$$
$$5x = 10 \quad \text{Add 2 to both sides.}$$
$$x = 2 \quad \text{Divide both sides by 5.}$$

We can find y by substituting 2 for x in either equation of the given system. Because $y = 3x - 2$ is already solved for y, it is easier to substitute in this equation.

$$\begin{aligned} y &= 3x - 2 \\ &= 3(2) - 2 \\ &= 6 - 2 \\ &= 4 \end{aligned}$$

The solution of the given system is $x = 2$ and $y = 4$, or just $(2, 4)$.

Check:

$$\begin{aligned} y &= 3x - 2 \\ 4 &\stackrel{?}{=} 3(2) - 2 \\ 4 &\stackrel{?}{=} 6 - 2 \\ 4 &= 4 \end{aligned} \qquad \begin{aligned} 2x + y &= 8 \\ 2(2) + 4 &\stackrel{?}{=} 8 \\ 4 + 4 &\stackrel{?}{=} 8 \\ 8 &= 8 \end{aligned}$$

Since the pair $x = 2$ and $y = 4$ is a solution, the lines represented by the equations of the given system intersect at the point $(2, 4)$. The equations of this system are independent, and the system is consistent.

To solve a system of equations in x and y by the substitution method, we follow these steps.

The Substitution Method

1. Solve one of the equations for x or y. (This step may not be necessary.)
2. Substitute the resulting expression for the variable obtained in Step 1 into the other equation, and solve that equation.
3. Find the value of the other variable by substituting the solution found in Step 2 into any equation containing both variables.
4. Check the solution in the equations of the original system.

EXAMPLE 1 Solve the system: $\begin{cases} 2x + y = -5 \\ 3x + 5y = -4 \end{cases}$.

Solution We solve one of the equations for one of its variables. Since the term y in the first equation has a coefficient of 1, we solve the first equation for y.

$$\begin{aligned} 2x + y &= -5 \\ y &= -5 - 2x && \text{Subtract } 2x \text{ from both sides.} \end{aligned}$$

We then substitute $-5 - 2x$ for y in the second equation and solve for x.

$$\begin{aligned} 3x + 5y &= -4 \\ 3x + 5(-5 - 2x) &= -4 \\ 3x - 25 - 10x &= -4 && \text{Remove parentheses.} \\ -7x - 25 &= -4 && \text{Combine like terms.} \\ -7x &= 21 && \text{Add 25 to both sides.} \\ x &= -3 && \text{Divide both sides by } -7. \end{aligned}$$

We can find y by substituting -3 for x in the equation $y = -5 - 2x$.

$$\begin{aligned} y &= -5 - 2x \\ &= -5 - 2(-3) \\ &= -5 + 6 \\ &= 1 \end{aligned}$$

The solution is $x = -3$ and $y = 1$, or just $(-3, 1)$.

Check:

$$\begin{aligned} 2x + y &= -5 \\ 2(-3) + 1 &\stackrel{?}{=} -5 \\ -6 + 1 &\stackrel{?}{=} -5 \\ -5 &= -5 \end{aligned} \qquad \begin{aligned} 3x + 5y &= -4 \\ 3(-3) + 5(1) &\stackrel{?}{=} -4 \\ -9 + 5 &\stackrel{?}{=} -4 \\ -4 &= -4 \end{aligned}$$

Self Check Solve the system: $\begin{cases} 2x - 3y = 13 \\ 3x + y = 3 \end{cases}$.

EXAMPLE 2 Solve the system: $\begin{cases} 2x + 3y = 5 \\ 3x + 2y = 0 \end{cases}$.

Solution We can solve the second equation for x:

$$\begin{aligned} 3x + 2y &= 0 \\ 3x &= -2y && \text{Subtract } 2y \text{ from both sides.} \\ x &= \frac{-2y}{3} && \text{Divide both sides by 3.} \end{aligned}$$

We then substitute $\frac{-2y}{3}$ for x in the other equation and solve for y.

$$\begin{aligned} 2x + 3y &= 5 \\ 2\left(\frac{-2y}{3}\right) + 3y &= 5 \\ \frac{-4y}{3} + 3y &= 5 && \text{Remove parentheses.} \\ 3\left(\frac{-4y}{3}\right) + 3(3y) &= 3(5) && \text{Multiply both sides by 3.} \\ -4y + 9y &= 15 && \text{Remove parentheses.} \\ 5y &= 15 && \text{Combine like terms.} \\ y &= 3 && \text{Divide both sides by 5.} \end{aligned}$$

We can find x by substituting 3 for y in the equation $x = \frac{-2y}{3}$.

$$\begin{aligned} x &= \frac{-2y}{3} \\ &= \frac{-2(3)}{3} \\ &= -2 \end{aligned}$$

Check the solution $(-2, 3)$ in each equation of the system.

Self Check Solve the system: $\begin{cases} 3x - 2y = -19 \\ 2x + 5y = 0 \end{cases}$.

EXAMPLE 3 Solve the system: $\begin{cases} 3(x - y) = 5 \\ x + 3 = -\frac{5}{2}y \end{cases}$.

Solution We begin by writing each equation in general form:

$$3(x - y) = 5 \qquad\qquad x + 3 = -\frac{5}{2}y$$

(1) $3x - 3y = 5$ $\qquad$ $2x + 6 = -5y$ Multiply both sides by 2.

(2) $2x + 5y = -6$ Add 5y and subtract 6 from both sides.

To solve the system formed by Equations 1 and 2, we first solve Equation 1 for x.

(1) $3x - 3y = 5$

$3x = 5 + 3y$ Add 3y to both sides.

(3) $x = \dfrac{5 + 3y}{3}$ Divide both sides by 3.

We then substitute $\frac{5 + 3y}{3}$ for x in Equation 2 and proceed as follows:

$2x + 5y = -6$

(2) $2\left(\dfrac{5 + 3y}{3}\right) + 5y = -6$

$2(5 + 3y) + 15y = -18$ Multiply both sides by 3.

$10 + 6y + 15y = -18$ Remove parentheses.

$10 + 21y = -18$ Combine like terms.

$21y = -28$ Subtract 10 from both sides.

$y = \dfrac{-28}{21}$ Divide both sides by 21.

$y = -\dfrac{4}{3}$ Simplify $\frac{-28}{21}$.

To find x, we substitute $-\frac{4}{3}$ for y in Equation 3 and simplify.

$$\begin{aligned} x &= \frac{5 + 3y}{3} \\ &= \frac{5 + 3\left(-\frac{4}{3}\right)}{3} \\ &= \frac{5 - 4}{3} \\ &= \frac{1}{3} \end{aligned}$$

Check the solution $\left(\frac{1}{3}, -\frac{4}{3}\right)$ in each equation.

Self Check Solve the system: $\begin{cases} 2(x + y) = -5 \\ x + 2 = -\frac{3}{5}y \end{cases}$.

Inconsistent Systems

EXAMPLE 4 Solve the system: $\begin{cases} x = 4(3 - y) \\ 2x = 4(3 - 2y) \end{cases}$.

Solution Since $x = 4(3 - y)$, we can substitute $4(3 - y)$ for x in the second equation and solve for y.

$$\begin{aligned} 2\mathbf{x} &= 4(3 - 2y) \\ 2[\mathbf{4(3 - y)}] &= 4(3 - 2y) \\ 8(3 - y) &= 4(3 - 2y) && 2 \cdot 4 = 8. \\ 24 - 8y &= 12 - 8y && \text{Remove parentheses.} \\ 24 &= 12 && \text{Add } 8y \text{ to both sides.} \end{aligned}$$

This impossible result indicates that the equations in this system are independent, but that the system is inconsistent. If each equation in this system were graphed, these graphs would be parallel lines. There are no solutions to this system.

Self Check Solve the system: $\begin{cases} 0.1x - 0.4 = 0.1y \\ -2y = 2(2 - x) \end{cases}$.

Dependent Equations

EXAMPLE 5 Solve the system: $\begin{cases} 3x = 4(6 - y) \\ 4y + 3x = 24 \end{cases}$.

Solution We can substitute $4(6 - y)$ for $3x$ in the second equation and proceed as follows:

$$\begin{aligned} 4y + \mathbf{3x} &= 24 \\ 4y + \mathbf{4(6 - y)} &= 24 \\ 4y + 24 - 4y &= 24 && \text{Remove parentheses.} \\ 24 &= 24 && \text{Combine like terms.} \end{aligned}$$

Although $24 = 24$ is true, we did not find y. This result indicates that the equations of this system are dependent. If either equation were graphed, the same line would result.

Because any ordered pair that satisfies one equation satisfies the other also, the system has infinitely many solutions. To find some of them, we substitute 8, 0, and 4 for x in either equation and solve for y. The pairs (8, 0), (0, 6), and (4, 3) are solutions.

Self Check Solve the system: $\begin{cases} 3y = -3(x + 4) \\ 3x + 3y = -12 \end{cases}$.

Self Check Answers

1. $(2, -3)$ **2.** $(-5, 2)$ **3.** $\left(-\frac{5}{4}, -\frac{5}{4}\right)$ **4.** no solution **5.** infinitely many solutions

Orals *Let $y = x + 1$. Find y after each quantity is substituted for x.*

1. $2z$

2. $z + 1$

3. $3t + 2$

4. $\frac{t}{3} + 3$

3.4 EXERCISES

REVIEW *Let $x = -2$ and $y = 3$ and evaluate each expression.*

1. $y^2 - x^2$

2. $-x^2 + y^3$

3. $\dfrac{3x - 2y}{2x + y}$

4. $-2x^2y^2$

5. $-x(3y - 4)$

6. $-2y(4x - y)$

VOCABULARY AND CONCEPTS *Fill in the blanks.*

7. We say the equation $y = 2x + 4$ is solved for __ or that y is expressed in ______ of x.

8. To check a solution of a system, we test whether the coordinates of the ordered pair ______ both equations.

9. Consider $2(x - 6) = 2x - 12$. The distributive property was applied to _______ parentheses.

10. In mathematics, to _________ means to replace an expression with one that is equivalent to it.

11. A system with dependent equations has ____________ solutions.

12. In the term y, the coefficient is understood to be __.

PRACTICE *Use substitution to solve each system. If the equations of a system are dependent or if a system is inconsistent, so indicate.*

13. $\begin{cases} y = 2x \\ x + y = 6 \end{cases}$

14. $\begin{cases} y = 3x \\ x + y = 4 \end{cases}$

15. $\begin{cases} y = 2x - 6 \\ 2x + y = 6 \end{cases}$

16. $\begin{cases} y = 2x - 9 \\ x + 3y = 8 \end{cases}$

17. $\begin{cases} y = 2x + 5 \\ x + 2y = -5 \end{cases}$

18. $\begin{cases} y = -2x \\ 3x + 2y = -1 \end{cases}$

19. $\begin{cases} 2a + 4b = -24 \\ a = 20 - 2b \end{cases}$

20. $\begin{cases} 3a + 6b = -15 \\ a = -2b - 5 \end{cases}$

21. $\begin{cases} 2a = 3b - 13 \\ b = 2a + 7 \end{cases}$

22. $\begin{cases} a = 3b - 1 \\ b = 2a + 2 \end{cases}$

23. $\begin{cases} 2x + y = 4 \\ 4x + y = 5 \end{cases}$

24. $\begin{cases} x + 3y = 3 \\ 2x + 3y = 4 \end{cases}$

25. $\begin{cases} r + 3s = 9 \\ 3r + 2s = 13 \end{cases}$

26. $\begin{cases} x - 2y = 2 \\ 2x + 3y = 11 \end{cases}$

27. $\begin{cases} 4x + 5y = 2 \\ 3x - y = 11 \end{cases}$

28. $\begin{cases} 5u + 3v = 5 \\ 4u - v = 4 \end{cases}$

29. $\begin{cases} 2x + y = 0 \\ 3x + 2y = 1 \end{cases}$

30. $\begin{cases} 3x - y = 7 \\ 2x + 3y = 1 \end{cases}$

31. $\begin{cases} 3x + 4y = -7 \\ 2y - x = -1 \end{cases}$

32. $\begin{cases} 4x + 5y = -2 \\ x + 2y = -2 \end{cases}$

33. $\begin{cases} 9x = 3y + 12 \\ 4 = 3x - y \end{cases}$

34. $\begin{cases} 8y = 15 - 4x \\ x + 2y = 4 \end{cases}$

35. $\begin{cases} 2x + 3y = 5 \\ 3x + 2y = 5 \end{cases}$

36. $\begin{cases} 3x - 2y = -1 \\ 2x + 3y = -5 \end{cases}$

37. $\begin{cases} a = \frac{3}{2}b + 5 \\ 2a - 3b = 8 \end{cases}$

38. $\begin{cases} x = \frac{1}{2}y + \frac{5}{4} \\ 4x - 2y = 5 \end{cases}$

39. $\begin{cases} 2x + 5y = -2 \\ 4x + 3y = 10 \end{cases}$

40. $\begin{cases} 3x + 4y = -6 \\ 2x - 3y = -4 \end{cases}$

41. $\begin{cases} 2x - 3y = -3 \\ 3x + 5y = -14 \end{cases}$

42. $\begin{cases} 4x - 5y = -12 \\ 5x - 2y = 2 \end{cases}$

43. $\begin{cases} 7x - 2y = -1 \\ -5x + 2y = -1 \end{cases}$

44. $\begin{cases} -8x + 3y = 22 \\ 4x + 3y = -2 \end{cases}$

45. $\begin{cases} 2a + 3b = 2 \\ 8a - 3b = 3 \end{cases}$

46. $\begin{cases} 3a - 2b = 0 \\ 9a + 4b = 5 \end{cases}$

47. $\begin{cases} y - x = 3x \\ 2(x + y) = 14 - y \end{cases}$

48. $\begin{cases} y + x = 2x + 2 \\ 2(3x - 2y) = 21 - y \end{cases}$

49. $\begin{cases} 3(x - 1) + 3 = 8 + 2y \\ 2(x + 1) = 4 + 3y \end{cases}$

50. $\begin{cases} 4(x - 2) = 19 - 5y \\ 3(x + 1) - 2y = 2y \end{cases}$

51. $\begin{cases} 6a = 5(3 + b + a) - a \\ 3(a - b) + 4b = 5(1 + b) \end{cases}$

52. $\begin{cases} 5(x + 1) + 7 = 7(y + 1) \\ 5(y + 1) = 6(1 + x) + 5 \end{cases}$

53. $\begin{cases} \frac{1}{2}x + \frac{1}{2}y = -1 \\ \frac{1}{3}x - \frac{1}{2}y = -4 \end{cases}$

54. $\begin{cases} \frac{2}{3}y + \frac{1}{5}z = 1 \\ \frac{1}{3}y - \frac{2}{5}z = 3 \end{cases}$

55. $\begin{cases} 5x = \frac{1}{2}y - 1 \\ \frac{1}{4}y = 10x - 1 \end{cases}$

56. $\begin{cases} \frac{2}{3}x = 1 - 2y \\ 2(5y - x) + 11 = 0 \end{cases}$

57. $\begin{cases} \frac{6x - 1}{3} - \frac{5}{3} = \frac{3y + 1}{2} \\ \frac{1 + 5y}{4} + \frac{x + 3}{4} = \frac{17}{2} \end{cases}$

58. $\begin{cases} \frac{5x - 2}{4} + \frac{1}{2} = \frac{3y + 2}{2} \\ \frac{7y + 3}{3} = \frac{x}{2} + \frac{7}{3} \end{cases}$

WRITING

59. Explain how to use substitution to solve a system of equations.

60. If the equations of a system are written in general form, why is it to your advantage to solve for a variable whose coefficient is 1?

SOMETHING TO THINK ABOUT

61. Could you use substitution to solve the system

$$\begin{cases} y = 2y + 4 \\ x = 3x - 5 \end{cases}$$

How would you solve it?

62. What are the advantages and disadvantages of
   a. the graphing method?
   b. the substitution method?

3.5 Solving Systems of Equations by Addition

- The Addition Method
- Inconsistent Systems
- Dependent Equations

Getting Ready *Add the left-hand sides and the right-hand sides of the equations in each system.*

1. $\begin{cases} 2x + 3y = 4 \\ 3x - 3y = 6 \end{cases}$

2. $\begin{cases} 4x - 2y = 1 \\ -4x + 3y = 5 \end{cases}$

3. $\begin{cases} 6x - 5y = 23 \\ -4x + 5y = 10 \end{cases}$

4. $\begin{cases} -5x + 6y = 18 \\ 5x + 12y = 10 \end{cases}$

We now consider a second method for solving systems of equations that will provide exact solutions. It is called the *addition method.*

The Addition Method

To solve the system

$$\begin{cases} x + y = 8 \\ x - y = -2 \end{cases}$$

by the addition method, we see that the coefficients of y are *opposites* and then add the left-hand sides and the right-hand sides of the equations to eliminate the variable y.

$$\begin{array}{r} x + y = 8 \\ \underline{x - y = -2} \end{array}$$

Equal quantities, $x - y$ and -2, are added to both sides of the equation $x + y = 8$. By the addition property of equality, the results will be equal.

Now, column by column, we add like terms.

Combine like terms.

$$\begin{array}{r} x + y = 8 \\ \underline{x - y = -2} \\ 2x \quad\;\; = 6 \end{array}$$

← Write each result here.

We can then solve the resulting equation for x.

$$2x = 6$$

$$x = 3 \quad \text{Divide both sides by 2.}$$

To find y, we substitute 3 for x in either equation of the system and solve it for y.

$$x + y = 8 \quad \text{The first equation of the system.}$$

$$3 + y = 8 \quad \text{Substitute 3 for } x.$$

$$y = 5 \quad \text{Subtract 3 from both sides.}$$

We check the solution by verifying that the pair (3, 5) satisfies each equation of the original system.

To solve a system of equations in x and y by the addition method, we follow these steps.

The Addition Method

1. If necessary, write both equations in general form: $Ax + By = C$.
2. If necessary, multiply one or both of the equations by nonzero quantities to make the coefficients of x (or the coefficients of y) opposites.
3. Add the equations to eliminate the term involving x (or y).
4. Solve the equation resulting from Step 3.
5. Find the value of the other variable by substituting the solution found in Step 4 into any equation containing both variables.
6. Check the solution in the equations of the original system.

EXAMPLE 1 Solve the system: $\begin{cases} 3y = 14 + x \\ x + 22 = 5y \end{cases}$.

Solution We can write the equations in the form

$$\begin{cases} -x + 3y = 14 \\ x - 5y = -22 \end{cases}$$

Because the coefficients of x are opposites, the terms involving x are eliminated when the equations are added. We solve the resulting equation for y.

$$\begin{aligned} -x + 3y &= 14 \\ x - 5y &= -22 \\ \hline -2y &= -8 \\ y &= 4 \quad \text{Divide both sides by } -2. \end{aligned}$$

To find x, we substitute 4 for y in either equation of the system. If we substitute 4 for y in the equation $-x + 3y = 14$, we have

$$\begin{aligned} -x + 3y &= 14 \\ -x + 3(4) &= 14 \\ -x + 12 &= 14 \quad \text{Simplify.} \\ -x &= 2 \quad \text{Subtract 12 from both sides.} \\ x &= -2 \quad \text{Divide both sides by } -1. \end{aligned}$$

Verify that $(-2, 4)$ satisfies each equation.

Self Check Solve the system: $\begin{cases} 3y = 7 - x \\ 2x - 3y = -22 \end{cases}$.

Sometimes we need to multiply both sides of one equation in a system by a number to make the coefficients of one of the variables opposites.

EXAMPLE 2 Solve the system: $\begin{cases} 3x + y = 7 \\ x + 2y = 4 \end{cases}$.

Solution If we add the equations as they are, neither variable will be eliminated. We must write the equations so that the coefficients of one of the variables are opposites. To eliminate x, we can multiply both sides of the second equation by -3 to get

$$\begin{cases} 3x + y = 7 \\ -3(x + 2y) = -3(4) \end{cases} \rightarrow \begin{cases} 3x + y = 7 \\ -3x - 6y = -12 \end{cases}$$

The coefficients of the terms $3x$ and $-3x$ are opposites. When the equations are added, x is eliminated.

$$\begin{aligned} 3x + y &= 7 \\ -3x - 6y &= -12 \\ \hline -5y &= -5 \\ y &= 1 \quad \text{Divide both sides by } -5. \end{aligned}$$

To find x, we substitute 1 for y in the equation $3x + y = 7$.

$$3x + y = 7$$
$$3x + (1) = 7 \quad \text{Substitute 1 for } y.$$
$$3x = 6 \quad \text{Subtract 1 from both sides.}$$
$$x = 2 \quad \text{Divide both sides by 3.}$$

Check the solution (2, 1) in the original system of equations.

Self Check Solve the system: $\begin{cases} 3x + 4y = 25 \\ 2x + y = 10 \end{cases}$.

In some instances, we must multiply both equations by nonzero quantities to make the coefficients of one of the variables opposites.

EXAMPLE 3 Solve the system: $\begin{cases} 2a - 5b = 10 \\ 3a - 2b = -7 \end{cases}$.

Solution The equations in the system must be written so that one of the variables will be eliminated when the equations are added.

To eliminate a, we can multiply the first equation by 3 and the second equation by -2 to get

$$\begin{cases} 3(2a - 5b) = 3(10) \\ -2(3a - 2b) = -2(-7) \end{cases} \rightarrow \begin{cases} 6a - 15b = 30 \\ -6a + 4b = 14 \end{cases}$$

When these equations are added, the terms $6a$ and $-6a$ are eliminated.

$$\begin{aligned} 6a - 15b &= 30 \\ -6a + 4b &= 14 \\ \hline -11b &= 44 \\ b &= -4 \quad \text{Divide both sides by } -11. \end{aligned}$$

To find a, we substitute -4 for b in the equation $2a - 5b = 10$.

$$2a - 5b = 10$$
$$2a - 5(-4) = 10 \quad \text{Substitute } -4 \text{ for } b.$$
$$2a + 20 = 10 \quad \text{Simplify.}$$
$$2a = -10 \quad \text{Subtract 20 from both sides.}$$
$$a = -5 \quad \text{Divide both sides by 2.}$$

Check the solution $(-5, -4)$ in the original equations.

Self Check Solve the system: $\begin{cases} 2a + 3b = 7 \\ 5a + 2b = 1 \end{cases}$.

Comment Note that a solution of Example 3 by the substitution method would involve fractions. In these cases, the addition method is usually easier.

EXAMPLE 4 Solve the system: $\begin{cases} \frac{5}{6}x + \frac{2}{3}y = \frac{7}{6} \\ \frac{10}{7}x - \frac{4}{9}y = \frac{17}{21} \end{cases}$.

Solution To clear the equations of fractions, we multiply both sides of the first equation by 6 and both sides of the second equation by 63. This gives the system

$$\begin{cases} 5x + 4y = 7 & (1) \\ 90x - 28y = 51 & (2) \end{cases}$$

We can solve for x by eliminating the terms involving y. To do so, we multiply Equation 1 by 7 and add the result to Equation 2.

$$\begin{aligned} 35x + 28y &= 49 \\ 90x - 28y &= 51 \\ \hline 125x \quad &= 100 \\ x &= \frac{100}{125} \quad \text{Divide both sides by 125.} \\ x &= \frac{4}{5} \quad \text{Simplify.} \end{aligned}$$

To solve for y, we substitute $\frac{4}{5}$ for x in Equation 1 and simplify.

$$\begin{aligned} 5x + 4y &= 7 \\ 5\left(\frac{4}{5}\right) + 4y &= 7 \\ 4 + 4y &= 7 \quad \text{Simplify.} \\ 4y &= 3 \quad \text{Subtract 4 from both sides.} \\ y &= \frac{3}{4} \quad \text{Divide both sides by 4.} \end{aligned}$$

Check the solution of $\left(\frac{4}{5}, \frac{3}{4}\right)$ in the original equations.

Self Check Solve the system: $\begin{cases} \frac{1}{3}x + \frac{1}{6}y = 1 \\ \frac{1}{2}x - \frac{1}{4}y = 0 \end{cases}$.

Inconsistent Systems

EXAMPLE 5 Solve the system: $\begin{cases} x - \frac{2y}{3} = \frac{8}{3} \\ -\frac{3x}{2} + y = -6 \end{cases}$.

Solution We can multiply both sides of the first equation by 3 and both sides of the second equation by 2 to clear the equations of fractions.

$$\begin{cases} 3\left(x - \frac{2y}{3}\right) = 3\left(\frac{8}{3}\right) \\ 2\left(-\frac{3x}{2} + y\right) = 2(-6) \end{cases} \rightarrow \begin{cases} 3x - 2y = 8 \\ -3x + 2y = -12 \end{cases}$$

We can add the resulting equations to eliminate the term involving x.

$$\begin{array}{r}3x - 2y = 8 \\ -3x + 2y = -12 \\ \hline 0 = -4\end{array}$$

Here, the terms involving both x and y drop out, and a false result is obtained. This shows that the equations of the system are independent, but the system itself is inconsistent. This system has no solution.

Self Check Solve the system: $\begin{cases} x - \dfrac{y}{3} = \dfrac{10}{3} \\ 3x - y = \dfrac{5}{2} \end{cases}$.

Dependent Equations

EXAMPLE 6 Solve the system: $\begin{cases} \dfrac{2x - 5y}{2} = \dfrac{19}{2} \\ -0.2x + 0.5y = -1.9 \end{cases}$.

Solution We can multiply both sides of the first equation by 2 to clear it of fractions and both sides of the second equation by 10 to clear it of decimals.

$$\begin{cases} \mathbf{2}\left(\dfrac{2x - 5y}{2}\right) = \mathbf{2}\left(\dfrac{19}{2}\right) \\ \mathbf{10}(-0.2x + 0.5y) = \mathbf{10}(-1.9) \end{cases} \rightarrow \begin{cases} 2x - 5y = 19 \\ -2x + 5y = -19 \end{cases}$$

We add the resulting equations to get

$$\begin{array}{r}2x - 5y = 19 \\ -2x + 5y = -19 \\ \hline 0 = 0\end{array}$$

As in Example 5, both x and y drop out. However, this time a true result is obtained. This shows that the equations are dependent and the system has infinitely many solutions. Any ordered pair that satisfies one equation satisfies the other also. Some solutions are $(2, -3)$, $(12, 1)$, and $(0, -\frac{19}{5})$.

Self Check Solve the system: $\begin{cases} \dfrac{3x + y}{6} = \dfrac{1}{3} \\ -0.3x - 0.1y = -0.2 \end{cases}$.

Self Check Answers

1. $(-5, 4)$ **2.** $(3, 4)$ **3.** $(-1, 3)$ **4.** $\left(\frac{3}{2}, 3\right)$ **5.** no solution **6.** infinitely many solutions

Orals *Use addition to solve each system for x.*

1. $\begin{cases} x + y = 1 \\ x - y = 1 \end{cases}$

2. $\begin{cases} 2x + y = 4 \\ x - y = 2 \end{cases}$

Use addition to solve each system for y.

3. $\begin{cases} -x + y = 3 \\ x + y = 3 \end{cases}$

4. $\begin{cases} x + 2y = 4 \\ -x - y = 1 \end{cases}$

3.5 EXERCISES

REVIEW *Solve each equation.*

1. $8(3x - 5) - 12 = 4(2x + 3)$

2. $5x - 13 = x - 1$

3. $x - 2 = \frac{x + 2}{3}$

4. $\frac{3}{2}(y + 4) = \frac{20 - y}{2}$

Solve each inequality and graph the solution.

5. $7x - 9 \leq 5$

6. $-2x + 6 > 16$

VOCABULARY AND CONCEPTS *Fill in the blanks.*

7. The numerical ________ of $-3x$ is -3.

8. The ________ of 4 is -4.

9. $Ax + By = C$ is the ______ form of the equation of a line.

10. When adding the equations

$$\begin{array}{r} 5x - 6y = 10 \\ \underline{-3x + 6y = 24} \end{array}$$

the variable y will be ________.

11. To clear the equation $\frac{2}{3}x + 4y = -\frac{4}{5}$ of fractions, we must multiply both sides by ___.

12. To solve the system

$$\begin{cases} 3x + 12y = 4 \\ 6x - 4y = 8 \end{cases}$$

we would multiply the first equation by ___ and add to eliminate the x.

PRACTICE *Use addition to solve each system.*

13. $\begin{cases} x + y = 5 \\ x - y = -3 \end{cases}$

14. $\begin{cases} x - y = 1 \\ x + y = 7 \end{cases}$

15. $\begin{cases} x - y = -5 \\ x + y = 1 \end{cases}$

16. $\begin{cases} x + y = 1 \\ x - y = 5 \end{cases}$

17. $\begin{cases} 2x + y = -1 \\ -2x + y = 3 \end{cases}$

18. $\begin{cases} 3x + y = -6 \\ x - y = -2 \end{cases}$

19. $\begin{cases} 2x - 3y = -11 \\ 3x + 3y = 21 \end{cases}$

20. $\begin{cases} 3x - 2y = 16 \\ -3x + 8y = -10 \end{cases}$

21. $\begin{cases} 2x + y = -2 \\ -2x - 3y = -6 \end{cases}$

22. $\begin{cases} 3x + 4y = 8 \\ 5x - 4y = 24 \end{cases}$

23. $\begin{cases} 4x + 3y = 24 \\ 4x - 3y = -24 \end{cases}$

24. $\begin{cases} 5x - 4y = 8 \\ -5x - 4y = 8 \end{cases}$

Use addition to solve each system. If the equations of a system are dependent or if a system is inconsistent, so indicate.

25. $\begin{cases} x + y = 5 \\ x + 2y = 8 \end{cases}$

26. $\begin{cases} x + 2y = 0 \\ x - y = -3 \end{cases}$

27. $\begin{cases} 2x + y = 4 \\ 2x + 3y = 0 \end{cases}$

28. $\begin{cases} 2x + 5y = -13 \\ 2x - 3y = -5 \end{cases}$

29. $\begin{cases} 3x + 29 = 5y \\ 4y - 34 = -3x \end{cases}$

30. $\begin{cases} 3x - 16 = 5y \\ 33 - 5y = 4x \end{cases}$

31. $\begin{cases} 2x = 3(y - 2) \\ 2(x + 4) = 3y \end{cases}$

32. $\begin{cases} 3(x - 2) = 4y \\ 2(2y + 3) = 3x \end{cases}$

33. $\begin{cases} -2(x + 1) = 3(y - 2) \\ 3(y + 2) = 6 - 2(x - 2) \end{cases}$

34. $\begin{cases} 5(x - 1) = 8 - 3(y + 2) \\ 4(x + 2) - 7 = 3(2 - y) \end{cases}$

35. $\begin{cases} 4(x + 1) = 17 - 3(y - 1) \\ 2(x + 2) + 3(y - 1) = 9 \end{cases}$

36. $\begin{cases} 3(x + 3) + 2(y - 4) = 5 \\ 3(x - 1) = -2(y + 2) \end{cases}$

37. $\begin{cases} 2x + y = 10 \\ x + 2y = 10 \end{cases}$

38. $\begin{cases} 3x + 2y = 0 \\ 2x - 3y = -13 \end{cases}$

39. $\begin{cases} 2x - y = 16 \\ 3x + 2y = 3 \end{cases}$

40. $\begin{cases} 3x + 4y = -17 \\ 4x - 3y = -6 \end{cases}$

41. $\begin{cases} 2x + 3y = 2 \\ 4x - 9y = -1 \end{cases}$

42. $\begin{cases} 4x + 5y = 2 \\ 16x - 15y = 1 \end{cases}$

43. $\begin{cases} 4x + 5y = -20 \\ 5x - 4y = -25 \end{cases}$

44. $\begin{cases} 3x - 5y = 4 \\ 7x + 3y = 68 \end{cases}$

45. $\begin{cases} 6x = -3y \\ 5y = 2x + 12 \end{cases}$

46. $\begin{cases} 3y = 4x \\ 5x = 4y - 2 \end{cases}$

47. $\begin{cases} 4(2x - y) = 18 \\ 3(x - 3) = 2y - 1 \end{cases}$

48. $\begin{cases} 2(2x + 3y) = 5 \\ 8x = 3(1 + 3y) \end{cases}$

49. $\begin{cases} 3(x - 2y) = 12 \\ x = 2(y + 2) \end{cases}$

50. $\begin{cases} 4x = 3(4 - y) \\ 3y = 4(2 - x) \end{cases}$

51. $\begin{cases} \frac{3}{5}x + \frac{4}{5}y = 1 \\ -\frac{1}{4}x + \frac{3}{8}y = 1 \end{cases}$

52. $\begin{cases} \frac{1}{2}x - \frac{1}{4}y = 1 \\ \frac{1}{3}x + y = 3 \end{cases}$

53. $\begin{cases} \frac{3}{5}x + y = 1 \\ \frac{4}{5}x - y = -1 \end{cases}$

54. $\begin{cases} \frac{1}{2}x + \frac{4}{7}y = -1 \\ 5x - \frac{4}{5}y = -10 \end{cases}$

55. $\begin{cases} \frac{x}{2} - \frac{y}{3} = -2 \\ \frac{2x - 3}{2} + \frac{6y + 1}{3} = \frac{17}{6} \end{cases}$

56. $\begin{cases} \frac{x + 2}{4} + \frac{y - 1}{3} = \frac{1}{12} \\ \frac{x + 4}{5} - \frac{y - 2}{2} = \frac{5}{2} \end{cases}$

57. $\begin{cases} \frac{x - 3}{2} + \frac{y + 5}{3} = \frac{11}{6} \\ \frac{x + 3}{3} - \frac{5}{12} = \frac{y + 3}{4} \end{cases}$

58. $\begin{cases} \frac{x + 2}{3} = \frac{3 - y}{2} \\ \frac{x + 3}{2} = \frac{2 - y}{3} \end{cases}$

WRITING

59. Why is it usually best to write the equations of a system in general form before using the addition method to solve it?

60. How would you decide whether to use substitution or addition to solve a system of equations?

SOMETHING TO THINK ABOUT

61. If possible, find a solution to the system

$$\begin{cases} x + y = 5 \\ x - y = -3 \\ 2x - y = -2 \end{cases}$$

62. If possible, find a solution to the system

$$\begin{cases} x + y = 5 \\ x - y = -3 \\ x - 2y = 0 \end{cases}$$

3.6 Applications of Systems of Equations

- Solving Problems with Two Variables

Getting Ready *Let x and y represent two numbers. Use an algebraic expression to denote each phrase.*

1. The sum of x and y
2. The difference when y is subtracted from x
3. The product of x and y
4. The quotient x divided by y
5. Give the formula for the area of a rectangle.
6. Give the formula for the perimeter of a rectangle.

We have previously set up equations involving one variable to solve problems. In this section, we consider ways to solve problems by using equations in two variables.

Solving Problems with Two Variables

The following steps are helpful when solving problems involving two unknown quantities.

Problem-Solving Strategy

1. Read the problem several times and *analyze* the facts. Occasionally, a sketch, chart, or diagram will help you visualize the facts of the problem.
2. Pick different variables to represent two unknown quantities. *Form two equations* involving each of the two variables. This will give a system of two equations in two variables.
3. *Solve the system* using the most convenient method: graphing, substitution, or addition.
4. *State the conclusion.*
5. *Check the solution* in the words of the problem.

EXAMPLE 1 **Farming** A farmer raises wheat and soybeans on 215 acres. If he wants to plant 31 more acres in wheat than in soybeans, how many acres of each should he plant?

Analyze the problem The farmer plants two fields, one in wheat and one in soybeans. We know that the number of acres of wheat planted plus the number of acres of soybeans planted will equal a total of 215 acres.

Form two equations If w represents the number of acres of wheat and s represents the number of acres of soybeans to be planted, we can form the two equations

The number of acres planted in wheat	plus	the number of acres planted in soybeans	equals	215 acres.
w	$+$	s	$=$	215

Since the farmer wants to plant 31 more acres in wheat than in soybeans, we have

The number of acres planted in wheat	minus	the number of acres planted in soybeans	equals	31 acres.
w	$-$	s	$=$	31

Solve the system We can now solve the system

$$\begin{cases} w + s = 215 & (1) \\ w - s = 31 & (2) \end{cases}$$

by the addition method.

$$\begin{aligned} w + s &= 215 \\ w - s &= 31 \\ \hline 2w \quad &= 246 \\ w &= 123 \quad \text{Divide both sides by 2.} \end{aligned}$$

To find s, we substitute 123 for w in Equation 1.

$$\begin{aligned} w + s &= 215 \\ 123 + s &= 215 \quad \text{Substitute 123 for } w. \\ s &= 92 \quad \text{Subtract 123 from both sides.} \end{aligned}$$

State the conclusion The farmer should plant 123 acres of wheat and 92 acres of soybeans.

Check the result The total acreage planted is $123 + 92$, or 215 acres. The area planted in wheat is 31 acres greater than that planted in soybeans, because $123 - 92 = 31$. The answers check.

EXAMPLE 2 **Lawn care** An installer of underground irrigation systems wants to cut a 20-foot length of plastic tubing into two pieces. The longer piece is to be 2 feet longer than twice the shorter piece. Find the length of each piece.

Analyze the problem Refer to Figure 3-27, which shows the pipe.

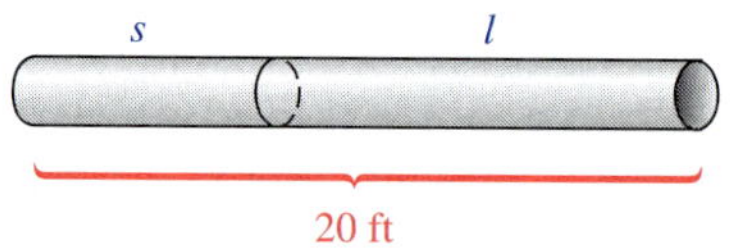

Figure 3-27

Form two equations We can let s represent the length of the shorter piece and l represent the length of the longer piece. Then we can form the equations

The length of the shorter piece	plus	the length of the longer piece	equals	20 feet.
s	$+$	l	$=$	20

Since the longer piece is 2 feet longer than twice the shorter piece, we have

The length of the longer piece	equals	2	times	the length of the shorter piece	plus	2 feet.
l	$=$	2	$\cdot$	s	$+$	2

Solve the system We can use the substitution method to solve the system

$$\begin{cases} s + l = 20 & (1) \\ l = 2s + 2 & (2) \end{cases}$$

$$\begin{aligned} s + (2s + 2) &= 20 && \text{Substitute } 2s + 2 \text{ for } l \text{ in Equation 1.} \\ 3s + 2 &= 20 && \text{Combine like terms.} \\ 3s &= 18 && \text{Subtract 2 from both sides.} \\ s &= 6 && \text{Divide both sides by 3.} \end{aligned}$$

State the conclusion The shorter piece should be 6 feet long. To find the length of the longer piece, we substitute 6 for s in Equation 1 and solve for l.

$$\begin{aligned} s + l &= 20 \\ 6 + l &= 20 && \text{Substitute 6 for } s. \\ l &= 14 && \text{Subtract 6 from both sides.} \end{aligned}$$

The longer piece should be 14 feet long.

Check the result The sum of 6 and 14 is 20. 14 is 2 more than twice 6. The answers check. ∎

EXAMPLE 3 **Gardening** Tom has 150 feet of fencing to enclose a rectangular garden. If the length is to be 5 feet less than 3 times the width, find the area of the garden.

Analyze the problem To find the area of a rectangle, we need to know its length and width. See Figure 3-28.

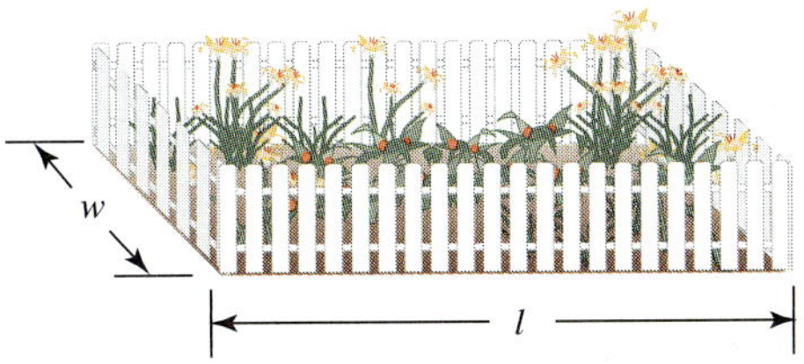

Figure 3-28

Form two equations We can let l represent the length of the garden and w represent the width. Since the perimeter of a rectangle is two lengths plus two widths, we can form the equations

2	times	the length of the garden	plus	2	times	the width of the garden	equals	150 feet.
2	$\cdot$	l	$+$	2	$\cdot$	w	$=$	150

Since the length is 5 feet less than 3 times the width,

The length of the garden	equals	3	times	the width of the garden	minus	5 feet.
l	$=$	3	$\cdot$	w	$-$	5

Solve the system We can use the substitution method to solve this system.

$$\begin{cases} 2l + 2w = 150 & (1) \\ l = 3w - 5 & (2) \end{cases}$$

$$\begin{aligned} 2(3w - 5) + 2w &= 150 && \text{Substitute } 3w - 5 \text{ for } l \text{ in Equation 1.} \\ 6w - 10 + 2w &= 150 && \text{Remove parentheses.} \\ 8w - 10 &= 150 && \text{Combine like terms.} \\ 8w &= 160 && \text{Add 10 to both sides.} \\ w &= 20 && \text{Divide both sides by 8.} \end{aligned}$$

The width of the garden is 20 feet. To find the length, we substitute 20 for w in Equation 2 and simplify.

$$\begin{aligned} l &= 3w - 5 \\ &= 3(20) - 5 && \text{Substitute 20 for } w. \\ &= 60 - 5 \\ &= 55 \end{aligned}$$

Since the dimensions of the rectangle are 55 feet by 20 feet, and the area of a rectangle is given by the formula

$$A = l \cdot w \qquad \text{Area = length times width.}$$

we have

$$\begin{aligned} A &= 55 \cdot 20 \\ &= 1{,}100 \end{aligned}$$

State the conclusion The garden covers an area of 1,100 square feet.

Check the result Because the dimensions of the garden are 55 feet by 20 feet, the perimeter is

$$\begin{aligned} P &= 2l + 2w \\ &= 2(55) + 2(20) && \text{Substitute for } l \text{ and } w. \\ &= 110 + 40 \\ &= 150 \end{aligned}$$

It is also true that 55 feet is 5 feet less than 3 times 20 feet. The answers check. ∎

EXAMPLE 4 **Manufacturing** The setup cost of a machine that mills brass plates is \$750. After setup, it costs \$0.25 to mill each plate. Management is considering the purchase of a larger machine that can produce the same plate at a cost of \$0.20 per plate. If the setup cost of the larger machine is \$1,200, how many plates would the company have to produce to make the purchase worthwhile?

Analyze the problem We begin by finding the number of plates (called the **break point**) that will cost equal amounts to produce on either machine.

Form two equations We can let c represent the cost of milling p plates. If we call the machine currently being used machine 1, and the new one machine 2, we can form the two equations

The cost of making p plates on machine 1	equals	the startup cost of machine 1	plus	the cost per plate on machine 1	times	the number of plates p to be made.
c	$=$	750	$+$	0.25	$\cdot$	p

The cost of making p plates on machine 2	equals	the startup cost of machine 2	plus	the cost per plate on machine 2	times	the number of plates p to be made.
c	$=$	1,200	$+$	0.20	$\cdot$	p

Solve the system Since the costs at the break point are equal, we can use the substitution method to solve the system

$$\begin{cases} c = 750 + 0.25p \\ c = 1{,}200 + 0.20p \end{cases}$$

$$\begin{aligned} 750 + 0.25p &= 1{,}200 + 0.20p && \text{Substitute } 750 + 0.25p \text{ for } c \text{ in the second equation.} \\ 0.25p &= 450 + 0.20p && \text{Subtract 750 from both sides.} \\ 0.05p &= 450 && \text{Subtract } 0.20p \text{ from both sides.} \\ p &= 9{,}000 && \text{Divide both sides by 0.05.} \end{aligned}$$

State the conclusion If 9,000 plates are milled, the cost will be the same on either machine. If more than 9,000 plates are milled, the cost will be cheaper on the newer machine, because it mills the plates less expensively than the smaller machine.

Check the solution Figure 3-29 verifies that the break point is 9,000 plates. It also interprets the solution graphically.

Current machine
$c = 750 + 0.25p$

p	c
0	750
1,000	1,000
5,000	2,000

New, larger machine
$c = 1{,}200 + 0.20p$

p	c
0	1,200
4,000	2,000
12,000	3,600

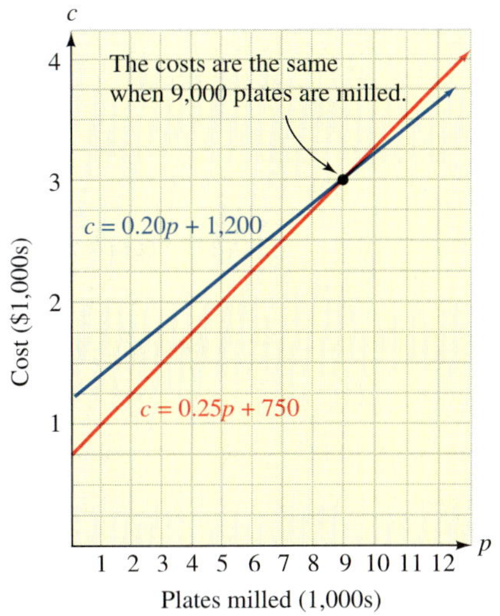

Figure 3-29

EXAMPLE 5 **Investing** Terri and Juan earned \$1,150 from a one-year investment of \$15,000. If Terri invested some of the money at 8% interest and Juan invested the rest at 7%, how much did each invest?

Analyze the problem We are told that Terri invested an unknown part of the \$15,000 at 8% and Juan invested the rest at 7%. Together, these investments earned \$1,150.

Form two equations We can let x represent the amount invested by Terri and y represent the amount of money invested by Juan. Because the total investment is \$15,000, we have

The amount invested by Terri	plus	the amount invested by Juan	equals	\$15,000.
x	$+$	y	$=$	15,000

Since the income on x dollars invested at 8% is $0.08x$, the income on y dollars invested at 7% is $0.07y$, and the combined income is \$1,150, we have

The income on the 8% investment	plus	the income on the 7% investment	equals	\$1,150.
$0.08x$	$+$	$0.07y$	$=$	1,150

Thus, we have the system

$$\begin{cases} x + y = 15{,}000 & (1) \\ 0.08x + 0.07y = 1{,}150 & (2) \end{cases}$$

Solve the system To solve the system, we use the addition method.

$$\begin{aligned} -8x - 8y &= -120{,}000 && \text{Multiply both sides of Equation 1 by } -8. \\ 8x + 7y &= 115{,}000 && \text{Multiply both sides of Equation 2 by 100.} \\ \hline -y &= -5{,}000 && \text{Add the equations together.} \\ y &= 5{,}000 && \text{Multiply both sides by } -1. \end{aligned}$$

To find x, we substitute 5,000 for y in Equation 1 and simplify.

$$\begin{aligned} x + y &= 15{,}000 \\ x + 5{,}000 &= 15{,}000 && \text{Substitute 5,000 for } y. \\ x &= 10{,}000 && \text{Subtract 5,000 from both sides.} \end{aligned}$$

State the conclusion Terri invested \$10,000, and Juan invested \$5,000.

Check the result

$$\begin{aligned} \$10{,}000 + \$5{,}000 &= \$15{,}000 && \text{The two investments total \$15,000.} \\ 0.08(\$10{,}000) &= \$800 && \text{Terri earned \$800.} \\ 0.07(\$5{,}000) &= \$350 && \text{Juan earned \$350.} \end{aligned}$$

The combined interest is $\$800 + \$350 = \$1{,}150$. The answers check. ■

EXAMPLE 6 **Boating** A boat traveled 30 kilometers downstream in 3 hours and made the return trip in 5 hours. Find the speed of the boat in still water.

Analyze the problem Traveling downstream, the speed of the boat will be faster than it would be in still water. Traveling upstream, the speed of the boat will be less than it would be in still water.

Form two equations We can let s represent the speed of the boat in still water and let c represent the speed of the current. Then the rate of speed of the boat while going downstream is $s + c$.

The rate of the boat while going upstream is $s - c$. We can organize the information of the problem as in Figure 3-30.

	Distance	=	Rate	·	Time
Downstream	30		$s + c$		3
Upstream	30		$s - c$		5

Figure 3-30

Because $d = r \cdot t$, the information in the table gives two equations in two variables.

$$\begin{cases} 30 = 3(s + c) \\ 30 = 5(s - c) \end{cases}$$

After removing parentheses and rearranging terms, we have

$$\begin{cases} 3s + 3c = 30 & (1) \\ 5s - 5c = 30 & (2) \end{cases}$$

Solve the system To solve this system by addition, we multiply Equation 1 by 5, multiply Equation 2 by 3, add the equations, and solve for s.

$$\begin{aligned} 15s + 15c &= 150 \\ \underline{15s - 15c} &\underline{= \ \ 90} \\ 30s \qquad &= 240 \\ s &= 8 \quad \text{Divide both sides by 30.} \end{aligned}$$

State the conclusion The speed of the boat in still water is 8 kilometers per hour.

Check the result We leave the check to the reader. ■

EXAMPLE 7 **Medical technology** A laboratory technician has one batch of antiseptic that is 40% alcohol and a second batch that is 60% alcohol. She would like to make 8 liters of solution that is 55% alcohol. How many liters of each batch should she use?

Analyze the problem Some 60% alcohol solution must be added to some 40% alcohol solution to make a 55% alcohol solution.

Form two equations We can let x represent the number of liters to be used from batch 1, let y represent the number of liters to be used from batch 2, and organize the information of the problem as in Figure 3-31.

	Fractional part that is alcohol	·	Number of liters of solution	=	Number of liters of alcohol
Batch 1	0.40		x		$0.40x$
Batch 2	0.60		y		$0.60y$
Mixture	0.55		8		$0.55(8)$

Figure 3-31

The information in Figure 3-31 provides two equations.

(1) $x + y = 8$ The number of liters of batch 1 plus the number of liters of batch 2 equals the total number of liters in the mixture.

(2) $0.40x + 0.60y = 0.55(8)$ The amount of alcohol in batch 1 plus the amount of alcohol in batch 2 equals the amount of alcohol in the mixture.

Solve the system We can use addition to solve this system.

$$\begin{aligned} -40x - 40y &= -320 && \text{Multiply both sides of Equation 1 by } -40.\\ 40x + 60y &= 440 && \text{Multiply both sides of Equation 2 by 100.}\\ \hline 20y &= 120 \\ y &= 6 && \text{Divide both sides by 20.} \end{aligned}$$

To find x, we substitute 6 for y in Equation 1 and simplify:

$$\begin{aligned} x + y &= 8 \\ x + 6 &= 8 && \text{Substitute 6 for } y. \\ x &= 2 && \text{Subtract 6 from both sides.} \end{aligned}$$

State the conclusion The technician should use 2 liters of the 40% solution and 6 liters of the 60% solution.

Check the result The check is left to the reader.

Orals *If x and y are integers, express each quantity.*

1. Twice x
2. One more than y
3. The sum of twice x and three times y.

If a book costs \$x and a calculator costs \$y, find

4. The cost of 3 books and 2 calculators
5. The cost of 4 books and 5 calculators

3.6 EXERCISES

REVIEW *Graph each inequality.*

1. $x < 4$
2. $x \geq -3$
3. $-1 < x \leq 2$
4. $-2 \leq x \leq 0$

Write each product using exponents.

5. $8 \cdot 8 \cdot 8 \cdot c$
6. $4(\pi)(r)(r)$
7. $a \cdot a \cdot b \cdot b$
8. $(-2)(-2)$

VOCABULARY AND CONCEPTS *Fill in the blanks.*

9. A ______ is a letter that stands for a number.
10. An ______ is a statement indicating that two quantities are equal.
11. $\begin{cases} a + b = 20 \\ a = 2b + 4 \end{cases}$ is a ______ of linear equations.
12. A ______ of a system of two linear equations satisfies both equations simultaneously.

PRACTICE *Use two equations in two variables to solve each problem.*

13. **Integer problem** One integer is twice another, and their sum is 96. Find the integers.
14. **Integer problem** The sum of two integers is 38, and their difference is 12. Find the integers.

15. Integer problem Three times one integer plus another integer is 29. If the first integer plus twice the second is 18, find the integers.

16. Integer problem Twice one integer plus another integer is 21. If the first integer plus 3 times the second is 33, find the integers.

APPLICATIONS *Use two equations in two variables to solve each problem.*

17. Raising livestock A rancher raises five times as many cows as horses. If he has 168 animals, how many cows does he have?

18. Grass seed mixture A landscaper used 100 pounds of grass seed containing twice as much bluegrass as rye. He added 15 more pounds of bluegrass to the mixture before seeding a lawn. How many pounds of bluegrass did he use?

19. Buying painting supplies Two partial receipts for paint supplies appear in the illustration. How much did each gallon of paint and each brush cost?

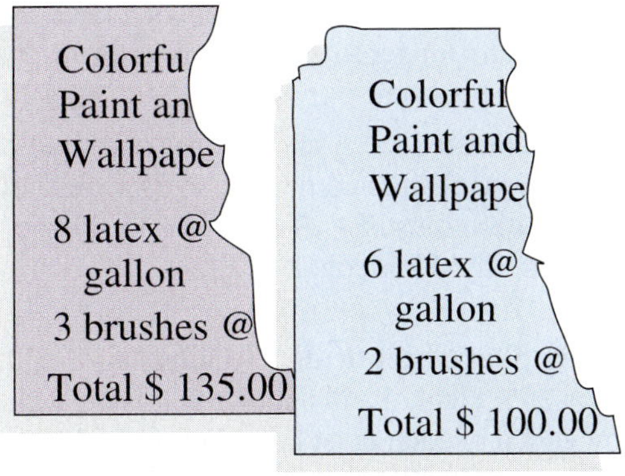

20. Buying baseball equipment One catcher's mitt and ten outfielder's gloves cost $239.50. How much does each cost if one catcher's mitt and five outfielder's gloves cost $134.50?

21. Buying contact lens cleaner Two bottles of contact lens cleaner and three bottles of soaking solution cost $29.40, and three bottles of cleaner and two bottles of soaking solution cost $28.60. Find the cost of each.

22. Buying clothes Two pairs of shoes and four pairs of socks cost $109, and three pairs of shoes and five pairs of socks cost $160. Find the cost of a pair of socks.

23. Cutting pipe A plumber wants to cut the pipe shown in the illustration into two pieces so that one piece is 5 feet longer than the other. How long should each piece be?

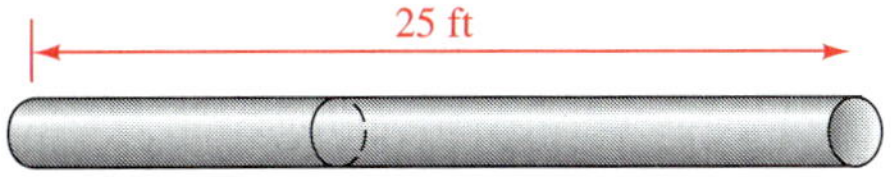

24. Cutting lumber A carpenter wants to cut a 20-foot board into two pieces so that one piece is 4 times as long as the other. How long should each piece be?

25. Splitting the lottery Chayla and Lena pool their resources to buy several lottery tickets. They win $250,000! They agree that Lena should get $50,000 more than Chayla, because she gave most of the money. How much will Chayla get?

26. Figuring inheritances In his will, a man left his older son $10,000 more than twice as much as he left his younger son. If the estate is worth $497,500, how much did the younger son get?

27. Television programming The producer of a 30-minute documentary about World War I divided it into two parts. Four times as much program time was devoted to the causes of the war as to the outcome. How long was each part of the documentary?

28. Government The salaries of the President and Vice President of the United States total $576,300 a year. If the President makes $203,700 more than the Vice President, find each of their salaries.

29. Causes of death In 1993, the number of Americans dying from cancer was six times the number that died from accidents. If the number of deaths from these two causes totaled 630,000, how many Americans died from each cause?

30. At the movies At an IMAX theater, the giant rectangular movie screen has a width 26 feet less than its length. If its perimeter is 332 feet, find the area of the screen.

31. Geometry The perimeter of the rectangle shown in the illustration is 110 feet. Find its dimensions.

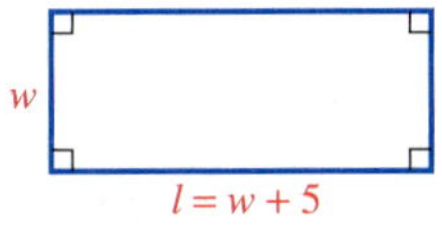

32. Geometry A rectangle is 3 times as long as it is wide, and its perimeter is 80 centimeters. Find its dimensions.

33. Geometry The length of a rectangle is 2 feet more than twice its width. If its perimeter is 34 feet, find its area.

34. Geometry A 50-meter path surrounds a rectangular garden. The width of the garden is two-thirds its length. Find its area.

35. Choosing a furnace A high-efficiency 90+ furnace costs \$2,250 and costs an average of \$412 per year to operate in Rockford, IL. An 80+ furnace costs only \$1,715 but costs \$466 per year to operate. Find the break point.

36. Making tires A company has two molds to form tires. One mold has a setup cost of \$600 and the other a setup cost of \$1,100. The cost to make each tire on the first machine is \$15, and the cost per tire on the second machine is \$13. Find the break point.

37. Choosing a furnace See Exercise 35. If you intended to live in a house for seven years, which furnace would you choose?

38. Making tires See Exercise 36. If you planned a production run of 500 tires, which mold would you use?

39. Investing money Bill invested some money at 5% annual interest, and Janette invested some at 7%. If their combined interest was \$310 on a total investment of \$5,000, how much did Bill invest?

40. Investing money Peter invested some money at 6% annual interest, and Martha invested some at 12%. If their combined investment was \$6,000 and their combined interest was \$540, how much money did Martha invest?

41. Buying tickets Students can buy tickets to a basketball game for \$1. The admission for nonstudents is \$2. If 350 tickets are sold and the total receipts are \$450, how many student tickets are sold?

42. Buying tickets If receipts for the movie advertised in the illustration were \$720 for an audience of 190 people, how many senior citizens attended?

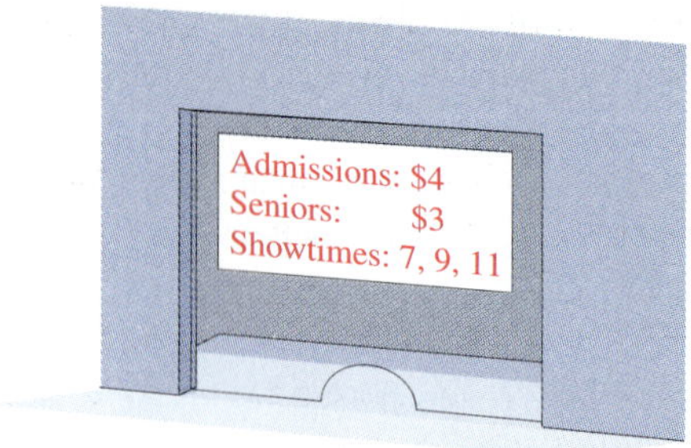

43. Boating A boat can travel 24 miles downstream in 2 hours and can make the return trip in 3 hours. Find the speed of the boat in still water.

44. Aviation With the wind, a plane can fly 3,000 miles in 5 hours. Against the same wind, the trip takes 6 hours. Find the airspeed of the plane (the speed in still air).

45. Aviation An airplane can fly downwind a distance of 600 miles in 2 hours. However, the return trip against the same wind takes 3 hours. Find the speed of the wind.

46. Finding the speed of a current It takes a motorboat 4 hours to travel 56 miles down a river, and it takes 3 hours longer to make the return trip. Find the speed of the current.

47. Mixing chemicals A chemist has one solution that is 40% alcohol and another that is 55% alcohol. How much of each must she use to make 15 liters of a solution that is 50% alcohol?

48. Mixing pharmaceuticals A nurse has a solution that is 25% alcohol and another that is 50% alcohol. How much of each must he use to make 20 liters of a solution that is 40% alcohol?

49. Mixing nuts A merchant wants to mix the peanuts with the cashews shown in the illustration to get 48 pounds of mixed nuts to sell at \$4 per pound. How many pounds of each should the merchant use?

50. **Mixing peanuts and candy** A merchant wants to mix peanuts worth \$3 per pound with jelly beans worth \$1.50 per pound to make 30 pounds of a mixture worth \$2.10 per pound. How many pounds of each should he use?

51. **Selling radios** An electronics store put two types of car radios on sale. One model sold for \$87, and the other sold for \$119. During the sale, the receipts for the 25 radios sold were \$2,495. How many of the less-expensive radios were sold?

52. **Selling ice cream** At a store, ice cream cones cost \$0.90 and sundaes cost \$1.65. One day, the receipts for a total of 148 cones and sundaes were \$180.45. How many cones were sold?

53. **Investing money** An investment of \$950 at one rate of interest and \$1,200 at a higher rate together generate an annual income of \$205.50. If the investment rates differ by 1%, find the lower rate. (*Hint:* Treat 1% as 0.01.)

54. **Selecting radio frequencies** In a radio, an inductor and a capacitor are used in a resonant circuit to select a wanted radio station at a frequency f and reject all others. The inductance L and the capacitance C determine the inductive reactance X_L and the capacitive reactance X_C of that circuit, where

$$X_L = 2\pi fL \qquad \text{and} \qquad X_C = \frac{1}{2\pi fC}$$

The radio station selected will be at the frequency f where $X_L = X_C$. Write the formula for f^2 in terms of L and C.

55. **Equilibrium price** The number of canoes sold at a marina depends on price. As the price gets higher, fewer canoes will be sold. The equation that relates the price of a canoe to the number sold is called a **demand equation.** Suppose that the demand equation for canoes is

$$p = -\frac{1}{2}q + 1,300$$

where p is the price and q is the number sold at that price.

The number of canoes produced also depends on price. As the price gets higher, more canoes will be manufactured. The equation that relates the number of canoes produced to the price is called a **supply equation.** Suppose that the supply equation for canoes is

$$p = \frac{1}{3}q + \frac{1,400}{3}$$

where p is the price and q is the number produced at that price. The **equilibrium price** is the price at which supply equals demand. Find the equilibrium price.

WRITING

56. Which problem in the preceding set did you find the hardest? Why?

57. Which problem in the preceding set did you find the easiest? Why?

SOMETHING TO THINK ABOUT

58. How many nails will balance one nut?

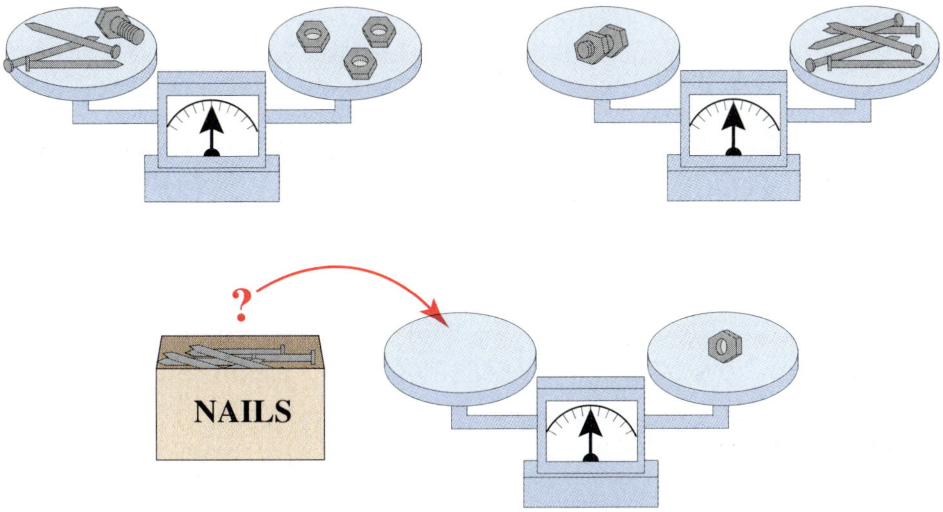

3.7 Systems of Linear Inequalities

- Solutions of Linear Inequalities
- Graphing Linear Inequalities
- An Application of Linear Inequalities
- Solving Systems of Linear Inequalities
- An Application of Systems of Linear Inequalities

Getting Ready *Graph* $y = \frac{1}{3}x + 3$ *and tell whether the given point lies on the line, above the line, or below the line.*

1. (0, 0) **2.** (0, 4) **3.** (2, 2) **4.** (6, 5)
5. (−3, 2) **6.** (6, 8) **7.** (−6, 0) **8.** (−9, 5)

We now discuss how to solve linear inequalities in two variables graphically. Then, we will show how to solve systems of inequalities.

Solutions of Linear Inequalities

A **linear inequality** in x and y is an inequality that can be written in one of the following forms:

$$Ax + By > C \qquad Ax + By < C \qquad Ax + By \geq C \qquad Ax + By \leq C$$

where A, B, and C are real numbers and A and B are not both 0. Some examples of linear inequalities are

$$2x - y > -3 \qquad y < 3 \qquad x + 4y \geq 6 \qquad x \leq -2$$

An ordered pair (x, y) is a solution of an inequality in x and y if a true statement results when the values of x and y are substituted into the inequality.

EXAMPLE 1 Determine whether each ordered pair is a solution of $y \geq x - 5$: **a.** (4, 2) and **b.** (0, −6).

Solution **a.** To determine whether (4, 2) is a solution, we substitute 4 for x and 2 for y.

$$y \geq x - 5$$
$$2 \geq 4 - 5$$
$$2 \geq -1$$

Since $2 \geq -1$ is a true inequality, (4, 2) is a solution.

b. To determine whether (0, −6) is a solution, we substitute 0 for x and −6 for y.

$$y \geq x - 5$$
$$-6 \geq 0 - 5$$
$$-6 \geq -5$$

Since $-6 \geq -5$ is a false inequality, $(0, -6)$ is not a solution.

Self Check Use the inequality in Example 1 and determine whether each ordered pair is a solution: **a.** $(8, 2)$ and **b.** $(-4, 3)$.

Graphing Linear Inequalities

The graph of $y = x - 5$ is a line consisting of the points whose coordinates satisfy the equation. The graph of the inequality $y \geq x - 5$ is not a line but rather an area bounded by a line, called a **half-plane.** The half-plane consists of the points whose coordinates satisfy the inequality.

EXAMPLE 2 Graph the inequality: $y \geq x - 5$.

Solution We begin by graphing the equation $y = x - 5$ as in Figure 3-32(a). Because the graph of $y \geq x - 5$ also indicates that y can be greater than $x - 5$, the coordinates of points other than those shown in Figure 3-32(a) satisfy the inequality. For example, the coordinates of the origin satisfy the inequality. We can verify this by letting x and y be 0 in the given inequality:

$$y \geq x - 5$$
$$0 \geq 0 - 5 \quad \text{Substitute 0 for } x \text{ and 0 for } y.$$
$$0 \geq -5$$

Because $0 \geq -5$ is true, the coordinates of the origin satisfy the original inequality. In fact, the coordinates of every point on the same side of the line as the origin satisfy the inequality. The graph of $y \geq x - 5$ is the half-plane that is shaded in Figure 3-32(b). Since the boundary line $y = x - 5$ is included, we draw it with a solid line.

$y = x - 5$

x	y	(x, y)
0	−5	(0, −5)
5	0	(5, 0)

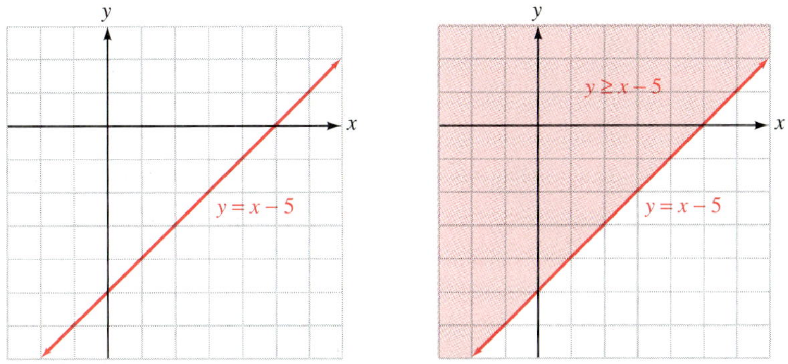

(a) (b)

Figure 3-32

Self Check Graph: $y \geq -x - 2$.

EXAMPLE 3 Graph: $x + 2y < 6$.

Solution We find the boundary by graphing the equation $x + 2y = 6$. Since the symbol $<$ does not include an $=$ sign, the points on the graph of $x + 2y = 6$ will not be a part of the graph. To show this, we draw the boundary line as a broken line. See Figure 3-33.

To determine which half-plane to shade, we substitute the coordinates of some point that lies on one side of the boundary line into $x + 2y < 6$. The origin is a convenient choice.

$$x + 2y < 6$$
$$0 + 2(0) < 6 \quad \text{Substitute 0 for } x \text{ and 0 for } y.$$
$$0 < 6$$

Since $0 < 6$ is true, we shade the side of the line that includes the origin. The graph is shown in Figure 3-33.

$x + 2y = 6$

x	y	(x, y)
0	3	(0, 3)
6	0	(6, 0)
4	1	(4, 1)

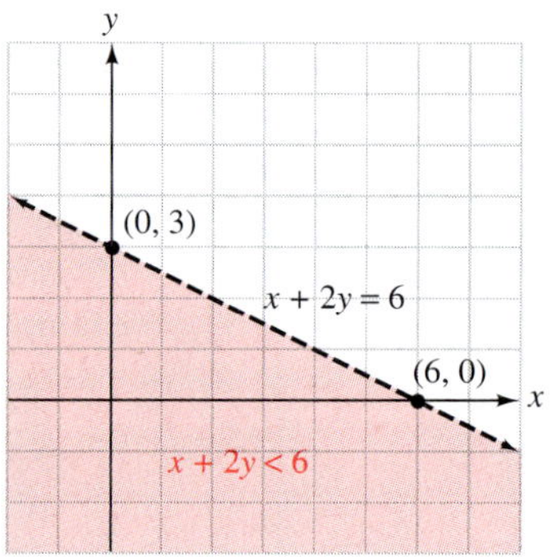

Figure 3-33

Self Check Graph: $2x - y < 4$.

EXAMPLE 4 Graph: $y > 2x$.

Solution To find the boundary line, we graph the equation $y = 2x$. Since the symbol $>$ does not include an equal sign, the points on the boundary are not a part of the graph of $y > 2x$. To show this, we draw the boundary as a broken line. See Figure 3-34(a).

To determine which half-plane to shade, we substitute the coordinates of some point that lies on one side of the boundary into $y > 2x$. Point $T(2, 0)$, for example, is below the boundary line. See Figure 3-34(a). To see if point $T(2, 0)$ satisfies $y > 2x$, we substitute 2 for x and 0 for y in the inequality.

$$y > 2x$$
$$0 > 2(2) \quad \text{Substitute 2 for } x \text{ and 0 for } y.$$
$$0 > 4$$

Since $0 > 4$ is false, the coordinates of point T do not satisfy the inequality, and point T is not on the side of the line we wish to shade. Instead, we shade the other side of the boundary line. The graph of the solution set of $y > 2x$ is shown in Figure 3-34(b).

$y = 2x$

x	y	(x, y)
0	0	(0, 0)
−1	−2	(−1, −2)
3	6	(3, 6)

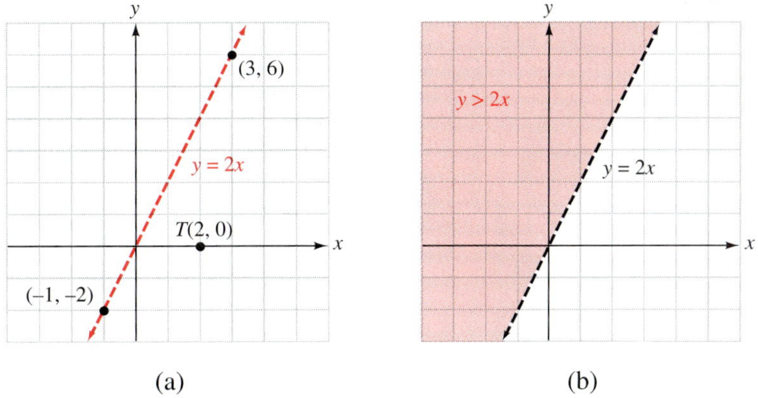

Figure 3-34

Self Check Graph: $y < 3x$.

An Application of Linear Inequalities

EXAMPLE 5 **Earning money** Chen has two part-time jobs, one paying \$5 per hour and the other paying \$6 per hour. He must earn at least \$120 per week to pay his expenses while attending college. Write an inequality that shows the various ways he can schedule his time to achieve his goal.

Solution If we let x represent the number of hours Chen works on the first job and y the number of hours he works on the second job, we have

The hourly rate on the first job	times	the hours worked on the first job	plus	the hourly rate on the second job	times	the hours worked on the second job.	is at least	\$120.
\$5	$\cdot$	x	$+$	\$6	$\cdot$	y	$\geq$	\$120

The graph of the inequality $5x + 6y \geq 120$ is shown in Figure 3-35. Any point in the shaded region indicates a possible way Chen can schedule his time and earn \$120 or more per week. For example, if he works 20 hours on the first job and 10 hours on the second job, he will earn

$$\begin{aligned} \$5(20) + \$6(10) &= \$100 + \$60 \\ &= \$160 \end{aligned}$$

Since Chen cannot work a negative number of hours, the graph in the figure has no meaning when either x or y is negative.

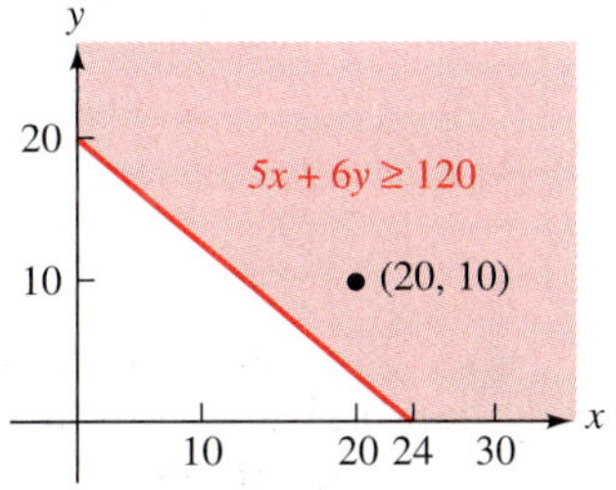

Figure 3-35

Solving Systems of Linear Inequalities

We have seen that the graph of a linear inequality in two variables is a half-plane. Therefore, we would expect the graph of a system of two linear inequalities to be two overlapping half-planes. For example, to solve the system

$$\begin{cases} x + y \geq 1 \\ x - y \geq 1 \end{cases}$$

we graph each inequality and then superimpose the graphs on one set of coordinate axes.

The graph of $x + y \geq 1$ includes the graph of the equation $x + y = 1$ and all points above it. Because the boundary line is included, we draw it with a solid line. See Figure 3-36(a).

The graph of $x - y \geq 1$ includes the graph of the equation $x - y = 1$ and all points below it. Because the boundary line is included, we draw it with a solid line. See Figure 3-36(b).

$x + y = 1$

x	y	(x, y)
0	1	(0, 1)
1	0	(1, 0)
2	−1	(2, −1)

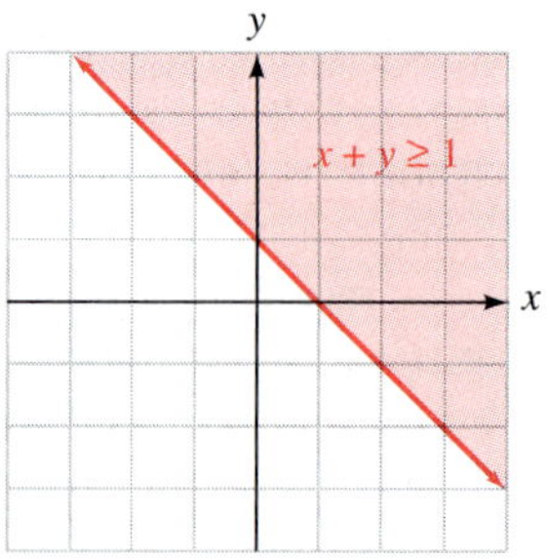

(a)

$x - y = 1$

x	y	(x, y)
0	−1	(0, −1)
1	0	(1, 0)
2	1	(2, 1)

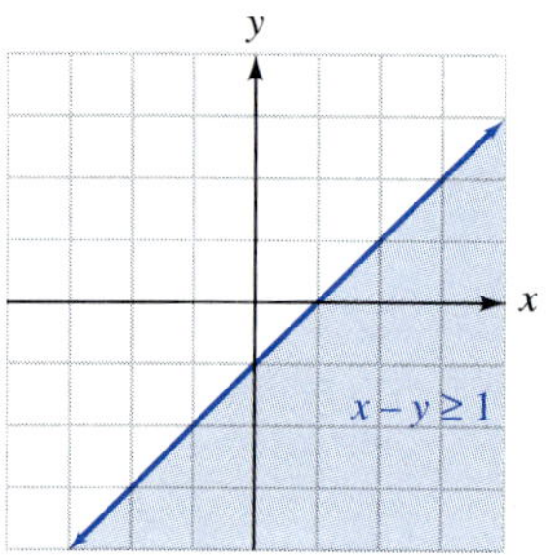

(b)

Figure 3-36

In Figure 3-37, we show the result when the graphs are superimposed on one coordinate system. The area that is shaded twice represents the set of solutions of the given system. Any point in the doubly shaded region has coordinates that satisfy both of the inequalities.

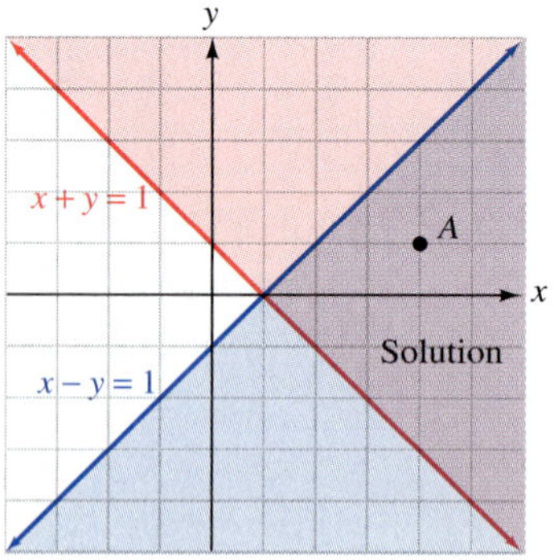

Figure 3-37

To see that this is true, we can pick a point, such as point A, that lies in the doubly shaded region and show that its coordinates satisfy both inequalities. Because point A has coordinates (4, 1), we have

$$\begin{aligned} x + y &\geq 1 \\ 4 + 1 &\geq 1 \\ 5 &\geq 1 \end{aligned} \qquad \begin{aligned} x - y &\geq 1 \\ 4 - 1 &\geq 1 \\ 3 &\geq 1 \end{aligned}$$

Since the coordinates of point A satisfy each inequality, point A is a solution. If we pick a point that is not in the doubly shaded region, its coordinates will not satisfy both of the inequalities.

In general, to solve systems of linear inequalities, we will take the following steps.

Solving Systems of Inequalities

1. Graph each inequality in the system on the same coordinate axes.
2. Find the region where the graphs overlap.
3. Pick a test point from the region to verify the solution.

EXAMPLE 6 Graph the solution set: $\begin{cases} 2x + y < 4 \\ -2x + y > 2 \end{cases}$.

Solution We graph each inequality on one set of coordinate axes, as in Figure 3-38.

- The graph of $2x + y < 4$ includes all points below the line $2x + y = 4$. Since the boundary is not included, we draw it as a broken line.
- The graph of $-2x + y > 2$ includes all points above the line $-2x + y = 2$. Since the boundary is not included, we draw it as a broken line.

The area that is shaded twice represents the set of solutions of the given system.

$2x + y = 4$

x	y	(x, y)
0	4	(0, 4)
1	2	(1, 2)
2	0	(2, 0)

$-2x + y = 2$

x	y	(x, y)
−1	0	(−1, 0)
0	2	(0, 2)
2	6	(2, 6)

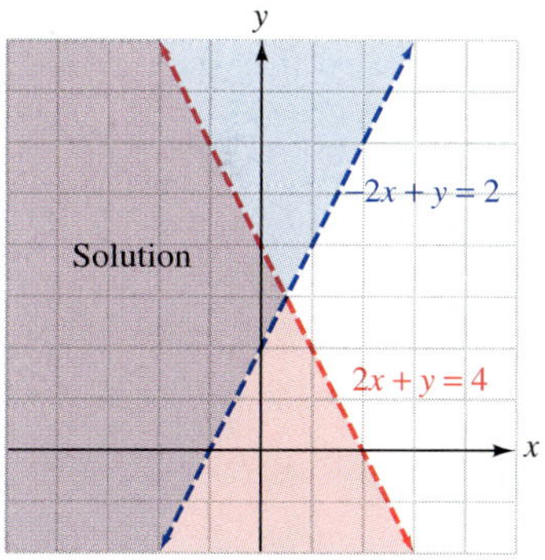

Figure 3-38

Pick a point in the doubly shaded region and show that it satisfies both inequalities.

Self Check Graph the solution set: $\begin{cases} x + 3y < 6 \\ -x + 3y < 6 \end{cases}$.

EXAMPLE 7 Graph the solution set: $\begin{cases} x \leq 2 \\ y > 3 \end{cases}$.

Solution We graph each inequality on one set of coordinate axes, as in Figure 3-39.

- The graph of $x \leq 2$ includes all points on the line $x = 2$ and all points to the left of the line. Since the boundary line is included, we draw it as a solid line.
- The graph $y > 3$ includes all points above the line $y = 3$. Since the boundary is not included, we draw it as a broken line.

The area that is shaded twice represents the set of solutions of the given system. Pick a point in the doubly shaded region and show that this is true.

$x = 2$

x	y	(x, y)
2	0	(2, 0)
2	2	(2, 2)
2	4	(2, 4)

$y = 3$

x	y	(x, y)
0	3	(0, 3)
1	3	(1, 3)
4	3	(4, 3)

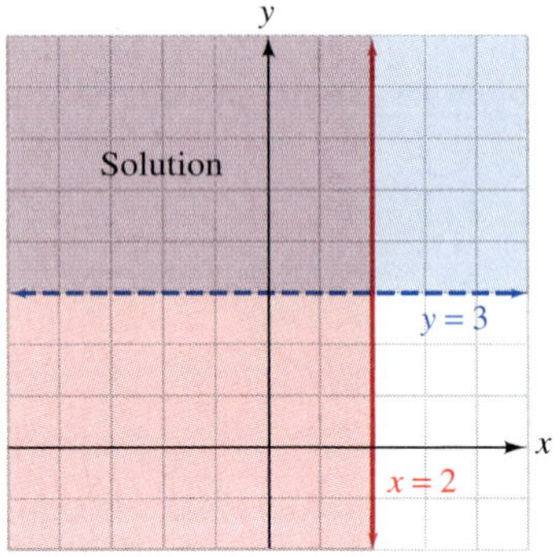

Figure 3-39

Self Check Graph the solution set: $\begin{cases} y \geq 1 \\ x > 2 \end{cases}$.

EXAMPLE 8 Graph the solution set: $\begin{cases} y < 3x - 1 \\ y \geq 3x + 1 \end{cases}$.

Solution We graph each inequality, as in Figure 3-40.

- The graph of $y < 3x - 1$ includes all of the points below the broken line $y = 3x - 1$.
- The graph of $y \geq 3x + 1$ includes all of the points on and above the solid line $y = 3x + 1$.

Since the graphs of these inequalities do not intersect, there are no solutions.

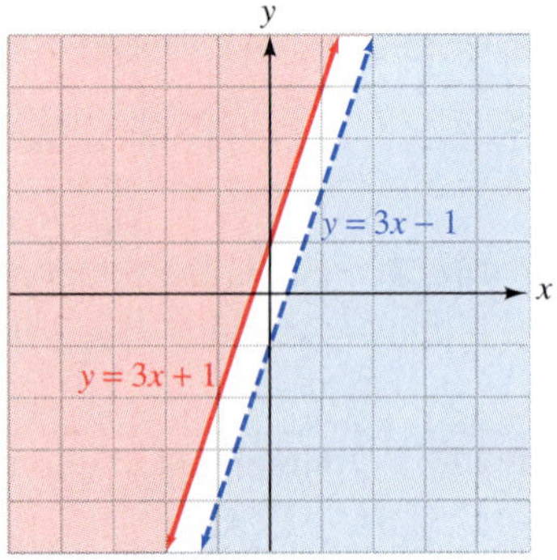

Figure 3-40

Self Check Graph the solution set: $\begin{cases} y \geq -\frac{1}{2}x + 1 \\ y \leq -\frac{1}{2}x - 1 \end{cases}$.

An Application of Systems of Linear Inequalities

EXAMPLE 9 **Landscaping** A man budgets from \$300 to \$600 for trees and bushes to landscape his yard. After shopping around, he finds that good trees cost \$150 and mature bushes cost \$75. What combinations of trees and bushes can he afford to buy?

Analyze the problem The man wants to spend *at least* \$300 but *not more than* \$600 for trees and bushes.

Form two inequalities We can let x represent the number of trees purchased and y the number of bushes purchased. We can then form the following system of inequalities.

The cost of a tree	times	the number of trees purchased	plus	the cost of a bush	times	the number of bushes purchased	should be at least	\$300.
\$150	$\cdot$	x	+	\$75	$\cdot$	y	$\geq$	\$300
The cost of a tree	times	the number of trees purchased	plus	the cost of a bush	times	the number of bushes purchased	should not be more than	\$600.
\$150	$\cdot$	x	+	\$75	$\cdot$	y	$\leq$	\$600

Solve the system We graph the system

$$\begin{cases} 150x + 75y \geq 300 \\ 150x + 75y \leq 600 \end{cases}$$

as in Figure 3-41. The coordinates of each point shown in the graph give a possible combination of the number of trees (x) and the number of bushes (y) that can be purchased. These possibilities are

(0, 4), (0, 5), (0, 6), (0, 7), (0, 8)
(1, 2), (1, 3), (1, 4), (1, 5), (1, 6)
(2, 0), (2, 1), (2, 2), (2, 3), (2, 4)
(3, 0), (3, 1), (3, 2), (4, 0)

Only these points can be used, because the man cannot buy part of a tree or part of a bush.

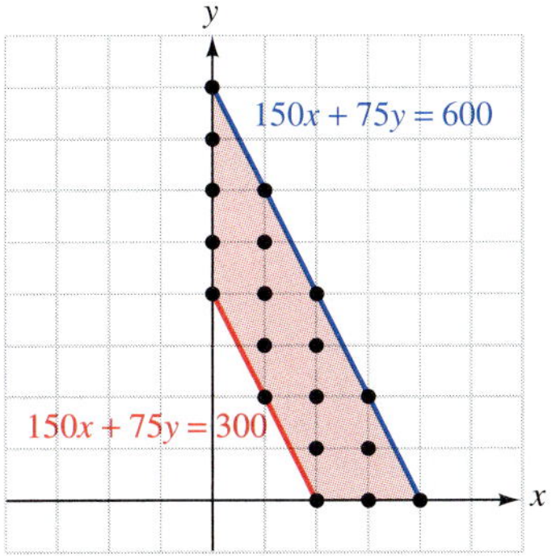

Figure 3-41

Self Check Answers

1. a. no, **b.** yes **2.**

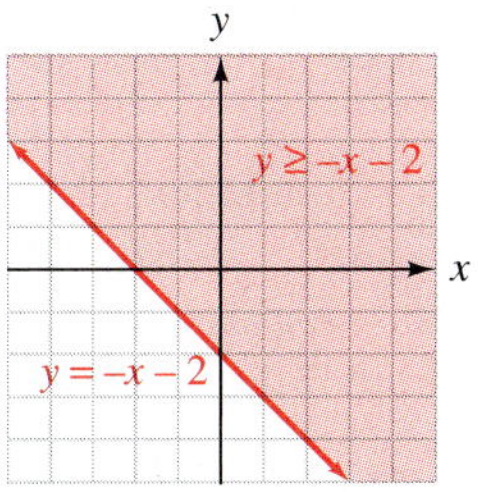

3.

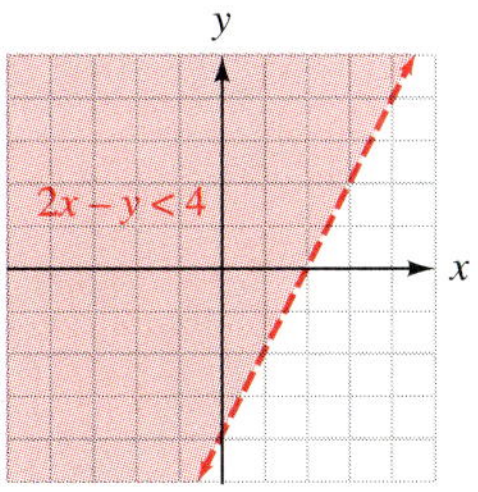

4.

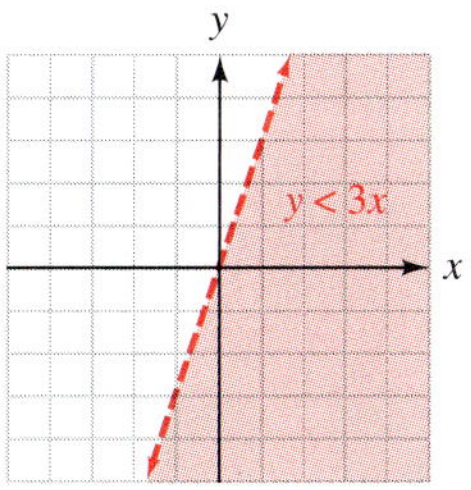

6.

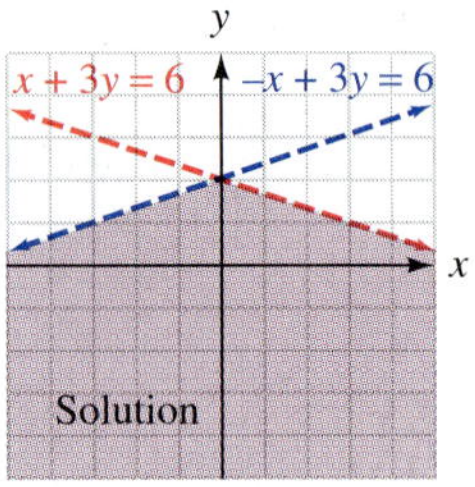

7.

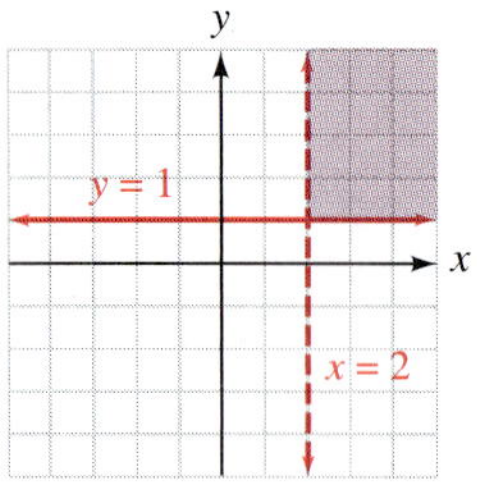

8. no solutions

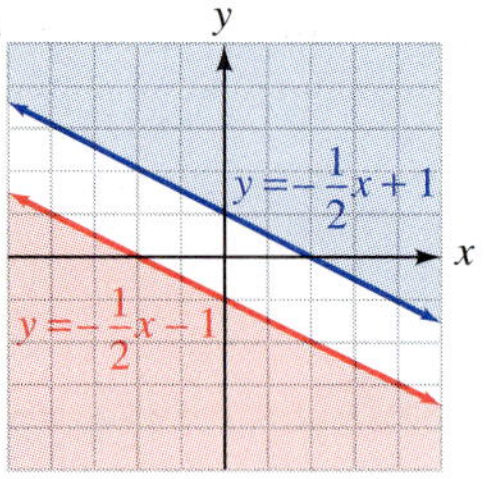

Orals *Tell whether the following coordinates satisfy* $y > 3x + 2$.

1. (0, 0) **2.** (5, 5) **3.** (−2, 4) **4.** (−3, −6)

Tell whether the following coordinates satisfy the inequality $y \le \frac{1}{2}x - 1$.

5. (0, 0) **6.** (2, 0) **7.** (4, 3) **8.** (−4, −3)

3.7 EXERCISES

REVIEW

1. Solve: $3x + 5 = 14$.

2. Solve: $2(x - 4) \le -12$.

3. Solve: $A = P + Prt$ for t.

4. Does the graph of $y = -x$ pass through the origin?

Simplify each expression.

5. $2a + 5(a - 3)$

6. $2t - 3(3 + t)$

7. $4(b - a) + 3b + 2a$

8. $3p + 2(q - p) + q$

VOCABULARY AND CONCEPTS *Fill in the blanks.*

9. $2x - y \le 4$ is a linear ________ in x and y.

10. The symbol $\le$ means ________ or _______.

11. In the accompanying graph, the line $2x - y = 4$ is the ________ of the graph $2x - y \le 4$.

12. In the accompanying graph, the line $2x - y = 4$ divides the rectangular coordinate system into two ________.

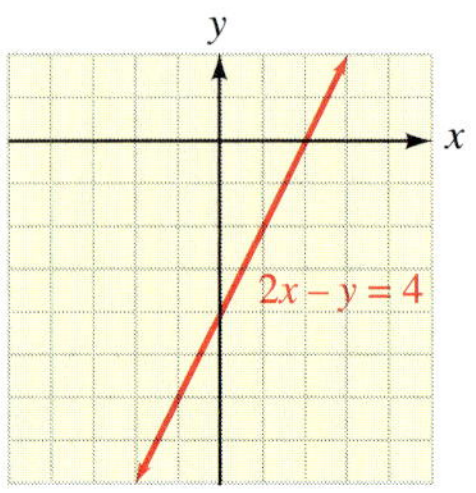

13. $\begin{cases} x + y > 2 \\ x + y < 4 \end{cases}$ is a system of linear ________.

14. The _______ of a system of linear inequalities are all the ordered pairs that make all of the inequalities of the system true at the same time.

15. Any point in the ___________ region of the graph of the solution of a system of two linear inequalities has coordinates that satisfy both inequalities of the system.

16. To graph a linear inequality such as $x + y > 2$, first graph the boundary with a dashed line. Then pick a test _____ to determine which half-plane to shade.

17. Tell whether each ordered pair is a solution of $5x - 3y \ge 0$.

a. (1, 1) **b.** (−2, −3)

c. (0, 0) **d.** $\left(\frac{1}{5}, \frac{4}{3}\right)$

18. Tell whether each ordered pair is a solution of $x + 4y < -1$.

a. (3, 1) **b.** (−2, 0)

c. (0.5, 0.2) **d.** $\left(-2, \frac{1}{4}\right)$

19. Tell whether the graph of each linear inequality includes the boundary line.

a. $y > -x$ **b.** $5x - 3y \le -2$

20. If a false statement results when the coordinates of a test point are substituted into a linear inequality, which half-plane should be shaded to represent the solution of the inequality?

PRACTICE *Complete the graph by shading the correct half-plane.*

21. $y \leq x + 2$

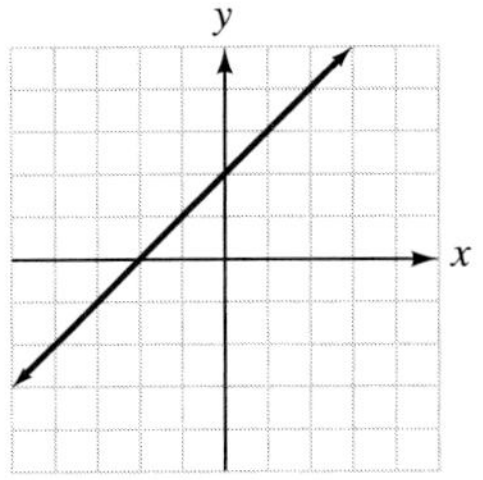

22. $y > x - 3$

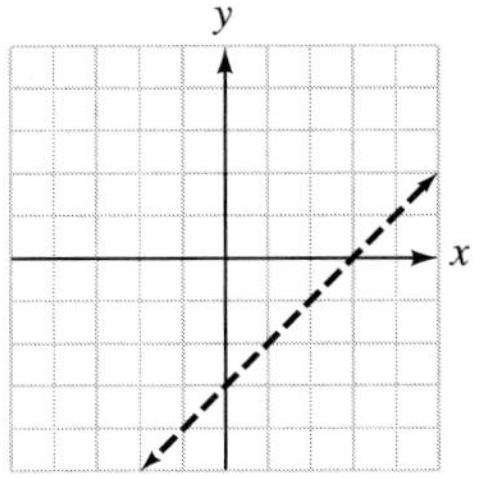

23. $y > 2x - 4$

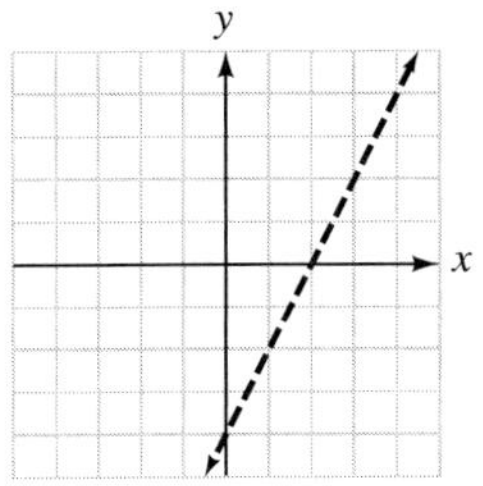

24. $y \leq -x + 1$

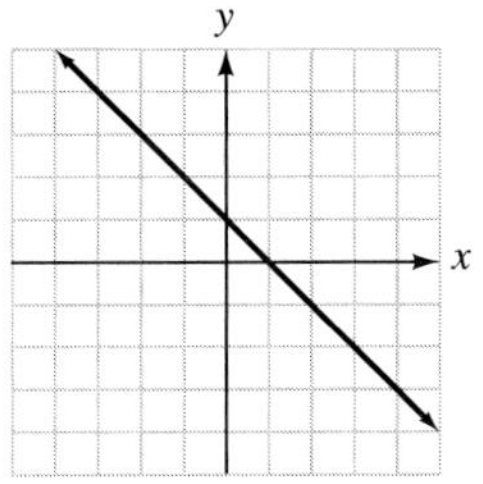

25. $x - 2y \leq 4$

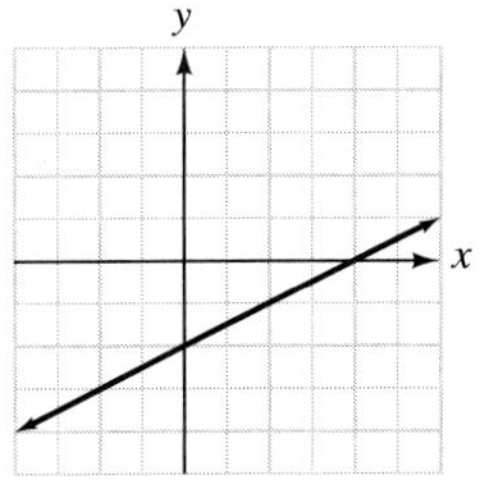

26. $3x + 2y \geq 12$

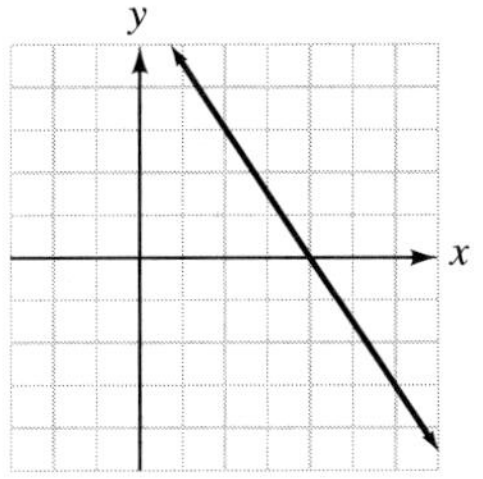

27. $y \leq 4x$

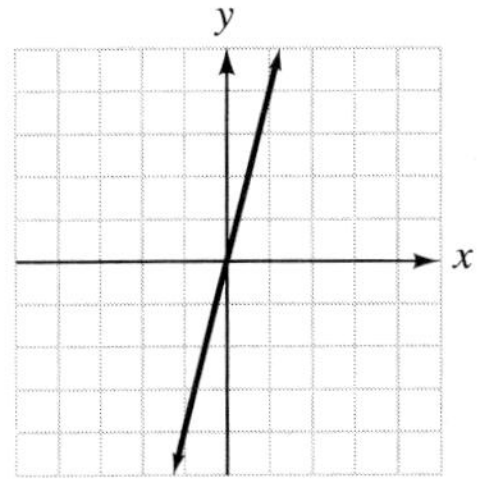

28. $y + 2x < 0$

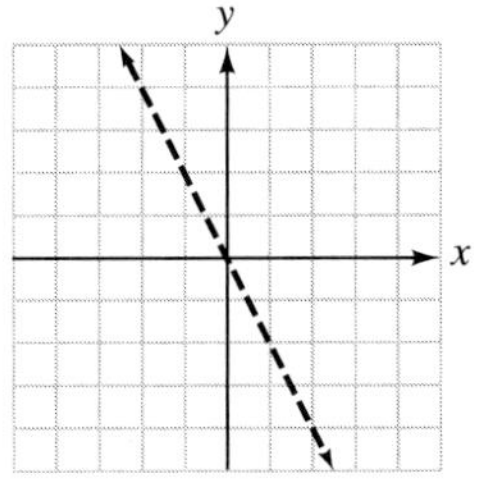

Graph each inequality.

29. $y \geq 3 - x$

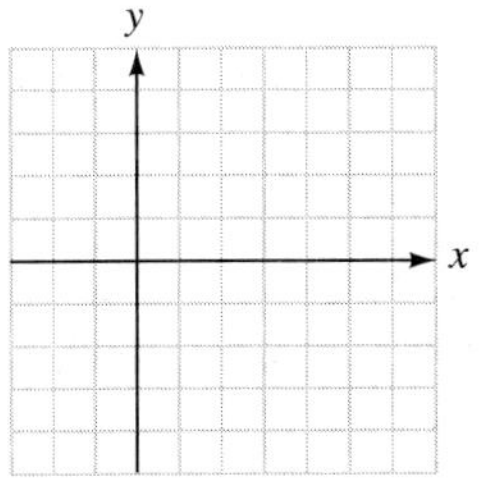

30. $y < 2 - x$

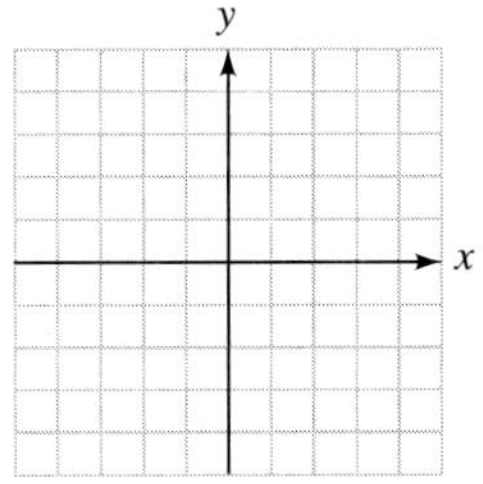

31. $y < 2 - 3x$

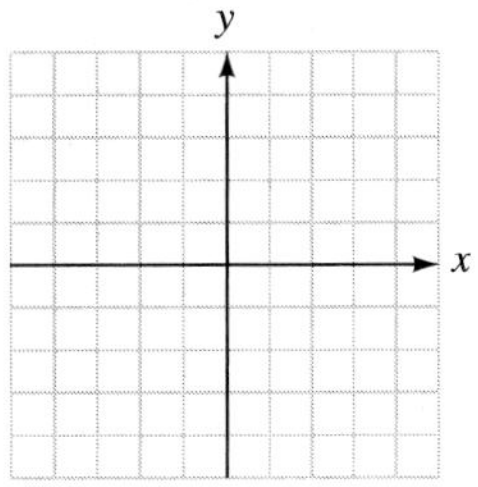

32. $y \geq 5 - 2x$

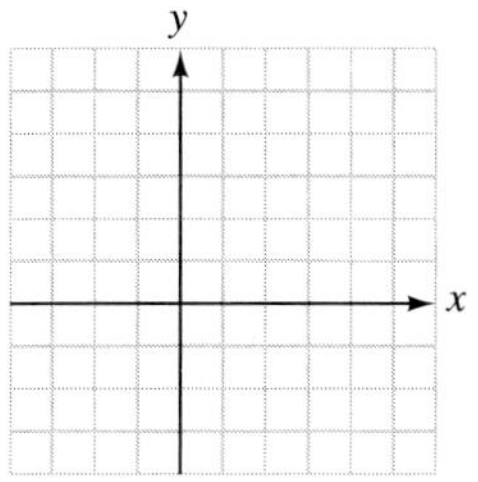

33. $y \geq 2x$

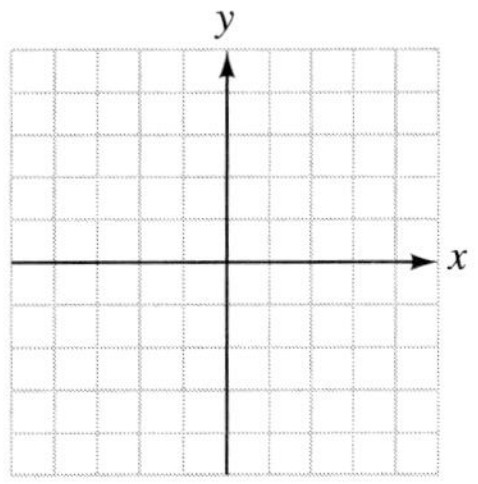

34. $y < 3x$

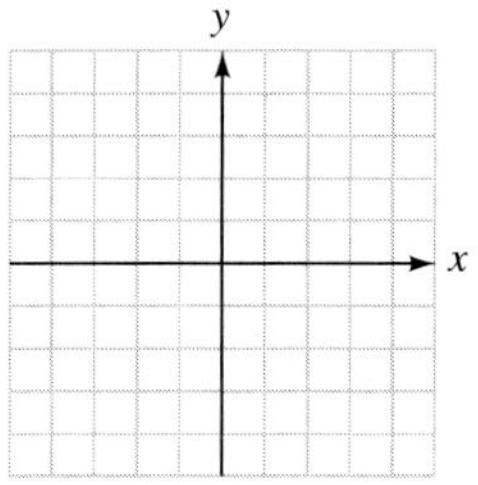

35. $2y - x < 8$

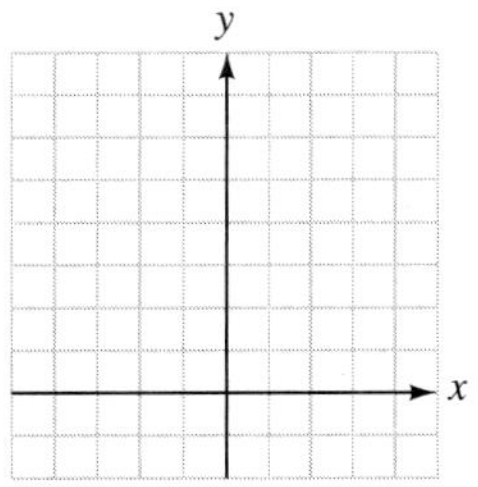

36. $y + 9x \geq 3$

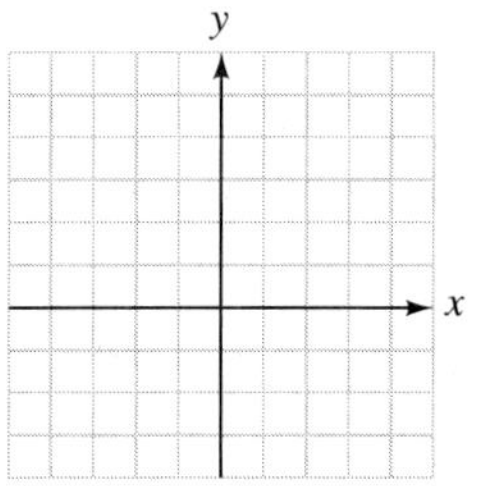

37. $3x - 4y > 12$

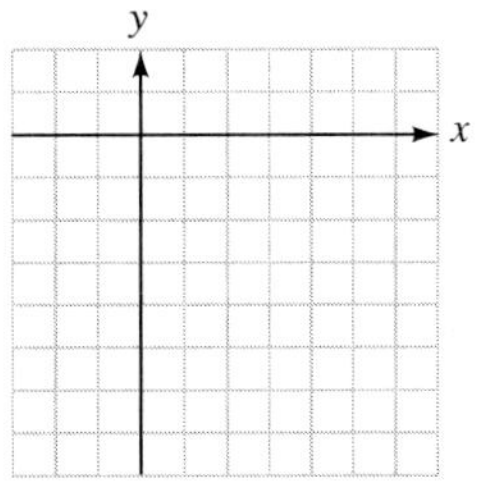

38. $4x + 3y \leq 12$

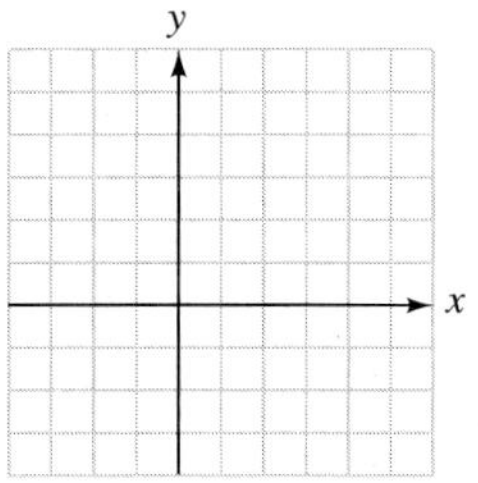

39. $5x + 4y \geq 20$

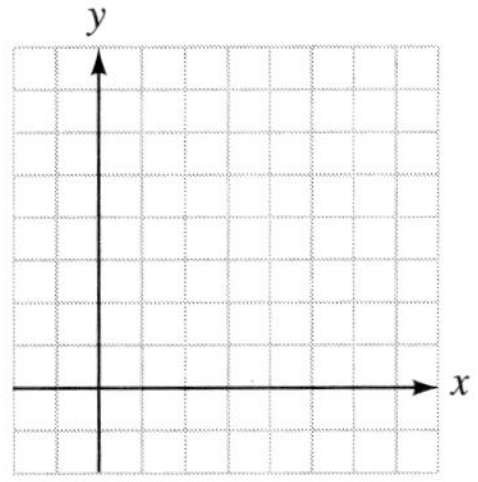

40. $7x - 2y < 21$

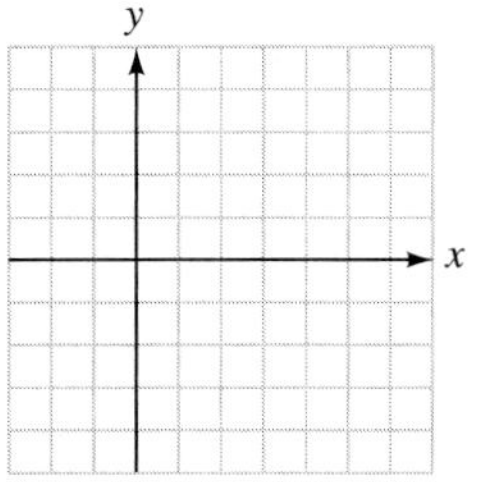

41. $x < 2$

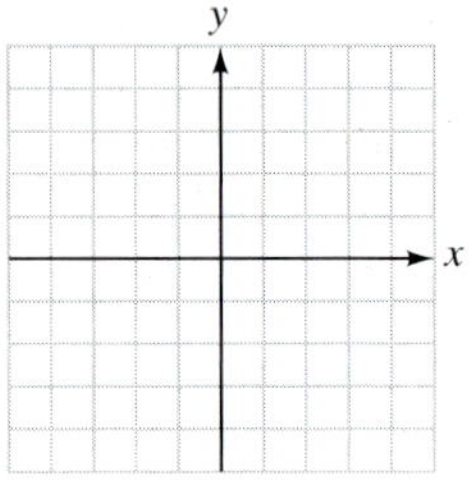

42. $y > -3$

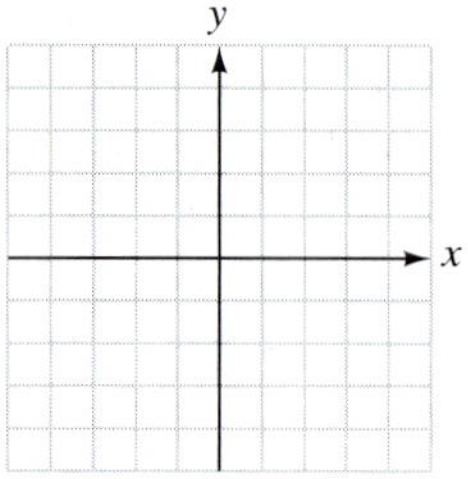

43. $y \leq 1$

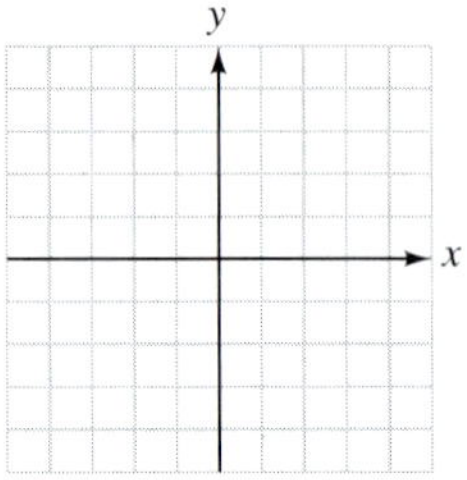

44. $x \geq -4$

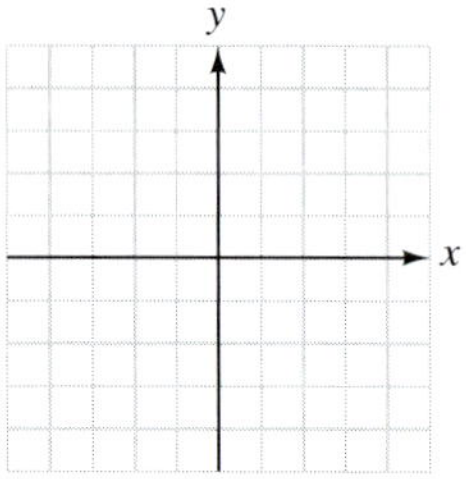

PRACTICE *Find the solution set of each system of inequalities, when possible.*

45. $\begin{cases} x + 2y \leq 3 \\ 2x - y \geq 1 \end{cases}$

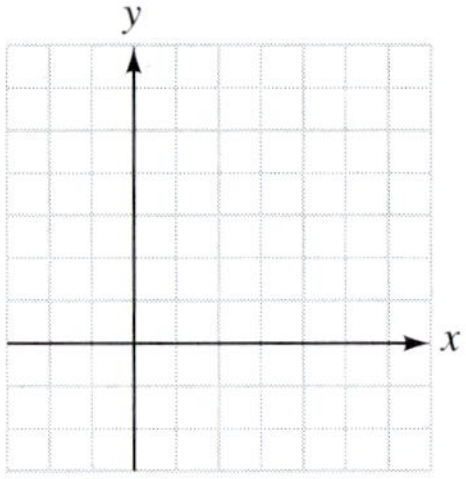

46. $\begin{cases} 2x + y \geq 3 \\ x - 2y \leq -1 \end{cases}$

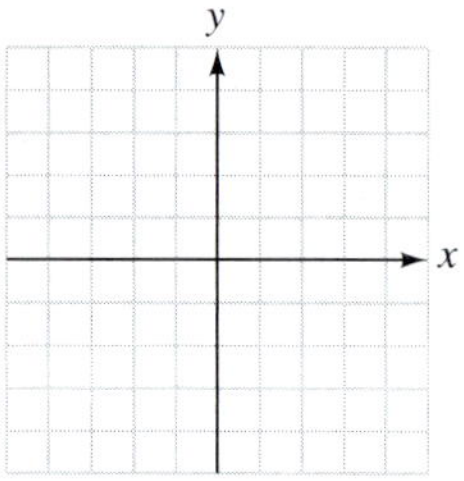

47. $\begin{cases} x + y < -1 \\ x - y > -1 \end{cases}$

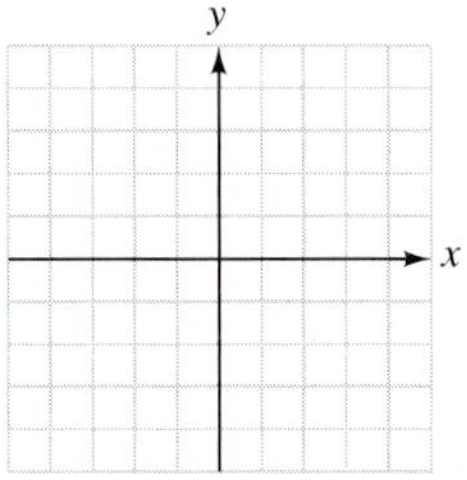

48. $\begin{cases} x + y > 2 \\ x - y < -2 \end{cases}$

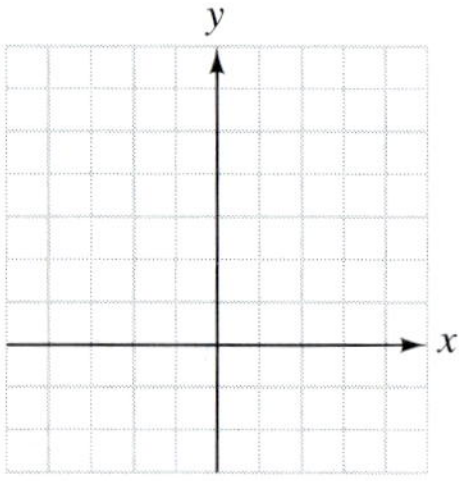

49. $\begin{cases} 2x - y < 4 \\ x + y \geq -1 \end{cases}$

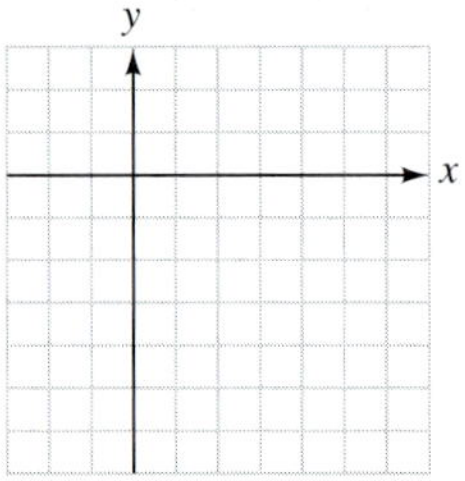

50. $\begin{cases} x - y \geq 5 \\ x + 2y < -4 \end{cases}$

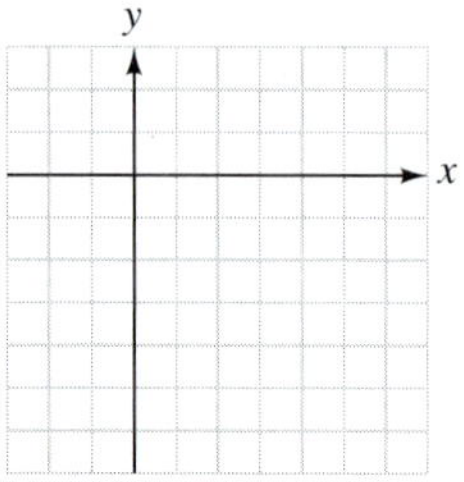

51. $\begin{cases} x > 2 \\ y \leq 3 \end{cases}$

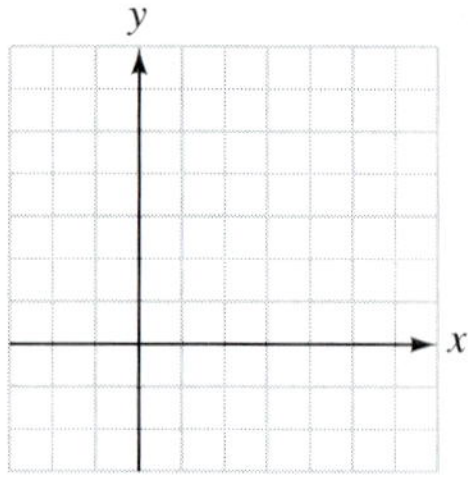

52. $\begin{cases} x \geq -1 \\ y > -2 \end{cases}$

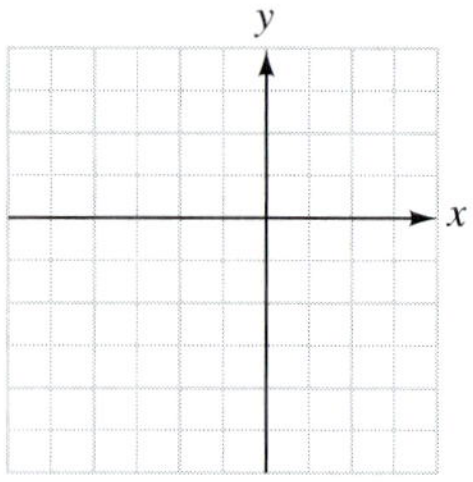

53. $\begin{cases} x + y < 1 \\ x + y > 3 \end{cases}$

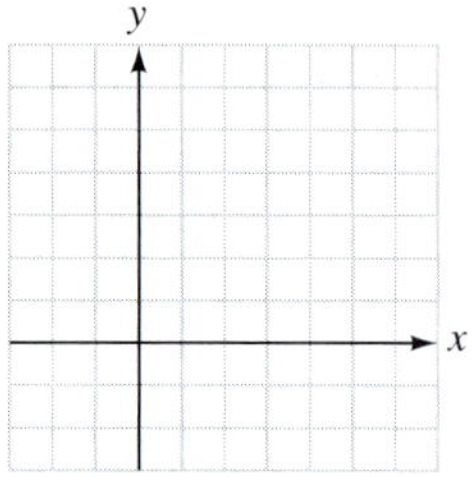

54. $\begin{cases} x \leq 0 \\ y < 0 \end{cases}$

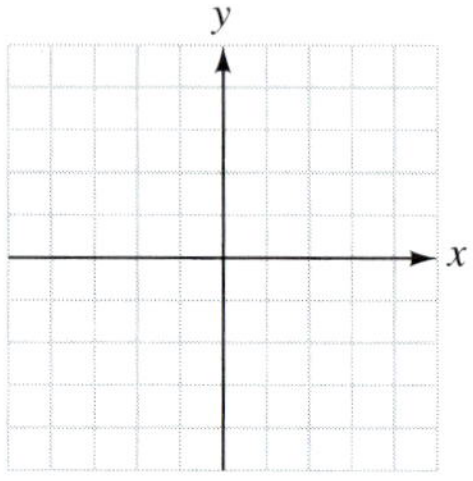

55. $\begin{cases} 3x + 4y > -7 \\ 2x - 3y \geq 1 \end{cases}$

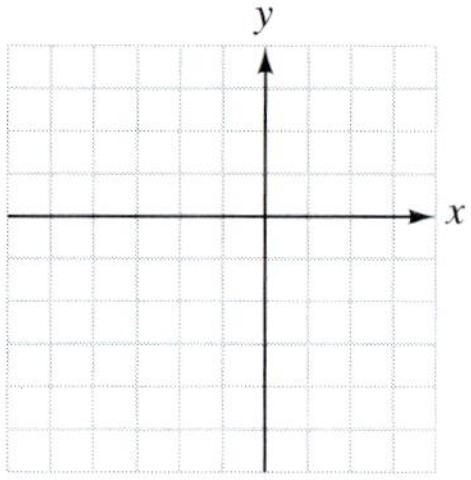

56. $\begin{cases} 3x + y \leq 1 \\ 4x - y > -8 \end{cases}$

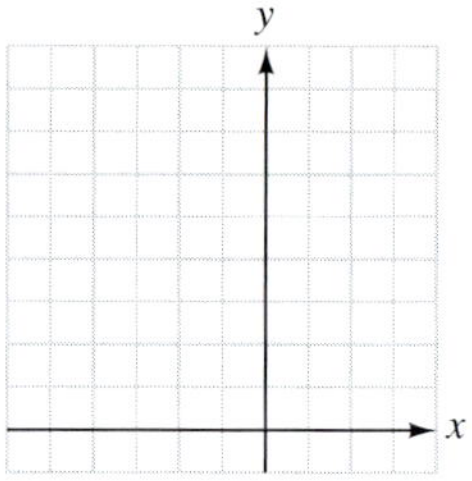

57. $\begin{cases} 2x - 4y > -6 \\ 3x + y \geq 5 \end{cases}$

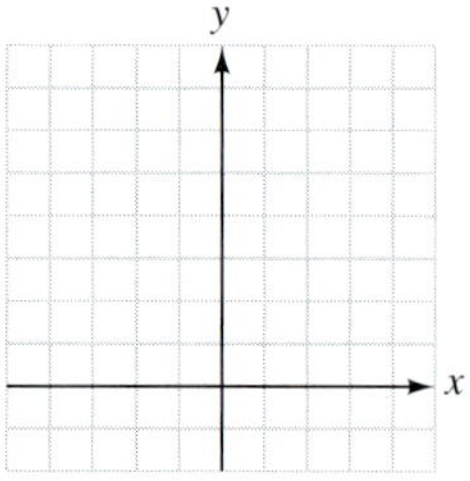

58. $\begin{cases} 2x - 3y < 0 \\ 2x + 3y \geq 12 \end{cases}$

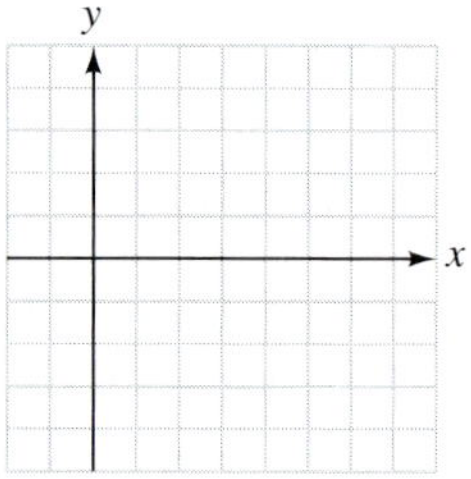

59. $\begin{cases} 3x - y \leq -4 \\ 3y > -2(x + 5) \end{cases}$

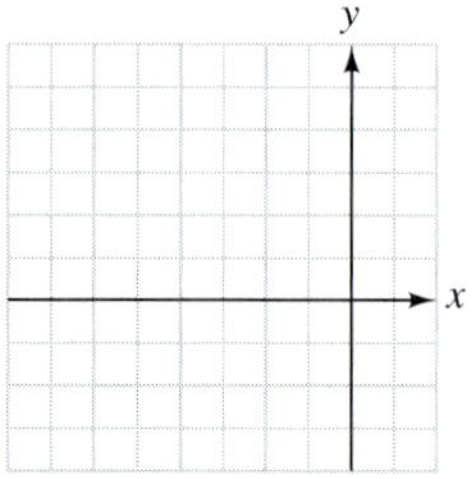

60. $\begin{cases} 3x + y < -2 \\ y > 3(1 - x) \end{cases}$

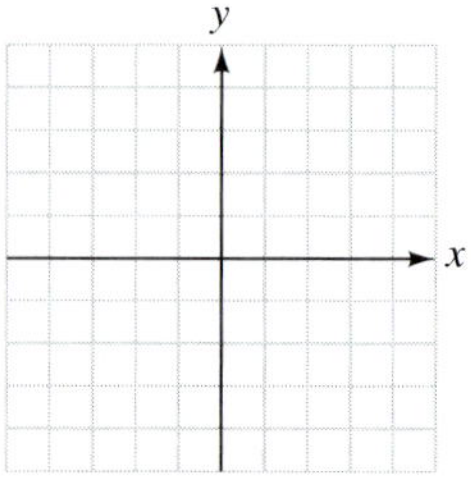

61. $\begin{cases} \dfrac{x}{2} + \dfrac{y}{3} \geq 2 \\ \dfrac{x}{2} - \dfrac{y}{2} < -1 \end{cases}$

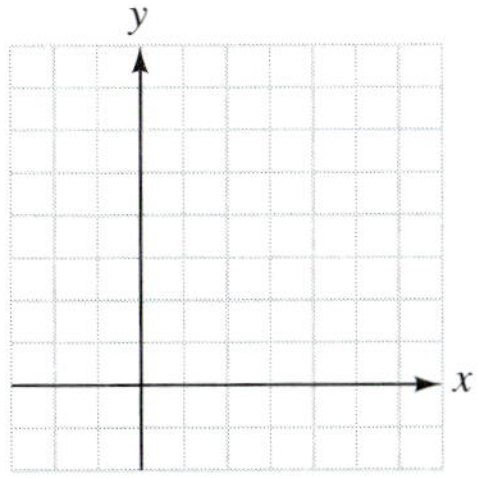

62. $\begin{cases} \dfrac{x}{3} - \dfrac{y}{2} < -3 \\ \dfrac{x}{3} + \dfrac{y}{2} > -1 \end{cases}$

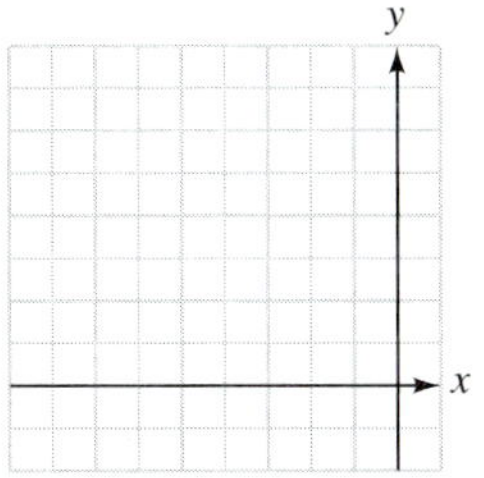

APPLICATIONS *Graph each inequality for nonnegative values of x and y. Then give some ordered pairs that satisfy the inequality.*

63. **Production planning** It costs a bakery \$3 to make a cake and \$4 to make a pie. Production costs cannot exceed \$120 per day. Find an inequality that shows the possible combinations of cakes (x) and pies (y) that can be made, and graph it in the illustration.

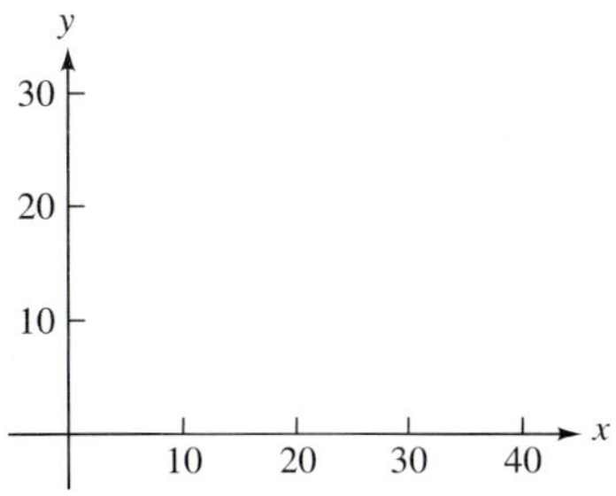

64. **Hiring baby sitters** Tomiko has a choice of two babysitters. Sitter 1 charges \$6 per hour, and sitter 2 charges \$7 per hour. Tomiko can afford no more than \$42 per week for sitters. Find an inequality that shows the possible ways that she can hire sitter 1 (x) and sitter 2 (y), and graph it in the illustration.

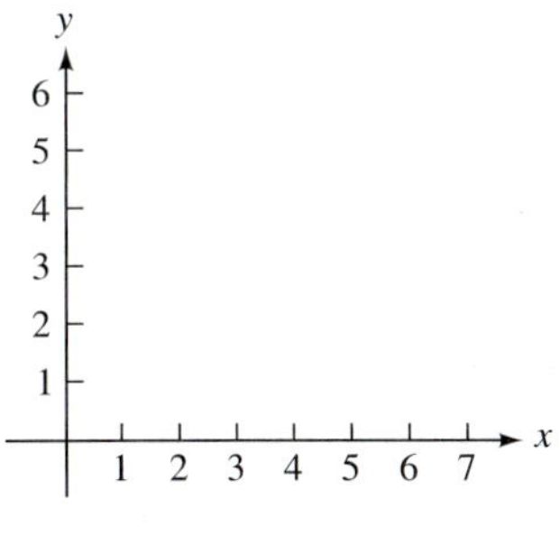

65. **Inventories** A clothing store advertises that it maintains an inventory of at least \$4,400 worth of men's jackets. A leather jacket costs \$100, and a nylon jacket costs \$88. Find an inequality that shows the possible ways that leather jackets (x) and nylon jackets (y) can be stocked, and graph it in the illustration.

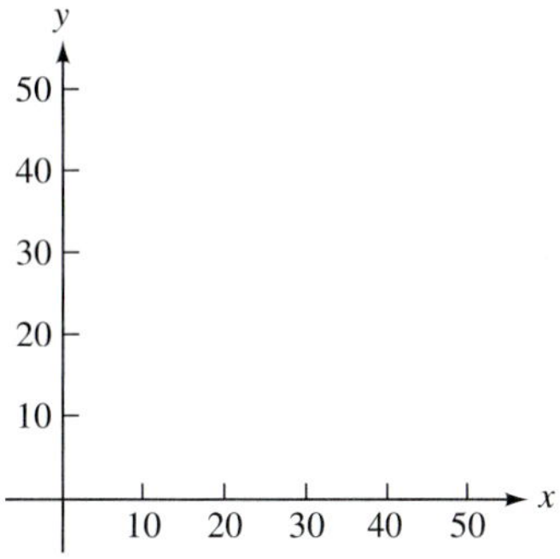

66. **Making sporting goods** To keep up with demand, a sporting goods manufacturer allocates at least 2,400 units of time per day to make baseballs and footballs. It takes 20 units of time to make a baseball and 30 units of time to make a football. Find an inequality that shows the possible ways to schedule the time to make baseballs (x) and footballs (y), and graph it in the illustration.

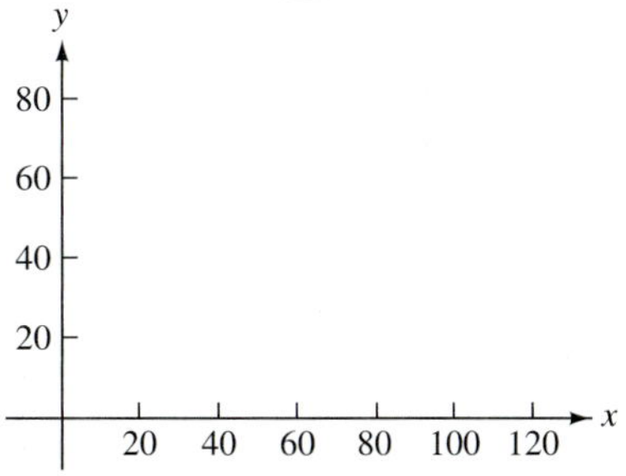

67. **Investing** Robert has up to \$8,000 to invest in two companies. Stock in Robotronics sells for \$40 per share, and stock in Macrocorp sells for \$50 per share. Find an inequality that shows the possible ways that he can buy shares of Robotronics (x) and Macrocorp (y), and graph it in the illustration on the next page.

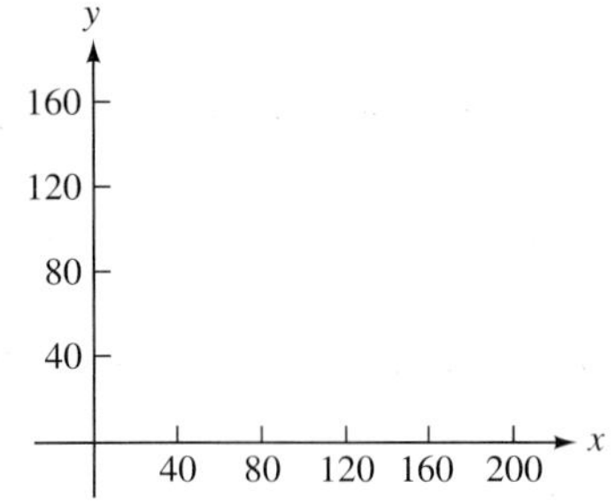

68. Buying tickets Tickets to the Rockford Rox baseball games cost \$6 for reserved seats and \$4 for general admission. Nightly receipts must average at least \$10,200 to meet expenses. Find an inequality that shows the possible ways that the Rox can sell reserved seats (x) and general admission tickets (y), and graph it in the illustration.

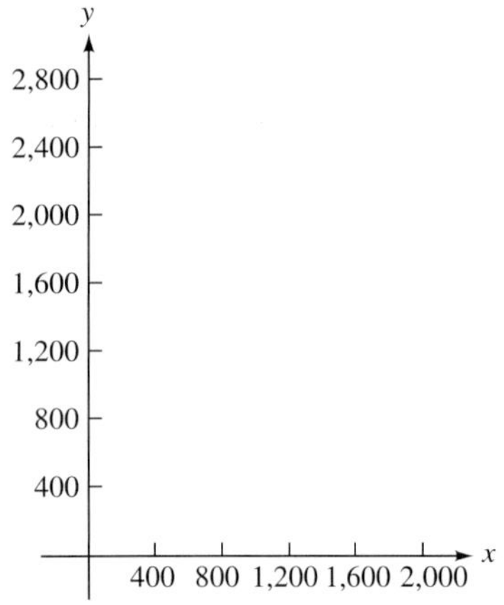

Graph each system of inequalities and give two possible solutions to each problem.

69. Buying compact discs Melodic Music has compact discs on sale for either \$10 or \$15. A customer wants to spend at least \$30 but no more than \$60 on CDs. Find a system of inequalities whose graph will show the possible combinations of \$10 CDs (x) and \$15 CDs (y) that the customer can buy, and graph it in the illustration.

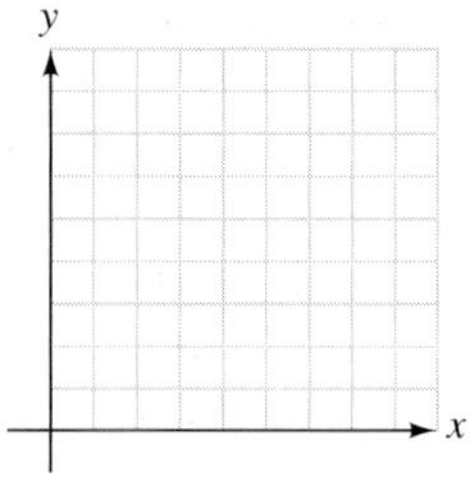

70. Buying boats Dry Boatworks wholesales aluminium boats for \$800 and fiberglass boats for \$600. Northland Marina wants to order at least \$2,400 but no more than \$4,800 worth of boats. Find a system of inequalities whose graph will show the possible combinations of aluminum boats (x) and fiberglass boats (y) that can be ordered, and graph it in the illustration.

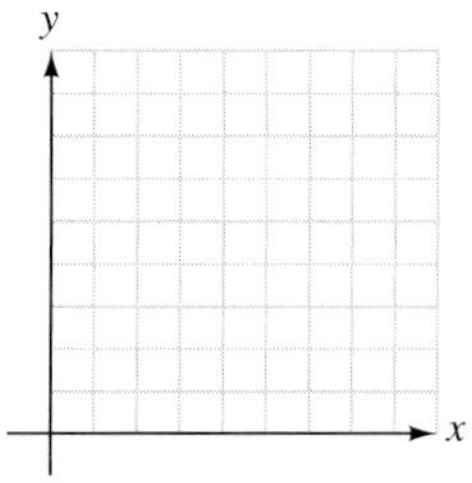

71. Buying furniture A distributor wholesales desk chairs for \$150 and side chairs for \$100. Best Furniture wants to order no more than \$900 worth of chairs and wants to order more side chairs than desk chairs. Find a system of inequalities whose graph will show the possible combinations of desk chairs (x) and side chairs (y) that can be ordered, and graph it in the illustration.

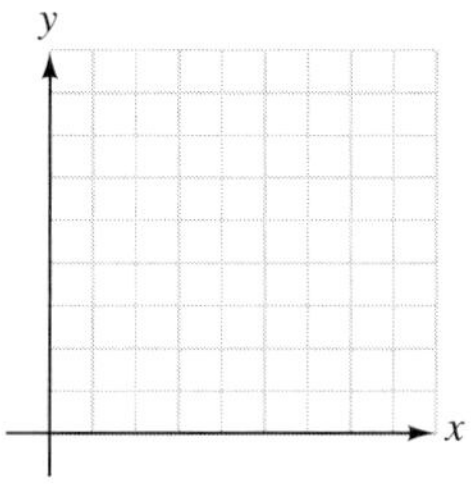

72. Ordering furnace equipment J. Bolden Heating Company wants to order no more than \$2,000 worth of electronic air cleaners and humidifiers from a wholesaler that charges \$500 for air cleaners and \$200 for humidifiers. Bolden wants more humidifiers than air cleaners. Find a system of inequalities whose graph will show the possible combinations of air cleaners (x) and humidifiers (y) that can be ordered, and graph it in the illustration.

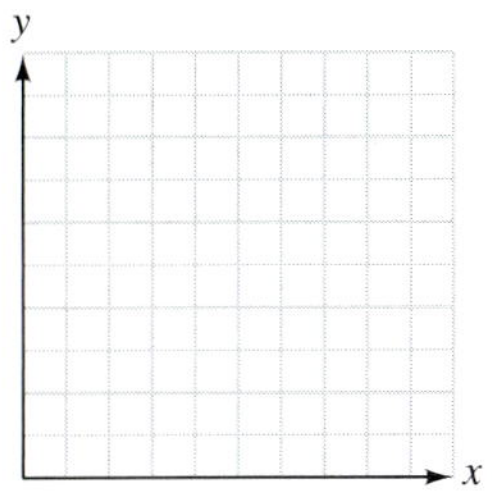

WRITING

73. Explain how to find the boundary for the graph of an inequality.

74. Explain how to decide which side of the boundary line to shade.

75. Explain how to use graphing to solve a system of inequalities.

76. Explain when a system of inequalities will have no solutions.

SOMETHING TO THINK ABOUT

77. What are some limitations of the graphing method for solving inequalities?

78. Graph $y = 3x + 1$, $y < 3x + 1$, and $y > 3x + 1$. What do you discover?

79. Can a system of inequalities have

a. no solutions?

b. exactly one solution?

c. infinitely many solutions?

80. Find a system of two inequalities that has a solution of (2, 0) but no solutions of the form (x, y) where $y < 0$.

Project

The graphing method of solving a system of equations is not as accurate as algebraic methods, and some systems are more difficult than others to solve accurately. For example, the two lines in Illustration 1(a) could be drawn carelessly, and the point of intersection would not be far from the correct location. If the lines in Illustration 1(b) were drawn carelessly, the point of intersection could move substantially from its correct location.

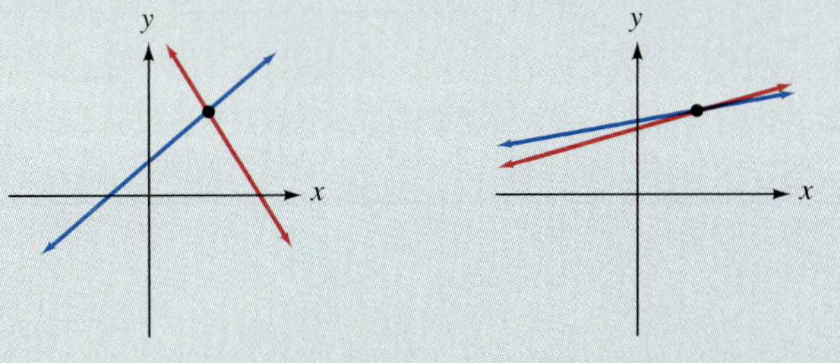

(a) (b)

Illustration 1

- Carefully solve each of these systems of equations graphically (by hand, not with a graphing caculator). Indicate your best estimate of the solution of each system.

$$\begin{cases} 2x - 4y = -7 \\ 4x + 2y = 11 \end{cases} \qquad \begin{cases} 5x - 4y = -1 \\ 12x - 10y = -3 \end{cases}$$

- Solve each system algebraically. How close were your graphical solutions to the actual solutions? Write a paragraph explaining any differences.
- Create a system of equations with the solutions $x = 3$, $y = 2$ for which an accurate solution could be obtained graphically.
- Create a system of equations with the solutions $x = 3$, $y = 2$ that is more difficult to solve accurately than the previous system, and write a paragraph explaining why.

Chapter Summary

CONCEPTS | **REVIEW EXERCISES**

3.1 The Rectangular Coordinate System

Any ordered pair of real numbers represents a point on the rectangular coordinate system.

Plot each point on the rectangular coordinate system in the illustration.

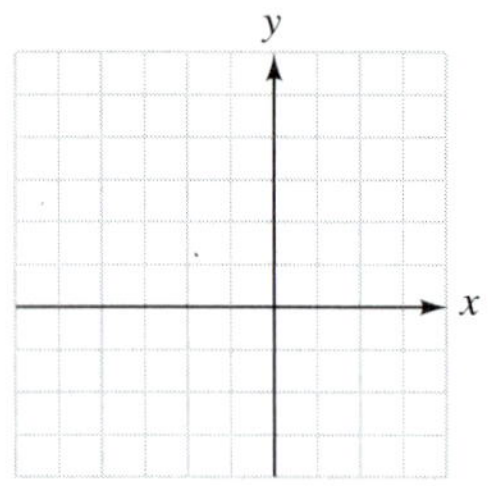

1. $A(1, 3)$ **2.** $B(1, -3)$

3. $C(-3, 1)$ **4.** $D(-3, -1)$

5. $E(0, 5)$ **6.** $F(-5, 0)$

Find the coordinates of each point in the illustration.

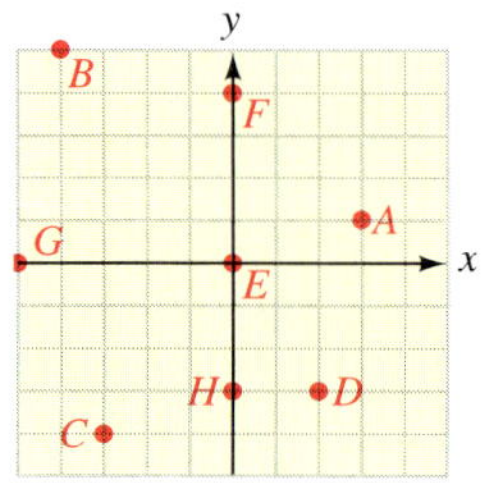

7. A **8.** B

9. C **10.** D

11. E **12.** F

13. G **14.** H

3.2 Graphing Linear Equations

An ordered pair of real numbers is a **solution** if it satisfies the equation.

Tell whether each pair satisfies the equation $3x - 4y = 12$.

15. $(2, 1)$ **16.** $\left(3, -\frac{3}{4}\right)$

To graph a linear equation,

1. Find three pairs (x, y) that satisfy the equation.
2. Plot each pair on the rectangular coordinate system.
3. Draw a line passing through the three points.

Graph each equation on a rectangular coordinate system.

17. $y = x - 5$

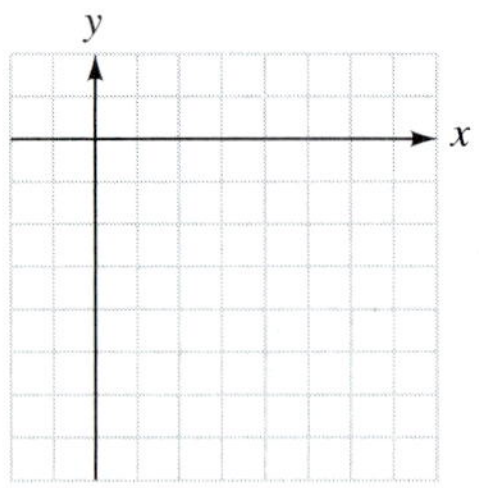

18. $y = 2x + 1$

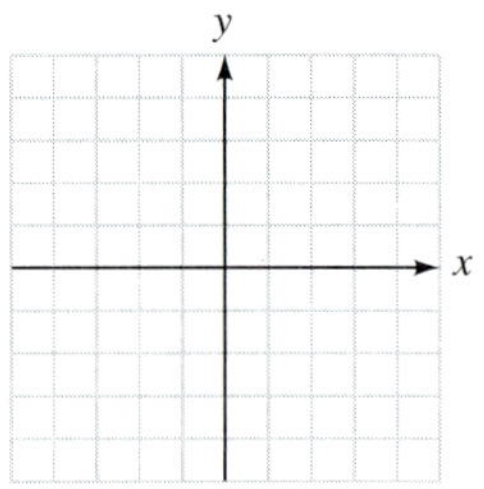

General form of an equation of a line:

$$Ax + By = C$$

(A and B are not both 0.)

19. $y = \dfrac{x}{2} + 2$

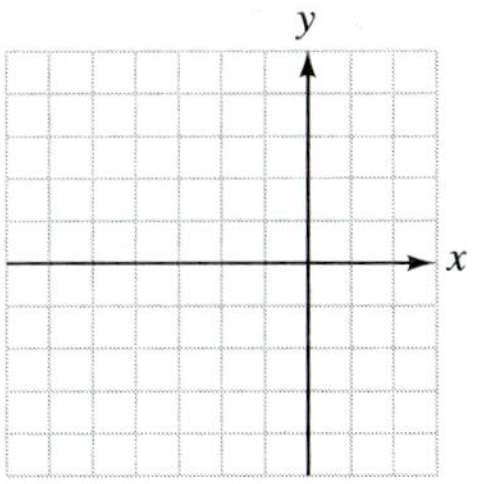

20. $y = 3$

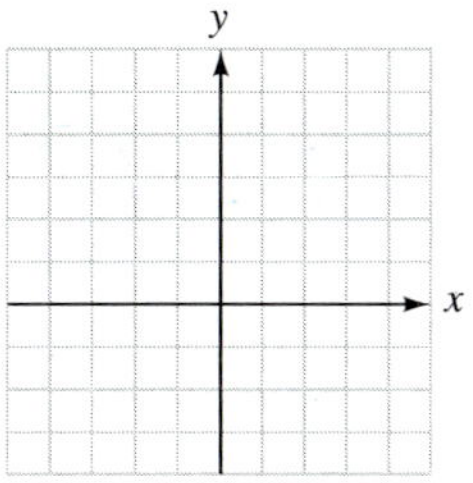

21. $x + y = 4$

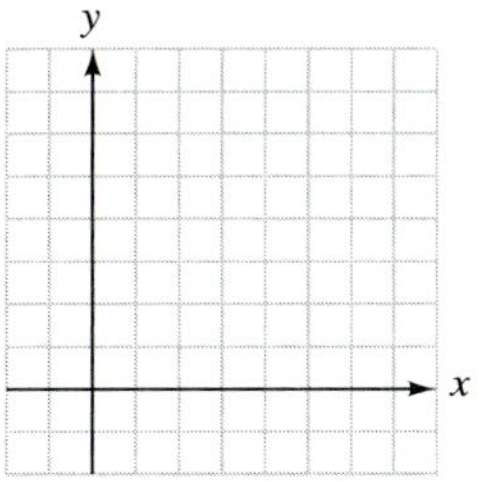

22. $x - y = -3$

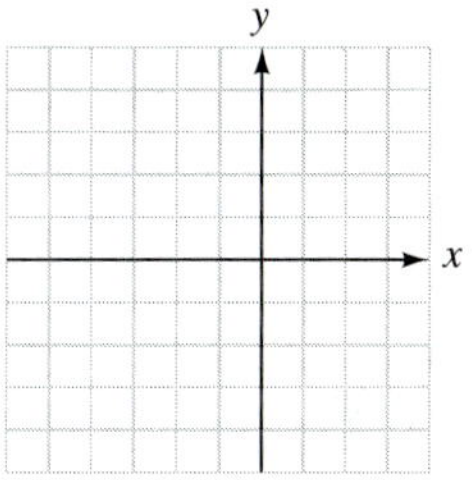

23. $3x + 5y = 15$

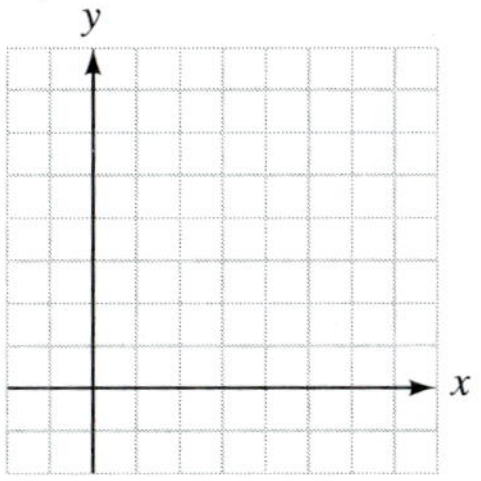

24. $7x - 4y = 28$

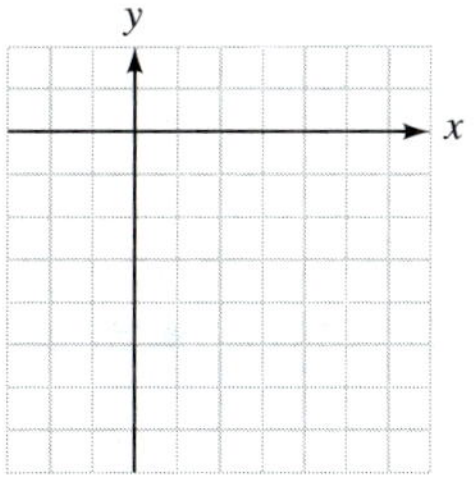

3.3 Solving Systems of Equations by Graphing

To solve a system of equations graphically, carefully graph each equation of the system. If the lines intersect, the coordinates of the point of intersection give the solution of the system.

Tell whether the ordered pair is a solution of the system.

25. $(1, 5), \begin{cases} 3x - y = -2 \\ 2x + 3y = 17 \end{cases}$

26. $(-2, 4), \begin{cases} 5x + 3y = 2 \\ -3x + 2y = 16 \end{cases}$

27. $\left(14, \dfrac{1}{2}\right), \begin{cases} 2x + 4y = 30 \\ \dfrac{x}{4} - y = 3 \end{cases}$

28. $\left(\dfrac{7}{2}, -\dfrac{2}{3}\right), \begin{cases} 4x - 6y = 18 \\ \dfrac{x}{3} + \dfrac{y}{2} = \dfrac{5}{6} \end{cases}$

Use the graphing method to solve each system.

29. $\begin{cases} x + y = 7 \\ 2x - y = 5 \end{cases}$

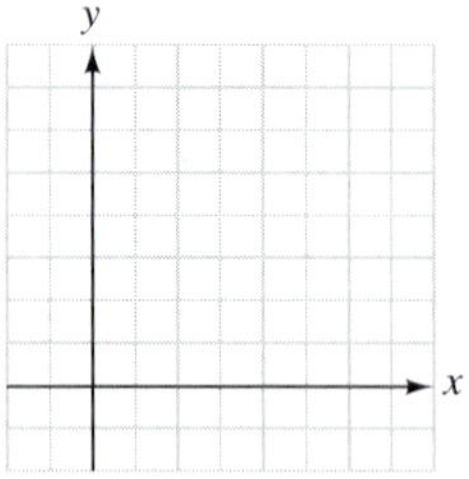

30. $\begin{cases} \dfrac{x}{3} + \dfrac{y}{5} = -1 \\ x - 3y = -3 \end{cases}$

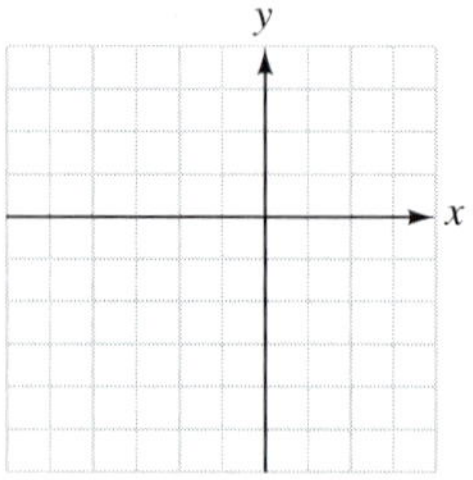

If a system of equations has infinitely many solutions, the equations of the system are called **dependent equations.**

If a system of equations has no solutions, it is called an **inconsistent system.**

31. $\begin{cases} 3x + 6y = 6 \\ x + 2y = 2 \end{cases}$

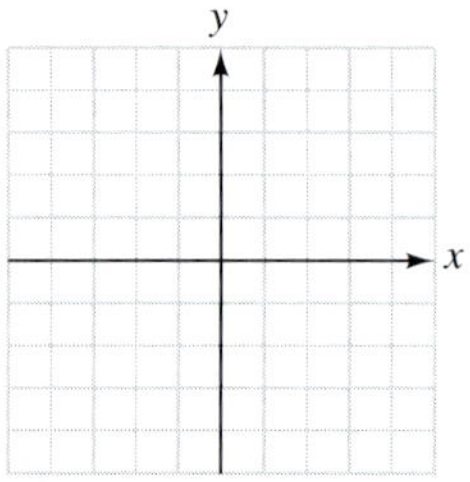

32. $\begin{cases} 6x + 3y = 12 \\ 2x + y = 2 \end{cases}$

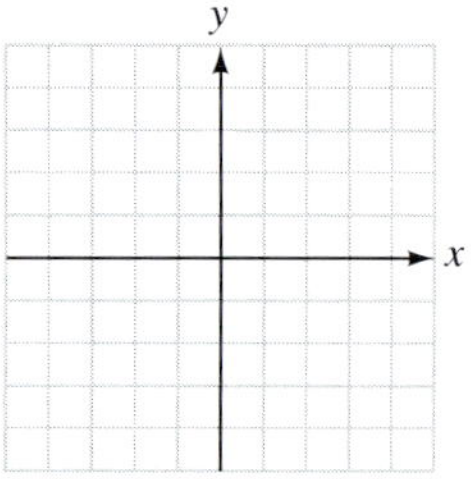

3.4 Solving Systems of Equations by Substitution

To solve a system of equations by substitution, solve one of the equations of the system for one of the variables, substitute the resulting expression into the other equation, and solve for the other variable.

Use substitution to solve each system.

33. $\begin{cases} x = 3y + 5 \\ 5x - 4y = 3 \end{cases}$

34. $\begin{cases} 3x - \dfrac{2y}{5} = 2(x - 2) \\ 2x - 3 = 3 - 2y \end{cases}$

35. $\begin{cases} 8x + 5y = 3 \\ 5x - 8y = 13 \end{cases}$

36. $\begin{cases} 6(x + 2) = y - 1 \\ 5(y - 1) = x + 2 \end{cases}$

3.5 Solving Systems of Equations by Addition

To solve a system of equations by addition, first multiply one or both of the equations by suitable constants, if necessary, to eliminate one of the variables when the equations are added. The equation that results can be solved for its single variable. Then substitute the value ob-

Use addition to solve each system.

37. $\begin{cases} 2x + y = 1 \\ 5x - y = 20 \end{cases}$

38. $\begin{cases} x + 8y = 7 \\ x - 4y = 1 \end{cases}$

39. $\begin{cases} 5x + y = 2 \\ 3x + 2y = 11 \end{cases}$

40. $\begin{cases} x + y = 3 \\ 3x = 2 - y \end{cases}$

41. $\begin{cases} 11x + 3y = 27 \\ 8x + 4y = 36 \end{cases}$

42. $\begin{cases} 9x + 3y = 5 \\ 3x = 4 - y \end{cases}$

tained back into one of the original equations and solve for the other variable.

43. $\begin{cases} 9x + 3y = 5 \\ 3x + y = \dfrac{5}{3} \end{cases}$

44. $\begin{cases} \dfrac{x}{3} + \dfrac{y+2}{2} = 1 \\ \dfrac{x+8}{8} + \dfrac{y-3}{3} = 0 \end{cases}$

3.6 Applications of Systems of Equations

Systems of equations are useful in solving many different types of problems.

45. Integer problem One number is 5 times another, and their sum is 18. Find the numbers.

46. Geometry The length of a rectangle is 3 times its width, and its perimeter is 24 feet. Find its dimensions.

47. Buying grapefruit A grapefruit costs 15 cents more than an orange. Together, they cost 85 cents. Find the cost of a grapefruit.

48. Utility bills A man's electric bill for January was $23 less than his gas bill. The two utilities cost him a total of $109. Find the amount of his gas bill.

49. Buying groceries Two gallons of milk and 3 dozen eggs cost $6.80. Three gallons of milk and 2 dozen eggs cost $7.35. How much does each gallon of milk cost?

50. Investing money Carlos invested part of $3,000 in a 10% certificate of deposit account and the rest in a 6% passbook account. If the total annual interest from both accounts is $270, how much did he invest at 6%?

3.7 Systems of Linear Inequalities

To graph a system of inequalities, first graph the individual inequalities of the system. The final solution, if one exists, is that region where all the individual graphs intersect.

Graph each inequality.

51. $y \geq x + 2$

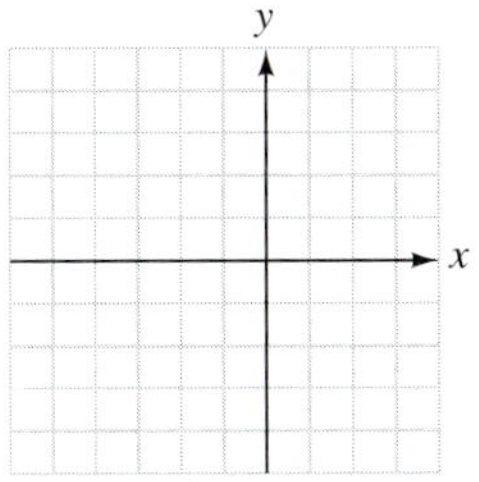

52. $x < 3$

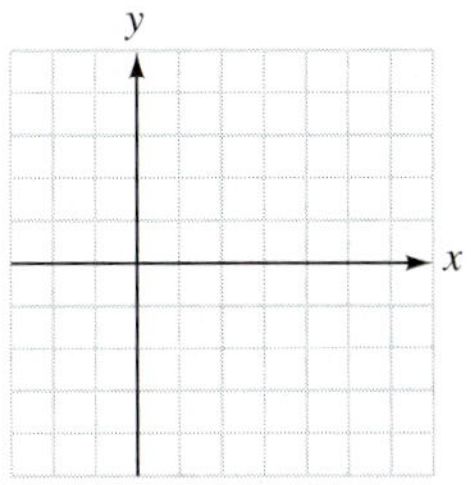

Solve each system of inequalities.

53. $\begin{cases} 5x + 3y < 15 \\ 3x - y > 3 \end{cases}$

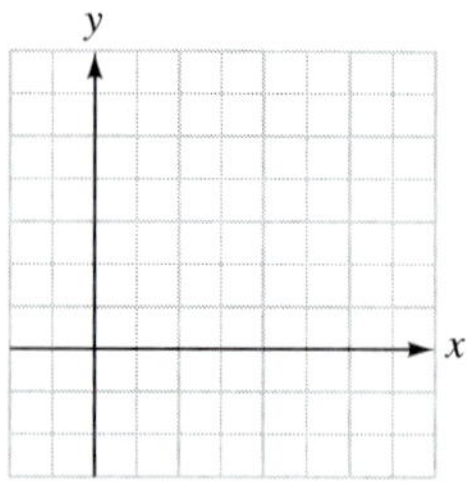

54. $\begin{cases} 5x - 3y \geq 5 \\ 3x + 2y \geq 3 \end{cases}$

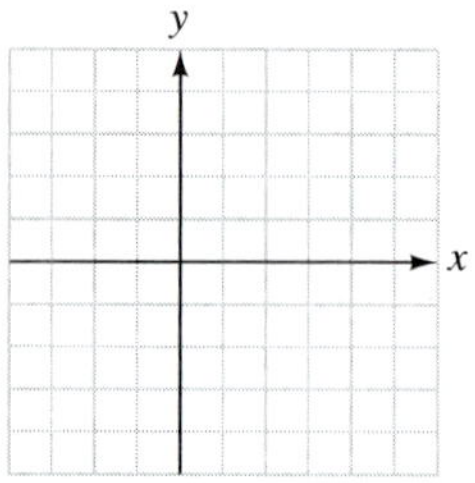

55. $\begin{cases} x \geq 3y \\ y < 3x \end{cases}$

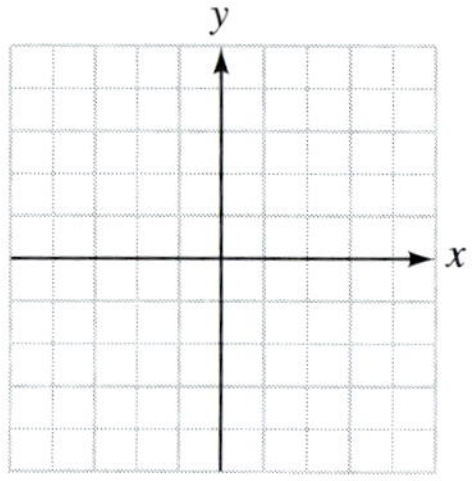

56. $\begin{cases} x \geq 0 \\ x \leq 3 \end{cases}$

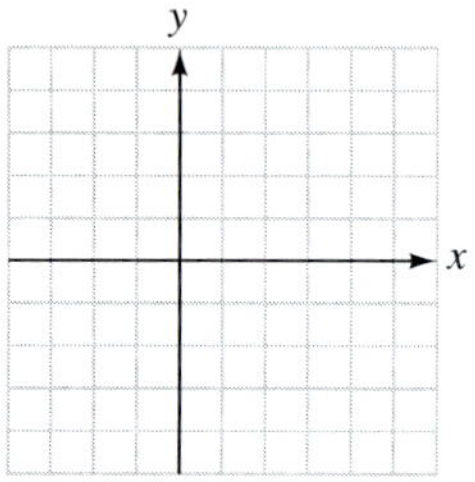

Chapter Test

Graph each equation.

1. $y = \dfrac{x}{2} + 1$

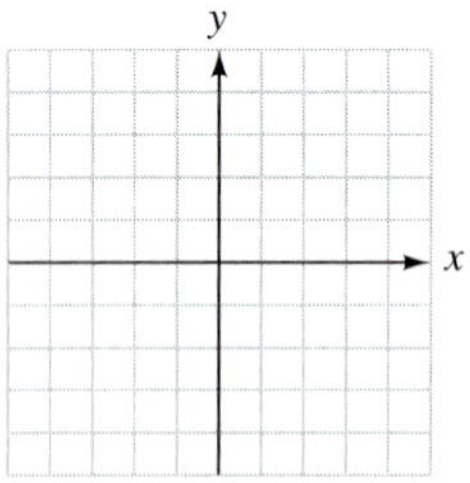

2. $2(x + 1) - y = 4$

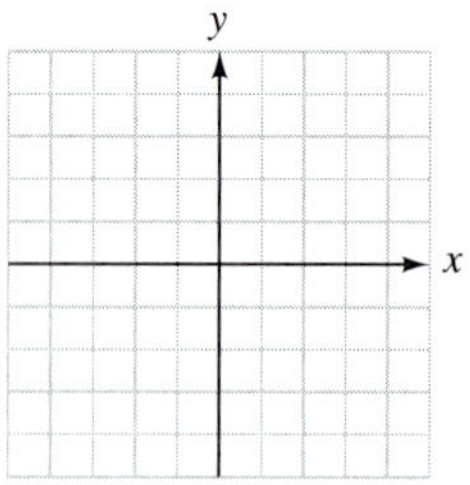

3. $x = 1$

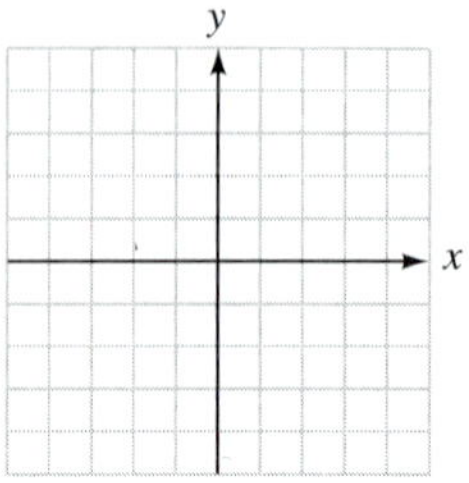

4. $2y = 8$

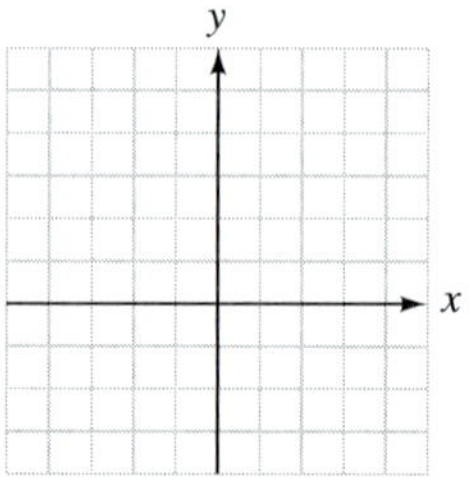

Tell whether the given ordered pair is a solution of the given system.

5. $(2, -3)$, $\begin{cases} 3x - 2y = 12 \\ 2x + 3y = -5 \end{cases}$

6. $(-2, -1)$, $\begin{cases} 4x + y = -9 \\ 2x - 3y = -7 \end{cases}$

Solve each system by graphing.

7. $\begin{cases} 3x + y = 7 \\ x - 2y = 0 \end{cases}$

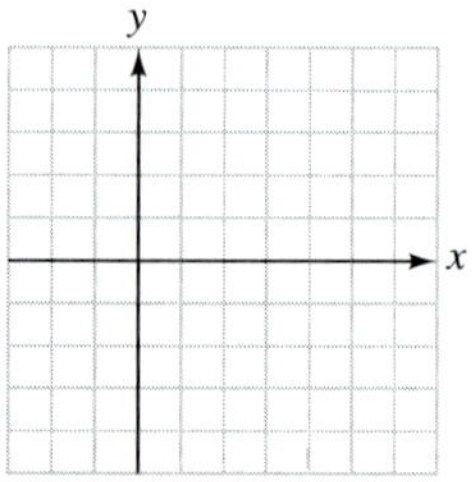

8. $\begin{cases} x + \dfrac{y}{2} = 1 \\ y = 1 - 3x \end{cases}$

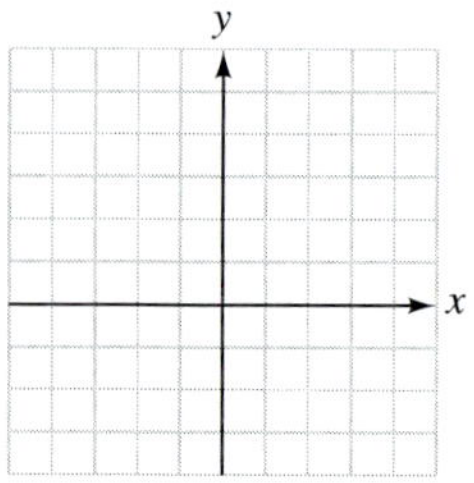

Solve each system by substitution.

9. $\begin{cases} y = x - 1 \\ x + y = -7 \end{cases}$

10. $\begin{cases} \dfrac{x}{6} + \dfrac{y}{10} = 3 \\ \dfrac{5x}{16} - \dfrac{3y}{16} = \dfrac{15}{8} \end{cases}$

Solve each system by addition.

11. $\begin{cases} 3x - y = 2 \\ 2x + y = 8 \end{cases}$

12. $\begin{cases} 4x + 3 = -3y \\ \dfrac{-x}{7} + \dfrac{4y}{21} = 1 \end{cases}$

Classify each system as consistent or inconsistent.

13. $\begin{cases} 2x + 3(y - 2) = 0 \\ -3y = 2(x - 4) \end{cases}$

14. $\begin{cases} \dfrac{x}{3} + y - 4 = 0 \\ -3y = x - 12 \end{cases}$

Use a system of equations in two variables to solve each problem.

15. The sum of two numbers is -18. One number is 2 greater than 3 times the other. Find the product of the numbers.

16. A woman invested some money at 8% and some at 9%. The interest on the combined investment of \$10,000 was \$840. How much was invested at 9%?

Solve each system of inequalities by graphing.

17. $\begin{cases} x + y < 3 \\ x - y < 1 \end{cases}$

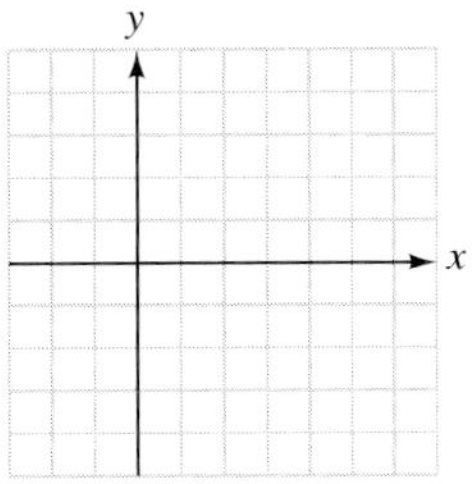

18. $\begin{cases} 2x + 3y \leq 6 \\ x \geq 2 \end{cases}$

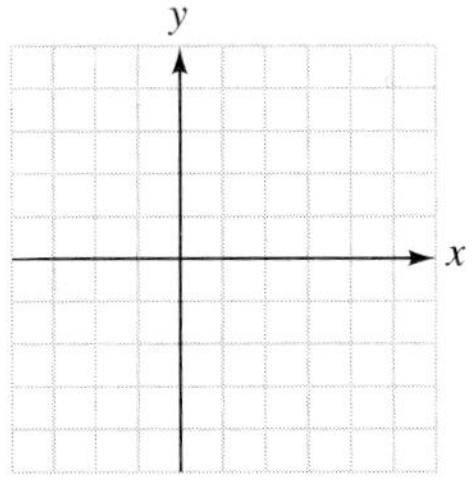

4 *Polynomials*

InfoTrac Project

Do a subject guide search on "HIV," choose the subdivision "prevention," and find the article "Success for AIDS prevention in the U.S. but catastrophe elsewhere." Write a summary of the article.

After rounding to the nearest hundred thousand, convert to scientific notation the figure representing the number of people who have escaped HIV infection because of prevention efforts in the U.S. Convert the amount of money saved to scientific notation and calculate the cost of treating one AIDS patient. Assuming that 4 million of the 7 million infected people in Asia will develop AIDS, use scientific notation to calculate the cost of treating them.

Complete this project after studying Section 4.3.

© Tom Stewart/CORBIS

Mathematics in Medicine

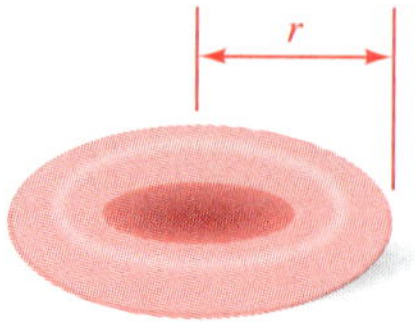

The red cells in human blood pick up oxygen in the lungs and carry it to all parts of the body. Each red cell is a tiny circular disc with an approximate radius of 0.00015 inch. Because the amount of oxygen carried depends on the surface area of the cells, and the cells are so tiny, a very great number is needed—about 25 trillion in an average adult.

What is the total surface area of all the red blood cells in the body of an average adult?

Exercise Set 4.3 Problems 49–50

In this chapter, we develop the rules for integer exponents and use them to express very large and small numbers in ***scientific notation.*** *We then discuss special algebraic expressions, called* ***polynomials,*** *and show how to add, subtract, multiply, and divide them.*

4.1 Natural-Number Exponents

- Exponents
- Powers of Expressions
- The Product Rule for Exponents
- The Power Rules for Exponents
- The Quotient Rule for Exponents

Getting Ready *Evaluate each expression.*

1. 2^3 **2.** 3^2 **3.** $3(2)$ **4.** $2(3)$

5. $2^3 + 2^2$ **6.** $2^3 \cdot 2^2$ **7.** $3^3 - 3^2$ **8.** $\frac{3^3}{3^2}$

In this section, we will revisit the topic of exponents. This time we will develop the basic rules used to manipulate exponential expressions.

Exponents

We have used natural-number exponents to indicate repeated multiplication. For example,

$$2^5 = 2 \cdot 2 \cdot 2 \cdot 2 \cdot 2 = 32 \qquad (-7)^3 = (-7)(-7)(-7) = -343$$
$$x^4 = x \cdot x \cdot x \cdot x \qquad -y^5 = -y \cdot y \cdot y \cdot y \cdot y$$

These examples suggest a definition for x^n, where n is a natural number.

Natural-Number Exponents

If n is a natural number, then

$$x^n = \overbrace{x \cdot x \cdot x \cdot \cdots \cdot x}^{n \text{ factors of } x}$$

In the exponential expression x^n, x is called the **base** and n is called the **exponent.** The entire expression is called a **power of x.**

$$\text{Base} \rightarrow x^n \leftarrow \text{Exponent}$$

Powers of Expressions

If an exponent is a natural number, it tells how many times its base is to be used as a factor. An exponent of 1 indicates that its base is to be used one time as a factor, an exponent of 2 indicates that its base is to be used two times as a factor, and so on.

$$3^1 = 3 \qquad (-y)^1 = -y \qquad (-4z)^2 = (-4z)(-4z) \quad \text{and} \quad (t^2)^3 = t^2 \cdot t^2 \cdot t^2$$

EXAMPLE 1 Show that **a.** -2^4 and **b.** $(-2)^4$ have different values.

Solution We find each power and show that the results are different.

$$\begin{aligned} -2^4 &= -(2^4) \\ &= -(2 \cdot 2 \cdot 2 \cdot 2) \\ &= -16 \end{aligned} \qquad\qquad \begin{aligned} (-2)^4 &= (-2)(-2)(-2)(-2) \\ &= 16 \end{aligned}$$

Since $-16 \neq 16$, it follows that $-2^4 \neq (-2)^4$.

Self Check Show that **a.** $(-4)^3$ and **b.** -4^3 have the same value.

EXAMPLE 2 Write each expression without exponents: **a.** r^3, **b.** $(-2s)^4$, and **c.** $\left(\frac{1}{3}ab\right)^5$.

Solution **a.** $r^3 = r \cdot r \cdot r$

b. $(-2s)^4 = (-2s)(-2s)(-2s)(-2s)$

c. $\left(\frac{1}{3}ab\right)^5 = \left(\frac{1}{3}ab\right)\left(\frac{1}{3}ab\right)\left(\frac{1}{3}ab\right)\left(\frac{1}{3}ab\right)\left(\frac{1}{3}ab\right)$

Self Check Write each expression without exponents: **a.** x^4 and **b.** $\left(-\frac{1}{2}xy\right)^3$.

The Product Rule for Exponents

To develop a rule for multiplying exponential expressions with the same base, we consider the product $x^2 \cdot x^3$. Since the expression x^2 means that x is to be used as a factor two times and the expression x^3 means that x is to be used as a factor three times, we have

$$\begin{aligned} x^2x^3 &= \overbrace{x \cdot x}^{\text{2 factors of } x} \cdot \overbrace{x \cdot x \cdot x}^{\text{3 factors of } x} \\ &= \overbrace{x \cdot x \cdot x \cdot x \cdot x}^{\text{5 factors of } x} \\ &= x^5 \end{aligned}$$

In general,

$$\begin{aligned} x^m \cdot x^n &= \overbrace{x \cdot x \cdot x \cdot \cdots \cdot x}^{m \text{ factors of } x} \cdot \overbrace{x \cdot x \cdot x \cdot \cdots \cdot x}^{n \text{ factors of } x} \\ &= \overbrace{x \cdot x \cdot x \cdot x \cdot x \cdot x \cdot \cdots \cdot x \cdot x \cdot x}^{m + n \text{ factors of } x} \\ &= x^{m+n} \end{aligned}$$

This discussion suggests the following rule: *To multiply two exponential expressions with the same base, keep the base and add the exponents.*

Product Rule for Exponents

If m and n are natural numbers, then

$$x^m x^n = x^{m+n}$$

EXAMPLE 3 Simplify each expression.

a. $x^3x^4 = x^{3+4}$ Keep the base and add the exponents.

$= x^7$ $3 + 4 = 7$.

b. $y^2y^4y = (y^2y^4)y$ Use the associative property to group y^2 and y^4 together.

$= (y^{2+4})y$ Keep the base and add the exponents.

$= y^6y$ $2 + 4 = 6$.

$= y^{6+1}$ Keep the base and add the exponents: $y = y^1$.

$= y^7$ $6 + 1 = 7$.

Self Check Simplify each expression: **a.** zz^3 and **b.** $x^2x^3x^6$.

EXAMPLE 4 Simplify: $(2y^3)(3y^2)$.

Solution $(2y^3)(3y^2) = 2(3)y^3y^2$ Use the commutative and associative properties to group the numbers together and the variables together.

$= 6y^{3+2}$ Multiply the coefficients. Keep the base and add the exponents.

$= 6y^5$ $3 + 2 = 5$.

Self Check Simplify: $(4x)(-3x^2)$.

Comment The product rule for exponents applies only to exponential expressions with the same base. An expression such as x^2y^3 cannot be simplified, because x^2 and y^3 have different bases.

The Power Rules for Exponents

To find another rule of exponents, we consider the expression $(x^3)^4$, which can be written as $x^3 \cdot x^3 \cdot x^3 \cdot x^3$. Because each of the four factors of x^3 contains three factors of x, there are $4 \cdot 3$ (or 12) factors of x. Thus, the expression can be written as x^{12}.

$$(x^3)^4 = x^3 \cdot x^3 \cdot x^3 \cdot x^3$$

$$= \overbrace{\underbrace{x \cdot x \cdot x}_{x^3} \cdot \underbrace{x \cdot x \cdot x}_{x^3} \cdot \underbrace{x \cdot x \cdot x}_{x^3} \cdot \underbrace{x \cdot x \cdot x}_{x^3}}^{12 \text{ factors of } x}$$

$$= x^{12}$$

In general,

$$(x^m)^n = \overbrace{x^m \cdot x^m \cdot x^m \cdot \cdots \cdot x^m}^{n \text{ factors of } x^m}$$
$$= \overbrace{x \cdot x \cdot x \cdot x \cdot x \cdot x \cdot x \cdot \cdots \cdot x}^{m \cdot n \text{ factors of } x}$$
$$= x^{m \cdot n}$$

This discussion suggests the following rule: *To raise an exponential expression to a power, keep the base and multiply the exponents.*

First Power Rule for Exponents

If m and n are natural numbers, then

$$(x^m)^n = x^{m \cdot n}$$

EXAMPLE 5 Write each expression using one exponent.

a. $(2^3)^7 = 2^{3 \cdot 7}$ Keep the base and multiply the exponents.
$= 2^{21}$ $3 \cdot 7 = 21$.

b. $(z^7)^7 = z^{7 \cdot 7}$ Keep the base and multiply the exponents.
$= z^{49}$ $7 \cdot 7 = 49$.

Self Check Write each expression using one exponent: **a.** $(y^5)^2$ and **b.** $(u^x)^y$.

In the next example, the product and power rules of exponents are both used.

EXAMPLE 6 Write each expression using one exponent.

a. $(x^2x^5)^2 = (x^7)^2$
$= x^{14}$

b. $(y^6y^2)^3 = (y^8)^3$
$= y^{24}$

c. $(z^2)^4(z^3)^3 = z^8z^9$
$= z^{17}$

d. $(x^3)^2(x^5x^2)^3 = x^6(x^7)^3$
$= x^6x^{21}$
$= x^{27}$

Self Check Write each expression using one exponent: **a.** $(a^4a^3)^3$ and **b.** $(a^3)^3(a^4)^2$.

To find more rules for exponents, we consider the expressions $(2x)^3$ and $\left(\frac{2}{x}\right)^3$.

$$(2x)^3 = (2x)(2x)(2x)$$
$$= (2 \cdot 2 \cdot 2)(x \cdot x \cdot x)$$
$$= 2^3x^3$$
$$= 8x^3$$

$$\left(\frac{2}{x}\right)^3 = \left(\frac{2}{x}\right)\left(\frac{2}{x}\right)\left(\frac{2}{x}\right) \quad (x \neq 0)$$
$$= \frac{2 \cdot 2 \cdot 2}{x \cdot x \cdot x}$$
$$= \frac{2^3}{x^3}$$
$$= \frac{8}{x^3}$$

These examples suggest the following rules: *To raise a product to a power, we raise each factor of the product to that power, and to raise a fraction to a power, we raise both the numerator and denominator to that power.*

More Power Rules for Exponents

If n is a natural number, then

$$(xy)^n = x^n y^n \qquad \text{and if } y \neq 0, \text{ then} \qquad \left(\frac{x}{y}\right)^n = \frac{x^n}{y^n}$$

EXAMPLE 7 Write each expression without parentheses. Assume there are no divisions by zero.

a. $(ab)^4 = a^4b^4$

b. $(3c)^3 = 3^3c^3$
$= 27c^3$

c. $(x^2y^3)^5 = (x^2)^5(y^3)^5$
$= x^{10}y^{15}$

d. $(-2x^3y)^2 = (-2)^2(x^3)^2y^2$
$= 4x^6y^2$

e. $\left(\frac{4}{k}\right)^3 = \frac{4^3}{k^3}$
$= \frac{64}{k^3}$

f. $\left(\frac{3x^2}{2y^3}\right)^5 = \frac{3^5(x^2)^5}{2^5(y^3)^5}$
$= \frac{243x^{10}}{32y^{15}}$

Self Check Write each expression without parentheses: **a.** $(3x^2y)^2$ and **b.** $\left(\frac{2x^3}{3y^2}\right)^4$.

The Quotient Rule for Exponents

To find a rule for dividing exponential expressions, we consider the fraction $\frac{4^5}{4^2}$, where the exponent in the numerator is greater than the exponent in the denominator. We can simplify the fraction as follows:

$$\frac{4^5}{4^2} = \frac{4 \cdot 4 \cdot 4 \cdot 4 \cdot 4}{4 \cdot 4}$$

$$= \frac{\overset{1}{\cancel{4}} \cdot \overset{1}{\cancel{4}} \cdot 4 \cdot 4 \cdot 4}{\underset{1}{\cancel{4}} \cdot \underset{1}{\cancel{4}}}$$

$$= 4^3$$

The result of 4^3 has a base of 4 and an exponent of $5 - 2$ (or 3). This suggests that *to divide exponential expressions with the same base, we keep the base and subtract the exponents.*

Quotient Rule for Exponents

If m and n are natural numbers, $m > n$, and $x \neq 0$, then

$$\frac{x^m}{x^n} = x^{m-n}$$

EXAMPLE 8 Simplify each expression. Assume there are no divisions by 0.

a. $\frac{x^4}{x^3} = x^{4-3}$
$= x^1$
$= x$

b. $\frac{8y^2y^6}{4y^3} = \frac{8y^8}{4y^3}$
$= \frac{8}{4}y^{8-3}$
$= 2y^5$

c. $\frac{a^3a^5a^7}{a^4a} = \frac{a^{15}}{a^5}$
$= a^{15-5}$
$= a^{10}$

d. $\frac{(a^3b^4)^2}{ab^5} = \frac{a^6b^8}{ab^5}$
$= a^{6-1}b^{8-5}$
$= a^5b^3$

Self Check Simplify: **a.** $\frac{a^5}{a^3}$, **b.** $\frac{6b^2b^3}{2b^4}$, and **c.** $\frac{(x^2y^3)^2}{x^3y^4}$.

We summarize the rules for positive exponents as follows.

Properties of Exponents

If n is a natural number, then

$$x^n = \underbrace{x \cdot x \cdot x \cdot \cdots \cdot x}_{n \text{ factors of } x}$$

If m and n are natural numbers and there are no divisions by 0, then

$$x^m x^n = x^{m+n} \qquad (x^m)^n = x^{m \cdot n} \qquad (xy)^n = x^n y^n \qquad \left(\frac{x}{y}\right)^n = \frac{x^n}{y^n}$$

$$\frac{x^m}{x^n} = x^{m-n} \quad \text{provided } m > n$$

Self Check Answers

1. a. -64, **b.** -64 **2. a.** $x \cdot x \cdot x \cdot x$, **b.** $\left(-\frac{1}{2}\right)\left(-\frac{1}{2}\right)\left(-\frac{1}{2}\right) \cdot x \cdot x \cdot x \cdot y \cdot y \cdot y$ **3. a.** z^4, **b.** x^{11} **4.** $-12x^3$ **5. a.** y^{10}, **b.** u^{xy} **6. a.** a^{21}, **b.** a^{17} **7. a.** $9x^4y^2$, **b.** $\frac{16x^{12}}{81y^8}$ **8. a.** a^2, **b.** $3b$, **c.** xy^2

Orals *Find the base and the exponent in each expression.*

1. x^3 **2.** 3^x **3.** ab^c **4.** $(ab)^c$

Evaluate each expression.

5. 6^2 **6.** $(-6)^2$ **7.** $2^3 + 1^3$ **8.** $(2 + 1)^3$

4.1 EXERCISES

REVIEW

1. Graph the real numbers -3, 0, 2, and $-\frac{3}{2}$ on a number line.

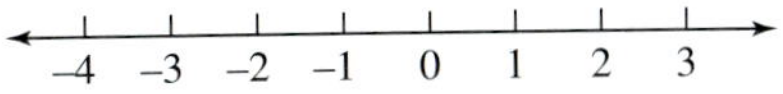

2. Graph the real numbers $-2 < x \leq 3$ on a number line.

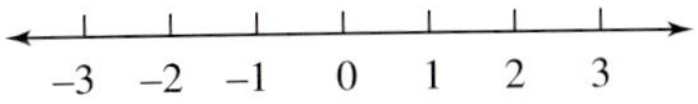

Write each algebraic expression as an English phrase.

3. $3(x + y)$

4. $3x + y$

Write each English phrase as an algebraic expression.

5. Three greater than the absolute value of twice x

6. The sum of the numbers y and z decreased by the sum of their squares

VOCABULARY AND CONCEPTS *Fill in the blanks.*

7. The base of the exponential expression $(-5)^3$ is ___. The exponent is ___.

8. The base of the exponential expression -5^3 is ___. The exponent is ___.

9. $(3x)^4$ means ______________.

10. Write $(-3y)(-3y)(-3y)$ as a power. ______

11. $y^5 =$ __________

12. $x^m x^n =$ ______

13. $(xy)^n =$ ____

14. $\left(\dfrac{a}{b}\right)^n =$ ____

15. $(a^b)^c =$ ____

16. $\dfrac{x^m}{x^n} =$ ______

17. The area of the square is $s \cdot s$. Why do you think the symbol s^2 is called "s squared"?

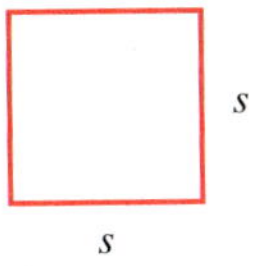

18. The volume of the cube is $s \cdot s \cdot s$. Why do you think the symbol s^3 is called "s cubed"?

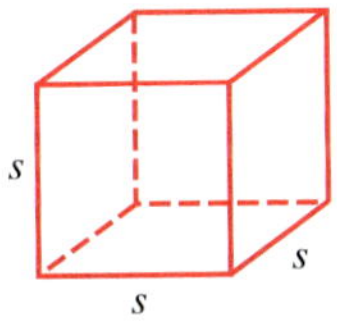

Identify the base and the exponent in each expression.

19. 4^3

20. $(-5)^2$

21. x^5

22. y^8

23. $(2y)^3$

24. $(-3x)^2$

25. $-x^4$

26. $(-x)^4$

27. x

28. $(xy)^3$

29. $2x^3$

30. $-3y^6$

Write each expression without using exponents.

31. 5^3

32. -4^5

33. x^7

34. $3x^3$

35. $-4x^5$

36. $(-2y)^4$

37. $(3t)^5$

38. a^3b^2

Write each expression using exponents.

39. $2 \cdot 2 \cdot 2$

40. $5 \cdot 5$

41. $x \cdot x \cdot x \cdot x$

42. $y \cdot y \cdot y \cdot y \cdot y \cdot y$

43. $(2x)(2x)(2x)$

44. $(-4y)(-4y)$

45. $-4t \cdot t \cdot t \cdot t$

46. $5 \cdot u \cdot u$

PRACTICE *Evaluate each expression.*

47. 5^4

48. $(-3)^3$

49. $2^2 + 3^2$

50. $2^3 - 2^2$

51. $5^4 - 4^3$

52. $2(4^3 + 3^2)$

53. $-5(3^4 + 4^3)$

54. $-5^2(4^3 - 2^6)$

Write each expression as an expression involving only one exponent.

55. x^4x^3

56. y^5y^2

57. x^5x^5

58. yy^3

59. tt^2

60. w^3w^5

61. $a^3a^4a^5$

62. $b^2b^3b^5$

63. $y^3(y^2y^4)$
64. $(y^4y)y^6$
65. $4x^2(3x^5)$
66. $-2y(y^3)$
67. $(-y^2)(4y^3)$
68. $(-4x^3)(-5x)$
69. $6x^3(-x^2)(-x^4)$
70. $-2x(-x^2)(-3x)$

71. $(3^2)^4$
72. $(4^3)^3$
73. $(y^5)^3$
74. $(b^3)^6$
75. $(a^3)^7$
76. $(b^2)^3$
77. $(x^2x^3)^5$
78. $(y^3y^4)^4$
79. $(3zz^2z^3)^5$
80. $(4t^3t^6t^2)^2$
81. $(x^5)^2(x^7)^3$
82. $(y^3y)^2(y^2)^2$
83. $(r^3r^2)^4(r^3r^5)^2$
84. $(s^2)^3(s^3)^2(s^4)^4$
85. $(s^3)^3(s^2)^2(s^5)^4$
86. $(yy^3)^3(y^2y^3)^4(y^3y^3)^2$

Write each expression without parentheses.

87. $(xy)^3$
88. $(uv^2)^4$
89. $(r^3s^2)^2$
90. $(a^3b^2)^3$
91. $(4ab^2)^2$
92. $(3x^2y)^3$
93. $(-2r^2s^3t)^3$
94. $(-3x^2y^4z)^2$
95. $\left(\frac{a}{b}\right)^3$
96. $\left(\frac{r^2}{s}\right)^4$
97. $\left(\frac{x^2}{y^3}\right)^5$
98. $\left(\frac{u^4}{v^2}\right)^6$
99. $\left(\frac{-2a}{b}\right)^5$
100. $\left(\frac{2t}{3}\right)^4$
101. $\left(\frac{b^2}{3a}\right)^3$
102. $\left(\frac{a^3b}{c^4}\right)^5$

Simplify each expression.

103. $\frac{x^5}{x^3}$
104. $\frac{a^6}{a^3}$
105. $\frac{y^3y^4}{yy^2}$
106. $\frac{b^4b^5}{b^2b^3}$
107. $\frac{12a^2a^3a^4}{4(a^4)^2}$
108. $\frac{16(aa^2)^3}{2a^2a^3}$
109. $\frac{(ab^2)^3}{(ab)^2}$
110. $\frac{(m^3n^4)^3}{(mn^2)^3}$
111. $\frac{20(r^4s^3)^4}{6(rs^3)^3}$
112. $\frac{15(x^2y^5)^5}{21(x^3y)^2}$
113. $\frac{17(x^4y^3)^8}{34(x^5y^2)^4}$
114. $\frac{35(r^3s^2)^2}{49r^2s^3}$

115. $\left(\frac{y^3y}{2yy^2}\right)^3$
116. $\left(\frac{3t^3t^4t^5}{4t^2t^6}\right)^3$
117. $\left(\frac{-2r^3r^3}{3r^4r}\right)^3$
118. $\left(\frac{-6y^4y^5}{5y^3y^5}\right)^2$

APPLICATIONS

119. **Bouncing balls** When a certain ball is dropped, it always rebounds to one-half of its previous height. If the ball is dropped from a height of 32 feet, explain why the expression $32(\frac{1}{2})^4$ represents the height of the ball on the fourth bounce. Find the height of the fourth bounce.

120. **Having babies** The probability that a couple will have n baby boys in a row is given by the formula $(\frac{1}{2})^n$. Find the probability that a couple will have four baby boys in a row.

121. **Investing** If an investment of $1,000 doubles every seven years, find the value of the investment after 28 years.

If P dollars are invested at a rate r, compounded annually, it will grow to A dollars in t years according to the formula

$$A = P(1 + r)^t$$

122. **Compound interest** How much will be in an account at the end of 2 years if $12,000 is invested at 5%, compounded annually?

123. **Compound interest** How much will be in an account at the end of 30 years if $8,000 is invested at 6%, compounded annually?

124. **Investing** Guess the answer to the following problem. Then use a calculator to find the correct answer. Were you close?
If the value of 1¢ is to double every day, what will the penny be worth after 31 days?

WRITING

125. Describe how you would multiply two exponential expressions with like bases.

126. Describe how you would divide two exponential expressions with like bases.

SOMETHING TO THINK ABOUT

127. Is the operation of raising to a power commutative? That is, is $a^b = b^a$? Explain.

128. Is the operation of raising to a power associative? That is, is $(a^b)^c = a^{(b^c)}$? Explain.

4.2 Zero and Negative-Integer Exponents

- Zero Exponents
- Negative-Integer Exponents
- Exponents with Variables

Getting Ready *Simplify by dividing out common factors.*

1. $\dfrac{3 \cdot 3 \cdot 3}{3 \cdot 3 \cdot 3 \cdot 3}$
2. $\dfrac{2yy}{2yyy}$
3. $\dfrac{3xx}{3xx}$
4. $\dfrac{xxy}{xxxyy}$

In the previous section, we discussed natural-number exponents. We now continue the discussion and include 0 and integer exponents.

Zero Exponents

When we discussed the quotient rule for exponents in the previous section, the exponent in the numerator was always greater than the exponent in the denominator. We now consider what happens when the exponents are equal.

If we apply the quotient rule to the fraction $\frac{5^3}{5^3}$, where the exponents in the numerator and denominator are equal, we obtain 5^0. However, because any nonzero number divided by itself equals 1, we also obtain 1.

$$\frac{5^3}{5^3} = 5^{3-3} = 5^0 \qquad \frac{5^3}{5^3} = \frac{\overset{1}{\cancel{5}} \cdot \overset{1}{\cancel{5}} \cdot \overset{1}{\cancel{5}}}{\underset{1}{\cancel{5}} \cdot \underset{1}{\cancel{5}} \cdot \underset{1}{\cancel{5}}} = 1$$

These are equal.

For this reason, we will define 5^0 to be equal to 1. In general, the following is true.

Zero Exponents

If x is any nonzero real number, then

$$x^0 = 1$$

Since $x \neq 0$, 0^0 is undefined.

EXAMPLE 1 Write each expression without exponents.

a. $\left(\dfrac{1}{13}\right)^0 = 1$

b. $\dfrac{x^5}{x^5} = x^{5-5} \quad (x \neq 0)$
$= x^0$
$= 1$

c. $3x^0 = 3(1)$
$= 3$

d. $(3x)^0 = 1$

e. $\dfrac{6^n}{6^n} = 6^{n-n}$
$= 6^0$
$= 1$

f. $\dfrac{y^m}{y^m} = y^{m-m} \quad (y \neq 0)$
$= y^0$
$= 1$

Parts c and d point out that $3x^0 \neq (3x)^0$.

Self Check Write each expression without exponents: **a.** $(-0.115)^0$, **b.** $\frac{4^2}{4^2}$, and **c.** $\frac{x^m}{x^m}$ $(x \neq 0)$.

Negative-Integer Exponents

If we apply the quotient rule to $\frac{6^2}{6^5}$, where the exponent in the numerator is less than the exponent in the denominator, we obtain 6^{-3}. However, by dividing out two factors of 6, we also obtain $\frac{1}{6^3}$.

$$\frac{6^2}{6^5} = 6^{2-5} = 6^{-3} \qquad \frac{6^2}{6^5} = \frac{\overset{1}{\cancel{6}} \cdot \overset{1}{\cancel{6}}}{\underset{1}{\cancel{6}} \cdot \underset{1}{\cancel{6}} \cdot 6 \cdot 6 \cdot 6} = \frac{1}{6^3}$$

These are equal.

For these reasons, we define 6^{-3} to be $\frac{1}{6^3}$. In general, the following is true.

Negative Exponents

If x is any nonzero number and n is a natural number, then

$$x^{-n} = \frac{1}{x^n}$$

EXAMPLE 2 Express each quantity without negative exponents or parentheses. Assume that no denominators are zero.

a. $3^{-5} = \frac{1}{3^5}$
$= \frac{1}{243}$

b. $x^{-4} = \frac{1}{x^4}$

c. $(2x)^{-2} = \frac{1}{(2x)^2}$
$= \frac{1}{4x^2}$

d. $2x^{-2} = 2\left(\frac{1}{x^2}\right)$
$= \frac{2}{x^2}$

e. $(-3a)^{-4} = \frac{1}{(-3a)^4}$
$= \frac{1}{81a^4}$

f. $(x^3x^2)^{-3} = (x^5)^{-3}$
$= \frac{1}{(x^5)^3}$
$= \frac{1}{x^{15}}$

Self Check Write each expression without negative exponents or parentheses: **a.** a^{-5}, **b.** $(3y)^{-3}$, and **c.** $(a^4a^3)^{-2}$.

Because of the definitions of negative and zero exponents, the product, power, and quotient rules are true for all integer exponents.

Properties of Exponents

If m and n are integers and there are no divisions by 0, then

$$x^m x^n = x^{m+n} \qquad (x^m)^n = x^{m \cdot n} \qquad (xy)^n = x^n y^n \qquad \left(\frac{x}{y}\right)^n = \frac{x^n}{y^n}$$

$$x^0 = 1 \quad (x \neq 0) \qquad x^{-n} = \frac{1}{x^n} \qquad \frac{x^m}{x^n} = x^{m-n}$$

EXAMPLE 3 Simplify and write the result without negative exponents. Assume that no denominators are zero.

a. $(x^{-3})^2 = x^{-6}$
$= \dfrac{1}{x^6}$

b. $\dfrac{x^3}{x^7} = x^{3-7}$
$= x^{-4}$
$= \dfrac{1}{x^4}$

c. $\dfrac{y^{-4}y^{-3}}{y^{-20}} = \dfrac{y^{-7}}{y^{-20}}$
$= y^{-7-(-20)}$
$= y^{-7+20}$
$= y^{13}$

d. $\dfrac{12a^3b^4}{4a^5b^2} = 3a^{3-5}b^{4-2}$
$= 3a^{-2}b^2$
$= \dfrac{3b^2}{a^2}$

e. $\left(-\dfrac{x^3y^2}{xy^{-3}}\right)^{-2} = (-x^{3-1}y^{2-(-3)})^{-2}$
$= (-x^2y^5)^{-2}$
$= \dfrac{1}{(-x^2y^5)^2}$
$= \dfrac{1}{x^4y^{10}}$

Self Check Simplify and write the result without negative exponents:
a. $(x^4)^{-3}$, **b.** $\dfrac{a^4}{a^8}$, **c.** $\dfrac{a^{-4}a^{-5}}{a^{-3}}$, and **d.** $\dfrac{20x^5y^3}{5x^3y^6}$.

Exponents with Variables

These properties of exponents are also true when the exponents are algebraic expressions.

EXAMPLE 4 Simplify each expression.

a. $x^{2m}x^{3m} = x^{2m+3m}$
$= x^{5m}$

b. $\dfrac{y^{2m}}{y^{4m}} = y^{2m-4m} \quad (y \neq 0)$
$= y^{-2m}$
$= \dfrac{1}{y^{2m}}$

c. $a^{2m-1}a^{2m} = a^{2m-1+2m}$
$= a^{4m-1}$

d. $(b^{m+1})^2 = b^{(m+1)2}$
$= b^{2m+2}$

Self Check Simplify: **a.** $z^{3n}z^{2n}$, **b.** $\frac{z^{3n}}{z^{5n}}$, and **c.** $(x^{m+2})^3$.

Accent on Technology — FINDING PRESENT VALUE

To find out how much money P must be invested at an annual rate i (expressed as a decimal) to have $\$A$ in n years, we use the formula $P = A(1 + i)^{-n}$. To find out how much we must invest at 6% to have $50,000 in 10 years, we substitute 50,000 for A, 0.06 (6%) for i, and 10 for n to get

$$P = A(1 + i)^{-n}$$
$$P = 50{,}000(1 + 0.06)^{-10}$$

To evaluate P with a calculator, we enter these numbers and press these keys:

(1 + .06) y^x 10 +/− × 50000 — On a calculator with a y^x and a +/− key.

50000 (1 + .06) ∧ ((−) 10) ENTER — On a graphing calculator.

Either way, we see that we must invest $27,919.74 to have $50,000 in 10 years.

Self Check Answers

1. a. 1, **b.** 1, **c.** 1 **2. a.** $\frac{1}{a^5}$, **b.** $\frac{1}{27y^3}$, **c.** $\frac{1}{a^{14}}$ **3. a.** $\frac{1}{x^{12}}$, **b.** $\frac{1}{a^4}$, **c.** $\frac{1}{a^6}$, **d.** $\frac{4x^2}{y^3}$ **4. a.** z^{5n}, **b.** $\frac{1}{z^{2n}}$, **c.** x^{3m+6}

Orals *Simplify each quantity.*

1. 2^{-1} **2.** 2^{-2} **3.** $\left(\frac{1}{2}\right)^{-1}$ **4.** $\left(\frac{7}{9}\right)^0$

5. $x^{-1}x^2$ **6.** $y^{-2}y^{-5}$ **7.** $\frac{x^5x^2}{x^7}$ **8.** $\left(\frac{x}{y}\right)^{-1}$

4.2 EXERCISES

REVIEW

1. If $a = -2$ and $b = 3$, evaluate $\frac{3a^2 + 4b + 8}{a + 2b^2}$.

2. Evaluate: $|-3 + 5 \cdot 2|$.

Solve each equation.

3. $5\left(x - \frac{1}{2}\right) = \frac{7}{2}$

4. $\frac{5(2 - x)}{6} = \frac{x + 6}{2}$

5. Solve $P = L + \frac{s}{f}i$ for s.

6. Solve $P = L + \frac{s}{f}i$ for i.

VOCABULARY AND CONCEPTS *Fill in the blanks.*

7. If x is any nonzero real number, then $x^0 =$ __.

8. If x is any nonzero real number, then $x^{-n} =$ __.

9. Since $\frac{6^4}{6^4} = 6^{4-4} = 6^0$ and $\frac{6^4}{6^4} = 1$, we define 6^0 to be __.

10. Since $\frac{8^3}{8^5} = 8^{3-5} = 8^{-2}$ and

$\frac{8^3}{8^5} = \frac{\not{8} \cdot \not{8} \cdot \not{8}}{\not{8} \cdot \not{8} \cdot \not{8} \cdot 8 \cdot 8} = \frac{1}{8^2}$, we define 8^{-2} to be ___.

PRACTICE *Simplify each expression. Write each answer without parentheses or negative exponents.*

11. $2^5 \cdot 2^{-2}$
12. $10^2 \cdot 10^{-4} \cdot 10^5$
13. $4^{-3} \cdot 4^{-2} \cdot 4^5$
14. $3^{-4} \cdot 3^5 \cdot 3^{-3}$
15. $\frac{3^5 \cdot 3^{-2}}{3^3}$
16. $\frac{6^2 \cdot 6^{-3}}{6^{-2}}$
17. $\frac{2^5 \cdot 2^7}{2^6 \cdot 2^{-3}}$
18. $\frac{5^{-2} \cdot 5^{-4}}{5^{-6}}$
19. $2x^0$
20. $(2x)^0$
21. $(-x)^0$
22. $-x^0$
23. $\left(\frac{a^2b^3}{ab^4}\right)^0$
24. $\frac{2}{3}\left(\frac{xyz}{x^2y}\right)^0$
25. $\frac{x^0 - 5x^0}{2x^0}$
26. $\frac{4a^0 + 2a^0}{3a^0}$
27. x^{-2}
28. y^{-3}
29. b^{-5}
30. c^{-4}
31. $(2y)^{-4}$
32. $(-3x)^{-1}$
33. $(ab^2)^{-3}$
34. $(m^2n^3)^{-2}$
35. $\frac{y^4}{y^5}$
36. $\frac{t^7}{t^{10}}$
37. $\frac{(r^2)^3}{(r^3)^4}$
38. $\frac{(b^3)^4}{(b^5)^4}$
39. $\frac{y^4y^3}{y^4y^{-2}}$
40. $\frac{x^{12}x^{-7}}{x^3x^4}$
41. $\frac{a^4a^{-2}}{a^2a^0}$
42. $\frac{b^0b^3}{b^{-3}b^4}$
43. $(ab^2)^{-2}$
44. $(c^2d^3)^{-2}$
45. $(x^2y)^{-3}$
46. $(-xy^2)^{-4}$
47. $(x^{-4}x^3)^3$
48. $(y^{-2}y)^3$
49. $(y^3y^{-2})^{-2}$
50. $(x^{-3}x^{-2})^2$
51. $(a^{-2}b^{-3})^{-4}$
52. $(y^{-3}z^5)^{-6}$
53. $(-2x^3y^{-2})^{-5}$
54. $(-3u^{-2}v^3)^{-3}$
55. $\left(\frac{a^3}{a^{-4}}\right)^2$
56. $\left(\frac{a^4}{a^{-3}}\right)^3$
57. $\left(\frac{b^5}{b^{-2}}\right)^{-2}$
58. $\left(\frac{b^{-2}}{b^3}\right)^{-3}$
59. $\left(\frac{4x^2}{3x^{-5}}\right)^4$
60. $\left(\frac{-3r^4r^{-3}}{r^{-3}r^7}\right)^3$
61. $\left(\frac{12y^3z^{-2}}{3y^{-4}z^3}\right)^2$
62. $\left(\frac{6xy^3}{3x^{-1}y}\right)^3$
63. $\left(\frac{2x^3y^{-2}}{4xy^2}\right)^7$
64. $\left(\frac{9u^2v^3}{18u^{-3}v}\right)^4$
65. $\left(\frac{14u^{-2}v^3}{21u^{-3}v}\right)^4$
66. $\left(\frac{-27u^{-5}v^{-3}w}{18u^3v^{-2}}\right)^4$
67. $\left(\frac{6a^2b^3}{2ab^2}\right)^{-2}$
68. $\left(\frac{15r^2s^{-2}t}{3r^{-3}s^3}\right)^{-3}$
69. $\left(\frac{18a^2b^3c^{-4}}{3a^{-1}b^2c}\right)^{-3}$
70. $\left(\frac{21x^{-2}y^2z^{-2}}{7x^3y^{-1}}\right)^{-2}$
71. $\frac{(2x^{-2}y)^{-3}}{(4x^2y^{-1})^3}$
72. $\frac{(ab^{-2}c)^2}{(a^{-2}b)^{-3}}$
73. $\frac{(17x^5y^{-5}z)^{-3}}{(17x^{-5}y^3z^2)^{-4}}$
74. $\frac{16(x^{-2}yz)^{-2}}{(2x^{-3}z^0)^4}$

Write each expression with a single exponent.

75. $x^{2m}x^m$
76. $y^{3m}y^{2m}$
77. $u^{2m}v^{3n}u^{3m}v^{-3n}$
78. $r^{2m}s^{-3}r^{3m}s^3$
79. $y^{3m+2}y^{-m}$
80. $x^{m+1}x^m$

81. $\dfrac{y^{3m}}{y^{2m}}$

82. $\dfrac{z^{4m}}{z^{2m}}$

83. $\dfrac{x^{3n}}{x^{6n}}$

84. $\dfrac{x^{m}}{x^{5m}}$

85. $(x^{m+1})^2$

86. $(y^2)^{m+1}$

87. $(x^{3-2n})^{-4}$

88. $(y^{1-n})^{-3}$

89. $(y^{2-n})^{-4}$

90. $(x^{3-4n})^{-2}$

APPLICATIONS

91. Present value How much money must be invested at 7% to have \$100,000 in 40 years?

92. Present value How much money must be invested at 8% to have \$100,000 in 40 years?

93. Present value How much money must be invested at 9% to have \$100,000 in 40 years?

94. Present value How much must be invested at 6% annual interest to have \$1,000,000 in 60 years?

95. Present value How much must be invested at 8% to have \$1,000,000 in 60 years?

96. Biology During bacterial reproduction, the time required for a population to double is called the **generation time.** If b bacteria are introduced into a medium, then after the generation time has elapsed, there will be $2b$ bacteria. After n generations, there will be $b \cdot 2^n$ bacteria. Give the meaning of this expression when $n = 0$.

WRITING

97. Tell how you would help a friend understand that 2^{-3} is not equal to -8.

98. Describe how you would verify on a calculator that

$$2^{-3} = \frac{1}{2^3}$$

SOMETHING TO THINK ABOUT

99. If a positive number x is raised to a negative power, is the result greater than, equal to, or less than x? Explore the possibilities.

100. We know that $x^{-n} = \dfrac{1}{x^n}$. Is it also true that $x^n = \dfrac{1}{x^{-n}}$? Explain.

4.3 Scientific Notation

- Scientific Notation
- Writing Numbers in Scientific Notation
- Changing from Scientific Notation to Standard Notation
- Using Scientific Notation to Simplify Computations

Getting Ready *Evaluate each expression.*

1. 10^2 **2.** 10^3 **3.** 10^1 **4.** 10^{-2}

5. $5(10^2)$ **6.** $8(10^3)$ **7.** $3(10^1)$ **8.** $7(10^{-2})$

We now use exponents to write very large and very small numbers in a compact form called *scientific notation.* In science, almost all large and small numbers are written in this form.

Scientific Notation

Scientists often deal with extremely large and extremely small numbers. For example,

- The distance from Earth to the sun is approximately 150,000,000 kilometers.
- Ultraviolet light emitted from a mercury arc has a wavelength of approximately 0.000025 centimeter.

The large number of zeros in these numbers makes them difficult to read and hard to remember. Scientific notation provides a compact way of writing large and small numbers.

Scientific Notation

A number is written in **scientific notation** if it is written as the product of a number between 1 (including 1) and 10 and an integer power of 10.

Each of the following numbers is written in scientific notation.

$$3.67 \times 10^{6} \qquad 2.24 \times 10^{-4} \quad \text{and} \quad 9.875 \times 10^{22}$$

Every number that is written in scientific notation has the following form:

An integer exponent

$$\_\_.\_\_ \times 10^{\_\_}$$

A decimal between 1 and 10

Writing Numbers in Scientific Notation

EXAMPLE 1 Change 150,000,000 to scientific notation

Solution We note that 1.5 lies between 1 and 10. To obtain 150,000,000, the decimal point in 1.5 must be moved eight places to the right. Because multiplying a number by 10 moves the decimal point one place to the right, we can accomplish this by multiplying 1.5 by 10 eight times.

1.5 0 0 0 0 0 0 0

8 places to the right

150,000,000 written in scientific notation is 1.5×10^{8}.

Self Check Change 93,000,000 to scientific notation.

EXAMPLE 2 Change 0.000025 to scientific notation.

Solution We note that 2.5 is between 1 and 10. To obtain 0.000025, the decimal point in 2.5 must be moved five places to the left. We can accomplish this by dividing 2.5 by 10^{5}, which is equivalent to multiplying 2.5 by $\frac{1}{10^{5}}$ (or by 10^{-5}).

0 0 0 0 2.5

5 places to the left

In scientific notation, 0.000025 is written 2.5×10^{-5}.

Self Check Write 0.00125 in scientific notation.

EXAMPLE 3 Write **a.** 235,000 and **b.** 0.00000235 in scientific notation.

Solution **a.** $235{,}000 = 2.35 \times 10^5$, because $2.35 \times 10^5 = 235{,}000$ and 2.35 is between 1 and 10.

b. $0.00000235 = 2.35 \times 10^{-6}$, because $2.35 \times 10^{-6} = 0.00000235$ and 2.35 is between 1 and 10.

Self Check Write **a.** 17,500 and **b.** 0.657 in scientific notation.

PERSPECTIVE

The Metric System

A common metric unit of length is the kilometer, which is 1,000 meters. Because 1,000 is 10^3, we can write $1 \text{ km} = 10^3 \text{ m}$. Similarly, 1 centimeter is one-hundredth of a meter: $1 \text{ cm} = 10^{-2} \text{ m}$. In the metric system, prefixes such as *kilo* and *centi* refer to powers of 10. Other prefixes are used in the metric system, as shown in the table.

To appreciate the magnitudes involved, consider these facts: Light, which travels 186,000 miles every second, will travel about one foot in one nanosecond. The distance to the nearest star is 43 petameters, and the diameter of an atom is about 10 nanometers. To measure some quantities, however, even these units are inadequate. The sun, for example, radiates 5×10^{26} watts. That's a lot of light bulbs!

Prefix	Symbol	Meaning	
peta	P	10^{15} =	1,000,000,000,000,000.
tera	T	10^{12} =	1,000,000,000,000.
giga	G	10^{9} =	1,000,000,000.
mega	M	10^{6} =	1,000,000.
kilo	k	10^{3} =	1,000.
deci	d	10^{-1} =	0.1
centi	c	10^{-2} =	0.01
milli	m	10^{-3} =	0.001
micro	μ	10^{-6} =	0.000 001
nano	n	10^{-9} =	0.000 000 001
pico	p	10^{-12} =	0.000 000 000 001
femto	f	10^{-15} =	0.000 000 000 000 001
atto	a	10^{-18} =	0.000 000 000 000 000 001

EXAMPLE 4 Write 432.0×10^5 in scientific notation.

Solution The number 432.0×10^5 is not written in scientific notation, because 432.0 is not a number between 1 and 10. To write the number in scientific notation, we proceed as follows:

$$432.0 \times 10^5 = 4.32 \times 10^2 \times 10^5 \quad \text{Write 432.0 in scientific notation.}$$
$$= 4.32 \times 10^7 \quad 10^2 \times 10^5 = 10^7.$$

Self Check Write 85×10^{-3} in scientific notation.

Changing from Scientific Notation to Standard Notation

We can change a number written in scientific notation to **standard notation.** For example, to write 9.3×10^7 in standard notation, we multiply 9.3 by 10^7.

$$\begin{aligned} 9.3 \times 10^7 &= 9.3 \times 10{,}000{,}000 \\ &= 93{,}000{,}000 \end{aligned}$$

EXAMPLE 5 Write **a.** 3.4×10^5 and **b.** 2.1×10^{-4} in standard notation.

Solution **a.** $3.4 \times 10^5 = 3.4 \times 100{,}000$
$= 340{,}000$

b. $$\begin{aligned} 2.1 \times 10^{-4} &= 2.1 \times \frac{1}{10^4} \\ &= 2.1 \times \frac{1}{10{,}000} \\ &= 0.00021 \end{aligned}$$

Self Check Write **a.** 4.76×10^5 and **b.** 9.8×10^{-3} in standard notation.

Each of the following numbers is written in both scientific and standard notation. In each case, the exponent gives the number of places that the decimal point moves, and the sign of the exponent indicates the direction that it moves.

$5.32 \times 10^5 = 5\,3\,2\,0\,0\,0.$ 5 places to the right.

$2.37 \times 10^6 = 2\,3\,7\,0\,0\,0\,0.$ 6 places to the right.

$8.95 \times 10^{-4} = 0\,.\,0\,0\,0\,8\,9\,5$ 4 places to the left.

$8.375 \times 10^{-3} = 0\,.\,0\,0\,8\,3\,7\,5$ 3 places to the left.

$9.77 \times 10^0 = 9.77$ No movement of the decimal point.

Using Scientific Notation to Simplify Computations

Another advantage of scientific notation becomes apparent when we simplify fractions such as

$$\frac{(0.0032)(25{,}000)}{0.00040}$$

that contain very large or very small numbers. Although we can simplify this fraction by using arithmetic, scientific notation provides an easier way. First, we write each number in scientific notation; then we do the arithmetic on the numbers and the exponential expressions separately. Finally, we write the result in standard form, if desired.

$$\begin{aligned}\frac{(0.0032)(25{,}000)}{0.00040} &= \frac{(3.2\times10^{-3})(2.5\times10^{4})}{4.0\times10^{-4}}\\ &= \frac{(3.2)(2.5)}{4.0}\times\frac{10^{-3}10^{4}}{10^{-4}}\\ &= \frac{8.0}{4.0}\times10^{-3+4-(-4)}\\ &= 2.0\times10^{5}\\ &= 200{,}000\end{aligned}$$

Accent on Technology — FINDING POWERS OF DECIMALS

To find the value of $(453.46)^5$, we can use a calculator and enter these numbers and press these keys:

453.46 y^x 5 = — On a calculator with a y^x key.

453.46 ∧ 5 ENTER — On a graphing calculator.

Either way, we have $(453.46)^5 = 1.917321395\times10^{13}$. Since this number is too large to show on the display, the calculator gives the result in scientific notation.

EXAMPLE 6 **Speed of light** In a vacuum, light travels 1 meter in approximately 0.000000003 second. How long does it take for light to travel 500 kilometers?

Solution Since 1 kilometer = 1,000 meters, the length of time for light to travel 500 kilometers (500 · 1,000) meters is given by

$$\begin{aligned}(0.000000003)(500)(1{,}000) &= (3\times10^{-9})(5\times10^{2})(1\times10^{3})\\ &= 3(5)\times10^{-9+2+3}\\ &= \mathbf{15}\times10^{-4}\\ &= \mathbf{1.5\times10^{1}}\times10^{-4}\\ &= 1.5\times10^{-3}\\ &= 0.0015\end{aligned}$$

Light travels 500 kilometers in approximately 0.0015 second. ∎

Self Check Answers

1. 9.3×10^{7} **2.** 1.25×10^{-3} **3. a.** 1.75×10^{4}, **b.** 6.57×10^{-1} **4.** 8.5×10^{-2} **5. a.** 476,000, **b.** 0.0098

Orals *Tell which number of each pair is the larger.*

1. 37.2 or 3.72×10^{2}

2. 37.2 or 3.72×10^{-1}

3. 3.72×10^{3} or 4.72×10^{3}

4. 3.72×10^{3} or 4.72×10^{2}

5. 3.72×10^{-1} or 4.72×10^{-2}

6. 3.72×10^{-3} or 2.72×10^{-2}

4.3 EXERCISES

REVIEW

1. If $y = -1$, find the value of $-5y^{55}$.
2. Evaluate $\dfrac{3a^2 - 2b}{2a + 2b}$ if $a = 4$ and $b = 3$.

Tell which property of real numbers justifies each statement.

3. $5 + z = z + 5$
4. $7(u + 3) = 7u + 7 \cdot 3$

Solve each equation.

5. $3(x - 4) - 6 = 0$
6. $8(3x - 5) - 4(2x + 3) = 12$

VOCABULARY AND CONCEPTS *Fill in the blanks.*

7. A number is written in ______________ when it is written as the product of a number between 1 (including 1) and 10 and an integer power of 10.
8. The number 125,000 is written in ________ notation.

PRACTICE *Write each number in scientific notation.*

9. 23,000
10. 4,750
11. 1,700,000
12. 290,000
13. 0.062
14. 0.00073
15. 0.0000051
16. 0.04
17. 42.5×10^2
18. 0.3×10^3
19. 0.25×10^{-2}
20. 25.2×10^{-3}

Write each number in standard notation.

21. 2.3×10^2
22. 3.75×10^4
23. 8.12×10^5
24. 1.2×10^3
25. 1.15×10^{-3}
26. 4.9×10^{-2}
27. 9.76×10^{-4}
28. 7.63×10^{-5}
29. 25×10^6
30. 0.07×10^3
31. 0.51×10^{-3}
32. 617×10^{-2}

Use scientific notation to simplify each expression. Give all answers in standard notation.

33. $(3.4 \times 10^2)(2.1 \times 10^3)$
34. $(4.1 \times 10^{-3})(3.4 \times 10^4)$
35. $\dfrac{9.3 \times 10^2}{3.1 \times 10^{-2}}$
36. $\dfrac{7.2 \times 10^6}{1.2 \times 10^8}$
37. $\dfrac{96{,}000}{(12{,}000)(0.00004)}$
38. $\dfrac{(0.48)(14{,}400{,}000)}{96{,}000{,}000}$

APPLICATIONS

39. **Distance to Alpha Centauri** The distance from Earth to the nearest star outside our solar system is approximately 25,700,000,000,000 miles. Write this number in scientific notation.
40. **Speed of sound** The speed of sound in air is 33,100 centimeters per second. Write this number in scientific notation.
41. **Distance to Mars** The distance from Mars to the sun is approximately 1.14×10^8 miles. Write this number in standard notation.
42. **Distance to Venus** The distance from Venus to the sun is approximately 6.7×10^7 miles. Write this number in standard notation.
43. **Length of one meter** One meter is approximately 0.00622 mile. Write this number in scientific notation.
44. **Angstrom** One angstrom is 1×10^{-7} millimeter. Write this number in standard notation.
45. **Distance between Mercury and the sun** The distance from Mercury to the sun is approximately 3.6×10^7 miles. Use scientific notation to express this distance in feet. (*Hint:* 5,280 feet = 1 mile.)
46. **Mass of a proton** The mass of one proton is approximately 1.7×10^{-24} gram. Use scientific notation to express the mass of 1 million protons.

47. **Speed of sound** The speed of sound in air is approximately 3.3×10^4 centimeters per second. Use scientific notation to express this speed in kilometers per second. (*Hint:* 100 centimeters = 1 meter and 1,000 meters = 1 kilometer.)

48. **Light year** One light year is approximately 5.87×10^{12} miles. Use scientific notation to express this distance in feet. (*Hint:* 5,280 feet = 1 mile.)

The bulk of the surface area of the red blood cell shown in the illustration is contained on its top and bottom. That area is $2\pi r^2$, twice the area of one circle. If there are N discs, their total surface area T will be N times the surface area of a single disc: $T = 2N\pi r^2$.

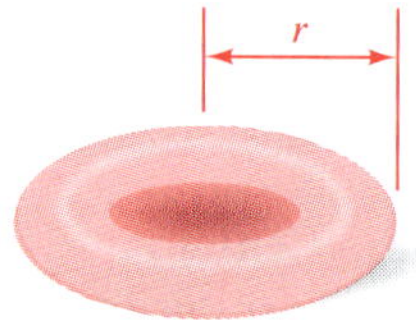

49. **Red blood cells** The red cells in human blood pick up oxygen in the lungs and carry it to all parts of the body. Each cell is a tiny circular disc with a radius of about 0.00015 in. Because the amount of oxygen carried depends on the surface area of the cells, and the cells are so tiny, a great number is needed—about 25 trillion in an average adult. Write these two numbers in scientific notation.

50. **Red blood cells** Find the total surface area of all the red blood cells in the body of an average adult. See Exercise 49.

WRITING

51. In what situations would scientific notation be more convenient than standard notation?

52. To multiply a number by a power of 10, we move the decimal point. Which way, and how far? Explain.

SOMETHING TO THINK ABOUT

53. Two positive numbers are written in scientific notation. How could you decide which is larger, without converting either to standard notation?

54. The product $1 \cdot 2 \cdot 3 \cdot 4 \cdot 5$, or 120, is called **5 factorial,** written 5!. Similarly, the number $6! = 6 \cdot 5 \cdot 4 \cdot 3 \cdot 2 \cdot 1 = 720$. Factorials get large very quickly. Calculate 30!, and write the number in standard notation. How large a factorial can you compute with a calculator?

4.4 Polynomials

- Polynomials
- Monomials, Binomials, and Trinomials
- Degree of a Polynomial
- Evaluating Polynomials
- Polynomial Functions
- Graphing Polynomial Functions

Getting Ready *Write each expression using exponents.*

1. $2xxyyy$
2. $3xyyy$
3. $2xx + 3yy$
4. $xxx + yyy$
5. $(3xxy)(2xyy)$
6. $(5xyzzz)(xyz)$
7. $3(5xy)\left(\frac{1}{3}xy\right)$
8. $(xy)(xz)(yz)(xyz)$

In algebra, exponential expressions are combined to form polynomials. In this section, we will introduce the topic of polynomials and graph some simple polynomial functions.

Polynomials

Recall that expressions such as

$$3x \qquad 4y^2 \qquad -8x^2y^3 \quad \text{and} \quad 25$$

with constant and/or variable factors are called **algebraic terms.** The numerical coefficients of the first three of these terms are 3, 4, and -8, respectively. Because $25 = 25x^0$, 25 is considered to be the numerical coefficient of the term 25.

Polynomials

> A **polynomial** is an algebraic expression that is the sum of one or more terms containing whole-number exponents on the variables.

Here are some examples of polynomials:

$$8xy^2t \qquad 3x + 2 \qquad 4y^2 - 2y + 3 \quad \text{and} \quad 3a - 4b - 4c + 8d$$

Comment The expression $2x^3 - 3y^{-2}$ is not a polynomial, because the second term contains a negative exponent on a variable base.

EXAMPLE 1 Tell whether each expression is a polynomial.

a. $x^2 + 2x + 1$ Yes.

b. $3x^{-1} - 2x - 3$ No. The first term has a negative exponent on a variable base.

c. $\frac{1}{2}x^3 - 2.3x + 5$ Yes.

Self Check Tell whether each expression is a polynomial: **a.** $3x^{-4} + 2x^2 - 3$ and **b.** $7.5x^3 - 4x^2 - 3x$.

Monomials, Binomials, and Trinomials

A polynomial with one term is called a **monomial.** A polynomial with two terms is called a **binomial.** A polynomial with three terms is called a **trinomial.** Here are some examples.

Monomials	Binomials	Trinomials
$5x^2y$	$3u^3 - 4u^2$	$-5t^2 + 4t + 3$
$-6x$	$18a^2b + 4ab$	$27x^3 - 6x - 2$
29	$-29z^{17} - 1$	$-32r^6 + 7y^3 - z$

EXAMPLE 2 Classify each polynomial as a monomial, a binomial, or a trinomial.

a. $5x^4 + 3x$ Since the polynomial has two terms, it is a binomial.

b. $7x^4 - 5x^3 - 2$ Since the polynomial has three terms, it is a trinomial.

c. $-5x^2y^3$ Since the polynomial has one term, it is a monomial.

Self Check Classify each polynomial as a monomial, a binomial, or a trinomial: **a.** $5x$, **b.** $-5x^2 + 2x - 5$, and **c.** $16x^2 - 9y^2$.

Degree of a Polynomial

The monomial $7x^6$ is called a **monomial of sixth degree** or a **monomial of degree 6,** because the variable x occurs as a factor six times. The monomial $3x^3y^4$ is a monomial of the seventh degree, because the variables x and y occur as factors a total of seven times. Other examples are

$-2x^3$ is a monomial of degree 3.

$47x^2y^3$ is a monomial of degree 5.

$18x^4y^2z^8$ is a monomial of degree 14.

8 is a monomial of degree 0, because $8 = 8x^0$.

These examples illustrate the following definition.

Degree of a Monomial

If a is a nonzero constant, the **degree of the monomial ax^n** is n.

The **degree of a monomial with several variables** is the sum of the exponents on those variables.

Comment Note that the degree of ax^n is not defined when $a = 0$. Since $ax^n = 0$ when $a = 0$, the constant 0 has no defined degree.

Because each term of a polynomial is a monomial, we define the degree of a polynomial by considering the degree of each of its terms.

Degree of a Polynomial

The **degree of a polynomial** is the same as the degree of its term with largest degree.

For example,

- $x^2 + 2x$ is a binomial of degree 2, because the degree of its first term is 2 and the degree of its other term is less than 2.
- $3x^3y^2 + 4x^4y^4 - 3x^3$ is a trinomial of degree 8, because the degree of its second term is 8 and the degree of each of its other terms is less than 8.
- $25x^4y^3z^7 - 15xy^8z^{10} - 32x^8y^8z^3 + 4$ is a polynomial of degree 19, because its second and third terms are of degree 19. Its other terms have degrees less than 19.

EXAMPLE 3 Find the degree of each polynomial.

a. $-4x^3 - 5x^2 + 3x$ 3, the degree of the term with largest degree.

b. $5x^4y^2 + 7xy^2 - 16x^3y^5$ 8, the degree of the last term.

c. $-17a^2b^3c^4 + 12a^3b^4c$ 9, the degree of the first term.

Self Check Find the degree of each polynomial: **a.** $15p^3q^4 - 25p^4q^2$ and **b.** $-14rs^3t^4 + 12r^3s^3t^3$.

Evaluating Polynomials

When a number is substituted for the variable in a polynomial, the polynomial takes on a numerical value. Finding that value is called **evaluating the polynomial.**

EXAMPLE 4 Evaluate the polynomial $3x^2 + 2$ when **a.** $x = 0$, **b.** $x = 2$, **c.** $x = -3$, and **d.** $x = -\frac{1}{5}$.

Solution

a. $$\begin{aligned} 3x^2 + 2 &= 3(0)^2 + 2 \\ &= 3(0) + 2 \\ &= 0 + 2 \\ &= 2 \end{aligned}$$

b. $$\begin{aligned} 3x^2 + 2 &= 3(2)^2 + 2 \\ &= 3(4) + 2 \\ &= 12 + 2 \\ &= 14 \end{aligned}$$

c. $$\begin{aligned} 3x^2 + 2 &= 3(-3)^2 + 2 \\ &= 3(9) + 2 \\ &= 27 + 2 \\ &= 29 \end{aligned}$$

d. $$\begin{aligned} 3x^2 + 2 &= 3\left(-\frac{1}{5}\right)^2 + 2 \\ &= 3\left(\frac{1}{25}\right) + 2 \\ &= \frac{3}{25} + \frac{50}{25} \\ &= \frac{53}{25} \end{aligned}$$

Self Check Evaluate $3x^2 + x - 2$ when **a.** $x = 2$ and **b.** $x = -1$.

When we evaluate a polynomial function for several values of its variable, we often write the results in a table.

EXAMPLE 5 Evaluate the polynomial $x^3 + 1$ when **a.** $x = -2$, **b.** $x = -1$, **c.** $x = 0$, **d.** $x = 1$, and **e.** $x = 2$. Write the results in a table.

Solution

	x	$x^3 + 1$	
a.	-2	-7	$x^3 + 1 = (-2)^3 + 1 = -7.$
b.	-1	0	$x^3 + 1 = (-1)^3 + 1 = 0.$
c.	0	1	$x^3 + 1 = (0)^3 + 1 = 1.$
d.	1	2	$x^3 + 1 = (1)^3 + 1 = 2.$
e.	2	9	$x^3 + 1 = (2)^3 + 1 = 9.$

Self Check Consider the polynomial $-x^3 + 1$. Complete the following table.

x	$-x^3 + 1$
-2	
-1	
0	
1	
2	

Polynomial Functions

Since the right-hand sides of the functions $y = f(x) = 2x - 3$, $y = f(x) = x^2$, and $y = f(t) = -16t^2 + 64t$ are polynomials, they are called **polynomial functions.** We can evaluate these functions at specific values of x by evaluating the polynomial on the right-hand side.

EXAMPLE 6 Evaluate $y = f(x) = 2x - 3$ for $x = -2$.

Solution To find $y = f(-2)$, we substitute -2 for x and evaluate y.

$$
\begin{aligned}
y &= 2x - 3 \\
y &= 2(-2) - 3 \\
&= -4 - 3 \\
&= -7
\end{aligned}
$$

Thus, $f(-2) = -7$.

Self Check Find $f(5)$.

Accent on Technology

HEIGHT OF A ROCKET

The height h (in feet) of a toy rocket launched straight up into the air with an initial velocity of 64 feet per second is given by the polynomial function

$$h = f(t) = -16t^2 + 64t$$

In this case, the height h is the dependent variable, and the time t is the independent variable. To find the height of the rocket 3.5 seconds after launch, we substitute 3.5 for t and evaluate h.

$$
\begin{aligned}
h &= -16t^2 + 64t \\
h &= -16(3.5)^2 + 64(3.5)
\end{aligned}
$$

To evaluate h using a calculator, we enter these numbers and press these keys:

16 [+/−] [×] 3.5 [x^2] [+] [(] 64 [×] 3.5 [)] [=] — On a calculator with a [+/−] key.

[(−)] 16 [×] 3.5 [x^2] [+] 64 [×] 3.5 [ENTER] — On a graphing calculator.

Either way, the display reads 28. After 3.5 seconds, the rocket will be 28 feet above the ground.

Graphing Polynomial Functions

We can graph polynomial functions as we graphed equations in Section 3.2. We make a table of values, plot points, and draw the line or curve that passes through those points.

In the next example, we graph the function $y = f(x) = 2x - 3$. Since its graph is a line, it is a **linear function.**

EXAMPLE 7 Graph: $y = f(x) = 2x - 3$.

Solution We substitute numbers for x, compute the corresponding values of $f(x)$, and list the results in a table, as in Figure 4-1. We then plot the pairs (x, y) and draw a line through the points, as shown in the figure. From the graph, we can see that x can be any value. This confirms that the domain is the set of all real numbers. We can also see that y can be any value. This confirms that the range is also the set of all real numbers.

$y = f(x) = 2x - 3$

x	y	(x, y)
−3	−9	(−3, −9)
−2	−7	(−2, −7)
−1	−5	(−1, −5)
0	−3	(0, −3)
1	−1	(1, −1)
2	1	(2, 1)
3	3	(3, 3)

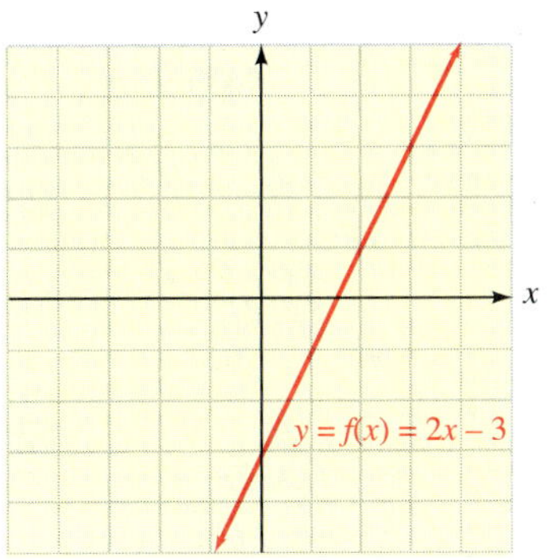

Figure 4-1

Self Check Graph $y = f(x) = \frac{1}{2}x + 3$ and tell whether it is a linear function.

In the next example, we graph the function $y = f(x) = x^2$, called the **squaring function.** Since the polynomial on the right-hand side is of second degree, we call this function a **quadratic function.**

EXAMPLE 8 Graph: $y = f(x) = x^2$.

Solution We substitute numbers for x, compute the corresponding values of $f(x)$, and list the results in a table, as in Figure 4-2. We then plot the pairs (x, y) and draw a smooth curve through the points, as shown in the figure. This curve is called a **parabola.** From the graph, we can see that x can be any value. This confirms that the domain is the set of all real numbers. We can also see that y is always a positive number or 0. This confirms that the range is the set of all real numbers such that $y \geq 0$.

$y = f(x) = x^2$

x	y	(x, y)
−3	9	(−3, 9)
−2	4	(−2, 4)
−1	1	(−1, 1)
0	0	(0, 0)
1	1	(1, 1)
2	4	(2, 4)
3	9	(3, 9)

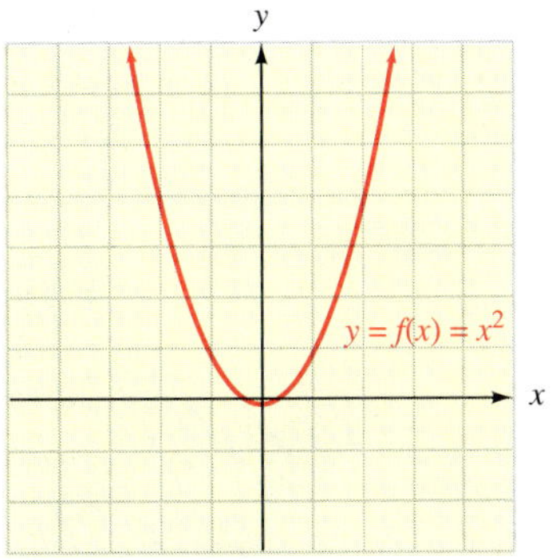

Figure 4-2

Self Check Graph $y = f(x) = x^2 - 3$ and compare the graph to the graph of $y = f(x) = x^2$ shown in Figure 4-2.

In the next example, we graph the function $y = f(x) = x^3$, called the **cubing function.**

EXAMPLE 9 Graph: $y = f(x) = x^3$.

Solution We substitute numbers for x, compute the corresponding values of $f(x)$, and list the results in a table, as in Figure 4-3. We then plot the pairs (x, y) and draw a smooth curve through the points, as shown in the figure.

$y = f(x) = x^3$

x	y	(x, y)
−2	−8	(−2, −8)
−1	−1	(−1, −1)
0	0	(0, 0)
1	1	(1, 1)
2	8	(2, 8)

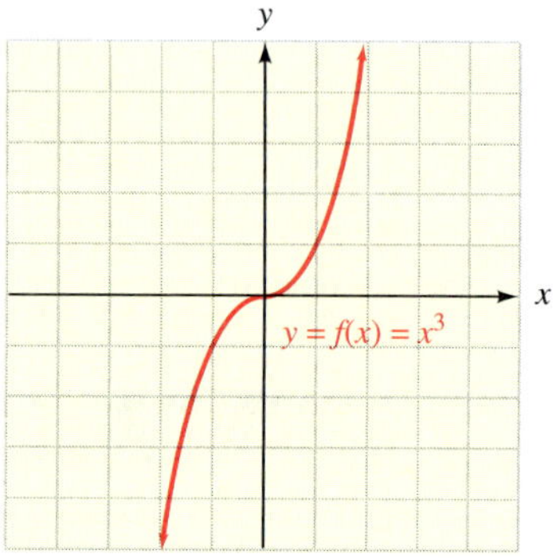

Figure 4-3

Self Check Graph $y = f(x) = x^3 + 3$ and compare the graph to the graph of $y = f(x) = x^3$ shown in Figure 4-3.

Accent on Technology GRAPHING POLYNOMIAL FUNCTIONS

It is possible to use a graphing calculator to generate tables and graphs for polynomial functions. For example, Figure 4-4 shows calculator tables and graphs of $y = f(x) = 2x - 3$, $y = f(x) = x^2$, and $y = f(x) = x^3$.

X	Y1	
−3	−9	
−2	−7	
−1	−5	
0	−3	
1	−1	
2	1	
3	3	

X = −3

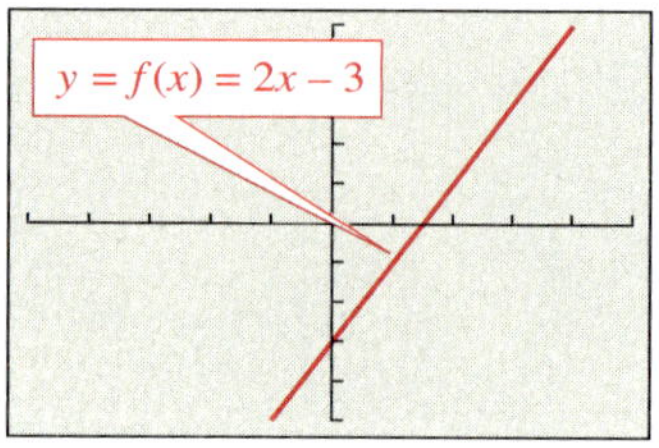

(a)

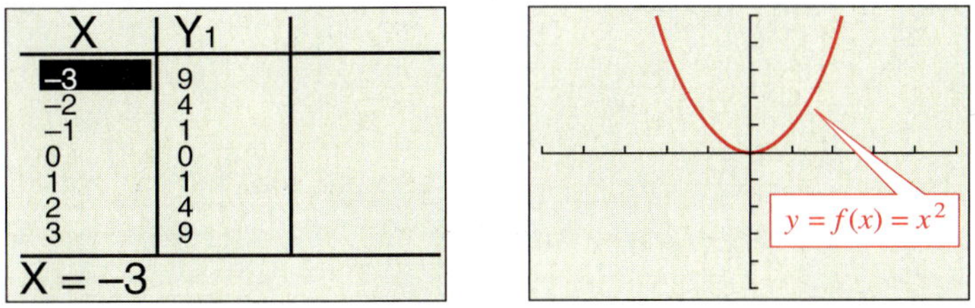

(b)

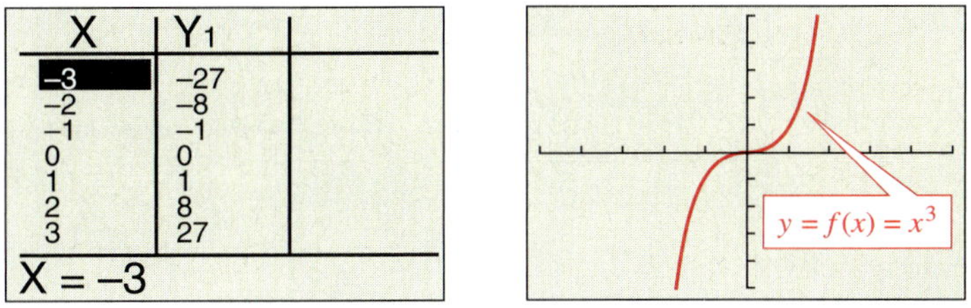

(c)

Figure 4-4

EXAMPLE 10 Graph: $y = f(x) = x^2 - 2x$.

Solution We substitute numbers for x, compute the corresponding values of $f(x)$, and list the results in a table, as in Figure 4-5. We then plot the pairs (x, y) and draw a smooth curve through the points, as shown in the figure.

$y = f(x) = x^2 - 2x$

x	y	(x, y)
-2	8	$(-2, 8)$
-1	3	$(-1, 3)$
0	0	$(0, 0)$
1	-1	$(1, -1)$
2	0	$(2, 0)$
3	3	$(3, 3)$
4	8	$(4, 8)$

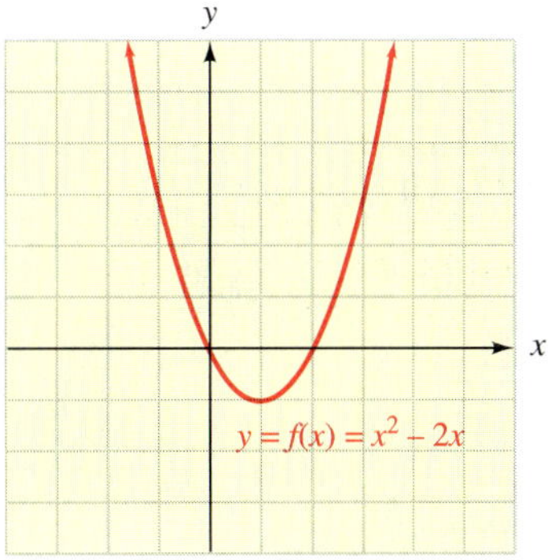

Figure 4-5

Self Check Use a graphing calculator to graph $y = f(x) = x^2 - 2x$.

Self Check Answers

1. a. no, **b.** yes **2. a.** monomial, **b.** trinomial, **c.** binomial **3. a.** 7, **b.** 9 **4. a.** 12, **b.** 0

5. 9, 2, 1, 0, −7 **6.** 7 **7.** a linear function

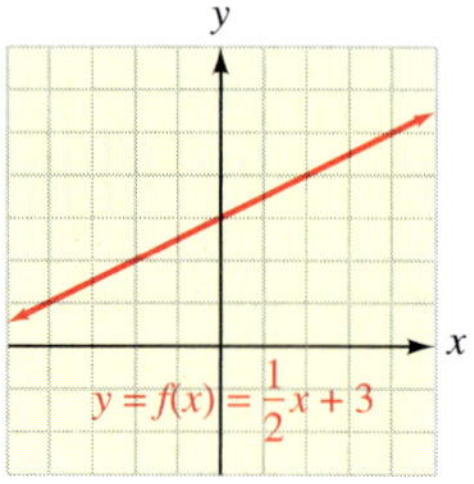

8. same shape but 3 units lower

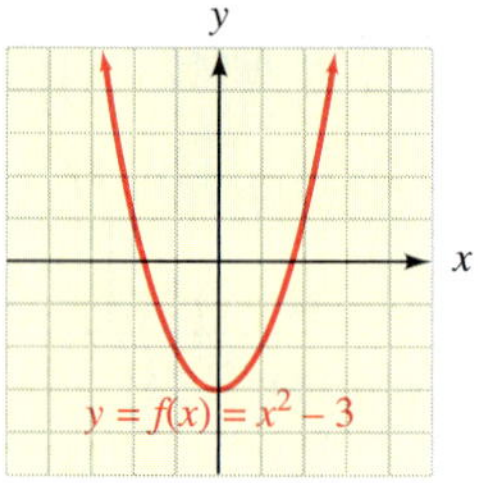

9. same shape but 3 units higher

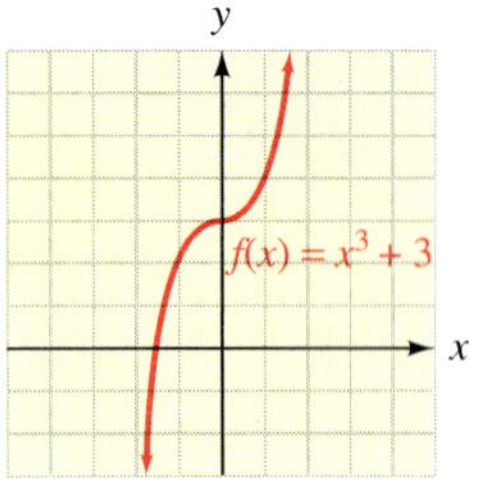

10.

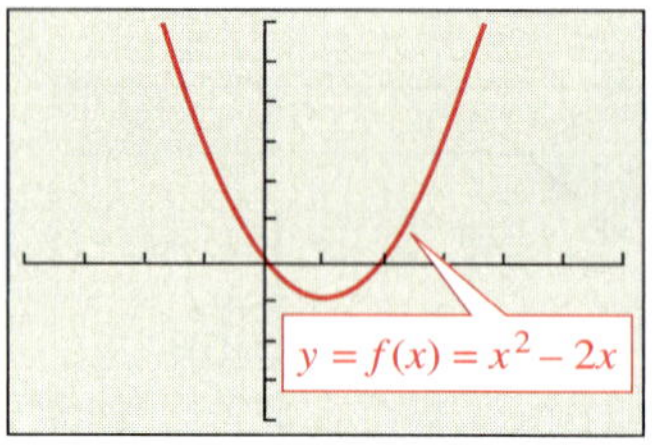

Orals *Give an example of a polynomial that is . . .*

1. a binomial
2. a monomial
3. a trinomial
4. not a monomial, a binomial, or a trinomial
5. of degree 3
6. of degree 1
7. of degree 0
8. has no defined degree

4.4 EXERCISES

REVIEW *Solve each equation.*

1. $5(u - 5) + 9 = 2(u + 4)$

2. $8(3a - 5) - 12 = 4(2a + 3)$

Solve each inequality and graph the solution set.

3. $-4(3y + 2) \leq 28$

4. $-5 < 3t + 4 \leq 13$

Write each expression without using parentheses or negative exponents.

5. $(x^2x^4)^3$

6. $(a^2)^3(a^3)^2$

7. $\left(\dfrac{y^2y^5}{y^4}\right)^3$

8. $\left(\dfrac{2t^3}{t}\right)^{-4}$

VOCABULARY AND CONCEPTS *Fill in the blanks.*

9. An expression with a constant and/or a variable is called an ________ term.

10. The numerical coefficient of the term $-25x^2y^3$ is ____.

11. A ________ is an algebraic expression that is the sum of one or more terms containing whole-number exponents on the variables.

12. A ________ is a polynomial with two terms.

13. A ________ is a polynomial with three terms.

14. A ________ is a polynomial with one term.

15. If $a \neq 0$, the ______ of ax^n is n.

16. The degree of a monomial with several variables is the _____ of the exponents on those variables.

17. Any equation in x and y where each input value x determines exactly one output value y is called a ________.

18. $f(x)$ is read as ______.

19. In a function, the set of all input values is called the ________.

20. In a function, the set of all output values is called the ______.

Tell whether each expression is a polynomial.

21. $x^3 - 5x^2 - 2$

22. $x^{-4} - 5x$

23. $\frac{1}{2}x^3 + 3$

24. $x^3 - 1$

Classify each polynomial as a monomial, a binomial, a trinomial, or none of these.

25. $3x + 7$

26. $3y - 5$

27. $3y^2 + 4y + 3$

28. $3xy$

29. $3z^2$

30. $3x^4 - 2x^3 + 3x - 1$

31. $5t - 32$

32. $9x^2y^3z^4$

33. $s^2 - 23s + 31$

34. $12x^3 - 12x^2 + 36x - 3$

35. $3x^5 - 2x^4 - 3x^3 + 17$

36. x^3

Give the degree of each polynomial.

37. $3x^4$

38. $3x^5 - 4x^2$

39. $-2x^2 + 3x^3$

40. $-5x^5 + 3x^2 - 3x$

41. $3x^2y^3 + 5x^3y^5$

42. $-2x^2y^3 + 4x^3y^2z$

43. $-5r^2s^2t - 3r^3st^2 + 3$

44. $4r^2s^3t^3 - 5r^2s^8$

45. $x^{12} + 3x^2y^3z^4$

46. 17^2x

47. 38

48. -25

PRACTICE *Evaluate* $5x - 3$ *for each value.*

49. $x = 2$

50. $x = 0$

51. $x = -1$

52. $x = -2$

Evaluate $-x^2 - 4$ *for each value.*

53. $x = 0$

54. $x = 1$

55. $x = -1$

56. $x = -2$

If $y = f(x) = 5x + 1$, *find each value.*

57. $f(0)$

58. $f(2)$

59. $f(-2)$

60. $f\left(\frac{2}{5}\right)$

If $y = f(x) = x^2 - 2x + 3$, *find each value.*

61. $f(0)$

62. $f(3)$

63. $f(-2)$

64. $f(-1)$

Complete each table.

65.

x	$x^2 - 3$
-2	
-1	
0	
1	
2	

66.

x	$-x^2 + 3$
-2	
-1	
0	
1	
2	

67.

x	$x^3 + 2$
-2	
-1	
0	
1	
2	

68.

x	$-x^3 + 2$
-2	
-1	
0	
1	
2	

Graph each polynomial function. Check your work with a graphing calculator.

69. $f(x) = x^2 - 1$

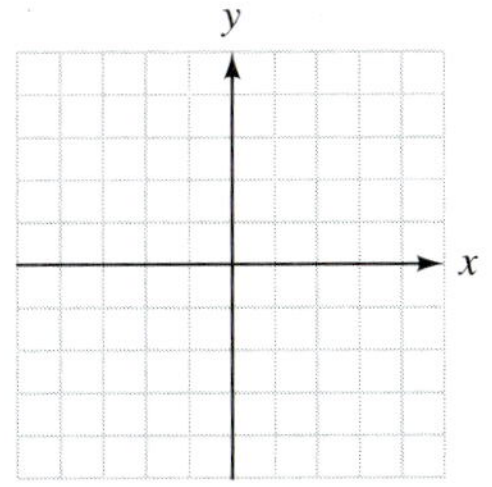

70. $f(x) = x^2 + 2$

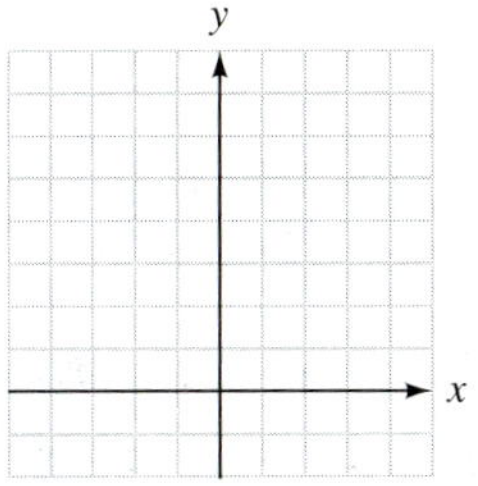

71. $f(x) = x^3 + 2$

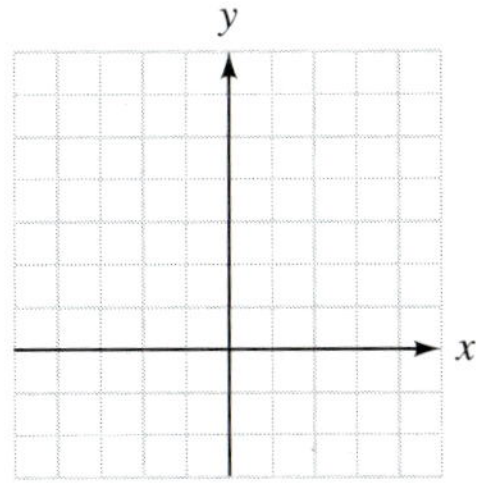

72. $f(x) = x^3 - 2$

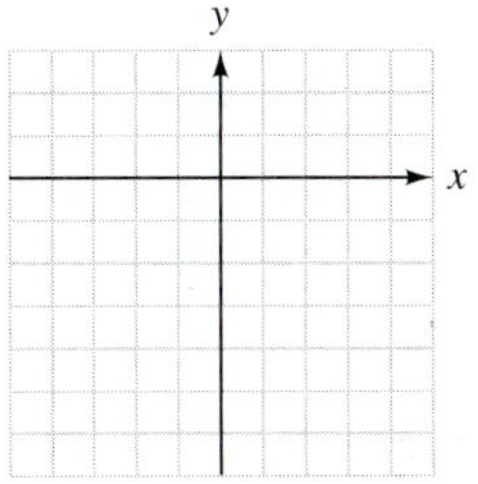

APPLICATIONS *Use a calculator to help solve each problem.*

73. **Height of a rocket** See the Accent on Technology section on page 254. Find the height of the rocket 2 seconds after launch.

74. Height of a rocket Again referring to page 254, make a table of values to find the rocket's height at various times. For what values of t will the height of the rocket be 0?

75. Computing revenue The revenue r (in dollars) that a manufacturer of desk chairs receives is given by the polynomial function

$$r = f(d) = -0.08d^2 + 100d$$

where d is the number of chairs manufactured. Find the revenue received when 815 chairs are manufactured.

76. Falling balloons Some students dropped balloons filled with water from a dormitory window. The height h (in feet) of the balloons t seconds after being dropped is given by the polynomial function

$$h = f(t) = -16t^2 + 12t + 20$$

How far above the ground is a balloon 1.5 seconds after being dropped?

77. Stopping distance The number of feet that a car travels before stopping depends on the driver's reaction time and the braking distance. For one driver, the stopping distance d is given by the function $d = f(v) = 0.04v^2 + 0.9v$, where v is the velocity of the car. Find the stopping distance when the driver is traveling at 30 mph.

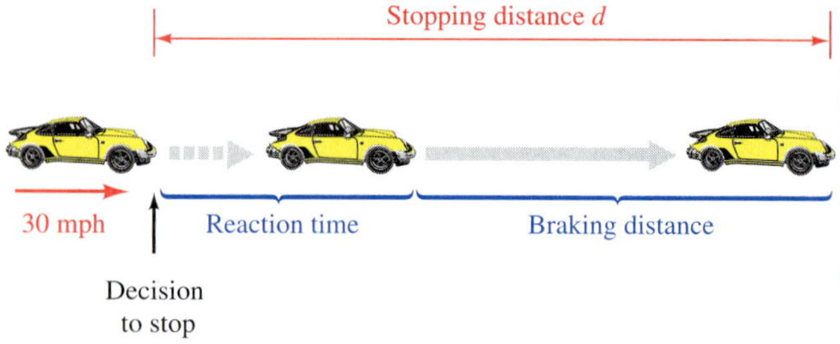

78. Stopping distance Find the stopping distance of the car discussed in Exercise 77 when the driver is going 70 mph.

WRITING

79. Describe how to determine the degree of a polynomial.

80. Describe how to classify a polynomial as a monomial, a binomial, a trinomial, or none of these.

SOMETHING TO THINK ABOUT

81. Find a polynomial whose value will be 1 if you substitute $\frac{3}{2}$ for x.

82. Graph the function $y = f(x) = -x^2$. What do you discover?

4.5 Adding and Subtracting Polynomials

- Adding Monomials
- Subtracting Monomials
- Adding Polynomials
- Subtracting Polynomials
- Adding and Subtracting Multiples of Polynomials
- An Application of Adding Polynomials

Getting Ready *Combine like terms and simplify, if possible.*

1. $3x + 2x$ **2.** $5y - 3y$ **3.** $19x + 6x$ **4.** $8z - 3z$
5. $9r + 3r$ **6.** $4r - 3s$ **7.** $7r - 7r$ **8.** $17r - 17r^2$

In this section, we will discuss how to add and subtract polynomials.

Adding Monomials

Recall that like terms have the same variables with the same exponents. For example,

$3xyz^2$ and $-2xyz^2$ are like terms.

$\frac{1}{2}ab^2c$ and $\frac{1}{3}a^2bd^2$ are unlike terms.

Also recall that to combine like terms, we add (or subtract) their coefficients and keep the same variables with the same exponents. For example,

$$2y + 5y = (2 + 5)y \qquad -3x^2 + 7x^2 = (-3 + 7)x^2$$
$$= 7y \qquad = 4x^2$$

Likewise,

$$4x^3y^2 + 9x^3y^2 = 13x^3y^2 \qquad 4r^2s^3t^4 + 7r^2s^3t^4 = 11r^2s^3t^4$$

These examples suggest that to add like monomials, we simply combine like terms.

EXAMPLE 1

a. $5xy^3 + 7xy^3 = 12xy^3$

b. $-7x^2y^2 + 6x^2y^2 + 3x^2y^2 = -x^2y^2 + 3x^2y^2$
$= 2x^2y^2$

c. $(2x^2)^2 + 81x^4 = 4x^4 + 81x^4$ $\quad (2x^2)^2 = (2x^2)(2x^2) = 4x^4.$
$= 85x^4$

Self Check Perform the following additions: **a.** $6a^3b^2 + 5a^3b^2$, **b.** $-2pq^2 + 5pq^2 + 8pq^2$, and **c.** $27x^6 + (2x^2)^3$.

Subtracting Monomials

To subtract one monomial from another, we add the negative of the monomial that is to be subtracted. In symbols, $x - y = x + (-y)$.

EXAMPLE 2

a. $8x^2 - 3x^2 = 8x^2 + (-3x^2)$
$= 5x^2$

b. $6x^3y^2 - 9x^3y^2 = 6x^3y^2 + (-9x^3y^2)$
$= -3x^3y^2$

c. $-3r^2st^3 - 5r^2st^3 = -3r^2st^3 + (-5r^2st^3)$
$= -8r^2st^3$

Self Check Find each difference: **a.** $12m^3 - 7m^3$ and **b.** $-4p^3q^2 - 8p^3q^2$.

Adding Polynomials

Because of the distributive property, we can remove parentheses enclosing several terms when the sign preceding the parentheses is +. We simply drop the parentheses.

$$\begin{aligned} +(3x^2 + 3x - 2) &= +1(3x^2 + 3x - 2) \\ &= 1(3x^2) + 1(3x) + 1(-2) \\ &= 3x^2 + 3x + (-2) \\ &= 3x^2 + 3x - 2 \end{aligned}$$

We can add polynomials by removing parentheses, if necessary, and then combining any like terms that are contained within the polynomials.

EXAMPLE 3 Add: $(3x^2 - 3x + 2) + (2x^2 + 7x - 4)$.

Solution

$$\begin{aligned} &(3x^2 - 3x + 2) + (2x^2 + 7x - 4) \\ &\quad = 3x^2 - 3x + 2 + 2x^2 + 7x - 4 \\ &\quad = 3x^2 + 2x^2 - 3x + 7x + 2 + (-4) \\ &\quad = 5x^2 + 4x - 2 \end{aligned}$$

Self Check Add: $(2a^2 - a + 4) + (5a^2 + 6a - 5)$.

Additions such as Example 3 are often written with like terms aligned vertically. We can then add column by column.

$$\begin{array}{r} 3x^2 - 3x + 2 \\ \underline{2x^2 + 7x - 4} \\ 5x^2 + 4x - 2 \end{array}$$

EXAMPLE 4 Add:

$$\begin{array}{rrr} 4x^2y & + 8x^2y^2 & - 3x^2y^3 \\ 3x^2y & - 8x^2y^2 & + 8x^2y^3 \\ \hline 7x^2y & & + 5x^2y^3 \end{array}$$

Self Check Add:

$$\begin{array}{rrr} 4pq^2 & + 6pq^3 & - 7pq^4 \\ \hline 2pq^2 & - 8pq^3 & + 9pq^4 \end{array}$$

Subtracting Polynomials

Because of the distributive property, we can remove parentheses enclosing several terms when the sign preceding the parentheses is $-$. We simply drop the negative sign and the parentheses, and *change the sign of every term within the parentheses.*

$$\begin{aligned} -(3x^2 + 3x - 2) &= -1(3x^2 + 3x - 2) \\ &= -1(3x^2) + (-1)(3x) + (-1)(-2) \\ &= -3x^2 + (-3x) + 2 \\ &= -3x^2 - 3x + 2 \end{aligned}$$

This suggests that the way to subtract polynomials is to remove parentheses and combine like terms.

EXAMPLE 5

a. $$\begin{aligned} (3x - 4) - (5x + 7) &= 3x - 4 - 5x - 7 \\ &= -2x - 11 \end{aligned}$$

b. $$\begin{aligned} (3x^2 - 4x - 6) - (2x^2 - 6x + 12) &= 3x^2 - 4x - 6 - 2x^2 + 6x - 12 \\ &= x^2 + 2x - 18 \end{aligned}$$

c. $$\begin{aligned} (-4rt^3 + 2r^2t^2) - (-3rt^3 + 2r^2t^2) &= -4rt^3 + 2r^2t^2 + 3rt^3 - 2r^2t^2 \\ &= -rt^3 \end{aligned}$$

Self Check Find the difference: $(-2a^2b + 5ab^2) - (-5a^2b - 7ab^2)$.

To subtract polynomials in vertical form, we add the negative of the **subtrahend** (the bottom polynomial) to the **minuend** (the top polynomial).

EXAMPLE 6 Subtract $3x^2y - 2xy^2$ from $2x^2y + 4xy^2$.

Solution We write the subtraction in vertical form, change the signs of the terms of the subtrahend, and add:

$$\begin{array}{r} 2x^2y + 4xy^2 \\ -\ \underline{3x^2y - 2xy^2} \end{array} \quad \rightarrow \quad \begin{array}{r} 2x^2y + 4xy^2 \\ +\ \underline{-3x^2y + 2xy^2} \\ -x^2y + 6xy^2 \end{array}$$

In horizontal form, the solution is

$$\begin{aligned} 2x^2y + 4xy^2 - (3x^2y - 2xy^2) &= 2x^2y + 4xy^2 - 3x^2y + 2xy^2 \\ &= -x^2y + 6xy^2 \end{aligned}$$

Self Check Find the difference:

$$\begin{array}{r} 5p^2q - 6pq + 7q \\ -\ \underline{2p^2q + 2pq - 8q} \end{array}$$

EXAMPLE 7 Subtract $6xy^2 + 4x^2y^2 - x^3y^2$ from $-2xy^2 - 3x^3y^2$.

Solution

$$\begin{array}{rrr} -2xy^2 & & -\ 3x^3y^2 \\ -\ \ \ 6xy^2 & +\ 4x^2y^2 & -\ \ x^3y^2 \\ \hline \end{array} \quad \rightarrow \quad \begin{array}{rrr} -2xy^2 & & -\ 3x^3y^2 \\ +\ -6xy^2 & -\ 4x^2y^2 & +\ \ x^3y^2 \\ \hline -8xy^2 & -\ 4x^2y^2 & -\ 2x^3y^2 \end{array}$$

In horizontal form, the solution is

$$\begin{aligned} &-2xy^2 - 3x^3y^2 - (6xy^2 + 4x^2y^2 - x^3y^2) \\ &\quad = -2xy^2 - 3x^3y^2 - 6xy^2 - 4x^2y^2 + x^3y^2 \\ &\quad = -8xy^2 - 4x^2y^2 - 2x^3y^2 \end{aligned}$$

Self Check Subtract $-2pq^2 - 2p^2q^2 + 3p^3q^2$ from $5pq^2 + 3p^2q^2 - p^3q^2$.

Adding and Subtracting Multiples of Polynomials

Because of the distributive property, we can remove parentheses enclosing several terms when a monomial precedes the parentheses. We simply multiply every term within the parentheses by that monomial. For example, to add $3(2x + 5)$ and $2(4x - 3)$, we proceed as follows:

$$\begin{aligned} 3(2x + 5) + 2(4x - 3) &= 6x + 15 + 8x - 6 \\ &= 6x + 8x + 15 - 6 && 15 + 8x = 8x + 15. \\ &= 14x + 9 && \text{Combine like terms.} \end{aligned}$$

EXAMPLE 8 **a.** $3(x^2 + 4x) + 2(x^2 - 4) = 3x^2 + 12x + 2x^2 - 8$
$= 5x^2 + 12x - 8$

b. $8(y^2 - 2y + 3) - 4(2y^2 + y - 3) = 8y^2 - 16y + 24 - 8y^2 - 4y + 12$
$= -20y + 36$

c. $-4(x^2y^2 - x^2y + 3x) - (x^2y^2 - 2x) + 3(x^2y^2 + 2x^2y)$
$= -4x^2y^2 + 4x^2y - 12x - x^2y^2 + 2x + 3x^2y^2 + 6x^2y$
$= -2x^2y^2 + 10x^2y - 10x$

Self Check Simplify: **a.** $2(a^3 - 3a) + 5(a^3 + 2a)$ and **b.** $5(x^2y + 2x^2) - (x^2y - 3x^2)$.

An Application of Adding Polynomials

EXAMPLE 9 **Property values** A house purchased for \$95,000 is expected to appreciate according to the formula $y = 2{,}500x + 95{,}000$, where y is the value of the house after x years. A second house purchased for \$125,000 is expected to appreciate according to the formula $y = 4{,}500x + 125{,}000$. Find one formula that will give the value of both properties after x years.

Solution The value of the first house after x years is given by the polynomial $2{,}500x + 95{,}000$. The value of the second house after x years is given by the polynomial $4{,}500x + 125{,}000$. The value of both houses will be the sum of these two polynomials.

$$2{,}500x + 95{,}000 + 4{,}500x + 125{,}000 = 7{,}000x + 220{,}000$$

The total value y of the properties is given by $y = 7{,}000x + 220{,}000$.

Self Check Answers

1. a. $11a^3b^2$, **b.** $11pq^2$, **c.** $35x^6$ **2. a.** $5m^3$, **b.** $-12p^3q^2$ **3.** $7a^2 + 5a - 1$ **4.** $6pq^2 - 2pq^3 + 2pq^4$ **5.** $3a^2b + 12ab^2$ **6.** $3p^2q - 8pq + 15q$ **7.** $7pq^2 + 5p^2q^2 - 4p^3q^2$ **8. a.** $7a^3 + 4a$, **b.** $4x^2y + 13x^2$

Orals *Simplify.*

1. $x^3 + 3x^3$
2. $3xy + xy$
3. $(x + 3y) - (x + y)$
4. $5(1 - x) + 3(x - 1)$
5. $(2x - y^2) - (2x + y^2)$
6. $5(x^2 + y) + (x^2 - y)$
7. $3x^2 + 2y + x^2 - y$
8. $2x^2y + y - (2x^2y - y)$

4.5 EXERCISES

REVIEW *Let $a = 3$, $b = -2$, $c = -1$, and $d = 2$. Evaluate each expression.*

1. $ab + cd$
2. $ad + bc$
3. $a(b + c)$
4. $d(b + a)$
5. Solve the inequality $-4(2x - 9) \geq 12$ and graph the solution set.
6. The **kinetic energy** of a moving object is given by the formula

$$K = \frac{mv^2}{2}$$

Solve the formula for m.

VOCABULARY AND CONCEPTS *Fill in the blanks.*

7. A ________ is a polynomial with one term.
8. If two polynomials are subtracted in vertical form, the bottom polynomial is called the ________, and the top polynomial is called the ________.
9. To add like monomials, add the numerical ________ and keep the ________.
10. $a - b = a +$ ____
11. To add two polynomials, combine any ________ contained in the polynomials.
12. To subtract polynomials, remove parentheses and combine ________.

Tell whether the terms are like or unlike terms. If they are like terms, add them.

13. $3y, 4y$
14. $3x^2, 5x^2$
15. $3x, 3y$
16. $3x^2, 6x$
17. $3x^3, 4x^3, 6x^3$
18. $-2y^4, -6y^4, 10y^4$
19. $-5x^3y^2, 13x^3y^2$
20. $23, 12x$
21. $-23t^6, 32t^6, 56t^6$
22. $32x^5y^3, -21x^5y^3, -11x^5y^3$
23. $-x^2y, xy, 3xy^2$
24. $4x^3y^2z, -6x^3y^2z, 2x^3y^2z$

PRACTICE *Simplify each expression.*

25. $4y + 5y$
26. $-2x + 3x$
27. $-8t^2 - 4t^2$
28. $15x^2 + 10x^2$
29. $32u^3 - 16u^3$
30. $25xy^2 - 7xy^2$
31. $18x^5y^2 - 11x^5y^2$
32. $17x^6y - 22x^6y$
33. $3rst + 4rst + 7rst$
34. $-2ab + 7ab - 3ab$
35. $-4a^2bc + 5a^2bc - 7a^2bc$
36. $(xy)^2 + 4x^2y^2 - 2x^2y^2$
37. $(3x)^2 - 4x^2 + 10x^2$
38. $(2x)^4 - (3x^2)^2$
39. $5x^2y^2 + 2(xy)^2 - (3x^2)y^2$
40. $-3x^3y^6 + 2(xy^2)^3 - (3x)^3y^6$
41. $(-3x^2y)^4 + (4x^4y^2)^2 - 2x^8y^4$
42. $5x^5y^{10} - (2xy^2)^5 + (3x)^5y^{10}$

Perform the operations and simplify.

43. $(3x + 7) + (4x - 3)$
44. $(2y - 3) + (4y + 7)$
45. $(4a + 3) - (2a - 4)$
46. $(5b - 7) - (3b + 5)$
47. $(2x + 3y) + (5x - 10y)$
48. $(5x - 8y) - (2x + 5y)$
49. $(-8x - 3y) - (11x + y)$
50. $(-4a + b) + (5a - b)$
51. $(3x^2 - 3x - 2) + (3x^2 + 4x - 3)$
52. $(3a^2 - 2a + 4) - (a^2 - 3a + 7)$
53. $(2b^2 + 3b - 5) - (2b^2 - 4b - 9)$
54. $(4c^2 + 3c - 2) + (3c^2 + 4c + 2)$
55. $(2x^2 - 3x + 1) - (4x^2 - 3x + 2) + (2x^2 + 3x + 2)$
56. $(-3z^2 - 4z + 7) + (2z^2 + 2z - 1) - (2z^2 - 3z + 7)$
57. $2(x + 3) + 3(x + 3)$
58. $5(x + y) + 7(x + y)$
59. $-8(x - y) + 11(x - y)$
60. $-4(a - b) - 5(a - b)$
61. $2(x^2 - 5x - 4) - 3(x^2 - 5x - 4) + 6(x^2 - 5x - 4)$
62. $7(x^2 + 3x + 1) + 9(x^2 + 3x + 1) - 5(x^2 + 3x + 1)$
63. Add: $\begin{array}{r} 3x^2 + 4x + 5 \\ \underline{2x^2 - 3x + 6} \end{array}$
64. Add: $\begin{array}{r} 2x^3 + 2x^2 - 3x + 5 \\ \underline{3x^3 - 4x^2 - x - 7} \end{array}$
65. Add: $\begin{array}{r} 2x^3 - 3x^2 + 4x - 7 \\ \underline{-9x^3 - 4x^2 - 5x + 6} \end{array}$
66. Add: $\begin{array}{r} -3x^3 + 4x^2 - 4x + 9 \\ \underline{2x^3 + 9x - 3} \end{array}$
67. Add: $\begin{array}{r} -3x^2y + 4xy + 25y^2 \\ \underline{5x^2y - 3xy - 12y^2} \end{array}$

68. Add:
$$\begin{array}{r} -6x^3z - 4x^2z^2 + 7z^3 \\ \underline{-7x^3z + 9x^2z^2 - 21z^3} \end{array}$$

69. Subtract:
$$\begin{array}{r} 3x^2 + 4x - 5 \\ \underline{-2x^2 - 2x + 3} \end{array}$$

70. Subtract:
$$\begin{array}{r} 3y^2 - 4y + 7 \\ \underline{6y^2 - 6y - 13} \end{array}$$

71. Subtract:
$$\begin{array}{r} 4x^3 + 4x^2 - 3x + 10 \\ \underline{5x^3 - 2x^2 - 4x - 4} \end{array}$$

72. Subtract:
$$\begin{array}{r} 3x^3 + 4x^2 + 7x + 12 \\ \underline{-4x^3 + 6x^2 + 9x - 3} \end{array}$$

73. Subtract:
$$\begin{array}{r} -2x^2y^2 - 4xy + 12y^2 \\ \underline{10x^2y^2 + 9xy - 24y^2} \end{array}$$

74. Subtract:
$$\begin{array}{r} 25x^3 - 45x^2z + 31xz^2 \\ \underline{12x^3 + 27x^2z - 17xz^2} \end{array}$$

75. Find the sum when $x^2 + x - 3$ is added to the sum of $2x^2 - 3x + 4$ and $3x^2 - 2$.

76. Find the sum when $3y^2 - 5y + 7$ is added to the sum of $-3y^2 - 7y + 4$ and $5y^2 + 5y - 7$.

77. Find the difference when $t^3 - 2t^2 + 2$ is subtracted from the sum of $3t^3 + t^2$ and $-t^3 + 6t - 3$.

78. Find the difference when $-3z^3 - 4z + 7$ is subtracted from the sum of $2z^2 + 3z - 7$ and $-4z^3 - 2z - 3$.

79. Find the sum when $3x^2 + 4x - 7$ is added to the sum of $-2x^2 - 7x + 1$ and $-4x^2 + 8x - 1$.

80. Find the difference when $32x^2 - 17x + 45$ is subtracted from the sum of $23x^2 - 12x - 7$ and $-11x^2 + 12x + 7$.

Simplify each expression.

81. $2(x + 3) + 4(x - 2)$

82. $3(y - 4) - 5(y + 3)$

83. $-2(x^2 + 7x - 1) - 3(x^2 - 2x + 7)$

84. $-5(y^2 - 2y - 6) + 6(2y^2 + 2y - 5)$

85. $2(2y^2 - 2y + 2) - 4(3y^2 - 4y - 1) + 4(y^3 - y^2 - y)$

86. $-4(z^2 - 5z) - 5(4z^2 - 1) + 6(2z - 3)$

87. $2(a^2b^2 - ab) - 3(ab + 2ab^2) + (b^2 - ab + a^2b^2)$

88. $3(xy^2 + y^2) - 2(xy^2 + 4y^2 + y^3) + 2(y^3 + y^2)$

89. $-4(x^2y^2 + xy^3 + xy^2z) - 2(x^2y^2 - 4xy^2z) - 2(8xy^3 - y)$

90. $-3(u^2v - uv^3 + uvw) + 4(uvw + w^2) - 3(w^2 + uvw)$

APPLICATIONS *Consider the following information: If a house was purchased for* \$105,000 *and is expected to appreciate* \$900 *per year, its value y after x years is given by the formula* $y = 900x + 105{,}000$.

91. Value of a house Find the expected value of the house in 10 years.

92. Value of a house A second house was purchased for \$120,000 and was expected to appreciate \$1,000 per year. Find a polynomial equation that will give the value of the house in x years.

93. Value of a house Find the value of the house discussed in Exercise 92 after 12 years.

94. Value of a house Find one polynomial equation that will give the combined value y of both houses after x years.

95. Value of two houses Find the value of the two houses after 20 years by

a. substituting 20 into the polynomial equations $y = 900x + 105{,}000$ and $y = 1{,}000x + 120{,}000$ and adding the results.

b. substituting into the result of Exercise 94.

96. Value of two houses Find the value of the two houses after 25 years by

a. substituting 25 into the polynomial equations $y = 900x + 105{,}000$ and $y = 1{,}000x + 120{,}000$ and adding the results.

b. substituting into the result of Exercise 94.

Consider the following information: A business bought two computers, one for \$6,600 and the other for \$9,200. The first computer is expected to depreciate \$1,100 per year and the second \$1,700 per year.

97. Value of a computer Write a polynomial equation that will give the value of the first computer after x years.

98. Value of a computer Write a polynomial equation that will give the value of the second computer after x years.

99. Value of two computers Find one polynomial equation that will give the value of both computers after x years.

100. Value of two computers In two ways, find the value of the computers after 3 years.

WRITING

101. How do you recognize like terms?

102. How do you add like terms?

SOMETHING TO THINK ABOUT

Let $P(x) = 3x - 5$. Find each value.

103. $P(x + h) + P(x)$

104. $P(x + h) - P(x)$

105. If $P(x) = x^{23} + 5x^2 + 73$ and $Q(x) = x^{23} + 4x^2 + 73$, find $P(7) - Q(7)$.

106. If two numbers written in scientific notation have the same power of 10, they can be added as similar terms:

$$2 \times 10^3 + 3 \times 10^3 = 5 \times 10^3$$

Without converting to standard form, how could you add

$$2 \times 10^3 + 3 \times 10^4$$

4.6 Multiplying Polynomials

- Multiplying Monomials
- Multiplying a Polynomial by a Monomial
- Multiplying a Binomial by a Binomial
- The FOIL Method
- Multiplying a Polynomial by a Binomial
- Multiplying Binomials to Solve Equations
- An Application of Multiplying Polynomials

Getting Ready *Simplify:*

1. $(2x)(3)$ **2.** $(3xxx)(x)$ **3.** $5x^2 \cdot x$ **4.** $8x^2x^3$

Use the distributive property to remove parentheses.

5. $3(x + 5)$ **6.** $-2(x + 5)$ **7.** $4(y - 3)$ **8.** $-2(y^2 - 3)$

We now discuss how to multiply polynomials. After introducing general methods for multiplication, we will introduce a special method, called the *FOIL method,* used for multiplying binomials.

Multiplying Monomials

We have previously multiplied monomials by other monomials. For example, to multiply $4x^2$ by $-2x^3$, we use the commutative and associative properties of multiplication to group the numerical factors together and the variable factors together. Then we multiply the numerical factors and multiply the variable factors.

$$4x^2(-2x^3) = 4(-2)x^2x^3$$
$$= -8x^5$$

This example suggests the following rule.

Multiplying Monomials

> To multiply two monomials, multiply the numerical factors and then multiply the variable factors.

EXAMPLE 1 Multiply: **a.** $3x^5(2x^5)$, **b.** $-2a^2b^3(5ab^2)$, and **c.** $-4y^5z^2(2y^3z^3)(3yz)$.

Solution **a.** $3x^5(2x^5) = 3(2)x^5x^5$
$$= 6x^{10}$$

b. $-2a^2b^3(5ab^2) = -2(5)a^2ab^3b^2$
$$= -10a^3b^5$$

c. $-4y^5z^2(2y^3z^3)(3yz) = -4(2)(3)y^5y^3yz^2z^3z$
$$= -24y^9z^6$$

Self Check Multiply: **a.** $(5a^2b^3)(6a^3b^4)$ and **b.** $(-15p^3q^2)(5p^3q^2)$.

Multiplying a Polynomial by a Monomial

To find the product of a monomial and a polynomial with more than one term, we use the distributive property. To multiply $2x + 4$ by $5x$, for example, we proceed as follows:

$5x(2x + 4) = 5x \cdot 2x + 5x \cdot 4$ Use the distributive property.
$= 10x^2 + 20x$ Multiply the monomials $5x \cdot 2x = 10x^2$ and $5x \cdot 4 = 20x$.

This example suggests the following rule.

Multiplying Polynomials by Monomials

> To multiply a polynomial with more than one term by a monomial, use the distributive property to remove parentheses and simplify.

EXAMPLE 2 Multiply: **a.** $3a^2(3a^2 - 5a)$ and **b.** $-2xz^2(2x - 3z + 2z^2)$.

Solution **a.** $3a^2(3a^2 - 5a) = 3a^2 \cdot 3a^2 - 3a^2 \cdot 5a$ Use the distributive property.
$= 9a^4 - 15a^3$ Multiply.

b. $-2xz^2(2x - 3z + 2z^2)$
$= -2xz^2 \cdot 2x - (-2xz^2) \cdot 3z + (-2xz^2) \cdot 2z^2$ Use the distributive property.
$= -4x^2z^2 - (-6xz^3) + (-4xz^4)$ Multiply.
$= -4x^2z^2 + 6xz^3 - 4xz^4$

Self Check Multiply: **a.** $2p^3(3p^2 - 5p)$ and **b.** $-5a^2b(3a + 2b - 4ab)$.

Multiplying a Binomial by a Binomial

To multiply two binomials, we must use the distributive property more than once. For example, to multiply $2a - 4$ by $3a + 5$, we proceed as follows.

$(2a - 4)(3a + 5) = (2a - 4) \cdot 3a + (2a - 4) \cdot 5$	Use the distributive property.
$= 3a(2a - 4) + 5(2a - 4)$	Use the commutative property of multiplication.
$= 3a \cdot 2a - 3a \cdot 4 + 5 \cdot 2a - 5 \cdot 4$	Use the distributive property.
$= 6a^2 - 12a + 10a - 20$	Do the multiplications.
$= 6a^2 - 2a - 20$	Combine like terms.

This example suggests the following rule.

Multiplying Two Binomials

> To multiply two binomials, multiply each term of one binomial by each term of the other binomial and combine like terms.

The FOIL Method

We can use a shortcut method, called the **FOIL** method, to multiply binomials. FOIL is an acronym for **F**irst terms, **O**uter terms, **I**nner terms, and **L**ast terms. To use this method to multiply $2a - 4$ by $3a + 5$, we

1. multiply the **F**irst terms $2a$ and $3a$ to obtain $6a^2$,
2. multiply the **O**uter terms $2a$ and 5 to obtain $10a$,
3. multiply the **I**nner terms -4 and $3a$ to obtain $-12a$, and
4. multiply the **L**ast terms -4 and 5 to obtain -20.

Then we simplify the resulting polynomial, if possible.

First terms Last terms

Inner terms Outer terms

$(2a - 4)(3a + 5) = 2a(3a) + 2a(5) + (-4)(3a) + (-4)(5)$	
$= 6a^2 + 10a - 12a - 20$	Simplify.
$= 6a^2 - 2a - 20$	Combine like terms.

EXAMPLE 3 Find each product:

F L I O

a. $(3x + 4)(2x - 3) = 3x(2x) + 3x(-3) + 4(2x) + 4(-3)$

$= 6x^2 - 9x + 8x - 12$

$= 6x^2 - x - 12$

b. $(2y - 7)(5y - 4) = 2y(5y) + 2y(-4) + (-7)(5y) + (-7)(-4)$

$= 10y^2 - 8y - 35y + 28$

$= 10y^2 - 43y + 28$

(F, O, I, L)

c. $(2r - 3s)(2r + t) = 2r(2r) + 2r(t) - 3s(2r) - 3s(t)$

$= 4r^2 + 2rt - 6rs - 3st$

(F, O, I, L)

Self Check Find each product: **a.** $(2a - 1)(3a + 2)$ and **b.** $(5y - 2z)(2y + 3z)$.

EXAMPLE 4 Simplify each expression:

a. $3(2x - 3)(x + 1)$

$= 3(2x^2 + 2x - 3x - 3)$ Multiply the binomials.

$= 3(2x^2 - x - 3)$ Combine like terms.

$= 6x^2 - 3x - 9$ Use the distributive property to remove parentheses.

b. $(x + 1)(x - 2) - 3x(x + 3)$

$= x^2 - 2x + x - 2 - 3x^2 - 9x$

$= -2x^2 - 10x - 2$ Combine like terms.

Self Check Simplify: $(x + 3)(2x - 1) + 2x(x - 1)$.

The products discussed in Example 5 are called **special products.**

EXAMPLE 5 Find each product:

a. $(x + y)^2 = (x + y)(x + y)$

$= x^2 + xy + xy + y^2$

$= x^2 + 2xy + y^2$

The square of the sum of two quantities has three terms: the square of the first quantity, plus twice the product of the quantities, plus the square of the second quantity.

b. $(x - y)^2 = (x - y)(x - y)$

$$= x^2 - xy - xy + y^2$$
$$= x^2 - 2xy + y^2$$

The square of the difference of two quantities has three terms: the square of the first quantity, minus twice the product of the quantities, plus the square of the second quantity.

c. $(x + y)(x - y) = x^2 - xy + xy - y^2$
$$= x^2 - y^2$$

The product of the sum and the difference of two quantities is a binomial. It is the product of the first quantities minus the product of the second quantities. Binomials that have the same terms, but different signs, are called **conjugate binomials.**

Self Check Find each product: **a.** $(p + 2)^2$, **b.** $(p - 2)^2$, and **c.** $(p + 2q)(p - 2q)$.

Because the products discussed in Example 5 occur so often, it is wise to learn their forms.

Special Products

$$(x + y)^2 = x^2 + 2xy + y^2$$
$$(x - y)^2 = x^2 - 2xy + y^2$$
$$(x + y)(x - y) = x^2 - y^2$$

Comment Note that $(x + y)^2 \neq x^2 + y^2$ and $(x - y)^2 \neq x^2 - y^2$.

Multiplying a Polynomial by a Binomial

We must use the distributive property more than once to multiply a polynomial by a binomial. For example, to multiply $3x^2 + 3x - 5$ by $2x + 3$, we proceed as follows:

$$\begin{aligned}(2x + 3)(3x^2 + 3x - 5) &= (2x + 3)3x^2 + (2x + 3)3x - (2x + 3)5 \\ &= 3x^2(2x + 3) + 3x(2x + 3) - 5(2x + 3) \\ &= 6x^3 + 9x^2 + 6x^2 + 9x - 10x - 15 \\ &= 6x^3 + 15x^2 - x - 15\end{aligned}$$

This example suggests the following rule.

Multiplying Polynomials

To multiply one polynomial by another, multiply each term of one polynomial by each term of the other polynomial and combine like terms.

It is often convenient to organize the work vertically.

EXAMPLE 6 **a.** Multiply:

$$\begin{array}{rrrrl} & 3a^2 - & 4a & + 7 & \\ & & 2a & + 5 & \\ \hline 2a(3a^2 - 4a + 7) \longrightarrow & 6a^3 - & 8a^2 & + 14a & \\ 5(3a^2 - 4a + 7) \longrightarrow & & + 15a^2 & - 20a & + 35 \\ \hline & 6a^3 + & 7a^2 & - 6a & + 35 \end{array}$$

b. Multiply:

$$\begin{array}{rrrrrl} & & 3y^2 - & 5y & + 4 & \\ & & - 4y^2 & - 3 & & \\ \hline -4y^2(3y^2 - 5y + 4) \longrightarrow & -12y^4 + & 20y^3 - & 16y^2 & & \\ -3(3y^2 - 5y + 4) \longrightarrow & & & - 9y^2 & + 15y & - 12 \\ \hline & -12y^4 + & 20y^3 - & 25y^2 & + 15y & - 12 \end{array}$$

Self Check Multiply: **a.** $(3x + 2)(2x^2 - 4x + 5)$ and **b.** $(-2x^2 + 3)(2x^2 - 4x - 1)$.

Multiplying Binomials to Solve Equations

To solve an equation such as $(x + 2)(x + 3) = x(x + 7)$, we can use the FOIL method to remove the parentheses on the left-hand side, use the distributive property to remove parentheses on the right-hand side, and proceed as follows:

$$(x + 2)(x + 3) = x(x + 7)$$
$$x^2 + 3x + 2x + 6 = x^2 + 7x$$
$$3x + 2x + 6 = 7x \quad \text{Subtract } x^2 \text{ from both sides.}$$
$$5x + 6 = 7x \quad \text{Combine like terms.}$$
$$6 = 2x \quad \text{Subtract } 5x \text{ from both sides.}$$
$$3 = x \quad \text{Divide both sides by 2.}$$

Check:

$$(x + 2)(x + 3) = x(x + 7)$$
$$(3 + 2)(3 + 3) \stackrel{?}{=} 3(3 + 7) \quad \text{Replace } x \text{ with 3.}$$
$$5(6) \stackrel{?}{=} 3(10) \quad \text{Do the additions within parentheses.}$$
$$30 = 30$$

EXAMPLE 7 Solve: $(x + 5)(x + 4) = (x + 9)(x + 10)$.

Solution We remove parentheses on both sides of the equation and proceed as follows:

$$(x + 5)(x + 4) = (x + 9)(x + 10)$$
$$x^2 + 4x + 5x + 20 = x^2 + 10x + 9x + 90$$
$$9x + 20 = 19x + 90 \quad \text{Subtract } x^2 \text{ from both sides and combine like terms.}$$
$$20 = 10x + 90 \quad \text{Subtract } 9x \text{ from both sides.}$$
$$-70 = 10x \quad \text{Subtract 90 from both sides.}$$
$$-7 = x \quad \text{Divide both sides by 10.}$$

Check: $(x + 5)(x + 4) = (x + 9)(x + 10)$

$(-7 + 5)(-7 + 4) \stackrel{?}{=} (-7 + 9)(-7 + 10)$ Replace x with -7.

$(-2)(-3) \stackrel{?}{=} (2)(3)$ Do the additions within parentheses.

$6 = 6$

Self Check Solve: $(x + 2)(x - 4) = (x + 6)(x - 3)$.

An Application of Multiplying Polynomials

EXAMPLE 8 **Dimensions of a painting** A square painting is surrounded by a border 2 inches wide. If the area of the border is 96 square inches, find the dimensions of the painting.

Analyze the problem Refer to Figure 4-6, which shows a square painting surrounded by a border 2 inches wide. We know that the area of this border is 96 square inches, and we are to find the dimensions of the painting.

Figure 4-6

Form an equation Let x represent the length of each side of the square painting. The outer rectangle is also a square, and its dimensions are $x + 4$ by $x + 4$ inches. Since the area of a square is the product of its length and width, the area of the larger square is $(x + 4)(x + 4)$, and the area of the painting is $x \cdot x$. If we subtract the area of the painting from the area of the larger square, the difference is 96 (the area of the border).

The area of the large square	minus	the area of the square painting	equals	the area of the border.
$(x + 4)(x + 4)$	$-$	$x \cdot x$	$=$	96

Solve the equation

$(x + 4)(x + 4) - x^2 = 96$ $x \cdot x = x^2$.

$x^2 + 8x + 16 - x^2 = 96$ $(x + 4)(x + 4) = x^2 + 8x + 16$.

$8x + 16 = 96$ Combine like terms.

$8x = 80$ Subtract 16 from both sides.

$x = 10$ Divide both sides by 8.

State the conclusion The dimensions of the painting are 10 inches by 10 inches.

Check the result Check the result.

Self Check Answers

1. a. $30a^5b^7$, **b.** $-75p^6q^4$ **2. a.** $6p^5 - 10p^4$, **b.** $-15a^3b - 10a^2b^2 + 20a^3b^2$ **3. a.** $6a^2 + a - 2$, **b.** $10y^2 + 11yz - 6z^2$ **4.** $4x^2 + 3x - 3$ **5. a.** $p^2 + 4p + 4$, **b.** $p^2 - 4p + 4$, **c.** $p^2 - 4q^2$ **6. a.** $6x^3 - 8x^2 + 7x + 10$, **b.** $-4x^4 + 8x^3 + 8x^2 - 12x - 3$ **7.** 2

Orals *Find each product.*

1. $2x^2(3x - 1)$
2. $5y(2y^2 - 3)$
3. $7xy(x + y)$
4. $-2y(2x - 3y)$
5. $(x + 3)(x + 2)$
6. $(x - 3)(x + 2)$
7. $(2x + 3)(x + 2)$
8. $(3x - 1)(3x + 1)$
9. $(x + 3)^2$
10. $(x - 5)^2$

4.6 EXERCISES

REVIEW *Tell which property of real numbers justifies each statement.*

1. $3(x + 5) = 3x + 3 \cdot 5$
2. $(x + 3) + y = x + (3 + y)$
3. $3(ab) = (ab)3$
4. $a + 0 = a$
5. Solve: $\frac{5}{3}(5y + 6) - 10 = 0$.
6. Solve $F = \frac{GMm}{d^2}$ for m.

VOCABULARY AND CONCEPTS *Fill in the blanks.*

7. A polynomial with one term is called a ________.
8. A ________ is a polynomial with two terms.
9. A polynomial with three terms is called a ________.
10. In the acronym FOIL, F stands for ________, O stands for ________, I stands for ________, and L stands for ________.

Consider the product $(2x + 5)(3x - 4)$.

11. The product of the first terms is ____.
12. The product of the outer terms is ____.
13. The product of the inner terms is ____.
14. The product of the last terms is ____.

PRACTICE *Find each product.*

15. $(3x^2)(4x^3)$
16. $(-2a^3)(3a^2)$
17. $(3b^2)(-2b)(4b^3)$
18. $(3y)(2y^2)(-y^4)$
19. $(2x^2y^3)(3x^3y^2)$
20. $(-x^3y^6z)(x^2y^2z^7)$
21. $(x^2y^5)(x^2z^5)(-3y^2z^3)$
22. $(-r^4st^2)(2r^2st)(rst)$
23. $(x^2y^3)^5$
24. $(a^3b^2c)^4$
25. $(a^3b^2c)(abc^3)^2$
26. $(xyz^3)(xy^2z^2)^3$
27. $3(x + 4)$
28. $-3(a - 2)$
29. $-4(t + 7)$
30. $6(s^2 - 3)$
31. $3x(x - 2)$
32. $4y(y + 5)$
33. $-2x^2(3x^2 - x)$
34. $4b^3(2b^2 - 2b)$
35. $3xy(x + y)$
36. $-4x^2(3x^2 - x)$
37. $2x^2(3x^2 + 4x - 7)$
38. $3y^3(2y^2 - 7y - 8)$
39. $\frac{1}{4}x^2(8x^5 - 4)$
40. $\frac{4}{3}a^2b(6a - 5b)$
41. $-\frac{2}{3}r^2t^2(9r - 3t)$
42. $-\frac{4}{5}p^2q(10p + 15q)$
43. $(3xy)(-2x^2y^3)(x + y)$
44. $(-2a^2b)(-3a^3b^2)(3a - 2b)$

45. $(a + 4)(a + 5)$

46. $(y - 3)(y + 5)$

47. $(3x - 2)(x + 4)$

48. $(t + 4)(2t - 3)$

49. $(2a + 4)(3a - 5)$

50. $(2b - 1)(3b + 4)$

51. $(3x - 5)(2x + 1)$

52. $(2y - 5)(3y + 7)$

53. $(x + 3)(2x - 3)$

54. $(2x + 3)(2x - 5)$

55. $(2s + 3t)(3s - t)$

56. $(3a - 2b)(4a + b)$

57. $(x + y)(x + z)$

58. $(a - b)(x + y)$

59. $(u + v)(u + 2t)$

60. $(x - 5y)(a + 2y)$

61. $(-2r - 3s)(2r + 7s)$

62. $(-4a + 3)(-2a - 3)$

63. $\begin{array}{r} 4x + 3 \\ \underline{x + 2} \end{array}$

64. $\begin{array}{r} 5r + 6 \\ \underline{2r - 1} \end{array}$

65. $\begin{array}{r} 4x - 2y \\ \underline{3x + 5y} \end{array}$

66. $\begin{array}{r} 5r + 6s \\ \underline{2r - s} \end{array}$

67. $\begin{array}{r} x^2 + x + 1 \\ \underline{x - 1} \end{array}$

68. $\begin{array}{r} 4x^2 - 2x + 1 \\ \underline{2x + 1} \end{array}$

69. $(2x + 1)(x^2 + 3x - 1)$

70. $(3x - 2)(2x^2 - x + 2)$

71. $(x + 4)(x + 4)$

72. $(a + 3)(a + 3)$

73. $(t - 3)(t - 3)$

74. $(z - 5)(z - 5)$

75. $(r + 4)(r - 4)$

76. $(b + 2)(b - 2)$

77. $(x + 5)^2$

78. $(y - 6)^2$

79. $(2s + 1)(2s + 1)$

80. $(3t - 2)(3t - 2)$

81. $(4x + 5)(4x - 5)$

82. $(5z + 1)(5z - 1)$

83. $(x - 2y)^2$

84. $(3a + 2b)^2$

85. $(2a - 3b)^2$

86. $(2x + 5y)^2$

87. $(4x + 5y)(4x - 5y)$

88. $(6p + 5q)(6p - 5q)$

89. $2(x - 4)(x + 1)$

90. $-3(2x + 3y)(3x - 4y)$

91. $3a(a + b)(a - b)$

92. $-2r(r + s)(r + s)$

93. $(4t + 3)(t^2 + 2t + 3)$

94. $(3x + y)(2x^2 - 3xy + y^2)$

95. $(-3x + y)(x^2 - 8xy + 16y^2)$

96. $(3x - y)(x^2 + 3xy - y^2)$

97. $(x - 2y)(x^2 + 2xy + 4y^2)$

98. $(2m + n)(4m^2 - 2mn + n^2)$

Simplify each expression.

99. $2t(t + 2) + 3t(t - 5)$

100. $3y(y + 2) + (y + 1)(y - 1)$

101. $3xy(x + y) - 2x(xy - x)$

102. $(a + b)(a - b) - (a + b)(a + b)$

103. $(x + y)(x - y) + x(x + y)$

104. $(2x - 1)(2x + 1) + x(2x + 1)$

105. $(x + 2)^2 - (x - 2)^2$

106. $(x - 3)^2 - (x + 3)^2$

107. $(2s - 3)(s + 2) + (3s + 1)(s - 3)$

108. $(3x + 4)(2x - 2) - (2x + 1)(x + 3)$

Solve each equation.

109. $(s - 4)(s + 1) = s^2 + 5$

110. $(y - 5)(y - 2) = y^2 - 4$

111. $z(z + 2) = (z + 4)(z - 4)$

112. $(z + 3)(z - 3) = z(z - 3)$

113. $(x + 4)(x - 4) = (x - 2)(x + 6)$

114. $(y - 1)(y + 6) = (y - 3)(y - 2) + 8$

115. $(a - 3)^2 = (a + 3)^2$

116. $(b + 2)^2 = (b - 1)^2$

117. $4 + (2y - 3)^2 = (2y - 1)(2y + 3)$

118. $7s^2 + (s - 3)(2s + 1) = (3s - 1)^2$

APPLICATIONS

119. Millstones The radius of one millstone in the illustration is 3 meters greater than the radius of the other, and their areas differ by 15π square meters. Find the radius of the larger millstone.

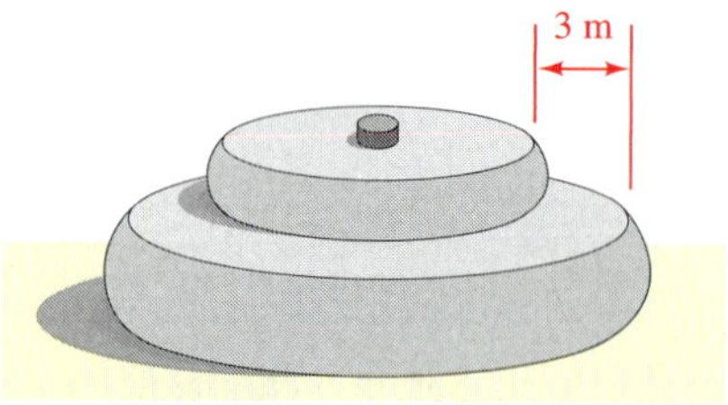

120. Bookbinding Two square sheets of cardboard used for making book covers differ in area by 44 square inches. An edge of the larger square is 2 inches greater than an edge of the smaller square. Find the length of an edge of the smaller square.

121. Baseball In major league baseball, the distance between bases is 30 feet greater than it is in softball. The bases in major league baseball mark the corners of a square that has an area 4,500 square feet greater than for softball. Find the distance between the bases in baseball.

122. Pulley design The radius of one pulley in the illustration is 1 inch greater than the radius of the second pulley, and their areas differ by 4π square inches. Find the radius of the smaller pulley.

WRITING

123. Describe the steps involved in finding the product of a binomial and its conjugate.

124. Writing the expression $(x + y)^2$ as $x^2 + y^2$ illustrates a common error. Explain.

SOMETHING TO THINK ABOUT

125. The area of the square in the illustration is the total of the areas of the four smaller regions. The picture illustrates the product $(x + y)^2$. Explain.

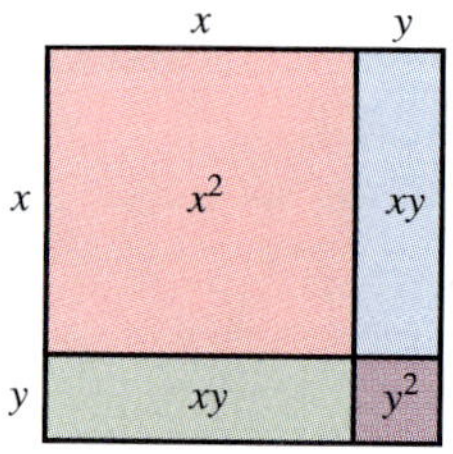

126. The illustration represents the product of two binomials. Explain.

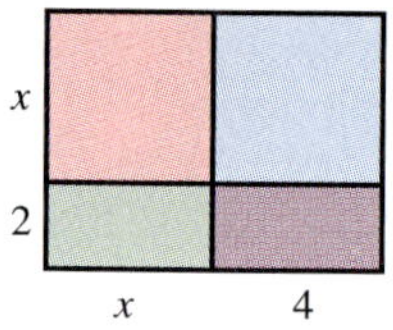

4.7 Dividing Polynomials by Monomials

- Dividing a Monomial by a Monomial
- Dividing a Polynomial by a Monomial
- An Application of Dividing a Polynomial by a Monomial

Getting Ready *Simplify each fraction.*

1. $\dfrac{4x^2y^3}{2xy}$ 2. $\dfrac{9xyz}{9xz}$ 3. $\dfrac{15x^2y}{10x}$ 4. $\dfrac{6x^2y}{6xy^2}$

5. $\dfrac{(2x^2)(5y^2)}{10xy}$ 6. $\dfrac{(5x^3y)(6xy^3)}{10x^4y^4}$

In this section, we will show how to divide polynomials by monomials. We will discuss how to divide polynomials by polynomials in the next section.

Dividing a Monomial by a Monomial

We have seen that dividing by a number is equivalent to multiplying by its reciprocal. For example, dividing the number 8 by 2 gives the same answer as multiplying 8 by $\frac{1}{2}$.

$$\frac{8}{2} = 4 \quad \text{and} \quad \frac{1}{2} \cdot 8 = 4$$

In general, the following is true.

Division

$$\frac{a}{b} = \frac{1}{b} \cdot a \quad (b \neq 0)$$

Recall that to simplify a fraction, we write both its numerator and denominator as the product of several factors and then divide out all common factors. For example,

$$\frac{20}{25} = \frac{4 \cdot 5}{5 \cdot 5} \qquad \text{Factor: } 20 = 4 \cdot 5 \text{ and } 25 = 5 \cdot 5.$$

$$= \frac{4 \cdot \overset{1}{\cancel{5}}}{\underset{1}{\cancel{5}} \cdot 5} \qquad \text{Divide out the common factor of 5.}$$

$$= \frac{4}{5} \qquad \tfrac{5}{5} = 1.$$

We can use the same method to simplify algebraic fractions that contain variables.

$$\frac{3p^2q}{6pq^3} = \frac{3 \cdot p \cdot p \cdot q}{2 \cdot 3 \cdot p \cdot q \cdot q \cdot q} \qquad \text{Factor: } p^2 = p \cdot p,\ 6 = 2 \cdot 3, \text{ and } q^3 = q \cdot q \cdot q.$$

$$= \frac{\overset{1}{\cancel{3}} \cdot \overset{1}{\cancel{p}} \cdot p \cdot \overset{1}{\cancel{q}}}{2 \cdot \underset{1}{\cancel{3}} \cdot \underset{1}{\cancel{p}} \cdot \underset{1}{\cancel{q}} \cdot q \cdot q} \qquad \text{Divide out the common factors of 3, } p\text{, and } q.$$

$$= \frac{p}{2q^2} \qquad \tfrac{3}{3} = 1,\ \tfrac{p}{p} = 1, \text{ and } \tfrac{q}{q} = 1.$$

To divide monomials, we can either use the previous method used for simplifying arithmetic fractions or use the rules of exponents.

EXAMPLE 1 Simplify: **a.** $\frac{x^2y}{xy^2}$ and **b.** $\frac{-8a^3b^2}{4ab^3}$.

Solution

Using Fractions

a. $$\frac{x^2y}{xy^2} = \frac{x \cdot x \cdot y}{x \cdot y \cdot y}$$

$$= \frac{\overset{1}{\cancel{x}} \cdot x \cdot \overset{1}{\cancel{y}}}{\underset{1}{\cancel{x}} \cdot y \cdot \underset{1}{\cancel{y}}}$$

$$= \frac{x}{y}$$

Using the Rules of Exponents

$$\frac{x^2y}{xy^2} = x^{2-1}y^{1-2}$$

$$= x^1y^{-1}$$

$$= \frac{x}{y}$$

b. $\frac{-8a^3b^2}{4ab^3} = \frac{-2 \cdot 4 \cdot a \cdot a \cdot a \cdot b \cdot b}{4 \cdot a \cdot b \cdot b \cdot b}$

$= \frac{-2 \cdot \overset{1}{\cancel{4}} \cdot \overset{1}{\cancel{a}} \cdot a \cdot a \cdot \overset{1}{\cancel{b}} \cdot \overset{1}{\cancel{b}}}{\underset{1}{\cancel{4}} \cdot \underset{1}{\cancel{a}} \cdot \underset{1}{\cancel{b}} \cdot \underset{1}{\cancel{b}} \cdot b}$

$= \frac{-2a^2}{b}$

$\frac{-8a^3b^2}{4ab^3} = \frac{(-1)2^3a^3b^2}{2^2ab^3}$

$= (-1)2^{3-2}a^{3-1}b^{2-3}$

$= (-1)2^1a^2b^{-1}$

$= \frac{-2a^2}{b}$

Self Check Simplify: $\frac{-5p^2q^3}{10pq^4}$.

Dividing a Polynomial by a Monomial

In Chapter 1, we saw that

$$\frac{a}{d} + \frac{b}{d} = \frac{a + b}{d}$$

Since this is true, we also have

$$\frac{a + b}{d} = \frac{a}{d} + \frac{b}{d}$$

This suggests that, to divide a polynomial by a monomial, we can divide each term in the polynomial by the monomial.

EXAMPLE 2 Simplify: $\frac{9x + 6y}{3xy}$.

Solution $\frac{9x + 6y}{3xy} = \frac{9x}{3xy} + \frac{6y}{3xy}$ Divide each term in the polynomial by the monomial.

$= \frac{3}{y} + \frac{2}{x}$ Simplify each fraction.

Self Check Simplify: $\frac{4a - 8b}{4ab}$.

EXAMPLE 3 Simplify: $\frac{6x^2y^2 + 4x^2y - 2xy}{2xy}$.

Solution $\frac{6x^2y^2 + 4x^2y - 2xy}{2xy}$

$= \frac{6x^2y^2}{2xy} + \frac{4x^2y}{2xy} - \frac{2xy}{2xy}$ Divide each term in the polynomial by the monomial.

$= 3xy + 2x - 1$ Simplify each fraction.

Self Check Simplify: $\frac{9a^2b - 6ab^2 + 3ab}{3ab}$.

EXAMPLE 4 Simplify: $\dfrac{12a^3b^2 - 4a^2b + a}{6a^2b^2}$.

Solution

$$\frac{12a^3b^2 - 4a^2b + a}{6a^2b^2}$$

$$= \frac{12a^3b^2}{6a^2b^2} - \frac{4a^2b}{6a^2b^2} + \frac{a}{6a^2b^2}$$ Divide each term in the polynomial by the monomial.

$$= 2a - \frac{2}{3b} + \frac{1}{6ab^2}$$ Simplify each fraction.

Self Check Simplify: $\dfrac{14p^3q + pq^2 - p}{7p^2q}$.

EXAMPLE 5 Simplify: $\dfrac{(x - y)^2 - (x + y)^2}{xy}$.

Solution

$$\frac{(x - y)^2 - (x + y)^2}{xy}$$

$$= \frac{x^2 - 2xy + y^2 - (x^2 + 2xy + y^2)}{xy}$$ Multiply the binomials in the numerator.

$$= \frac{x^2 - 2xy + y^2 - x^2 - 2xy - y^2}{xy}$$ Remove parentheses.

$$= \frac{-4xy}{xy}$$ Combine like terms.

$$= -4$$ Divide out xy.

Self Check Simplify: $\dfrac{(x + y)^2 - (x - y)^2}{xy}$.

An Application of Dividing a Polynomial by a Monomial

The cross-sectional area of the trapezoidal drainage ditch shown in Figure 4-7 is given by the formula $A = \frac{1}{2}h(B + b)$, where B and b are its bases and h is its height. To solve the formula for b, we proceed as follows.

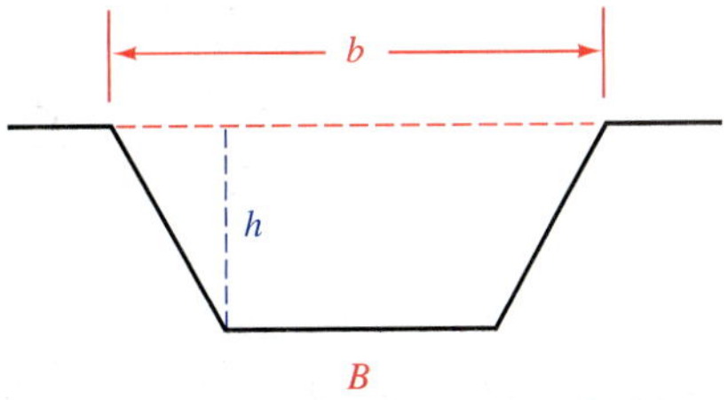

Figure 4-7

$$A = \frac{1}{2}h(B + b)$$

$$2A = 2 \cdot \frac{1}{2}h(B + b) \quad \text{Multiply both sides by 2.}$$

$$2A = h(B + b) \quad \text{Simplify: } 2 \cdot \tfrac{1}{2} = \tfrac{2}{2} = 1.$$

$$2A = hB + hb \quad \text{Use the distributive property to remove parentheses.}$$

$$2A - hB = hB - hB + hb \quad \text{Subtract } hB \text{ from both sides.}$$

$$2A - hB = hb \quad \text{Combine like terms: } hB - hB = 0.$$

$$\frac{2A - hB}{h} = \frac{hb}{h} \quad \text{Divide both sides by } h.$$

$$\frac{2A - hB}{h} = b \quad \frac{hb}{h} = b.$$

EXAMPLE 6 Another student worked the previous problem in a different way and got a result of $b = \frac{2A}{h} - B$. Is this result correct?

Solution To show that this result is correct, we must show that $\frac{2A - hB}{h} = \frac{2A}{h} - B$. We can do this by dividing $2A - hB$ by h.

$$\frac{2A - hB}{h} = \frac{2A}{h} - \frac{hB}{h} \quad \text{Divide each term in the polynomial by the monomial.}$$

$$= \frac{2A}{h} - B \quad \text{Simplify: } \tfrac{hB}{h} = B.$$

The results are the same.

Self Check Suppose another student got $2A - B$. Is this result correct?

Self Check Answers

1. $-\frac{p}{2q}$ **2.** $\frac{1}{b} - \frac{2}{a}$ **3.** $3a - 2b + 1$ **4.** $2p + \frac{q}{7p} - \frac{1}{7pq}$ **5.** 4 **6.** no

Orals *Simplify each fraction.*

1. $\frac{4x^3y}{2xy}$

2. $\frac{6x^3y^2}{3x^3y}$

3. $\frac{35ab^2c^3}{7abc}$

4. $\frac{-14p^2q^5}{7pq^4}$

5. $\frac{(x + y) + (x - y)}{2x}$

6. $\frac{(2x^2 - z) + (x^2 + z)}{x}$

4.7 EXERCISES

REVIEW *Identify each polynomial as a monomial, a binomial, a trinomial, or none of these.*

1. $5a^2b + 2ab^2$
2. $-3x^3y$
3. $-2x^3 + 3x^2 - 4x + 12$
4. $17t^2 - 15t + 27$
5. Find the degree of the trinomial $3x^2 - 2x + 4$.
6. What is the numerical coefficient of the second term of the trinomial $-7t^2 - 5t + 17$?

VOCABULARY AND CONCEPTS *Fill in the blanks.*

7. A ________ is an algebraic expression in which the exponents on the variables are whole numbers.
8. A ________ is a polynomial with one algebraic term.
9. A binomial is a polynomial with ____ terms.
10. A trinomial is a polynomial with _____ terms.
11. $\frac{1}{b} \cdot a = \frac{}{—}$
12. $\frac{15x - 6y}{6xy} = \frac{}{——} \cdot (15x - 6y)$

PRACTICE *Simplify each fraction.*

13. $\frac{5}{15}$
14. $\frac{64}{128}$
15. $\frac{-125}{75}$
16. $\frac{-98}{21}$
17. $\frac{120}{160}$
18. $\frac{70}{420}$
19. $\frac{-3,612}{-3,612}$
20. $\frac{-288}{-112}$
21. $\frac{-90}{360}$
22. $\frac{8,423}{-8,423}$
23. $\frac{5,880}{2,660}$
24. $\frac{-762}{366}$

Perform each division by simplifying each fraction. Write all answers without using negative or zero exponents.

25. $\frac{xy}{yz}$
26. $\frac{a^2b}{ab^2}$
27. $\frac{r^3s^2}{rs^3}$
28. $\frac{y^4z^3}{y^2z^2}$
29. $\frac{8x^3y^2}{4xy^3}$
30. $\frac{-3y^3z}{6yz^2}$
31. $\frac{12u^5v}{-4u^2v^3}$
32. $\frac{16rst^2}{-8rst^3}$
33. $\frac{-16r^3y^2}{-4r^2y^4}$
34. $\frac{35xyz^2}{-7x^2yz}$
35. $\frac{-65rs^2t}{15r^2s^3t}$
36. $\frac{112u^3z^6}{-42u^3z^6}$
37. $\frac{x^2x^3}{xy^6}$
38. $\frac{(xy)^2}{x^2y^3}$
39. $\frac{(a^3b^4)^3}{ab^4}$
40. $\frac{(a^2b^3)^3}{a^6b^6}$
41. $\frac{15(r^2s^3)^2}{-5(rs^5)^3}$
42. $\frac{-5(a^2b)^3}{10(ab^2)^3}$
43. $\frac{-32(x^3y)^3}{128(x^2y^2)^3}$
44. $\frac{68(a^6b^7)^2}{-96(abc^2)^3}$
45. $\frac{(5a^2b)^3}{(2a^2b^2)^3}$
46. $\frac{-(4x^3y^3)^2}{(x^2y^4)^8}$
47. $\frac{-(3x^3y^4)^3}{-(9x^4y^5)^2}$
48. $\frac{(2r^3s^2t)^2}{-(4r^2s^2t^2)^2}$
49. $\frac{(a^2a^3)^4}{(a^4)^3}$
50. $\frac{(b^3b^4)^5}{(bb^2)^2}$
51. $\frac{(z^3z^{-4})^3}{(z^{-3})^2}$
52. $\frac{(t^{-3}t^5)}{(t^2)^{-3}}$

Perform each division.

53. $\frac{6x + 9y}{3xy}$
54. $\frac{8x + 12y}{4xy}$
55. $\frac{5x - 10y}{25xy}$
56. $\frac{2x - 32}{16x}$
57. $\frac{3x^2 + 6y^3}{3x^2y^2}$
58. $\frac{4a^2 - 9b^2}{12ab}$
59. $\frac{15a^3b^2 - 10a^2b^3}{5a^2b^2}$
60. $\frac{9a^4b^3 - 16a^3b^4}{12a^2b}$
61. $\frac{4x - 2y + 8z}{4xy}$
62. $\frac{5a^2 + 10b^2 - 15ab}{5ab}$

63. $\dfrac{12x^3y^2 - 8x^2y - 4x}{4xy}$

64. $\dfrac{12a^2b^2 - 8a^2b - 4ab}{4ab}$

65. $\dfrac{-25x^2y + 30xy^2 - 5xy}{-5xy}$

66. $\dfrac{-30a^2b^2 - 15a^2b - 10ab^2}{-10ab}$

Simplify each numerator and perform the division.

67. $\dfrac{5x(4x - 2y)}{2y}$

68. $\dfrac{9y^2(x^2 - 3xy)}{3x^2}$

69. $\dfrac{(-2x)^3 + (3x^2)^2}{6x^2}$

70. $\dfrac{(-3x^2y)^3 + (3xy^2)^3}{27x^3y^4}$

71. $\dfrac{4x^2y^2 - 2(x^2y^2 + xy)}{2xy}$

72. $\dfrac{-5a^3b - 5a(ab^2 - a^2b)}{10a^2b^2}$

73. $\dfrac{(3x - y)(2x - 3y)}{6xy}$

74. $\dfrac{(2m - n)(3m - 2n)}{-3m^2n^2}$

75. $\dfrac{(a + b)^2 - (a - b)^2}{2ab}$

76. $\dfrac{(x - y)^2 + (x + y)^2}{2x^2y^2}$

APPLICATIONS

77. **Reconciling formulas** Are the formulas the same?

$$l = \frac{P - 2w}{2} \quad \text{and} \quad l = \frac{P}{2} - w$$

78. **Reconciling formulas** Are the formulas the same?

$$r = \frac{G + 2b}{2b} \quad \text{and} \quad r = \frac{G}{2b} + b$$

79. **Phone bills** On a phone bill, the following formulas are given to compute the average cost per minute of x minutes of phone usage. Are they equivalent?

$$C = \frac{0.15x + 12}{x} \quad \text{and} \quad C = 0.15 + \frac{12}{x}$$

80. **Electric bills** On an electric bill, the following formulas are given to compute the average cost of x kwh of electricity. Are they equivalent?

$$C = \frac{0.08x + 5}{x} \quad \text{and} \quad C = 0.08x + \frac{5}{x}$$

WRITING

81. Describe how you would simplify the fraction

$$\frac{4x^2y + 8xy^2}{4xy}$$

82. A fellow student attempts to simplify the fraction $\frac{3x + 5}{x + 5}$ by dividing out the $x + 5$:

$$\frac{3x + 5}{x + 5} = \frac{3\cancel{x + 5}}{\cancel{x + 5}} = 3$$

What would you say to him?

SOMETHING TO THINK ABOUT

83. If $x = 501$, evaluate $\dfrac{x^{500} - x^{499}}{x^{499}}$.

84. An exercise reads as follows:

Simplify: $\dfrac{3x^3y + 6xy^2}{3xy^3}$

It contains a misprint: one mistyped letter or digit. The correct answer is $\frac{x^2}{y} + 2$. Fix the exercise.

4.8 Dividing Polynomials by Polynomials

- Dividing Polynomials by Polynomials
- Writing Powers in Descending Order
- The Case of the Missing Terms

Getting Ready *Divide:*

1. $12\overline{)156}$
2. $17\overline{)357}$
3. $13\overline{)247}$
4. $19\overline{)247}$

We now complete our work on operations on polynomials by considering how to divide one polynomial by another.

Dividing Polynomials by Polynomials

To divide one polynomial by another, we use a method similar to long division in arithmetic. We will illustrate the method with several examples.

EXAMPLE 1 Divide $x^2 + 5x + 6$ by $x + 2$.

Solution Here the divisor is $x + 2$, and the dividend is $x^2 + 5x + 6$.

Step 1:

$$\begin{array}{r} x\phantom{{}+5x+6} \\ x+2\,\overline{)\,x^2+5x+6} \end{array}$$

How many times does x divide x^2? $x^2/x = x$. Place the x above the division symbol.

Step 2:

$$\begin{array}{r} x\phantom{{}+5x+6} \\ x+2\,\overline{)\,x^2+5x+6} \\ \underline{x^2+2x}\phantom{{}+6} \end{array}$$

Multiply each item in the divisor by x. Place the product under $x^2 + 5x$ and draw a line.

Step 3:

$$\begin{array}{r} x\phantom{{}+5x+6} \\ x+2\,\overline{)\,x^2+5x+6} \\ \underline{x^2+2x}\phantom{{}+6} \\ 3x+6 \end{array}$$

Subtract $x^2 + 2x$ from $x^2 + 5x$ by adding the negative of $x^2 + 2x$ to $x^2 + 5x$.

Bring down the 6.

Step 4:

$$\begin{array}{r} x+3\phantom{{}+6} \\ x+2\,\overline{)\,x^2+5x+6} \\ \underline{x^2+2x}\phantom{{}+6} \\ 3x+6 \end{array}$$

How many times does x divide $3x$? $3x/x = +3$. Place the $+3$ above the division symbol.

Step 5:

$$\begin{array}{r} x+3\phantom{{}+6} \\ x+2\,\overline{)\,x^2+5x+6} \\ \underline{x^2+2x}\phantom{{}+6} \\ 3x+6 \\ \underline{3x+6} \end{array}$$

Multiply each term in the divisor by 3. Place the product under the $3x + 6$ and draw a line.

Step 6:

$$\begin{array}{r} x+3\phantom{{}+6} \\ x+2\,\overline{)\,x^2+5x+6} \\ \underline{x^2+2x}\phantom{{}+6} \\ 3x+6 \\ \underline{3x+6} \\ 0 \end{array}$$

Subtract $3x + 6$ from $3x + 6$ by adding the negative of $3x + 6$.

The quotient is $x + 3$, and the remainder is 0.

Step 7: Check by verifying that $x + 2$ times $x + 3$ is $x^2 + 5x + 6$.

$$\begin{aligned}(x + 2)(x + 3) &= x^2 + 3x + 2x + 6 \\ &= x^2 + 5x + 6\end{aligned}$$

Self Check Divide $x^2 + 7x + 12$ by $x + 3$.

EXAMPLE 2 Divide: $\dfrac{6x^2 - 7x - 2}{2x - 1}$.

Solution Here the divisor is $2x - 1$, and the dividend is $6x^2 - 7x - 2$.

Step 1:

$$\begin{array}{r} 3x\phantom{{}-7x-2} \\ 2x-1\overline{)6x^2-7x-2} \end{array}$$

How many times does $2x$ divide $6x^2$? $6x^2/2x = 3x$. Place the $3x$ above the division symbol.

Step 2:

$$\begin{array}{r} 3x\phantom{{}-7x-2} \\ 2x-1\overline{)6x^2-7x-2} \\ \underline{6x^2-3x}\phantom{{}-2} \end{array}$$

Multiply each term in the divisor by $3x$. Place the product under $6x^2 - 7x$ and draw a line.

Step 3:

$$\begin{array}{r} 3x\phantom{{}-7x-2} \\ 2x-1\overline{)6x^2-7x-2} \\ \underline{6x^2-3x}\phantom{{}-2} \\ -4x-2 \end{array}$$

Subtract $6x^2 - 3x$ from $6x^2 - 7x$ by adding the negative of $6x^2 - 3x$ to $6x^2 - 7x$.

Bring down the -2.

Step 4:

$$\begin{array}{r} 3x-2\phantom{{}-7x} \\ 2x-1\overline{)6x^2-7x-2} \\ \underline{6x^2-3x}\phantom{{}-2} \\ -4x-2 \end{array}$$

How many times does $2x$ divide $-4x$? $-4x/2x = -2$. Place the -2 above the division symbol.

Step 5:

$$\begin{array}{r} 3x-2\phantom{{}-7x} \\ 2x-1\overline{)6x^2-7x-2} \\ \underline{6x^2-3x}\phantom{{}-2} \\ -4x-2 \\ \underline{-4x+2} \end{array}$$

Multiply each term in the divisor by -2. Place the product under the $-4x - 2$ and draw a line.

Step 6:

$$\begin{array}{r} 3x-2\phantom{{}-7x} \\ 2x-1\overline{)6x^2-7x-2} \\ \underline{6x^2-3x}\phantom{{}-2} \\ -4x-2 \\ \underline{-4x+2} \\ -4 \end{array}$$

Subtract $-4x + 2$ from $-4x - 2$ by adding the negative of $-4x + 2$.

Here the quotient is $3x - 2$, and the remainder is -4. It is common to write the answer in quotient + $\frac{\text{remainder}}{\text{divisor}}$ form:

$$3x - 2 + \frac{-4}{2x - 1}$$

where the fraction $\dfrac{-4}{2x - 1}$ is formed by dividing the remainder by the divisor.

Step 7: To check the answer, we multiply $3x - 2 + \frac{-4}{2x-1}$ by $2x - 1$. The product should be the dividend.

$$\begin{aligned}(2x-1)\left(3x - 2 + \frac{-4}{2x-1}\right) &= (2x-1)(3x-2) + (2x-1)\left(\frac{-4}{2x-1}\right)\\ &= (2x-1)(3x-2) - 4\\ &= 6x^2 - 4x - 3x + 2 - 4\\ &= 6x^2 - 7x - 2\end{aligned}$$

Self Check Divide: $\dfrac{8x^2 + 6x - 3}{2x + 3}$.

Writing Powers in Descending Order

The division method works best when exponents of the terms in the divisor and the dividend are written in descending order. This means that the term involving the highest power of x appears first, the term involving the second-highest power of x appears second, and so on. For example, the terms in

$$3x^3 + 2x^2 - 7x + 5 \qquad 5 = 5^0.$$

have their exponents written in descending order.

If the powers in the dividend or divisor are not in descending order, we can use the commutative property of addition to write them that way.

EXAMPLE 3 Divide $4x^2 + 2x^3 + 12 - 2x$ by $x + 3$.

Solution We write the dividend so that the exponents are in descending order and divide.

$$\begin{array}{r} 2x^2 - 2x + 4 \\ x + 3\overline{)2x^3 + 4x^2 - 2x + 12} \\ \underline{2x^3 + 6x^2} \qquad\qquad\quad \\ -2x^2 - 2x \qquad\; \\ \underline{-2x^2 - 6x} \qquad\; \\ +4x + 12 \\ \underline{+4x + 12} \\ 0 \end{array}$$

Check: $(x + 3)(2x^2 - 2x + 4) = 2x^3 - 2x^2 + 4x + 6x^2 - 6x + 12$
$= 2x^3 + 4x^2 - 2x + 12$

Self Check Divide $x^2 - 10x + 6x^3 + 4$ by $2x - 1$.

The Case of the Missing Terms

When we write the terms of a dividend in descending powers of x, we may notice that some powers of x are missing. For example, in the dividend of

$$x + 1\overline{)3x^4 - 7x^2 - 3x + 15}$$

the term involving x^3 is missing. When this happens, we should either write the term with a coefficient of 0 or leave a blank space for it. In this case, we would write the dividend as

$$3x^4 + 0x^3 - 7x^2 - 3x + 15 \qquad \text{or} \qquad 3x^4 \qquad\quad - 7x^2 - 3x + 15$$

EXAMPLE 4 Divide: $\dfrac{x^2 - 4}{x + 2}$.

Solution Since $x^2 - 4$ does not have a term involving x, we must either include the term $0x$ or leave a space for it.

$$\begin{array}{r@{}l}
 & x - 2 \\
x + 2 & \overline{)x^2 + 0x - 4} \\
 & \underline{x^2 + 2x} \\
 & \; - 2x - 4 \\
 & \; \underline{- 2x - 4} \\
 & \;0
\end{array}$$

Check: $(x + 2)(x - 2) = x^2 - 2x + 2x - 4$
$\qquad\qquad\qquad\qquad\quad = x^2 - 4$

Self Check Divide: $\dfrac{x^2 - 9}{x - 3}$.

EXAMPLE 5 Divide: $x^3 + y^3$ by $x + y$.

Solution We write $x^3 + y^3$, leaving spaces for the missing terms, and proceed as follows.

$$\begin{array}{r@{}l}
 & x^2 - xy + y^2 \\
x + y & \overline{)x^3 \qquad\qquad\qquad\quad + y^3} \\
 & \underline{x^3 + x^2y} \\
 & \; - x^2y \\
 & \; \underline{- x^2y - xy^2} \\
 & \; + xy^2 + y^3 \\
 & \;\underline{xy^2 + y^3} \\
 & 0
\end{array}$$

Check: $(x + y)(x^2 - xy + y^2) = x^3 - x^2y + xy^2 + x^2y - xy^2 + y^3$
$\qquad\qquad\qquad\qquad\qquad\qquad\quad = x^3 + y^3$

Self Check Divide: $x^3 - y^3$ by $x - y$.

Self Check Answers

1. $x + 4$ **2.** $4x - 3 + \dfrac{6}{2x + 3}$ **3.** $3x^2 + 2x - 4$ **4.** $x + 3$ **5.** $x^2 + xy + y^2$

Orals *Divide and give the answer in* quotient $+ \frac{\text{remainder}}{\text{divisor}}$ *form.*

1. $x\overline{)2x + 3}$
2. $x\overline{)3x - 5}$
3. $x + 1\overline{)2x + 3}$
4. $x + 1\overline{)3x + 5}$
5. $x + 1\overline{)x^2 + x}$
6. $x + 2\overline{)x^2 + 2x}$

4.8 EXERCISES

REVIEW

1. List the composite numbers between 20 and 30.
2. Graph the set of prime numbers between 10 and 20 on a number line.

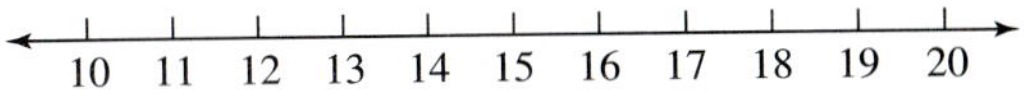

Let $a = -2$ and $b = 3$. Evaluate each expression.

3. $|a - b|$
4. $|a + b|$
5. $-|a^2 - b^2|$
6. $a - |-b|$

Simplify each expression.

7. $3(2x^2 - 4x + 5) + 2(x^2 + 3x - 7)$
8. $-2(y^3 + 2y^2 - y) - 3(3y^3 + y)$

VOCABULARY AND CONCEPTS *Fill in the blanks.*

9. In the division $x + 1\overline{)x^2 + 2x + 1}$, $x + 1$ is called the ______, and $x^2 + 2x + 1$ is called the ______.
10. The answer to a division problem is called the ______.
11. If a division does not come out even, the leftover part is called a ______.
12. The exponents in $2x^4 + 3x^3 + 4x^2 - 7x - 2$ are said to be written in ______ order.

Write each polynomial with the powers in descending order.

13. $4x^3 + 7x - 2x^2 + 6$
14. $5x^2 + 7x^3 - 3x - 9$
15. $9x + 2x^2 - x^3 + 6x^4$
16. $7x^5 + x^3 - x^2 + 2x^4$

Identify the missing terms in each polynomial.

17. $5x^4 + 2x^2 - 1$
18. $-3x^5 - 2x^3 + 4x - 6$

PRACTICE *Perform each division.*

19. Divide $x^2 + 4x + 4$ by $x + 2$.
20. Divide $x^2 - 5x + 6$ by $x - 2$.
21. Divide $y^2 + 13y + 12$ by $y + 1$.
22. Divide $z^2 - 7z + 12$ by $z - 3$.
23. Divide $a^2 + 2ab + b^2$ by $a + b$.
24. Divide $a^2 - 2ab + b^2$ by $a - b$.
25. $\dfrac{6a^2 + 5a - 6}{2a + 3}$
26. $\dfrac{8a^2 + 2a - 3}{2a - 1}$
27. $\dfrac{3b^2 + 11b + 6}{3b + 2}$
28. $\dfrac{3b^2 - 5b + 2}{3b - 2}$
29. $\dfrac{2x^2 - 7xy + 3y^2}{2x - y}$
30. $\dfrac{3x^2 + 5xy - 2y^2}{x + 2y}$

Write the powers of x in descending order (if necessary) and perform each division.

31. $5x + 3\overline{)11x + 10x^2 + 3}$
32. $2x - 7\overline{)-x - 21 + 2x^2}$
33. $4 + 2x\overline{)-10x - 28 + 2x^2}$
34. $1 + 3x\overline{)9x^2 + 1 + 6x}$
35. $2x - y\overline{)xy - 2y^2 + 6x^2}$
36. $2y + x\overline{)3xy + 2x^2 - 2y^2}$
37. $x + 3y\overline{)2x^2 - 3y^2 + 5xy}$
38. $2x - 3y\overline{)2x^2 - 3y^2 - xy}$
39. $3x - 2y\overline{)-10y^2 + 13xy + 3x^2}$
40. $2x + 3y\overline{)-12y^2 + 10x^2 + 7xy}$

41. $4x + y\overline{)-19xy + 4x^2 - 5y^2}$

42. $x - 4y\overline{)5x^2 - 4y^2 - 19xy}$

43. $2x + 3\overline{)2x^3 + 7x^2 + 4x - 3}$

44. $2x - 1\overline{)2x^3 - 3x^2 + 5x - 2}$

45. $3x + 2\overline{)6x^3 + 10x^2 + 7x + 2}$

46. $4x + 3\overline{)4x^3 - 5x^2 - 2x + 3}$

47. $2x + y\overline{)2x^3 + 3x^2y + 3xy^2 + y^3}$

48. $3x - 2y\overline{)6x^3 - x^2y + 4xy^2 - 4y^3}$

Perform each division. If there is a remainder, leave the answer in quotient + $\frac{\text{remainder}}{\text{divisor}}$ *form.*

49. $\dfrac{2x^2 + 5x + 2}{2x + 3}$

50. $\dfrac{3x^2 - 8x + 3}{3x - 2}$

51. $\dfrac{4x^2 + 6x - 1}{2x + 1}$

52. $\dfrac{6x^2 - 11x + 2}{3x - 1}$

53. $\dfrac{x^3 + 3x^2 + 3x + 1}{x + 1}$

54. $\dfrac{x^3 + 6x^2 + 12x + 8}{x + 2}$

55. $\dfrac{2x^3 + 7x^2 + 4x + 3}{2x + 3}$

56. $\dfrac{6x^3 + x^2 + 2x + 1}{3x - 1}$

57. $\dfrac{2x^3 + 4x^2 - 2x + 3}{x - 2}$

58. $\dfrac{3y^3 - 4y^2 + 2y + 3}{y + 3}$

59. $\dfrac{x^2 - 1}{x - 1}$

60. $\dfrac{x^2 - 9}{x + 3}$

61. $\dfrac{4x^2 - 9}{2x + 3}$

62. $\dfrac{25x^2 - 16}{5x - 4}$

63. $\dfrac{x^3 + 1}{x + 1}$

64. $\dfrac{x^3 - 8}{x - 2}$

65. $\dfrac{a^3 + a}{a + 3}$

66. $\dfrac{y^3 - 50}{y - 5}$

67. $3x - 4\overline{)15x^3 - 23x^2 + 16x}$

68. $2y + 3\overline{)21y^2 + 6y^3 - 20}$

WRITING

69. Distinguish among *dividend, divisor, quotient,* and *remainder.*

70. How would you check the results of a division?

SOMETHING TO THINK ABOUT

71. What's wrong here?

$$
\begin{array}{r}
x + 1 \\
x - 2\overline{)x^2 + 3x - 2} \\
\underline{x^2 - 2x} \phantom{{}- 2} \\
x - 2 \\
\underline{x - 2} \\
0
\end{array}
$$

72. What's wrong here?

$$
\begin{array}{r}
3x \phantom{{}+ 10x + 7} \\
x + 2\overline{)3x^2 + 10x + 7} \\
\underline{3x^2 + 9x} \phantom{{}+ 7} \\
x + 7
\end{array}
$$

The quotient is $3x$ and the remainder is $x + 7$.

Project

To discover a pattern in the behavior of polynomials, consider the polynomial $2x^2 - 3x - 5$. First, evaluate the polynomial at $x = 1$ and $x = 3$. Then divide the polynomial by $x - 1$ and again by $x - 3$.

1. What do you notice about the remainders of these divisions?

2. Try others. For example, evaluate the polynomial at $x = 2$ and then divide by $x - 2$.

3. Can you make the pattern hold when you evaluate the polynomial at $x = -2$?

4. Does the pattern hold for other polynomials? Try some polynomials of your own, experiment, and report your conclusions.

Chapter Summary

CONCEPTS	REVIEW EXERCISES
4.1	**Natural-Number Exponents**
If n is a natural number, then $x^n = \underbrace{x \cdot x \cdot x \cdot \cdots \cdot x}_{n \text{ factors of } x}$	Write each expression without exponents. **1.** $(-3x)^4$ **2.** $\left(\frac{1}{2}pq\right)^3$ Evaluate each expression. **3.** 5^3 **4.** 3^5 **5.** $(-8)^2$ **6.** -8^2 **7.** $3^2 + 2^2$ **8.** $(3 + 2)^2$
If m and n are integers, then $x^m x^n = x^{m+n}$ $(x^m)^n = x^{m \cdot n}$ $(xy)^n = x^n y^n$ $\left(\frac{x}{y}\right)^n = \frac{x^n}{y^n} \quad (y \neq 0)$ $\frac{x^m}{x^n} = x^{m-n} \quad (x \neq 0)$	Perform the operations and simplify. **9.** x^3x^2 **10.** x^2x^7 **11.** $(y^7)^3$ **12.** $(x^{21})^2$ **13.** $(ab)^3$ **14.** $(3x)^4$ **15.** $b^3b^4b^5$ **16.** $-z^2(z^3y^2)$ **17.** $(16s)^2s$ **18.** $-3y(y^5)$ **19.** $(x^2x^3)^3$ **20.** $(2x^2y)^2$ **21.** $\frac{x^7}{x^3}$ **22.** $\left(\frac{x^2y}{xy^2}\right)^2$ **23.** $\frac{8(y^2x)^2}{4(yx^2)^2}$ **24.** $\frac{(5y^2z^3)^3}{25(yz)^5}$
4.2	**Zero and Negative-Integer Exponents**
$x^0 = 1 \quad (x \neq 0)$ $x^{-n} = \frac{1}{x^n} \quad (x \neq 0)$	Write each expression without negative exponents or parentheses. **25.** x^0 **26.** $(3x^2y^2)^0$ **27.** $(3x^0)^2$ **28.** $(3x^2y^0)^2$ **29.** x^{-3} **30.** $x^{-2}x^3$ **31.** y^4y^{-3} **32.** $\frac{x^3}{x^{-7}}$

33. $(x^{-3}x^{4})^{-2}$

34. $(a^{-2}b)^{-3}$

35. $\left(\frac{x^2}{x}\right)^{-5}$

36. $\left(\frac{15z^4}{5z^3}\right)^{-2}$

4.3 Scientific Notation

A number is written in scientific notation if it is written as the product of a number between 1 (including 1) and 10 and an integer power of 10.

Write each number in scientific notation.

37. 728

38. 9,370

39. 0.0136

40. 0.00942

41. 7.73

42. 753×10^3

43. 0.018×10^{-2}

44. 600×10^2

Write each number in standard notation.

45. 7.26×10^5

46. 3.91×10^{-4}

47. 2.68×10^0

48. 5.76×10^1

49. 739×10^{-2}

50. 0.437×10^{-3}

51. $\frac{(0.00012)(0.00004)}{0.00000016}$

52. $\frac{(4{,}800)(20{,}000)}{600{,}000}$

4.4 Polynomials

Find the degree of each polynomial and classify it as a monomial, a binomial, or a trinomial.

53. $13x^7$

54. $5^3x + x^2$

55. $-3x^5 + x - 1$

56. $9xy + 21x^3y^2$

When a number is substituted for the variable in a polynomial, the polynomial takes on a numerical value.

Evaluate $3x + 2$ for each value of x.

57. $x = 3$

58. $x = 0$

59. $x = -2$

60. $x = \frac{2}{3}$

Evaluate $5x^4 - x$ for each value of x.

61. $x = 3$

62. $x = 0$

63. $x = -2$

64. $x = -0.3$

Any equation in x and y where each value of x determines a single value of y is a **function.** We say that y is a function of x.

If $y = f(x) = x^2 - 4$, find each value.

65. $f(0)$

66. $f(5)$

67. $f(-2)$

68. $f\left(\frac{1}{2}\right)$

Graph each polynomial function.

69. $y = f(x) = x^2 - 5$

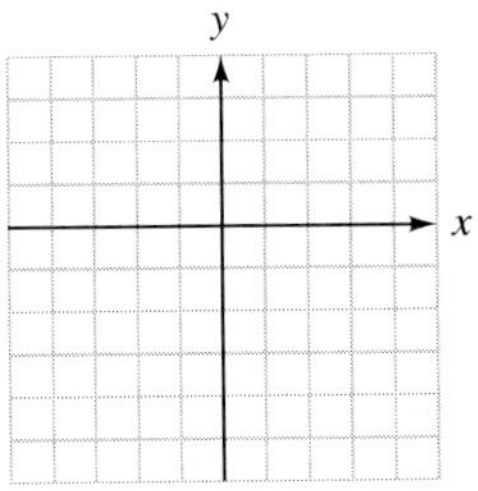

70. $y = f(x) = x^3 - 2$

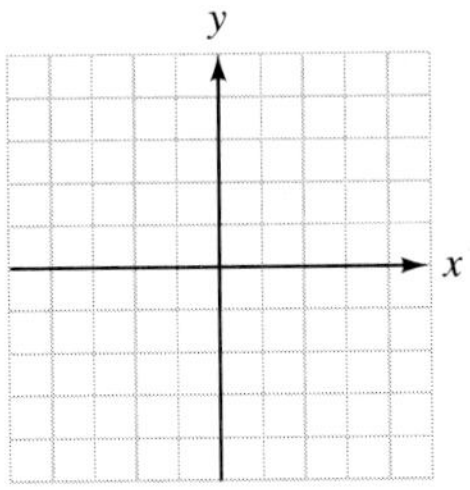

4.5 Adding and Subtracting Polynomials

When adding or subtracting polynomials, combine like terms by adding or subtracting the numerical coefficients and using the same variables and the same exponents.

Simplify each expression, if possible.

71. $3x + 5x - x$

72. $3x + 2y$

73. $(xy)^2 + 3x^2y^2$

74. $-2x^2yz + 3yx^2z$

75. $(3x^2 + 2x) + (5x^2 - 8x)$

76. $(7a^2 + 2a - 5) - (3a^2 - 2a + 1)$

77. $3(9x^2 + 3x + 7) - 2(11x^2 - 5x + 9)$

78. $4(4x^3 + 2x^2 - 3x - 8) - 5(2x^3 - 3x + 8)$

4.6 Multiplying Polynomials

To multiply two monomials, first multiply the numerical factors and then multiply the variable factors using the properties of exponents.

Find each product.

79. $(2x^2y^3)(5xy^2)$

80. $(xyz^3)(x^3z)^2$

To multiply a polynomial with more than one term by a monomial, multiply each term of the polynomial by the monomial and simplify.

Find each product.

81. $5(x + 3)$

82. $3(2x + 4)$

83. $x^2(3x^2 - 5)$

84. $2y^2(y^2 + 5y)$

85. $-x^2y(y^2 - xy)$

86. $-3xy(xy - x)$

To multiply two binomials, use the FOIL method.

Find each product.

87. $(x + 3)(x + 2)$

88. $(2x + 1)(x - 1)$

89. $(3a - 3)(2a + 2)$

90. $6(a - 1)(a + 1)$

91. $(a - b)(2a + b)$

92. $(3x - y)(2x + y)$

Special products:

$(x + y)^2 = x^2 + 2xy + y^2$

$(x - y)^2 = x^2 - 2xy + y^2$

$(x + y)(x - y) = x^2 - y^2$

Find each product.

93. $(x + 3)(x + 3)$

94. $(x + 5)(x - 5)$

95. $(y - 2)(y + 2)$

96. $(x + 4)^2$

97. $(x - 3)^2$

98. $(y - 1)^2$

99. $(2y + 1)^2$

100. $(y^2 + 1)(y^2 - 1)$

To multiply one polynomial by another, multiply each term of one polynomial by each term of the other polynomial, and simplify.

Find each product.

101. $(3x + 1)(x^2 + 2x + 1)$

102. $(2a - 3)(4a^2 + 6a + 9)$

Solve each equation.

103. $x^2 + 3 = x(x + 3)$

104. $x^2 + x = (x + 1)(x + 2)$

105. $(x + 2)(x - 5) = (x - 4)(x - 1)$

106. $(x - 1)(x - 2) = (x - 3)(x + 1)$

107. $x^2 + x(x + 2) = x(2x + 1) + 1$

108. $(x + 5)(3x + 1) = x^2 + (2x - 1)(x - 5)$

4.7 Dividing Polynomials by Monomials

To divide a polynomial by a monomial, divide each term in the polynomial by the monomial.

Perform each division.

109. $\dfrac{3x + 6y}{2xy}$

110. $\dfrac{14xy - 21x}{7xy}$

111. $\dfrac{15a^2bc + 20ab^2c - 25abc^2}{-5abc}$

112. $\dfrac{(x + y)^2 + (x - y)^2}{-2xy}$

4.8	Dividing Polynomials by Polynomials
Use long division to divide one polynomial by another.	Perform each division. **113.** $x + 2\overline{)x^2 + 3x + 5}$ **114.** $x - 1\overline{)x^2 - 6x + 5}$ **115.** $x + 3\overline{)2x^2 + 7x + 3}$ **116.** $3x - 1\overline{)3x^2 + 14x - 2}$ **117.** $2x - 1\overline{)6x^3 + x^2 + 1}$ **118.** $3x + 1\overline{)-13x - 4 + 9x^3}$

Chapter Test

1. Use exponents to rewrite $2xxxyyyy$.

2. Evaluate: $3^2 + 5^3$

Write each expression as an expression containing only one exponent.

3. $y^2(yy^3)$

4. $(-3b^2)(2b^3)(-b^2)$

5. $(2x^3)^5(x^2)^3$

6. $(2rr^2r^3)^3$

Simplify each expression. Write answers without using parentheses or negative exponents.

7. $3x^0$

8. $2y^{-5}y^2$

9. $\dfrac{y^2}{yy^{-2}}$

10. $\left(\dfrac{a^2b^{-1}}{4a^3b^{-2}}\right)^{-3}$

11. Write 28,000 in scientific notation.

12. Write 0.0025 in scientific notation.

13. Write 7.4×10^3 in standard notation.

14. Write 9.3×10^{-5} in standard notation.

15. Classify $3x^2 + 2$ as a monomial, a binomial, or a trinomial.

16. Find the degree of the polynomial $3x^2y^3z^4 + 2x^3y^2z - 5x^2y^3z^5$.

17. Evaluate $x^2 + x - 2$ when $x = -2$.

18. Graph the polynomial function $y = f(x) = x^2 + 2$.

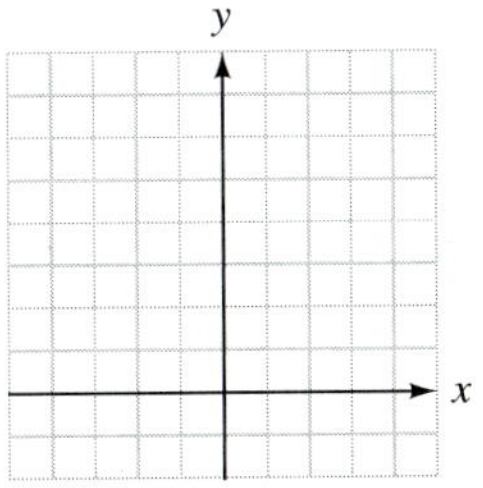

19. Simplify: $-6(x - y) + 2(x + y) - 3(x + 2y)$.

20. Simplify:
$-2(x^2 + 3x - 1) - 3(x^2 - x + 2) + 5(x^2 + 2)$.

21. Add:
$$\begin{array}{r} 3x^3 + 4x^2 - x - 7 \\ \underline{2x^3 - 2x^2 + 3x + 2} \end{array}$$

22. Subtract:
$$\begin{array}{r} 2x^2 - 7x + 3 \\ \underline{3x^2 - 2x - 1} \end{array}$$

Find each product.

23. $(-2x^3)(2x^2y)$

24. $3y^2(y^2 - 2y + 3)$

25. $(2x - 5)(3x + 4)$

26. $(2x - 3)(x^2 - 2x + 4)$

27. Solve: $(a + 2)^2 = (a - 3)^2$.

28. Simplify: $\dfrac{8x^2y^3z^4}{16x^3y^2z^4}$.

29. Simplify: $\dfrac{6a^2 - 12b^2}{24ab}$.

30. Divide: $2x + 3\overline{)2x^2 - x - 6}$.

CUMULATIVE REVIEW EXERCISES

Evaluate each expression. Assume that $x = 2$ and $y = -5$.

1. $5 + 3 \cdot 2$

2. $3 \cdot 5^2 - 4$

3. $\dfrac{3x - y}{xy}$

4. $\dfrac{x^2 - y^2}{x + y}$

Solve each equation.

5. $\dfrac{4}{5}x + 6 = 18$

6. $x - 2 = \dfrac{x + 2}{3}$

7. $2(5x + 2) = 3(3x - 2)$

8. $4(y + 1) = -2(4 - y)$

Graph the solution of each inequality.

9. $5x - 3 > 7$

10. $7x - 9 < 5$

11. $-2 < -x + 3 < 5$

12. $0 \le \dfrac{4 - x}{3} \le 2$

Solve each formula for the indicated variable.

13. $A = p + prt$, for r

14. $A = \dfrac{1}{2}bh$, for h

Graph each equation.

15. $3x - 4y = 12$

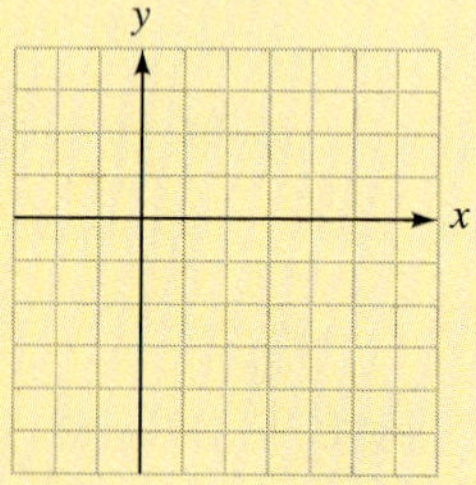

16. $y - 2 = \dfrac{1}{2}(x - 4)$

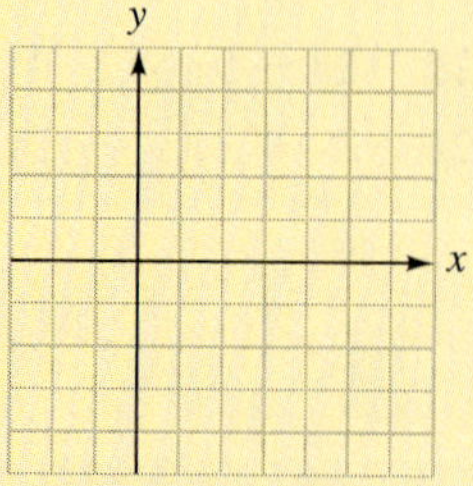

Let $y = f(x) = 5x - 2$ and find each value.

17. $f(0)$

18. $f(3)$

19. $f(-2)$

20. $f\left(\dfrac{1}{5}\right)$

Write each expression as an expression using only one exponent.

21. $(y^3y^5)y^6$

22. $\dfrac{x^3y^4}{x^2y^3}$

23. $\dfrac{a^4b^{-3}}{a^{-3}b^3}$

24. $\left(\dfrac{-x^{-2}y^3}{x^{-3}y^2}\right)^2$

Perform each operation.

25. $(3x^2 + 2x - 7) - (2x^2 - 2x + 7)$

26. $(3x - 7)(2x + 8)$

27. $(x - 2)(x^2 + 2x + 4)$

28. $x - 3\overline{)2x^2 - 5x - 3}$

29. Astronomy The **parsec,** a unit of distance used in astronomy, is 3×10^{16} meters. The distance from Earth to Betelgeuse, a star in the constellation Orion, is 1.6×10^2 parsecs. Use scientific notation to express this distance in meters.

30. Surface area The total surface area A of a box with dimensions l, w, and d is given by the formula

$$A = 2lw + 2wd + 2ld$$

If $A = 202$ square inches, $l = 9$ inches, and $w = 5$ inches, find d.

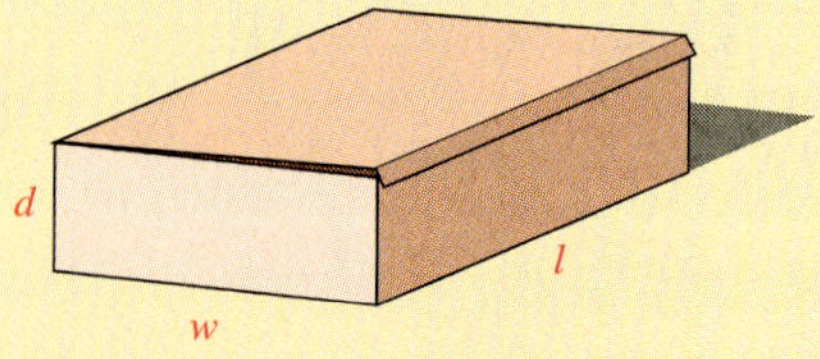

31. Concentric circles The area of the ring between the two concentric circles of radius r and R is given by the formula

$$A = \pi(R + r)(R - r)$$

If $r = 3$ inches and $R = 17$ inches, find A to the nearest tenth.

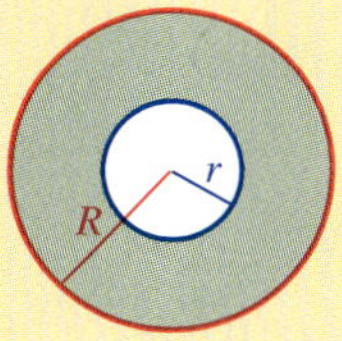

32. Employee discounts Employees at an appliance store can purchase merchandise at 25% less than the regular price. An employee buys a TV set for \$414.72, including 8% sales tax. Find the regular price of the TV.

5 Factoring Polynomials

InfoTrac Project

Do a subject guide search on "arches," choose "View periodical references," and find the article "Another grand arch for Paris." Write a summary of the article.

The area of a cross section of the arch is 7,700 square meters. In the following equation, x represents $1/10$ of the width of the arch. Using factoring, solve the equation for x.

$$2x^2 - 3x - 77 = 0$$

Use your answer to find the width of the arch and then find its height.

Complete this project after studying Section 5.8.

©AFP/CORBIS

Mathematics in Ecology

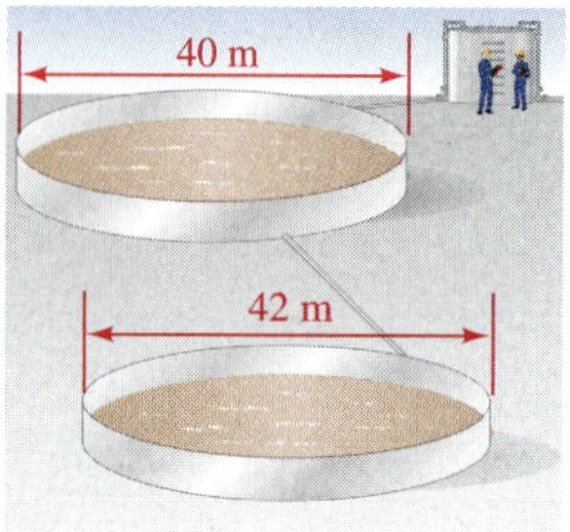

Many types of bacteria cannot survive in air. In one step in waste treatment, sewage is exposed to the air by placing it in large, shallow, circular aeration pools. One sewage processing plant has two such pools, with diameters of 40 and 42 meters. To meet new clean-water standards, the plant must double its capacity by building another pool.

How large a pool should the engineering department specify to double the plant's capacity in this phase of sewage treatment?

Exercise Set 5.8
Problems 35–36

In this chapter, we will reverse the operation of multiplying polynomials and show what polynomials were used to find a given product. We also will use this skill to solve many equations and, in the next chapter, to simplify algebraic fractions.

5.1 Factoring Out the Greatest Common Factor; Factoring by Grouping

- **Factoring Natural Numbers**
- **Factoring Monomials**
- **Factoring Out a Common Monomial**
- **Factoring Out a Negative Factor**
- **Factoring Out a Polynomial**
- **Factoring by Grouping**

Getting Ready *Simplify each expression by removing parentheses.*

1. $5(x + 3)$ **2.** $7(y - 8)$ **3.** $x(3x - 2)$ **4.** $y(5y + 9)$

5. $3(x + y) + a(x + y)$ **6.** $x(y + 1) + 5(y + 1)$

7. $5(x + 1) - y(x + 1)$ **8.** $x(x + 2) - y(x + 2)$

In this chapter, we shall reverse the operation of multiplication and show how to find the factors of a known product. The process of finding the individual factors of a product is called **factoring.**

Factoring Natural Numbers

Because 4 divides 12 exactly, 4 is called a **factor** of 12. The numbers 1, 2, 3, 4, 6, and 12 are the natural-number factors of 12, because each one divides 12 exactly. Recall that a natural number greater than 1 whose only factors are 1 and the number itself is called a **prime number.** For example, 19 is a prime number, because

1. 19 is a natural number greater than 1, and
2. The only two natural number factors of 19 are 1 and 19.

The prime numbers less than 50 are

2, 3, 5, 7, 11, 13, 17, 19, 23, 29, 31, 37, 41, 43, and 47

A natural number is said to be in **prime-factored form** if it is written as the product of factors that are prime numbers.

To find the prime-factored form of a natural number, we can use a **factoring tree.** For example, to find the prime-factored form of 60, we proceed as follows:

Solution 1

1. Start with 60.
2. Factor 60 as $6 \cdot 10$.
3. Factor 6 and 10.

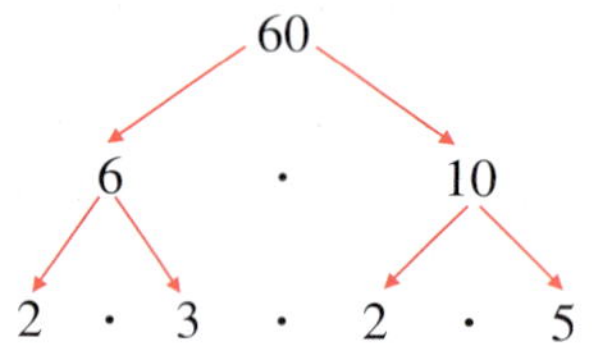

Solution 2

1. Start with 60.
2. Factor 60 as $4 \cdot 15$.
3. Factor 4 and 15.

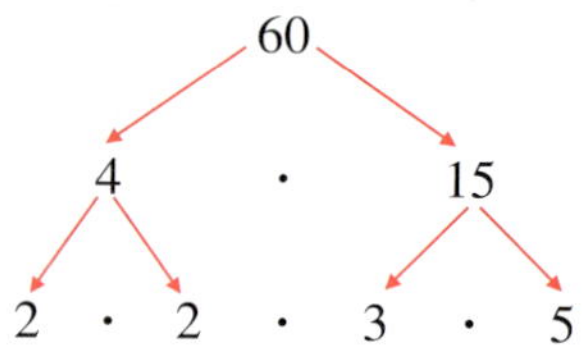

We stop when only prime numbers appear. In either case, the prime factors of 60 are $2 \cdot 2 \cdot 3 \cdot 5$. Thus, the prime-factored form of 60 is $2^2 \cdot 3 \cdot 5$. This illustrates the **fundamental theorem of arithmetic,** which states that there is only one prime factorization for any natural number greater than 1.

The right-hand sides of the equations

$$42 = 2 \cdot 3 \cdot 7$$
$$60 = 2^2 \cdot 3 \cdot 5$$
$$90 = 2 \cdot 3^2 \cdot 5$$

show the prime-factored forms (or **prime factorizations**) of 42, 60, and 90. The largest natural number that divides each of these numbers is called their **greatest common factor (GCF).** The GCF of 42, 60, and 90 is 6, because 6 is the largest natural number that divides each of these numbers:

$$\frac{42}{6} = 7 \qquad \frac{60}{6} = 10 \qquad \text{and} \qquad \frac{90}{6} = 15$$

Factoring Monomials

Algebraic monomials can also have a greatest common factor. The right-hand sides of the equations

$$6a^2b^3 = 2 \cdot 3 \cdot a \cdot a \cdot b \cdot b \cdot b$$
$$4a^3b^2 = 2 \cdot 2 \cdot a \cdot a \cdot a \cdot b \cdot b$$
$$18a^2b = 2 \cdot 3 \cdot 3 \cdot a \cdot a \cdot b$$

show the prime factorizations of $6a^2b^3$, $4a^3b^2$, and $18a^2b$. Since all three of these monomials have one factor of 2, two factors of a, and one factor of b, the GCF is

$$2 \cdot a \cdot a \cdot b \quad \text{or} \quad 2a^2b$$

To find the GCF of several monomials, we follow these steps.

Finding the Greatest Common Factor (GCF)

1. Find the prime factorization of each monomial.
2. List each common factor the least number of times it appears in any one monomial.
3. Find the product of the factors found in the list to obtain the GCF.

PERSPECTIVE

Much of the mathematics that we have inherited from earlier times is the result of teamwork. In a battle early in the 12th century, control of the Spanish city of Toledo was taken from the Mohammedans, who had ruled there for four centuries. Libraries in this great city contained many books written in Arabic, full of knowledge that was unknown in Europe.

The Archbishop of Toledo wanted to share this knowledge with the rest of the world. He knew that these books should be translated into Latin, the universal language of scholarship. But what European scholar could read Arabic? The citizens of Toledo knew both Arabic and Spanish, and most scholars of Europe could read Spanish.

Teamwork saved the day. A citizen of Toledo read the Arabic text aloud, in Spanish. The scholars listened to the Spanish version and wrote it down in Latin. One of these scholars was an Englishman, Robert of Chester. It was he who translated al-Khowarazmi's book, *Ihm al-jabr wa'l muqabalah,* the beginning of the subject we now know as algebra.

EXAMPLE 1 Find the GCF of $10x^3y^2$, $60x^2y$, and $30xy^2$.

Solution **1.** Find the prime factorization of each monomial.

$$10x^3y^2 = 2 \cdot 5 \cdot x \cdot x \cdot x \cdot y \cdot y$$
$$60x^2y = 2 \cdot 2 \cdot 3 \cdot 5 \cdot x \cdot x \cdot y$$
$$30xy^2 = 2 \cdot 3 \cdot 5 \cdot x \cdot y \cdot y$$

2. List each common factor the least number of times it appears in any one monomial: 2, 5, x, and y.

3. Find the product of the factors in the list:

$$2 \cdot 5 \cdot x \cdot y = 10xy$$

Self Check Find the GCF of $20a^2b^3$, $12ab^4$, and $8a^3b^2$.

Factoring Out a Common Monomial

Recall that the distributive property provides a way to multiply a polynomial by a monomial. For example,

$$\begin{aligned} 3x^2(2x - 3y) &= 3x^2 \cdot 2x - 3x^2 \cdot 3y \\ &= 6x^3 - 9x^2y \end{aligned}$$

To reverse this process and factor the product $6x^3 - 9x^2y$, we can find the GCF of each monomial (which is $3x^2$) and then use the distributive property in reverse.

$$\begin{aligned} 6x^3 - 9x^2y &= 3x^2 \cdot 2x - 3x^2 \cdot 3y \\ &= 3x^2(2x - 3y) \end{aligned}$$

This process is called **factoring out the greatest common factor.**

EXAMPLE 2 Factor: $12y^2 + 20y$.

Solution To find the GCF, we find the prime factorization of $12y^2$ and $20y$.

$$\left.\begin{aligned} 12y^2 &= 2 \cdot 2 \cdot 3 \cdot y \cdot y \\ 20y &= 2 \cdot 2 \cdot 5 \cdot y \end{aligned}\right\} \text{GCF} = 4y$$

We can use the distributive property to factor out the GCF of $4y$.

$$\begin{aligned} 12y^2 + 20y &= 4y \cdot 3y + 4y \cdot 5 \\ &= 4y(3y + 5) \end{aligned}$$

Check by verifying that $4y(3y + 5) = 12y^2 + 20y$.

Self Check Factor: $15x^3 - 20x^2$.

EXAMPLE 3 Factor: $35a^3b^2 - 14a^2b^3$.

Solution To find the GCF, we find the prime factorization of $35a^3b^2$ and $-14a^2b^3$.

$$\left.\begin{aligned} 35a^3b^2 &= 5 \cdot 7 \cdot a \cdot a \cdot a \cdot b \cdot b \\ -14a^2b^3 &= -2 \cdot 7 \cdot a \cdot a \cdot b \cdot b \cdot b \end{aligned}\right\} \text{GCF} = 7a^2b^2$$

We factor out the GCF of $7a^2b^2$.

$$\begin{aligned} 35a^3b^2 - 14a^2b^3 &= 7a^2b^2 \cdot 5a - 7a^2b^2 \cdot 2b \\ &= 7a^2b^2(5a - 2b) \end{aligned}$$

Check: $7a^2b^2(5a - 2b) = 35a^3b^2 - 14a^2b^3$.

Self Check Factor: $40x^2y^3 + 15x^3y^2$.

EXAMPLE 4 Factor: $a^2b^2 - ab$.

Solution We factor out the GCF, which is ab.

$$\begin{aligned} a^2b^2 - ab &= ab \cdot ab - ab \cdot 1 \\ &= ab(ab - 1) \end{aligned}$$

Comment The last term of $a^2b^2 - ab$ has an implied coefficient of 1. When ab is factored out, we must write the coefficient of 1.

Check: $ab(ab - 1) = a^2b^2 - ab$.

Self Check Factor: $x^3y^5 + x^2y^3$.

EXAMPLE 5 Factor: $12x^3y^2z + 6x^2yz - 3xz$.

Solution We factor out the GCF, which is $3xz$.

$$\begin{aligned} 12x^3y^2z + 6x^2yz - 3xz &= 3xz \cdot 4x^2y^2 + 3xz \cdot 2xy - 3xz \cdot 1 \\ &= 3xz(4x^2y^2 + 2xy - 1) \end{aligned}$$

Check: $3xz(4x^2y^2 + 2xy - 1) = 12x^3y^2z + 6x^2yz - 3xz$

Self Check Factor: $6ab^2c - 12a^2bc + 3ab$.

Factoring Out a Negative Factor

It is often useful to factor out the negative of a monomial.

EXAMPLE 6 Factor -1 out of $-a^3 + 2a^2 - 4$.

Solution

$$\begin{aligned} &-a^3 + 2a^2 - 4 \\ &= (-1)a^3 + (-1)(-2a^2) + (-1)4 && (-1)(-2a^2) = +2a^2. \\ &= -1(a^3 - 2a^2 + 4) && \text{Factor out } -1. \\ &= -(a^3 - 2a^2 + 4) && \text{The coefficient of 1 need not be written.} \end{aligned}$$

Check: $-(a^3 - 2a^2 + 4) = -a^3 + 2a^2 - 4$

Self Check Factor -1 out of $-b^4 - 3b^2 + 2$.

EXAMPLE 7 Factor out the negative of the GCF: $-18a^2b + 6ab^2 - 12a^2b^2$.

Solution The GCF is $6ab$. To factor out its negative, we factor out $-6ab$.

$$\begin{aligned} -18a^2b + 6ab^2 - 12a^2b^2 &= (-6ab)3a - (-6ab)b + (-6ab)2ab \\ &= -6ab(3a - b + 2ab) \end{aligned}$$

Check: $-6ab(3a - b + 2ab) = -18a^2b + 6ab^2 - 12a^2b^2$

Self Check Factor out the negative of the GCF: $-25xy^2 - 15x^2y + 30x^2y^2$.

Factoring Out a Polynomial

If the GCF of several terms is a polynomial, we can factor out the common polynomial factor. For example, since $a + b$ is a common factor of $(a + b)x$ and $(a + b)y$, we can factor out the $a + b$.

$$(a + b)x + (a + b)y = (a + b)(x + y)$$

We can check by verifying that $(a + b)(x + y) = (a + b)x + (a + b)y$.

EXAMPLE 8 Factor $a + 3$ out of $(a + 3) + (a + 3)^2$.

Solution Recall that $a + 3$ is equal to $(a + 3)1$ and that $(a + 3)^2$ is equal to $(a + 3)(a + 3)$. We can factor out $a + 3$ and simplify.

$$\begin{aligned} (a + 3) + (a + 3)^2 &= (a + 3)1 + (a + 3)(a + 3) \\ &= (a + 3)[1 + (a + 3)] \\ &= (a + 3)(a + 4) \end{aligned}$$

Self Check Factor out $y + 2$: $(y + 2)^2 - 3(y + 2)$.

EXAMPLE 9 Factor: $6a^2b^2(x + 2y) - 9ab(x + 2y)$.

Solution The GCF of $6a^2b^2$ and $9ab$ is $3ab$. We can factor out this GCF as well as $(x + 2y)$.

$$\begin{aligned} &6a^2b^2(x + 2y) - 9ab(x + 2y) \\ &= 3ab \cdot 2ab(x + 2y) - 3ab \cdot 3(x + 2y) \\ &= 3ab(x + 2y)(2ab - 3) && \text{Factor out } 3ab(x + 2y). \end{aligned}$$

Self Check Factor: $4p^3q^2(2a + b) + 8p^2q^3(2a + b)$.

Factoring by Grouping

Suppose we wish to factor

$$ax + ay + cx + cy$$

Although no factor is common to all four terms, there is a common factor of a in $ax + ay$ and a common factor of c in $cx + cy$. We can factor out the a and the c to obtain

$$\begin{aligned} ax + ay + cx + cy &= a(x + y) + c(x + y) \\ &= (x + y)(a + c) \qquad \text{Factor out } x + y. \end{aligned}$$

We can check the result by multiplication.

$$\begin{aligned} (x + y)(a + c) &= ax + cx + ay + cy \\ &= ax + ay + cx + cy \end{aligned}$$

Thus, $ax + ay + cx + cy$ factors as $(x + y)(a + c)$. This type of factoring is called **factoring by grouping.**

EXAMPLE 10 Factor: $2c + 2d - cd - d^2$.

Solution

$$\begin{aligned} 2c + 2d - cd - d^2 &= 2(c + d) - d(c + d) \qquad \text{Factor out 2 from } 2c + 2d \text{ and } -d \text{ from } -cd - d^2. \\ &= (c + d)(2 - d) \qquad \text{Factor out } c + d. \end{aligned}$$

Check:

$$\begin{aligned} (c + d)(2 - d) &= 2c - cd + 2d - d^2 \\ &= 2c + 2d - cd - d^2 \end{aligned}$$

Self Check Factor: $3a + 3b - ac - bc$.

EXAMPLE 11 Factor: $x^2y - ax - xy + a$.

Solution

$$\begin{aligned} x^2y - ax - xy + a &= x(xy - a) - 1(xy - a) \qquad \text{Factor out } x \text{ from } x^2y - ax \text{ and } -1 \text{ from } -xy + a. \\ &= (xy - a)(x - 1) \qquad \text{Factor out } xy - a. \end{aligned}$$

Check by multiplication.

Self Check Factor: $pq^2 + tq + 2pq + 2t$.

Comment When factoring expressions such as those in the previous two examples, don't think that $2(c + d) - d(c + d)$ or $x(xy - a) - 1(xy - a)$ are in factored form. To be in factored form, the final result must be a product.

EXAMPLE 12 Factor: **a.** $a(c - d) + b(d - c)$ and **b.** $ac + bd - ad - bc$.

Solution

a.

$$\begin{aligned} a(c - d) + b(d - c) &= a(c - d) - b(-d + c) \qquad \text{Factor } -1 \text{ from } d - c. \\ &= a(c - d) - b(c - d) \qquad -d + c = c - d. \\ &= (c - d)(a - b) \qquad \text{Factor out } c - d. \end{aligned}$$

b. In this example, we cannot factor anything from the first two terms or the last two terms. However, if we rearrange the terms, the factoring is routine:

$$ac + bd - ad - bc = ac - ad + bd - bc \quad bd - ad = -ad + bd.$$
$$= a(c - d) + b(d - c) \quad \text{Factor } a \text{ from } ac - ad \text{ and } b \text{ from } bd - bc.$$
$$= (c - d)(a - b) \quad \text{See part a.}$$

Self Check Factor: $ax - by - ay + bx$.

Self Check Answers

1. $4ab^2$ **2.** $5x^2(3x - 4)$ **3.** $5x^2y^2(8y + 3x)$ **4.** $x^2y^3(xy^2 + 1)$ **5.** $3ab(2bc - 4ac + 1)$ **6.** $-(b^4 + 3b^2 - 2)$ **7.** $-5xy(5y + 3x - 6xy)$ **8.** $(y + 2)(y - 1)$ **9.** $4p^2q^2(2a + b)(p + 2q)$ **10.** $(a + b)(3 - c)$ **11.** $(pq + t)(q + 2)$ **12.** $(a + b)(x - y)$

Orals *Find the prime factorization of each number.*

1. 36 **2.** 27 **3.** 81 **4.** 45

Find the greatest common factor:

5. 3, 6, and 9
6. $3a^2b$, $6ab$, and $9ab^2$
7. $a(x + 3)$ and $3(x + 3)$
8. $5(a - 1)$ and $xy(a - 1)$

Factor out the greatest common factor:

9. $15xy + 10$
10. $15xy + 10xy^2$
11. $a(x + 3) - 3(x + 3)$
12. $b(x - 2) + (x - 2)$

5.1 EXERCISES

REVIEW *Solve each equation and check all solutions.*

1. $3x - 2(x + 1) = 5$
2. $5(y - 1) + 1 = y$
3. $\dfrac{2x - 7}{5} = 3$
4. $2x - \dfrac{x}{2} = 5x$

VOCABULARY AND CONCEPTS *Fill in the blanks.*

5. A natural number greater than 1 whose only factors are 1 and itself is called a ______ number.

6. If a natural number is written as the product of prime numbers, it is written in ____________ form.

7. The GCF of several natural numbers is the ______ number that divides each of the numbers.

8. The GCF of $6a^3b^3 - 3a^2b^2$ is ______.

9. The GCF of $x(a + b) - y(a + b)$ is ______.

10. Check the results of a factoring problem by ____________.

PRACTICE *Find the prime factorization of each number.*

11. 12
12. 24
13. 15
14. 20
15. 40
16. 62
17. 98
18. 112
19. 225
20. 144
21. 288
22. 968

Complete each factorization.

23. $4a + 12 = \square(a + 3)$
24. $3t - 27 = 3(t - \square)$
25. $r^4 + r^2 = r^2(\square + 1)$
26. $a^3 - a^2 = \square(a - 1)$
27. $4y^2 + 8y - 2xy = 2y(2y + \square - \square)$
28. $3x^2 - 6xy + 9xy^2 = \square(\square - 2y + 3y^2)$

29. $a(x + y) + b(x + y) = (x + y)\underline{\qquad}$

30. $p(m - n) - q(m - n) = \underline{\qquad}(p - q)$

31. $(r - s)p - (r - s)q = (r - s)\underline{\qquad}$

32. $ax + bx + ap + bp = x\underline{\qquad} + p\underline{\qquad}$

$= \underline{\qquad}(x + p)$

Factor out the greatest common factor.

33. $3x + 6$

34. $2y - 10$

35. $xy - xz$

36. $uv + ut$

37. $t^3 + 2t^2$

38. $b^3 - 3b^2$

39. $r^4 - r^2$

40. $a^3 + a^2$

41. $a^3b^3z^3 - a^2b^3z^2$

42. $r^3s^6t^9 + r^2s^2t^2$

43. $24x^2y^3z^4 + 8xy^2z^3$

44. $3x^2y^3 - 9x^4y^3z$

45. $12uvw^3 - 18uv^2w^2$

46. $14xyz - 16x^2y^2z$

47. $3x + 3y - 6z$

48. $2x - 4y + 8z$

49. $ab + ac - ad$

50. $rs - rt + ru$

51. $4y^2 + 8y - 2xy$

52. $3x^2 - 6xy + 9xy^2$

53. $12r^2 - 3rs + 9r^2s^2$

54. $6a^2 - 12a^3b + 36ab$

55. $abx - ab^2x + abx^2$

56. $a^2b^2x^2 + a^3b^2x^2 - a^3b^3x^3$

57. $4x^2y^2z^2 - 6xy^2z^2 + 12xyz^2$

58. $32xyz + 48x^2yz + 36xy^2z$

59. $70a^3b^2c^2 + 49a^2b^3c^3 - 21a^2b^2c^2$

60. $8a^2b^2 - 24ab^2c + 9b^2c^2$

Factor out −1 from each polynomial.

61. $-a - b$

62. $-x - 2y$

63. $-2x + 5y$

64. $-3x + 8z$

65. $-2a + 3b$

66. $-2x + 5y$

67. $-3m - 4n + 1$

68. $-3r + 2s - 3$

69. $-3xy + 2z + 5w$

70. $-4ab + 3c - 5d$

71. $-3ab - 5ac + 9bc$

72. $-6yz + 12xz - 5xy$

Factor out the greatest common factor, including −1.

73. $-3x^2y - 6xy^2$

74. $-4a^2b^2 + 6ab^2$

75. $-4a^2b^3 + 12a^3b^2$

76. $-25x^4y^3z^2 + 30x^2y^3z^4$

77. $-4a^2b^2c^2 + 14a^2b^2c - 10ab^2c^2$

78. $-10x^4y^3z^2 + 8x^3y^2z - 20x^2y$

79. $-14a^6b^6 + 49a^2b^3 - 21ab$

80. $-35r^9s^9t^9 + 25r^6s^6t^6 + 75r^3s^3t^3$

81. $-5a^2b^3c + 15a^3b^4c^2 - 25a^4b^3c$

82. $-7x^5y^4z^3 + 49x^5y^5z^4 - 21x^6y^4z^3$

Factor each expression.

83. $(x + y)2 + (x + y)b$

84. $(a - b)c + (a - b)d$

85. $3(x + y) - a(x + y)$

86. $x(y + 1) - 5(y + 1)$

87. $3(r - 2s) - x(r - 2s)$

88. $x(a + 2b) + y(a + 2b)$

89. $(x - 3)^2 + (x - 3)$

90. $(3t + 5)^2 - (3t + 5)$

91. $2x(a^2 + b) + 2y(a^2 + b)$

92. $3x(c - 3d) + 6y(c - 3d)$

93. $3x^2(r + 3s) - 6y^2(r + 3s)$

94. $9a^2b^2(3x - 2y) - 6ab(3x - 2y)$

95. $3x(a + b + c) - 2y(a + b + c)$

96. $2m(a - 2b + 3c) - 21xy(a - 2b + 3c)$

97. $14x^2y(r + 2s - t) - 21xy(r + 2s - t)$

98. $15xy^3(2x - y + 3z) + 25xy^2(2x - y + 3z)$

99. $(x + 3)(x + 1) - y(x + 1)$

100. $x(x^2 + 2) - y(x^2 + 2)$

101. $(3x - y)(x^2 - 2) + (x^2 - 2)$

102. $(x - 5y)(a + 2) - (x - 5y)$

103. $2x + 2y + ax + ay$

104. $bx + bz + 5x + 5z$

105. $7r + 7s - kr - ks$

106. $9p - 9q + mp - mq$

107. $xr + xs + yr + ys$

108. $pm - pn + qm - qn$

109. $2ax + 2bx + 3a + 3b$

110. $3xy + 3xz - 5y - 5z$

111. $2ab + 2ac + 3b + 3c$

112. $3ac + a + 3bc + b$

113. $2x^2 + 2xy - 3x - 3y$

114. $3ab + 9a - 2b - 6$

115. $3tv - 9tw + uv - 3uw$

116. $ce - 2cf + 3de - 6df$

117. $9mp + 3mq - 3np - nq$

118. $ax + bx - a - b$

119. $mp - np - m + n$

120. $6x^2u - 3x^2v + 2yu - yv$

121. $x(a - b) + y(b - a)$

122. $p(m - n) - q(n - m)$

Factor each expression. Factor out all common factors first, if they exist.

123. $ax^3 + bx^3 + 2ax^2y + 2bx^2y$

124. $x^3y^2 - 2x^2y^2 + 3xy^2 - 6y^2$

125. $4a^2b + 12a^2 - 8ab - 24a$

126. $-4abc - 4ac^2 + 2bc + 2c^2$

127. $x^3 + 2x^2 + x + 2$

128. $y^3 - 3y^2 - 5y + 15$

129. $x^3y - x^2y - xy^2 + y^2$

130. $2x^3z - 4x^2z + 32xz - 64z$

Factor each expression completely. You may have to rearrange some terms first.

131. $2r - bs - 2s + br$

132. $5x + ry + rx + 5y$

133. $ax + by + bx + ay$

134. $mr + ns + ms + nr$

135. $ac + bd - ad - bc$

136. $sx - ry + rx - sy$

137. $ar^2 - brs + ars - br^2$

138. $a^2bc + a^2c + abc + ac$

139. $ba + 3 + a + 3b$

140. $xy + 7 + y + 7x$

141. $pr + qs - ps - qr$

142. $ac - bd - ad + bc$

WRITING

143. When we add $5x$ and $7x$, we combine like terms: $5x + 7x = 12x$. Explain how this is related to factoring out a common factor.

144. Explain how you would factor $x(a - b) + y(b - a)$.

SOMETHING TO THINK ABOUT

145. Think of two positive integers. Divide their product by their greatest common factor. Why do you think the result is called the **lowest common multiple** of the two integers? (*Hint:* The **multiples** of an integer such as 5 are 5, 10, 15, 20, 25, 30, and so on.)

146. Two integers are **relatively prime** if their greatest common factor is 1. For example, 6 and 25 are relatively prime, but 6 and 15 are not. If the greatest common factor of three integers is 1, must any two of them be relatively prime? Explain.

147. Factor $ax + ay + bx + by$ by grouping the first two terms and the last two terms. Then rearrange the terms as $ax + bx + ay + by$, and factor again by grouping the first two and the last two. Do the results agree?

148. Factor $2xy + 2xz - 3y - 3z$ by grouping in two different ways.

5.2 Factoring the Difference of Two Squares

▮ Factoring the Difference of Two Squares ▮ Multistep Factoring

Getting Ready *Multiply the binomials.*

1. $(a + b)(a - b)$

2. $(2r + s)(2r - s)$

3. $(3x + 2y)(3x - 2y)$

4. $(4x^2 + 3)(4x^2 - 3)$

Whenever we multiply a binomial of the form $x + y$ by a binomial of the form $x - y$, we obtain a binomial of the form $x^2 - y^2$.

$$\begin{aligned}(x + y)(x - y) &= x^2 - xy + xy - y^2 \\ &= x^2 - y^2\end{aligned}$$

In this section, we will show how to reverse the multiplication process and factor binomials such as $x^2 - y^2$.

Factoring the Difference of Two Squares

The binomial $x^2 - y^2$ is called the **difference of two squares,** because x^2 is the square of x and y^2 is the square of y. The difference of the squares of two quantities always factors into the sum of those two quantities multiplied by the difference of those two quantities.

Factoring the Difference of Two Squares

$$x^2 - y^2 = (x + y)(x - y)$$

If we think of the difference of two squares as the square of a **F**irst quantity minus the square of a **L**ast quantity, we have the formula

$$F^2 - L^2 = (F + L)(F - L)$$

and we say, *To factor the square of a First quantity minus the square of a Last quantity, we multiply the First plus the Last by the First minus the Last.*

To factor $x^2 - 9$, we note that it can be written in the form $x^2 - 3^2$ and use the formula for factoring the difference of two squares:

$$\begin{aligned}&\mathbf{F}^2 - \mathbf{L}^2 = (\mathbf{F} + \mathbf{L})(\mathbf{F} - \mathbf{L}) \\ &\downarrow \quad\ \ \downarrow \qquad\ \ \downarrow \quad\ \downarrow \ \ \downarrow \quad\ \downarrow \\ &x^2 - 3^2 = (x + 3)(x - 3)\end{aligned}$$

We can check by verifying that $(x + 3)(x - 3) = x^2 - 9$.

To factor the difference of two squares, it is helpful to know the integers that are perfect squares. The number 400, for example, is a perfect square, because $20^2 = 400$. The perfect integer squares less than 400 are

1, 4, 9, 16, 25, 36, 49, 64, 81, 100, 121, 144, 169, 196, 225, 256, 289, 324, 361

Expressions containing variables such as x^4y^2 are also perfect squares, because they can be written as the square of a quantity:

$$x^4y^2 = (x^2y)^2$$

EXAMPLE 1 Factor: $25x^2 - 49$.

Solution We can write $25x^2 - 49$ in the form $(5x)^2 - 7^2$ and use the formula for factoring the difference of two squares:

$$\begin{array}{ccccccc} \mathbf{F}^2 & - & \mathbf{L}^2 & = & (\mathbf{F} + \mathbf{L})(\mathbf{F} - \mathbf{L}) \\ \downarrow & & \downarrow & & \downarrow \quad \downarrow \quad \downarrow \quad \downarrow \\ (\mathbf{5x})^2 & - & \mathbf{7}^2 & = & (\mathbf{5x} + \mathbf{7})(\mathbf{5x} - \mathbf{7}) \end{array}$$

Substitute $5x$ for F and 7 for L.

We can check by multiplying $5x + 7$ and $5x - 7$.

$$\begin{aligned} (5x + 7)(5x - 7) &= 25x^2 - 35x + 35x - 49 \\ &= 25x^2 - 49 \end{aligned}$$

Self Check Factor: $16a^2 - 81$.

EXAMPLE 2 Factor: $4y^4 - 25z^2$.

Solution We can write $4y^4 - 25z^2$ in the form $(2y^2)^2 - (5z)^2$ and use the formula for factoring the difference of two squares:

$$\begin{array}{ccccccc} \mathbf{F}^2 & - & \mathbf{L}^2 & = & (\mathbf{F} + \mathbf{L})(\mathbf{F} - \mathbf{L}) \\ \downarrow & & \downarrow & & \downarrow \quad \downarrow \quad \downarrow \quad \downarrow \\ (\mathbf{2y^2})^2 & - & (\mathbf{5z})^2 & = & (\mathbf{2y^2} + \mathbf{5z})(\mathbf{2y^2} - \mathbf{5z}) \end{array}$$

Check by multiplication.

Self Check Factor: $9m^2 - 64n^4$.

Multistep Factoring

We can often factor out a greatest common factor before factoring the difference of two squares. To factor $8x^2 - 32$, for example, we factor out the GCF of 8 and then factor the resulting difference of two squares.

$$\begin{aligned} 8x^2 - 32 &= 8(x^2 - 4) && \text{Factor out 8.} \\ &= 8(x^2 - 2^2) && \text{Write 4 as } 2^2. \\ &= 8(x + 2)(x - 2) && \text{Factor the difference of two squares.} \end{aligned}$$

We can check by multiplication:

$$\begin{aligned} 8(x + 2)(x - 2) &= 8(x^2 - 4) \\ &= 8x^2 - 32 \end{aligned}$$

EXAMPLE 3 Factor: $2a^2x^3y - 8b^2xy$.

Solution We factor out the GCF of $2xy$ and then factor the resulting difference of two squares.

$$\begin{aligned} & 2a^2x^3y - 8b^2xy \\ &= 2xy \cdot a^2x^2 - 2xy \cdot 4b^2 && \text{The GCF is } 2xy. \\ &= 2xy(a^2x^2 - 4b^2) && \text{Factor out } 2xy. \\ &= 2xy[(ax)^2 - (2b)^2] && \text{Write } a^2x^2 \text{ as } (ax)^2 \text{ and } 4b^2 \text{ as } (2b)^2. \\ &= 2xy(ax + 2b)(ax - 2b) && \text{Factor the difference of two squares.} \end{aligned}$$

Check by multiplication.

Self Check Factor: $2p^2q^2s - 18r^2s$.

Sometimes we must factor a difference of two squares more than once to factor a polynomial. For example, the binomial $625a^4 - 81b^4$ can be written in the form $(25a^2)^2 - (9b^2)^2$, which factors as

$$\begin{aligned} 625a^4 - 81b^4 &= (25a^2)^2 - (9b^2)^2 \\ &= (25a^2 + 9b^2)(25a^2 - 9b^2) \end{aligned}$$

Since the factor $25a^2 - 9b^2$ can be written in the form $(5a)^2 - (3b)^2$, it is the difference of two squares and can be factored as $(5a + 3b)(5a - 3b)$. Thus,

$$625a^4 - 81b^4 = (25a^2 + 9b^2)(5a + 3b)(5a - 3b)$$

Comment The binomial $25a^2 + 9b^2$ is the **sum of two squares,** because it can be written in the form $(5a)^2 + (3b)^2$. If we are limited to integer coefficients, binomials that are the sum of two squares cannot be factored. Polynomials that do not factor over the integers are called **prime polynomials.**

EXAMPLE 4 Factor: $2x^4y - 32y$.

Solution

$$\begin{aligned} 2x^4y - 32y &= 2y \cdot x^4 - 2y \cdot 16 \\ &= 2y(x^4 - 16) && \text{Factor out the GCF of } 2y. \\ &= 2y\,(x^2 + 4)(x^2 - 4) && \text{Factor } x^4 - 16. \\ &= 2y(x^2 + 4)(x + 2)(x - 2) && \text{Factor } x^2 - 4. \text{ Note that } x^2 + 4 \text{ does not factor.} \end{aligned}$$

Check by multiplication.

Self Check Factor: $48a^5 - 3ab^4$.

Example 5 requires the techniques of factoring out a common factor, factoring by grouping, and factoring the difference of two squares.

EXAMPLE 5 Factor: $2x^3 - 8x + 2yx^2 - 8y$.

Solution

$$\begin{aligned} 2x^3 - 8x + 2yx^2 - 8y &= 2(x^3 - 4x + yx^2 - 4y) && \text{Factor out 2.} \\ &= 2[x(x^2 - 4) + y(x^2 - 4)] && \text{Factor out } x \text{ from } x^3 - 4x \text{ and } y \text{ from } yx^2 - 4y. \\ &= 2[(x^2 - 4)(x + y)] && \text{Factor out } x^2 - 4. \\ &= 2(x + 2)(x - 2)(x + y) && \text{Factor } x^2 - 4. \end{aligned}$$

Check by multiplication.

Self Check Factor: $3a^3 - 12a + 3a^2b - 12b$.

Comment To *factor* an expression means to factor the expression *completely*.

Self Check Answers

1. $(4a + 9)(4a - 9)$ 2. $(3m + 8n^2)(3m - 8n^2)$ 3. $2s(pq + 3r)(pq - 3r)$ 4. $3a(4a^2 + b^2)(2a + b)(2a - b)$
5. $3(a + 2)(a - 2)(a + b)$

Orals *Factor each binomial.*

1. $x^2 - 9$
2. $y^2 - 36$
3. $z^2 - 4$
4. $p^2 - q^2$
5. $25 - t^2$
6. $36 - r^2$
7. $100 - y^2$
8. $100 - y^4$

5.2 EXERCISES

REVIEW

1. In the study of the flow of fluids, Bernoulli's law is given by the equation

$$\frac{p}{w} + \frac{v^2}{2g} + h = k$$

Solve the equation for p.

2. Solve Bernoulli's law for h. (See Exercise 1.)

VOCABULARY AND CONCEPTS *Fill in the blanks.*

3. A binomial of the form $a^2 - b^2$ is called the ____________.
4. A binomial of the form $a^2 + b^2$ is called the ____________.
5. $p^2 - q^2 = (p + q)$______
6. The ____ of two squares cannot be factored by using only integer coefficients.

Complete each factorization.

7. $x^2 - 9 = (x + 3)$______
8. $p^2 - q^2 =$ ______$(p - q)$
9. $4m^2 - 9n^2 = (2m + 3n)$________
10. $16p^2 - 25q^2 =$ ________$(4p - 5q)$

PRACTICE *Factor each expression, if possible.*

11. $x^2 - 16$
12. $x^2 - 25$
13. $y^2 - 49$
14. $y^2 - 81$
15. $4y^2 - 49$
16. $9z^2 - 4$
17. $9x^2 - y^2$
18. $4x^2 - z^2$
19. $25t^2 - 36u^2$
20. $49u^2 - 64v^2$
21. $16a^2 - 25b^2$
22. $36a^2 - 121b^2$
23. $a^2 + b^2$
24. $121a^2 - 144b^2$
25. $a^4 - 4b^2$
26. $9y^2 + 16z^2$
27. $49y^2 - 225z^4$
28. $25x^2 + 36y^2$
29. $196x^4 - 169y^2$
30. $144a^4 + 169b^4$
31. $8x^2 - 32y^2$
32. $2a^2 - 200b^2$
33. $2a^2 - 8y^2$
34. $32x^2 - 8y^2$
35. $3r^2 - 12s^2$
36. $45u^2 - 20v^2$

37. $x^3 - xy^2$

38. $a^2b - b^3$

39. $4a^2x - 9b^2x$

40. $4b^2y - 16c^2y$

41. $3m^3 - 3mn^2$

42. $2p^2q - 2q^3$

43. $4x^4 - x^2y^2$

44. $9xy^2 - 4xy^4$

45. $2a^3b - 242ab^3$

46. $50c^4d^2 - 8c^2d^4$

47. $x^4 - 81$

48. $y^4 - 625$

49. $a^4 - 16$

50. $b^4 - 256$

51. $a^4 - b^4$

52. $m^4 - 16n^4$

53. $81r^4 - 256s^4$

54. $x^8 - y^4$

55. $a^4 - b^8$

56. $16y^8 - 81z^4$

57. $x^8 - y^8$

58. $x^8y^8 - 1$

59. $2x^4 - 2y^4$

60. $a^5 - ab^4$

61. $a^4b - b^5$

62. $m^5 - 16mn^4$

63. $48m^4n - 243n^5$

64. $2x^4y - 512y^5$

65. $3a^5y + 6ay^5$

66. $2p^{10}q - 32p^2q^5$

67. $3a^{10} - 3a^2b^4$

68. $2x^9y + 2xy^9$

69. $2x^8y^2 - 32y^6$

70. $3a^8 - 243a^4b^8$

71. $a^6b^2 - a^2b^6c^4$

72. $a^2b^3c^4 - a^2b^3d^4$

73. $a^2b^7 - 625a^2b^3$

74. $16x^3y^4z - 81x^3y^4z^5$

75. $243r^5s - 48rs^5$

76. $1{,}024m^5n - 324mn^5$

77. $16(x - y)^2 - 9$

78. $9(x + 1)^2 - y^2$

79. $a^3 - 9a + 3a^2 - 27$

80. $b^3 - 25b - 2b^2 + 50$

81. $y^3 - 16y - 3y^2 + 48$

82. $a^3 - 49a + 2a^2 - 98$

83. $3x^3 - 12x + 3x^2 - 12$

84. $2x^3 - 18x - 6x^2 + 54$

85. $3m^3 - 3mn^2 + 3am^2 - 3an^2$

86. $ax^3 - axy^2 - bx^3 + bxy^2$

87. $2m^3n^2 - 32mn^2 + 8m^2 - 128$

88. $2x^3y + 4x^2y - 98xy - 196y$

WRITING

89. Explain how to factor the difference of two squares.

90. Explain why $x^4 - y^4$ is not completely factored as $(x^2 + y^2)(x^2 - y^2)$.

SOMETHING TO THINK ABOUT

91. It is easy to multiply 399 by 401 without a calculator: The product is $400^2 - 1$, or 159,999. Explain.

92. Use the method in the previous exercise to find $498 \cdot 502$ without a calculator.

5.3 Factoring Trinomials with Lead Coefficients of 1

- Factoring Trinomials of the Form $x^2 + bx + c$
- Factoring Out -1
- Prime Trinomials
- Multistep Factoring
- Factoring Trinomials by Grouping
- Factoring Perfect-Square Trinomials

Getting Ready *Multiply the binomials.*

1. $(x + 6)(x + 6)$ **2.** $(y - 7)(y - 7)$ **3.** $(a - 3)(a - 3)$

4. $(x + 4)(x + 5)$ **5.** $(r - 2)(r - 5)$ **6.** $(m + 3)(m - 7)$

7. $(a - 3b)(a + 4b)$ **8.** $(u - 3v)(u - 5v)$ **9.** $(x + 4y)(x - 6y)$

We now discuss how to factor trinomials of the form $x^2 + bx + c$, where the coefficient of x^2 is 1 and there are no common factors.

Factoring Trinomials of the Form $x^2 + bx + c$

The product of two binomials is often a trinomial. For example,

$$(x + 3)(x + 3) = x^2 + 6x + 9 \quad \text{and} \quad (x - 4y)(x - 4y) = x^2 - 8xy + 16y^2$$

For this reason, we should not be surprised that many trinomials factor into the product of two binomials. To develop a method for factoring trinomials, we multiply $(x + a)$ and $(x + b)$.

$$(x + a)(x + b) = x^2 + bx + ax + ab \qquad \text{Use the FOIL method.}$$
$$= x^2 + ax + bx + ab \qquad \text{Write } bx + ax \text{ as } ax + bx.$$
$$= x^2 + (a + b)x + ab \qquad \text{Factor } x \text{ out of } ax + bx.$$

From the result, we can see that

- the coefficient of the middle term is the sum of a and b, and
- the last term is the product of a and b

We can use these facts to factor trinomials with lead coefficients of 1.

EXAMPLE 1 Factor: $x^2 + 5x + 6$.

Solution To factor this trinomial, we will write it as the product of two binomials. Since the first term of the trinomial is x^2, the first term of each binomial factor must be x. To fill in the following blanks, we must find two integers whose product is $+6$ and whose sum is $+5$.

$$x^2 + 5x + 6 = (x \quad\quad)(x \quad\quad)$$

The positive factorizations of 6 and the sums of the factors are shown in the following table.

Product of the factors	Sum of the factors
$1(6) = 6$	$1 + 6 = 7$
$2(3) = 6$	$2 + 3 = 5$

The last row contains the integers +2 and +3, whose product is +6 and whose sum is +5. So we can fill in the blanks with +2 and +3.

$$x^2 + 5x + 6 = (x + 2)(x + 3)$$

To check the result, we verify that $(x + 2)$ times $(x + 3)$ is $x^2 + 5x + 6$.

$$\begin{aligned}(x + 2)(x + 3) &= x^2 + 3x + 2x + 2 \cdot 3 \\ &= x^2 + 5x + 6\end{aligned}$$

Self Check Factor: $y^2 + 5y + 4$.

Comment In Example 1, the factors can be written in either order. An equivalent factorization is $x^2 + 5x + 6 = (x + 3)(x + 2)$.

EXAMPLE 2 Factor: $y^2 - 7y + 12$.

Solution Since the first term of the trinomial is y^2, the first term of each binomial factor must be y. To fill in the following blanks, we must find two integers whose product is +12 and whose sum is −7.

$$y^2 - 7y + 12 = (y \quad)(y \quad)$$

The factorizations of 12 and the sums of the factors are shown in the table.

Product of the factors	Sum of the factors
$1(12) = 12$	$1 + 12 = 13$
$2(6) = 12$	$2 + 6 = 8$
$3(4) = 12$	$3 + 4 = 7$
$-1(-12) = 12$	$-1 + (-12) = -13$
$-2(-6) = 12$	$-2 + (-6) = -8$
$-3(-4) = 12$	$-3 + (-4) = -7$

The last row contains the integers −3 and −4, whose product is +12 and whose sum is −7. So we can fill in the blanks with −3 and −4.

$$y^2 - 7y + 12 = (y - 3)(y - 4)$$

To check the result, we verify that $(y - 3)$ times $(y - 4)$ is $y^2 - 7y + 12$.

$$\begin{aligned}(y - 4)(y - 3) &= y^2 - 3y - 4y + 12 \\ &= y^2 - 7y + 12\end{aligned}$$

Self Check Factor: $p^2 - 5p + 6$.

EXAMPLE 3 Factor: $a^2 + 2a - 15$.

Solution Since the first term is a^2, the first term of each binomial factor must be a. To fill in the blanks, we must find two integers whose product is −15 and whose sum is +2.

$$a^2 + 2a - 15 = (a \quad)(a \quad)$$

The factorizations of −15 and the sums of the factors are shown in the table.

Product of the factors	Sum of the factors
$1(-15) = -15$	$1 + (-15) = -14$
$3(-5) = -15$	$3 + (-5) = -2$
$5(-3) = -15$	$5 + (-3) = 2$
$15(-1) = -15$	$15 + (-1) = 14$

The third row contains the integers $+5$ and -3, whose product is -15 and whose sum is $+2$. So we can fill in the blanks with $+5$ and -3.

$$a^2 + 2a - 15 = (a + 5)(a - 3)$$

Check: $(a + 5)(a - 3) = a^2 - 3a + 5a - 15$
$= a^2 + 2a - 15$

Self Check Factor: $p^2 + 3p - 18$.

EXAMPLE 4 Factor: $z^2 - 4z - 21$.

Solution Since the first term is z^2, the first term of each binomial factor must be z. To fill in the blanks, we must find two integers whose product is -21 and whose sum is -4.

$$z^2 - 4z - 21 = (z \quad\quad)(z \quad\quad)$$

The factorizations of -21 and the sums of the factors are shown in the table.

Product of the factors	Sum of the factors
$1(-21) = -21$	$1 + (-21) = -20$
$3(-7) = -21$	$3 + (-7) = -4$
$7(-3) = -21$	$7 + (-3) = 4$
$21(-1) = -21$	$21 + (-1) = 20$

The second row contains the integers $+3$ and -7, whose product is -21 and whose sum is -4. So we can fill in the blanks with $+3$ and -7.

$$z^2 - 4z - 21 = (z + 3)(z - 7)$$

Check: $(z + 3)(z - 7) = z^2 - 7z + 3z - 21$
$= z^2 - 4z - 21$

Self Check Factor: $q^2 - 2q - 24$.

The next example has two variables.

EXAMPLE 5 Factor: $x^2 + xy - 6y^2$.

Solution Since the first term is x^2, the first term of each binomial factor must be x. Since the last term is $-6y^2$, the second term of each binomial factor has a factor of y. To fill in the blanks, we must find coefficients whose product is -6 that will give a middle term of xy.

$$x^2 + xy - 6y^2 = (x \;\square\; y)(x \;\square\; y)$$

The factorizations of -6 and the sums of the factors are shown in the table.

Product of the factors	Sum of the factors
$1(-6) = -6$	$1 + (-6) = -5$
$2(-3) = -6$	$2 + (-3) = -1$
$3(-2) = -6$	$3 + (-2) = 1$
$6(-1) = -6$	$6 + (-1) = 5$

The third row contains the integers 3 and -2. These are the only integers whose product is -6 and will give the correct middle term of xy. So we can fill in the blanks with 3 and -2.

$$x^2 + xy - 6y^2 = (x + 3y)(x - 2y)$$

Check: $(x + 3y)(x - 2y) = x^2 - 2xy + 3xy - 6y^2$
$= x^2 + xy - 6y^2$

Self Check Factor: $a^2 + ab - 12b^2$.

Factoring Out −1

When the coefficient of the first term is -1, we begin by factoring out -1.

EXAMPLE 6 Factor: $-x^2 + 2x + 15$.

Solution We factor out -1 and then factor the trinomial.

$$-x^2 + 2x + 15 = -(x^2 - 2x - 15) \quad \text{Factor out } -1.$$
$$= -(x - 5)(x + 3) \quad \text{Factor } x^2 - 2x - 15.$$

Check: $-(x - 5)(x + 3) = -(x^2 + 3x - 5x - 15)$
$= -(x^2 - 2x - 15)$
$= -x^2 + 2x + 15$

Self Check Factor: $-x^2 + 11x - 18$.

Prime Trinomials

If a trinomial cannot be factored using only integer coefficients, it is called a **prime polynomial.**

EXAMPLE 7 Factor: $x^2 + 2x + 3$, if possible.

Solution To factor the trinomial, we must find two integers whose product is $+3$ and whose sum is $+2$. The possible factorizations of 3 and the sums of the factors are shown in the table.

Product of the factors	Sum of the factors
$1(3) = 3$	$1 + 3 = 4$
$-1(-3) = 3$	$-1 + (-3) = -4$

Since two integers whose product is $+3$ and whose sum is $+2$ do not exist, $x^2 + 2x + 3$ cannot be factored. It is a prime trinomial.

Self Check Factor: $x^2 - 4x + 6$, if possible.

Multistep Factoring

The following examples require more than one step.

EXAMPLE 8 Factor: $-3ax^2 + 9a - 6ax$.

Solution We write the trinomial in descending powers of x and factor out the common factor of $-3a$.

$$\begin{aligned} -3ax^2 + 9a - 6ax &= -3ax^2 - 6ax + 9a \\ &= -3a(x^2 + 2x - 3) \end{aligned}$$

Finally, we factor the trinomial $x^2 + 2x - 3$.

$$-3ax^2 + 9a - 6ax = -3a(x + 3)(x - 1)$$

Check:
$$\begin{aligned} -3a(x + 3)(x - 1) &= -3a(x^2 + 2x - 3) \\ &= -3ax^2 - 6ax + 9a \\ &= -3ax^2 + 9a - 6ax \end{aligned}$$

Self Check Factor: $-2pq^2 + 6p - 4pq$.

EXAMPLE 9 Factor: $m^2 - 2mn + n^2 - 64a^2$.

Solution We group the first three terms together and factor the resulting trinomial.

$$\begin{aligned} m^2 - 2mn + n^2 - 64a^2 &= (m - n)(m - n) - 64a^2 \\ &= (m - n)^2 - (8a)^2 \end{aligned}$$

Then we factor the resulting difference of two squares:

$$\begin{aligned} m^2 - 2mn + n^2 - 64a^2 &= (m - n)^2 - (8a)^2 \\ &= (m - n + 8a)(m - n - 8a) \end{aligned}$$

Self Check Factor: $p^2 + 4pq + 4q^2 - 25y^2$.

Factoring Trinomials by Grouping

An alternate way of factoring trinomials of the form $x^2 + bx + c$ uses the technique of factoring by grouping. For example, to factor $x^2 + x - 12$ by grouping, we proceed as follows:

1. Determine the value of c: $c = -12$. This number is called the **key number.**
2. Find two factors of the key number -12 whose sum is $b = 1$. Two such factors are 4 and -3.

$$4(-3) = -12 \quad \text{and} \quad 4 + (-3) = 1$$

3. Use the factors 4 and -3 as coefficients of two terms to be placed between x^2 and -12.

$$x^2 + x - 12 = x^2 + 4x - 3x - 12$$

4. Factor the right-hand side of the previous equation by grouping.

$$x^2 + 4x - 3x - 12 = x(x + 4) - 3(x + 4)$$ Factor x out of $x^2 + 4x$ and -3 out of $-3x - 12$.

$$= (x + 4)(x - 3)$$ Factor out $x + 4$.

We can check this factorization by multiplication.

EXAMPLE 10 Factor: $y^2 + 7y + 10$.

Solution To factor this trinomial by grouping, we note that it is written in the form $y^2 + by + c$, with $b = 7$ and $c = 10$.

First, we identify c to be 10. Then, we find two factors of 10 whose sum is $b = 7$. Two such factors are 2 and 5. We use these factors as coefficients of two terms to be placed between y^2 and 10.

$$y^2 + 7y + 10 = y^2 + 2y + 5y + 10$$

Finally, we factor the right-hand side of the previous equation by grouping

$$y^2 + 2y + 5y + 10 = y(y + 2) + 5(y + 2)$$ Factor out y from $y^2 + 2y$ and factor out 5 from $5y + 10$.

$$= (y + 2)(y + 5)$$ Factor out $y + 2$.

Self Check Use grouping to factor $p^2 - 7p + 12$.

EXAMPLE 11 Factor: $z^2 - 4z - 21$.

Solution This is the trinomial of Example 4. To factor it by grouping, we note that the trinomial is written in the form $z^2 + bz + c$, with $b = -4$ and $c = -21$.

First, we identify c to be -21. Then we find two factors of -21 whose sum is $b = -4$. Two such factors are 3 and -7. We use these factors as coefficients of two terms to be placed between z^2 and -21:

$$z^2 - 4z - 21 = z^2 + 3z - 7z - 21$$

Finally, we factor the right-hand side of the previous equation by grouping.

$$z^2 + 3z - 7z - 21 = z(z + 3) - 7(z + 3)$$ Factor out z from $z^2 + 3z$ and factor out -7 from $-7z - 21$.

$$= (z + 3)(z - 7)$$ Factor out $z + 3$.

Self Check Use grouping to factor $a^2 + 2a - 15$. This is the trinomial of Example 3.

Factoring Perfect-Square Trinomials

We have discussed the following special product formulas used to square binomials.

Special Product Formulas

$$(x + y)^2 = x^2 + 2xy + y^2$$
$$(x - y)^2 = x^2 - 2xy + y^2$$

These formulas can be used in reverse order to factor special trinomials called **perfect-square trinomials.**

(1) $$x^2 + 2xy + y^2 = (x + y)^2$$
(2) $$x^2 - 2xy + y^2 = (x - y)^2$$

In words, Formula 1 states that *if a trinomial is the square of one quantity, plus twice the product of two quantities, plus the square of the second quantity, it factors into the square of the sum of the quantities.*

Formula 2 states that *if a trinomial is the square of one quantity, minus twice the product of two quantities, plus the square of the second quantity, it factors into the square of the difference of the quantities.*

The trinomials on the left-hand sides of the previous equations are perfect-square trinomials, because they are the results of squaring a binomial. Although we can factor perfect-square trinomials by using the techniques discussed earlier in this section, we can usually factor them by inspecting their terms. For example, $x^2 + 8x + 16$ is a perfect-square trinomial, because

- The first term x^2 is the square of x.
- The last term 16 is the square of 4.
- The middle term $8x$ is twice the product of x and 4.

Thus,

$$x^2 + 8x + 16 = x^2 + 2(x)(4) + 4^2$$
$$= (x + 4)^2$$

EXAMPLE 12 Factor: $x^2 - 10x + 25$.

Solution $x^2 - 10x + 25$ is a perfect-square trinomial, because

- The first term x^2 is the square of x.
- The last term 25 is the square of 5.
- The middle term $-10x$ is the negative of twice the product of x and 5.

Thus,

$$x^2 - 10x + 25 = x^2 - 2(x)(5) + 5^2$$
$$= (x - 5)^2$$

Self Check Factor: $x^2 + 10x + 25$.

Self Check Answers

1. $(y + 1)(y + 4)$ **2.** $(p - 3)(p - 2)$ **3.** $(p + 6)(p - 3)$ **4.** $(q + 4)(q - 6)$ **5.** $(a - 3b)(a + 4b)$ **6.** $-(x - 9)(x - 2)$ **7.** It is prime. **8.** $-2p(q + 3)(q - 1)$ **9.** $(p + 2q + 5y)(p + 2q - 5y)$ **10.** $(p - 4)(p - 3)$ **11.** $(a + 5)(a - 3)$ **12.** $(x + 5)^2$

Orals *Finish factoring each problem.*

1. $x^2 + 5x + 4 = (x + 1)(x + \square)$
2. $x^2 - 5x + 6 = (x \square 2)(x \square 3)$
3. $x^2 + x - 6 = (x \square 2)(x + \square)$
4. $x^2 - x - 6 = (x \square 3)(x + \square)$
5. $x^2 + 5x - 6 = (x + \square)(x - \square)$
6. $x^2 - 7x + 6 = (x - \square)(x - \square)$

5.3 EXERCISES

REVIEW *Graph the solution of each inequality on a number line.*

1. $x - 3 > 5$
2. $x + 4 \leq 3$
3. $-3x - 5 \geq 4$
4. $2x - 3 < 7$
5. $\frac{3(x - 1)}{4} < 12$
6. $\frac{-2(x + 3)}{3} \geq 9$
7. $-2 < x \leq 4$
8. $-5 \leq x + 1 < 5$

VOCABULARY AND CONCEPTS *Complete each formula.*

9. $x^2 + 2xy + y^2 =$ ______
10. $x^2 - 2xy + y^2 =$ ______

Complete each factorization.

11. $y^2 + 6y + 8 = (y + \square)(y + \square)$
12. $z^2 - 3z - 10 = (z + \square)(z - \square)$
13. $x^2 - xy - 2y^2 = (x + \square)(x - \square)$
14. $a^2 + ab - 6b^2 = (a + \square)(a - \square)$

Factor each trinomial, if possible. Check each result.

15. $x^2 + 3x + 2$
16. $y^2 + 4y + 3$
17. $z^2 + 12z + 11$
18. $x^2 + 7x + 10$
19. $a^2 - 4a - 5$
20. $b^2 + 6b - 7$
21. $t^2 - 9t + 14$
22. $c^2 - 9c + 8$
23. $u^2 + 10u + 15$
24. $v^2 + 9v + 15$
25. $y^2 - y - 30$
26. $x^2 - 3x - 40$
27. $a^2 + 6a - 16$
28. $x^2 + 5x - 24$
29. $t^2 - 5t - 50$
30. $a^2 - 10a - 39$
31. $r^2 - 9r - 12$
32. $s^2 + 11s - 26$
33. $y^2 + 2yz + z^2$
34. $r^2 - 2rs + 4s^2$
35. $x^2 + 4xy + 4y^2$
36. $a^2 + 10ab + 9b^2$
37. $m^2 + 3mn - 10n^2$
38. $m^2 - mn - 12n^2$
39. $a^2 - 4ab - 12b^2$
40. $p^2 + pq - 6q^2$

41. $u^2 + 2uv - 15v^2$

42. $m^2 - 11mn + 10n^2$

Factor each trinomial. Factor out −1 first.

43. $-x^2 - 7x - 10$

44. $-x^2 + 9x - 20$

45. $-y^2 - 2y + 15$

46. $-y^2 - 3y + 18$

47. $-t^2 - 15t + 34$

48. $-t^2 - t + 30$

49. $-r^2 + 14r - 40$

50. $-r^2 + 14r - 45$

51. $-a^2 - 4ab - 3b^2$

52. $-a^2 - 6ab - 5b^2$

53. $-x^2 + 6xy + 7y^2$

54. $-x^2 - 10xy + 11y^2$

Write each trinomial in descending powers of one variable, and then factor it.

55. $4 - 5x + x^2$

56. $y^2 + 5 + 6y$

57. $10y + 9 + y^2$

58. $x^2 - 13 - 12x$

59. $c^2 - 5 + 4c$

60. $b^2 - 6 - 5b$

61. $-r^2 + 2s^2 + rs$

62. $u^2 - 3v^2 + 2uv$

63. $4rx + r^2 + 3x^2$

64. $-a^2 + 5b^2 + 4ab$

65. $-3ab + a^2 + 2b^2$

66. $-13yz + y^2 - 14z^2$

Completely factor each trinomial. Factor out any common monomials first (including −1, if necessary).

67. $2x^2 + 10x + 12$

68. $3y^2 - 21y + 18$

69. $3y^3 + 6y^2 + 3y$

70. $4x^4 + 16x^3 + 16x^2$

71. $-5a^2 + 25a - 30$

72. $-2b^2 + 20b - 18$

73. $3z^2 - 15tz + 12t^2$

74. $5m^2 + 45mn - 50n^2$

75. $12xy + 4x^2y - 72y$

76. $48xy + 6xy^2 + 96x$

77. $-4x^2y - 4x^3 + 24xy^2$

78. $3x^2y^3 + 3x^3y^2 - 6xy^4$

Completely factor each expression.

79. $x^2 + 4x + 4 - y^2$

80. $p^2 - 2p + 1 - q^2$

81. $b^2 - 6b + 9 - c^2$

82. $m^2 + 8m + 16 - n^2$

83. $a^2 + 2ab + b^2 - 4$

84. $a^2 + 6a + 9 - b^2$

85. $b^2 - y^2 - 4y - 4$

86. $c^2 - a^2 + 8a - 16$

87. $x^2 + 6x + 9$

88. $x^2 + 10x + 25$

89. $y^2 - 8y + 16$

90. $z^2 - 2z + 1$

91. $t^2 + 20t + 100$

92. $r^2 + 24r + 144$

93. $u^2 - 18u + 81$

94. $v^2 - 14v + 49$

95. $x^2 + 4xy + 4y^2$

96. $a^2 + 6ab + 9b^2$

97. $r^2 - 10rs + 25s^2$

98. $m^2 - 12mn + 36n^2$

WRITING

99. Explain how you would write a trinomial in descending order.

100. Explain how to use the FOIL method to check the factoring of a trinomial.

SOMETHING TO THINK ABOUT

101. Two students factor $2x^2 + 20x + 42$ and get two different answers: $(2x + 6)(x + 7)$, and $(x + 3)(2x + 14)$. Do both answers check? Why don't they agree? Is either completely correct?

102. Find the error:

$x = y$	
$x^2 = xy$	Multiply both sides by x.
$x^2 - y^2 = xy - y^2$	Subtract y^2 from both sides.
$(x + y)(x - y) = y(x - y)$	Factor.
$x + y = y$	Divide both sides by $(x - y)$.
$y + y = y$	Substitute y for its equal, x.
$2y = y$	Combine like terms.
$2 = 1$	Divide both sides by y.

5.4 Factoring General Trinomials

- Factoring Trinomials of the Form $ax^2 + bx + c$
- Factoring Trinomials by Grouping
- Factoring Perfect-Square Trinomials

Getting Ready *Multiply and combine like terms.*

1. $(2x + 1)(3x + 2)$ **2.** $(3y - 2)(2y - 5)$ **3.** $(4t - 3)(2t + 3)$

4. $(2r + 5)(2r - 3)$ **5.** $(2m - 3)(3m - 2)$ **6.** $(4a + 3)(4a + 1)$

In the previous section, we saw how to factor trinomials whose leading coefficients are 1. We now show how to factor trinomials whose lead coefficients are other than 1.

Factoring Trinomials of the Form $ax^2 + bx + c$

We must consider more combinations of factors when we factor trinomials with lead coefficients other than 1.

EXAMPLE 1 Factor: $2x^2 + 5x + 3$.

Solution Since the first term is $2x^2$, the first terms of the binomial factors must be $2x$ and x. To fill in the blanks, we must find two factors of $+3$ that will give a middle term of $+5x$.

$$(2x \quad)(x \quad)$$

Since the sign of each term of the trinomial is $+$, we need to consider only positive factors of the last term (3). Since the positive factors of 3 are 1 and 3, there are two possible factorizations.

$$(2x + 1)(x + 3) \quad \text{or} \quad (2x + 3)(x + 1)$$

The first possibility is incorrect, because it gives a middle term of $7x$. The second possibility is correct, because it gives a middle term of $5x$. Thus,

$$2x^2 + 5x + 3 = (2x + 3)(x + 1)$$

Check by multiplication.

Self Check Factor: $3x^2 + 7x + 2$.

EXAMPLE 2 Factor: $6x^2 - 17x + 5$.

Solution Since the first term is $6x^2$, the first terms of the binomial factors must be $6x$ and x or $3x$ and $2x$. To fill in the blanks, we must find two factors of $+5$ that will give a middle term of $-17x$.

$$(6x \quad)(x \quad) \quad \text{or} \quad (3x \quad)(2x \quad)$$

Since the sign of the third term is + and the sign of the middle term is −, we need to consider only negative factors of the last term (5). Since the negative factors of 5 are −1 and −5, there are four possible factorizations.

$$(6x - 1)(x - 5) \qquad (6x - 5)(x - 1)$$
$$(3x - 1)(2x - 5) \qquad (3x - 5)(2x - 1)$$

Only the possibility printed in color gives the correct middle term of $-17x$. Thus,

$$6x^2 - 17x + 5 = (3x - 1)(2x - 5)$$

Check by multiplication.

Self Check Factor: $6x^2 - 7x + 2$.

EXAMPLE 3 Factor: $3y^2 - 4y - 4$.

Solution Since the first term is $3y^2$, the first terms of the binomial factors must be $3y$ and y. To fill in the blanks, we must find two factors of -4 that will give a middle term of $-4y$.

$$(3y \quad\quad)(y \quad\quad)$$

Since the sign of the third term is −, the signs inside the binomial factors will be different. Because the factors of the last term (4) are 1, 2, and 4, there are six possibilities to consider.

$$(3y + 1)(y - 4) \qquad (3y + 4)(y - 1)$$
$$(3y - 1)(y + 4) \qquad (3y - 4)(y + 1)$$
$$(3y - 2)(y + 2) \qquad (3y + 2)(y - 2)$$

Only the possibility printed in color gives the correct middle term of $-4y$. Thus,

$$3y^2 - 4y - 4 = (3y + 2)(y - 2)$$

Check by multiplication.

Self Check Factor: $5a^2 - 7a - 6$.

EXAMPLE 4 Factor: $6b^2 + 7b - 20$.

Solution Since the first term is $6b^2$, the first terms of the binomial factors must be $6b$ and b or $3b$ and $2b$. To fill in the blanks, we must find two factors of -20 that will give a middle term of $+7b$.

$$(6b \quad\quad)(b \quad\quad) \quad \text{or} \quad (3b \quad\quad)(2b \quad\quad)$$

Since the sign of the third term is −, the signs inside the binomial factors will be different. Because the factors of the last term (20) are 1, 2, 4, 5, 10, and 20, there are many possible combinations for the last terms. We must try to find one that will give a last term of -20 and a sum of the products of the outer terms and inner terms of $+7b$.

If we pick factors of $6b$ and b for the first terms and -5 and $+4$ for the last terms, we have

$$(6b - 5)(b + 4)$$

$$\begin{array}{r} -5b \\ \underline{24b} \\ 19b \end{array}$$

which gives a wrong middle term of $19b$.

If we pick factors of $3b$ and $2b$ for the first terms and -4 and $+5$ for the last terms, we have

$$(3b - 4)(2b + 5)$$

$$\begin{array}{r} -8b \\ \underline{15b} \\ 7b \end{array}$$

which gives the correct middle term of $+7b$ and the correct last term of -20. Thus,

$$6b^2 + 7b - 20 = (3b - 4)(2b + 5)$$

Check by multiplication.

Self Check Factor: $4x^2 + 4x - 3$.

The next example has two variables.

EXAMPLE 5 Factor: $2x^2 + 7xy + 6y^2$.

Solution Since the first term is $2x^2$, the first terms of the binomial factors must be $2x$ and x. To fill in the blanks, we must find two factors of $6y^2$ that will give a middle term of $+7xy$.

$$(2x \quad\quad)(x \quad\quad)$$

Since the sign of each term is +, the signs inside the binomial factors will be +. The possible factors of the last term $(6y^2)$ are y, $2y$, $3y$, and $6y$. We must try to find one that will give a last term of $+6y^2$ and a sum of the products of the outer terms and inner terms of $+7xy$.

If we pick factors of $6y$ and y, we have

$$(2x + y)(x + 6y)$$

$$\begin{array}{r} xy \\ \underline{12xy} \\ 13xy \end{array}$$

which gives a wrong middle term of $13xy$.

If we pick factors of $3y$ and $2y$, we have

$$(2x + 3y)(x + 2y)$$

$$\begin{array}{r} 3xy \\ \underline{4xy} \\ 7xy \end{array}$$

which gives a correct middle term of $7xy$. Thus,

$$2x^2 + 7xy + 6y^2 = (2x + 3y)(x + 2y)$$

Check by multiplication.

Self Check Factor: $4x^2 + 8xy + 3y^2$.

Because some guesswork is often necessary, it is difficult to give specific rules for factoring trinomials. However, the following hints are often helpful.

Factoring General Trinomials

1. Write the trinomial in descending powers of one variable.
2. Factor out any GCF (including -1 if that is necessary to make the coefficient of the first term positive).
3. If the sign of the third term is $+$, the signs between the terms of the binomial factors are the same as the sign of the middle term. If the sign of the third term is $-$, the signs between the terms of the binomial factors are opposite.
4. Try combinations of first terms and last terms until you find one that works, or until you exhaust all the possibilities. If no combination works, the trinomial is prime.
5. Check the factorization by multiplication.

EXAMPLE 6 Factor: $2x^2y - 8x^3 + 3xy^2$.

Solution **Step 1:** Write the trinomial in descending powers of x.

$$-8x^3 + 2x^2y + 3xy^2$$

Step 2: Factor out the negative of the GCF, which is $-x$.

$$-8x^3 + 2x^2y + 3xy^2 = -x(8x^2 - 2xy - 3y^2)$$

Step 3: Because the sign of the third term of the trinomial factor is $-$, the signs within its binomial factors will be opposites.

Step 4: Find the binomial factors of the trinomial.

$$\begin{aligned} -8x^3 + 2x^2y + 3xy^2 &= -x(8x^2 - 2xy - 3y^2) \\ &= -x(2x + y)(4x - 3y) \end{aligned}$$

Step 5: Check by multiplication.

$$\begin{aligned} -x(2x + y)(4x - 3y) &= -x(8x^2 - 6xy + 4xy - 3y^2) \\ &= -x(8x^2 - 2xy - 3y^2) \\ &= -8x^3 + 2x^2y + 3xy^2 \\ &= 2x^2y - 8x^3 + 3xy^2 \end{aligned}$$

Self Check Factor: $12y - 2y^3 - 2y^2$.

Factoring Trinomials by Grouping

An alternate way of factoring trinomials of the form $ax^2 + bx + c$ uses the technique of factoring by grouping. For example, to factor $6x^2 - 17x + 5$ (Example 2) by grouping, we proceed as follows:

1. Determine the product ac: $6(5) = 30$. This number is called the **key number.**
2. Find two factors of the key number 30 whose sum is $b = -17$. Two such factors are -15 and -2.

   $$-15(-2) = 30 \quad \text{and} \quad -15 + (-2) = -17$$

3. Use -15 and -2 as coefficients of two terms to be placed between $6x^2$ and 5.

   $$6x^2 - 17x + 5 = 6x^2 - 15x - 2x + 5$$

4. Factor the right-hand side of the previous equation by grouping.

   $$\begin{aligned} 6x^2 - 15x - 2x + 5 &= 3x(2x - 5) - 1(2x - 5) && \text{Factor out } 3x \text{ from } 6x^2 - 15x \text{ and } -1 \text{ from } -2x + 5. \\ &= (2x - 5)(3x - 1) && \text{Factor out } 2x - 5. \end{aligned}$$

We can verify this factorization by multiplication.

EXAMPLE 7 Factor: $4y^2 + 12y + 5$.

Solution To factor this trinomial by grouping, we note that it is written in the form $ay^2 + by + c$, with $a = 4$, $b = 12$, and $c = 5$. Since $a = 4$ and $c = 5$, we have $ac = 20$.

We now find two factors of 20 whose sum is $b = 12$. Two such factors are 10 and 2. We use these factors as coefficients of two terms to be placed between $4y^2$ and 5.

$$4y^2 + 12y + 5 = 4y^2 + 10y + 2y + 5$$

Finally, we factor the right-hand side of the previous equation by grouping

$$\begin{aligned} 4y^2 + 10y + 2y + 5 &= 2y(2y + 5) + (2y + 5) && \text{Factor out } 2y \text{ from } 4y^2 + 10y. \\ &= 2y(2y + 5) + 1 \cdot (2y + 5) && (2y + 5) = 1 \cdot (2y + 5). \\ &= (2y + 5)(2y + 1) && \text{Factor out } 2y + 5. \end{aligned}$$

Self Check Use grouping to factor $2p^2 - 7p + 3$.

EXAMPLE 8 Factor: $6b^2 + 7b - 20$.

Solution This is the trinomial of Example 4. Since $a = 6$ and $c = -20$ in the trinomial, $ac = -120$. We now find two factors of -120 whose sum is $+7$. Two such factors are 15 and -8. We use these factors as coefficients of two terms to be placed between $6b^2$ and -20.

$$6b^2 + 7b - 20 = 6b^2 + 15b - 8b - 20$$

Finally, we factor the right-hand side of the previous equation by grouping.

$$6b^2 + 15b - 8b - 20 = 3b(2b + 5) - 4(2b + 5)$$ Factor out $3b$ from $6b^2 + 15b$ and -4 from $-8b - 20$.

$$= (2b + 5)(3b - 4)$$ Factor out $2b + 5$.

Self Check Factor: $3y^2 - 4y - 4$ (Example 3).

Factoring Perfect-Square Trinomials

As before, we can factor perfect-square trinomials by inspection.

EXAMPLE 9 Factor: $4x^2 - 20x + 25$.

Solution $4x^2 - 20x + 25$ is a perfect-square trinomial, because

- The first term $4x^2$ is the square of $2x$: $(2x)^2 = 4x^2$.
- The last term 25 is the square of 5: $5^2 = 25$.
- The middle term $-20x$ is the negative of twice the product of $2x$ and 5.

Thus,

$$\begin{aligned} 4x^2 - 20x + 25 &= (2x)^2 - 2(2x)(5) + 5^2 \\ &= (2x - 5)^2 \end{aligned}$$

Self Check Factor: $9x^2 - 12x + 4$.

The next examples combine several factoring techniques.

EXAMPLE 10 Factor: $4x^2 - 4xy + y^2 - 9$.

Solution

$$4x^2 - 4xy + y^2 - 9$$
$$= (4x^2 - 4xy + y^2) - 9$$ Group the first three terms.
$$= (2x - y)^2 - 9$$ Factor the perfect-square trinomial.
$$= [(2x - y) + 3][(2x - y) - 3]$$ Factor the difference of two squares.
$$= (2x - y + 3)(2x - y - 3)$$ Remove the inner parentheses.

Check by multiplication.

Self Check Factor: $x^2 + 4x + 4 - y^2$.

EXAMPLE 11 Factor: $9 - 4x^2 - 4xy - y^2$.

Solution

$$9 - 4x^2 - 4xy - y^2 = 9 - (4x^2 + 4xy + y^2)$$ Factor -1 from the trinomial.
$$= 9 - (2x + y)(2x + y)$$ Factor the perfect-square trinomial.

$= 9 - (2x + y)^2$ — $(2x + y)(2x + y) = (2x + y)^2$.

$= [3 + (2x + y)][3 - (2x + y)]$ — Factor the difference of two squares.

$= (3 + 2x + y)(3 - 2x - y)$ — Remove parentheses.

Check by multiplication.

Self Check Factor: $16 - a^2 - 2a - 1$.

Self Check Answers

1. $(3x + 1)(x + 2)$ 2. $(3x - 2)(2x - 1)$ 3. $(5a + 3)(a - 2)$ 4. $(2x + 3)(2x - 1)$ 5. $(2x + 3y)(2x + y)$ 6. $-2y(y + 3)(y - 2)$ 7. $(2p - 1)(p - 3)$ 8. $(3y + 2)(y - 2)$ 9. $(3x - 2)^2$ 10. $(x + 2 + y)(x + 2 - y)$ 11. $(a + 5)(3 - a)$

Orals *Finish factoring each problem.*

1. $2x^2 + 5x + 3 = (\quad x + \quad)(x + 1)$
2. $6x^2 + 5x + 1 = (\quad x + 1)(3x + 1)$
3. $6x^2 + 5x - 1 = (x \quad 1)(6x \quad 1)$
4. $6x^2 + x - 1 = (2x \quad 1)(3x \quad 1)$
5. $4x^2 + 4x - 3 = (2x + \quad)(2x - \quad)$
6. $4x^2 - x - 3 = (4x + \quad)(x - \quad)$

5.4 EXERCISES

REVIEW

1. The nth term l of an arithmetic sequence is

$$l = f + (n - 1)d$$

where f is the first term and d is the common difference. Remove the parentheses and solve for n.

2. The sum S of n consecutive terms of an arithmetic sequence is

$$S = \frac{n}{2}(f + l)$$

where f is the first term and l is the nth term. Solve for f.

VOCABULARY AND CONCEPTS *Fill in the blanks.*

3. To factor a general trinomial, first write the trinomial in ________ powers of one variable.
4. If the sign of the first and third terms of a trinomial are +, the signs within the binomial factors are ________ as the sign of the middle term.
5. If the sign of the first term of a trinomial is + and the sign of the third term is −, the signs within the binomial factors are ________.
6. Always check factorizations by ________.

Complete each factorization.

7. $6x^2 + x - 2 = (3x + \quad)(2x - \quad)$
8. $15x^2 - 7x - 4 = (5x - \quad)(3x + \quad)$
9. $12x^2 - 7xy + y^2 = (3x - \quad)(4x - \quad)$
10. $6x^2 + 5xy - 6y^2 = (2x + \quad)(3x - \quad)$

PRACTICE *Factor each trinomial.*

11. $2x^2 - 3x + 1$
12. $2y^2 - 7y + 3$
13. $3a^2 + 13a + 4$
14. $2b^2 + 7b + 6$
15. $4z^2 + 13z + 3$
16. $4t^2 - 4t + 1$
17. $6y^2 + 7y + 2$
18. $4x^2 + 8x + 3$
19. $6x^2 - 7x + 2$
20. $4z^2 - 9z + 2$
21. $3a^2 - 4a - 4$
22. $8u^2 - 2u - 15$
23. $2x^2 - 3x - 2$
24. $12y^2 - y - 1$
25. $2m^2 + 5m - 12$
26. $10u^2 - 13u - 3$

27. $10y^2 - 3y - 1$

28. $6m^2 + 19m + 3$

29. $12y^2 - 5y - 2$

30. $10x^2 + 21x - 10$

31. $5t^2 + 13t + 6$

32. $16y^2 + 10y + 1$

33. $16m^2 - 14m + 3$

34. $16x^2 + 16x + 3$

35. $3x^2 - 4xy + y^2$

36. $2x^2 + 3xy + y^2$

37. $2u^2 + uv - 3v^2$

38. $2u^2 + 3uv - 2v^2$

39. $4a^2 - 4ab + b^2$

40. $2b^2 - 5bc + 2c^2$

41. $6r^2 + rs - 2s^2$

42. $3m^2 + 5mn + 2n^2$

43. $4x^2 + 8xy + 3y^2$

44. $4b^2 + 15bc - 4c^2$

45. $4a^2 - 15ab + 9b^2$

46. $12x^2 + 5xy - 3y^2$

Write the terms of each trinomial in descending powers of one variable. Then factor the trinomial, if possible.

47. $-13x + 3x^2 - 10$

48. $-14 + 3a^2 - a$

49. $15 + 8a^2 - 26a$

50. $16 - 40a + 25a^2$

51. $12y^2 + 12 - 25y$

52. $12t^2 - 1 - 4t$

53. $3x^2 + 6 + x$

54. $25 + 2u^2 + 3u$

55. $2a^2 + 3b^2 + 5ab$

56. $11uv + 3u^2 + 6v^2$

57. $pq + 6p^2 - q^2$

58. $-11mn + 12m^2 + 2n^2$

59. $b^2 + 4a^2 + 16ab$

60. $3b^2 + 3a^2 - ab$

61. $12x^2 + 10y^2 - 23xy$

62. $5ab + 25a^2 - 2b^2$

Factor each polynomial.

63. $4x^2 + 10x - 6$

64. $9x^2 + 21x - 18$

65. $y^3 + 13y^2 + 12y$

66. $2xy^2 + 8xy - 24x$

67. $6x^3 - 15x^2 - 9x$

68. $9y^3 + 3y^2 - 6y$

69. $30r^5 + 63r^4 - 30r^3$

70. $6s^5 - 26s^4 - 20s^3$

71. $4a^2 - 4ab - 8b^2$

72. $6x^2 + 3xy - 18y^2$

73. $8x^2 - 12xy - 8y^2$

74. $24a^2 + 14ab + 2b^2$

75. $-16m^3n - 20m^2n^2 - 6mn^3$

76. $-84x^4 - 100x^3y - 24x^2y^2$

77. $-28u^3v^3 + 26u^2v^4 - 6uv^5$

78. $-16x^4y^3 + 30x^3y^4 + 4x^2y^5$

79. $4x^2 + 12x + 9$

80. $4x^2 - 4x + 1$

81. $9x^2 + 12x + 4$

82. $4x^2 - 20x + 25$

83. $16x^2 - 8xy + y^2$

84. $25x^2 + 20xy + 4y^2$

85. $4x^2 + 4xy + y^2 - 16$

86. $9x^2 - 6x + 1 - d^2$

87. $9 - a^2 - 4ab - 4b^2$

88. $25 - 9a^2 + 6ac - c^2$

89. $9p^2 + 1 + 6p - q^2$

90. $16m^2 - 24m - n^2 + 9$

WRITING

91. Describe an organized approach to finding all of the possibilities when you attempt to factor $12x^2 - 4x + 9$.

92. Explain how to determine whether a trinomial is prime.

SOMETHING TO THINK ABOUT

93. For what values of b will the trinomial $6x^2 + bx + 6$ be factorable?

94. For what values of b will the trinomial $5y^2 - by - 3$ be factorable?

5.5 Factoring the Sum and Difference of Two Cubes

- Factoring the Sum of Two Cubes
- Factoring the Difference of Two Cubes
- Multistep Factoring

Getting Ready *Find each product.*

1. $(x - 3)(x^2 + 3x + 9)$
2. $(x + 2)(x^2 - 2x + 4)$
3. $(y + 4)(y^2 - 4y + 16)$
4. $(r - 5)(r^2 + 5r + 25)$
5. $(a - b)(a^2 + ab + b^2)$
6. $(a + b)(a^2 - ab + b^2)$

Recall that the difference of the squares of two quantities factors into the product of two binomials. One binomial is the sum of the quantities, and the other is the difference of the quantities.

$$x^2 - y^2 = (x + y)(x - y) \quad \text{or} \quad F^2 - L^2 = (F + L)(F - L)$$

In this section, we will discuss similar formulas for factoring the sum of two cubes and the difference of two cubes.

Factoring the Sum of Two Cubes

To discover the formula for factoring the sum of two cubes, we find the following product:

$$(x + y)(x^2 - xy + y^2) = (x + y)x^2 - (x + y)xy + (x + y)y^2 \quad \text{Use the distributive property.}$$
$$= x^3 + x^2y - x^2y - xy^2 + xy^2 + y^3$$
$$= x^3 + y^3$$

This result justifies the formula for factoring the **sum of two cubes.**

Factoring the Sum of Two Cubes

$$x^3 + y^3 = (x + y)(x^2 - xy + y^2)$$

If we think of the sum of two cubes as the cube of a **F**irst quantity plus the cube of a **L**ast quantity, we have the formula

$$F^3 + L^3 = (F + L)(F^2 - FL + L^2)$$

In words, we say, *To factor the cube of a* ***First*** *quantity plus the cube of a* ***Last*** *quantity, we multiply the* ***First*** *plus the* ***Last*** *by*

- *the* ***First*** *squared*
- *minus the* ***First*** *times the* ***Last***
- *plus the* ***Last*** *squared.*

To factor the sum of two cubes, it is helpful to know the cubes of the numbers from 1 to 10:

1, 8, 27, 64, 125, 216, 343, 512, 729, 1,000

Expressions containing variables such as x^6y^3 are also perfect cubes, because they can be written as the cube of a quantity:

$$x^6y^3 = (x^2y)^3$$

EXAMPLE 1 Factor: $x^3 + 8$.

Solution The binomial $x^3 + 8$ is the sum of two cubes, because

$$x^3 + 8 = x^3 + 2^3$$

Thus, $x^3 + 8$ factors as $(x + 2)$ times the trinomial $x^2 - 2x + 2^2$.

$$\begin{array}{l} \mathbf{F}^3 + \mathbf{L}^3 = (\mathbf{F} + \mathbf{L})(\mathbf{F}^2 - \mathbf{F}\ \mathbf{L} + \mathbf{L}^2) \\ \downarrow \\ \mathbf{x}^3 + \mathbf{2}^3 = (\mathbf{x} + \mathbf{2})(\mathbf{x}^2 - \mathbf{x} \cdot \mathbf{2} + \mathbf{2}^2) \\ \qquad\quad = (x + 2)(x^2 - 2x + 4) \end{array}$$

Check by multiplication.

$$\begin{aligned} (x + 2)(x^2 - 2x + 4) &= (x + 2)x^2 - (x + 2)2x + (x + 2)4 \\ &= x^3 + 2x^2 - 2x^2 - 4x + 4x + 8 \\ &= x^3 + 8 \end{aligned}$$

Self Check Factor: $p^3 + 64$.

EXAMPLE 2 Factor: $8b^3 + 27c^3$.

Solution The binomial $8b^3 + 27c^3$ is the sum of two cubes, because

$$8b^3 + 27c^3 = (2b)^3 + (3c)^3$$

Thus, $8b^3 + 27c^3$ factors as $(2b + 3c)$ times the trinomial $(2b)^2 - (2b)(3c) + (3c)^2$.

$$\begin{array}{l} \mathbf{F}^3 + \mathbf{L}^3 = (\mathbf{F} + \mathbf{L})(\mathbf{F}^2 - \mathbf{F}\ \mathbf{L} + \mathbf{L}^2) \\ \downarrow \\ (\mathbf{2b})^3 + (\mathbf{3c})^3 = (\mathbf{2b} + \mathbf{3c})[(\mathbf{2b})^2 - (\mathbf{2b})(\mathbf{3c}) + (\mathbf{3c})^2] \\ \qquad\qquad\quad = (2b + 3c)(4b^2 - 6bc + 9c^2) \end{array}$$

Check by multiplication.

$$\begin{aligned} &(2b + 3c)(4b^2 - 6bc + 9c^2) \\ &\quad = (2b + 3c)4b^2 - (2b + 3c)6bc + (2b + 3c)9c^2 \\ &\quad = 8b^3 + 12b^2c - 12b^2c - 18bc^2 + 18bc^2 + 27c^3 \\ &\quad = 8b^3 + 27c^3 \end{aligned}$$

Self Check Factor: $1{,}000p^3 + q^3$.

Factoring the Difference of Two Cubes

To discover the formula for factoring the difference of two cubes, we find the following product:

$$(x - y)(x^2 + xy + y^2) = (x - y)x^2 + (x - y)xy + (x - y)y^2$$ Use the distributive property.

$$= x^3 - x^2y + x^2y - xy^2 + xy^2 - y^3$$
$$= x^3 - y^3$$

This result justifies the formula for factoring the **difference of two cubes.**

Factoring the Difference of Two Cubes

$$x^3 - y^3 = (x - y)(x^2 + xy + y^2)$$

If we think of the difference of two cubes as the cube of a **F**irst quantity minus the cube of a **L**ast quantity, we have the formula

$$\text{F}^3 - \text{L}^3 = (\text{F} - \text{L})(\text{F}^2 + \text{FL} + \text{L}^2)$$

In words, we say, *To factor the cube of a **F**irst quantity minus the cube of a **L**ast quantity, we multiply the **F**irst minus the **L**ast by*

- *the **F**irst squared*
- *plus the **F**irst times the **L**ast*
- *plus the **L**ast squared.*

EXAMPLE 3 Factor: $a^3 - 64b^3$.

Solution The binomial $a^3 - 64b^3$ is the difference of two cubes.

$$a^3 - 64b^3 = a^3 - (4b)^3$$

Thus, its factors are the difference $a - 4b$ and the trinomial $a^2 + a(4b) + (4b)^2$.

$$\text{F}^3 - \text{L}^3 = (\text{F} - \text{L})(\text{F}^2 + \text{F}\,\text{L} + \text{L}^2)$$
$$\downarrow \quad \downarrow \quad \downarrow \quad \downarrow \quad \downarrow \quad \downarrow \quad \downarrow \quad \downarrow$$
$$a^3 - (4b)^3 = (a - 4b)[a^2 + a(4b) + (4b)^2]$$
$$= (a - 4b)(a^2 + 4ab + 16b^2)$$

Check by multiplication.

$$(a - 4b)(a^2 + 4ab + 16b^2)$$
$$= (a - 4b)a^2 + (a - 4b)4ab + (a - 4b)16b^2$$
$$= a^3 - 4a^2b + 4a^2b - 16ab^2 + 16ab^2 - 64b^3$$
$$= a^3 - 64b^3$$

Self Check Factor: $27p^3 - 8$.

Multistep Factoring

Sometimes we must factor out a greatest common factor before factoring a sum or difference of two cubes.

EXAMPLE 4 Factor: $-2t^5 + 128t^2$.

Solution

$$-2t^5 + 128t^2 = -2t^2(t^3 - 64) \quad \text{Factor out } -2t^2.$$
$$= -2t^2(t - 4)(t^2 + 4t + 16) \quad \text{Factor } t^3 - 64.$$

Check by multiplication.

Self Check Factor: $-3p^4 + 81p$.

EXAMPLE 5 Factor: $x^6 - 64$.

Solution The binomial $x^6 - 64$ is both the difference of two squares and the difference of two cubes. Since it is easier to factor the difference of two squares first, the expression factors into the product of a sum and a difference.

$$x^6 - 64 = (x^3)^2 - 8^2$$
$$= (x^3 + 8)(x^3 - 8)$$

Because $x^3 + 8$ is the sum of two cubes and $x^3 - 8$ is the difference of two cubes, each of these binomials can be factored.

$$x^6 - 64 = (x^3 + 8)(x^3 - 8)$$
$$= (x + 2)(x^2 - 2x + 4)(x - 2)(x^2 + 2x + 4)$$

Check by multiplication.

Self Check Factor: $a^6 - 1$.

Self Check Answers

1. $(p + 4)(p^2 - 4p + 16)$ **2.** $(10p + q)(100p^2 - 10pq + q^2)$ **3.** $(3p - 2)(9p^2 + 6p + 4)$
4. $-3p(p - 3)(p^2 + 3p + 9)$ **5.** $(a + 1)(a^2 - a + 1)(a - 1)(a^2 + a + 1)$

Orals *Factor each sum or difference of two cubes.*

1. $x^3 - y^3$
2. $x^3 + y^3$
3. $a^3 + 8$
4. $b^3 - 27$
5. $1 + 8x^3$
6. $8 - r^3$
7. $x^3y^3 + 1$
8. $125 - 8t^3$

5.5 EXERCISES

REVIEW

1. The length of one fermi is 1×10^{-13} centimeter, approximately the radius of a proton. Express this number in standard notation.

2. In the 14th century, the Black Plague killed about 25,000,000 people, which was 25% of the population of Europe. Find the population at that time, expressed in scientific notation.

VOCABULARY AND CONCEPTS *Complete each formula.*

3. $x^3 + y^3 = (x + y)$ ____________

4. $x^3 - y^3 = (x - y)$ ____________

PRACTICE *Factor each expression.*

5. $y^3 + 1$

6. $x^3 - 8$

7. $a^3 - 27$

8. $b^3 + 125$

9. $8 + x^3$

10. $27 - y^3$

11. $s^3 - t^3$
12. $8u^3 + w^3$
13. $27x^3 + y^3$
14. $x^3 - 27y^3$
15. $a^3 + 8b^3$
16. $27a^3 - b^3$
17. $64x^3 - 27$
18. $27x^3 + 125$
19. $27x^3 - 125y^3$
20. $64x^3 + 27y^3$
21. $a^6 - b^3$
22. $a^3 + b^6$
23. $x^9 + y^6$
24. $x^3 - y^9$

Factor each expression. Factor out any greatest common factors first.

25. $2x^3 + 54$
26. $2x^3 - 2$
27. $-x^3 + 216$
28. $-x^3 - 125$
29. $64m^3x - 8n^3x$
30. $16r^4 + 128rs^3$
31. $x^4y + 216xy^4$
32. $16a^5 - 54a^2b^3$
33. $81r^4s^2 - 24rs^5$
34. $4m^5n + 500m^2n^4$
35. $125a^6b^2 + 64a^3b^5$
36. $216a^4b^4 - 1{,}000ab^7$
37. $y^7z - yz^4$
38. $x^{10}y^2 - xy^5$
39. $2mp^4 + 16mpq^3$
40. $24m^5n - 3m^2n^4$

Factor each expression completely. Factor a difference of two squares first.

41. $x^6 - 1$
42. $x^6 - y^6$
43. $x^{12} - y^6$
44. $a^{12} - 64$

Factor each expression completely.

45. $3(x^3 + y^3) - z(x^3 + y^3)$

46. $x(8a^3 - b^3) + 4(8a^3 - b^3)$

47. $(m^3 + 8n^3) + (m^3x + 8n^3x)$

48. $(a^3x + b^3x) - (a^3y + b^3y)$

49. $(a^4 + 27a) - (a^3b + 27b)$

50. $(x^4 + xy^3) - (x^3y + y^4)$

51. $y^3(y^2 - 1) - 27(y^2 - 1)$

52. $z^3(y^2 - 4) + 8(y^2 - 4)$

WRITING

53. Explain how to factor $a^3 + b^3$.
54. Explain the difference between $x^3 - y^3$ and $(x - y)^3$.

SOMETHING TO THINK ABOUT

55. Let $a = 11$ and $b = 7$. Use a calculator to verify that

$$a^3 - b^3 = (a - b)(a^2 + ab + b^2)$$

56. Let $p = 5$ and $q = -2$. Use a calculator to verify that

$$p^3 + q^3 = (p + q)(p^2 - pq + q^2)$$

5.6 Summary of Factoring Techniques

Identifying Factoring Types

Getting Ready *Factor each polynomial.*

1. $3ax^2 + 3a^2x$

2. $x^2 - 9y^2$

3. $x^3 - 8$
4. $2x^2 - 8$
5. $x^2 - 3x - 10$
6. $6x^2 - 13x + 6$
7. $6x^2 - 14x + 4$
8. $ax^2 + bx^2 - ay^2 - by^2$

In this brief section, we will discuss ways to approach a randomly chosen factoring problem.

Identifying Factoring Types

Suppose we wish to factor the trinomial

$$x^4y + 7x^3y - 18x^2y$$

We begin by attempting to identify the problem type. The first type we look for is **factoring out a common factor.** Because the trinomial has a common factor of x^2y, we factor it out:

$$x^4y + 7x^3y - 18x^2y = x^2y(x^2 + 7x - 18)$$

We can factor the remaining trinomial $x^2 + 7x - 18$ as $(x + 9)(x - 2)$. Thus,

$$\begin{aligned} x^4y + 7x^3y - 18x^2y &= x^2y(x^2 + 7x - 18) \\ &= x^2y(x + 9)(x - 2) \end{aligned}$$

To identify the type of factoring problem, we follow these steps.

Factoring a Polynomial

1. Factor out all common factors.
2. If an expression has two terms, check to see if the problem type is
   a. the **difference of two squares:** $x^2 - y^2 = (x + y)(x - y)$
   b. the **sum of two cubes:** $x^3 + y^3 = (x + y)(x^2 - xy + y^2)$
   c. the **difference of two cubes:** $x^3 - y^3 = (x - y)(x^2 + xy + y^2)$
3. If an expression has three terms, check to see if it is a **perfect trinomial square:**

   $$x^2 + 2xy + y^2 = (x + y)(x + y)$$
   $$x^2 - 2xy + y^2 = (x - y)(x - y)$$

   If the trinomial is not a perfect trinomial square, attempt to factor the trinomial as a **general trinomial.**
4. If an expression has four terms, try to factor the expression by **grouping.**
5. Continue factoring until each individual factor is prime.
6. Check the results by multiplying.

EXAMPLE 1 Factor: $x^5y^2 - xy^6$.

Solution We begin by factoring out the common factor of xy^2.

$$x^5y^2 - xy^6 = xy^2(x^4 - y^4)$$

Since the expression $x^4 - y^4$ has two terms, we check to see whether it is the difference of two squares, which it is. As the difference of two squares, it factors as $(x^2 + y^2)(x^2 - y^2)$.

$$\begin{aligned} x^5y^2 - xy^6 &= xy^2(x^4 - y^4) \\ &= xy^2(x^2 + y^2)(x^2 - y^2) \end{aligned}$$

The binomial $x^2 + y^2$ is the sum of two squares and cannot be factored. However, $x^2 - y^2$ is the difference of two squares and factors as $(x + y)(x - y)$.

$$\begin{aligned} x^5y^2 - xy^6 &= xy^2(x^4 - y^4) \\ &= xy^2(x^2 + y^2)(x^2 - y^2) \\ &= xy^2(x^2 + y^2)(x + y)(x - y) \end{aligned}$$

Since each individual factor is prime, the given expression is in completely factored form.

Self Check Factor: $-a^5b + ab^5$.

EXAMPLE 2 Factor: $x^6 - x^4y^2 - x^3y^3 + xy^5$.

Solution We begin by factoring out the common factor of x.

$$x^6 - x^4y^2 - x^3y^3 + xy^5 = x(x^5 - x^3y^2 - x^2y^3 + y^5)$$

Since $x^5 - x^3y^2 - x^2y^3 + y^5$ has four terms, we try factoring it by grouping:

$$\begin{aligned} x^6 - x^4y^2 - x^3y^3 + xy^5 &= x(x^5 - x^3y^2 - x^2y^3 + y^5) \\ &= x[x^3(x^2 - y^2) - y^3(x^2 - y^2)] \\ &= x(x^2 - y^2)(x^3 - y^3) \qquad \text{Factor out } x^2 - y^2. \end{aligned}$$

Finally, we factor the difference of two squares and the difference of two cubes:

$$x^6 - x^4y^2 - x^3y^3 + xy^5 = x(x + y)(x - y)(x - y)(x^2 + xy + y^2)$$

Since each factor is prime, the given expression is in prime-factored form.

Self Check Factor: $2a^5 - 2a^2b^3 - 8a^3 + 8b^3$.

Self Check Answers

1. $-ab(a^2 + b^2)(a + b)(a - b)$ **2.** $2(a + 2)(a - 2)(a - b)(a^2 + ab + b^2)$

Orals *Indicate which factoring technique you would use first, if any.*

1. $2x^2 - 4x$

2. $16 - 25y^2$

3. $125 + r^3s^3$

4. $ax + ay - x - y$

5. $x^2 + 4$

6. $8x^2 - 50$

7. $25r^2 - s^4$

8. $8a^3 - 27b^3$

5.6 EXERCISES

REVIEW *Solve each equation, if possible.*

1. $2(t - 5) + t = 3(2 - t)$
2. $5 + 3(2x - 1) = 2(4 + 3x) - 24$
3. $5 - 3(t + 1) = t + 2$
4. $4m - 3 = -2(m + 1) - 3$

VOCABULARY AND CONCEPTS *Fill in the blanks.*

5. The first step in any factoring problem is to factor out all common ______, if possible.
6. If a polynomial has two terms, check to see if it is the ______________, the sum of two cubes, or the ________ of two cubes.
7. If a polynomial has three terms, try to factor it as the product of two ________.
8. If a polynomial has four or more terms, try factoring by ________.

PRACTICE *Factor each expression.*

9. $6x + 3$
10. $x^2 - 9$
11. $x^2 - 6x - 7$
12. $a^3 + b^3$
13. $6t^2 + 7t - 3$
14. $3rs^2 - 6r^2st$
15. $4x^2 - 25$
16. $ac + ad + bc + bd$
17. $t^2 - 2t + 1$
18. $6p^2 - 3p - 2$
19. $a^3 - 8$
20. $2x^2 - 32$
21. $x^2y^2 - 2x^2 - y^2 + 2$
22. $a^2c + a^2d^2 + bc + bd^2$
23. $70p^4q^3 - 35p^4q^2 + 49p^5q^2$
24. $a^2 + 2ab + b^2 - y^2$
25. $2ab^2 + 8ab - 24a$
26. $t^4 - 16$
27. $-8p^3q^7 - 4p^2q^3$
28. $8m^2n^3 - 24mn^4$
29. $4a^2 - 4ab + b^2 - 9$
30. $3rs + 6r^2 - 18s^2$
31. $x^2 + 7x + 1$
32. $3a^3 + 24b^3$
33. $-2x^5 + 128x^2$
34. $16 - 40z + 25z^2$
35. $14t^3 - 40t^2 + 6t^4$
36. $6x^2 + 7x - 20$
37. $a^2(x - a) - b^2(x - a)$
38. $5x^3y^3z^4 + 25x^2y^3z^2 - 35x^3y^2z^5$
39. $8p^6 - 27q^6$
40. $2c^2 - 5cd - 3d^2$
41. $125p^3 - 64y^3$
42. $8a^2x^3y - 2b^2xy$
43. $-16x^4y^2z + 24x^5y^3z^4 - 15x^2y^3z^7$
44. $2ac + 4ad + bc + 2bd$
45. $81p^4 - 16q^4$
46. $6x^2 - x - 16$
47. $4x^2 + 9y^2$
48. $30a^4 + 5a^3 - 200a^2$
49. $54x^3 + 250y^6$
50. $6a^3 + 35a^2 - 6a$
51. $10r^2 - 13r - 4$
52. $4x^2 + 4x + 1 - y^2$
53. $21t^3 - 10t^2 + t$
54. $16x^2 - 40x^3 + 25x^4$
55. $x^5 - x^3y^2 + x^2y^3 - y^5$
56. $a^3x^3 - a^3y^3 + b^3x^3 - b^3y^3$
57. $2a^2c - 2b^2c + 4a^2d - 4b^2d$
58. $3a^2x^2 + 6a^2x + 3a^2 - 3b^2$

WRITING

59. Explain how to identify the type of factoring required to factor a polynomial.
60. Which factoring technique do you find most difficult? Why?

SOMETHING TO THINK ABOUT

61. Write $x^6 - y^6$ as $(x^3)^2 - (y^3)^2$, factor it as the difference of two squares, and show that you get

$$(x + y)(x^2 - xy + y^2)(x - y)(x^2 + xy + y^2)$$

Write $x^6 - y^6$ as $(x^2)^3 - (y^2)^3$, factor it as the difference of two cubes, and show that you get

$$(x + y)(x - y)(x^4 + x^2y^2 + y^4)$$

62. Verify that the results of Exercise 61 agree by showing the parts in color agree. Which do you think is completely factored?

5.7 Solving Equations by Factoring

- Quadratic Equations
- The Zero-Factor Property
- Solving Quadratic Equations by Factoring

Getting Ready *Solve each equation.*

1. $x + 3 = 4$ **2.** $y - 8 = 5$ **3.** $3x - 2 = 7$ **4.** $5y + 9 = 19$

In this section, we will learn how to use factoring to solve many equations that contain second-degree polynomials. These equations are called *quadratic equations.*

Quadratic Equations

Equations such as

$$3x + 2 = 0 \quad \text{and} \quad 9x - 6 = 0$$

that contain first-degree polynomials are called **linear equations.** Equations such as

$$9x^2 - 6x = 0 \quad \text{and} \quad 3x^2 + 4x - 7 = 0$$

that contain second-degree polynomials are called **quadratic equations.**

Quadratic Equations

A **quadratic equation** is an equation of the form

$$ax^2 + bx + c = 0$$

where a, b, and c are real numbers, and $a \neq 0$.

The Zero-Factor Property

Many quadratic equations can be solved by factoring. For example, to solve the quadratic equation

$$x^2 + 5x - 6 = 0$$

we begin by factoring the trinomial and writing the equation as

(1) $$(x + 6)(x - 1) = 0$$

This equation indicates that the product of two quantities is 0. However, if the product of two quantities is 0, then at least one of those quantities must be 0. This fact is called the **zero-factor property.**

Zero-Factor Property

Suppose a and b represent two real numbers. Then

If $ab = 0$, then $a = 0$ or $b = 0$.

By applying the zero factor property to Equation 1, we have

$$x + 6 = 0 \qquad \text{or} \qquad x - 1 = 0$$

We can solve each of these linear equations to get

$$x = -6 \qquad \text{or} \qquad x = 1$$

To check, we substitute -6 for x, and then 1 for x in the original equation and simplify.

For x = −6

$$\begin{aligned} x^2 + 5x - 6 &= 0 \\ (-6)^2 + 5(-6) - 6 &\stackrel{?}{=} 0 \\ 36 - 30 - 6 &\stackrel{?}{=} 0 \\ 36 - 36 &\stackrel{?}{=} 0 \\ 0 &= 0 \end{aligned}$$

For x = 1

$$\begin{aligned} x^2 + 5x - 6 &= 0 \\ (1)^2 + 5(1) - 6 &\stackrel{?}{=} 0 \\ 1 + 5 - 6 &\stackrel{?}{=} 0 \\ 0 &= 0 \end{aligned}$$

Both solutions check.

Solving Quadratic Equations by Factoring

The quadratic equations $9x^2 - 6x = 0$ and $4x^2 - 36 = 0$ are each missing a term. The first equation is missing the constant term, and the second equation is missing the term involving x. These types of equations can often be solved by factoring.

EXAMPLE 1 Solve: $9x^2 - 6x = 0$.

Solution We begin by factoring the left-hand side of the equation.

$$9x^2 - 6x = 0$$

$$3x(3x - 2) = 0 \qquad \text{Factor out the common factor of } 3x.$$

By the zero-factor property, we have

$$3x = 0 \qquad \text{or} \qquad 3x - 2 = 0$$

We can solve each of these equations to get

$$x = 0 \qquad \text{or} \qquad x = \frac{2}{3}$$

Check: We substitute these results for x in the original equation and simplify.

For $x = 0$

$$9x^2 - 6x = 0$$
$$9(0)^2 - 6(0) \stackrel{?}{=} 0$$
$$0 - 0 \stackrel{?}{=} 0$$
$$0 = 0$$

For $x = \frac{2}{3}$

$$9x^2 - 6x = 0$$
$$9\left(\frac{2}{3}\right)^2 - 6\left(\frac{2}{3}\right) \stackrel{?}{=} 0$$
$$9\left(\frac{4}{9}\right) - 6\left(\frac{2}{3}\right) \stackrel{?}{=} 0$$
$$4 - 4 \stackrel{?}{=} 0$$
$$0 = 0$$

Both solutions check.

Self Check Solve: $5y^2 + 10y = 0$.

EXAMPLE 2 Solve: $4x^2 - 36 = 0$.

Solution To make the numbers smaller, we divide both sides of the equation by 4. Then we proceed as follows:

$$4x^2 - 36 = 0$$
$$x^2 - 9 = 0 \quad \text{Divide both sides by 4.}$$
$$(x + 3)(x - 3) = 0 \quad \text{Factor } x^2 - 9.$$
$$x + 3 = 0 \quad \text{or} \quad x - 3 = 0 \quad \text{Set each factor equal to 0.}$$
$$x = -3 \quad | \quad x = 3 \quad \text{Solve each linear equation.}$$

Check each solution.

For $x = -3$

$$4x^2 - 36 = 0$$
$$4(-3)^2 - 36 \stackrel{?}{=} 0$$
$$4(9) - 36 \stackrel{?}{=} 0$$
$$0 = 0$$

For $x = 3$

$$4x^2 - 36 = 0$$
$$4(3)^2 - 36 \stackrel{?}{=} 0$$
$$4(9) - 36 \stackrel{?}{=} 0$$
$$0 = 0$$

Both solutions check.

Self Check Solve: $9p^2 - 64 = 0$.

In the next example, we solve an equation whose polynomial is a trinomial.

EXAMPLE 3 Solve: $x^2 - 3x - 18 = 0$.

Solution

$$x^2 - 3x - 18 = 0$$
$$(x + 3)(x - 6) = 0 \quad \text{Factor } x^2 - 3x - 18.$$
$$x + 3 = 0 \quad \text{or} \quad x - 6 = 0 \quad \text{Set each factor equal to 0.}$$
$$x = -3 \quad | \quad x = 6 \quad \text{Solve each linear equation.}$$

Check each solution.

Self Check Solve: $x^2 + 3x - 18 = 0$.

EXAMPLE 4 Solve: $2x^2 + 3x = 2$.

Solution We write the equation in the form $ax^2 + bx + c = 0$ and solve for x.

$$2x^2 + 3x = 2$$

$$2x^2 + 3x - 2 = 0 \quad \text{Subtract 2 from both sides.}$$

$$(2x - 1)(x + 2) = 0 \quad \text{Factor } 2x^2 + 3x - 2.$$

$$2x - 1 = 0 \quad \text{or} \quad x + 2 = 0 \quad \text{Set each factor equal to 0.}$$

$$2x = 1 \quad \Big| \quad x = -2 \quad \text{Solve each linear equation.}$$

$$x = \frac{1}{2}$$

Check each solution.

Self Check Solve: $3x^2 - 5x - 2 = 0$.

EXAMPLE 5 Solve: $(x - 2)(x^2 - 7x + 6) = 0$.

Solution We begin by factoring the quadratic trinomial.

$$(x - 2)(x^2 - 7x + 6) = 0$$

$$(x - 2)(x - 6)(x - 1) = 0 \quad \text{Factor } x^2 - 7x + 6.$$

If the product of these three quantities is 0, then at least one of the quantities must be 0.

$$x - 2 = 0 \quad \text{or} \quad x - 6 = 0 \quad \text{or} \quad x - 1 = 0$$

$$x = 2 \quad \Big| \quad x = 6 \quad \Big| \quad x = 1$$

Check each solution.

Self Check Solve: $(x + 3)(x^2 + 7x - 8) = 0$.

EXAMPLE 6 Solve: $x^3 - 2x^2 - 63x = 0$.

Solution We begin by completely factoring the left-hand side.

$$x^3 - 2x^2 - 63x = 0$$

$$x(x^2 - 2x - 63) = 0 \quad \text{Factor out } x.$$

$$x(x + 7)(x - 9) = 0 \quad \text{Factor the trinomial.}$$

$$x = 0 \quad \text{or} \quad x + 7 = 0 \quad \text{or} \quad x - 9 = 0 \quad \text{Set each factor equal to 0.}$$

$$\Big| \quad x = -7 \quad \Big| \quad x = 9 \quad \text{Solve each linear equation.}$$

The solutions are 0, −7, and 9. Check each one.

Self Check Solve: $x^3 - x^2 - 2x = 0$.

EXAMPLE 7 Solve: $6x^3 + 12x = 17x^2$.

Solution To get all of the terms on the left-hand side, we subtract $17x^2$ from both sides. Then we proceed as follows:

$$6x^3 + 12x = 17x^2$$

$$6x^3 - 17x^2 + 12x = 0 \qquad \text{Subtract } 17x^2 \text{ from both sides.}$$

$$x(6x^2 - 17x + 12) = 0 \qquad \text{Factor out } x.$$

$$x(2x - 3)(3x - 4) = 0 \qquad \text{Factor } 6x^2 - 17x + 12.$$

$$x = 0 \quad \text{or} \quad 2x - 3 = 0 \quad \text{or} \quad 3x - 4 = 0 \qquad \text{Set each factor equal to 0.}$$

$$x = 0 \quad \Big| \quad 2x = 3 \quad \Big| \quad 3x = 4 \qquad \text{Solve the linear equations.}$$

$$x = \frac{3}{2} \quad \Big| \quad x = \frac{4}{3}$$

Check each solution.

Self Check Solve: $6x^3 + 7x^2 = 5x$.

Self Check Answers

1. $0, -2$ **2.** $\frac{8}{3}, -\frac{8}{3}$ **3.** $3, -6$ **4.** $2, -\frac{1}{3}$ **5.** $1, -3, -8$ **6.** $0, 2, -1$ **7.** $0, \frac{1}{2}, -\frac{5}{3}$

Orals *Solve each equation.*

1. $(x - 8)(x - 7) = 0$
2. $(x + 9)(x - 2) = 0$
3. $x^2 + 7x = 0$
4. $x^2 - 12x = 0$
5. $x^2 - 2x + 1 = 0$
6. $x^2 + x - 20 = 0$

5.7 EXERCISES

REVIEW *Simplify each expression and write all results without using negative exponents.*

1. $u^3u^2u^4$

2. $\frac{y^6}{y^8}$

3. $\frac{a^3b^4}{a^2b^5}$

4. $(3x^5)^0$

VOCABULARY AND CONCEPTS *Fill in the blanks.*

5. An equation of the form $ax^2 + bx + c = 0$, where $a \neq 0$, is called a ________ equation.

If $ab = 0$, then $a =$ __ or $b =$ __.

...atic equation contains a ______-degree

8. If the product of three numbers is 0, then at least one of the numbers must be __.

PRACTICE *Solve each equation.*

9. $(x - 2)(x + 3) = 0$

10. $(x - 3)(x - 2) = 0$

11. $(x - 4)(x + 1) = 0$

12. $(x + 5)(x + 2) = 0$

13. $(2x - 5)(3x + 6) = 0$

14. $(3x - 4)(x + 1) = 0$

15. $(x - 1)(x + 2)(x - 3) = 0$

16. $(x + 2)(x + 3)(x - 4) = 0$

17. $x^2 - 3x = 0$

18. $x^2 + 5x = 0$

19. $2x^2 - 5x = 0$
20. $5x^2 + 7x = 0$
21. $x^2 - 7x = 0$
22. $x^2 + 5x = 0$
23. $3x^2 + 8x = 0$
24. $5x^2 - x = 0$
25. $8x^2 - 16x = 0$
26. $15x^2 - 20x = 0$
27. $10x^2 + 2x = 0$
28. $5x^2 + x = 0$
29. $x^2 - 25 = 0$
30. $x^2 - 36 = 0$
31. $y^2 - 49 = 0$
32. $x^2 - 121 = 0$
33. $4x^2 - 1 = 0$
34. $9y^2 - 1 = 0$
35. $9y^2 - 4 = 0$
36. $16z^2 - 25 = 0$
37. $x^2 = 49$
38. $z^2 = 25$
39. $4x^2 = 81$
40. $9y^2 = 64$
41. $x^2 - 13x + 12 = 0$
42. $x^2 + 7x + 6 = 0$
43. $x^2 - 2x - 15 = 0$
44. $x^2 - x - 20 = 0$
45. $x^2 - 4x - 21 = 0$
46. $x^2 + 2x - 15 = 0$
47. $x^2 + 8 - 9x = 0$
48. $45 + x^2 - 14x = 0$
49. $a^2 + 8a = -15$
50. $a^2 - a = 56$
51. $2y - 8 = -y^2$
52. $-3y + 18 = y^2$
53. $2x^2 - 5x + 2 = 0$
54. $2x^2 + x - 3 = 0$
55. $6x^2 + x - 2 = 0$
56. $12x^2 + 5x - 3 = 0$
57. $5p^2 - 6p + 1 = 0$
58. $6q^2 - 5q + 1 = 0$
59. $14m^2 + 23m + 3 = 0$
60. $35n^2 - 34n + 8 = 0$
61. $(x - 1)(x^2 + 5x + 6) = 0$
62. $(x - 2)(x^2 - 8x + 7) = 0$
63. $(x + 3)(x^2 + 2x - 15) = 0$
64. $(x + 4)(x^2 - 2x - 15) = 0$
65. $(x - 5)(2x^2 + x - 3) = 0$
66. $(a + 1)(6a^2 + a - 2) = 0$
67. $(p^2 - 81)(p + 2) = 0$
68. $(4q^2 - 49)(q - 7) = 0$
69. $x^3 + 3x^2 + 2x = 0$
70. $x^3 - 7x^2 + 10x = 0$
71. $x^3 - 27x - 6x^2 = 0$
72. $x^3 - 22x - 9x^2 = 0$
73. $3x^2 - 8x = 3$
74. $2x^2 - 11x = 21$
75. $15x^2 - 2 = 7x$
76. $8x^2 + 10x = 3$
77. $x(6x + 5) = 6$
78. $x(2x - 3) = 14$
79. $(x + 1)(8x + 1) = 18x$
80. $4x(3x + 2) = x + 12$
81. $2x(3x^2 + 10x) = -6x$
82. $2x^3 = 2x(x + 2)$
83. $x^3 + 7x^2 = x^2 - 9x$
84. $x^2(x + 10) = 2x(x - 8)$

WRITING

85. If the product of several numbers is 0, at least one of the numbers is 0. Explain why.
86. Explain the error in this solution.

$$5x^2 + 2x = 10$$
$$x(5x + 2) = 10$$
$$x = 10 \quad \text{or} \quad 5x + 2 = 10$$
$$5x = 8$$
$$x = \frac{8}{5}$$

SOMETHING TO THINK ABOUT

87. Explain how you would factor

$$3a + 3b + 3c - ax - bx - cx$$

88. Explain how you would factor

$$9 - a^2 - 4ab - 4b^2$$

89. Solve in two ways: $3a^2 + 9a - 2a - 6 = 0$
90. Solve in two ways:

$$p^2 - 2p + p - 2 = 0$$

5.8 Problem Solving

- Integer Problems
- Ballistics
- Geometry Problems

Getting Ready

1. One side of a square is s inches long. Find an expression that represents its area.

2. The length of a rectangle is 4 centimeters more than twice the width. If w represents the width, find an expression that represents the length.

3. If x represents the smaller of two consecutive integers, find an expression that represents their product.

4. The length of a rectangle is 3 inches greater than the width. If w represents the width of the rectangle, find an expression that represents the area.

Finally, we can use the methods for solving quadratic equations discussed in the previous section to solve problems.

Integer Problems

EXAMPLE 1 One negative integer is 5 less than another, and their product is 84. Find the integers.

Analyze the problem Let x represent the larger number. Then $x - 5$ represents the smaller number. We know that the product of the negative integers is 84.

Form an equation Since their product is 84, we can form the equation $x(x - 5) = 84$.

Solve the equation To solve the equation, we proceed as follows.

$$\begin{aligned} x(x-5) &= 84 \\ x^2 - 5x &= 84 && \text{Remove parentheses.} \\ x^2 - 5x - 84 &= 0 && \text{Subtract 84 from both sides.} \\ (x-12)(x+7) &= 0 && \text{Factor.} \\ x - 12 = 0 \quad \text{or} \quad x + 7 &= 0 && \text{Set each factor equal to 0.} \\ x = 12 \quad | \quad x &= -7 && \text{Solve each linear equation.} \end{aligned}$$

State the conclusion Since we need two negative numbers, we discard the result $x = 12$. The two negative integers are

$$x = -7 \qquad \text{and} \qquad \begin{aligned} x - 5 &= -7 - 5 \\ &= -12 \end{aligned}$$

Check the result The number -12 is five less than -7, and $(-12)(-7) = 84$. ∎

Ballistics

EXAMPLE 2 If an object is thrown straight up into the air with an initial velocity of 112 feet per second, its height after t seconds is given by the formula

$$h = 112t - 16t^2$$

where h represents the height of the object in feet. After this object has been thrown, in how many seconds will it hit the ground?

Analyze the problem Before the object is thrown, its height above the ground is 0. When it is thrown, it will go up and then come down. When it hits the ground, its height will again be 0.

Form an equation Thus, we set h equal to 0 in the formula $h = 112t - 16t^2$ to form the equation $0 = 112t - 16t^2$.

$$h = 112t - 16t^2$$
$$0 = 112t - 16t^2$$

Solve the equation We solve the equation as follows.

$$0 = 112t - 16t^2$$
$$0 = 16t(7 - t) \quad \text{Factor out } 16t.$$
$$16t = 0 \quad \text{or} \quad 7 - t = 0 \quad \text{Set each factor equal to 0.}$$
$$t = 0 \quad \Big| \quad t = 7 \quad \text{Solve each linear equation.}$$

State the conclusion When $t = 0$, the object's height above the ground is 0 feet, because it has just been released. When $t = 7$, the height is again 0 feet. The object has hit the ground. The solution is 7 seconds.

Check the result.

Geometry Problems

Recall that the area of a rectangle is given by the formula

$$A = lw$$

where A represents the area, l the length, and w the width of the rectangle. The perimeter of a rectangle is given by the formula

$$P = 2l + 2w$$

where P represents the perimeter of the rectangle, l the length, and w the width.

EXAMPLE 3 Assume that the rectangle in Figure 5-1 has an area of 52 square centimeters and that its length is 1 centimeter more than 3 times its width. Find the perimeter of the rectangle.

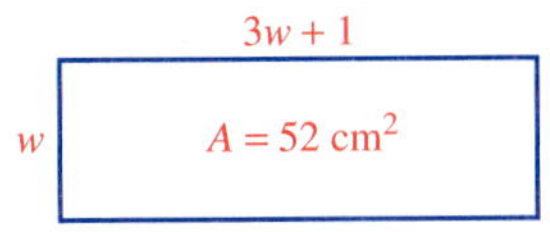

Figure 5-1

Analyze the problem Let w represent the width of the rectangle. Then $3w + 1$ represents its length. Its area is 52 square centimeters. We can use this fact to find the values of its width and length. Then we can find the perimeter.

Form and solve an equation Because the area is 52 square centimeters, we substitute 52 for A and $3w + 1$ for l in the formula $A = lw$ and solve for w.

$$A = lw$$
$$52 = (3w + 1)w$$
$$52 = 3w^2 + w \quad \text{Remove parentheses.}$$
$$0 = 3w^2 + w - 52 \quad \text{Subtract 52 from both sides.}$$
$$0 = (3w + 13)(w - 4) \quad \text{Factor.}$$
$$3w + 13 = 0 \quad \text{or} \quad w - 4 = 0 \quad \text{Set each factor equal to 0.}$$
$$3w = -13 \quad \Big| \quad w = 4 \quad \text{Solve each linear equation.}$$
$$w = -\frac{13}{3}$$

Because the width of a rectangle cannot be negative, we discard the result $w = -\frac{13}{3}$. Thus, the width of the rectangle is 4, and the length is given by

$$3w + 1 = 3(4) + 1$$
$$= 12 + 1$$
$$= 13$$

The dimensions of the rectangle are 4 centimeters by 13 centimeters. We find the perimeter by substituting 13 for l and 4 for w in the formula for the perimeter.

$$P = 2l + 2w$$
$$= 2(13) + 2(4)$$
$$= 26 + 8$$
$$= 34$$

State the conclusion The perimeter of the rectangle is 34 centimeters.

Check the result A rectangle with dimensions of 13 centimeters by 4 centimeters does have an area of 52 square centimeters, and the length is 1 centimeter more than 3 times the width. A rectangle with these dimensions has a perimeter of 34 centimeters. ∎

EXAMPLE 4 The triangle in Figure 5-2 has an area of 10 square centimeters and a height that is 3 centimeters less than twice the length of its base. Find the length of the base and the height of the triangle.

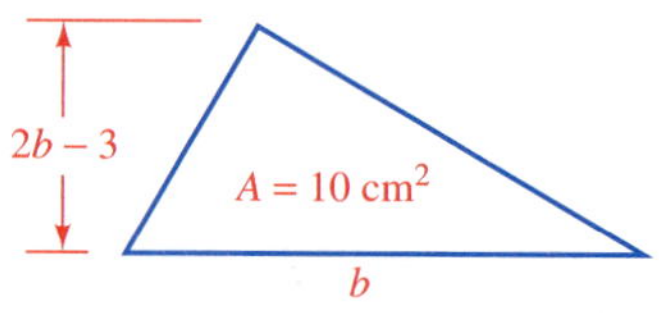

Figure 5-2

Analyze the problem Let b represent the length of the base of the triangle. Then $2b - 3$ represents the height. Because the area is 10 square centimeters, we can substitute 10 for A and $2b - 3$ for h in the formula $A = \frac{1}{2}bh$ and solve for b.

Form and solve an equation

$$A = \frac{1}{2}bh$$

$$10 = \frac{1}{2}b(2b - 3)$$

$$20 = b(2b - 3) \quad \text{Multiply both sides by 2.}$$

$$20 = 2b^2 - 3b \quad \text{Remove parentheses.}$$

$$0 = 2b^2 - 3b - 20 \quad \text{Subtract 20 from both sides.}$$

$$0 = (2b + 5)(b - 4) \quad \text{Factor.}$$

$$2b + 5 = 0 \quad \text{or} \quad b - 4 = 0 \quad \text{Set both factors equal to 0.}$$

$$2b = -5 \quad \bigg| \quad b = 4 \quad \text{Solve each linear equation.}$$

$$b = -\frac{5}{2}$$

State the conclusion Because a triangle cannot have a negative number for the length of its base, we discard the result $b = -\frac{5}{2}$. The length of the base of the triangle is 4 centimeters. Its height is $2(4) - 3$, or 5 centimeters.

Check the result If the base of the triangle has a length of 4 centimeters and the height of the triangle is 5 centimeters, its height is 3 centimeters less than twice the length of its base. Its area is 10 centimeters.

$$A = \frac{1}{2}bh$$

$$= \frac{1}{2}(4)(5)$$

$$= 2(5)$$

$$= 10$$

Orals *Give the formula for . . .*

1. The area of a rectangle

2. The area of a triangle

3. The area of a square

4. The volume of a rectangular solid

5. The perimeter of a rectangle

6. The perimeter of a square

5.8 EXERCISES

REVIEW *Solve each equation.*

1. $-2(5z + 2) = 3(2 - 3z)$

2. $3(2a - 1) - 9 = 2a$

3. A rectangle is 3 times as long as it is wide, and its perimeter is 120 centimeters. Find its area.

4. A woman invested \$15,000, part at 7% annual interest and part at 8% annual interest. If she receives \$1,100 interest per year, how much did she invest at 7%?

VOCABULARY AND CONCEPTS *Fill in the blanks.*

5. The first step in the problem-solving process is to _______ the problem.

6. The last step in the problem-solving process is to _____________.

PRACTICE *Solve each problem.*

7. Integer problem One positive integer is 2 more than another. Their product is 35. Find the integers.

8. Integer problem One positive integer is 5 less than 4 times another. Their product is 21. Find the integers.

9. Integer problem If 4 is added to the square of a composite integer, the result is 5 less than 10 times that integer. Find the integer.

10. Integer problem If 3 times the square of a certain natural number is added to the number itself, the result is 14. Find the number.

APPLICATIONS

An object has been thrown straight up into the air. The formula $h = vt - 16t^2$ *gives the height h of the object above the ground after t seconds when it is thrown upward with an initial velocity v.*

11. Time of flight After how many seconds will an object hit the ground if it was thrown with a velocity of 144 feet per second?

12. Time of flight After how many seconds will an object hit the ground if it was thrown with a velocity of 160 feet per second?

13. Ballistics If a cannonball is fired with an upward velocity of 220 feet per second, at what times will it be at a height of 600 feet?

14. Ballistics A cannonball's initial upward velocity is 128 feet per second. At what times will it be 192 feet above the ground?

15. Exhibition diving At a resort, tourists watch swimmers dive from a cliff to the water 64 feet below. A diver's height h above the water t seconds after diving is given by $h = -16t^2 + 64$. How long does a dive last?

16. Forensic medicine The kinetic energy E of a moving object is given by $E = \frac{1}{2}mv^2$, where m is the mass of the object (in kilograms) and v is the object's velocity (in meters per second). Kinetic energy is measured in joules. By the damage done to a victim, a police pathologist determines that the energy of a 3-kilogram mass at impact was 54 joules. Find the velocity at impact.

In Exercises 17–18, note that in the triangle $y^2 = h^2 + x^2$.

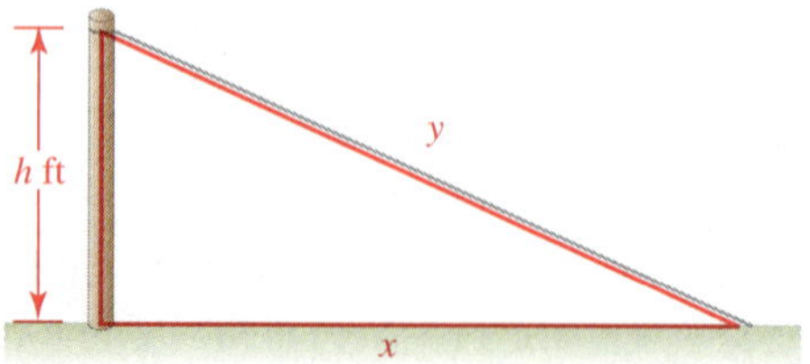

17. Ropes course A camper slides down the cable of a high-adventure ropes course to the ground as shown in the illustration. At what height did the camper start his slide?

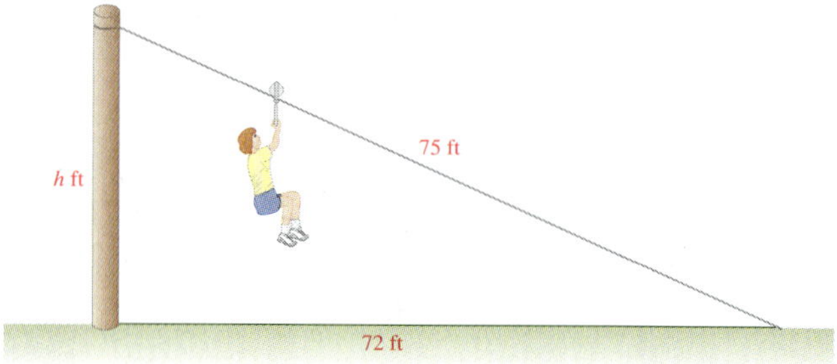

18. Ropes course If the pole and the landing area discussed in Exercise 17 are 24 feet apart and the high end of the cable is 7 feet, how long is the cable?

19. Insulation The area of the rectangular slab of foam insulation is 36 square meters. Find the dimensions of the slab.

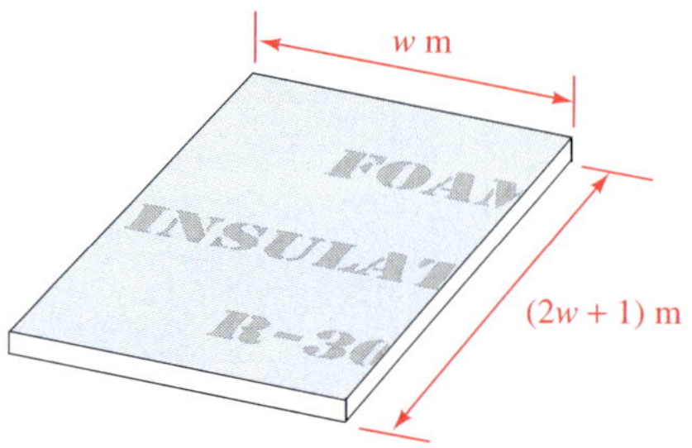

20. Shipping pallets The length of a rectangular shipping pallet is 2 feet less than 3 times its width. Its area is 21 square feet. Find the dimensions of the pallet.

21. Carpentry A room containing 143 square feet is 2 feet longer than it is wide. How long a crown molding is needed to trim the perimeter of the ceiling?

22. Designing a tent The length of the base of the triangular sheet of canvas above the door of the tent is 2 feet more than twice its height. The area is 30 square feet. Find the height and the length of the base of the triangle.

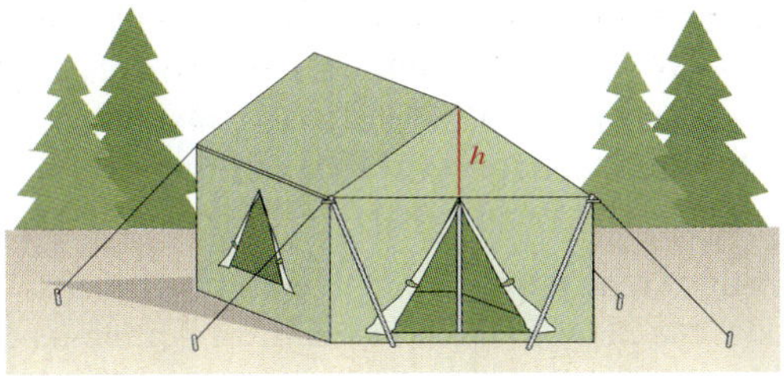

23. Dimensions of a triangle The height of a triangle is 2 inches less than 5 times the length of its base. The area is 36 square inches. Find the length of the base and the height of the triangle.

24. Area of a triangle The base of a triangle is numerically 3 less than its area, and the height is numerically 6 less than its area. Find the area of the triangle.

25. Area of a triangle The length of the base and the height of a triangle are numerically equal. Their sum is 6 less than the number of units in the area of the triangle. Find the area of the triangle.

26. Dimensions of a parallelogram The formula for the area of a parallelogram is $A = bh$. The area of the parallelogram in the illustration is 200 square centimeters. If its base is twice its height, how long is the base?

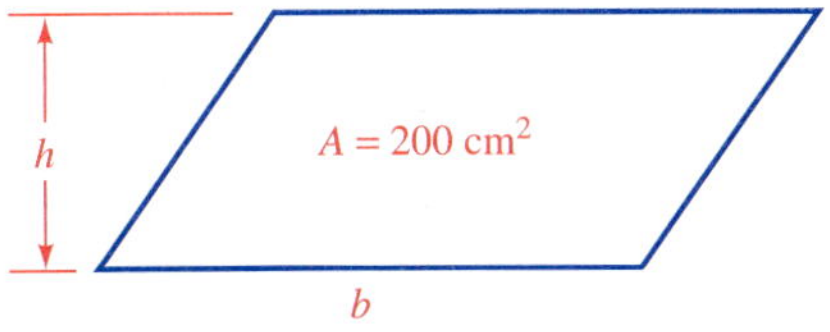

27. Swimming pool borders The owners of the rectangular swimming pool want to surround the pool with a crushed-stone border of uniform width. They have enough stone to cover 74 square meters. How wide should they make the border? (*Hint:* The area of the larger rectangle minus the area of the smaller is the area of the border.)

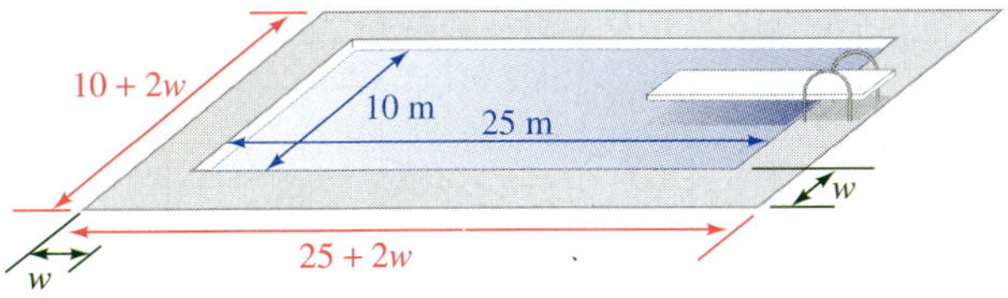

28. House construction The formula for the area of a trapezoid is $A = \frac{h(B + b)}{2}$. The area of the trapezoidal truss in the illustration is 24 square meters. Find the height of the trapezoid if one base is 8 meters and the other base is the same as the height.

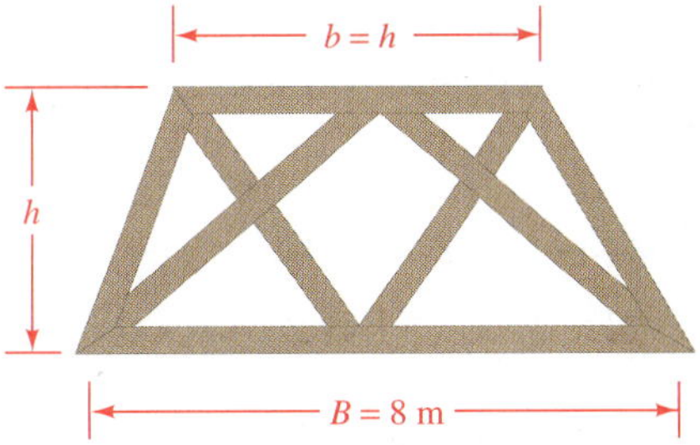

29. Volume of a solid The volume of a rectangular solid is given by the formula $V = lwh$, where l is the length, w is the width, and h is the height. The volume of the rectangular solid in the illustration is 210 cubic centimeters. Find the width of the rectangular solid if its length is 10 centimeters and its height is 1 centimeter longer than twice its width.

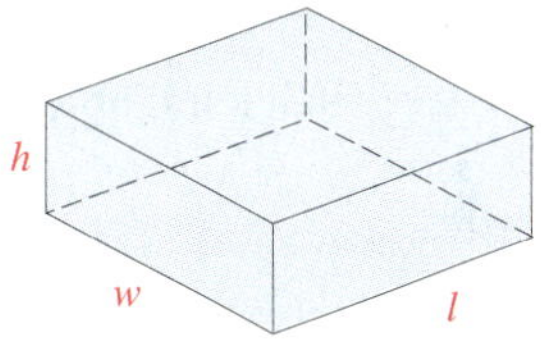

30. Volume of a pyramid The volume of a pyramid is given by the formula $V = \frac{Bh}{3}$, where B is the area of its base and h is its height. The volume of the pyramid in the illustration is 192 cubic centimeters. Find the dimensions of its rectangular base if one edge of the base is 2 centimeters longer than the other, and the height of the pyramid is 12 centimeters.

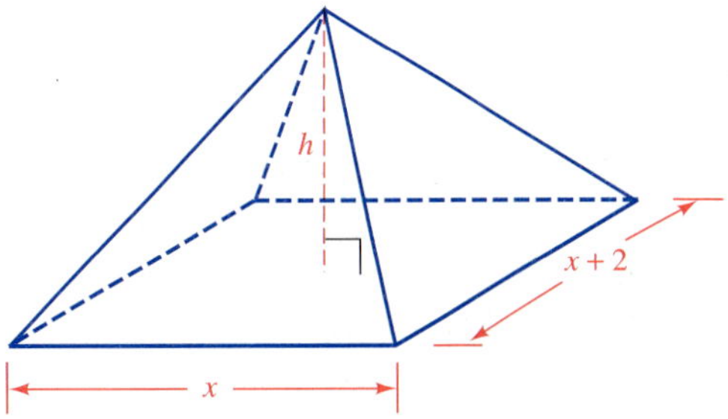

31. Volume of a pyramid The volume of a pyramid is 84 cubic centimeters. Its height is 9 centimeters, and one side of its rectangular base is 3 centimeters shorter than the other. Find the dimensions of its base. (See Exercise 30.)

32. Volume of a solid The volume of a rectangular solid is 72 cubic centimeters. Its height is 4 centimeters, and its width is 3 centimeters shorter than its length. Find the sum of its length and width. (See Exercise 29.)

33. Telephone connections The number of connections C that can be made among n telephones is given by the formula

$$C = \frac{1}{2}(n^2 - n)$$

How many telephones are needed to make 66 connections?

34. Football schedules If each of t teams in a high school football league plays every other team in the league once, the total number T of games played is given by the formula

$$T = \frac{t(t - 1)}{2}$$

If the season is 10 games long, how many teams are in the league?

35. Sewage treatment In one step in waste treatment, sewage is exposed to air by placing it in circular aeration pools. One sewage processing plant has two such pools, with diameters of 40 and 42 meters. Find the combined area of the pools.

36. Sewage treatment To meet new clean-water standards, the plant in Exercise 35 must double its capacity by building another pool.

Find the radius of the circular pool that the engineering department should specify to double the plant's capacity.

WRITING

37. Explain the steps you would use to set up and solve an application problem.

38. Explain how you should check the solution to an application problem.

SOMETHING TO THINK ABOUT

39. Here is an easy-sounding problem:
The length of a rectangle is 2 feet greater than the width, and the area is 18 square feet. Find the width of the rectangle.
Set up the equation. Can you solve it? Why not?

40. Does the equation in Exercise 39 have a solution, even if you can't find it? If it does, find an estimate of the solution.

Project

Because the length of each side of the largest square in Figure 5-3 is $x + y$, its area is $(x + y)^2$. This area is also the sum of four smaller areas, which illustrates the fractorization

$$x^2 + 2xy + y^2 = (x + y)^2$$

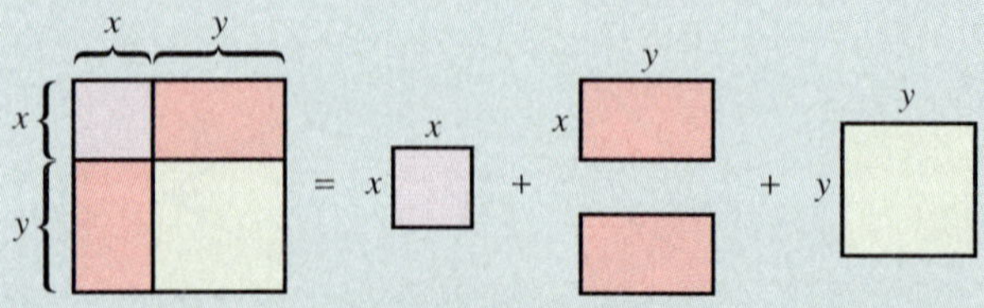

Figure 5-3

What factorization is illustrated by each of the following figures?

1.
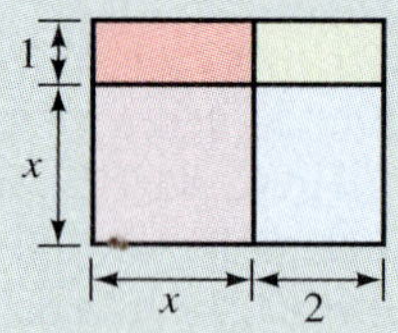

2.
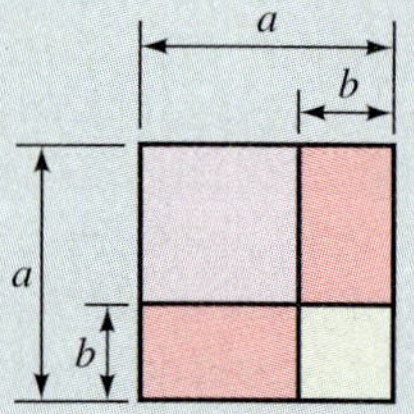

3.
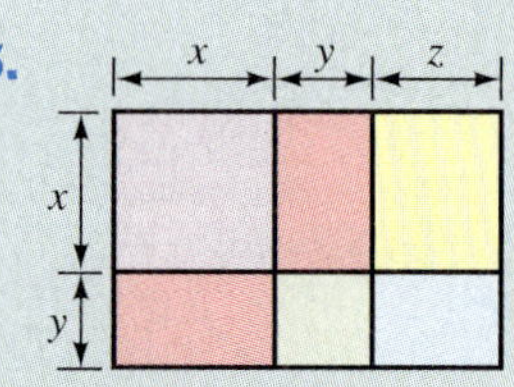

4.
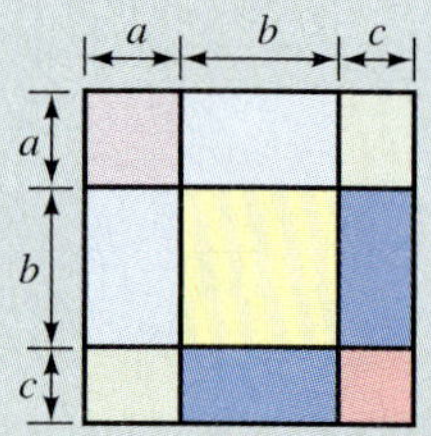

5. Factor the expression

$$a^2 + ac + 2a + ab + bc + 2b$$

and draw a figure that illustrates the factorization.

6. Verify the factorization

$$x^3 + 3x^2y + 3xy^2 + y^3 = (x + y)^3$$

Hint: Expand the right-hand side:

$$(x + y)^3 = (x + y)(x + y)(x + y)$$

Then draw a figure that illustrates the factorization.

Chapter Summary

CONCEPTS | **REVIEW EXERCISES**

5.1 Factoring Out the Greatest Common Factor; Factoring by Grouping

A natural number is in **prime-factored form** if it is written as the product of prime-number factors.

The **greatest common factor (GCF)** of several monomials is found by taking each common prime factor the fewest number of times it appears in any one monomial.

Find the prime factorization of each number.

1. 35
2. 45
3. 96
4. 102
5. 87
6. 99
7. 2,050
8. 4,096

Factor each expression completely.

9. $3x + 9y$
10. $5ax^2 + 15a$
11. $7x^2 + 14x$
12. $3x^2 - 3x$
13. $2x^3 + 4x^2 - 8x$
14. $ax + ay - az$
15. $ax + ay - a$
16. $x^2yz + xy^2z$

If a polynomial has four terms, consider factoring it by grouping.

Factor each polynomial.

17. $(x + y)a + (x + y)b$
18. $(x + y)^2 + (x + y)$
19. $2x^2(x + 2) + 6x(x + 2)$
20. $3x(y + z) - 9x(y + z)^2$
21. $3p + 9q + ap + 3aq$
22. $ar - 2as + 7r - 14s$
23. $x^2 + ax + bx + ab$

24. $xy + 2x - 2y - 4$

25. $xa + yb + ya + xb$

5.2 Factoring the Difference of Two Squares

To factor the difference of two squares, use the pattern

$$x^2 - y^2 = (x + y)(x - y)$$

Factor each expression.

26. $x^2 - 9$

27. $x^2y^2 - 16$

28. $(x + 2)^2 - y^2$

29. $z^2 - (x + y)^2$

30. $6x^2y - 24y^3$

31. $(x + y)^2 - z^2$

5.3–5.4 Factoring Trinomials

Factor trinomials using these steps:

1. Write the trinomial with the exponents of one variable in descending order.
2. Factor out any greatest common factor (including -1 if that is necessary to make the coefficient of the first term positive).
3. If the sign of the third term is $+$, the signs between the terms of the binomial factors are the same as the sign of the trinomial's second term. If the sign of the third term is $-$, the signs between the terms of the binomials are opposite.
4. Try various combinations of first terms and last terms until you find the one that works. If none work, the trinomial is prime.
5. Check by multiplication.

Factor each polynomial.

32. $x^2 + 10x + 21$

33. $x^2 + 4x - 21$

34. $x^2 + 2x - 24$

35. $x^2 - 4x - 12$

All

Factor each polynomial.

36. $2x^2 - 5x - 3$

37. $3x^2 - 14x - 5$

38. $6x^2 + 7x - 3$

39. $6x^2 + 3x - 3$

40. $6x^3 + 17x^2 - 3x$

41. $4x^3 - 5x^2 - 6x$

5.5 Factoring the Sum and Difference of Two Cubes

The sum and difference of two cubes factor according to the patterns

$$x^3 + y^3 = (x + y)(x^2 - xy + y^2)$$

$$x^3 - y^3 = (x - y)(x^2 + xy + y^2)$$

Factor each polynomial.

42. $c^3 - 27$

43. $d^3 + 8$

44. $2x^3 + 54$

45. $2ab^4 - 2ab$

5.6 Summary of Factoring Techniques

Factoring polynomials:

1. Factor out all common factors.
2. If an expression has two terms, check to see if it is
   - **a.** the **difference of two squares:** $a^2 - b^2 = (a + b)(a - b)$
   - **b.** the **sum of two cubes:** $a^3 + b^3 = (a + b)(a^2 - ab + b^2)$
   - **c.** the **difference of two cubes:** $a^3 - b^3 = (a - b)(a^2 + ab + b^2)$
3. If an expression has three terms, check to see if it is a **perfect trinomial square:**

   $a^2 + 2ab + b^2 = (a + b)(a + b)$

   $a^2 - 2ab + b^2 = (a - b)(a - b)$

   If the trinomial is not a trinomial square, attempt to factor it as a **general trinomial.**
4. If an expression has four or more terms, factor it by **grouping.**

Factor each polynomial

46. $3x^2y - xy^2 - 6xy + 2y^2$

47. $5x^2 + 10x - 15xy - 30y$

48. $2a^2x + 2abx + a^3 + a^2b$

49. $x^2 + 2ax + a^2 - y^2$

50. $x^2 - 4 + bx + 2b$

51. $ax^6 - ay^6$

5.7 Solving Equations by Factoring

Zero factor property:
If a and b represent two real numbers and if $ab = 0$, then $a = 0$ or $b = 0$.

Solve each equation.

52. $x^2 + 2x = 0$

53. $2x^2 - 6x = 0$

54. $3x^2 = 2x$

55. $5x^2 + 25x = 0$

56. $x^2 - 9 = 0$

57. $x^2 - 25 = 0$

58. $a^2 - 7a + 12 = 0$

59. $x^2 - 2x - 15 = 0$

60. $2x - x^2 + 24 = 0$

61. $16 + x^2 - 10x = 0$

62. $2x^2 - 5x - 3 = 0$

63. $2x^2 + x - 3 = 0$

64. $4x^2 = 1$

65. $9x^2 = 4$

66. $x^3 - 7x^2 + 12x = 0$

67. $x^3 + 5x^2 + 6x = 0$

68. $2x^3 + 5x^2 = 3x$

69. $3x^3 - 2x = x^2$

5.8 Problem Solving

70. Number problem The sum of two numbers is 12, and their product is 35. Find the numbers.

71. Number problem If 3 times the square of a positive number is added to 5 times the number, the result is 2. Find the number.

72. Dimensions of a rectangle A rectangle is 2 feet longer than it is wide, and its area is 48 square feet. Find its dimensions.

73. Gardening A rectangular flower bed is 3 feet longer than twice its width, and its area is 27 square feet. Find its dimensions.

74. Geometry A rectangle is 3 feet longer than it is wide. Its area is numerically equal to its perimeter. Find its dimensions.

Chapter Test

1. Find the prime factorization of 196.

2. Find the prime factorization of 111.

Factor out the greatest common factor.

3. $60ab^2c^3 + 30a^3b^2c - 25a$

4. $3x^2(a + b) - 6xy(a + b)$

Factor each expression.

5. $ax + ay + bx + by$

6. $x^2 - 25$

7. $3a^2 - 27b^2$

8. $16x^4 - 81y^4$

9. $x^2 + 4x + 3$

10. $x^2 - 9x - 22$

11. $x^2 + 10xy + 9y^2$

12. $6x^2 - 30xy + 24y^2$

13. $3x^2 + 13x + 4$

14. $2a^2 + 5a - 12$

15. $2x^2 + 3xy - 2y^2$

16. $12 - 25x + 12x^2$

17. $12a^2 + 6ab - 36b^2$

18. $x^3 - 64$

19. $216 + 8a^3$

20. $x^9z^3 - y^3z^6$

Solve each equation.

21. $x^2 + 3x = 0$

22. $2x^2 + 5x + 3 = 0$

23. $9y^2 - 81 = 0$

24. $-3(y - 6) + 2 = y^2 + 2$

25. $10x^2 - 13x = 9$

26. $10x^2 - x = 9$

27. $10x^2 + 43x = 9$

28. $10x^2 - 89x = 9$

29. Cannon fire A cannonball is fired straight up into the air with a velocity of 192 feet per second. In how many seconds will it hit the ground? (Its height above the ground is given by the formula $h = vt - 16t^2$, where v is the velocity and t is the time in seconds.)

30. Base of a triangle The base of a triangle with an area of 40 square meters is 2 meters longer than it is high. Find the base of the triangle.

6 Proportion and Rational Expressions

©Wayne Eastep/Getty Images

InfoTrac project

Do a subject guide search on "volume," and under "volume (cubic content)," choose "periodicals." Find the article "Rules of thumb: oil flow, line volume, linefill in barrels, small hole leakage." In the formula for V (speed), "*d* sup 2" means d^2." What is the speed of oil moving through a 17-inch pipe with a throughput of 225,000 barrels of oil per day? Using the formula for V (speed), solve for Q.

Two pipes are emptying into a large oil tanker at the same time. One pipe has an inner diameter of 14 inches and the other has an inner diameter of 15 inches. The oil flows through the 15-inch pipe 1.5 mph faster than through the 14-inch pipe. If the two pipes together have a throughput of 1,006,111 barrels per day, find the speed of the oil flowing through the 14-inch pipe.

Complete this project after studying Section 6.8.

Mathematics in Food Preparation

A recipe for chocolate chip cookies calls for $1\frac{1}{4}$ cups of flour and 1 cup of sugar. The recipe will make $3\frac{1}{2}$ dozen cookies. How many cups of flour will be needed to make 12 dozen cookies?

Exercise Set 6.2 Problem 51

In Chapter 6, we discuss ratio, proportion, and rational expressions, the fractions of algebra. After learning how to simplify, add, subtract, multiply, and divide rational expressions, we will solve equations and solve problems that involve rational expressions.

6.1 Ratios

▮ Ratios ▮ Unit Costs ▮ Rates

Getting Ready *Simplify each fraction.*

1. $\frac{2}{4}$ **2.** $\frac{8}{12}$ **3.** $-\frac{20}{25}$ **4.** $\frac{-45}{81}$

In this section, we will discuss ratios, unit costs, and rates. These ideas are important in many areas of everyday life.

Ratios

Ratios appear often in real-life situations. For example,

- To prepare fuel for a Lawnboy lawnmower, gasoline must be mixed with oil in the ratio of 50 to 1.
- To make 14-karat jewelry, gold is mixed with other metals in the ratio of 14 to 10.
- At Rock Valley College, the ratio of students to faculty is 16 to 1.

Ratios give us a way to compare numerical quantities.

Ratios

A **ratio** of two numbers is their indicated quotient. In symbols,

If a and b are two numbers, the ratio of a to b is $\frac{a}{b}$.

Comment The denominator b cannot be 0 in the fraction $\frac{a}{b}$, but b can be 0 in the ratio $\frac{a}{b}$. For example, the ratio of women to men on a women's softball team could be 25 to 0. However, these applications are rare.

Some examples of ratios are

$$\frac{7}{9}, \quad \frac{21}{27}, \quad \text{and} \quad \frac{2{,}290}{1{,}317}$$

- The fraction $\frac{7}{9}$ can be read as "the ratio of 7 to 9."
- The fraction $\frac{21}{27}$ can be read as "the ratio of 21 to 27."
- The ratio $\frac{2,290}{1,317}$ can be read as "the ratio of 2,290 to 1,317."

Because $\frac{7}{9}$ and $\frac{21}{27}$ represent equal numbers, they are **equal ratios.**

EXAMPLE 1 Express each phrase as a ratio in lowest terms: **a.** the ratio of 15 to 12 and **b.** the ratio of 0.3 to 1.2.

Solution **a.** The ratio of 15 to 12 can be written as the fraction $\frac{15}{12}$. After simplifying, the ratio is $\frac{5}{4}$.

b. The ratio of 0.3 to 1.2 can be written as the fraction $\frac{0.3}{1.2}$. We can simplify this fraction as follows:

$$\frac{0.3}{1.2} = \frac{0.3 \cdot \mathbf{10}}{1.2 \cdot \mathbf{10}}$$ To clear the decimal, multiply both numerator and denominator by 10.

$$= \frac{3}{12}$$ Multiply: $0.3 \cdot 10 = 3$ and $1.2 \cdot 10 = 12$.

$$= \frac{1}{4}$$ Simplify the fraction: $\frac{3}{12} = \frac{\overset{1}{\cancel{3}} \cdot 1}{\underset{1}{\cancel{3}} \cdot 4} = \frac{1}{4}$.

Self Check Express each ratio in lowest terms: **a.** the ratio of 8 to 12 and **b.** the ratio of 3.2 to 16.

EXAMPLE 2 Express each phrase as a ratio in lowest terms: **a.** the ratio of 3 meters to 8 meters and **b.** the ratio of 4 ounces to 1 pound.

Solution **a.** The ratio of 3 meters to 8 meters can be written as the fraction $\frac{3 \text{ meters}}{8 \text{ meters}}$, or just $\frac{3}{8}$.

b. Ratios should be expressed in the same units. Since there are 16 ounces in 1 pound, the proper ratio is $\frac{4 \text{ ounces}}{16 \text{ ounces}}$, which simplifies to $\frac{1}{4}$.

Self Check Express each ratio in lowest terms: **a.** the ratio of 8 ounces to 2 pounds and **b.** the ratio of 1 foot to 2 yards. (*Hint:* 3 feet = 1 yard.)

EXAMPLE 3 At a college, there are 2,772 students and 154 faculty members. Write a fraction in simplified form that expresses the ratio of students per faculty member.

Solution The ratio of students to faculty is 2,772 to 154. We can write this ratio as the fraction $\frac{2,772}{154}$ and simplify it.

$$\frac{2,772}{154} = \frac{18 \cdot \overset{1}{\cancel{154}}}{1 \cdot \underset{1}{\cancel{154}}}$$

$$= \frac{18}{1}$$ $\frac{154}{154} = 1$.

The ratio of students to faculty is 18 to 1.

Self Check In a college graduating class, 224 students out of 632 went on to graduate school. Write a fraction in simplified form that expresses the ratio of the number of students going on to the number in the graduating class.

Unit Costs

The *unit cost* of an item is the ratio of its cost to its quantity. For example, the unit cost (the cost per pound) of 5 pounds of coffee priced at \$20.75 is given by

$$\frac{\$20.75}{5\text{ pounds}} = \frac{\$2{,}075}{500\text{ pounds}} \qquad \text{To eliminate the decimal, multiply numerator and denominator by 100.}$$

$$= \$4.15\text{ per pound} \qquad \$2{,}070 \div 500 = \$4.15.$$

The unit cost is \$4.15 per pound.

EXAMPLE 4 Olives come packaged in a 12-ounce jar, which sells for \$3.09, or in a 6-ounce jar, which sells for \$1.53. Which is the better buy?

Solution To find the better buy, we must find each unit cost. The unit cost of the 12-ounce jar is

$$\frac{\$3.09}{12\text{ ounces}} = \frac{309¢}{12\text{ ounces}} \qquad \text{Change \$3.09 to 309 cents.}$$

$$= 25.75¢\text{ per ounce}$$

The unit cost of the 6-ounce jar is

$$\frac{\$1.53}{6\text{ ounces}} = \frac{153¢}{6\text{ ounces}} \qquad \text{Change \$1.53 to 153 cents.}$$

$$= 25.5¢\text{ per ounce}$$

Since the unit cost is less when olives are packaged in 6-ounce jars, that is the better buy.

Self Check A fast-food restaurant sells a 12-ounce soft drink for 79¢ and a 16-ounce soft drink for 99¢. Which is the better buy?

Rates

When ratios are used to compare quantities with different units, they are often called *rates*. For example, if we drive 413 miles in 7 hours, the average rate of speed is the quotient of the miles driven to the length of time of the trip.

$$\text{Average rate of speed} = \frac{413\text{ miles}}{7\text{ hours}} = \frac{59\text{ miles}}{1\text{ hour}} \qquad \frac{413}{7} = \frac{\cancel{7} \cdot 59}{\cancel{7} \cdot 1} = \frac{59}{1}.$$

The rate $\frac{59\text{ miles}}{1\text{ hour}}$ can be expressed in any of the following forms:

$59\,\frac{\text{miles}}{\text{hour}}$, 59 miles per hour, 59 miles/hour, or 59 mph

EXAMPLE 5 Find the hourly rate of pay for a student who earns \$370 for working 40 hours.

Solution We can write the rate of pay as

$$\text{Rate of pay} = \frac{\$370}{40 \text{ hours}}$$

and simplify by dividing 370 by 40.

$$\text{Rate of pay} = 9.25\,\frac{\text{dollars}}{\text{hour}}$$

The rate is \$9.25 per hour.

Self Check Lawanda earns \$316 per 40-hour week managing a dress shop. Find her hourly rate of pay.

EXAMPLE 6 One household used 813.75 kilowatt hours of electricity during a 31-day period. Find the rate of energy consumption in kilowatt hours per day.

Solution We can write the rate of energy consumption as

$$\text{Rate of energy consumption} = \frac{813.75 \text{ kilowatt hours}}{31 \text{ days}}$$

and simplify by dividing 813.75 by 31.

$$\text{Rate of energy consumption} = 26.25\,\frac{\text{kilowatt hours}}{\text{day}}$$

The rate of consumption in 26.25 kilowatt hours per day.

Self Check To heat a house for 30 days a furnace burned 72 therms of natural gas. Find the rate of gas consumption in therms per day.

EXAMPLE 7 A textbook costs \$49.22, including sales tax. If the tax was \$3.22, find the sales tax rate.

Solution Since the tax was \$3.22, the cost of the book alone was

$$\$49.22 - \$3.22 = \$46.00$$

We can write the sales tax rate as

$$\begin{aligned}\text{Sales tax rate} &= \frac{\text{amount of sales tax}}{\text{cost of the book, without tax}}\\ &= \frac{\$3.22}{\$46}\end{aligned}$$

and simplify by dividing 3.22 by 46.

$$\text{Sales tax rate} = \frac{3.22}{46} = 0.07$$

The tax rate is 0.07, or 7%.

Self Check A sport coat costs $160.50, including sales tax. If the cost of the coat without tax is $150, find the sales tax rate.

Accent on Technology: COMPUTING GAS MILEAGE

A man drove a total of 775 miles. Along the way, he stopped for gas three times, pumping 10.5, 11.3, and 8.75 gallons of gas. He started with the tank half-full and ended with the tank half-full. To find how many miles he got per gallon, we need to divide the total distance by the total number of gallons of gas consumed.

$$\frac{775}{10.5 + 11.3 + 8.75} \quad \begin{array}{l} \leftarrow \text{Total distance} \\ \leftarrow \text{Total number of gallons consumed} \end{array}$$

We can make this calculation by entering these numbers and pressing these keys.

775 ÷ (10.5 + 11.3 + 8.75) = On a scientific calculator.

775 ÷ (10.5 + 11.3 + 8.75) ENTER On a graphing calculator.

Either way, the display will read 25.36824877. To the nearest one-hundredth, he got 25.37 mpg.

Self Check Answers

1. a. $\frac{2}{3}$, **b.** $\frac{1}{5}$ **2. a.** $\frac{1}{4}$, **b.** $\frac{1}{6}$ **3.** $\frac{28}{79}$ **4.** the 16-oz drink **5.** $7.90 per hour **6.** 2.4 therms per day **7.** 7%

Orals *Express as a ratio in lowest terms.*

1. 5 to 7 **2.** 50 to 1 **3.** 3 to 9 **4.** 7 to 10

6.1 EXERCISES

REVIEW *Solve each equation.*

1. $2x + 4 = 38$

2. $\frac{x}{2} - 4 = 38$

3. $3(x + 2) = 24$

4. $\frac{x - 6}{3} = 20$

Factor each expression.

5. $2x + 6$

6. $x^2 - 49$

7. $2x^2 - x - 6$

8. $x^3 + 27$

VOCABULARY AND CONCEPTS *Fill in the blanks.*

9. A ratio is a _______ of two numbers.

10. The _______ of an item is the quotient of its cost to its quantity.

11. The ratios $\frac{2}{3}$ and $\frac{4}{6}$ are _____ ratios.

12. The quotient $\frac{500 \text{ miles}}{15 \text{ hours}}$ is called a ____.

13. Give three examples of ratios that you have encountered this past week.

14. Suppose that a basketball player made 8 free throws out of 12 tries. The ratio of $\frac{8}{12}$ can be simplified as $\frac{2}{3}$. Interpret this result.

PRACTICE *Express each phrase as a ratio in lowest terms.*

15. 5 to 7
16. 3 to 5
17. 17 to 34
18. 19 to 38
19. 22 to 33
20. 14 to 21
21. 7 to 24.5
22. 0.65 to 0.15
23. 4 ounces to 12 ounces
24. 3 inches to 15 inches
25. 12 minutes to 1 hour
26. 8 ounces to 1 pound
27. 3 days to 1 week
28. 4 inches to 2 yards
29. 18 months to 2 years
30. 8 feet to 4 yards

Refer to the monthly family budget shown in the table. Give each ratio in lowest terms.

Item	Amount
Rent	\$750
Food	\$652
Gas and electric	\$188
Phone	\$125
Entertainment	\$110

31. Find the total amount of the budget.
32. Find the ratio of the amount budgeted for rent to the total budget.
33. Find the ratio of the amount budgeted for entertainment to the total budget.
34. Find the ratio of the amount budgeted for phone to the amount budgeted for entertainment.

Refer to the tax deductions listed in the table. Give each ratio in lowest terms.

Item	Amount
Medical	\$ 995
Real estate tax	\$1,245
Contributions	\$1,680
Mortgage interest	\$4,580
Union dues	\$ 225

35. Find the total amount of deductions.
36. Find the ratio of real estate tax deductions to the total deductions.
37. Find the ratio of the contributions to the total deductions.
38. Find the ratio of the mortgage interest deduction to the union dues deduction.

APPLICATIONS *Express each result in lowest terms. Use a calculator when it is helpful.*

39. **Faculty-to-student ratio** At a college, there are 125 faculty members and 2,000 students. Find the faculty-to-student ratio.
40. **Ratio of men to women** In a state senate, there are 94 men and 24 women. Find the ratio of men to women.
41. **Unit cost of gasoline** A driver pumped 17 gallons of gasoline into his tank at a cost of \$30.09. Write a quotient of dollars to gallons, and give the unit cost of gasoline.
42. **Unit cost of grass seed** A 50-pound bag of grass seed costs \$222.50. Write a quotient of dollars to pounds, and give the unit cost of grass seed.
43. **Unit cost of cranberry juice** A 12-ounce can of cranberry juice sells for 84¢. Give the unit cost in cents per ounce.
44. **Unit cost of beans** A 24-ounce package of green beans sells for \$1.29. Give the unit cost in cents per ounce.
45. **Comparative shopping** A 6-ounce can of orange juice sells for 89¢, and an 8-ounce can sells for \$1.19. Which is the better buy?
46. **Comparing speeds** A car travels 345 miles in 6 hours, and a truck travels 376 miles in 6.2 hours. Which vehicle travels faster?
47. **Comparing reading speeds** One seventh-grader read a 54-page book in 40 minutes, and another read an 80-page book in 62 minutes. If the books were equally difficult, which student read faster?
48. **Comparative shopping** A 30-pound bag of fertilizer costs \$12.25, and an 80-pound bag costs \$30.25. Which is the better buy?

In this proportion, the product of the extremes is equal to the product of the means.

$1 \cdot 6 = 6$ and $2 \cdot 3 = 6$

This illustrates a fundamental property of proportions.

Fundamental Property of Proportions

In any proportion, the product of the extremes is equal to the product of the means.

In the proportion $\frac{a}{b} = \frac{c}{d}$, a and d are the extremes, and b and c are the means. We can show that the product of the extremes (ad) is equal to the product of the means (bc) by multiplying both sides of the proportion by bd and observing that $ad = bc$.

$$\frac{a}{b} = \frac{c}{d}$$

$$\frac{bd}{1} \cdot \frac{a}{b} = \frac{bd}{1} \cdot \frac{c}{d}$$ To eliminate the fractions, multiply both sides by $\frac{bd}{1}$.

$$\frac{abd}{b} = \frac{bcd}{d}$$ Multiply the numerators and multiply the denominators.

$$ad = bc$$ Divide out the common factors: $\frac{b}{b} = 1$ and $\frac{d}{d} = 1$.

Since $ad = bc$, the product of the extremes equals the product of the means.

To determine whether an equation is a proportion, we can check to see whether the product of the extremes is equal to the product of the means.

EXAMPLE 1 Determine whether each equation is a proportion: **a.** $\frac{3}{7} = \frac{9}{21}$ and **b.** $\frac{8}{3} = \frac{13}{5}$.

Solution In each case, we check to see whether the product of the extremes is equal to the product of the means.

a. The product of the extremes is $3 \cdot 21 = 63$. The product of the means is $7 \cdot 9 = 63$. Since the products are equal, the equation is a proportion: $\frac{3}{7} = \frac{9}{21}$.

b. The product of the extremes is $8 \cdot 5 = 40$. The product of the means is $3 \cdot 13 = 39$. Since the products are not equal, the equation is not a proportion: $\frac{8}{3} \neq \frac{13}{5}$.

Self Check Determine whether the equation is a proportion: $\frac{6}{13} = \frac{24}{53}$.

When two pairs of numbers such as 2, 3 and 8, 12 form a proportion, we say that they are **proportional.** To show that 2, 3, 8, and 12 are proportional, we check to see whether the equation

$$\frac{2}{3} = \frac{8}{12}$$

is a proportion. To do so, we find the product of the extremes and the product of the means:

$2 \cdot 12 = 24$ $3 \cdot 8 = 24$

Since the products are equal, the equation is a proportion, and the numbers are proportional.

EXAMPLE 2 Determine whether 3, 7, 36, and 91 are proportional.

Solution We check to see whether $\frac{3}{7} = \frac{36}{91}$ is a proportion by finding two products:

$$3 \cdot 91 = 273 \quad \text{The product of the extremes.}$$
$$7 \cdot 36 = 252 \quad \text{The product of the means.}$$

Since the products are not equal, the numbers are not proportional.

Self Check Determine whether 6, 11, 54, and 99 are proportional.

Solving Proportions

Suppose that we know three terms in the proportion

$$\frac{x}{5} = \frac{24}{20}$$

To find the unknown term, we multiply the extremes and multiply the means, set them equal, and solve for x:

$$\frac{x}{5} = \frac{24}{20}$$

$$20x = 5 \cdot 24 \quad \text{In a proportion, the product of the extremes is equal to the product of the means.}$$

$$20x = 120 \quad \text{Multiply: } 5 \cdot 24 = 120.$$

$$\frac{20x}{\mathbf{20}} = \frac{120}{\mathbf{20}} \quad \text{To undo the multiplication by 20, divide both sides by 20.}$$

$$x = 6 \quad \text{Simplify: } \tfrac{20}{20} = 1 \text{ and } \tfrac{120}{20} = 6.$$

The first term is 6.

EXAMPLE 3 Solve: $\frac{12}{18} = \frac{3}{x}$.

Solution

$$\frac{12}{18} = \frac{3}{x}$$

$$12 \cdot x = 18 \cdot 3 \quad \text{In a proportion, the product of the extremes equals the product of the means.}$$

$$12x = 54 \quad \text{Multiply: } 18 \cdot 3 = 54.$$

$$\frac{12x}{\mathbf{12}} = \frac{54}{\mathbf{12}} \quad \text{To undo the multiplication by 12, divide both sides by 12.}$$

$$x = \frac{9}{2} \quad \text{Simplify: } \tfrac{12}{12} = 1 \text{ and } \tfrac{54}{12} = \tfrac{9}{2}.$$

Thus, $x = \frac{9}{2}$.

Self Check Solve: $\frac{15}{x} = \frac{25}{40}$.

EXAMPLE 4 Find the third term of the proportion $\frac{3.5}{7.2} = \frac{x}{15.84}$.

Solution

$$\frac{3.5}{7.2} = \frac{x}{15.84}$$

$3.5(15.84) = 7.2x$ — In a proportion, the product of the extremes equals the product of the means.

$55.44 = 7.2x$ — Multiply: $3.5 \cdot 15.84 = 55.44$.

$\frac{55.44}{7.2} = \frac{7.2x}{7.2}$ — To undo the multiplication by 7.2, divide both sides by 7.2.

$7.7 = x$ — Simplify: $\frac{55.44}{7.2} = 7.7$ and $\frac{7.2}{7.2} = 1$.

The third term is 7.7.

Self Check Find the second term of the proportion $\frac{6.7}{x} = \frac{33.5}{38}$.

Accent on Technology

SOLVING EQUATIONS WITH A CALCULATOR

To solve the equation in Example 4 with a calculator, we can proceed as follows.

$$\frac{3.5}{7.2} = \frac{x}{15.84}$$

$\frac{3.5(15.84)}{7.2} = x$ — Multiply both sides by 15.84.

We can find x by entering these numbers and pressing these keys.

3.5 × 15.84 ÷ 7.2 = — On a scientific calculator.

3.5 × 15.84 ÷ 7.2 ENTER — On a graphing calculator.

Either way, the display will read 7.7. Thus, $x = 7.7$.

EXAMPLE 5 Solve: $\frac{2x + 1}{4} = \frac{10}{8}$.

Solution

$$\frac{2x + 1}{4} = \frac{10}{8}$$

$8(2x + 1) = 40$ — In a proportion, the product of the extremes equals the product of the means.

$16x + 8 = 40$ — Use the distributive property to remove parentheses.

$16x + 8 - 8 = 40 - 8$ — To undo the addition of 8, subtract 8 from both sides.

$16x = 32$ — Simplify: $8 - 8 = 0$ and $40 - 8 = 32$.

$\frac{16x}{16} = \frac{32}{16}$ — To undo the multiplication by 16, divide both sides by 16.

$x = 2$ — Simplify: $\frac{16}{16} = 1$ and $\frac{32}{16} = 2$.

Thus, $x = 2$.

Self Check Solve: $\dfrac{3x - 1}{2} = \dfrac{12.5}{5}$.

Problem Solving

When solving application problems, we often set two unit costs or two rates equal to form a proportion.

EXAMPLE 6 If 6 apples cost \$1.38, how much will 16 apples cost?

Solution Let c represent the cost of 16 apples. The ratios of the numbers of apples to their costs are equal.

6 apples is to \$1.38 as 16 apples is to \$$c$.

$$\begin{array}{r} \text{6 apples} \rightarrow \\ \text{Cost of 6 apples} \rightarrow \end{array} \frac{6}{1.38} = \frac{16}{c} \begin{array}{l} \leftarrow \text{16 apples} \\ \leftarrow \text{Cost of 16 apples} \end{array}$$

$6 \cdot c = 1.38(16)$ — In a proportion, the product of the extremes is equal to the product of the means.

$6c = 22.08$ — Do the multiplication: $1.38 \cdot 16 = 22.08$.

$\dfrac{6c}{6} = \dfrac{22.08}{6}$ — To undo the multiplication by 6, divide both sides by 6.

$c = 3.68$ — Simplify: $\frac{6}{6} = 1$ and $\frac{22.08}{6} = 3.68$.

Sixteen apples will cost \$3.68.

Self Check If 9 tickets to a concert cost \$112.50, how much will 15 tickets cost?

EXAMPLE 7 A solution contains 2 quarts of antifreeze and 5 quarts of water. How many quarts of antifreeze must be mixed with 18 quarts of water to have the same concentration?

Solution Let q represent the number of quarts of antifreeze to be mixed with the water. The ratios of the quarts of antifreeze to the quarts of water are equal.

2 quarts antifreeze is to 5 quarts water as q quarts antifreeze is to 18 quarts water.

$$\begin{array}{r} \text{2 quarts antifreeze} \rightarrow \\ \text{5 quarts water} \rightarrow \end{array} \frac{2}{5} = \frac{q}{18} \begin{array}{l} \leftarrow q \text{ quarts of antifreeze} \\ \leftarrow \text{18 quarts of water} \end{array}$$

$2 \cdot 18 = 5q$ — In a proportion, the product of the extremes is equal to the product of the means.

$36 = 5q$ — Do the multiplication: $2 \cdot 18 = 36$.

$\dfrac{36}{5} = \dfrac{5q}{5}$ — To undo the multiplication by 5, divide both sides by 5.

$\dfrac{36}{5} = q$ — Simplify: $\frac{5}{5} = 1$.

The mixture should contain $\frac{36}{5}$ or 7.2 quarts of antifreeze.

Self Check A solution should contain 2 ounces of alcohol for every 7 ounces of water. How much alcohol should be added to 20 ounces of water to get the proper concentration?

EXAMPLE 8 A recipe for rhubarb cake calls for $1\frac{1}{4}$ cups of sugar for every $2\frac{1}{2}$ cups of flour. How many cups of flour are needed if the baker intends to use 3 cups of sugar?

Solution Let f represent the number of cups of flour to be mixed with the sugar. The ratios of the cups of sugar to the cups of flour are equal.

$1\frac{1}{4}$ cups sugar is to $2\frac{1}{2}$ cups flour as 3 cups sugar is to f cups flour.

$1\frac{1}{4}$ cups sugar → $\frac{1\frac{1}{4}}{2\frac{1}{2}} = \frac{3}{f}$ ← 3 cups sugar
$2\frac{1}{2}$ cups flour → ← f cups flour

$$\frac{1.25}{2.5} = \frac{3}{f}$$ Change the fractions to decimals.

$$1.25f = 2.5 \cdot 3$$ In a proportion, the product of the extremes is equal to the product of the means.

$$1.25f = 7.5$$ Do the multiplication: $2.5 \cdot 3 = 7.5$.

$$\frac{1.25f}{\mathbf{1.25}} = \frac{7.5}{\mathbf{1.25}}$$ To undo the multiplication by 1.25, divide both sides by 1.25.

$$f = 6$$ Divide: $\frac{1.25}{1.25} = 1$ and $\frac{7.5}{1.25} = 6$.

The baker should use 6 cups of flour.

Self Check How many cups of sugar will be needed to make several cakes that will require a total of 25 cups of flour?

EXAMPLE 9 In a manufacturing process, 15 parts out of 90 were found to be defective. How many defective parts will be expected in a run of 120 parts?

Solution Let d represent the expected number of defective parts. In each run, the ratio of the defective parts to the total number of parts should be the same.

15 defective parts is to 90 as d defective parts is to 120.

15 defective parts → $\frac{15}{90} = \frac{d}{120}$ ← d defective parts
90 parts → ← 120 parts

$$15 \cdot 120 = 90d$$ In a proportion, the product of the extremes is equal to the product of the means.

$$1{,}800 = 90d$$ Do the multiplication: $15 \cdot 120 = 1{,}800$.

$$\frac{1{,}800}{\mathbf{90}} = \frac{90d}{\mathbf{90}}$$ To undo the multiplication by 90, divide both sides by 90.

$$20 = d$$ Divide: $\frac{1{,}800}{90} = 20$ and $\frac{90}{90} = 1$.

The expected number of defective parts is 20.

Self Check How many defective parts will be expected in a run of 3,000 parts?

Similar Triangles

If two angles of one triangle have the same measure as two angles of a second triangle, the triangles will have the same shape. Triangles with the same shape are called **similar triangles.** In Figure 6-1, $\triangle ABC \sim \triangle DEF$ (read the symbol $\sim$ as "is similar to").

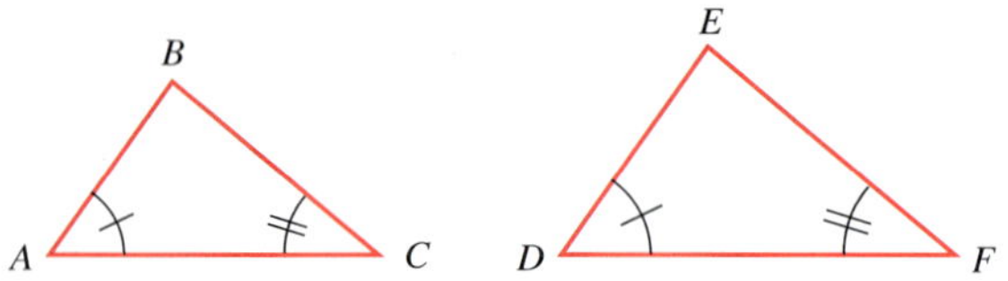

Figure 6-1

Property of Similar Triangles

If two triangles are similar, all pairs of corresponding sides are in proportion.

In the similar triangles shown in Figure 6-1, the following proportions are true.

$$\frac{\overline{AB}}{\overline{DE}} = \frac{\overline{BC}}{\overline{EF}}, \quad \frac{\overline{BC}}{\overline{EF}} = \frac{\overline{CA}}{\overline{FD}}, \quad \text{and} \quad \frac{\overline{CA}}{\overline{FD}} = \frac{\overline{AB}}{\overline{DE}}$$

EXAMPLE 10 A tree casts a shadow 18 feet long at the same time as a woman 5 feet tall casts a shadow that is 1.5 feet long. Find the height of the tree.

Solution Figure 6-2 shows the triangles determined by the tree and its shadow and the woman and her shadow.

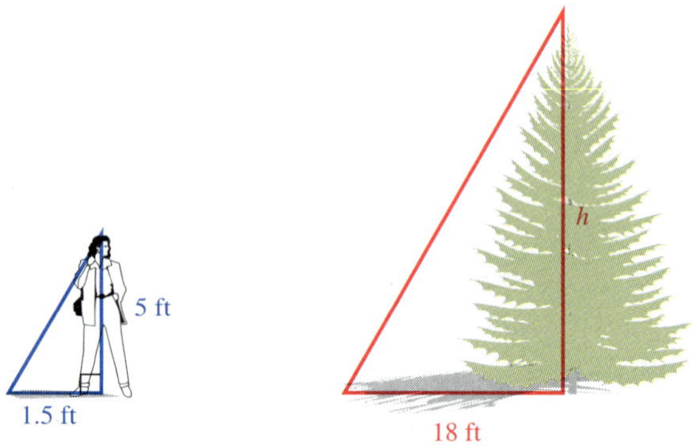

Figure 6-2

Since the triangles have the same shape, they are similar, and the lengths of their corresponding sides are in proportion. If we let h represent the height of the tree, we can find h by solving the following proportion.

$$\frac{h}{5} = \frac{18}{1.5}$$ $\frac{\text{Height of the tree}}{\text{Height of the woman}} = \frac{\text{Shadow of the tree}}{\text{Shadow of the woman}}$.

$$1.5h = 5(18)$$ In a proportion, the product of the extremes is equal to the product of the means.

$$h = 60$$ To undo the multiplication by 1.5, divide both sides by 1.5 and simplify.

The tree is 60 feet tall.

Self Check Find the height of the tree if the woman is 5 feet 6 inches tall and her shadow is still 1.5 feet long.

Self Check Answers

1. no **2.** yes **3.** 24 **4.** 7.6 **5.** 2 **6.** \$187.50 **7.** $\frac{40}{7}$ oz **8.** 12.5 cups **9.** 500 **10.** 66 ft

Orals *Which expressions are proportions?*

1. $\frac{3}{5} = \frac{6}{10}$ **2.** $\frac{1}{2} = \frac{1}{3}$ **3.** $\frac{1}{2} + \frac{1}{4}$ **4.** $\frac{1}{x} = \frac{2}{2x}$

6.2 EXERCISES

REVIEW

1. Change $\frac{9}{10}$ to a percent.
2. Change $\frac{7}{8}$ to a percent.
3. Change $33\frac{1}{3}\%$ to a fraction.
4. Change 75% to fraction.
5. Find 30% of 1,600.
6. Find $\frac{1}{2}\%$ of 520.
7. **Shopping** If Maria bought a dress for 25% off the original price of \$98, how much did the dress cost?
8. **Shopping** Bill purchased a shirt on sale for \$17.50. Find the original cost of the shirt if it was marked down 30%.

VOCABULARY AND CONCEPTS *Fill in the blanks.*

9. A ________ is a statement that two ______ are equal.
10. The first and fourth terms of a proportion are called the ________ of the proportion.
11. The second and third terms of a proportion are called the _______ of the proportion.
12. When two pairs of numbers form a proportion, we say that the numbers are ___________.
13. If two triangles have the same ______, they are said to be similar.
14. If two triangles are similar, the lengths of their corresponding sides are in __________.
15. The equation $\frac{a}{b} = \frac{c}{d}$ is a proportion if the product ___ is equal to the product ___.
16. If $3 \cdot 10 = 17 \cdot x$, then ______ is a proportion. (Note that answers may differ.)
17. Read $\triangle ABC$ as ________ ABC.
18. The symbol ~ is read as _________.

PRACTICE *Tell whether each statement is a proportion.*

19. $\frac{9}{7} = \frac{81}{70}$ **20.** $\frac{5}{2} = \frac{20}{8}$

21. $\frac{-7}{3} = \frac{14}{-6}$ **22.** $\frac{13}{-19} = \frac{-65}{95}$

23. $\frac{9}{19} = \frac{38}{80}$ **24.** $\frac{40}{29} = \frac{29}{22}$

25. $\frac{10.4}{3.6} = \frac{41.6}{14.4}$ **26.** $\frac{13.23}{3.45} = \frac{39.96}{11.35}$

Solve for the variable in each proportion.

27. $\frac{2}{3} = \frac{x}{6}$ **28.** $\frac{3}{6} = \frac{x}{8}$

29. $\frac{5}{10} = \frac{3}{c}$ **30.** $\frac{7}{14} = \frac{2}{b}$

31. $\frac{-6}{x} = \frac{8}{4}$ **32.** $\frac{4}{x} = \frac{2}{8}$

33. $\frac{x}{3} = \frac{9}{3}$ **34.** $\frac{x}{2} = \frac{-18}{6}$

35. $\frac{x+1}{5} = \frac{3}{15}$ **36.** $\frac{x-1}{7} = \frac{2}{21}$

37. $\frac{x+3}{12} = \frac{-7}{6}$ **38.** $\frac{x+7}{-4} = \frac{3}{12}$

39. $\frac{4-x}{13} = \frac{11}{26}$ **40.** $\frac{5-x}{17} = \frac{13}{34}$

41. $\frac{2x + 1}{18} = \frac{14}{3}$

42. $\frac{2x - 1}{18} = \frac{9}{54}$

43. $\frac{3p - 2}{12} = \frac{p + 1}{3}$

44. $\frac{12}{m} = \frac{18}{m + 2}$

APPLICATIONS *Set up and solve a proportion.*

45. **Grocery shopping** If 3 pints of yogurt cost $1, how much will 51 pints cost?

46. **Shopping for clothes** If shirts are on sale at two for $25, how much will 5 shirts cost?

47. **Gardening** Garden seed is on sale at 3 packets for 50¢. How much will 39 packets cost?

48. **Cooking** A recipe for spaghetti sauce requires four 16-ounce bottles of catsup to make two gallons of sauce. How many bottles of catsup are needed to make 10 gallons of sauce?

49. **Mixing perfume** A perfume is to be mixed in the ratio of 3 drops of pure essence to 7 drops of alcohol. How many drops of pure essence should be mixed with 56 drops of alcohol?

50. **Making cologne** A cologne can be made by mixing 2 drops of pure essence with 5 drops of distilled water. How many drops of water should be used with 15 drops of pure essence?

51. **Making cookies** A recipe for chocolate chip cookies calls for $1\frac{1}{4}$ cups of flour and 1 cup of sugar. The recipe will make $3\frac{1}{2}$ dozen cookies. How many cups of flour will be needed to make 12 dozen cookies?

52. **Making brownies** A recipe for brownies calls for 4 eggs and $1\frac{1}{2}$ cups of flour. If the recipe makes 15 brownies, how many cups of flour will be needed to make 130 brownies?

53. **Quality control** In a manufacturing process, 95% of the parts made are to be within specifications. How many defective parts would be expected in a run of 940 pieces?

54. **Quality control** Out of a sample of 500 men's shirts, 17 were rejected because of crooked collars. How many crooked collars would you expect to find in a run of 15,000 shirts?

55. **Gas consumption** If a car can travel 42 miles on 1 gallon of gas, how much gas will it need to travel 315 miles?

56. **Gas consumption** If a truck gets 12 mpg, how far can it go on 17 gallons of gas?

57. **Computing paychecks** Chen earns $412 for a 40-hour week. If he missed 10 hours of work last week, how much did he get paid?

58. **Model railroading** An HO-scale model railroad engine is 9 inches long. If HO scale is 87 feet to 1 foot, how long is a real engine?

59. **Model railroading** An N-scale model railroad caboose is 3.5 inches long. If N scale is 169 feet to 1 foot, how long is a real caboose?

60. **Model houses** A model house is built to a scale of 1 inch to 8 inches. If a model house is 36 inches wide, how wide is the real house?

61. **Staffing** A school board determined that there should be 3 teachers for every 50 students. How many teachers are needed for an enrollment of 2,700 students?

62. **Drafting** In a scale drawing, a 280-foot antenna tower is drawn 7 inches high. The building next to it is drawn 2 inches high. How tall is the actual building?

63. **Mixing fuel** The instructions on a can of oil intended to be added to lawnmower gasoline read:

Recommended	Gasoline	Oil
50 to 1	6 gal	16 oz

Are these instructions correct? (*Hint:* There are 128 ounces in 1 gallon.)

64. **Mixing fuel** See Exercise 63. How much oil should be mixed with 28 gallons of gas?

Use similar triangles to solve each problem.

65. **Height of a tree** A tree casts a shadow of 26 feet at the same time as a 6-foot man casts a shadow of 4 feet. Find the height of the tree.

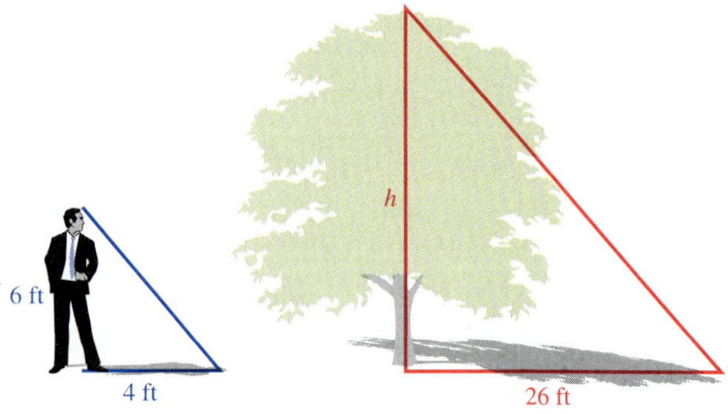

66. Height of a flagpole A man places a mirror on the ground and sees the reflection of the top of a flagpole, as in the illustration. The two triangles in the illustration are similar. Find the height h of the flagpole.

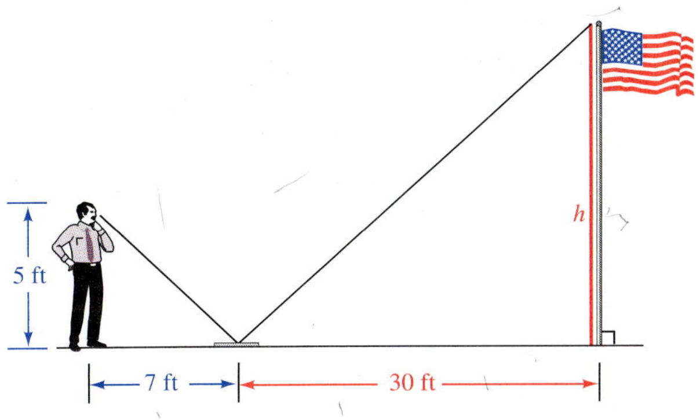

67. Width of a river Use the dimensions in the illustration to find w, the width of the river. The two triangles in the illustration are similar.

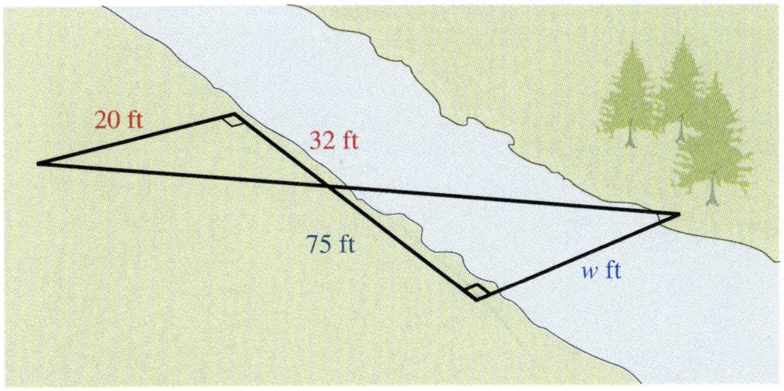

68. Flight path An airplane ascends 100 feet as it flies a horizontal distance of 1,000 feet. How much altitude will it gain as it flies a horizontal distance of 1 mile? (*Hint:* 5,280 feet = 1 mile.)

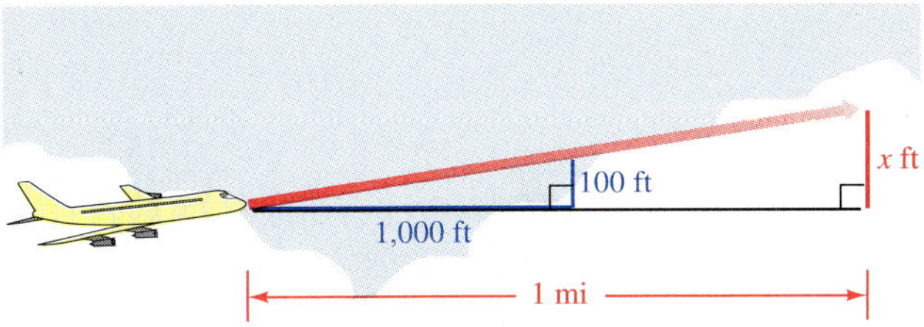

69. Flight path An airplane descends 1,350 feet as it flies a horizontal distance of 1 mile. How much altitude is lost as it flies a horizontal distance of 5 miles?

70. Ski runs A ski course falls 100 feet in every 300 feet of horizontal run. If the total horizontal run is $\frac{1}{2}$ mile, find the height of the hill.

71. Mountain travel A road ascends 750 feet in every 2,500 feet of travel. By how much will the road rise in a trip of 10 miles?

72. Photo enlargements The 3-by-5 photo in the illustration is to be blown up to the larger size. Find x.

WRITING

73. Explain the difference between a ratio and a proportion.

74. Explain how to tell whether the equation $\frac{3.2}{3.7} = \frac{5.44}{6.29}$ is a proportion.

SOMETHING TO THINK ABOUT

75. Verify that $\frac{3}{5} = \frac{12}{20} = \frac{3 + 12}{5 + 20}$. Is the following rule always true?

$$\frac{a}{b} = \frac{c}{d} = \frac{a + c}{b + d}$$

76. Verify that since $\frac{3}{5} = \frac{9}{15}$, then $\frac{3 + 5}{5} = \frac{9 + 15}{15}$. Is the following rule always true?

$$\text{If } \frac{a}{b} = \frac{c}{d}, \text{ then } \frac{a + b}{b} = \frac{c + d}{d}.$$

6.3 Simplifying Rational Expressions

- Simplifying Rational Expressions
- Division by 1
- Dividing Polynomials That Are Negatives

Getting Ready *Simplify.*

1. $\frac{12}{16}$ 2. $\frac{16}{8}$ 3. $\frac{25}{55}$ 4. $\frac{36}{72}$

Fractions such as $\frac{1}{2}$ and $\frac{3}{4}$ that are the quotient of two integers are *rational numbers.* Expressions such as

$$\frac{a}{a+2} \quad \text{and} \quad \frac{5x^2+3}{x^2+x-12}$$

where the numerators and denominators are polynomials, are called **rational expressions.** Since rational expressions indicate division, we must exclude any values of the variable that will make the denominator equal to 0. For example, a cannot be -2 in the rational expression

$$\frac{a}{a+2}$$

because the denominator will be 0:

$$\frac{a}{a+2} = \frac{-2}{-2+2} = \frac{-2}{0}$$

When the denominator of a rational expression is 0, we say that the expression is undefined.

EXAMPLE 1 Find all values of x such that the following rational expression is undefined.

$$\frac{5x^2+3}{x^2+x-12}$$

Solution To find the values of x that make the rational expression undefined, we set its denominator equal to 0 and solve for x.

$$x^2 + x - 12 = 0$$

$$(x+4)(x-3) = 0 \qquad \text{Factor the trinomial.}$$

$$x + 4 = 0 \quad \text{or} \quad x - 3 = 0 \qquad \text{Set each factor equal to 0.}$$

$$x = -4 \quad | \quad x = 3 \qquad \text{Solve each equation.}$$

We can check by substituting 3 and -4 for x and verifying that the rational expression is undefined.

For $x = 3$

$$\frac{5x^2 + 3}{x^2 + x - 12} = \frac{5(3)^2 + 3}{3^2 + 3 - 12} = \frac{5(9) + 3}{9 + 3 - 12} = \frac{45 + 3}{12 - 12} = \frac{48}{0}$$

For $x = -4$

$$\frac{5x^2 + 3}{x^2 + x - 12} = \frac{5(-4)^2 + 3}{(-4)^2 + (-4) - 12} = \frac{5(16) + 3}{16 - 4 - 12} = \frac{80 + 3}{12 - 12} = \frac{83}{0}$$

Since the denominator is 0 when $x = 3$ or $x = -4$, the rational expression $\frac{5x^2 + 3}{x^2 + x - 12}$ is undefined at these values. ∎

Simplifying Rational Expressions

We have seen that a fraction can be simplified by dividing out common factors shared by its numerator and denominator. For example,

$$\frac{18}{30} = \frac{3 \cdot 6}{5 \cdot 6} = \frac{3 \cdot \overset{1}{\cancel{6}}}{5 \cdot \underset{1}{\cancel{6}}} = \frac{3}{5} \qquad -\frac{6}{15} = -\frac{3 \cdot 2}{3 \cdot 5} = -\frac{\overset{1}{\cancel{3}} \cdot 2}{\underset{1}{\cancel{3}} \cdot 5} = -\frac{2}{5}$$

These examples illustrate the **fundamental property of fractions,** first discussed in Section 1.2.

The Fundamental Property of Fractions

If a, b, and x are real numbers, then

$$\frac{a \cdot x}{b \cdot x} = \frac{a}{b} \qquad (b \neq 0 \text{ and } x \neq 0)$$

Since rational expressions are fractions, we can use the fundamental property of fractions to simplify rational expressions. We factor the numerator and denominator of the rational expression and divide out all common factors. When all common factors have been divided out, we say that the rational expression has been **written in lowest terms.** To simplify a rational expression **means to write it in lowest terms.**

EXAMPLE 2 Simplify: $\frac{21x^2y}{14xy^2}$.

Solution

$$\frac{21x^2y}{14xy^2} = \frac{3 \cdot 7 \cdot x \cdot x \cdot y}{2 \cdot 7 \cdot x \cdot y \cdot y}$$ Factor the numerator and denominator.

$$= \frac{3 \cdot \overset{1}{\cancel{7}} \cdot \overset{1}{\cancel{x}} \cdot x \cdot \overset{1}{\cancel{y}}}{2 \cdot \underset{1}{\cancel{7}} \cdot \underset{1}{\cancel{x}} \cdot y \cdot \underset{1}{\cancel{y}}}$$ Divide out the common factors of 7, x, and y.

$$= \frac{3x}{2y}$$

This rational expression can also be simplified by using the rules of exponents:

$$\frac{21x^2y}{14xy^2} = \frac{3 \cdot 7}{2 \cdot 7}x^{2-1}y^{1-2} \quad \frac{x^2}{x} = x^{2-1}; \frac{y}{y^2} = y^{1-2}.$$

$$= \frac{3}{2}xy^{-1} \quad 2 - 1 = 1; 1 - 2 = -1.$$

$$= \frac{3}{2} \cdot \frac{x}{y} \quad y^{-1} = \frac{1}{y}.$$

$$= \frac{3x}{2y} \quad \text{Multiply.}$$

Self Check Simplify: $\frac{32a^3b^2}{24ab^4}$.

PERSPECTIVE

The fraction $\frac{8}{4}$ is equal to 2, because $4 \cdot 2 = 8$. The expression $\frac{8}{0}$ is undefined, because there is no number x for which $0 \cdot x = 8$. The expression $\frac{0}{0}$ presents a different problem, however, because $\frac{0}{0}$ seems to equal any number. For example, $\frac{0}{0} = 17$, because $0 \cdot 17 = 0$. Similarly, $\frac{0}{0} = \pi$, because $0 \cdot \pi = 0$. Since "no answer" and "any answer" are both unacceptable, division by 0 is not allowed.

Although $\frac{0}{0}$ represents many numbers, there is often one best answer. In the 17th century, mathematicians such as Sir Isaac Newton (1642–1727) and Gottfried Wilhelm von Leibniz (1646–1716) began to look more closely at expressions related to the fraction $\frac{0}{0}$. They discovered that under certain conditions, there was one best answer. Expressions related to $\frac{0}{0}$ are called **indeterminate forms.** One of these expressions, called a **derivative,** is the foundation of **calculus,** an important area of mathematics discovered independently by both Newton and Leibniz.

EXAMPLE 3 Simplify: $\frac{x^2 + 3x}{3x + 9}$.

Solution

$$\frac{x^2 + 3x}{3x + 9} = \frac{x(x + 3)}{3(x + 3)} \quad \text{Factor the numerator and the denominator.}$$

$$= \frac{x\overset{1}{\cancel{(x + 3)}}}{3\underset{1}{\cancel{(x + 3)}}} \quad \text{Divide out the common factor of } x + 3.$$

$$= \frac{x}{3}$$

Self Check Simplify: $\frac{x^2 - 5x}{5x - 25}$.

Division by 1

Any number divided by 1 remains unchanged. For example,

$$\frac{37}{1} = 37, \quad \frac{5x}{1} = 5x, \quad \text{and} \quad \frac{3x + y}{1} = 3x + y$$

In general, for any real number a, the following is true.

Division by 1

$$\frac{a}{1} = a$$

EXAMPLE 4 Simplify: $\dfrac{x^3 + x^2}{x + 1}$.

Solution

$$\begin{aligned} \frac{x^3 + x^2}{x + 1} &= \frac{x^2(x + 1)}{x + 1} && \text{Factor the numerator.} \\ &= \frac{x^2\overset{1}{\cancel{(x + 1)}}}{\underset{1}{\cancel{x + 1}}} && \text{Divide out the common factor of } x + 1. \\ &= \frac{x^2}{1} \\ &= x^2 && \text{Denominators of 1 need not be written.} \end{aligned}$$

Self Check Simplify: $\dfrac{x^2 - x}{x - 1}$. ▮

Dividing Polynomials That Are Negatives

If the terms of two polynomials are the same, except for sign, the polynomials are called **negatives** of each other. For example,

$x - y$ and $-x + y$ are negatives,
$2a - 1$ and $-2a + 1$ are negatives, and
$3x^2 - 2x + 5$ and $-3x^2 + 2x - 5$ are negatives.

Example 5 shows why the quotient of two binomials that are negatives is always -1.

EXAMPLE 5 Simplify: **a.** $\dfrac{x - y}{y - x}$ and **b.** $\dfrac{2a - 1}{1 - 2a}$.

Solution We can rearrange terms in each numerator, factor out -1, and proceed as follows:

a.
$$\begin{aligned} \frac{x - y}{y - x} &= \frac{-y + x}{y - x} \\ &= \frac{-(y - x)}{y - x} \\ &= \frac{-\overset{1}{\cancel{(y - x)}}}{\underset{1}{\cancel{y - x}}} \\ &= -1 \end{aligned}$$

b.
$$\begin{aligned} \frac{2a - 1}{1 - 2a} &= \frac{-1 + 2a}{1 - 2a} \\ &= \frac{-(1 - 2a)}{1 - 2a} \\ &= \frac{-\overset{1}{\cancel{(1 - 2a)}}}{\underset{1}{\cancel{1 - 2a}}} \\ &= -1 \end{aligned}$$

Self Check Simplify: $\dfrac{3p - 2q}{2q - 3p}$.

The previous example suggests this important result.

Division of Negatives The quotient of any nonzero expression and its negative is -1.

EXAMPLE 6 Simplify: $\dfrac{x^2 + 13x + 12}{x^2 - 144}$.

Solution

$$\frac{x^2 + 13x + 12}{x^2 - 144} = \frac{(x + 1)(x + 12)}{(x + 12)(x - 12)}$$ Factor the numerator and denominator.

$$= \frac{(x + 1)\overset{1}{\cancel{(x + 12)}}}{\underset{1}{\cancel{(x + 12)}}(x - 12)}$$ Divide out the common factor of $x + 12$.

$$= \frac{x + 1}{x - 12}$$

Self Check Simplify: $\dfrac{x^2 - 9}{x^3 - 3x^2}$.

Comment Remember that only *factors* common to the *entire numerator* and *entire denominator* can be divided out. *Terms* that are common to the numerator and denominator *cannot* be divided out. For example, consider the correct simplification

$$\frac{5 + 8}{5} = \frac{13}{5}$$

It would be incorrect to divide out the common *term* of 5 in the above simplification. Doing so gives an incorrect answer.

$$\frac{5 + 8}{5} \neq \frac{\overset{1}{\cancel{5}} + 8}{\underset{1}{\cancel{5}}} = \frac{1 + 8}{1} = 9$$

EXAMPLE 7 Simplify: $\dfrac{5(x + 3) - 5}{7(x + 3) - 7}$.

Solution We cannot divide out $x + 3$, because it is not a factor of the entire numerator, nor is it a factor of the entire denominator. Instead, we simplify the numerator and denominator, factor them, and divide out any common factors.

$$\frac{5(x+3)-5}{7(x+3)-7} = \frac{5x+15-5}{7x+21-7} \quad \text{Remove parentheses.}$$

$$= \frac{5x+10}{7x+14} \quad \text{Combine like terms.}$$

$$= \frac{5(x+2)}{7(x+2)} \quad \text{Factor the numerator and denominator.}$$

$$= \frac{5\overset{1}{\cancel{(x+2)}}}{7\underset{1}{\cancel{(x+2)}}} \quad \text{Divide out the common factor of } x+2.$$

$$= \frac{5}{7}$$

Self Check Simplify: $\frac{4(x-2)+4}{3(x-2)+3}$.

EXAMPLE 8 Simplify: $\frac{x(x+3)-3(x-1)}{x^2+3}$.

Solution

$$\frac{x(x+3)-3(x-1)}{x^2+3} = \frac{x^2+3x-3x+3}{x^2+3} \quad \text{Remove parentheses in the numerator.}$$

$$= \frac{x^2+3}{x^2+3} \quad \text{Combine like terms in the numerator.}$$

$$= \frac{\overset{1}{\cancel{x^2+3}}}{\underset{1}{\cancel{x^2+3}}} \quad \text{Divide out the common factor of } x^2+3.$$

$$= 1$$

Self Check Simplify: $\frac{a(a+2)-2(a-1)}{a^2+2}$.

Sometimes rational expressions do not simplify. For example, to attempt to simplify

$$\frac{x^2+x-2}{x^2+x}$$

we factor the numerator and denominator.

$$\frac{x^2+x-2}{x^2+x} = \frac{(x+2)(x-1)}{x(x+1)}$$

Because there are no factors common to the numerator and denominator, this rational expression is already in lowest terms.

EXAMPLE 9 Simplify: $\frac{x^3+8}{x^2+ax+2x+2a}$.

Solution

$$\frac{x^3+8}{x^2+ax+2x+2a} = \frac{(x+2)(x^2-2x+4)}{x(x+a)+2(x+a)}$$ Factor the numerator and begin to factor the denominator.

$$= \frac{(x+2)(x^2-2x+4)}{(x+a)(x+2)}$$ Finish factoring the denominator.

$$= \frac{\overset{1}{\cancel{(x+2)}}(x^2-2x+4)}{(x+a)\underset{1}{\cancel{(x+2)}}}$$ Divide out the common factor of $x + 2$.

$$= \frac{x^2-2x+4}{x+a}$$

Self Check Simplify: $\dfrac{ab+3a-2b-6}{a^3-8}$.

Self Check Answers

2. $\frac{4a^2}{3b^2}$ **3.** $\frac{x}{5}$ **4.** x **5.** -1 **6.** $\frac{x+3}{x^2}$ **7.** $\frac{4}{3}$ **8.** 1 **9.** $\frac{b+3}{a^2+2a+4}$

Orals *Simplify each rational expression.*

1. $\frac{14}{21}$ **2.** $\frac{34}{17}$ **3.** $\frac{xyz}{wxy}$ **4.** $\frac{8x^2}{4x}$

5. $\frac{6x^2y}{6xy^2}$ **6.** $\frac{x^2y^3}{x^2y^4}$ **7.** $\frac{x+y}{y+x}$ **8.** $\frac{x-y}{y-x}$

6.3 EXERCISES

REVIEW

1. State the associative property of addition.

2. State the distributive property.

3. What is the additive identity?

4. What is the multiplicative identity?

5. Find the additive inverse of $-\frac{5}{3}$.

6. Find the multiplicative inverse of $-\frac{5}{3}$.

VOCABULARY AND CONCEPTS *Fill in the blanks.*

7. In a fraction, the part above the fraction bar is called the ________.

8. In a fraction, the part below the fraction bar is called the ________.

9. The denominator of a fraction cannot be __.

10. A fraction that has polynomials in its numerator and denominator is called a ________ expression.

11. $x - 2$ and $2 - x$ are called ________ of each other.

12. To *simplify* a rational expression means to write it in ______ terms.

13. The fundamental property of fractions states that $\frac{ac}{bc} = \frac{\_\_}{}$.

14. Any number x divided by 1 is __.

15. To simplify a rational expression, we ______ the numerator and denominator and divide out ________ factors.

16. A rational expression cannot be simplified when it is written in __________.

PRACTICE *Find all values of the variable for which the following rational expressions are undefined.*

17. $\frac{2y+1}{y-2}$ **18.** $\frac{3a^2+5a}{3a-2}$

19. $\frac{3x-13}{x^2-x-2}$ **20.** $\frac{2p^2+5p}{6p^2-p-1}$

21. $\dfrac{2m^2 + 5m}{2m^2 - m - 3}$

22. $\dfrac{5q^2 - 3}{6q^2 - q - 2}$

Write each expression in lowest terms. If it is already in lowest terms, so indicate. Assume that no denominators are 0.

23. $\dfrac{8}{10}$

24. $\dfrac{16}{28}$

25. $\dfrac{28}{35}$

26. $\dfrac{14}{20}$

27. $\dfrac{8}{52}$

28. $\dfrac{15}{21}$

29. $\dfrac{10}{45}$

30. $\dfrac{21}{35}$

31. $\dfrac{-18}{54}$

32. $\dfrac{16}{40}$

33. $\dfrac{4x}{2}$

34. $\dfrac{2x}{4}$

35. $\dfrac{-6x}{18}$

36. $\dfrac{-25y}{5}$

37. $\dfrac{45}{9a}$

38. $\dfrac{48}{16y}$

39. $\dfrac{7 + 3}{5z}$

40. $\dfrac{(3 - 18)k}{25}$

41. $\dfrac{(3 + 4)a}{24 - 3}$

42. $\dfrac{x + x}{2}$

43. $\dfrac{2x}{3x}$

44. $\dfrac{5y}{7y}$

45. $\dfrac{6x^2}{4x^2}$

46. $\dfrac{9xy}{6xy}$

47. $\dfrac{2x^2}{3y}$

48. $\dfrac{5y^2}{2y^2}$

49. $\dfrac{15x^2y}{5xy^2}$

50. $\dfrac{12xz}{4xz^2}$

51. $\dfrac{28x}{32y}$

52. $\dfrac{14xz^2}{7x^2z^2}$

53. $\dfrac{x + 3}{3(x + 3)}$

54. $\dfrac{2(x + 7)}{x + 7}$

55. $\dfrac{5x + 35}{x + 7}$

56. $\dfrac{x - 9}{3x - 27}$

57. $\dfrac{x^2 + 3x}{2x + 6}$

58. $\dfrac{xz - 2x}{yz - 2y}$

59. $\dfrac{15x - 3x^2}{25y - 5xy}$

60. $\dfrac{3y + xy}{3x + xy}$

61. $\dfrac{6a - 6b + 6c}{9a - 9b + 9c}$

62. $\dfrac{3a - 3b - 6}{2a - 2b - 4}$

63. $\dfrac{x - 7}{7 - x}$

64. $\dfrac{d - c}{c - d}$

65. $\dfrac{6x - 3y}{3y - 6x}$

66. $\dfrac{3c - 4d}{4c - 3d}$

67. $\dfrac{a + b - c}{c - a - b}$

68. $\dfrac{x - y - z}{z + y - x}$

69. $\dfrac{x^2 + 3x + 2}{x^2 + x - 2}$

70. $\dfrac{x^2 + x - 6}{x^2 - x - 2}$

71. $\dfrac{x^2 - 8x + 15}{x^2 - x - 6}$

72. $\dfrac{x^2 - 6x - 7}{x^2 + 8x + 7}$

73. $\dfrac{2x^2 - 8x}{x^2 - 6x + 8}$

74. $\dfrac{3y^2 - 15y}{y^2 - 3y - 10}$

75. $\dfrac{xy + 2x^2}{2xy + y^2}$

76. $\dfrac{3x + 3y}{x^2 + xy}$

77. $\dfrac{x^2 + 3x + 2}{x^3 + x^2}$

78. $\dfrac{6x^2 - 13x + 6}{3x^2 + x - 2}$

79. $\dfrac{x^2 - 8x + 16}{x^2 - 16}$

80. $\dfrac{3x + 15}{x^2 - 25}$

81. $\dfrac{2x^2 - 8}{x^2 - 3x + 2}$

82. $\dfrac{3x^2 - 27}{x^2 + 3x - 18}$

83. $\dfrac{x^2 - 2x - 15}{x^2 + 2x - 15}$

84. $\dfrac{x^2 + 4x - 77}{x^2 - 4x - 21}$

85. $\dfrac{x^2 - 3(2x - 3)}{9 - x^2}$

86. $\dfrac{x(x - 8) + 16}{16 - x^2}$

87. $\dfrac{4(x+3)+4}{3(x+2)+6}$

88. $\dfrac{4+2(x-5)}{3x-5(x-2)}$

89. $\dfrac{x^2-9}{(2x+3)-(x+6)}$

90. $\dfrac{x^2+5x+4}{2(x+3)-(x+2)}$

91. $\dfrac{x^3+1}{x^2-x+1}$

92. $\dfrac{x^3-1}{x^2+x+1}$

93. $\dfrac{2a^3-16}{2a^2+4a+8}$

94. $\dfrac{3y^3+81}{y^2-3y+9}$

95. $\dfrac{ab+b+2a+2}{ab+a+b+1}$

96. $\dfrac{xy+2y+3x+6}{x^2+5x+6}$

97. $\dfrac{xy+3y+3x+9}{x^2-9}$

98. $\dfrac{ab+b^2+2a+2b}{a^2+2a+ab+2b}$

WRITING

99. Explain why $\dfrac{x-7}{7-x}=-1$.

100. Explain why $\dfrac{x+7}{7+x}=1$.

SOMETHING TO THINK ABOUT

101. Exercise 85 has two possible answers: $\dfrac{3-x}{3+x}$ and $-\dfrac{x-3}{x+3}$. Why is either answer correct?

102. Find two different-looking but correct answers for the following problem.

Simplify: $\dfrac{y^2+5(2y+5)}{25-y^2}$.

6.4 Multiplying and Dividing Rational Expressions

- Multiplying Rational Expressions
- Multiplying a Rational Expression by a Polynomial
- Dividing Rational Expressions
- Dividing a Rational Expression by a Polynomial
- Combined Operations

Getting Ready *Multiply or divide the fractions and simplify.*

1. $\dfrac{3}{7}\cdot\dfrac{14}{9}$
2. $\dfrac{21}{15}\cdot\dfrac{10}{3}$
3. $\dfrac{19}{38}\cdot 6$
4. $42\cdot\dfrac{3}{21}$
5. $\dfrac{4}{9}\div\dfrac{8}{45}$
6. $\dfrac{11}{7}\div\dfrac{22}{14}$
7. $\dfrac{75}{12}\div\dfrac{50}{6}$
8. $\dfrac{13}{5}\div\dfrac{26}{20}$

Just like arithmetic fractions, rational expressions can be multiplied, divided, added, and subtracted. In this section, we will show how to multiply and divide rational expressions, the fractions of algebra.

Multiplying Rational Expressions

Recall that to multiply fractions, we multiply their numerators and multiply their denominators. For example, to find the product of $\frac{4}{7}$ and $\frac{3}{5}$, we proceed as follows.

$$\frac{4}{7} \cdot \frac{3}{5} = \frac{4 \cdot 3}{7 \cdot 5}$$ Multiply the numerators and multiply the denominators.

$$= \frac{12}{35}$$ $4 \cdot 3 = 12$ and $7 \cdot 5 = 35$.

This suggests the rule for multiplying rational expressions.

Multiplying Rational Expressions

If a, b, c, and d are real numbers and $b \neq 0$ and $d \neq 0$, then

$$\frac{a}{b} \cdot \frac{c}{d} = \frac{ac}{bd}$$

EXAMPLE 1 Multiply: **a.** $\frac{1}{3} \cdot \frac{2}{5}$, **b.** $\frac{7}{9} \cdot \frac{-5}{3x}$, **c.** $\frac{x^2}{2} \cdot \frac{3}{y^2}$, and **d.** $\frac{t+1}{t} \cdot \frac{t-1}{t-2}$.

Solution **a.** $\frac{1}{3} \cdot \frac{2}{5} = \frac{1 \cdot 2}{3 \cdot 5}$

$$= \frac{2}{15}$$

b. $\frac{7}{9} \cdot \frac{-5}{3x} = \frac{7(-5)}{9 \cdot 3x}$

$$= \frac{-35}{27x}$$

c. $\frac{x^2}{2} \cdot \frac{3}{y^2} = \frac{x^2 \cdot 3}{2 \cdot y^2}$

$$= \frac{3x^2}{2y^2}$$

d. $\frac{t+1}{t} \cdot \frac{t-1}{t-2} = \frac{(t+1)(t-1)}{t(t-2)}$

Self Check Multiply: $\frac{3x}{4} \cdot \frac{p-3}{y}$.

EXAMPLE 2 Multiply: $\frac{35x^2y}{7y^2z} \cdot \frac{z}{5xy}$.

Solution

$$\frac{35x^2y}{7y^2z} \cdot \frac{z}{5xy} = \frac{35x^2y \cdot z}{7y^2z \cdot 5xy}$$ Multiply the numerators and multiply the denominators.

$$= \frac{5 \cdot 7 \cdot x \cdot x \cdot y \cdot z}{7 \cdot y \cdot y \cdot z \cdot 5 \cdot x \cdot y}$$ Factor.

$$= \frac{\overset{1}{\cancel{5}} \cdot \overset{1}{\cancel{7}} \cdot \overset{1}{\cancel{x}} \cdot x \cdot \overset{1}{\cancel{y}} \cdot \overset{1}{\cancel{z}}}{\underset{1}{\cancel{7}} \cdot \underset{1}{\cancel{y}} \cdot y \cdot \underset{1}{\cancel{z}} \cdot \underset{1}{\cancel{5}} \cdot \underset{1}{\cancel{x}} \cdot y}$$ Divide out common factors.

$$= \frac{x}{y^2}$$

Self Check Multiply: $\frac{a^2b^2}{2a} \cdot \frac{9a^3}{3b^3}$.

EXAMPLE 3 Multiply: $\frac{x^2 - x}{2x + 4} \cdot \frac{x + 2}{x}$.

Solution

$$\frac{x^2 - x}{2x + 4} \cdot \frac{x + 2}{x} = \frac{(x^2 - x)(x + 2)}{(2x + 4)(x)}$$ Multiply the numerators and multiply the denominators.

$$= \frac{x(x - 1)(x + 2)}{2(x + 2)x}$$ Factor.

$$= \frac{\overset{1}{\cancel{x}}(x - 1)\overset{1}{\cancel{(x + 2)}}}{2\underset{1}{\cancel{(x + 2)}}\underset{1}{\cancel{x}}}$$ Divide out common factors.

$$= \frac{x - 1}{2}$$

Self Check Multiply: $\frac{x^2 + x}{3x + 6} \cdot \frac{x + 2}{x + 1}$.

EXAMPLE 4 Multiply: $\frac{x^2 - 3x}{x^2 - x - 6}$ and $\frac{x^2 + x - 2}{x^2 - x}$.

Solution

$$\frac{x^2 - 3x}{x^2 - x - 6} \cdot \frac{x^2 + x - 2}{x^2 - x}$$

$$= \frac{(x^2 - 3x)(x^2 + x - 2)}{(x^2 - x - 6)(x^2 - x)}$$ Multiply the numerators and multiply the denominators.

$$= \frac{x(x - 3)(x + 2)(x - 1)}{(x + 2)(x - 3)x(x - 1)}$$ Factor.

$$= \frac{\overset{1}{\cancel{x}}\overset{1}{\cancel{(x - 3)}}\overset{1}{\cancel{(x + 2)}}\overset{1}{\cancel{(x - 1)}}}{\underset{1}{\cancel{(x + 2)}}\underset{1}{\cancel{(x - 3)}}\underset{1}{\cancel{x}}\underset{1}{\cancel{(x - 1)}}}$$ Divide out common factors.

$$= 1$$

Self Check Multiply: $\frac{a^2 + a}{a^2 - 4} \cdot \frac{a^2 - a - 2}{a^2 + 2a + 1}$.

Multiplying a Rational Expression by a Polynomial

Since any number divided by 1 remains unchanged, we can write any polynomial as a rational expression by inserting a denominator of 1.

EXAMPLE 5 Multiply: $\frac{x^2 + x}{x^2 + 8x + 7} \cdot x + 7$.

Solution

$$\frac{x^2+x}{x^2+8x+7}\cdot(x+7) = \frac{x^2+x}{x^2+8x+7}\cdot\frac{x+7}{1}$$ Write $x + 7$ as a fraction with a denominator of 1.

$$= \frac{x(x+1)(x+7)}{(x+1)(x+7)1}$$ Multiply the fractions and factor where possible.

$$= \frac{x\overset{1}{\cancel{(x+1)}}\overset{1}{\cancel{(x+7)}}}{1\underset{1}{\cancel{(x+1)}}\underset{1}{\cancel{(x+7)}}}$$ Divide out all common factors.

$$= x$$

Self Check Multiply: $a - 7 \cdot \dfrac{a^2 - a}{a^2 - 8a + 7}$.

Dividing Rational Expressions

Recall that division by a nonzero number is equivalent to multiplying by the reciprocal of that number. Thus, to divide two fractions, we can invert the divisor (the fraction following the ÷ sign) and multiply. For example, to divide $\frac{4}{7}$ by $\frac{3}{5}$, we proceed as follows:

$$\frac{4}{7} \div \frac{3}{5} = \frac{4}{7}\cdot\frac{5}{3}$$ Invert $\frac{3}{5}$ and change the division to a multiplication.

$$= \frac{20}{21}$$ Multiply the numerators and multiply the denominators.

This suggests the rule for dividing rational expressions.

Dividing Rational Expressions

If a is a real number and b, c, and d are nonzero real numbers, then

$$\frac{a}{b} \div \frac{c}{d} = \frac{a}{b}\cdot\frac{d}{c} = \frac{ad}{bc}$$

EXAMPLE 6 Divide: **a.** $\dfrac{7}{13} \div \dfrac{21}{26}$ and **b.** $\dfrac{-9x}{35y} \div \dfrac{15x^2}{14}$.

Solution **a.**

$$\frac{7}{13} \div \frac{21}{26} = \frac{7}{13}\cdot\frac{26}{21}$$ Invert the divisor and multiply.

$$= \frac{7\cdot 2\cdot 13}{13\cdot 3\cdot 7}$$ Multiply the fractions and factor where possible.

$$= \frac{\overset{1}{\cancel{7}}\cdot 2\cdot \overset{1}{\cancel{13}}}{\underset{1}{\cancel{13}}\cdot 3\cdot \underset{1}{\cancel{7}}}$$ Divide out common factors.

$$= \frac{2}{3}$$

b. $\frac{-9x}{35y} \div \frac{15x^2}{14} = \frac{-9x}{35y} \cdot \frac{14}{15x^2}$ Invert the divisor and multiply.

$= \frac{-3 \cdot 3 \cdot x \cdot 2 \cdot 7}{5 \cdot 7 \cdot y \cdot 3 \cdot 5 \cdot x \cdot x}$ Multiply the fractions and factor where possible.

$= \frac{-3 \cdot \overset{1}{\cancel{3}} \cdot \overset{1}{\cancel{x}} \cdot 2 \cdot \overset{1}{\cancel{7}}}{5 \cdot \underset{1}{\cancel{7}} \cdot y \cdot \underset{1}{\cancel{3}} \cdot 5 \cdot \underset{1}{\cancel{x}} \cdot x}$ Divide out common factors.

$= -\frac{6}{25xy}$ Multiply the remaining factors.

Self Check Divide: $\frac{-8a}{3b} \div \frac{16a^2}{9b^2}$.

EXAMPLE 7 Divide: $\frac{x^2 + x}{3x - 15} \div \frac{x^2 + 2x + 1}{6x - 30}$.

Solution

$\frac{x^2 + x}{3x - 15} \div \frac{x^2 + 2x + 1}{6x - 30}$

$= \frac{x^2 + x}{3x - 15} \cdot \frac{6x - 30}{x^2 + 2x + 1}$ Invert the divisor and multiply.

$= \frac{x(x + 1) \cdot 2 \cdot 3(x - 5)}{3(x - 5)(x + 1)(x + 1)}$ Multiply the fractions and factor.

$= \frac{x\overset{1}{\cancel{(x + 1)}} \cdot 2 \cdot \overset{1}{\cancel{3}}\overset{1}{\cancel{(x - 5)}}}{\underset{1}{\cancel{3}}\underset{1}{\cancel{(x - 5)}}\underset{1}{\cancel{(x + 1)}}(x + 1)}$ Divide out all common factors.

$= \frac{2x}{x + 1}$

Self Check Divide: $\frac{a^2 - 1}{a^2 + 4a + 3} \div \frac{a - 1}{a^2 + 2a - 3}$.

Dividing a Rational Expression by a Polynomial

To divide a rational expression by a polynomial, we write the polynomial as a rational expression by inserting a denominator of 1 and then divide the expressions.

EXAMPLE 8 Divide: $\frac{2x^2 - 3x - 2}{2x + 1} \div (4 - x^2)$.

Solution

$$\frac{2x^2 - 3x - 2}{2x + 1} \div (4 - x^2)$$

$$= \frac{2x^2 - 3x - 2}{2x + 1} \div \frac{4 - x^2}{1}$$ Write $4 - x^2$ as a fraction with a denominator of 1.

$$= \frac{2x^2 - 3x - 2}{2x + 1} \cdot \frac{1}{4 - x^2}$$ Invert the divisor and multiply.

$$= \frac{(2x + 1)(x - 2) \cdot 1}{(2x + 1)(2 + x)(2 - x)}$$ Multiply the fractions and factor where possible.

$$= \frac{\overset{1}{\cancel{(2x + 1)}}\overset{-1}{\cancel{(x - 2)}} \cdot 1}{\underset{1}{\cancel{(2x + 1)}}(2 + x)\underset{1}{\cancel{(2 - x)}}}$$ Divide out common factors: $\frac{x - 2}{2 - x} = -1$.

$$= \frac{-1}{2 + x}$$

$$= -\frac{1}{2 + x}$$

Self Check Divide: $(b - a) \div \dfrac{a^2 - b^2}{a^2 + ab}$. ∎

Combined Operations

Unless parentheses indicate otherwise, we perform multiplications and divisions in order from left to right.

EXAMPLE 9 Simplify: $\dfrac{x^2 - x - 6}{x - 2} \div \dfrac{x^2 - 4x}{x^2 - x - 2} \cdot \dfrac{x - 4}{x^2 + x}$.

Solution Since there are no parentheses to indicate otherwise, we perform the division first.

$$\frac{x^2 - x - 6}{x - 2} \div \frac{x^2 - 4x}{x^2 - x - 2} \cdot \frac{x - 4}{x^2 + x}$$

$$= \frac{x^2 - x - 6}{x - 2} \cdot \frac{x^2 - x - 2}{x^2 - 4x} \cdot \frac{x - 4}{x^2 + x}$$ Invert the divisor and multiply.

$$= \frac{(x + 2)(x - 3)(x + 1)(x - 2)(x - 4)}{(x - 2)x(x - 4)x(x + 1)}$$ Multiply the fractions and factor.

$$= \frac{(x + 2)(x - 3)\overset{1}{\cancel{(x + 1)}}\overset{1}{\cancel{(x - 2)}}\overset{1}{\cancel{(x - 4)}}}{\underset{1}{\cancel{(x - 2)}}x\underset{1}{\cancel{(x - 4)}}x\underset{1}{\cancel{(x + 1)}}}$$ Divide out all common factors.

$$= \frac{(x + 2)(x - 3)}{x^2}$$

Self Check Simplify: $\dfrac{a^2 + ab}{ab - b^2} \cdot \dfrac{a^2 - b^2}{a^2 + ab} \div \dfrac{a + b}{b}$. ∎

EXAMPLE 10 Simplify: $\frac{x^2 + 6x + 9}{x^2 - 2x}\left(\frac{x^2 - 4}{x^2 + 3x} \div \frac{x + 2}{x}\right)$.

Solution We perform the division within the parentheses first.

$$\frac{x^2 + 6x + 9}{x^2 - 2x}\left(\frac{x^2 - 4}{x^2 + 3x} \div \frac{x + 2}{x}\right)$$

$$= \frac{x^2 + 6x + 9}{x^2 - 2x}\left(\frac{x^2 - 4}{x^2 + 3x} \cdot \frac{x}{x + 2}\right)$$ Invert the divisor and multiply.

$$= \frac{(x + 3)(x + 3)(x + 2)(x - 2)x}{x(x - 2)x(x + 3)(x + 2)}$$ Multiply the fractions and factor where possible.

$$= \frac{\overset{1}{\cancel{(x + 3)}}(x + 3)\overset{1}{\cancel{(x + 2)}}\overset{1}{\cancel{(x - 2)}}\overset{1}{\cancel{x}}}{\underset{1}{\cancel{x}}\underset{1}{\cancel{(x - 2)}}x\underset{1}{\cancel{(x + 3)}}\underset{1}{\cancel{(x + 2)}}}$$ Divide out all common factors.

$$= \frac{x + 3}{x}$$

Self Check Simplify: $\frac{x^2 - 2x}{x^2 + 6x + 9} \div \left(\frac{x^2 - 4}{x^2 + 3x} \cdot \frac{x}{x + 2}\right)$.

Self Check Answers

1. $\frac{3x(p - 3)}{4y}$ 2. $\frac{3a^4}{2b}$ 3. $\frac{x}{3}$ 4. $\frac{a}{a + 2}$ 5. a 6. $-\frac{3b}{2a}$ 7. $a - 1$ 8. $-a$ 9. 1 10. $\frac{x}{x + 3}$

Orals *Perform the operations and simplify.*

1. $\frac{x}{2} \cdot \frac{3}{x}$
2. $\frac{x + 1}{5} \cdot \frac{7}{x + 1}$
3. $\frac{5}{x + 7} \cdot (x + 7)$
4. $\frac{3}{7} \div \frac{3}{7}$
5. $\frac{3}{4} \div 3$
6. $(x + 1) \div \frac{x + 1}{x}$

6.4 EXERCISES

REVIEW *Simplify each expression. Write all answers without using negative exponents.*

1. $2x^3y^2(-3x^2y^4z)$
2. $\frac{8x^4y^5}{-2x^3y^2}$
3. $(3y)^{-4}$
4. $(a^{-2}a)^{-3}$
5. $\frac{x^{3m}}{x^{4m}}$
6. $(3x^2y^3)^0$

Perform the operations and simplify.

7. $-4(y^3 - 4y^2 + 3y - 2) + 6(-2y^2 + 4) - 4(-2y^3 - y)$
8. $y - 5\overline{)5y^3 - 3y^2 + 4y - 1}$

VOCABULARY AND CONCEPTS *Fill in the blanks.*

9. In a fraction, the part above the fraction bar is called the ________.
10. In a fraction, the part below the fraction bar is called the ________.

11. To multiply fractions, we multiply their ________ and multiply their ________.

12. $\frac{a}{b} \cdot \frac{c}{d} = \overline{\quad}$

13. To write a polynomial in fractional form, we insert a denominator of __.

14. $\frac{a}{b} \div \frac{c}{d} = \frac{a}{b} \cdot \overline{\quad}$

15. To divide two fractions, invert the ______ and ______.

16. Unless parentheses indicate otherwise, do multiplications and divisions in order from ___ to ____.

PRACTICE *Perform the multiplications. Simplify the answers, if possible.*

17. $\frac{5}{7} \cdot \frac{9}{13}$

18. $\frac{2}{7} \cdot \frac{5}{11}$

19. $\frac{25}{35} \cdot \frac{-21}{55}$

20. $-\frac{27}{24} \cdot \left(-\frac{56}{35}\right)$

21. $\frac{2}{3} \cdot \frac{15}{2} \cdot \frac{1}{7}$

22. $\frac{2}{5} \cdot \frac{10}{9} \cdot \frac{3}{2}$

23. $\frac{3x}{y} \cdot \frac{y}{2}$

24. $\frac{2y}{z} \cdot \frac{z}{3}$

25. $\frac{5y}{7} \cdot \frac{7x}{5z}$

26. $\frac{4x}{3y} \cdot \frac{3y}{7x}$

27. $\frac{7z}{9z} \cdot \frac{4z}{2z}$

28. $\frac{8z}{2x} \cdot \frac{16x}{3x}$

29. $\frac{2x^2y}{3xy} \cdot \frac{3xy^2}{2}$

30. $\frac{2x^2z}{z} \cdot \frac{5x}{z}$

31. $\frac{8x^2y^2}{4x^2} \cdot \frac{2xy}{2y}$

32. $\frac{9x^2y}{3x} \cdot \frac{3xy}{3y}$

33. $\frac{-2xy}{x^2} \cdot \frac{3xy}{2}$

34. $\frac{-3x}{x^2} \cdot \frac{2xz}{3}$

35. $\frac{ab^2}{a^2b} \cdot \frac{b^2c^2}{abc} \cdot \frac{abc^2}{a^3c^2}$

36. $\frac{x^3y}{z} \cdot \frac{xz^3}{x^2y^2} \cdot \frac{yz}{xyz}$

37. $\frac{10r^2st^3}{6rs^2} \cdot \frac{3r^3t}{2rst} \cdot \frac{2s^3t^4}{5s^2t^3}$

38. $\frac{3a^3b}{25cd^3} \cdot \frac{-5cd^2}{6ab} \cdot \frac{10abc^2}{2bc^2d}$

39. $\frac{z+7}{7} \cdot \frac{z+2}{z}$

40. $\frac{a-3}{a} \cdot \frac{a+3}{5}$

41. $\frac{x-2}{2} \cdot \frac{2x}{x-2}$

42. $\frac{y+3}{y} \cdot \frac{3y}{y+3}$

43. $\frac{x+5}{5} \cdot \frac{x}{x+5}$

44. $\frac{y-9}{y+9} \cdot \frac{y}{9}$

45. $\frac{(x+1)^2}{x+1} \cdot \frac{x+2}{x+1}$

46. $\frac{(y-3)^2}{y-3} \cdot \frac{y-3}{y-3}$

47. $\frac{2x+6}{x+3} \cdot \frac{3}{4x}$

48. $\frac{3y-9}{y-3} \cdot \frac{y}{3y^2}$

49. $\frac{x^2-x}{x} \cdot \frac{3x-6}{3x-3}$

50. $\frac{5z-10}{z+2} \cdot \frac{3}{3z-6}$

51. $\frac{7y-14}{y-2} \cdot \frac{x^2}{7x}$

52. $\frac{y^2+3y}{9} \cdot \frac{3x}{y+3}$

53. $\frac{x^2+x-6}{5x} \cdot \frac{5x-10}{x+3}$

54. $\frac{z^2+4z-5}{5z-5} \cdot \frac{5z}{z+5}$

55. $\frac{m^2-2m-3}{2m+4} \cdot \frac{m^2-4}{m^2+3m+2}$

56. $\frac{p^2-p-6}{3p-9} \cdot \frac{p^2-9}{p^2+6p+9}$

57. $\frac{x^2+7xy+12y^2}{x^2+2xy-8y^2} \cdot \frac{x^2-xy-2y^2}{x^2+4xy+3y^2}$

58. $\frac{m^2+9mn+20n^2}{m^2-25n^2} \cdot \frac{m^2-9mn+20n^2}{m^2-16n^2}$

59. $\frac{ax+bx+ay+by}{x^3-y^3} \cdot \frac{x^2+xy+y^2}{ax+bx}$

60. $\frac{a^2-ab+b^2}{a^3+b^3} \cdot \frac{ac+ad+bc+bd}{c^2-d^2}$

61. $\frac{x^2-y^2}{y^2-xy} \cdot \frac{yx^3-y^4}{ax+ay+bx+by}$

62. $\frac{xw-xz+wy-yz}{x^2+2xy+y^2} \cdot \frac{x^3-y^3}{z^2-w^2}$

63. $\frac{abc^2}{a+1} \cdot \frac{c}{a^2b^2} \cdot \frac{a^2+a}{ac}$

64. $\frac{x^3yz^2}{4x+8} \cdot \frac{x^2-4}{2x^2y^2z^2} \cdot \frac{8yz}{x-2}$

65. $\frac{3x^2+5x+2}{x^2-9} \cdot \frac{x-3}{x^2-4} \cdot \frac{x^2+5x+6}{6x+4}$

66. $\frac{x^2-25}{3x+6} \cdot \frac{x^2+x-2}{2x+10} \cdot \frac{6x}{3x^2-18x+15}$

Perform each division. Simplify answers when possible.

67. $\frac{1}{3} \div \frac{1}{2}$

68. $\frac{3}{4} \div \frac{1}{3}$

69. $\frac{21}{14} \div \frac{5}{2}$

70. $\frac{14}{3} \div \frac{10}{3}$

71. $\frac{2}{y} \div \frac{4}{3}$

72. $\frac{3}{a} \div \frac{a}{9}$

73. $\frac{3x}{2} \div \frac{x}{2}$

74. $\frac{y}{6} \div \frac{2}{3y}$

75. $\frac{3x}{y} \div \frac{2x}{4}$

76. $\frac{3y}{8} \div \frac{2y}{4y}$

77. $\frac{4x}{3x} \div \frac{2y}{9y}$

78. $\frac{14}{7y} \div \frac{10}{5z}$

79. $\frac{x^2}{3} \div \frac{2x}{4}$

80. $\frac{z^2}{z} \div \frac{z}{3z}$

81. $\frac{x^2y}{3xy} \div \frac{xy^2}{6y}$

82. $\frac{2xz}{z} \div \frac{4x^2}{z^2}$

83. $\frac{x+2}{3x} \div \frac{x+2}{2}$

84. $\frac{z-3}{3z} \div \frac{z+3}{z}$

85. $\frac{(z-2)^2}{3z^2} \div \frac{z-2}{6z}$

86. $\frac{(x+7)^2}{x+7} \div \frac{(x-3)^2}{x+7}$

87. $\frac{(z-7)^2}{z+2} \div \frac{z(z-7)}{5z^2}$

88. $\frac{y(y+2)}{y^2(y-3)} \div \frac{y^2(y+2)}{(y-3)^2}$

89. $\frac{x^2-4}{3x+6} \div \frac{x-2}{x+2}$

90. $\frac{x^2-9}{5x+15} \div \frac{x-3}{x+3}$

91. $\frac{x^2-1}{3x-3} \div \frac{x+1}{3}$

92. $\frac{x^2-16}{x-4} \div \frac{3x+12}{x}$

93. $\frac{5x^2+13x-6}{x+3} \div \frac{5x^2-17x+6}{x-2}$

94. $\frac{x^2-x-6}{2x^2+9x+10} \div \frac{x^2-25}{2x^2+15x+25}$

95. $\frac{2x^2+8x-42}{x-3} \div \frac{2x^2+14x}{x^2+5x}$

96. $\frac{x^2-2x-35}{3x^2+27x} \div \frac{x^2+7x+10}{6x^2+12x}$

97. $\frac{ab+4a+2b+8}{b^2+4b+16} \div \frac{b^2-16}{b^3-64}$

98. $\frac{r^3-s^3}{r^2-s^2} \div \frac{r^2+rs+s^2}{mr+ms+nr+ns}$

99. $\frac{p^3-p^2q+pq^2}{mp-mq+np-nq} \div \frac{q^3+p^3}{q^2-p^2}$

100. $\frac{s^3-r^3}{r^2+rs+s^2} \div \frac{pr-ps-qr+qs}{q^2-p^2}$

Perform the operations.

101. $\frac{x}{3} \cdot \frac{9}{4} \div \frac{x^2}{6}$

102. $\frac{y^2}{2} \div \frac{4}{y} \cdot \frac{y^2}{8}$

103. $\frac{x^2}{18} \div \frac{x^3}{6} \div \frac{12}{x^2}$

104. $\frac{y^3}{3y} \cdot \frac{3y^2}{4} \div \frac{15}{20}$

105. $\frac{x^2-1}{x^2-9} \cdot \frac{x+3}{x+2} \div \frac{5}{x+2}$

106. $\frac{2}{3x-3} \div \frac{2x+2}{x-1} \cdot \frac{5}{x+1}$

107. $\frac{x^2-4}{2x+6} \div \frac{x+2}{4} \cdot \frac{x+3}{x-2}$

108. $\frac{x^2-5x}{x+1} \cdot \frac{x+1}{x^2+3x} \div \frac{x-5}{x-3}$

109. $\frac{x-x^2}{x^2-4}\left(\frac{2x+4}{x+2} \div \frac{5}{x+2}\right)$

110. $\frac{2}{3x-3} \div \left(\frac{2x+2}{x-1} \cdot \frac{5}{x+1}\right)$

111. $\frac{y^2}{x+1} \cdot \frac{x^2+2x+1}{x^2-1} \div \frac{3y}{xy-y}$

112. $\dfrac{x^2 - y^2}{x^4 - x^3} \div \dfrac{x - y}{x^2} \div \dfrac{x^2 + 2xy + y^2}{x + y}$

113. $\dfrac{x^2 + x - 6}{x^2 - 4} \cdot \dfrac{x^2 + 2x}{x - 2} \div \dfrac{x^2 + 3x}{x + 2}$

114. $\dfrac{x^2 - x - 6}{x^2 + 6x - 7} \cdot \dfrac{x^2 + x - 2}{x^2 + 2x} \div \dfrac{x^2 + 7x}{x^2 - 3x}$

WRITING

115. Explain how to multiply two fractions and how to simplify the result.

116. Explain why any mathematical expression can be written as a fraction.

117. To divide fractions, you must first know how to multiply fractions. Explain.

118. Explain how to do the division $\dfrac{a}{b} \div \dfrac{c}{d} \div \dfrac{e}{f}$.

SOMETHING TO THINK ABOUT

119. Let x equal a number of your choosing. Without simplifying first, use a calculator to evaluate

$$\frac{x^2 + x - 6}{x^2 + 3x} \cdot \frac{x^2}{x - 2}$$

Try again, with a different value of x. If you were to simplify the expression, what do you think you would get?

120. Simplify the expression in Exercise 119 to determine whether your answer was correct.

6.5 Adding and Subtracting Rational Expressions

- Adding Rational Expressions with Like Denominators
- Subtracting Rational Expressions with Like Denominators
- Combined Operations
- The LCD
- Adding Rational Expressions with Unlike Denominators
- Subtracting Rational Expressions with Unlike Denominators
- Combined Operations

Getting Ready *Add or subtract the fractions and simplify.*

1. $\dfrac{1}{5} + \dfrac{3}{5}$ **2.** $\dfrac{3}{7} + \dfrac{4}{7}$ **3.** $\dfrac{3}{8} + \dfrac{4}{8}$ **4.** $\dfrac{18}{19} + \dfrac{20}{19}$

5. $\dfrac{5}{9} - \dfrac{4}{9}$ **6.** $\dfrac{7}{12} - \dfrac{1}{12}$ **7.** $\dfrac{7}{13} - \dfrac{9}{13}$ **8.** $\dfrac{20}{10} - \dfrac{7}{10}$

We now discuss how to add and subtract rational expressions.

Adding Rational Expressions with Like Denominators

To add rational expressions with a common denominator, we add their numerators and keep the common denominator. For example,

$$\frac{2x}{7} + \frac{3x}{7} = \frac{2x + 3x}{7} \quad \text{Add the numerators and keep the common denominator.}$$

$$= \frac{5x}{7} \quad 2x + 3x = 5x.$$

In general, we have the following result.

Adding Rational Expressions with Like Denominators

If a, b, and d represent real numbers, then

$$\frac{a}{d} + \frac{b}{d} = \frac{a + b}{d} \quad (d \neq 0)$$

EXAMPLE 1 Perform each addition.

a. $\frac{xy}{8z} + \frac{3xy}{8z} = \frac{xy + 3xy}{8z}$ Add the numerators and keep the common denominator.

$= \frac{4xy}{8z}$ Combine like terms.

$= \frac{xy}{2z}$ $\frac{4xy}{8z} = \frac{4xy}{4 \cdot 2z} = \frac{xy}{2z}$, because $\frac{4}{4} = 1$.

b. $\frac{3x + y}{5x} + \frac{x + y}{5x} = \frac{3x + y + x + y}{5x}$ Add the numerators and keep the common denominator.

$= \frac{4x + 2y}{5x}$ Combine like terms.

Self Check Add: **a.** $\frac{x}{7} + \frac{y}{7}$, and **b.** $\frac{3x}{7y} + \frac{4x}{7y}$.

EXAMPLE 2 Add: $\frac{3x + 21}{5x + 10} + \frac{8x + 1}{5x + 10}$.

Solution Since the rational expressions have the same denominator, we add their numerators and keep the common denominator.

$\frac{3x + 21}{5x + 10} + \frac{8x + 1}{5x + 10} = \frac{3x + 21 + 8x + 1}{5x + 10}$ Add the fractions.

$= \frac{11x + 22}{5x + 10}$ Combine like terms.

$= \frac{11\overset{1}{\cancel{(x + 2)}}}{5\underset{1}{\cancel{(x + 2)}}}$ Factor and divide out the common factor of $x + 2$.

$= \frac{11}{5}$

Self Check Add: $\frac{x + 4}{6x - 12} + \frac{x - 8}{6x - 12}$.

Subtracting Rational Expressions with Like Denominators

To subtract rational expressions with a common denominator, we subtract their numerators and keep the common denominator.

Subtracting Rational Expressions with Like Denominators

If a, b, and d represent real numbers, then

$$\frac{a}{d} - \frac{b}{d} = \frac{a-b}{d} \quad (d \neq 0)$$

EXAMPLE 3 Subtract: **a.** $\frac{5x}{3} - \frac{2x}{3}$ and **b.** $\frac{5x+1}{x-3} - \frac{4x-2}{x-3}$.

Solution In each part, the rational expressions have the same denominator. To subtract them, we subtract their numerators and keep the common denominator.

a.
$$\frac{5x}{3} - \frac{2x}{3} = \frac{5x-2x}{3} \quad \text{Subtract the numerators and keep the common denominator.}$$
$$= \frac{3x}{3} \quad \text{Combine like terms.}$$
$$= \frac{x}{1} \quad \tfrac{3}{3} = 1.$$
$$= x \quad \text{Denominators of 1 need not be written.}$$

b.
$$\frac{5x+1}{x-3} - \frac{4x-2}{x-3} = \frac{(5x+1)-(4x-2)}{x-3} \quad \text{Subtract the numerators and keep the common denominator.}$$
$$= \frac{5x+1-4x+2}{x-3} \quad \text{Remove parentheses.}$$
$$= \frac{x+3}{x-3} \quad \text{Combine like terms.}$$

Self Check Subtract: $\frac{2y+1}{y+5} - \frac{y-4}{y+5}$.

Combined Operations

To add and/or subtract three or more rational expressions, we follow the rules for order of operations.

EXAMPLE 4 Simplify: $\frac{3x+1}{x-7} - \frac{5x+2}{x-7} + \frac{2x+1}{x-7}$.

Solution This example involves both addition and subtraction of rational expressions. Unless parentheses indicate otherwise, we do additions and subtractions from left to right.

$$\frac{3x+1}{x-7} - \frac{5x+2}{x-7} + \frac{2x+1}{x-7}$$

$$= \frac{(3x+1) - (5x+2) + (2x+1)}{x-7}$$ Combine the numerators and keep the common denominator.

$$= \frac{3x+1-5x-2+2x+1}{x-7}$$ Remove parentheses.

$$= \frac{0}{x-7}$$ Combine like terms.

$$= 0$$ Simplify.

Self Check Simplify: $\frac{2a-3}{a-5} + \frac{3a+2}{a-5} - \frac{24}{a-5}$.

Example 4 illustrates that if the numerator of a rational expression is 0 and the denominator is not, the value of the expression is 0.

The LCD

Since the denominators of the fractions in the addition $\frac{4}{7} + \frac{3}{5}$ are different, we cannot add the fractions in their present form.

four-sevenths + three-fifths

Different denominators

To add these fractions, we need to find a common denominator. The smallest common denominator (called the **least** or **lowest common denominator**) is the easiest one to use.

Least Common Denominator

The **least common denominator (LCD)** for a set of fractions is the smallest number that each denominator will divide exactly.

In the addition $\frac{4}{7} + \frac{3}{5}$, the denominators are 7 and 5. The smallest number that 7 and 5 will divide exactly is 35. This is the LCD. We now build each fraction into a fraction with a denominator of 35.

$$\frac{4}{7} + \frac{3}{5} = \frac{4 \cdot 5}{7 \cdot 5} + \frac{3 \cdot 7}{5 \cdot 7}$$ Multiply numerator and denominator of $\frac{4}{7}$ by 5, and multiply numerator and denominator of $\frac{3}{5}$ by 7.

$$= \frac{20}{35} + \frac{21}{35}$$ Do the multiplications.

Now that the fractions have a common denominator, we can add them.

$$\frac{20}{35} + \frac{21}{35} = \frac{20+21}{35} = \frac{41}{35}$$

EXAMPLE 5 Change **a.** $\frac{1}{2y}$, **b.** $\frac{3y}{5}$, and **c.** $\frac{7x}{10y}$ into rational expressions with a denominator of $30y$.

Solution To build each rational expression, we multiply the numerator and denominator by what it takes to make the denominator $30y$.

a. $\frac{1}{2y} = \frac{1 \cdot 15}{2y \cdot 15} = \frac{15}{30y}$

b. $\frac{3y}{5} = \frac{3y \cdot 6y}{5 \cdot 6y} = \frac{18y^2}{30y}$

c. $\frac{7x}{10y} = \frac{7x \cdot 3}{10y \cdot 3} = \frac{21x}{30y}$

Self Check Change $\frac{5a}{6b}$ into a rational expression with a denominator of $30ab$.

There is a process that we can use to find the least common denominator of several rational expressions.

Finding the Least Common Denominator (LCD)

1. List the different denominators that appear in the rational expressions.
2. Completely factor each denominator.
3. Form a product using each different factor obtained in Step 2. Use each different factor the *greatest* number of times it appears in any one factorization. The product formed by multiplying these factors is the LCD.

EXAMPLE 6 Find the LCD of $\frac{5a}{24b}$, $\frac{11a}{18b}$, and $\frac{35a}{36b}$.

Solution We list and factor each denominator into the product of prime numbers.

$$24b = 2 \cdot 2 \cdot 2 \cdot 3 \cdot b = 2^3 \cdot 3 \cdot b$$
$$18b = 2 \cdot 3 \cdot 3 \cdot b = 2 \cdot 3^2 \cdot b$$
$$36b = 2 \cdot 2 \cdot 3 \cdot 3 \cdot b = 2^2 \cdot 3^2 \cdot b$$

We then form a product with factors of 2, 3, and b. To find the LCD, we use each of these factors the greatest number of times it appears in any one factorization. We use 2 three times, because it appears three times as a factor of 24. We use 3 twice, because it occurs twice as a factor of 18 and 36. We use b once.

$$\begin{aligned} \text{LCD} &= 2 \cdot 2 \cdot 2 \cdot 3 \cdot 3 \cdot b \\ &= 8 \cdot 9 \cdot b \\ &= 72b \end{aligned}$$

Self Check Find the LCD of $\frac{3y}{28z}$ and $\frac{5x}{21z}$.

Adding Rational Expressions with Unlike Denominators

The following steps summarize how to add rational expressions that have unlike denominators.

Adding Rational Expressions with Unlike Denominators

To add rational expressions with unlike denominators,

1. Find the LCD.
2. Write each fraction as a fraction with a denominator that is the LCD.
3. Add the resulting fractions and simplify the result, if possible.

To add $\frac{4x}{7}$ and $\frac{3x}{5}$, we first find the LCD, which is 35. We then build the rational expressions so that each one has a denominator of 35. Finally, we add the results.

$$\frac{4x}{7} + \frac{3x}{5} = \frac{4x \cdot 5}{7 \cdot 5} + \frac{3x \cdot 7}{5 \cdot 7}$$ Multiply numerator and denominator of $\frac{4x}{7}$ by 5 and numerator and denominator of $\frac{3x}{5}$ by 7.

$$= \frac{20x}{35} + \frac{21x}{35}$$ Do the multiplications.

$$= \frac{41x}{35}$$ Add the numerators and keep the common denominator.

EXAMPLE 7 Add: $\frac{5a}{24b}$, $\frac{11a}{18b}$, and $\frac{35a}{36b}$.

Solution In Example 6, we saw that the LCD of these rational expressions is $2 \cdot 2 \cdot 2 \cdot 3 \cdot 3 \cdot b = 72b$. To add the rational expressions, we first factor each denominator:

$$\frac{5a}{24b} + \frac{11a}{18b} + \frac{35a}{36b} = \frac{5a}{2 \cdot 2 \cdot 2 \cdot 3 \cdot b} + \frac{11a}{2 \cdot 3 \cdot 3 \cdot b} + \frac{35a}{2 \cdot 2 \cdot 3 \cdot 3 \cdot b}$$

In each resulting expression, we multiply the numerator and the denominator by whatever it takes to build the denominator to the lowest common denominator of $2 \cdot 2 \cdot 2 \cdot 3 \cdot 3 \cdot b$.

$$= \frac{5a \cdot 3}{2 \cdot 2 \cdot 2 \cdot 3 \cdot b \cdot 3} + \frac{11a \cdot 2 \cdot 2}{2 \cdot 3 \cdot 3 \cdot b \cdot 2 \cdot 2} + \frac{35a \cdot 2}{2 \cdot 2 \cdot 3 \cdot 3 \cdot b \cdot 2}$$

$$= \frac{15a + 44a + 70a}{72b}$$ Do the multiplications.

$$= \frac{129a}{72b}$$ Add the fractions.

$$= \frac{43a}{24b}$$ Simplify.

Self Check Add: $\frac{3y}{28z} + \frac{5x}{21z}$.

EXAMPLE 8 Add: $\frac{5y}{14x} + \frac{2y}{21x}$.

Solution We first find the LCD.

$$\left.\begin{aligned} 14x &= 2 \cdot 7 \cdot x \\ 21x &= 3 \cdot 7 \cdot x \end{aligned}\right\} \text{LCD} = 2 \cdot 3 \cdot 7 \cdot x = 42x$$

We then build the rational expressions so that each one has a denominator of $42x$.

$$\frac{5y}{14x} + \frac{2y}{21x} = \frac{5y \cdot 3}{14x \cdot 3} + \frac{2y \cdot 2}{21x \cdot 2}$$ Multiply the numerator and denominator of $\frac{5y}{14x}$ by 3 and those of $\frac{2y}{21x}$ by 2.

$$= \frac{15y}{42x} + \frac{4y}{42x}$$ Do the multiplications.

$$= \frac{19y}{42x}$$ Add the fractions.

Self Check Add: $\frac{3y}{4x} + \frac{2y}{3x}$.

EXAMPLE 9 Add: $\frac{1}{x} + \frac{x}{y}$.

Solution By inspection, the LCD is xy.

$$\frac{1}{x} + \frac{x}{y} = \frac{1(y)}{x(y)} + \frac{(x)x}{(x)y}$$ Build the fractions to get the common denominator of xy.

$$= \frac{y}{xy} + \frac{x^2}{xy}$$ Do the multiplications.

$$= \frac{y + x^2}{xy}$$ Add the fractions.

Self Check Add: $\frac{a}{b} + \frac{3}{a}$.

Subtracting Rational Expressions with Unlike Denominators

To subtract rational expressions with unlike denominators, we first change them into expressions with the same denominator.

EXAMPLE 10 Subtract: $\frac{x}{x+1} - \frac{3}{x}$.

Solution By inspection, the least common denominator is $(x + 1)x$.

$$\frac{x}{x+1} - \frac{3}{x} = \frac{x(x)}{(x+1)x} - \frac{3(x+1)}{x(x+1)}$$ Build the fractions to get the common denominator.

$$= \frac{x(x) - 3(x+1)}{(x+1)x}$$ Subtract the numerators and keep the common denominator.

$$= \frac{x^2 - 3x - 3}{(x+1)x}$$ Do the multiplications in the numerator.

Self Check Subtract: $\frac{a}{a-1} - \frac{5}{b}$.

EXAMPLE 11 Subtract: $\dfrac{a}{a-1} - \dfrac{2}{a^2-1}$.

Solution We factor $a^2 - 1$ and discover that the LCD is $(a + 1)(a - 1)$.

$$\frac{a}{a-1} - \frac{2}{a^2-1}$$

$$= \frac{a(a+1)}{(a-1)(a+1)} - \frac{2}{(a+1)(a-1)}$$ Build the first fraction and factor the denominator of the second fraction.

$$= \frac{a(a+1)-2}{(a-1)(a+1)}$$ Subtract the numerators and keep the common denominator.

$$= \frac{a^2+a-2}{(a-1)(a+1)}$$ Remove parentheses.

$$= \frac{(a+2)\overset{1}{\cancel{(a-1)}}}{\underset{1}{\cancel{(a-1)}}(a+1)}$$ Factor and divide out the common factor of $a - 1$.

$$= \frac{a+2}{a+1}$$ Simplify.

Self Check Subtract: $\dfrac{b}{b+1} - \dfrac{3}{b^2-1}$. ▮

EXAMPLE 12 Subtract: $\dfrac{3}{x-y} - \dfrac{x}{y-x}$.

Solution We note that the second denominator is the negative of the first. So we can multiply the numerator and denominator of the second fraction by -1 to get

$$\frac{3}{x-y} - \frac{x}{y-x} = \frac{3}{x-y} - \frac{-1x}{-1(y-x)}$$ Multiply numerator and denominator by -1.

$$= \frac{3}{x-y} - \frac{-x}{-y+x}$$ Remove parentheses.

$$= \frac{3}{x-y} - \frac{-x}{x-y}$$ $-y + x = x - y$.

$$= \frac{3-(-x)}{x-y}$$ Subtract the numerators and keep the common denominator.

$$= \frac{3+x}{x-y}$$ $-(-x) = x$.

Self Check Subtract: $\dfrac{5}{a-b} - \dfrac{2}{b-a}$. ▮

Combined Operations

To add and/or subtract three or more rational expressions, we follow the rules for the order of operations.

EXAMPLE 13 Perform the operations: $\frac{3}{x^2y} + \frac{2}{xy} - \frac{1}{xy^2}$.

Solution Find the least common denominator.

$$\left.\begin{aligned} x^2y &= x \cdot x \cdot y \\ xy &= x \cdot y \\ xy^2 &= x \cdot y \cdot y \end{aligned}\right\} \quad \text{Factor each denominator.}$$

In any one of these denominators, the factor x occurs at most twice, and the factor y occurs at most twice. Thus,

$$\begin{aligned} \text{LCD} &= x \cdot x \cdot y \cdot y \\ &= x^2y^2 \end{aligned}$$

We build each rational expression into an expression with a denominator of x^2y^2.

$$\frac{3}{x^2y} + \frac{2}{xy} - \frac{1}{xy^2}$$

$$= \frac{3 \cdot y}{x \cdot x \cdot y \cdot y} + \frac{2 \cdot x \cdot y}{x \cdot y \cdot x \cdot y} - \frac{1 \cdot x}{x \cdot y \cdot y \cdot x}$$ Factor each denominator and build each fraction.

$$= \frac{3y + 2xy - x}{x^2y^2}$$ Do the multiplications and combine the numerators.

Self Check Combine: $\frac{5}{ab^2} - \frac{b}{a} + \frac{a}{b}$.

EXAMPLE 14 Perform the operations: $\frac{3}{x^2 - y^2} + \frac{2}{x - y} - \frac{1}{x + y}$.

Solution Find the least common denominator.

$$\left.\begin{aligned} x^2 - y^2 &= (x - y)(x + y) \\ x - y &= x - y \\ x + y &= x + y \end{aligned}\right\} \quad \text{Factor each denominator, where possible.}$$

Since the least common denominator is $(x - y)(x + y)$, we build each fraction into a new fraction with that common denominator.

$$\frac{3}{x^2 - y^2} + \frac{2}{x - y} - \frac{1}{x + y}$$

$$= \frac{3}{(x - y)(x + y)} + \frac{2}{x - y} - \frac{1}{x + y}$$ Factor.

$$= \frac{3}{(x - y)(x + y)} + \frac{2(x + y)}{(x - y)(x + y)} - \frac{1(x - y)}{(x + y)(x - y)}$$ Build each fraction to get a common denominator.

$$= \frac{3 + 2(x + y) - 1(x - y)}{(x - y)(x + y)}$$ Combine the numerators and keep the common denominator.

$$= \frac{3 + 2x + 2y - x + y}{(x - y)(x + y)}$$ Remove parentheses.

$$= \frac{3 + x + 3y}{(x - y)(x + y)}$$ Combine like terms.

Self Check Combine: $\frac{5}{a^2 - b^2} - \frac{3}{a + b} + \frac{4}{a - b}$

Self Check Answers

1. a. $\frac{x + y}{7}$, b. $\frac{x}{y}$ 2. $\frac{1}{3}$ 3. 1 4. 5 5. $\frac{25a^2}{30ab}$ 6. $84z$ 7. $\frac{9y + 20x}{84z}$ 8. $\frac{17y}{12x}$ 9. $\frac{a^2 + 3b}{ab}$ 10. $\frac{ab - 5a + 5}{(a - 1)b}$ 11. $\frac{b^2 - b - 3}{(b + 1)(b - 1)}$ 12. $\frac{7}{a - b}$ 13. $\frac{5 - b^3 + a^2b}{ab^2}$ 14. $\frac{a + 7b + 5}{(a + b)(a - b)}$

Orals *Indicate whether the expressions are equal.*

1. $\frac{1}{2}, \frac{6}{12}$
2. $\frac{3}{8}, \frac{15}{40}$
3. $\frac{7}{9}, \frac{14}{27}$
4. $\frac{5}{10}, \frac{15}{30}$
5. $\frac{x}{3}, \frac{3x}{9}$
6. $\frac{5}{3}, \frac{5x}{3y}$
7. $\frac{5}{3}, \frac{5x}{3x}$
8. $\frac{5y}{10}, \frac{y}{2}$

6.5 EXERCISES

REVIEW *Write each number in prime-factored form.*

1. 49
2. 64
3. 136
4. 242
5. 102
6. 315
7. 144
8. 145

VOCABULARY AND CONCEPTS *Fill in the blanks.*

9. The _____ for a set of fractions is the smallest number that each denominator divides exactly.
10. When we multiply the numerator and denominator of a fraction by some number to get a common denominator, we say that we are _______ the fraction.
11. To add two fractions with like denominators, we add their _________ and keep the _______________.
12. To subtract two fractions with ______ denominators, we need to find a common denominator.

PRACTICE *Perform the operations. Simplify answers, if possible.*

13. $\frac{1}{3} + \frac{1}{3}$
14. $\frac{3}{4} + \frac{3}{4}$
15. $\frac{2}{9} + \frac{1}{9}$
16. $\frac{5}{7} + \frac{9}{7}$
17. $\frac{2x}{y} + \frac{2x}{y}$
18. $\frac{4y}{3x} + \frac{2y}{3x}$
19. $\frac{4}{7y} + \frac{10}{7y}$
20. $\frac{x^2}{4y} + \frac{x^2}{4y}$
21. $\frac{y + 2}{5z} + \frac{y + 4}{5z}$
22. $\frac{x + 3}{x^2} + \frac{x + 5}{x^2}$
23. $\frac{3x - 5}{x - 2} + \frac{6x - 13}{x - 2}$
24. $\frac{8x - 7}{x + 3} + \frac{2x + 37}{x + 3}$
25. $\frac{5}{7} - \frac{4}{7}$
26. $\frac{5}{9} - \frac{3}{9}$
27. $\frac{35}{72} - \frac{44}{72}$
28. $\frac{35}{99} - \frac{13}{99}$
29. $\frac{2x}{y} - \frac{x}{y}$
30. $\frac{7y}{5} - \frac{4y}{5}$

31. $\frac{9y}{3x} - \frac{6y}{3x}$

32. $\frac{5r^2}{2r} - \frac{r^2}{2r}$

33. $\frac{6x - 5}{3xy} - \frac{3x - 5}{3xy}$

34. $\frac{7x + 7}{5y} - \frac{2x + 7}{5y}$

35. $\frac{3y - 2}{y + 3} - \frac{2y - 5}{y + 3}$

36. $\frac{5x + 8}{x + 5} - \frac{3x - 2}{x + 5}$

37. $\frac{13x}{15} + \frac{12x}{15} - \frac{5x}{15}$

38. $\frac{13y}{32} + \frac{13y}{32} - \frac{10y}{32}$

39. $\frac{x}{3y} + \frac{2x}{3y} - \frac{x}{3y}$

40. $\frac{5y}{8x} + \frac{4y}{8x} - \frac{y}{8x}$

41. $\frac{3x}{y + 2} - \frac{3y}{y + 2} + \frac{x + y}{y + 2}$

42. $\frac{3y}{x - 5} + \frac{x}{x - 5} - \frac{y - x}{x - 5}$

43. $\frac{x + 1}{x - 2} - \frac{2(x - 3)}{x - 2} + \frac{3(x + 1)}{x - 2}$

44. $\frac{3xy}{x - y} - \frac{x(3y - x)}{x - y} - \frac{x(x - y)}{x - y}$

Build each fraction into an equivalent fraction with the indicated denominator.

45. $\frac{25}{4}$; 20

46. $\frac{5}{y}$; xy

47. $\frac{8}{x}$; x^2y

48. $\frac{7}{y}$; xy^2

49. $\frac{3x}{x + 1}$; $(x + 1)^2$

50. $\frac{5y}{y - 2}$; $(y - 2)^2$

51. $\frac{2y}{x}$; $x^2 + x$

52. $\frac{3x}{y}$; $y^2 - y$

53. $\frac{z}{z - 1}$; $z^2 - 1$

54. $\frac{y}{y + 2}$; $y^2 - 4$

55. $\frac{2}{x + 1}$; $x^2 + 3x + 2$

56. $\frac{3}{x - 1}$; $x^2 + x - 2$

Several denominators are given. Find the LCD.

57. $2x, 6x$

58. $3y, 9y$

59. $3x, 6y, 9xy$

60. $2x^2, 6y, 3xy$

61. $x^2 - 1, x + 1$

62. $y^2 - 9, y - 3$

63. $x^2 + 6x, x + 6, x$

64. $xy^2 - xy, xy, y - 1$

65. $x^2 - 4x - 5, x^2 - 25$

66. $x^2 - x - 6, x^2 - 9$

Perform the operations. Simplify answers, if possible.

67. $\frac{1}{2} + \frac{2}{3}$

68. $\frac{2}{3} - \frac{5}{6}$

69. $\frac{2y}{9} + \frac{y}{3}$

70. $\frac{8a}{15} - \frac{5a}{12}$

71. $\frac{21x}{14} - \frac{5x}{21}$

72. $\frac{7y}{6} + \frac{10y}{9}$

73. $\frac{4x}{3} + \frac{2x}{y}$

74. $\frac{2y}{5x} - \frac{y}{2}$

75. $\frac{2}{x} - 3x$

76. $14 + \frac{10}{y^2}$

77. $\frac{y + 2}{5y} + \frac{y + 4}{15y}$

78. $\frac{x + 3}{x^2} + \frac{x + 5}{2x}$

79. $\frac{x + 5}{xy} - \frac{x - 1}{x^2y}$

80. $\frac{y - 7}{y^2} - \frac{y + 7}{2y}$

81. $\frac{x}{x + 1} + \frac{x - 1}{x}$

82. $\frac{3x}{xy} + \frac{x + 1}{y - 1}$

83. $\frac{x - 1}{x} + \frac{y + 1}{y}$

84. $\frac{a + 2}{b} + \frac{b - 2}{a}$

85. $\dfrac{x}{x-2} + \dfrac{4+2x}{x^2-4}$

86. $\dfrac{y}{y+3} - \dfrac{2y-6}{y^2-9}$

87. $\dfrac{x+1}{x-1} + \dfrac{x-1}{x+1}$

88. $\dfrac{2x}{x+2} + \dfrac{x+1}{x-3}$

89. $\dfrac{2x+2}{x-2} - \dfrac{2x}{2-x}$

90. $\dfrac{y+3}{y-1} - \dfrac{y+4}{1-y}$

91. $\dfrac{2x}{x^2-3x+2} + \dfrac{2x}{x-1} - \dfrac{x}{x-2}$

92. $\dfrac{4a}{a-2} - \dfrac{3a}{a-3} + \dfrac{4a}{a^2-5a+6}$

93. $\dfrac{2x}{x-1} + \dfrac{3x}{x+1} - \dfrac{x+3}{x^2-1}$

94. $\dfrac{a}{a-1} - \dfrac{2}{a+2} + \dfrac{3(a-2)}{a^2+a-2}$

95. $\dfrac{x+1}{2x+4} - \dfrac{x^2}{2x^2-8}$

96. $\dfrac{x+1}{x+2} - \dfrac{x^2+1}{x^2-x-6}$

WRITING

97. Explain how to add rational expressions with the same denominator.

98. Explain how to subtract rational expressions with the same denominator.

99. Explain how to find a lowest common denominator.

100. Explain how to add two rational expressions with different denominators.

SOMETHING TO THINK ABOUT

101. Find the error:

$$\frac{2x+3}{x+5} - \frac{x+2}{x+5} = \frac{2x+3-x+2}{x+5} = \frac{x+5}{x+5} = 1$$

102. Find the error:

$$\frac{5x-4}{y} + \frac{x}{y} = \frac{5x-4+x}{y+y} = \frac{6x-4}{2y} = \frac{3x-2}{y}$$

Show that each formula is true.

103. $\dfrac{a}{b} + \dfrac{c}{d} = \dfrac{ad+bc}{bd}$

104. $\dfrac{a}{b} - \dfrac{c}{d} = \dfrac{ad-bc}{bd}$

6.6 Complex Fractions

- Simplifying Complex Fractions
- Simplifying Fractions with Terms Containing Negative Exponents

Getting Ready *Use the distributive property to remove parentheses, and simplify.*

1. $3\left(1+\dfrac{1}{3}\right)$
2. $10\left(\dfrac{1}{5}-2\right)$
3. $4\left(\dfrac{3}{2}+\dfrac{1}{4}\right)$
4. $14\left(\dfrac{3}{7}-1\right)$
5. $x\left(\dfrac{3}{x}+3\right)$
6. $y\left(\dfrac{2}{y}-1\right)$
7. $4x\left(3-\dfrac{1}{2x}\right)$
8. $6xy\left(\dfrac{1}{2x}+\dfrac{1}{3y}\right)$

In this section, we consider fractions that contain fractions. These complicated fractions are called *complex fractions.*

Simplifying Complex Fractions

Fractions such as

$$\frac{\frac{1}{3}}{4}, \quad \frac{\frac{5}{3}}{\frac{2}{9}}, \quad \frac{x + \frac{1}{2}}{3 - x}, \quad \text{and} \quad \frac{\frac{x+1}{2}}{x + \frac{1}{x}}$$

that contain fractions in their numerators or denominators are called **complex fractions.** Complex fractions can often be simplified. For example, we can simplify

$$\frac{\frac{5x}{3}}{\frac{2y}{9}}$$

by doing the division:

$$\frac{\frac{5x}{3}}{\frac{2y}{9}} = \frac{5x}{3} \div \frac{2y}{9} = \frac{5x}{3} \cdot \frac{9}{2y} = \frac{5x \cdot 3 \cdot \overset{1}{\cancel{3}}}{\underset{1}{\cancel{3}} \cdot 2y} = \frac{15x}{2y}$$

There are two methods that we can use to simplify complex fractions.

Simplifying Complex Fractions

> ***Method 1***
>
> Write the numerator and the denominator of the complex fraction as single fractions. Then divide the fractions and simplify.
>
> ***Method 2***
>
> Multiply the numerator and denominator of the complex fraction by the LCD of the fractions in its numerator and denominator. Then simplify the results, if possible.

To simplify $\frac{\frac{3x}{5} + 1}{2 - \frac{x}{5}}$ using Method 1, we proceed as follows:

$$\frac{\frac{3x}{5} + 1}{2 - \frac{x}{5}} = \frac{\frac{3x}{5} + \frac{5}{5}}{\frac{10}{5} - \frac{x}{5}}$$

Change 1 to $\frac{5}{5}$ and 2 to $\frac{10}{5}$.

$$= \frac{\frac{3x + 5}{5}}{\frac{10 - x}{5}}$$

Add the fractions in the numerator and subtract the fractions in the denominator.

$$= \frac{3x+5}{5} \div \frac{10-x}{5}$$ Write the complex fraction as an equivalent division problem.

$$= \frac{3x+5}{5} \cdot \frac{5}{10-x}$$ Invert the divisor and multiply.

$$= \frac{(3x+5)5}{5(10-x)}$$ Multiply the fractions.

$$= \frac{3x+5}{10-x}$$ Divide out the common factor of 5: $\frac{5}{5} = 1$.

To use Method 2, we proceed as follows:

$$\frac{\frac{3x}{5}+1}{2-\frac{x}{5}} = \frac{5\left(\frac{3x}{5}+1\right)}{5\left(2-\frac{x}{5}\right)}$$ Multiply both numerator and denominator by 5, the LCD of $\frac{3x}{5}$ and $\frac{x}{5}$.

$$= \frac{5 \cdot \frac{3x}{5} + 5 \cdot 1}{5 \cdot 2 - 5 \cdot \frac{x}{5}}$$ Remove parentheses.

$$= \frac{3x+5}{10-x}$$ Do the multiplications.

In this example, Method 2 is easier than Method 1. With practice, you will be able to see which method is best to use in any given situation.

EXAMPLE 1 Simplify: $\frac{\frac{x}{3}}{\frac{y}{3}}$.

Solution

Method 1

$$\frac{\frac{x}{3}}{\frac{y}{3}} = \frac{x}{3} \div \frac{y}{3}$$

$$= \frac{x}{3} \cdot \frac{3}{y}$$

$$= \frac{3x}{3y}$$

$$= \frac{x}{y}$$

Method 2

$$\frac{\frac{x}{3}}{\frac{y}{3}} = \frac{3\left(\frac{x}{3}\right)}{3\left(\frac{y}{3}\right)}$$

$$= \frac{\frac{x}{1}}{\frac{y}{1}}$$

$$= \frac{x}{y}$$

Self Check Simplify: $\frac{\frac{a}{4}}{\frac{5}{b}}$.

EXAMPLE 2 Simplify: $\dfrac{\dfrac{x}{x+1}}{\dfrac{y}{x}}$.

Solution

Method 1

$$\frac{\dfrac{x}{x+1}}{\dfrac{y}{x}} = \frac{x}{x+1} \div \frac{y}{x}$$

$$= \frac{x}{x+1} \cdot \frac{x}{y}$$

$$= \frac{x^2}{y(x+1)}$$

Method 2

$$\frac{\dfrac{x}{x+1}}{\dfrac{y}{x}} = \frac{x(x+1)\left(\dfrac{x}{x+1}\right)}{x(x+1)\left(\dfrac{y}{x}\right)}$$

$$= \frac{\dfrac{x^2}{1}}{\dfrac{y(x+1)}{1}}$$

$$= \frac{x^2}{y(x+1)}$$

Self Check Simplify: $\dfrac{\dfrac{x}{y}}{\dfrac{x}{y+1}}$.

EXAMPLE 3 Simplify: $\dfrac{1+\dfrac{1}{x}}{1-\dfrac{1}{x}}$.

Solution

Method 1

$$\frac{1+\dfrac{1}{x}}{1-\dfrac{1}{x}} = \frac{\dfrac{x}{x}+\dfrac{1}{x}}{\dfrac{x}{x}-\dfrac{1}{x}}$$

$$= \frac{\dfrac{x+1}{x}}{\dfrac{x-1}{x}}$$

$$= \frac{x+1}{x} \div \frac{x-1}{x}$$

$$= \frac{x+1}{x} \cdot \frac{x}{x-1}$$

$$= \frac{(x+1)x}{x(x-1)}$$

$$= \frac{x+1}{x-1}$$

Method 2

$$\frac{1+\dfrac{1}{x}}{1-\dfrac{1}{x}} = \frac{x\left(1+\dfrac{1}{x}\right)}{x\left(1-\dfrac{1}{x}\right)}$$

$$= \frac{x+1}{x-1}$$

Self Check Simplify: $\dfrac{\frac{1}{x}+1}{\frac{1}{x}-1}$.

EXAMPLE 4 Simplify: $\dfrac{1}{1+\frac{1}{x+1}}$.

Solution We use Method 2.

$$\frac{1}{1+\frac{1}{x+1}} = \frac{(x+1)\cdot 1}{(x+1)\left(1+\frac{1}{x+1}\right)}$$ Multiply numerator and denominator by $x+1$.

$$= \frac{x+1}{(x+1)1+1}$$ Simplify.

$$= \frac{x+1}{x+2}$$ Simplify.

Self Check Simplify: $\dfrac{2}{\frac{1}{x+2}-2}$.

Simplifying Fractions with Terms Containing Negative Exponents

Many fractions with terms containing negative exponents are complex fractions in disguise.

EXAMPLE 5 Simplify: $\dfrac{x^{-1}+y^{-2}}{x^{-2}-y^{-1}}$.

Solution Write the fraction as a complex fraction and simplify:

$$\frac{x^{-1}+y^{-2}}{x^{-2}-y^{-1}} = \frac{\frac{1}{x}+\frac{1}{y^2}}{\frac{1}{x^2}-\frac{1}{y}}$$

$$= \frac{x^2y^2\left(\frac{1}{x}+\frac{1}{y^2}\right)}{x^2y^2\left(\frac{1}{x^2}-\frac{1}{y}\right)}$$ Multiply numerator and denominator by x^2y^2.

$$= \frac{xy^2+x^2}{y^2-x^2y}$$ Remove parentheses.

$$= \frac{x(y^2+x)}{y(y-x^2)}$$ Attempt to simplify the fraction by factoring the numerator and denominator.

The result cannot be simplified.

Self Check Simplify: $\dfrac{x^{-2} - y^{-1}}{x^{-1} + y^{-2}}$.

Self Check Answers

1. $\dfrac{ab}{20}$ **2.** $\dfrac{y + 1}{y}$ **3.** $\dfrac{1 + x}{1 - x}$ **4.** $\dfrac{2(x + 2)}{-2x - 3}$ **5.** $\dfrac{y(y - x^2)}{x(y^2 + x)}$

Orals *Simplify each complex fraction.*

1. $\dfrac{\frac{2}{3}}{\frac{1}{2}}$ **2.** $\dfrac{2}{\frac{1}{2}}$ **3.** $\dfrac{\frac{1}{2}}{2}$ **4.** $\dfrac{1 + \frac{1}{2}}{\frac{1}{2}}$

6.6 EXERCISES

REVIEW *Write each expression as an expression involving only one exponent.*

1. $t^3t^4t^2$ **2.** $(a^0a^2)^3$

3. $-2r(r^3)^2$ **4.** $(s^3)^2(s^4)^0$

Write each expression without parentheses or negative exponents.

5. $\left(\dfrac{3r}{4r^3}\right)^4$ **6.** $\left(\dfrac{12y^{-3}}{3y^2}\right)^{-2}$

7. $\left(\dfrac{6r^{-2}}{2r^3}\right)^{-2}$ **8.** $\left(\dfrac{4x^3}{5x^{-3}}\right)^{-2}$

VOCABULARY AND CONCEPTS *Fill in the blanks.*

9. If a fraction has a fraction in its numerator or denominator, it is called a ______________.

10. The denominator of the complex fraction

$$\dfrac{\frac{3}{x} + \frac{x}{y}}{\frac{1}{x} + 2} \text{ is } \_\_\_\_\_.$$

11. In Method 1, we write the numerator and denominator of a complex fraction as _____ fractions and then _____.

12. In Method 2, we multiply the numerator and denominator of the complex fraction by the _____ of the fractions in its numerator and denominator.

PRACTICE *Simplify each complex fraction.*

13. $\dfrac{\frac{2}{3}}{\frac{3}{4}}$

14. $\dfrac{\frac{3}{5}}{\frac{2}{7}}$

15. $\dfrac{\frac{4}{5}}{\frac{32}{15}}$

16. $\dfrac{\frac{7}{8}}{\frac{49}{4}}$

17. $\dfrac{\frac{2}{3} + 1}{\frac{1}{3} + 1}$

18. $\dfrac{\frac{3}{5} - 2}{\frac{2}{5} - 2}$

19. $\dfrac{\frac{1}{2} + \frac{3}{4}}{\frac{3}{2} + \frac{1}{4}}$

20. $\dfrac{\frac{2}{3} - \frac{5}{2}}{\frac{2}{3} - \frac{3}{2}}$

21. $\dfrac{\frac{x}{y}}{\frac{1}{x}}$

22. $\dfrac{\frac{y}{x}}{\frac{x}{xy}}$

23. $\dfrac{\frac{5t^2}{9x^2}}{\frac{3t}{x^2t}}$

24. $\dfrac{\frac{5w^2}{4tz}}{\frac{15wt}{z^2}}$

25. $\dfrac{\frac{1}{x} - 3}{\frac{5}{x} + 2}$

26. $\dfrac{\frac{1}{y} + 3}{\frac{3}{y} - 2}$

27. $\dfrac{\frac{2}{x} + 2}{\frac{4}{x} + 2}$

28. $\dfrac{\frac{3}{x} - 3}{\frac{9}{x} - 3}$

29. $\dfrac{\frac{3y}{x} - y}{y - \frac{y}{x}}$

30. $\dfrac{\frac{y}{x} + 3y}{y + \frac{2y}{x}}$

31. $\dfrac{\frac{1}{x+1}}{1 + \frac{1}{x+1}}$

32. $\dfrac{\frac{1}{x-1}}{1 - \frac{1}{x-1}}$

33. $\dfrac{\frac{x}{x+2}}{\frac{x}{x+2} + x}$

34. $\dfrac{\frac{2}{x-2}}{\frac{2}{x-2} - 1}$

35. $\dfrac{1}{\frac{1}{x} + \frac{1}{y}}$

36. $\dfrac{1}{\frac{b}{a} - \frac{a}{b}}$

37. $\dfrac{\frac{2}{x}}{\frac{2}{y} - \frac{4}{x}}$

38. $\dfrac{\frac{2y}{3}}{\frac{2y}{3} - \frac{8}{y}}$

39. $\dfrac{3 + \frac{3}{x-1}}{3 - \frac{3}{x}}$

40. $\dfrac{2 - \frac{2}{x+1}}{2 + \frac{2}{x}}$

41. $\dfrac{\frac{3}{x} + \frac{4}{x+1}}{\frac{2}{x+1} - \frac{3}{x}}$

42. $\dfrac{\frac{5}{y-3} - \frac{2}{y}}{\frac{1}{y} + \frac{2}{y-3}}$

43. $\dfrac{\frac{2}{x} - \frac{3}{x+1}}{\frac{2}{x+1} - \frac{3}{x}}$

44. $\dfrac{\frac{5}{y} + \frac{4}{y+1}}{\frac{4}{y} - \frac{5}{y+1}}$

45. $\dfrac{\frac{1}{y^2+y} - \frac{1}{xy+x}}{\frac{1}{xy+x} - \frac{1}{y^2+y}}$

46. $\dfrac{\frac{2}{b^2-1} - \frac{3}{ab-a}}{\frac{3}{ab-a} - \frac{2}{b^2-1}}$

47. $\dfrac{x^{-2}}{y^{-1}}$

48. $\dfrac{a^{-4}}{b^{-2}}$

49. $\dfrac{1 + x^{-1}}{x^{-1} - 1}$

50. $\dfrac{y^{-2} + 1}{y^{-2} - 1}$

51. $\dfrac{a^{-2} + a}{a + 1}$

52. $\dfrac{t - t^{-2}}{1 - t^{-1}}$

53. $\dfrac{2x^{-1} + 4x^{-2}}{2x^{-2} + x^{-1}}$

54. $\dfrac{x^{-2} - 3x^{-3}}{3x^{-2} - 9x^{-3}}$

55. $\dfrac{1 - 25y^{-2}}{1 + 10y^{-1} + 25y^{-2}}$

56. $\dfrac{1 - 9x^{-2}}{1 - 6x^{-1} + 9x^{-2}}$

WRITING

57. Explain how to use Method 1 to simplify

$$\dfrac{1 + \frac{1}{x}}{3 - \frac{1}{x}}$$

58. Explain how to use Method 2 to simplify the expression in Exercise 57.

SOMETHING TO THINK ABOUT

59. Simplify each complex fraction:

$$\frac{1}{1+1}, \quad \frac{1}{1+\frac{1}{2}}, \quad \frac{1}{1+\frac{1}{1+\frac{1}{2}}}, \quad \frac{1}{1+\frac{1}{1+\frac{1}{1+\frac{1}{2}}}}$$

60. In Exercise 59, what is the pattern in the numerators and denominators of the four answers? What would be the next answer?

6.7 Solving Equations That Contain Rational Expressions

- Solving Equations That Contain Rational Expressions
- Extraneous Solutions
- Formulas

Getting Ready *Simplify.*

1. $3\left(x + \frac{1}{3}\right)$ **2.** $8\left(x - \frac{1}{8}\right)$ **3.** $x\left(\frac{3}{x} + 2\right)$ **4.** $3y\left(\frac{1}{3} - \frac{2}{y}\right)$

5. $6x\left(\frac{5}{2x} + \frac{2}{3x}\right)$ **6.** $9x\left(\frac{7}{9} + \frac{2}{3x}\right)$

7. $(y - 1)\left(\frac{1}{y - 1} + 1\right)$ **8.** $(x + 2)\left(3 - \frac{1}{x + 2}\right)$

We will now use our knowledge of rational expressions to solve equations that contain rational expressions with variables in their denominators. To do so, we will use new equation-solving methods that sometimes lead to false solutions. For this reason, it is very important to check all answers.

Solving Equations That Contain Rational Expressions

To solve equations containing rational expressions, it is usually best to eliminate them. To do so, we multiply both sides of the equation by the LCD of the rational expressions that appear in the equation. For example, to solve $\frac{x}{3} + 1 = \frac{x}{6}$, we multiply both sides of the equation by 6:

$$\frac{x}{3} + 1 = \frac{x}{6}$$

$$6\left(\frac{x}{3} + 1\right) = 6\left(\frac{x}{6}\right)$$

We then use the distributive property to remove parentheses, simplify, and solve the resulting equation for x.

$$6 \cdot \frac{x}{3} + 6 \cdot 1 = 6 \cdot \frac{x}{6}$$

$$2x + 6 = x$$

$$x + 6 = 0 \quad \text{Subtract } x \text{ from both sides.}$$

$$x = -6 \quad \text{Subtract 6 from both sides.}$$

Check: $\frac{x}{3} + 1 = \frac{x}{6}$

$$\frac{-6}{3} + 1 \stackrel{?}{=} \frac{-6}{6} \quad \text{Substitute } -6 \text{ for } x.$$

$$-2 + 1 \stackrel{?}{=} -1 \quad \text{Simplify.}$$
$$-1 = -1$$

EXAMPLE 1 Solve: $\frac{4}{x} + 1 = \frac{6}{x}$.

Solution To clear the equation of rational expressions, we multiply both sides by the LCD of $\frac{4}{x}$ and $\frac{6}{x}$, which is x.

$$\frac{4}{x} + 1 = \frac{6}{x}$$
$$x\left(\frac{4}{x} + 1\right) = x\left(\frac{6}{x}\right) \quad \text{Multiply both sides by } x.$$
$$x \cdot \frac{4}{x} + x \cdot 1 = x \cdot \frac{6}{x} \quad \text{Remove parentheses.}$$
$$4 + x = 6 \quad \text{Simplify.}$$
$$x = 2 \quad \text{Subtract 4 from both sides.}$$

Check:
$$\frac{4}{x} + 1 = \frac{6}{x}$$
$$\frac{4}{2} + 1 \stackrel{?}{=} \frac{6}{2} \quad \text{Substitute 2 for } x.$$
$$2 + 1 \stackrel{?}{=} 3 \quad \text{Simplify.}$$
$$3 = 3$$

Self Check Solve: $\frac{6}{x} - 1 = \frac{3}{x}$.

Extraneous Solutions

If we multiply both sides of an equation by an expression that involves a variable, as we did in Example 1, we *must* check the apparent solutions. The next example shows why.

EXAMPLE 2 Solve: $\frac{x+3}{x-1} = \frac{4}{x-1}$.

Solution To clear the equation of rational expressions, we multiply both sides by $x - 1$, the LCD of the fractions contained in the equation.

$$\frac{x+3}{x-1} = \frac{4}{x-1}$$
$$(x-1)\frac{x+3}{x-1} = (x-1)\frac{4}{x-1} \quad \text{Multiply both sides by } x - 1.$$
$$x + 3 = 4 \quad \text{Simplify.}$$
$$x = 1 \quad \text{Subtract 3 from both sides.}$$

Because both sides were multiplied by an expression containing a variable, we must check the apparent solution.

$$\frac{x+3}{x-1} = \frac{4}{x-1}$$

$$\frac{1+3}{1-1} \stackrel{?}{=} \frac{4}{1-1} \quad \text{Substitute 1 for } x.$$

$$\frac{4}{0} = \frac{4}{0} \quad \text{Division by 0 is undefined.}$$

Since zeros appear in the denominators, the fractions are undefined. Thus, 1 is a false solution, and the equation has no solutions. Such false solutions are often called **extraneous solutions.**

Self Check Solve: $\frac{x+5}{x-2} = \frac{7}{x-2}$.

The first two examples suggest the steps to follow when solving equations that contain rational expressions.

Solving Equations Containing Rational Expressions

1. Find any restrictions on the variable. Remember that the denominator of a fraction cannot be 0.
2. Multiply both sides of the equation by the LCD of the rational expressions appearing in the equation to clear the equation of fractions.
3. Solve the resulting equation.
4. Check the solutions to determine any extraneous roots.

EXAMPLE 3 Solve: $\frac{3x+1}{x+1} - 2 = \frac{3(x-3)}{x+1}$.

Solution Since the denominator $x + 1$ cannot be 0, x cannot be -1. To clear the equation of rational expressions, we multiply both sides by $x + 1$, the LCD of the rational expressions contained in the equation. We can then solve the resulting equation.

$$\frac{3x+1}{x+1} - 2 = \frac{3(x-3)}{x+1}$$

$$(x+1)\left[\frac{3x+1}{x+1} - 2\right] = (x+1)\left[\frac{3(x-3)}{x+1}\right]$$

$$3x + 1 - 2(x+1) = 3(x-3) \quad \text{Use the distributive property to remove parentheses.}$$

$$3x + 1 - 2x - 2 = 3x - 9 \quad \text{Remove parentheses.}$$

$$x - 1 = 3x - 9 \quad \text{Combine like terms.}$$

$$-2x = -8 \quad \text{On both sides, subtract } 3x \text{ and add 1.}$$

$$x = 4 \quad \text{Divide both sides by } -2.$$

Check:

$$\frac{3x + 1}{x + 1} - 2 = \frac{3(x - 3)}{x + 1}$$

$$\frac{3(4) + 1}{4 + 1} - 2 \stackrel{?}{=} \frac{3(4 - 3)}{4 + 1} \quad \text{Substitute 4 for } x.$$

$$\frac{13}{5} - \frac{10}{5} \stackrel{?}{=} \frac{3(1)}{5}$$

$$\frac{3}{5} = \frac{3}{5}$$

Self Check Solve: $\frac{12}{x + 1} - 5 = \frac{2}{x + 1}$.

Many times, we will have to factor a denominator to find the LCD.

EXAMPLE 4 Solve: $\frac{x + 2}{x + 3} + \frac{1}{x^2 + 2x - 3} = 1$.

Solution To find any restricted values of x and the LCD, we must factor the second denominator.

$$\frac{x + 2}{x + 3} + \frac{1}{x^2 + 2x - 3} = 1$$

$$\frac{x + 2}{x + 3} + \frac{1}{(x + 3)(x - 1)} = 1 \quad \text{Factor } x^2 + 2x - 3.$$

Since $x + 3$ and $x - 1$ cannot be 0, x cannot be -3 or 1. To clear the equation of rational expressions, we multiply both sides by $(x + 3)(x - 1)$, the LCD of the fractions contained in the equation.

$$(x + 3)(x - 1)\left[\frac{x + 2}{x + 3} + \frac{1}{(x + 3)(x - 1)}\right] = (x + 3)(x - 1)1 \quad \text{Multiply both sides by } (x + 3)(x - 1).$$

$$(x + 3)(x - 1)\frac{x + 2}{x + 3} + (x + 3)(x - 1)\frac{1}{(x + 3)(x - 1)} = (x + 3)(x - 1)1 \quad \text{Remove brackets.}$$

$$(x - 1)(x + 2) + 1 = (x + 3)(x - 1) \quad \text{Simplify.}$$

$$x^2 + x - 2 + 1 = x^2 + 2x - 3 \quad \text{Remove parentheses.}$$

$$x - 2 + 1 = 2x - 3 \quad \text{Subtract } x^2 \text{ from both sides.}$$

$$x - 1 = 2x - 3 \quad \text{Combine like terms.}$$

$$-x - 1 = -3 \quad \text{Subtract } 2x \text{ from both sides.}$$

$$-x = -2 \quad \text{Add 1 to both sides.}$$

$$x = 2 \quad \text{Divide both sides by } -1.$$

Verify that 2 is a solution of the given equation.

EXAMPLE 5 Solve: $\frac{4}{5} + y = \frac{4y - 50}{5y - 25}$.

Solution Since $5y - 25$ cannot be 0, y cannot be 5.

$$\frac{4}{5} + y = \frac{4y - 50}{5y - 25}$$

$$\frac{4}{5} + y = \frac{4y - 50}{5(y - 5)} \qquad \text{Factor } 5y - 25.$$

$$5(y - 5)\left[\frac{4}{5} + y\right] = 5(y - 5)\left[\frac{4y - 50}{5(y - 5)}\right] \qquad \text{Multiply both sides by } 5(y - 5).$$

$$4(y - 5) + 5y(y - 5) = 4y - 50 \qquad \text{Remove brackets.}$$

$$4y - 20 + 5y^2 - 25y = 4y - 50 \qquad \text{Remove parentheses.}$$

$$5y^2 - 25y - 20 = -50 \qquad \text{Subtract } 4y \text{ from both sides and rearrange terms.}$$

$$5y^2 - 25y + 30 = 0 \qquad \text{Add 50 to both sides.}$$

$$y^2 - 5y + 6 = 0 \qquad \text{Divide both sides by 5.}$$

$$(y - 3)(y - 2) = 0 \qquad \text{Factor } y^2 - 5y + 6.$$

$$y - 3 = 0 \quad \text{or} \quad y - 2 = 0 \qquad \text{Set each factor equal to 0.}$$

$$y = 3 \quad | \quad y = 2$$

Verify that 3 and 2 both satisfy the original equation.

Self Check Solve: $\frac{x - 6}{3x - 9} - \frac{1}{3} = \frac{x}{2}$.

Formulas

Many formulas are equations that contain fractions.

EXAMPLE 6 The formula $\frac{1}{r} = \frac{1}{r_1} + \frac{1}{r_2}$ is used in electronics to calculate parallel resistances. Solve the formula for r.

Solution We eliminate the denominators by multiplying both sides by the LCD, which is rr_1r_2.

$$\frac{1}{r} = \frac{1}{r_1} + \frac{1}{r_2}$$

$$rr_1r_2\left(\frac{1}{r}\right) = rr_1r_2\left(\frac{1}{r_1} + \frac{1}{r_2}\right) \qquad \text{Multiply both sides by } rr_1r_2.$$

$$\frac{rr_1r_2}{r} = \frac{rr_1r_2}{r_1} + \frac{rr_1r_2}{r_2} \qquad \text{Remove parentheses.}$$

$$r_1r_2 = rr_2 + rr_1 \qquad \text{Simplify.}$$

$$r_1r_2 = r(r_2 + r_1) \qquad \text{Factor out } r.$$

$$\frac{r_1r_2}{r_2 + r_1} = r \qquad \text{Divide both sides by } r_2 + r_1.$$

or

$$r = \frac{r_1r_2}{r_2 + r_1}$$

Self Check Solve the formula for r_1.

Self Check Answers

1. 3 **2.** 2 is extraneous. **3.** 1 **5.** 1, 2 **6.** $r_1 = \frac{rr_2}{r_2 - r}$

Orals *Indicate your first step when solving each equation.*

1. $\frac{x-3}{5} = \frac{x}{2}$

2. $\frac{1}{x-1} = \frac{8}{x}$

3. $\frac{y}{9} + 5 = \frac{y+1}{3}$

4. $\frac{5x-8}{3} + 3x = \frac{x}{5}$

6.7 EXERCISES

REVIEW *Factor each expression.*

1. $x^2 + 4x$

2. $x^2 - 16y^2$

3. $2x^2 + x - 3$

4. $6a^2 - 5a - 6$

5. $x^4 - 16$

6. $4x^2 + 10x - 6$

VOCABULARY AND CONCEPTS *Fill in the blanks.*

7. False solutions that result from multiplying both sides of an equation by a variable are called ________ solutions.

8. If the product of two numbers is 1, the numbers are called ________.

9. To clear an equation of rational expressions, we multiply both sides by the _____ of the expressions in the equation.

10. If you multiply both sides of an equation by an expression that involves a variable, you must _____ the solution.

11. To clear the equation $\frac{1}{x} + \frac{2}{y} = 5$ of denominators, we multiply both sides by ___.

12. To clear the equation $\frac{x}{x-2} - \frac{x}{x-1} = 5$ of denominators, we multiply both sides by ________.

PRACTICE *Solve each equation and check the solution. If an equation has no solution, so indicate.*

13. $\frac{x}{2} + 4 = \frac{3x}{2}$

14. $\frac{y}{3} + 6 = \frac{4y}{3}$

15. $\frac{2y}{5} - 8 = \frac{4y}{5}$

16. $\frac{3x}{4} - 6 = \frac{x}{4}$

17. $\frac{x}{3} + 1 = \frac{x}{2}$

18. $\frac{x}{2} - 3 = \frac{x}{5}$

19. $\frac{x}{5} - \frac{x}{3} = -8$

20. $\frac{2}{3} + \frac{x}{4} = 7$

21. $\frac{3a}{2} + \frac{a}{3} = -22$

22. $\frac{x}{2} + x = \frac{9}{2}$

23. $\frac{x-3}{3} + 2x = -1$

24. $\frac{x+2}{2} - 3x = x + 8$

25. $\frac{z-3}{2} = z + 2$

26. $\frac{b+2}{3} = b - 2$

27. $\frac{5(x+1)}{8} = x + 1$

28. $\frac{3(x-1)}{2} + 2 = x$

29. $\frac{c-4}{4} = \frac{c+4}{8}$

30. $\frac{t+3}{2} = \frac{t-3}{3}$

31. $\frac{x+1}{3} + \frac{x-1}{5} = \frac{2}{15}$

32. $\frac{y-5}{7} + \frac{y-7}{5} = \frac{-2}{5}$

33. $\frac{3x-1}{6} - \frac{x+3}{2} = \frac{3x+4}{3}$

34. $\frac{2x+3}{3} + \frac{3x-4}{6} = \frac{x-2}{2}$

35. $\frac{3}{x} + 2 = 3$

36. $\frac{2}{x} + 9 = 11$

37. $\frac{5}{a} - \frac{4}{a} = 8 + \frac{1}{a}$

38. $\frac{11}{b} + \frac{13}{b} = 12$

39. $\dfrac{2}{y+1} + 5 = \dfrac{12}{y+1}$

40. $\dfrac{1}{t-3} = \dfrac{-2}{t-3} + 1$

41. $\dfrac{1}{x-1} + \dfrac{3}{x-1} = 1$

42. $\dfrac{3}{p+6} - 2 = \dfrac{7}{p+6}$

43. $\dfrac{a^2}{a+2} - \dfrac{4}{a+2} = a$

44. $\dfrac{z^2}{z+1} + 2 = \dfrac{1}{z+1}$

45. $\dfrac{x}{x-5} - \dfrac{5}{x-5} = 3$

46. $\dfrac{3}{y-2} + 1 = \dfrac{3}{y-2}$

47. $\dfrac{3r}{2} - \dfrac{3}{r} = \dfrac{3r}{2} + 3$

48. $\dfrac{2p}{3} - \dfrac{1}{p} = \dfrac{2p-1}{3}$

49. $\dfrac{1}{3} + \dfrac{2}{x-3} = 1$

50. $\dfrac{3}{5} + \dfrac{7}{x+2} = 2$

51. $\dfrac{u}{u-1} + \dfrac{1}{u} = \dfrac{u^2+1}{u^2-u}$

52. $\dfrac{v}{v+2} + \dfrac{1}{v-1} = 1$

53. $\dfrac{3}{x-2} + \dfrac{1}{x} = \dfrac{2(3x+2)}{x^2-2x}$

54. $\dfrac{5}{x} + \dfrac{3}{x+2} = \dfrac{-6}{x(x+2)}$

55. $\dfrac{7}{q^2-q-2} + \dfrac{1}{q+1} = \dfrac{3}{q-2}$

56. $\dfrac{-5}{s^2+s-2} + \dfrac{3}{s+2} = \dfrac{1}{s-1}$

57. $\dfrac{3y}{3y-6} + \dfrac{8}{y^2-4} = \dfrac{2y}{2y+4}$

58. $\dfrac{x-3}{4x-4} + \dfrac{1}{9} = \dfrac{x-5}{6x-6}$

59. $y + \dfrac{2}{3} = \dfrac{2y-12}{3y-9}$

60. $y + \dfrac{3}{4} = \dfrac{3y-50}{4y-24}$

61. $\dfrac{5}{4y+12} - \dfrac{3}{4} = \dfrac{5}{4y+12} - \dfrac{y}{4}$

62. $\dfrac{3}{5x-20} + \dfrac{4}{5} = \dfrac{3}{5x-20} - \dfrac{x}{5}$

63. $\dfrac{x}{x-1} - \dfrac{12}{x^2-x} = \dfrac{-1}{x-1}$

64. $1 - \dfrac{3}{b} = \dfrac{-8b}{b^2+3b}$

65. $\dfrac{z-4}{z-3} = \dfrac{z+2}{z+1}$

66. $\dfrac{a+2}{a+8} = \dfrac{a-3}{a-2}$

67. $\dfrac{n}{n^2-9} + \dfrac{n+8}{n+3} = \dfrac{n-8}{n-3}$

68. $\dfrac{x-3}{x-2} - \dfrac{1}{x} = \dfrac{x-3}{x}$

69. $\dfrac{b+2}{b+3} + 1 = \dfrac{-7}{b-5}$

70. $\dfrac{x-4}{x-3} + \dfrac{x-2}{x-3} = x - 3$

71. Solve the formula $\dfrac{1}{a} + \dfrac{1}{b} = 1$ for a.

72. Solve the formula $\dfrac{1}{a} - \dfrac{1}{b} = 1$ for b.

73. **Optics** The focal length f of a lens is given by the formula

$$\frac{1}{f} = \frac{1}{d_1} + \frac{1}{d_2}$$

where d_1 is the distance from the object to the lens and d_2 is the distance from the lens to the image.

Solve the formula for f.

74. Solve the formula in Exercise 73 for d_1.

WRITING

75. Explain how you would decide what to do first when you solve an equation that involves fractions.

76. Explain why it is important to check your solutions to an equation that contains fractions with variables in the denominator.

SOMETHING TO THINK ABOUT

77. What number is equal to its own reciprocal?

78. Solve: $x^{-2} + x^{-1} = 0$.

6.8 Applications of Equations That Contain Rational Expressions

- Problem Solving

Getting Ready

1. If it takes 5 hours to fill a pool, what part could be filled in 1 hour?
2. \$$x$ is invested at 5% annual interest. Write an expression for the interest earned in one year.
3. Write an expression for the amount of an investment that earns \$$y$ interest in one year at 5%.
4. Express how long it takes to travel y miles at 52 mph.

In this final section, we will consider problems whose solutions depend on solving equations containing rational expressions.

Problem Solving

EXAMPLE 1 **Number problem** If the same number is added to both the numerator and denominator of the fraction $\frac{3}{5}$, the result is $\frac{4}{5}$. Find the number.

Analyze the problem We are asked to find a number. If we add it to both the numerator and denominator of the fraction $\frac{3}{5}$, we will get $\frac{4}{5}$.

Form an equation Let n represent the unknown number and add n to both the numerator and denominator of $\frac{3}{5}$. Then set the result equal to $\frac{4}{5}$ to get the equation

$$\frac{3+n}{5+n} = \frac{4}{5}$$

Solve the equation To solve the equation, we proceed as follows:

$$\frac{3+n}{5+n} = \frac{4}{5}$$

$$\mathbf{5(5+n)}\frac{3+n}{5+n} = \mathbf{5(5+n)}\frac{4}{5} \quad \text{Multiply both sides by } 5(5+n).$$

$$5(3+n) = (5+n)4 \quad \text{Simplify.}$$

$$15 + 5n = 20 + 4n \quad \text{Use the distributive property to remove parentheses.}$$

$$5n = 5 + 4n \quad \text{Subtract 15 from both sides.}$$

$$n = 5 \quad \text{Subtract } 4n \text{ from both sides.}$$

State the conclusion The number is 5.

Check the result Add 5 to both the numerator and denominator of $\frac{3}{5}$ and get

$$\frac{3+5}{5+5} = \frac{8}{10} = \frac{4}{5}$$

The result checks.

EXAMPLE 2 **Draining an oil tank** An inlet pipe can fill an oil tank in 7 days, and a second inlet pipe can fill the same tank in 9 days. If both pipes are used, how long will it take to fill the tank?

Analyze the problem The key is to note what each pipe can do in 1 day. If you add what the first pipe can do in 1 day to what the second pipe can do in 1 day, the sum is what they can do together in 1 day. Since the first pipe can fill the tank in 7 days, it can do $\frac{1}{7}$ of the job in 1 day. Since the second pipe can fill the tank in 9 days, it can do $\frac{1}{9}$ of the job in 1 day. If it takes x days for both pipes to fill the tank, together they can do $\frac{1}{x}$ of the job in 1 day.

Form an equation Let x represent the number of days it will take to fill the tank if both inlet pipes are used. Then we form the equation

What the first inlet pipe can do in 1 day	plus	what the second inlet pipe can do in 1 day	equals	what they can do together in 1 day.
$\frac{1}{7}$	$+$	$\frac{1}{9}$	$=$	$\frac{1}{x}$

Solve the equation To solve the equation, we proceed as follows:

$$\frac{1}{7} + \frac{1}{9} = \frac{1}{x}$$

$$\mathbf{63x}\left(\frac{1}{7} + \frac{1}{9}\right) = \mathbf{63x}\left(\frac{1}{x}\right) \quad \text{Multiply both sides by } 63x.$$

$$9x + 7x = 63 \quad \text{Use the distributive property to remove parentheses and simplify.}$$

$$16x = 63 \quad \text{Combine like terms.}$$

$$x = \frac{63}{16} \quad \text{Divide both sides by 16.}$$

State the conclusion It will take $\frac{63}{16}$ or $3\frac{15}{16}$ days for both inlet pipes to fill the tank.

Check the result In $\frac{63}{16}$ days, the first pipe fills $\frac{1}{7}\left(\frac{63}{16}\right)$ of the tank, and the second pipe fills $\frac{1}{9}\left(\frac{63}{16}\right)$ of the tank. The sum of these efforts, $\frac{9}{16} + \frac{7}{16}$, is equal to one full tank. ∎

EXAMPLE 3 **Track and field** A coach can run 10 miles in the same amount of time that her best student athlete can run 12 miles. If the student can run 1 mph faster than the coach, how fast can the student run?

Analyze the problem This is a uniform motion problem, which is based on the formula $d = rt$, where d is the distance traveled, r is the rate, and t is the time. If we solve this formula for t, we obtain

$$t = \frac{d}{r}$$

If the coach runs 10 miles at some unknown rate of r mph, it will take $\frac{10}{r}$ hours. If the student runs 12 miles at some unknown rate of $(r + 1)$ mph, it will take $\frac{12}{r+1}$ hours. We can organize the information of the problem as in Table 6-2.

	d	$=$	r	$\cdot$	t
Student	12		$r + 1$		$\frac{12}{r+1}$
Coach	10		r		$\frac{10}{r}$

Table 6-2

Because the times are given to be equal, we know that $\frac{12}{r+1} = \frac{10}{r}$.

Form an equation Let r be the rate that the coach can run. Then $r + 1$ is the rate that the student can run. We can form the equation.

The time it takes the student to run 12 miles	equals	the time it takes the coach to run 10 miles.
$\frac{12}{r+1}$	$=$	$\frac{10}{r}$

Solve the equation We can solve the equation as follows:

$$\frac{12}{r+1} = \frac{10}{r}$$

$$r(r+1)\frac{12}{r+1} = r(r+1)\frac{10}{r} \quad \text{Multiply both sides by } r(r+1).$$

$$12r = 10(r+1) \quad \text{Simplify.}$$

$$12r = 10r + 10 \quad \text{Use the distributive property to remove parentheses.}$$

$$2r = 10 \quad \text{Subtract } 10r \text{ from both sides.}$$

$$r = 5 \quad \text{Divide both sides by 2.}$$

State the conclusion The coach can run 5 mph. The student, running 1 mph faster, can run 6 mph.

Check the result Verify that this result checks. ∎

EXAMPLE 4 **Comparing investments** At one bank, a sum of money invested for one year will earn \$96 interest. If invested in bonds, that same money would earn \$108, because the interest rate paid by the bonds is 1% greater than that paid by the bank. Find the bank's rate of interest.

Analyze the problem This interest problem is based on the formula $i = pr$, where i is the interest, p is the principal (the amount invested), and r is the annual rate of interest. If we solve this formula for p, we obtain

$$p = \frac{i}{r}$$

If we let r represent the bank's rate of interest, then $r + 0.01$ represents the rate paid by the bonds. If an investment at a bank earns \$96 interest at some unknown rate r, the principal invested is $\frac{96}{r}$. If an investment in bonds earns \$108 interest at some unknown rate $(r + 0.01)$, the principal invested is $\frac{108}{r + 0.01}$. We can organize the information of the problem as in Table 6-3.

	Interest	= Principal	· Rate
Bank	96	$\frac{96}{r}$	r
Bonds	108	$\frac{108}{r+0.01}$	$r + 0.01$

Table 6-3

Form an equation Because the same principal would be invested in either account, we can set up the following equation:

$$\frac{96}{r} = \frac{108}{r + 0.01}$$

Solve the equation We can solve the equation as follows:

$$\frac{96}{r} = \frac{108}{r + 0.01}$$

$$\mathbf{r(r + 0.01)} \cdot \frac{96}{r} = \frac{108}{r + 0.01} \cdot \mathbf{r(r + 0.01)} \quad \text{Multiply both sides by } r(r + 0.01).$$

$$96(r + 0.01) = 108r$$

$$96r + 0.96 = 108r \quad \text{Remove parentheses.}$$

$$0.96 = 12r \quad \text{Subtract } 96r \text{ from both sides.}$$

$$0.08 = r \quad \text{Divide both sides by 12.}$$

State the conclusion The bank's interest rate is 0.08, or 8%. The bonds pay 9% interest, a rate 1% greater than that paid by the bank.

Check the results Verify that these rates check.

Orals

1. What is the formula that relates the principal p that is invested, the earned interest i, and the rate r for 1 year?
2. What is the formula that relates the distance d traveled at a speed r, for a time t?
3. What is the formula that relates the cost C of purchasing q items that cost $\$d$ each?

6.8 EXERCISES

REVIEW *Solve each equation.*

1. $x^2 - 5x - 6 = 0$
2. $x^2 - 25 = 0$
3. $(t + 2)(t^2 + 7t + 12) = 0$
4. $2(y - 4) = -y^2$
5. $y^3 - y^2 = 0$
6. $5a^3 - 125a = 0$
7. $(x^2 - 1)(x^2 - 4) = 0$
8. $6t^3 + 35t^2 = 6t$

VOCABULARY AND CONCEPTS

9. List the five steps used in problem solving.
10. Write 6% as a decimal.

PRACTICE

11. **Number problem** If the denominator of $\frac{3}{4}$ is increased by a number and the numerator is doubled, the result is 1. Find the number.

12. Number problem If a number is added to the numerator of $\frac{7}{8}$ and the same number is subtracted from the denominator, the result is 2. Find the number.

13. Number problem If a number is added to the numerator of $\frac{3}{4}$ and twice as much is added to the denominator, the result is $\frac{4}{7}$. Find the number.

14. Number problem If a number is added to the numerator of $\frac{5}{7}$ and twice as much is subtracted from the denominator, the result is 8. Find the number.

15. Number problem The sum of a number and its reciprocal is $\frac{13}{6}$. Find the numbers.

16. Number problem The sum of the reciprocals of two consecutive even integers is $\frac{7}{24}$. Find the integers.

APPLICATIONS

17. Filling a pool An inlet pipe can fill an empty swimming pool in 5 hours, and another inlet pipe can fill the pool in 4 hours. How long will it take both pipes to fill the pool?

18. Filling a pool One inlet pipe can fill an empty pool in 4 hours, and a drain can empty the pool in 8 hours. How long will it take the pipe to fill the pool if the drain is left open?

19. Roofing a house A homeowner estimates that it will take 7 days to roof his house. A professional roofer estimates that he could roof the house in 4 days. How long will it take if the homeowner helps the roofer?

20. Sewage treatment A sludge pool is filled by two inlet pipes. One pipe can fill the pool in 15 days and the other pipe can fill it in 21 days. However, if no sewage is added, waste removal will empty the pool in 36 days. How long will it take the two inlet pipes to fill an empty pool?

21. Touring A tourist can bicycle 28 miles in the same time as he can walk 8 miles. If he can ride 10 mph faster than he can walk, how much time should he allow to walk a 30-mile trail? (*Hint:* How fast can he walk?)

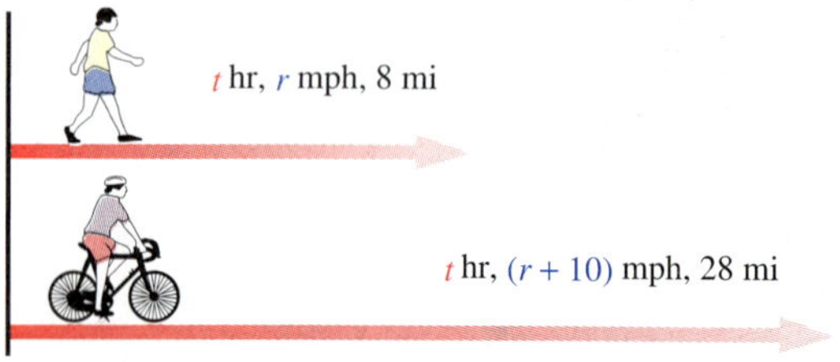

22. Comparing travel A plane can fly 300 miles in the same time as it takes a car to go 120 miles. If the car travels 90 mph slower than the plane, find the speed of the plane.

23. Boating A boat that can travel 18 mph in still water can travel 22 miles downstream in the same amount of time that it can travel 14 miles upstream. Find the speed of the current in the river.

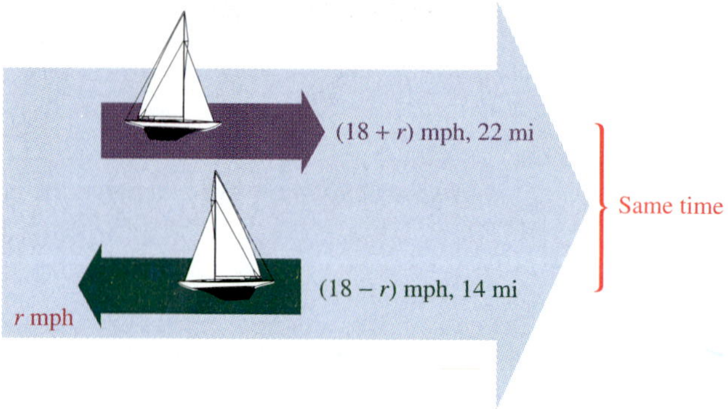

24. Wind speed A plane can fly 300 miles downwind in the same amount of time as it can travel 210 miles upwind. Find the velocity of the wind if the plane can fly 255 mph in still air.

25. Comparing investments Two certificates of deposit pay interest at rates that differ by 1%. Money invested for one year in the first CD earns \$175 interest. The same principal invested in the other CD earns \$200. Find the two rates of interest.

26. Comparing interest rates Two bond funds pay interest at rates that differ by 2%. Money invested for one year in the first fund earns \$315 interest. The same amount invested in the other fund earns \$385. Find the lower rate of interest.

27. Sharing costs Some office workers bought a \$35 gift for their boss. If there had been two more employees to contribute, everyone's cost would have been \$2 less. How many workers contributed to the gift?

28. Sales A dealer bought some radios for a total of \$1,200. She gave away 6 radios as gifts, sold each of the rest for \$10 more than she paid for each radio, and broke even. How many radios did she buy?

29. Sales A bookstore can purchase several calculators for a total cost of \$120. If each calculator cost \$1 less, the bookstore could purchase 10 additional calculators at the same total cost. How many calculators can be purchased at the regular price?

30. **Furnace repair** A repairman purchased several furnace-blower motors for a total cost of \$210. If his cost per motor had been \$5 less, he could have purchased 1 additional motor. How many motors did he buy at the regular rate?

31. **River tours** A river boat tour begins by going 60 miles upstream against a 5 mph current. Then the boat turns around and returns with the current. What still-water speed should the captain use to complete the tour in 5 hours?

32. **Travel time** A company president flew 680 miles in a corporate jet but returned in a smaller plane that could fly only half as fast. If the total travel time was 6 hours, find the speeds of the planes.

WRITING

33. The key to solving shared work problems is to ask, "How much of the job could be done in 1 unit of time?" Explain.

34. It is difficult to check the solution of a shared work problem. Explain how you could decide if the answer is at least reasonable.

SOMETHING TO THINK ABOUT

35. Create a problem, involving either investment income or shared work, that can be solved by an equation that contains rational expressions.

36. Solve the problem you created in Exercise 35.

Projects

1. If the sides of two similar triangles are in the ratio of 1 to 1, the triangles are said to be **congruent.** Congruent triangles have the same shape and the same size (area).

   a. Draw several triangles with sides of length 1, 1.5, and 2 inches. Are the triangles all congruent? What general rule could you make?

   b. Draw several triangles with the dimensions shown in the illustration. Are the triangles all congruent? What general rule could you make?

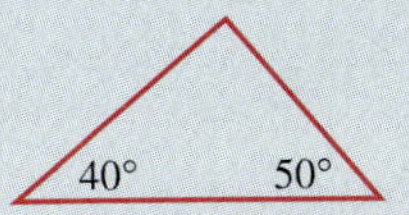

   c. Draw several triangles with the dimensions shown in the illustration. Are the triangles all congruent? What general rule could you make?

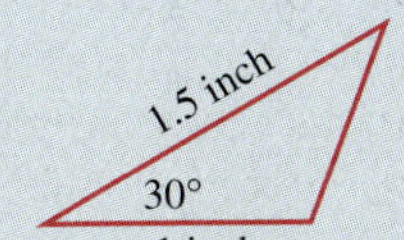

2. If y is equal to a polynomial divided by a polynomial, we call the resulting function a **rational function.** The simplest of these functions is defined by the equation $y = \frac{1}{x}$. Since the denominator of this fraction cannot be 0, the domain of the function is the set of real numbers, except 0. Since a fraction with a numerator of 1 cannot be 0, the range is also the set of real numbers, except 0.

   Construct the graph of this function by making a table of values containing at least eight ordered pairs, plotting the ordered pairs, and joining the points with two curves. Then graph each of the following rational functions and find each one's domain and range.

   a. $y = \dfrac{2}{x}$ b. $y = -\dfrac{1}{x}$

   c. $y = -\dfrac{1}{x+1}$ d. $y = \dfrac{2}{x-1}$

3. Suppose that the cost of telephone service is \$6 per month plus 5¢ per call. If n represents the number of calls made one month, the cost C of phone service that month will be given by $C = 0.05n + 6$. If we divide the total cost C by the number of calls n, we will obtain the average cost per call, which we will denote as $\bar{c}$.

$$(1)\quad \bar{c} = \frac{C}{n} = \frac{0.05n + 6}{n}$$

$\bar{c}$ is the average cost per call, C is the total monthly cost, and n is the number of phone calls made that month.

Use the rational function in Equation 1 to find the average monthly cost per call when

a. 5 calls were made and **b.** 25 calls were made.

Assume that a phone company charges \$15 a month and 5¢ per phone call and answer the following questions.

a. Write a function that will give the cost C per month for making n phone calls.

b. Write a function that will give the average cost $\bar{c}$ per call during the month.

c. Find the cost if 45 phone calls were made during the month.

d. Find the average cost per call if 45 calls were made that month.

Chapter Summary

CONCEPTS	REVIEW EXERCISES
6.1	**Ratios**
A **ratio** is the comparison of two numbers by their indicated quotient.	Write each ratio as a fraction in lowest terms. **1.** 3 to 6 **2.** $12x$ to $15x$ **3.** 2 feet to 1 yard **4.** 5 pints to 3 quarts
The **unit cost** of an item is the ratio of its cost to its quantity.	**5.** If three pounds of coffee cost \$8.79, find the unit cost (the cost per pound).
Rates are ratios that are used to compare quantities with different units.	**6.** If a factory used 2,275 kwh of electricity in February, what was the rate of energy consumption in kwh per week?
6.2	**Proportions and Similar Triangles**
A **proportion** is a statement that two ratios are equal.	Determine whether the following equations are proportions. **7.** $\frac{4}{7} = \frac{20}{34}$ **8.** $\frac{5}{7} = \frac{30}{42}$
In any proportion, the product of the extremes is equal to the product of the means.	Solve each proportion. **9.** $\frac{3}{x} = \frac{6}{9}$ **10.** $\frac{x}{3} = \frac{x}{5}$ **11.** $\frac{x-2}{5} = \frac{x}{7}$ **12.** $\frac{4x-1}{18} = \frac{x}{6}$
The measures of corresponding sides of similar triangles are in proportion.	**13.** A telephone pole casts a shadow 12 feet long at the same time that a man 6 feet tall casts a shadow of 3.6 feet. How tall is the pole?

6.3 Simplifying Rational Expressions

If b and c are not 0, then

$$\frac{a}{b} = \frac{a \cdot c}{b \cdot c}$$

$$\frac{a}{1} = a$$

$\frac{a}{0}$ is undefined.

14. For what values of x is the rational expression undefined?

$$\frac{x + 4}{(x + 3)(x - 3)}$$

Write each fraction in lowest terms.

15. $\frac{10}{25}$

16. $-\frac{12}{18}$

17. $-\frac{51}{153}$

18. $\frac{105}{45}$

19. $\frac{3x^2}{6x^3}$

20. $\frac{5xy^2}{2x^2y^2}$

21. $\frac{x^2}{x^2 + x}$

22. $\frac{x + 2}{x^2 + 2x}$

23. $\frac{6xy}{3xy}$

24. $\frac{8x^2y}{2x(4xy)}$

25. $\frac{3p - 2}{2 - 3p}$

26. $\frac{x^2 - x - 56}{x^2 - 5x - 24}$

27. $\frac{2x^2 - 16x}{2x^2 - 18x + 16}$

28. $\frac{a^2 + 2a + ab + 2b}{a^2 + 2ab + b^2}$

6.4 Multiplying and Dividing Rational Expressions

$$\frac{a}{b} \cdot \frac{c}{d} = \frac{a \cdot c}{b \cdot d} \quad (b, d \neq 0)$$

Perform each multiplication and simplify.

29. $\frac{3xy}{2x} \cdot \frac{4x}{2y^2}$

30. $\frac{3x}{x^2 - x} \cdot \frac{2x - 2}{x^2}$

31. $\frac{x^2 + 3x + 2}{x^2 + 2x} \cdot \frac{x}{x + 1}$

32. $\frac{x^2 + x}{3x - 15} \cdot \frac{6x - 30}{x^2 + 2x + 1}$

$$\frac{a}{b} \div \frac{c}{d} = \frac{a}{b} \cdot \frac{d}{c} \quad (b, c, d \neq 0)$$

Perform each division and simplify.

33. $\frac{3x^2}{5x^2y} \div \frac{6x}{15xy^2}$

34. $\frac{x^2 + 5x}{x^2 + 4x - 5} \div \frac{x^2}{x - 1}$

35. $\frac{x^2 - x - 6}{2x - 1} \div \frac{x^2 - 2x - 3}{2x^2 + x - 1}$

36. $\dfrac{x^2 - 3x}{x^2 - x - 6} \div \dfrac{x^2 - x}{x^2 + x - 2}$

37. $\dfrac{x^2 + 4x + 4}{x^2 + x - 6}\left(\dfrac{x - 2}{x - 1} \div \dfrac{x + 2}{x^2 + 2x - 3}\right)$

6.5 Adding and Subtracting Rational Expressions

$\dfrac{a}{d} + \dfrac{b}{d} = \dfrac{a + b}{d} \quad (d \neq 0)$

$\dfrac{a}{d} - \dfrac{b}{d} = \dfrac{a - b}{d} \quad (d \neq 0)$

To add or subtract rational expressions with unlike denominators, first find the LCD of the expressions. Then express each one in equivalent form with this LCD. Finally, add or subtract the expressions.

Perform each operation. Simplify all answers.

38. $\dfrac{x}{x + y} + \dfrac{y}{x + y}$

39. $\dfrac{3x}{x - 7} - \dfrac{x - 2}{x - 7}$

40. $\dfrac{x}{x - 1} + \dfrac{1}{x}$

41. $\dfrac{1}{7} - \dfrac{1}{x}$

42. $\dfrac{3}{x + 1} - \dfrac{2}{x}$

43. $\dfrac{x + 2}{2x} - \dfrac{2 - x}{x^2}$

44. $\dfrac{x}{x + 2} + \dfrac{3}{x} - \dfrac{4}{x^2 + 2x}$

45. $\dfrac{2}{x - 1} - \dfrac{3}{x + 1} + \dfrac{x - 5}{x^2 - 1}$

6.6 Complex Fractions

To simplify a complex fraction, use either of these methods:

1. Write the numerator and denominator of the complex fraction as single fractions, do the division of the fractions, and simplify.
2. Multiply both the numerator and the denominator of the complex fraction by the LCD of the fractions that appear in the numerator and the denominator; then simplify.

Simplify each complex fraction.

46. $\dfrac{\frac{3}{2}}{\frac{2}{3}}$

47. $\dfrac{\frac{3}{2} + 1}{\frac{2}{3} + 1}$

48. $\dfrac{\frac{1}{x} + 1}{\frac{1}{x} - 1}$

49. $\dfrac{1 + \frac{3}{x}}{2 - \frac{1}{x^2}}$

50. $\dfrac{\frac{2}{x - 1} + \frac{x - 1}{x + 1}}{\frac{1}{x^2 - 1}}$

51. $\dfrac{\frac{a}{b} + c}{\frac{b}{a} + c}$

6.7 Solving Equations That Contain Rational Expressions

To solve an equation that contains rational expressions, change it to another equation without rational expressions. Do so by multiplying both sides by the LCD of the rational expressions. Check all solutions.

Solve each equation and check all answers.

52. $\frac{3}{x} = \frac{2}{x - 1}$

53. $\frac{5}{x + 4} = \frac{3}{x + 2}$

54. $\frac{2}{3x} + \frac{1}{x} = \frac{5}{9}$

55. $\frac{2x}{x + 4} = \frac{3}{x - 1}$

56. $\frac{2}{x - 1} + \frac{3}{x + 4} = \frac{-5}{x^2 + 3x - 4}$

57. $\frac{4}{x + 2} - \frac{3}{x + 3} = \frac{6}{x^2 + 5x + 6}$

58. Solve for r_1: $\frac{1}{r} = \frac{1}{r_1} + \frac{1}{r_2}$.

6.8 Applications of Equations That Contain Rational Expressions

59. The efficiency E of a Carnot engine is given by the formula

$$E = 1 - \frac{T_2}{T_1}$$

Solve the formula for T_1.

60. Radioactive tracers are used for diagnostic work in nuclear medicine. The **effective half-life** H of a radioactive material in a biological organism is given by the formula

$$H = \frac{RB}{R + B}$$

where R is the radioactive half-life and B is the biological half-life of the tracer. Solve the formula for R.

61. Pumping a basement If one pump can empty a flooded basement in 18 hours and a second pump can empty the basement in 20 hours, how long will it take to empty the basement when both pumps are used?

62. Painting houses If a homeowner can paint a house in 14 days and a professional painter can paint it in 10 days, how long will it take if they work together?

63. Exercise A jogger can bicycle 30 miles in the same time as he can jog 10 miles. If he can ride 10 mph faster than he can jog, how fast can he jog?

64. Wind speed A plane can fly 400 miles downwind in the same amount of time as it can travel 320 miles upwind. If the plane can fly at 360 mph in still air, find the velocity of the wind.

Chapter Test

1. Express as a ratio in lowest terms: 6 feet to 3 yards.

2. Is the equation $\frac{3xy}{5xy} = \frac{3xt}{5xt}$ a proportion?

3. Solve for y: $\frac{y}{y-1} = \frac{y-2}{y}$.

4. A tree casts a shadow that is 30 feet long when a 6-foot-tall man casts a shadow that is 4 feet long. How tall is the tree?

5. Simplify: $\frac{48x^2y}{54xy^2}$.

6. Simplify: $\frac{2x^2 - x - 3}{4x^2 - 9}$.

7. Simplify: $\frac{3(x+2) - 3}{2x - 4 - (x - 5)}$.

8. Multiply and simplify: $\frac{12x^2y}{15xyz} \cdot \frac{25y^2z}{16xt}$.

9. Multiply and simplify: $\frac{x^2 + 3x + 2}{3x + 9} \cdot \frac{x + 3}{x^2 - 4}$.

10. Divide and simplify: $\frac{8x^2y}{25xt} \div \frac{16x^2y^3}{30xyt^3}$.

11. Divide and simplify: $\frac{x^2 - x}{3x^2 + 6x} \div \frac{3x - 3}{3x^3 + 6x^2}$.

12. Simplify: $\frac{x^2 + xy}{x - y} \cdot \frac{x^2 - y^2}{x^2 - 2x} \div \frac{x^2 + 2xy + y^2}{x^2 - 4}$.

13. Add: $\frac{5x - 4}{x - 1} + \frac{5x + 3}{x - 1}$.

14. Subtract: $\frac{3y + 7}{2y + 3} - \frac{3(y - 2)}{2y + 3}$.

15. Add: $\frac{x + 1}{x} + \frac{x - 1}{x + 1}$.

16. Subtract: $\frac{5x}{x - 2} - 3$.

17. Simplify: $\frac{\frac{8x^2}{xy^3}}{\frac{4y^3}{x^2y^3}}$.

18. Simplify: $\frac{1 + \frac{y}{x}}{\frac{y}{x} - 1}$.

19. Solve for x: $\frac{x}{10} - \frac{1}{2} = \frac{x}{5}$.

20. Solve for x: $3x - \frac{2(x + 3)}{3} = 16 - \frac{x + 2}{2}$.

21. Solve for x: $\frac{7}{x + 4} - \frac{1}{2} = \frac{3}{x + 4}$.

22. Solve for B: $H = \frac{RB}{R + B}$.

23. **Cleaning highways** One highway worker could pick up all the trash on a strip of highway in 7 hours, and his helper could pick up the trash in 9 hours. How long will it take them if they work together?

24. **Boating** A boat can motor 28 miles downstream in the same amount of time as it can motor 18 miles upstream. Find the speed of the current if the boat can motor at 23 mph in still water.

25. **Flight path** A plane drops 575 feet as it flies a horizontal distance of $\frac{1}{2}$ mile. How much altitude will it lose as it flies a horizontal distance of 7 miles?

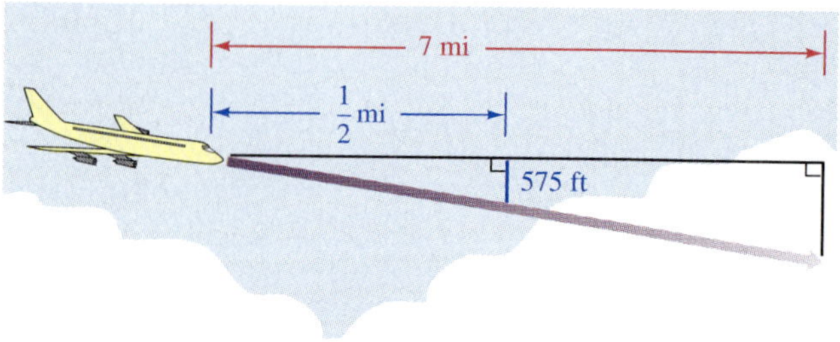

CUMULATIVE REVIEW EXERCISES

Simplify each expression.

1. x^2x^5
2. $(x^2)^5$
3. $\frac{x^5}{x^2}$
4. $(3x^5)^0$
5. $(3x^2 - 2x) + (6x^3 - 3x^2 - 1)$
6. $(4x^3 - 2x) - (2x^3 - 2x^2 - 3x + 1)$
7. $3(5x^2 - 4x + 3) + 2(-x^2 + 2x - 4)$
8. $4(3x^2 - 4x - 1) - 2(-2x^2 + 4x - 3)$

Perform each operation.

9. $(3x^3y^2)(-4x^2y^3)$
10. $-5x^2(7x^3 - 2x^2 - 2)$
11. $(3x + 1)(2x + 4)$
12. $(5x - 4y)(3x + 2y)$
13. $x + 3\overline{)x^2 + 7x + 12}$
14. $2x - 3\overline{)2x^3 - x^2 - x - 3}$

Factor each expression.

15. $3x^2y - 6xy^2$
16. $3(a + b) + x(a + b)$
17. $2a + 2b + ab + b^2$
18. $25p^4 - 16q^2$
19. $x^2 - 11x - 12$
20. $x^2 - xy - 6y^2$
21. $6a^2 - 7a - 20$
22. $8m^2 - 10mn - 3n^2$
23. $p^3 - 27q^3$
24. $8r^3 + 64s^3$

Solve each equation.

25. $\frac{4}{5}x + 6 = 8$
26. $5 - \frac{x + 2}{3} = 7 - x$
27. $6x^2 - x - 2 = 0$
28. $5x^2 = 10x$
29. $x^2 + 3x + 2 = 0$
30. $2y^2 + 5y - 12 = 0$

Solve each inequality and graph the solution set.

31. $5x - 3 > 7$
32. $7x - 9 < 5$
33. $-2 < -x + 3 < 5$
34. $0 \leq \frac{4 - x}{3} \leq 2$

Graph each equation.

35. Graph: $4x - 3y = 12$.

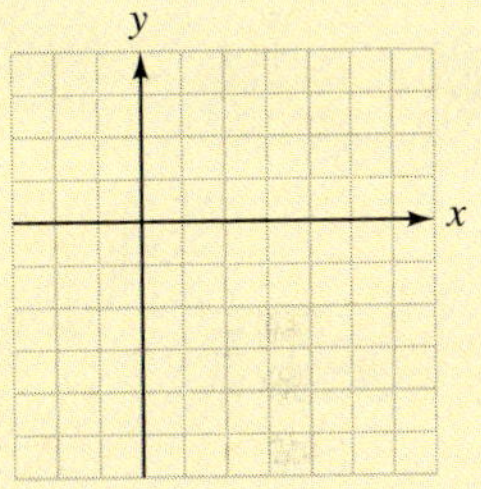

36. Graph: $3x + 4y = 4y + 12$.

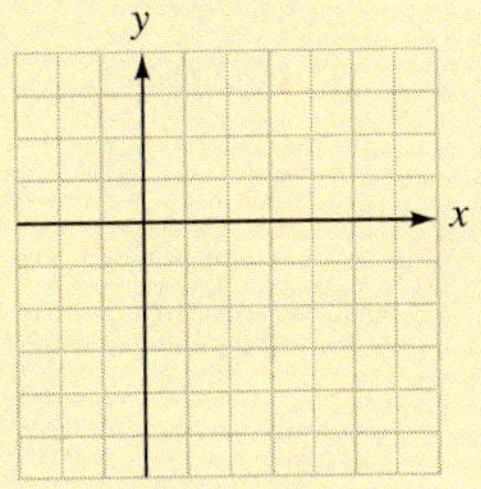

If $f(x) = 2x^2 - 3$, find each value.

37. $f(0)$
38. $f(3)$
39. $f(-2)$
40. $f(2x)$

Simplify each fraction.

41. $\frac{x^2 + 2x + 1}{x^2 - 1}$
42. $\frac{x^2 + 2x - 15}{x^2 + 3x - 10}$

Perform the operation(s) and simplify when possible.

43. $\dfrac{x^2 + x - 6}{5x - 5} \cdot \dfrac{5x - 10}{x + 3}$

44. $\dfrac{p^2 - p - 6}{3p - 9} \div \dfrac{p^2 + 6p + 9}{p^2 - 9}$

45. $\dfrac{3x}{x + 2} + \dfrac{5x}{x + 2} - \dfrac{7x - 2}{x + 2}$

46. $\dfrac{x - 1}{x + 1} + \dfrac{x + 1}{x - 1}$

47. $\dfrac{a + 1}{2a + 4} - \dfrac{a^2}{2a^2 - 8}$

48. $\dfrac{\frac{1}{x} + \frac{1}{y}}{\frac{1}{x} - \frac{1}{y}}$

9 Radicals and Rational Exponents

© Royalty-Free/CORBIS

InfoTrac

Do a subject guide search on "alternating current." Click on "View periodical references," and find the article "AC fundamentals" from *EC&M Electrical Construction & Maintenance,* Jan 1, 2003, v102. In an alternating current circuit, voltage, current, and impedance are represented by complex numbers. Using the formula found in the article, find the voltage (in volts) of a circuit with a current of $5 - 6.4i$ amperes and an impedance of $3.2 + 4.25i$ ohms. Now find the current of a circuit with voltage of $6.25 - 4.00i$ volts and an impedance of $5.3 + 6.75i$ ohms.

Complete this project after studying Section 9.7.

Mathematics in Photography

Many cameras have an adjustable lens opening, called an **aperture,** that controls the amount of light passing through the circular lens. Various lenses—wide-angle, close-up, and telephoto—are distinguished by their **focal lengths.**

The ***f*-number** of a lens is its focal length divided by the diameter of its aperture.

$$f\text{-number} = \frac{f}{d}$$

f is the focal length, and d is the diameter of the aperture.

Example 5
Section 9.5

For example, a lens with a focal length of 12 centimeters and an aperture 6 centimeters in diameter has an f-number of $\frac{12}{6}$ and is called an $f/2$ lens.

If the area of an aperture is reduced to admit half as much light as an $f/2$ lens, the f-number of the lens will change. Find the new f-number.

In this chapter, we will reverse the squaring process and learn how to find square roots of numbers. We will also learn how to find other roots of numbers and discuss complex numbers.

9.1 Radical Expressions

- Square Roots
- Square Roots of Expressions with Variables
- Cube Roots
- *n*th Roots
- Square Root and Cube Root Functions
- Standard Deviation

Getting Ready *Find each power.*

1. 0^2 **2.** 4^2 **3.** $(-4)^2$ **4.** -4^2

5. $\left(\frac{2}{5}\right)^3$ **6.** $\left(-\frac{3}{4}\right)^4$ **7.** $(7xy)^2$ **8.** $(7xy)^3$

In this section, we will discuss square roots and other roots of algebraic expressions. We will also consider their related functions.

Square Roots

When solving problems, we must often find what number must be squared to obtain a second number a. If such a number can be found, it is called a **square root** of a. For example,

- 0 is a square root of 0, because $0^2 = 0$.
- 4 is a square root of 16, because $4^2 = 16$.
- -4 is a square root of 16, because $(-4)^2 = 16$.
- $7xy$ is a square root of $49x^2y^2$, because $(7xy)^2 = 49x^2y^2$.
- $-7xy$ is a square root of $49x^2y^2$, because $(-7xy)^2 = 49x^2y^2$.

All positive numbers have two real-number square roots: one that is positive and one that is negative.

EXAMPLE 1 Find the two square roots of 121.

Solution The two square roots of 121 are 11 and -11, because

$$11^2 = 121 \qquad \text{and} \qquad (-11)^2 = 121$$

Self Check Find the square roots of 144.

To express square roots, we use the symbol $\sqrt{\ }$, called a **radical sign.** For example,

$\sqrt{121} = 11$ Read as "The positive square root of 121 is 11."

$-\sqrt{121} = -11$ Read as "The negative square root of 121 is -11."

The number under the radical sign is called a **radicand.**

Square Root of *a*

If $a > 0$, $\sqrt{a}$ is the positive number whose square is a. In symbols,

$$(\sqrt{a})^2 = a$$

The positive number $\sqrt{a}$ is called the **principal square root of *a*.** The principal square root of 0 is 0: $\sqrt{0} = 0$.

Comment Remember that the principal square root of a positive number is always positive. Although 5 and -5 are both square roots of 25, only 5 is the principal square root. The radical expression $\sqrt{25}$ represents 5. The radical expression $-\sqrt{25}$ represents -5.

Because of the previous definition, the square root of any number squared is that number. For example,

$$(\sqrt{10})^2 = \sqrt{10} \cdot \sqrt{10} = 10 \qquad (\sqrt{a})^2 = \sqrt{a} \cdot \sqrt{a} = a.$$

Comment These examples suggest that if any number a can be factored into two equal factors, either of those factors is a square root of a.

EXAMPLE 2 Simplify each radical:

a. $\sqrt{1} = 1$

b. $\sqrt{81} = 9$

c. $-\sqrt{81} = -9$

d. $-\sqrt{225} = -15$

e. $\sqrt{\dfrac{1}{4}} = \dfrac{1}{2}$

f. $-\sqrt{\dfrac{16}{121}} = -\dfrac{4}{11}$

g. $\sqrt{0.04} = 0.2$

h. $-\sqrt{0.0009} = -0.03$

Self Check Simplify: **a.** $-\sqrt{49}$ and **b.** $\sqrt{\dfrac{25}{49}}$.

Numbers such as 1, 4, 9, 16, 49, and 1,600 are called **integer squares,** because each one is the square of an integer. The square root of every integer square is a rational number.

$\sqrt{1} = 1, \quad \sqrt{4} = 2, \quad \sqrt{9} = 3, \quad \sqrt{16} = 4, \quad \sqrt{49} = 7, \quad \sqrt{1{,}600} = 40$

The square roots of many positive integers are not rational numbers. For example, $\sqrt{11}$ is an **irrational number.** To find an approximate value of $\sqrt{11}$ with a calculator, we enter these numbers and press these keys.

11 2nd $\sqrt{\ }$ — On a scientific calculator.

2nd $\sqrt{\ }$ 11 ENTER — On a graphing calculator.

Either way, we will see that

$$\sqrt{11} \approx 3.31662479$$

Square roots of negative numbers are not real numbers. For example, $\sqrt{-9}$ is not a real number, because no real number squared equals -9. Square roots of negative numbers come from a set called **imaginary numbers,** which we will discuss later in this chapter.

Square Roots of Expressions with Variables

If $x \neq 0$, the positive number x^2 has x and $-x$ for its two square roots. To denote the positive square root of $\sqrt{x^2}$, we must know whether x is positive or negative.

If $x > 0$, we can write

$\sqrt{x^2} = x$ — $\sqrt{x^2}$ represents the positive square root of x^2, which is x.

If x is negative, then $-x > 0$, and we can write

$\sqrt{x^2} = -x$ — $\sqrt{x^2}$ represents the positive square root of x^2, which is $-x$.

If we don't know whether x is positive or negative, we must use absolute value symbols to guarantee that $\sqrt{x^2}$ is positive.

Definition of $\sqrt{x^2}$

If x can be any real number then

$$\sqrt{x^2} = |x|$$

EXAMPLE 3 Simplify each expression. Assume that x can be any real number.

a. $\sqrt{16x^2} = \sqrt{(4x)^2}$ — Write $16x^2$ as $(4x)^2$.

$= |4x|$ — Because $(|4x|)^2 = 16x^2$. Since x could be negative, absolute value symbols are needed.

$= 4|x|$ — Since 4 is a positive constant in the product $4x$, we can write it outside the absolute value symbols.

b. $\sqrt{x^2 + 2x + 1}$

$= \sqrt{(x + 1)^2}$ — Factor $x^2 + 2x + 1$.

$= |x + 1|$ — Because $(x + 1)^2 = x^2 + 2x + 1$. Since $x + 1$ can be negative (for example, when $x = -5$), absolute value symbols are needed.

c. $\sqrt{x^4} = x^2$ — Because $(x^2)^2 = x^4$. Since $x^2 \geq 0$, no absolute value symbols are needed.

Self Check Simplify: **a.** $\sqrt{25a^2}$ and **b.** $\sqrt{16a^4}$.

Cube Roots

The **cube root of x** is any number whose cube is x. For example,

4 is a cube root of 64, because $4^3 = 64$.
$3x^2y$ is a cube root of $27x^6y^3$, because $(3x^2y)^3 = 27x^6y^3$.
$-2y$ is a cube root of $-8y^3$, because $(-2y)^3 = -8y^3$.

Cube Roots

The **cube root of a** is denoted as $\sqrt[3]{a}$ and is the number whose cube is a. In symbols,

$$\left(\sqrt[3]{a}\right)^3 = a$$

Comment The previous definition implies that if a can be factored into three equal factors, any one of those factors is a cube root of a.

We note that 64 has two real-number square roots, 8 and -8. However, 64 has only one real-number cube root, 4, because 4 is the only real number whose cube is 64. Since every real number has exactly one real cube root, it is unnecessary to use absolute value symbols when simplifying cube roots.

Definition of $\sqrt[3]{x^3}$

If x is any real number, then

$$\sqrt[3]{x^3} = x$$

EXAMPLE 4 Simplify each radical:

a. $\sqrt[3]{125} = 5$ Because $5^3 = 5 \cdot 5 \cdot 5 = 125$.

b. $\sqrt[3]{\frac{1}{8}} = \frac{1}{2}$ Because $\left(\frac{1}{2}\right)^3 = \frac{1}{2} \cdot \frac{1}{2} \cdot \frac{1}{2} = \frac{1}{8}$.

c. $\sqrt[3]{-27x^3} = -3x$ Because $(-3x)^3 = (-3x)(-3x)(-3x) = -27x^3$.

d. $\sqrt[3]{-\frac{8a^3}{27b^3}} = -\frac{2a}{3b}$ Because $\left(-\frac{2a}{3b}\right)^3 = \left(-\frac{2a}{3b}\right)\left(-\frac{2a}{3b}\right)\left(-\frac{2a}{3b}\right) = -\frac{8a^3}{27b^3}$.

e. $\sqrt[3]{0.216x^3y^6} = 0.6xy^2$ Because $(0.6xy^2)^3 = (0.6xy^2)(0.6xy^2)(0.6xy^2) = 0.216x^3y^6$.

Self Check Simplify: **a.** $\sqrt[3]{1{,}000}$, **b.** $\sqrt[3]{\frac{1}{27}}$, and **c.** $\sqrt[3]{125a^3}$.

*n*th Roots

Just as there are square roots and cube roots, there are fourth roots, fifth roots, sixth roots, and so on.

When n is an odd natural number greater than 1, $\sqrt[n]{x}$ represents an **odd root.** Since every real number has only one real nth root when n is odd, we don't need to use absolute value symbols when finding odd roots. For example,

$\sqrt[5]{243} = 3$ because $3^5 = 243$

$\sqrt[7]{-128x^7} = -2x$ because $(-2x)^7 = -128x^7$

When n is an even natural number greater than 1, $\sqrt[n]{x}$ represents an **even root.** In this case, there will be one positive and one negative real nth root. For example, the two real sixth roots of 729 are 3 and -3, because $3^6 = 729$ and $(-3)^6 = 729$. When finding even roots, we use absolute value symbols to guarantee that the principal nth root is positive.

$\sqrt[4]{(-3)^4} = |-3| = 3$ — Because $3^4 = (-3)^4$. We could also simplify this as follows: $\sqrt[4]{(-3)^4} = \sqrt[4]{81} = 3$.

$\sqrt[6]{729x^6} = |3x| = 3|x|$ — Because $(3|x|)^6 = 729x^6$. The absolute value symbols guarantee that the sixth root is positive.

In the radical $\sqrt[n]{x}$, n is called the **index** (or **order**) of the radical. When the index is 2, the radical is a square root, and we usually do not write the index.

$$\sqrt[2]{x} = \sqrt{x}$$

Comment When n is an even number greater than 1, and $x < 0$, $\sqrt[n]{x}$ is not a real number. For example, $\sqrt[4]{-81}$ is not a real number, because no real number raised to the 4th power is -81.

EXAMPLE 5 Simplify each radical:

a. $\sqrt[4]{625} = 5$, because $5^4 = 625$ — Read $\sqrt[4]{625}$ as "the fourth root of 625."

b. $\sqrt[5]{-32} = -2$, because $(-2)^5 = -32$ — Read $\sqrt[5]{-32}$ as "the fifth root of -32."

c. $\sqrt[6]{\frac{1}{64}} = \frac{1}{2}$, because $\left(\frac{1}{2}\right)^6 = \frac{1}{64}$ — Read $\sqrt[6]{\frac{1}{64}}$ as "the sixth root of $\frac{1}{64}$."

d. $\sqrt[7]{10^7} = 10$, because $10^7 = 10^7$ — Read $\sqrt[7]{10^7}$ as "the seventh root of 10^7."

Self Check Simplify: **a.** $\sqrt[4]{\frac{1}{81}}$ and **b.** $\sqrt[5]{10^5}$.

When finding the nth root of an nth power, we can use the following rules.

Definition of $\sqrt[n]{a^n}$

If n is an odd natural number greater than 1, then $\sqrt[n]{a^n} = a$.

If n is an even natural number, then $\sqrt[n]{a^n} = |a|$.

EXAMPLE 6 Simplify each radical. Assume that x can be any real number.

Solution

a. $\sqrt[5]{x^5} = x$ — Since n is odd, absolute value symbols aren't needed.

b. $\sqrt[4]{16x^4} = |2x| = 2|x|$ — Since n is even and x can be negative, absolute value symbols are needed to guarantee that the result is positive.

c. $\sqrt[6]{(x+4)^6} = |x+4|$ — Absolute value symbols are needed to guarantee that the result is positive.

d. $\sqrt[3]{(x+1)^3} = x + 1$ — Since n is odd, absolute value symbols aren't needed.

e. $\sqrt{(x^2 + 4x + 4)^2} = \sqrt{[(x+2)^2]^2}$ — Factor $x^2 + 4x + 4$.

$= \sqrt{(x+2)^4}$

$= (x+2)^2$ — Since $(x+2)^2$ is always positive, absolute value symbols aren't needed.

Self Check Simplify: **a.** $\sqrt[4]{16a^4}$ and **b.** $\sqrt[5]{(a+5)^5}$.

We summarize the possibilities for $\sqrt[n]{x}$ as follows.

Definition for $\sqrt[n]{x}$

If n is a natural number greater than 1 and x is a real number, then

If $x > 0$, then $\sqrt[n]{x}$ is the positive number such that $\left(\sqrt[n]{x}\right)^n = x$.

If $x = 0$, then $\sqrt[n]{x} = 0$.

If $x < 0$ $\begin{cases} \text{and } n \text{ is odd, then } \sqrt[n]{x} \text{ is the real number such that } \left(\sqrt[n]{x}\right)^n = x. \\ \text{and } n \text{ is even, then } \sqrt[n]{x} \text{ is not a real number.} \end{cases}$

Square Root and Cube Root Functions

Since there is one principal square root for every nonnegative real number x, the equation $f(x) = \sqrt{x}$ determines a function, called the **square root function.**

EXAMPLE 7 Graph $f(x) = \sqrt{x}$ and find its domain and range.

Solution We can make a table of values and plot points to get the graph shown in Figure 9-1(a), or we can use a graphing calculator with window settings of $[-1, 9]$ for x and $[-2, 5]$ for y to get the graph shown in Figure 9-1(b) Since the equation defines a function, its graph passes the vertical line test.

$f(x) = \sqrt{x}$

x	$f(x)$	$(x, f(x))$
0	0	(0, 0)
1	1	(1, 1)
4	2	(4, 2)
9	3	(9, 3)

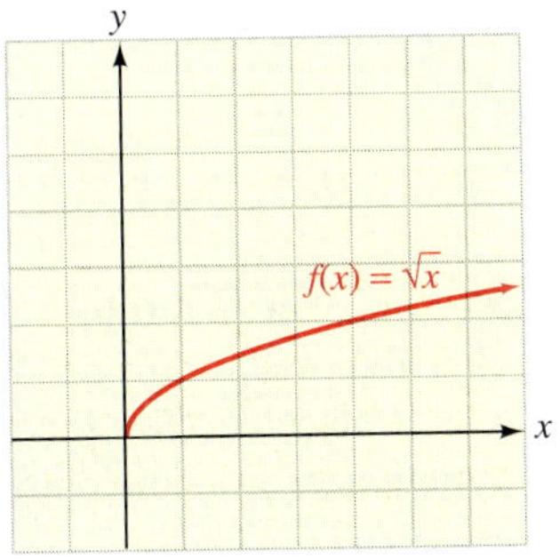

(a)

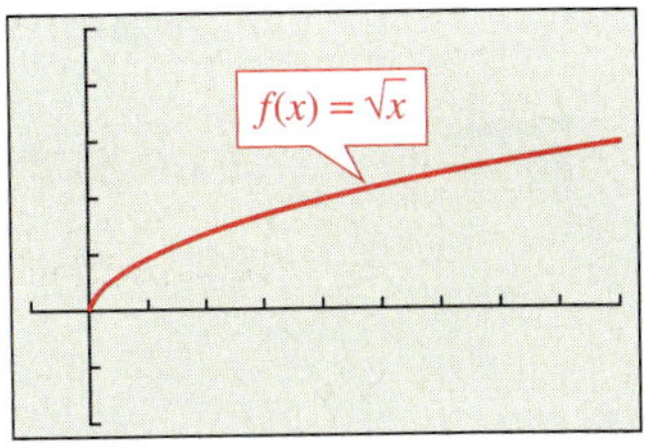

(b)

Figure 9-1

From either graph, we can see that the domain and the range are the set of nonnegative real numbers, which is the interval $[0, \infty)$.

Self Check Graph $f(x) = \sqrt{x} + 2$ and compare it to the graph of $f(x) = \sqrt{x}$.

The graphs of many functions are translations or reflections of the square root function. For example, if $k > 0$,

- The graph of $f(x) = \sqrt{x} + k$ is the graph of $f(x) = \sqrt{x}$ translated k units up.
- The graph of $f(x) = \sqrt{x} - k$ is the graph of $f(x) = \sqrt{x}$ translated k units down.
- The graph of $f(x) = \sqrt{x + k}$ is the graph of $f(x) = \sqrt{x}$ translated k units to the left.

- The graph of $f(x) = \sqrt{x - k}$ is the graph of $f(x) = \sqrt{x}$ translated k units to the right.
- The graph of $f(x) = -\sqrt{x}$ is the graph of $f(x) = \sqrt{x}$ reflected about the x-axis.

EXAMPLE 8 Graph $f(x) = -\sqrt{x + 4} - 2$ and find its domain and range.

Solution This graph will be the reflection of $f(x) = \sqrt{x}$ about the x-axis, translated 4 units to the left and 2 units down. See Figure 9-2(a). We can confirm this graph by using a graphing calculator with window settings of $[-5, 6]$ for x and $[-6, 2]$ for y to get the graph shown in Figure 9-2(b).

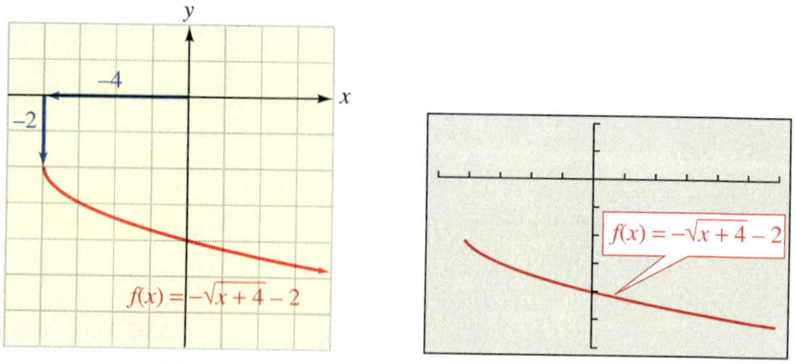

(a) (b)

Figure 9-2

From either graph, we can see that the domain is the interval $[-4, \infty)$ and that the range is the interval $(-\infty, -2]$.

Self Check Graph: $f(x) = \sqrt{x - 2} - 4$.

EXAMPLE 9 **Period of a pendulum** The **period of a pendulum** is the time required for the pendulum to swing back and forth to complete one cycle. (See Figure 9-3.) The period t (in seconds) is a function of the pendulum's length l, which is defined by the formula

$$t = f(l) = 2\pi\sqrt{\frac{l}{32}}$$

Find the period of a pendulum that is 5 feet long.

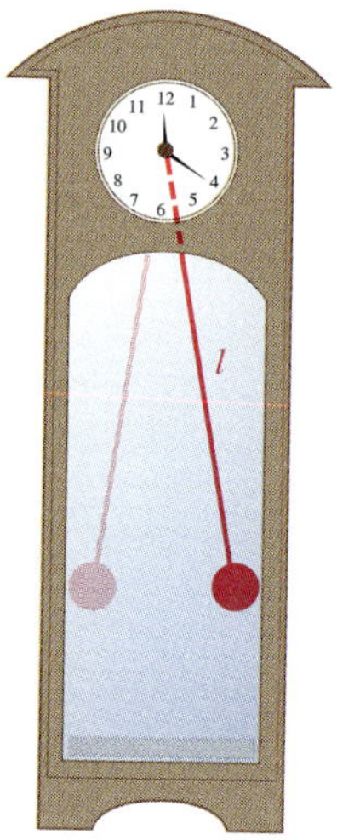

Figure 9-3

Solution We substitute 5 for l in the formula and simplify.

$$t = 2\pi\sqrt{\frac{l}{32}}$$

$$t = 2\pi\sqrt{\frac{5}{32}}$$

$$\approx 2.483647066 \quad \text{Use a calculator.}$$

To the nearest tenth, the period is 2.5 seconds.

Self Check To the nearest hundredth, find the period of a pendulum that is 3 feet long.

Accent on Technology — FINDING THE PERIOD OF A PENDULUM

To solve Example 9 with a graphing calculator with window settings of [−2, 10] for x and [−2, 10] for y, we graph the function $f(x) = 2\pi\sqrt{\frac{x}{32}}$, as in Figure 9-4(a). We then trace and move the cursor toward an x value of 5 until we see the coordinates shown in Figure 9-4(b). The period is given by the y-value shown in the screen. By zooming in, we can get better results.

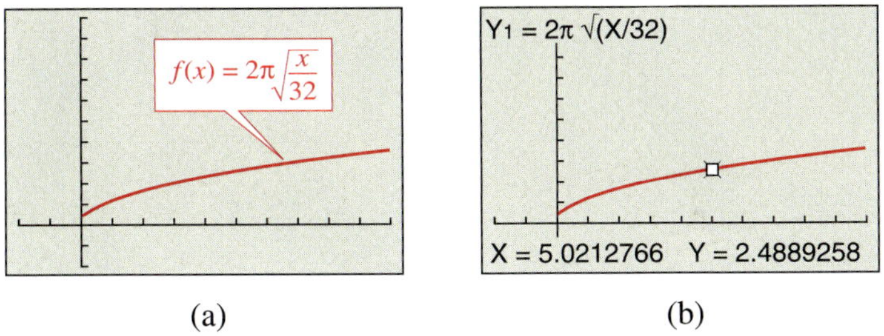

Figure 9-4

The equation $f(x) = \sqrt[3]{x}$ defines a **cube root function.** From the graph shown in Figure 9-5(a), we can see that the domain and range of the function $f(x) = \sqrt[3]{x}$ are the set of real numbers. Note that the graph of $f(x) = \sqrt[3]{x}$ passes the vertical line test. Figures 9-5(b) and 9-5(c) show several translations of the cube root function.

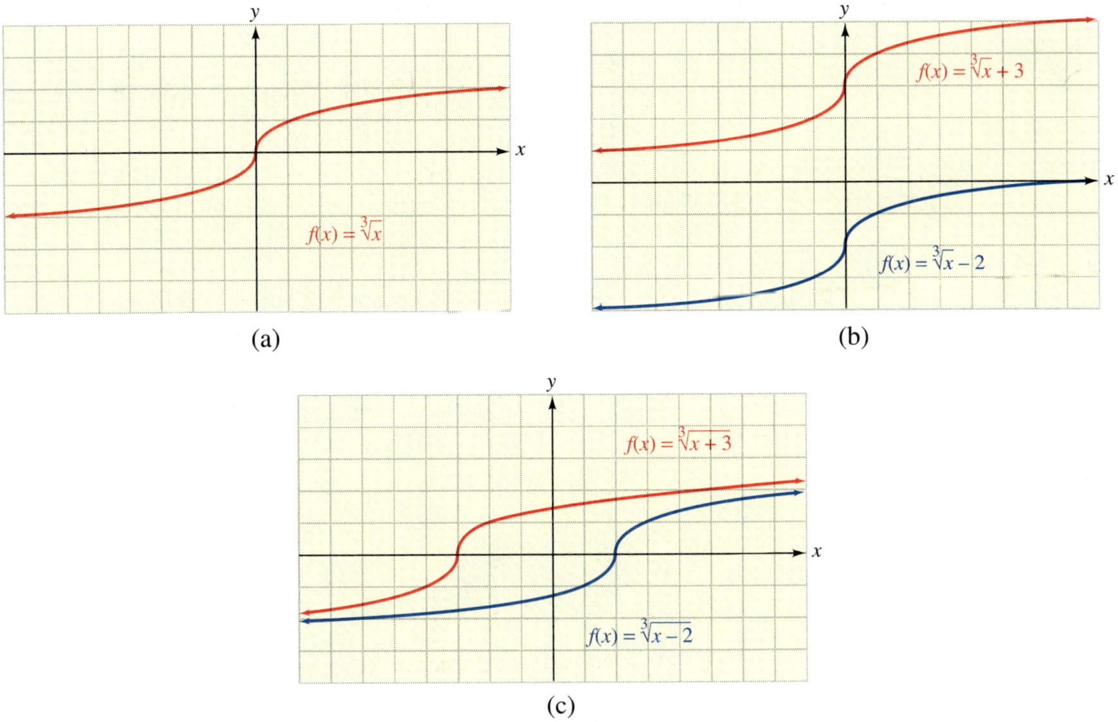

Figure 9-5

Standard Deviation

In statistics, the **standard deviation** is used to tell which of a set of distributions is the most variable. To see how to compute the standard deviation of a distribution, we consider the distribution 4, 5, 5, 8, 13 and construct the following table.

Original terms	Mean of the distribution	Differences (original term minus mean)	Squares of the differences from the mean
4	7	−3	9
5	7	−2	4
5	7	−2	4
8	7	1	1
13	7	6	36

The population **standard deviation** of the distribution is the positive square root of the mean of the numbers shown in column 4 of the table.

$$\text{Standard deviation} = \sqrt{\frac{\text{sum of the squares of the differences from the mean}}{\text{number of differences}}}$$

$$= \sqrt{\frac{9 + 4 + 4 + 1 + 36}{5}}$$

$$= \sqrt{\frac{54}{5}}$$

$$\approx 3.286335345 \quad \text{Use a calculator.}$$

To the nearest hundredth, the standard deviation of the given distribution is 3.29.

The symbol for the population standard deviation is σ, the lowercase Greek letter *sigma.*

EXAMPLE 10 Which of the following distributions has the most variability: **a.** 3, 5, 7, 8, 12 or **b.** 1, 4, 6, 11?

Solution We compute the standard deviation of each distribution.

a.

Original terms	Mean of the distribution	Differences (original term minus mean)	Squares of the differences from the mean
3	7	−4	16
5	7	−2	4
7	7	0	0
8	7	1	1
12	7	5	25

$$\sigma = \sqrt{\frac{16 + 4 + 0 + 1 + 25}{5}} = \sqrt{\frac{46}{5}} \approx 3.03$$

b.

Original terms	Mean of the distribution	Differences (original term minus mean)	Squares of the differences from the mean
1	5.5	−4.5	20.25
4	5.5	−1.5	2.25
6	5.5	0.5	0.25
11	5.5	5.5	30.25

$$\sigma = \sqrt{\frac{20.25 + 2.25 + 0.25 + 30.25}{4}} = \sqrt{\frac{53}{4}} \approx 3.64$$

Since the standard deviation for the second distribution is greater than the standard deviation for the first distribution, the second distribution has the greater variability.

Self Check Answers

1. 12, −12 **2. a.** −7, **b.** $\frac{5}{7}$ **3. a.** $5|a|$, **b.** $4a^2$ **4. a.** 10, **b.** $\frac{1}{3}$, **c.** $5a$ **5. a.** $\frac{1}{3}$, **b.** 10 **6. a.** $2|a|$, **b.** $a + 5$ **7.** It is 2 units higher.

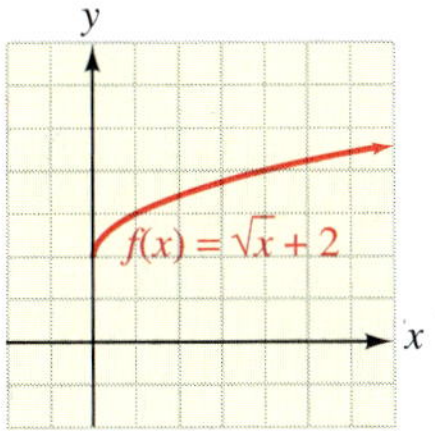

8.

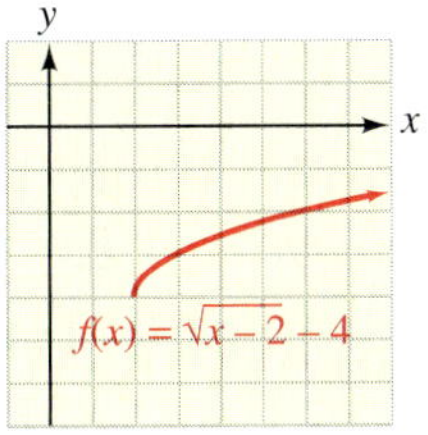

9. 1.92 sec

Orals *Simplify each radical, if possible.*

1. $\sqrt{9}$ **2.** $-\sqrt{16}$ **3.** $\sqrt[3]{-8}$ **4.** $\sqrt[5]{32}$

5. $\sqrt{64x^2}$ **6.** $\sqrt[3]{-27x^3}$

7. $\sqrt{-3}$ **8.** $\sqrt[4]{(x+1)^8}$

9.1 EXERCISES

REVIEW *Simplify each rational expression.*

1. $\frac{x^2 + 7x + 12}{x^2 - 16}$

2. $\frac{a^3 - b^3}{b^2 - a^2}$

Perform the operations.

3. $\frac{x^2 - x - 6}{x^2 - 2x - 3} \cdot \frac{x^2 - 1}{x^2 + x - 2}$

4. $\frac{x^2 - 3x - 4}{x^2 - 5x + 6} \div \frac{x^2 - 2x - 3}{x^2 - x - 2}$

5. $\frac{3}{m + 1} + \frac{3m}{m - 1}$

6. $\frac{2x + 3}{3x - 1} - \frac{x - 4}{2x + 1}$

VOCABULARY AND CONCEPTS *Fill in the blanks.*

7. $5x^2$ is the square root of $25x^4$, because ______ $= 25x^4$.

8. 6 is a square root of 36 because ______.

9. The principal square root of x ($x > 0$) is the ______ square root of x.

10. $\sqrt{x^2} =$ ___.

11. The graph of $f(x) = \sqrt{x} + 3$ is the graph of $f(x) = \sqrt{x}$ translated __ units __.
12. The graph of $f(x) = \sqrt{x + 5}$ is the graph of $f(x) = \sqrt{x}$ translated __ units to the __.
13. $\left(\sqrt[3]{x}\right)^3 =$ __
14. $\sqrt[3]{x^3} =$ __
15. When n is an odd number greater than 1, $\sqrt[n]{x}$ represents an ____ root.
16. When n is a positive ____ number, $\sqrt[n]{x}$ represents an even root.
17. $\sqrt{0} =$ __
18. The ______ deviation of a set of numbers is the positive square root of the mean of the squares of the differences of the numbers from the mean.

Identify the radicand in each expression.

19. $\sqrt{3x^2}$
20. $5\sqrt{x}$
21. $ab^2\sqrt{a^2 + b^3}$
22. $\frac{1}{2}x\sqrt{\frac{x}{y}}$

PRACTICE *Find each square root, if possible.*

23. $\sqrt{121}$
24. $\sqrt{144}$
25. $-\sqrt{64}$
26. $-\sqrt{1}$
27. $\sqrt{\frac{1}{9}}$
28. $-\sqrt{\frac{4}{25}}$
29. $-\sqrt{\frac{25}{49}}$
30. $\sqrt{\frac{49}{81}}$
31. $\sqrt{-25}$
32. $\sqrt{0.25}$
33. $\sqrt{0.16}$
34. $\sqrt{-49}$
35. $\sqrt{(-4)^2}$
36. $\sqrt{(-9)^2}$
37. $\sqrt{-36}$
38. $-\sqrt{-4}$

Use a calculator to find each square root. Give the answer to four decimal places.

39. $\sqrt{12}$
40. $\sqrt{340}$
41. $\sqrt{679.25}$
42. $\sqrt{0.0063}$

Find each square root. Assume that all variables are unrestricted, and use absolute value symbols when necessary.

43. $\sqrt{4x^2}$
44. $\sqrt{16y^4}$
45. $\sqrt{9a^4}$
46. $\sqrt{16b^2}$
47. $\sqrt{(t + 5)^2}$
48. $\sqrt{(a + 6)^2}$
49. $\sqrt{(-5b)^2}$
50. $\sqrt{(-8c)^2}$
51. $\sqrt{a^2 + 6a + 9}$
52. $\sqrt{x^2 + 10x + 25}$
53. $\sqrt{t^2 + 24t + 144}$
54. $\sqrt{m^2 + 30m + 225}$

Simplify each cube root.

55. $\sqrt[3]{1}$
56. $\sqrt[3]{-8}$
57. $\sqrt[3]{-125}$
58. $\sqrt[3]{512}$
59. $\sqrt[3]{-\frac{8}{27}}$
60. $\sqrt[3]{\frac{125}{216}}$
61. $\sqrt[3]{0.064}$
62. $\sqrt[3]{0.001}$
63. $\sqrt[3]{8a^3}$
64. $\sqrt[3]{-27x^6}$
65. $\sqrt[3]{-1{,}000p^3q^3}$
66. $\sqrt[3]{343a^6b^3}$
67. $\sqrt[3]{-\frac{1}{8}m^6n^3}$
68. $\sqrt[3]{\frac{27}{1{,}000}a^6b^6}$
69. $\sqrt[3]{0.008z^9}$
70. $\sqrt[3]{0.064s^9t^6}$

Simplify each radical, if possible.

71. $\sqrt[4]{81}$
72. $\sqrt[6]{64}$
73. $-\sqrt[5]{243}$
74. $-\sqrt[4]{625}$
75. $\sqrt[5]{-32}$
76. $\sqrt[6]{729}$
77. $\sqrt[4]{\frac{16}{625}}$
78. $\sqrt[5]{-\frac{243}{32}}$
79. $-\sqrt[5]{-\frac{1}{32}}$
80. $\sqrt[6]{-729}$
81. $\sqrt[4]{-256}$
82. $-\sqrt[4]{\frac{81}{256}}$

Simplify each radical. Assume that all variables are unrestricted, and use absolute value symbols where necessary.

83. $\sqrt[4]{16x^4}$
84. $\sqrt[5]{32a^5}$
85. $\sqrt[3]{8a^3}$
86. $\sqrt[6]{64x^6}$
87. $\sqrt[4]{\frac{1}{16}x^4}$
88. $\sqrt[4]{\frac{1}{81}x^8}$
89. $\sqrt[4]{x^{12}}$
90. $\sqrt[8]{x^{24}}$
91. $\sqrt[5]{-x^5}$
92. $\sqrt[3]{-x^6}$
93. $\sqrt[3]{-27a^6}$
94. $\sqrt[5]{-32x^5}$

Simplify each radical. Assume that all variables are unrestricted, and use absolute value symbols when necessary.

95. $\sqrt[25]{(x+2)^{25}}$

96. $\sqrt[44]{(x+4)^{44}}$

97. $\sqrt[8]{0.00000001x^{16}y^8}$

98. $\sqrt[5]{0.00032x^{10}y^5}$

Find each value given that $f(x) = \sqrt{x-4}$.

99. $f(4)$

100. $f(8)$

101. $f(20)$

102. $f(29)$

Find each value given that $g(x) = \sqrt{x-8}$

103. $g(9)$

104. $g(17)$

105. $g(8.25)$

106. $g(8.64)$

Find each value given that $f(x) = \sqrt{x^2+1}$. *Give each answer to four decimal places.*

107. $f(4)$

108. $f(6)$

109. $f(2.35)$

110. $f(21.57)$

Graph each function and find its domain and range.

111. $f(x) = \sqrt{x+4}$

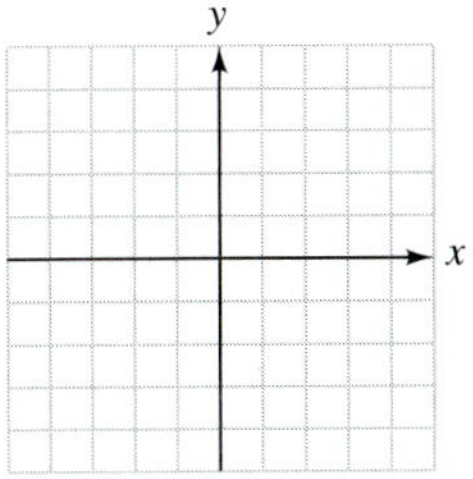

112. $f(x) = -\sqrt{x-2}$

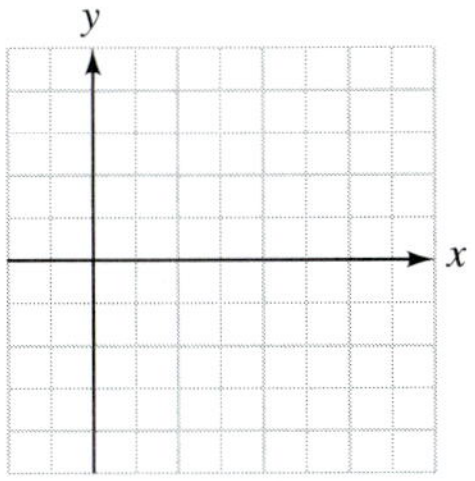

113. $f(x) = -\sqrt{x} - 3$

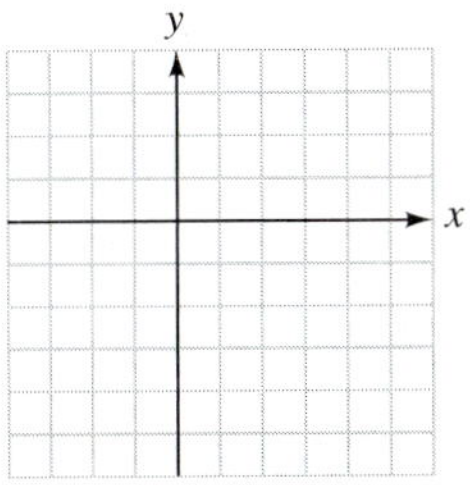

114. $f(x) = \sqrt[3]{x} - 1$

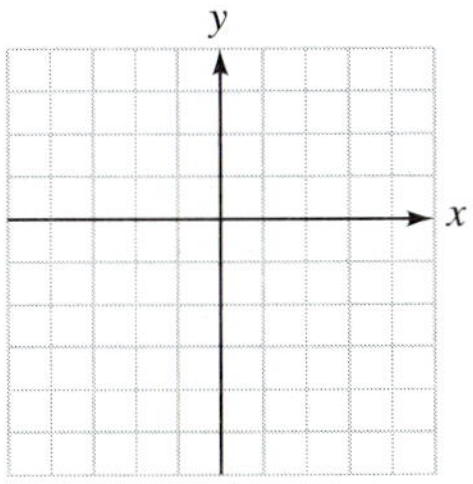

115. Find the standard deviation of the following distribution to the nearest hundredth: 2, 5, 5, 6, 7.

116. Find the standard deviation of the following distribution to the nearest hundredth: 3, 6, 7, 9, 11, 12.

117. **Statistics** In statistics, the formula

$$s_{\bar{x}} = \frac{s}{\sqrt{N}}$$

gives an estimate of the standard error of the mean. Find $s_{\bar{x}}$ to four decimal places when $s = 65$ and $N = 30$.

118. **Statistics** In statistics, the formula

$$\sigma_{\bar{x}} = \frac{\sigma}{\sqrt{N}}$$

gives the standard deviation of means of samples of size N. Find $\sigma_{\bar{x}}$ to four decimal places when $\sigma = 12.7$ and $N = 32$.

APPLICATIONS *Use a calculator to solve each problem.*

119. **Radius of a circle** The radius r of a circle is given by the formula $r = \sqrt{\frac{A}{\pi}}$, where A is its area. Find the radius of a circle whose area is 9π square units.

120. **Diagonal of a baseball diamond** The diagonal d of a square is given by the formula $d = \sqrt{2s^2}$, where s is the length of each side. Find the diagonal of the baseball diamond.

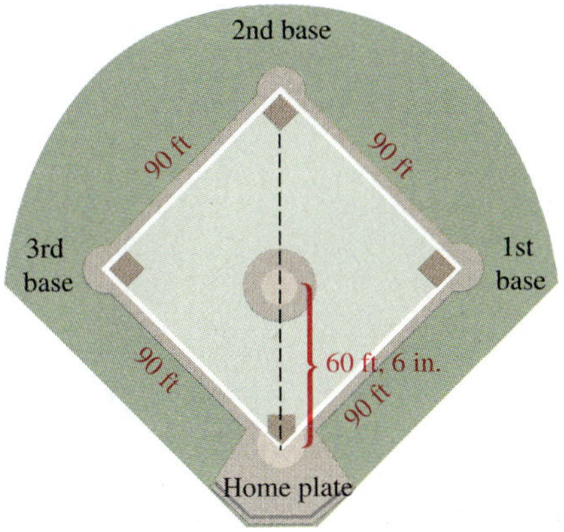

121. **Falling objects** The time t (in seconds) that it will take for an object to fall a distance of s feet is given by the formula

$$t = \frac{\sqrt{s}}{4}$$

If a stone is dropped down a 256-foot well, how long will it take it to hit bottom?

122. Law enforcement Police sometimes use the formula $s = k\sqrt{l}$ to estimate the speed s (in mph) of a car involved in an accident. In this formula, l is the length of the skid in feet, and k is a constant depending on the condition of the pavement. For wet pavement, $k \approx 3.24$. How fast was a car going if its skid was 400 feet on wet pavement?

123. Electronics When the resistance in a circuit is 18 ohms, the current I (measured in amperes) and the power P (measured in watts) are related by the formula

$$I = \sqrt{\frac{P}{18}}$$

Find the current used by an electrical appliance that is rated at 980 watts.

124. Medicine The approximate pulse rate p (in beats per minute) of an adult who is t inches tall is given by the formula

$$p = \frac{590}{\sqrt{t}}$$

Find the approximate pulse rate of an adult who is 71 inches tall.

WRITING

125. If x is any real number, then $\sqrt{x^2} = x$ is not correct. Explain.

126. If x is any real number, then $\sqrt[3]{x^3} = |x|$ is not correct. Explain.

SOMETHING TO THINK ABOUT

127. Is $\sqrt{x^2 - 4x + 4} = x - 2$? What are the exceptions?

128. When is $\sqrt{x^2} \neq x$?

9.2 Applications of Radicals

- The Pythagorean Theorem
- The Distance Formula

Getting Ready *Evaluate each expression.*

1. $3^2 + 4^2$

2. $5^2 + 12^2$

3. $(5 - 2)^2 + (2 + 1)^2$

4. $(111 - 21)^2 + (60 - 4)^2$

In this section, we will discuss the Pythagorean theorem, a theorem that shows the relationship of the sides of a right triangle. We will then use this theorem to develop a formula that gives the distance between two points on the coordinate plane.

The Pythagorean Theorem

If we know the lengths of two legs of a right triangle, we can find the length of the **hypotenuse** (the side opposite the 90° angle) by using the **Pythagorean theorem.**

Pythagorean Theorem

If a and b are the lengths of two legs of a right triangle and c is the length of the hypotenuse, then

$$a^2 + b^2 = c^2$$

In words, the Pythagorean theorem says,

In any right triangle, the square of the hypotenuse is equal to the sum of the squares of the two legs.

Since the lengths of the sides of a triangle are positive numbers, we can use the **square root property of equality** and the Pythagorean theorem to find the length of the third side of any right triangle when the measures of two sides are given.

Square Root Property of Equality

Let a and b represent positive numbers,

if $a = b$, then $\sqrt{a} = \sqrt{b}$.

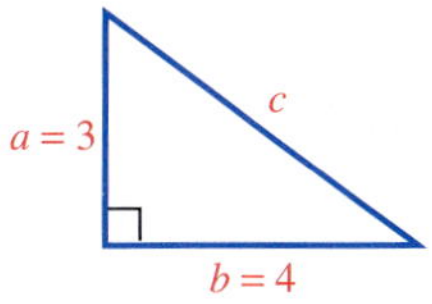

Figure 9-6

For example, suppose the right triangle shown in Figure 9-6 has legs of length 3 and 4 units. To find the length of the hypotenuse, we use the Pythagorean theorem.

$$a^2 + b^2 = c^2$$
$$3^2 + 4^2 = c^2$$
$$9 + 16 = c^2$$
$$25 = c^2$$
$$\sqrt{25} = \sqrt{c^2}$$ Use the square root property and take the positive square root of both sides.
$$5 = c$$

The length of the hypotenuse is 5 units.

EXAMPLE 1 **Fighting fires** To fight a forest fire, the forestry department plans to clear a rectangular fire break around the fire, as shown in Figure 9-7. Crews are equipped with mobile communications with a 3,000-yard range. Can crews at points A and B remain in radio contact?

Solution Points A, B, and C form a right triangle. To find the distance c from point A to point B, we can use the Pythagorean theorem, substituting 2,400 for a and 1,000 for b and solving for c.

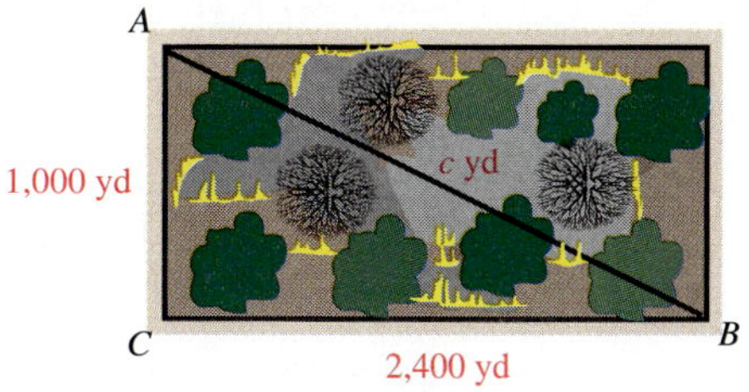

Figure 9-7

$$a^2 + b^2 = c^2$$
$$2{,}400^2 + 1{,}000^2 = c^2$$
$$5{,}760{,}000 + 1{,}000{,}000 = c^2$$
$$6{,}760{,}000 = c^2$$
$$\sqrt{6{,}760{,}000} = \sqrt{c^2}$$ Take the positive square root of both sides.
$$2{,}600 = c$$ Use a calculator to find the square root.

The two crews are 2,600 yards apart. Because this distance is less than the range of the radios, they can communicate.

Self Check Can the crews communicate if $b = 1{,}500$ yards?

PERSPECTIVE

Pythagoras was a teacher. Although it was unusual at that time, his classes were coeducational. He and his followers formed a secret society with two rules: Membership was for life, and members could not reveal the secrets they knew.

Much of their teaching was good mathematics, but some ideas were strange. To them, numbers were sacred. Because beans were used as counters to represent numbers, Pythagoreans refused to eat beans. They also believed that the *only* numbers were the whole numbers. To them, fractions were not numbers; $\frac{2}{3}$ was just a way of comparing the whole numbers 2 and 3. They believed that whole numbers were the building blocks of the universe. The basic Pythagorean doctrine was, "All things are number," and they meant *whole* number.

The Pythagorean theorem was an important discovery of the Pythagorean school, yet it caused some controversy. The right triangle in the illustration has two legs of length 1. By the Pythagorean theorem, the length of the hypotenuse is $\sqrt{2}$. One of their own group, Hippasus of Metapontum, discovered that $\sqrt{2}$ is an irrational number: There are *no* whole numbers a and b that make the fraction $\frac{a}{b}$ exactly equal to $\sqrt{2}$. This discovery was not appreciated by the other Pythagoreans. How could everything in the universe be described with whole numbers, when the side of this simple triangle couldn't? The Pythagoreans had a choice. Either revise and expand their beliefs, or cling to the old. According to legend, the group was at sea at the time of the discovery. Rather than upset the system, they threw Hippasus overboard.

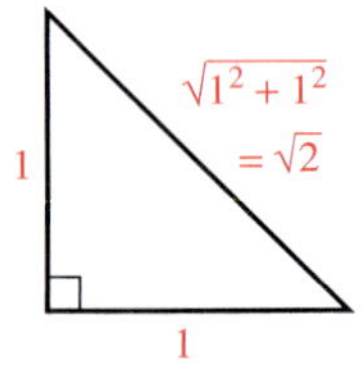

The Distance Formula

We can use the Pythagorean theorem to develop a formula to find the distance between any two points that are graphed on a rectangular coordinate system.

To find the distance d between points P and Q shown in Figure 9-8, we construct the right triangle PRQ. The distance between P and R is $|x_2 - x_1|$, and the distance between R and Q is $|y_2 - y_1|$. We apply the Pythagorean theorem to the right triangle PRQ to get

$$[d(PQ)]^2 = |x_2 - x_1|^2 + |y_2 - y_1|^2$$ Read $d(PQ)$ as "the distance between P and Q."

$$= (x_2 - x_1)^2 + (y_2 - y_1)^2$$ Because $|x_2 - x_1|^2 = (x_2 - x_1)^2$ and $|y_2 - y_1|^2 = (y_2 - y_1)^2$.

or

(1) $$d(PQ) = \sqrt{(x_2 - x_1)^2 + (y_2 - y_1)^2}$$

Equation 1 is called the **distance formula.**

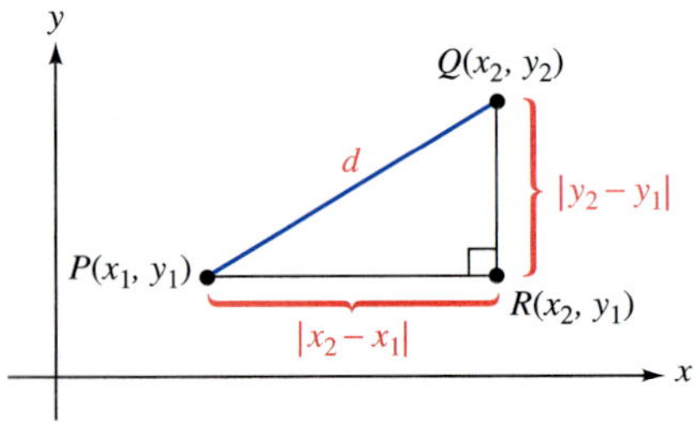

Figure 9-8

Distance Formula

The distance between two points $P(x_1, y_1)$ and $Q(x_2, y_2)$ is given by the formula

$$d(PQ) = \sqrt{(x_2 - x_1)^2 + (y_2 - y_1)^2}$$

EXAMPLE 2 Find the distance between points $(-2, 3)$ and $(4, -5)$.

Solution To find the distance, we can use the distance formula by substituting 4 for x_2, -2 for x_1, -5 for y_2, and 3 for y_1.

$$\begin{aligned} d(PQ) &= \sqrt{(x_2 - x_1)^2 + (y_2 - y_1)^2} \\ &= \sqrt{[4 - (-2)]^2 + (-5 - 3)^2} \\ &= \sqrt{(4 + 2)^2 + (-5 - 3)^2} \\ &= \sqrt{6^2 + (-8)^2} \\ &= \sqrt{36 + 64} \\ &= \sqrt{100} \\ &= 10 \end{aligned}$$

The distance between P and Q is 10 units.

Self Check Find the distance between $P(-2, -2)$ and $Q(3, 10)$.

EXAMPLE 3 **Building a freeway** In a city, streets run north and south, and avenues run east and west. Streets and avenues are 850 feet apart. The city plans to construct a straight freeway from the intersection of 25th Street and 8th Avenue to the intersection of 115th Street and 64th Avenue. How long will the freeway be?

Solution We can represent the roads by the coordinate system in Figure 9-9, where the units on each axis represent 850 feet. We represent the end of the freeway at 25th Street and 8th Avenue by the point $(x_1, y_1) = (25, 8)$. The other end is $(x_2, y_2) = (115, 64)$.

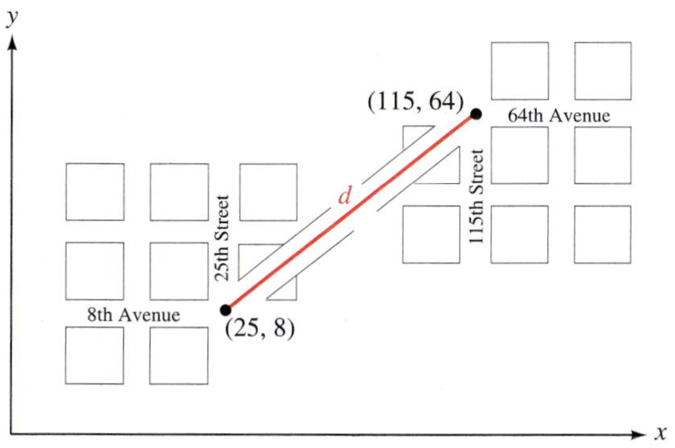

Figure 9-9

We can use the distance formula to find the length of the freeway.

$$d = \sqrt{(x_2 - x_1)^2 + (y_2 - y_1)^2}$$
$$d = \sqrt{(115 - 25)^2 + (64 - 8)^2}$$
$$= \sqrt{90^2 + 56^2}$$
$$= \sqrt{8{,}100 + 3{,}136}$$
$$= \sqrt{11{,}236}$$
$$= 106 \quad \text{Use a calculator.}$$

Because each unit is 850 feet, the length of the freeway is 106(850) = 90,100 feet. Since 5,280 feet = 1 mile, we can divide 90,100 by 5,280 to convert 90,100 feet to 17.064394 miles. Thus, the freeway will be about 17 miles long.

EXAMPLE 4 **Bowling** The velocity, v, of an object after it has fallen d feet is given by the equation $v^2 = 64d$. If an inexperienced bowler lofts the ball 4 feet, with what velocity does it strike the alley?

Solution We find the velocity by substituting 4 for d in the equation $v^2 = 64d$ and solving for v.

$$v^2 = 64d$$
$$v^2 = 64(4)$$
$$v^2 = 256$$
$$v = \sqrt{256} \quad \text{Take the positive square root of both sides.}$$
$$= 16$$

The ball strikes the alley with a velocity of 16 feet per second.

Self Check Answers

1. yes **2.** 13

Orals *Evaluate each expression.*

1. $\sqrt{25}$ **2.** $\sqrt{100}$ **3.** $\sqrt{169}$

4. $\sqrt{3^2 + 4^2}$ **5.** $\sqrt{8^2 + 6^2}$ **6.** $\sqrt{5^2 + 12^2}$

7. $\sqrt{5^2 - 3^2}$ **8.** $\sqrt{5^2 - 4^2}$ **9.** $\sqrt{169 - 12^2}$

9.2 EXERCISES

REVIEW *Find each product.*

1. $(4x + 2)(3x - 5)$

2. $(3y - 5)(2y + 3)$

3. $(5t + 4s)(3t - 2s)$

4. $(4r - 3)(2r^2 + 3r - 4)$

VOCABULARY AND CONCEPTS *Fill in the blanks.*

5. In a right triangle, the side opposite the 90° angle is called the ________.

6. In a right triangle, the two shorter sides are called ____.

7. If a and b are the lengths of two legs of a right triangle and c is the length of the hypotenuse, then ________.

8. In any right triangle, the square of the hypotenuse is equal to the ____ of the squares of the two ____.

9. With the ______ formula, we can find the distance between two points on a rectangular coordinate system.

10. $d(PQ) =$ ______________

PRACTICE *The lengths of two sides of the right triangle ABC shown in the illustration are given. Find the length of the missing side.*

11. $a = 6$ ft and $b = 8$ ft

12. $a = 10$ cm and $c = 26$ cm

13. $b = 18$ m and $c = 82$ m

14. $b = 7$ ft and $c = 25$ ft

15. $a = 14$ in. and $c = 50$ in.

16. $a = 8$ cm and $b = 15$ cm

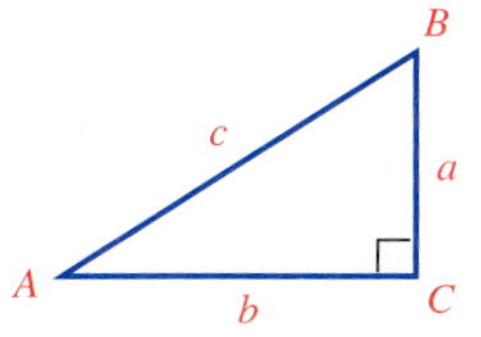

APPLICATIONS *Give each answer to the nearest tenth.*

17. Geometry Find the length of the diagonal of one of the faces of the cube.

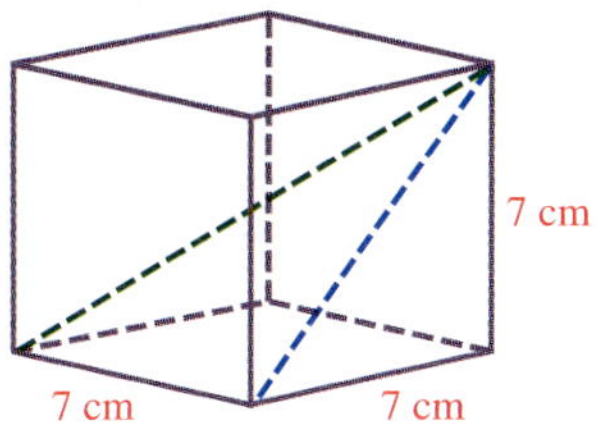

18. Geometry Find the length of the diagonal of the cube shown in the illustration above.

Find the distance between the given points. If an answer is not exact, use a calculator and give the answer to the nearest tenth.

19. $(0, 0), (3, -4)$

20. $(0, 0), (-6, 8)$

21. $(2, 4), (5, 8)$

22. $(5, 9), (8, 13)$

23. $(-2, -8), (3, 4)$

24. $(-5, -2), (7, 3)$

25. $(6, 8), (12, 16)$

26. $(10, 4), (2, -2)$

27. $(-3, 5), (-5, -5)$

28. $(2, -3), (4, -8)$

29. Geometry Show that the point $(5, 1)$ is equidistant from points $(7, 0)$ and $(3, 0)$.

30. Geometry Show that a triangle with vertices at $(2, 3)$, $(-3, 4)$, and $(1, -2)$ is a right triangle. (*Hint:* If the Pythagorean theorem holds, the triangle is a right triangle.)

31. Geometry Show that a triangle with vertices at $(-2, 4)$, $(2, 8)$, and $(6, 4)$ is isosceles.

32. Geometry Show that a triangle with vertices at $(-2, 13)$, $(-8, 9)$, and $(-2, 5)$ is isosceles.

APPLICATIONS

33. Sailing Refer to the sailboat in the illustration. How long must a rope be to fasten the top of the mast to the bow?

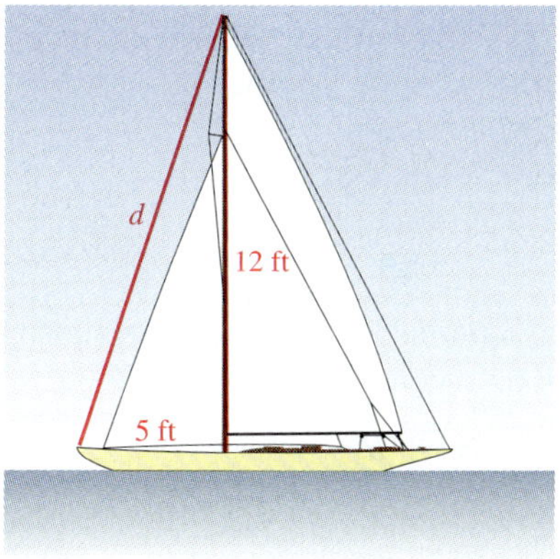

34. Carpentry The gable end of the roof shown is divided in half by a vertical brace. Find the distance from eaves to peak.

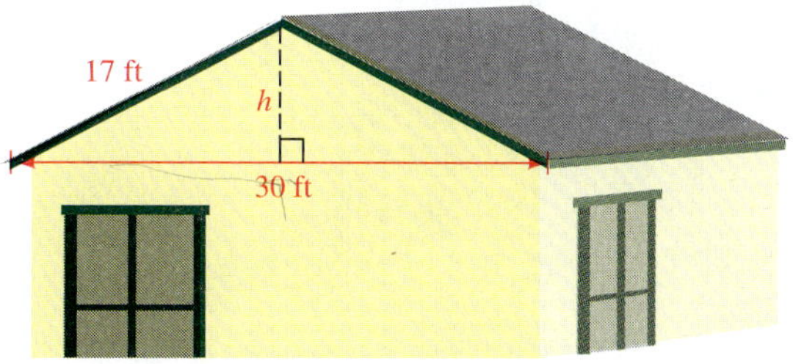

In Exercises 35–38 on the next page, use a calculator. The baseball diamond is a square, 90 feet on a side.

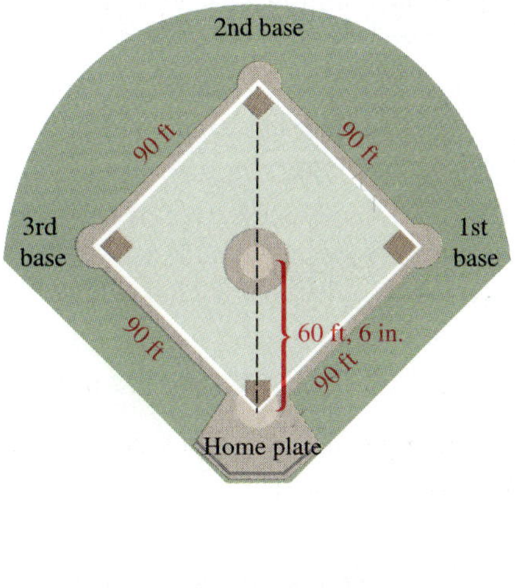

35. Baseball How far must a catcher throw the ball to throw out a runner stealing second base?

36. Baseball In baseball, the pitcher's mound is 60 feet, 6 inches from home plate. How far from the mound is second base?

37. Baseball If the third baseman fields a ground ball 10 feet directly behind third base, how far must he throw the ball to throw a runner out at first base?

38. Baseball The shortstop fields a grounder at a point one-third of the way from second base to third base. How far will he have to throw the ball to make an out at first base?

Use a calculator.

39. Packing a tennis racket The diagonal d of a rectangular box with dimensions $a \times b \times c$ is given by

$$d = \sqrt{a^2 + b^2 + c^2}$$

Will the racket shown below fit in the shipping carton?

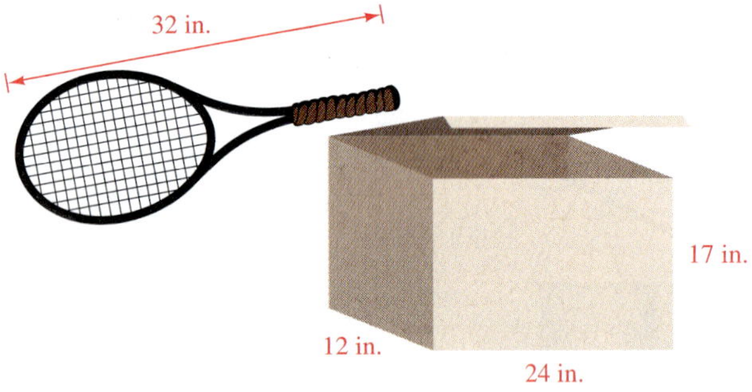

40. Shipping packages A delivery service won't accept a package for shipping if any dimension exceeds 21 inches. An archaeologist wants to ship a 36-inch femur bone. Will it fit in a 3-inch-tall box that has a 21-inch-square base?

41. Shipping packages Can the archaeologist in Exercise 40 ship the femur bone in a cubical box 21 inches on an edge?

42. Reach of a ladder The base of the 37-foot ladder in the illustration is 9 feet from the wall. Will the top reach a window ledge that is 35 feet above the ground?

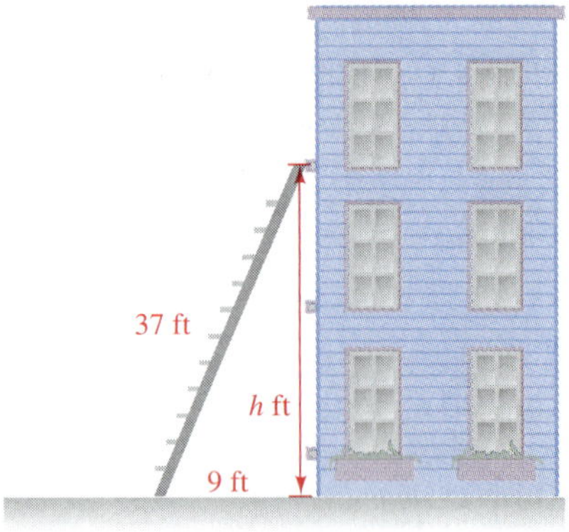

43. Telephone service The telephone cable in the illustration currently runs from A to B to C to D. How much cable is required to run from A to D directly?

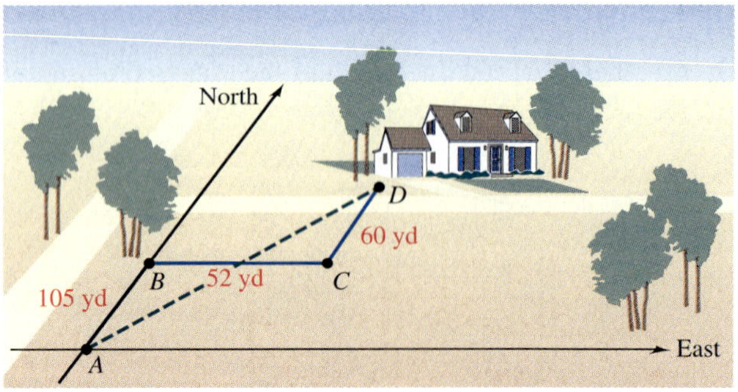

44. Electric service The power company routes its lines as shown in the illustration. How much wire could be saved by going directly from A to E?

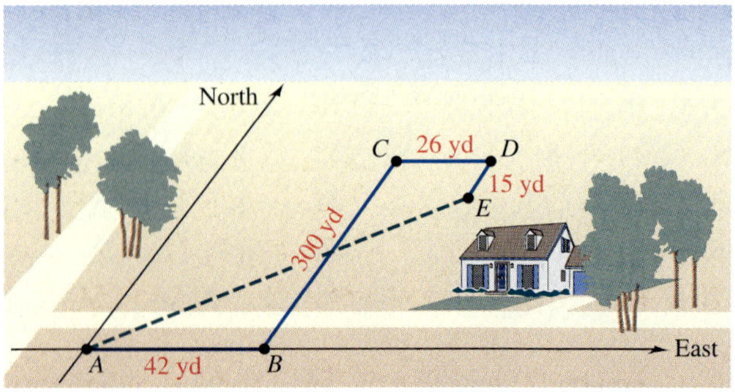

45. Supporting a weight A weight placed on the tight wire pulls the center down 1 foot. By how much is the wire stretched? Round the answer to the nearest hundredth of a foot.

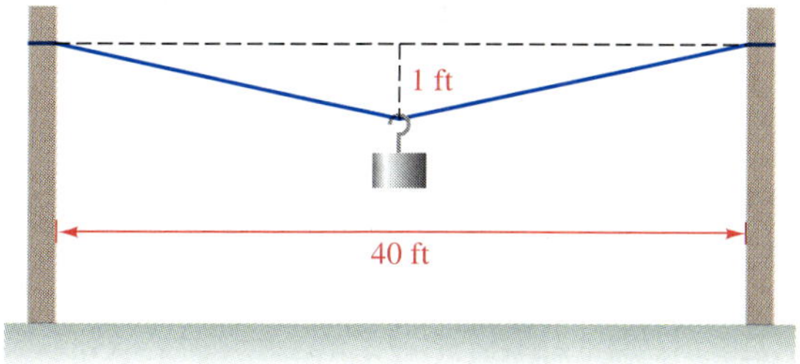

46. Geometry The side, s, of a square with area A square feet is given by the formula $s = \sqrt{A}$. Find the perimeter of a square with an area of 49 square feet.

47. Surface area of a cube The total surface area, A, of a cube is related to its volume, V, by the formula $A = 6\sqrt[3]{V^2}$. Find the surface area of a cube with a volume of 8 cubic centimeters?

48. Area of many cubes A grain of table salt is a cube with a volume of approximately 6×10^{-6} cubic in., and there are about 1.5 million grains of salt in one cup. Find the total surface area of the salt in one cup. (See Exercise 47.)

WRITING

49. State the Pythagorean theorem.

50. Explain the distance formula.

SOMETHING TO THINK ABOUT

51. The formula

$$I = \frac{703w}{h^2}$$

(where w is weight in pounds and h is height in inches) can be used to estimate body mass index, I. The scale shown in the table can be used to judge a person's risk of heart attack. A girl weighing 104 pounds is 54.1 inches tall. Find her estimated body mass index.

20–26	normal
27–29	higher risk
30 and above	very high risk

52. What is the risk of a heart attack for a man who is 6 feet tall and weighs 220 pounds?

9.3 Rational Exponents

- Rational Exponents
- Exponential Expressions with Variables in Their Bases
- Fractional Exponents with Numerators Other Than 1
- Negative Fractional Exponents
- Simplifying Radical Expressions

Getting Ready *Simplify each expression.*

1. x^3x^4 **2.** $(a^3)^4$ **3.** $\frac{a^8}{a^4}$ **4.** a^0

5. x^{-4} **6.** $(ab^2)^3$ **7.** $\left(\frac{b^2}{c^3}\right)^3$ **8.** $(a^2a^3)^2$

Rational Exponents

We have seen that positive integer exponents indicate the number of times that a base is to be used as a factor in a product. For example, x^5 means that x is to be used as a factor five times.

$$x^5 = \overbrace{x \cdot x \cdot x \cdot x \cdot x}^{\text{5 factors of } x}$$

Furthermore, we recall the following properties of exponents.

Rules of Exponents

If there are no divisions by 0, then for all integers m and n,

1. $x^m x^n = x^{m+n}$
2. $(x^m)^n = x^{mn}$
3. $(xy)^n = x^n y^n$
4. $\left(\frac{x}{y}\right)^n = \frac{x^n}{y^n}$
5. $x^0 = 1 \quad (x \neq 0)$
6. $x^{-n} = \frac{1}{x^n}$
7. $\frac{x^m}{x^n} = x^{m-n}$
8. $\left(\frac{x}{y}\right)^{-n} = \left(\frac{y}{x}\right)^n$

To show how to raise bases to fractional powers, we consider the expression $10^{1/2}$. Since fractional exponents must obey the same rules as integer exponents, the square of $10^{1/2}$ is equal to 10.

$$\begin{aligned} (10^{1/2})^2 &= 10^{(1/2)2} && \text{Keep the base and multiply the exponents.} \\ &= 10^1 && \tfrac{1}{2} \cdot 2 = 1. \\ &= 10 && 10^1 = 10. \end{aligned}$$

However, we have seen that

$$\left(\sqrt{10}\right)^2 = 10$$

Since $(10^{1/2})^2$ and $\left(\sqrt{10}\right)^2$ both equal 10, we define $10^{1/2}$ to be $\sqrt{10}$. Likewise, we define

$$10^{1/3} \text{ to be } \sqrt[3]{10} \qquad \text{and} \qquad 10^{1/4} \text{ to be } \sqrt[4]{10}$$

Rational Exponents

If n is a natural number greater than 1, and $\sqrt[n]{x}$ is a real number, then

$$x^{1/n} = \sqrt[n]{x}$$

EXAMPLE 1 Simplify each expression:

a. $9^{1/2} = \sqrt{9} = 3$

b. $-\left(\frac{16}{9}\right)^{1/2} = -\sqrt{\frac{16}{9}} = -\frac{4}{3}$

c. $(-64)^{1/3} = \sqrt[3]{-64} = -4$

d. $16^{1/4} = \sqrt[4]{16} = 2$

e. $\left(\frac{1}{32}\right)^{1/5} = \sqrt[5]{\frac{1}{32}} = \frac{1}{2}$

f. $0^{1/8} = \sqrt[8]{0} = 0$

g. $-(32x^5)^{1/5} = -\sqrt[5]{32x^5} = -2x$

h. $(xyz)^{1/4} = \sqrt[4]{xyz}$

Self Check Assume that $x > 0$. Simplify: **a.** $16^{1/2}$, **b.** $\left(\frac{27}{8}\right)^{1/3}$, and **c.** $-(16x^4)^{1/4}$.

EXAMPLE 2 Write each radical using a fractional exponent: **a.** $\sqrt[4]{5xyz}$ and **b.** $\sqrt[5]{\frac{xy^2}{15}}$.

Solution **a.** $\sqrt[4]{5xyz} = (5xyz)^{1/4}$ **b.** $\sqrt[5]{\frac{xy^2}{15}} = \left(\frac{xy^2}{15}\right)^{1/5}$

Self Check Write the radical using a fractional exponent: $\sqrt[6]{4ab}$.

Exponential Expressions with Variables in Their Bases

As with radicals, when n is even in the expression $x^{1/n}$ $(n > 1)$, there are two real nth roots and we must use absolute value symbols to guarantee that the simplified result is positive.

When n is odd, there is only one real nth root, and we don't need to use absolute value symbols.

When n is even and x is negative, the expression $x^{1/n}$ is not a real number.

EXAMPLE 3 Assume that all variables can be any real number, and simplify each expression.

a. $(-27x^3)^{1/3} = -3x$ — Because $(-3x)^3 = -27x^3$. Since n is odd, no absolute value symbols are needed.

b. $(49x^2)^{1/2} = |7x|$ — Because $(|7x|)^2 = 49x^2$. Since $7x$ can be negative, absolute value symbols are needed.

$= 7|x|$

c. $(256a^8)^{1/8} = 2|a|$ — Because $(2|a|)^8 = 256a^8$. Since a can be any real number, $2a$ can be negative. Thus, absolute value symbols are needed.

d. $[(y + 1)^2]^{1/2} = |y + 1|$ — Because $|y + 1|^2 = (y + 1)^2$. Since y can be any real number, $y + 1$ can be negative, and the absolute value symbols are needed.

e. $(25b^4)^{1/2} = 5b^2$ — Because $(5b^2)^2 = 25b^4$. Since $b^2 \geq 0$, no absolute value symbols are needed.

f. $(-256x^4)^{1/4}$ is not a real number. — Because no real number raised to the 4th power is $-256x^4$.

Self Check Simplify each expression: **a.** $(625a^4)^{1/4}$ and **b.** $(b^4)^{1/2}$.

We summarize the cases as follows.

Summary of the Definitions of $x^{1/n}$

If n is a natural number greater than 1 and x is a real number, then

If $x > 0$, then $x^{1/n}$ is the positive number such that $(x^{1/n})^n = x$.

If $x = 0$, then $x^{1/n} = 0$.

If $x < 0$ $\begin{cases} \text{and } n \text{ is odd, then } x^{1/n} \text{ is the real number such that } (x^{1/n})^n = x. \\ \text{and } n \text{ is even, then } x^{1/n} \text{ is not a real number.} \end{cases}$

Fractional Exponents with Numerators Other Than 1

We can extend the definition of $x^{1/n}$ to include fractional exponents with numerators other than 1. For example, since $4^{3/2}$ can be written as $(4^{1/2})^3$, we have

$$4^{3/2} = (4^{1/2})^3 = \left(\sqrt{4}\right)^3 = 2^3 = 8$$

Thus, we can simplify $4^{3/2}$ by cubing the square root of 4. We can also simplify $4^{3/2}$ by taking the square root of 4 cubed.

$$4^{3/2} = (4^3)^{1/2} = 64^{1/2} = \sqrt{64} = 8$$

In general, we have the following rule.

Changing from Rational Exponents to Radicals

If m and n are positive integers, $x \geq 0$, and $\frac{m}{n}$ is in simplified form, then

$$x^{m/n} = \left(\sqrt[n]{x}\right)^m = \sqrt[n]{x^m}$$

Because of the previous definition, we can interpret $x^{m/n}$ in two ways:

1. $x^{m/n}$ means the mth power of the nth root of x.
2. $x^{m/n}$ means the nth root of the mth power of x.

EXAMPLE 4 Simplify each expression:

a. $27^{2/3} = \left(\sqrt[3]{27}\right)^2 = 3^2 = 9$ or $27^{2/3} = \sqrt[3]{27^2} = \sqrt[3]{729} = 9$

b. $\left(\frac{1}{16}\right)^{3/4} = \left(\sqrt[4]{\frac{1}{16}}\right)^3 = \left(\frac{1}{2}\right)^3 = \frac{1}{8}$ or $\left(\frac{1}{16}\right)^{3/4} = \sqrt[4]{\left(\frac{1}{16}\right)^3} = \sqrt[4]{\frac{1}{4{,}096}} = \frac{1}{8}$

c. $(-8x^3)^{4/3} = \left(\sqrt[3]{-8x^3}\right)^4 = (-2x)^4 = 16x^4$ or $(-8x^3)^{4/3} = \sqrt[3]{(-8x^3)^4} = \sqrt[3]{4{,}096x^{12}} = 16x^4$

Self Check Simplify: **a.** $16^{3/2}$ and **b.** $(-27x^6)^{2/3}$.

Comment To avoid large numbers, it is usually better to find the root of the base first, as shown in Example 4.

Negative Fractional Exponents

To be consistent with the definition of negative integer exponents, we define $x^{-m/n}$ as follows.

Definition of $x^{-m/n}$

If m and n are positive integers, $\frac{m}{n}$ is in simplified form, and $x^{1/n}$ is a real number, then

$$x^{-m/n} = \frac{1}{x^{m/n}} \quad \text{and} \quad \frac{1}{x^{-m/n}} = x^{m/n} \quad (x \neq 0)$$

EXAMPLE 5 Write each expression without negative exponents, if possible.

a. $64^{-1/2} = \frac{1}{64^{1/2}}$

$= \frac{1}{8}$

b. $16^{-3/2} = \frac{1}{16^{3/2}}$

$= \frac{1}{(16^{1/2})^3}$

$= \frac{1}{64}$ $\quad (16^{1/2})^3 = 4^3 = 64.$

c. $(-32x^5)^{-2/5} = \frac{1}{(-32x^5)^{2/5}} \quad (x \neq 0)$

$= \frac{1}{[(-32x^5)^{1/5}]^2}$

$= \frac{1}{(-2x)^2}$

$= \frac{1}{4x^2}$

d. $(-16)^{-3/4}$ is not a real number, because $(-16)^{1/4}$ is not a real number.

Self Check Write each expression without negative exponents: **a.** $25^{-3/2}$ and **b.** $(-27a^3)^{-2/3}$.

Comment By definition, 0^0 is undefined. A base of 0 raised to a negative power is also undefined, because 0^{-2} would equal $\frac{1}{0^2}$, which is undefined since we cannot divide by 0.

We can use the laws of exponents to simplify many expressions with fractional exponents.

EXAMPLE 6 Write all answers without negative exponents. Assume that all variables represent positive numbers. Thus, no absolute value symbols are necessary.

a. $5^{2/7}5^{3/7} = 5^{2/7+3/7}$ $\quad$ Use the rule $x^m x^n = x^{m+n}$.

$= 5^{5/7}$ $\quad$ Add: $\frac{2}{7} + \frac{3}{7} = \frac{5}{7}$.

b. $(5^{2/7})^3 = 5^{(2/7)(3)}$ $\quad$ Use the rule $(x^m)^n = x^{mn}$.

$= 5^{6/7}$ $\quad$ Multiply: $\frac{2}{7}(3) = \frac{2}{7}\left(\frac{3}{1}\right) = \frac{6}{7}$.

c. $(a^{2/3}b^{1/2})^6 = (a^{2/3})^6(b^{1/2})^6$ $\quad$ Use the rule $(xy)^n = x^n y^n$.

$= a^{12/3}b^{6/2}$ $\quad$ Use the rule $(x^m)^n = x^{mn}$ twice.

$= a^4b^3$ $\quad$ Simplify the exponents.

d. $\dfrac{a^{8/3}a^{1/3}}{a^2} = a^{8/3+1/3-2}$ Use the rules $x^m x^n = x^{m+n}$ and $\dfrac{x^m}{x^n} = x^{m-n}$.

$= a^{8/3+1/3-6/3}$ $2 = \frac{6}{3}$.

$= a^{3/3}$ $\frac{8}{3} + \frac{1}{3} - \frac{6}{3} = \frac{3}{3}$.

$= a$ $\frac{3}{3} = 1$.

Self Check Simplify: **a.** $(x^{1/3}y^{3/2})^6$ and **b.** $\dfrac{x^{5/3}x^{2/3}}{x^{1/3}}$.

EXAMPLE 7 Assume that all variables represent positive numbers, and perform the operations. Write all answers without negative exponents.

a. $a^{4/5}(a^{1/5} + a^{3/5}) = a^{4/5}a^{1/5} + a^{4/5}a^{3/5}$ Use the distributive property.

$= a^{4/5+1/5} + a^{4/5+3/5}$ Use the rule $x^m x^n = x^{m+n}$.

$= a^{5/5} + a^{7/5}$ Simplify the exponents.

$= a + a^{7/5}$

Comment Note that $a + a^{7/5} \neq a^{1+7/5}$. The expression $a + a^{7/5}$ cannot be simplified, because a and $a^{7/5}$ are not like terms.

b. $x^{1/2}(x^{-1/2} + x^{1/2}) = x^{1/2}x^{-1/2} + x^{1/2}x^{1/2}$ Use the distributive property.

$= x^{1/2-1/2} + x^{1/2+1/2}$ Use the rule $x^m x^n = x^{m+n}$.

$= x^0 + x^1$ Simplify.

$= 1 + x$ $x^0 = 1$.

c. $(x^{2/3} + 1)(x^{2/3} - 1) = x^{4/3} - x^{2/3} + x^{2/3} - 1$ Use the FOIL method.

$= x^{4/3} - 1$ Combine like terms.

d. $(x^{1/2} + y^{1/2})^2 = (x^{1/2} + y^{1/2})(x^{1/2} + y^{1/2})$ Use the FOIL method.

$= x + 2x^{1/2}y^{1/2} + y$

Simplifying Radical Expressions

We can simplify many radical expressions by using the following steps.

Using Fractional Exponents to Simplify Radicals

1. Change the radical expression into an exponential expression with rational exponents.
2. Simplify the rational exponents.
3. Change the exponential expression back into a radical.

EXAMPLE 8 Simplify: **a.** $\sqrt[4]{3^2}$, **b.** $\sqrt[8]{x^6}$, and **c.** $\sqrt[9]{27x^6y^3}$.

Solution **a.**

$\sqrt[4]{3^2} = (3^2)^{1/4}$ Change the radical to an exponential expression.

$= 3^{2/4}$ Use the rule $(x^m)^n = x^{mn}$.

$= 3^{1/2}$ $\frac{2}{4} = \frac{1}{2}$.

$= \sqrt{3}$ Change back to radical notation.

b.

$\sqrt[8]{x^6} = (x^6)^{1/8}$ Change the radical to an exponential expression.

$= x^{6/8}$ Use the rule $(x^m)^n = x^{mn}$.

$= x^{3/4}$ $\frac{6}{8} = \frac{3}{4}$.

$= (x^3)^{1/4}$ $\frac{3}{4} = 3\left(\frac{1}{4}\right)$.

$= \sqrt[4]{x^3}$ Change back to radical notation.

c.

$\sqrt[9]{27x^6y^3} = (3^3x^6y^3)^{1/9}$ Write 27 as 3^3 and change the radical to an exponential expression.

$= 3^{3/9}x^{6/9}y^{3/9}$ Raise each factor to the $\frac{1}{9}$ power by multiplying the fractional exponents.

$= 3^{1/3}x^{2/3}y^{1/3}$ Simplify each fractional exponent.

$= (3x^2y)^{1/3}$ Use the rule $(xy)^n = x^ny^n$.

$= \sqrt[3]{3x^2y}$ Change back to radical notation.

Self Check Simplify: **a.** $\sqrt[6]{3^3}$ and **b.** $\sqrt[4]{64x^2y^2}$.

Self Check Answers

1. a. 4, **b.** $\frac{3}{2}$, **c.** $-2x$ **2.** $(4ab)^{1/6}$ **3. a.** $5|a|$, **b.** b^2, **4. a.** 64, **b.** $9x^4$ **5. a.** $\frac{1}{125}$, **b.** $\frac{1}{9a^2}$ **6. a.** x^2y^9 **b.** x^2 **8. a.** $\sqrt{3}$ **b.** $\sqrt{8xy}$

Orals *Simplify each expression.*

1. $4^{1/2}$ **2.** $9^{1/2}$ **3.** $27^{1/3}$ **4.** $1^{1/4}$

5. $4^{3/2}$ **6.** $8^{2/3}$ **7.** $\left(\frac{1}{4}\right)^{1/2}$ **8.** $\left(\frac{1}{4}\right)^{-1/2}$

9. $(8x^3)^{1/3}$ **10.** $(16x^8)^{1/4}$

9.3 EXERCISES

REVIEW *Solve each inequality.*

1. $5x - 4 < 11$

2. $2(3t - 5) \geq 8$

3. $\frac{4}{5}(r - 3) > \frac{2}{3}(r + 2)$

4. $-4 < 2x - 4 \leq 8$

5. How much water must be added to 5 pints of a 20% alcohol solution to dilute it to a 15% solution?

6. A grocer bought some boxes of apples for \$70. However, 4 boxes were spoiled. The grocer sold the remaining boxes at a profit of \$2 each. How many boxes did the grocer sell if she managed to break even?

VOCABULARY AND CONCEPTS *Fill in the blanks.*

7. $a^4 =$ ________ **8.** $a^ma^n =$ ____

9. $(a^m)^n =$ ____ **10.** $(ab)^n =$ ____

11. $\left(\frac{a}{b}\right)^n = \frac{\quad}{\quad}$

12. $a^0 =$ __, provided $a \neq$ __.

13. $a^{-n} = \frac{\quad}{\quad}$, provided $a \neq$ __.

14. $\frac{a^m}{a^n} =$ ____

15. $\left(\frac{a}{b}\right)^{-n} =$ ____

16. $x^{1/n} =$ ____

17. $(x^n)^{1/n} =$ ____, provided n is even.

18. $x^{m/n} = \sqrt[n]{x^m} =$ ______

PRACTICE *Change each expression into radical notation.*

19. $7^{1/3}$

20. $26^{1/2}$

21. $8^{1/5}$

22. $13^{1/7}$

23. $(3x)^{1/4}$

24. $(4ab)^{1/6}$

25. $\left(\frac{1}{2}x^3y\right)^{1/4}$

26. $\left(\frac{3}{4}a^2b^2\right)^{1/5}$

27. $(4a^2b^3)^{1/5}$

28. $(5pq^2)^{1/3}$

29. $(x^2 + y^2)^{1/2}$

30. $(x^3 + y^3)^{1/3}$

Change each radical to an exponential expression.

31. $\sqrt{11}$

32. $\sqrt[3]{12}$

33. $\sqrt[4]{3a}$

34. $\sqrt[7]{12xy}$

35. $3\sqrt[5]{a}$

36. $4\sqrt[3]{p}$

37. $\sqrt[6]{\frac{1}{7}abc}$

38. $\sqrt[7]{\frac{3}{8}p^2q}$

39. $\sqrt[5]{\frac{1}{2}mn}$

40. $\sqrt[8]{\frac{2}{7}p^2q}$

41. $\sqrt[3]{a^2 - b^2}$

42. $\sqrt{x^2 + y^2}$

Simplify each expression, if possible. Assume that all variables are unrestricted, and use absolute value symbols when necessary.

43. $4^{1/2}$

44. $64^{1/2}$

45. $27^{1/3}$

46. $125^{1/3}$

47. $16^{1/4}$

48. $625^{1/4}$

49. $32^{1/5}$

50. $0^{1/5}$

51. $\left(\frac{1}{4}\right)^{1/2}$

52. $\left(\frac{1}{16}\right)^{1/2}$

53. $\left(\frac{1}{8}\right)^{1/3}$

54. $\left(\frac{1}{16}\right)^{1/4}$

55. $-16^{1/4}$

56. $-125^{1/3}$

57. $(-27)^{1/3}$

58. $(-125)^{1/3}$

59. $(-64)^{1/2}$

60. $(-243)^{1/5}$

61. $0^{1/3}$

62. $(-216)^{1/2}$

63. $(25y^2)^{1/2}$

64. $(-27x^3)^{1/3}$

65. $(16x^4)^{1/4}$

66. $(-16x^4)^{1/2}$

67. $(243x^5)^{1/5}$

68. $[(x + 1)^4]^{1/4}$

69. $(-64x^8)^{1/4}$

70. $[(x + 5)^3]^{1/3}$

71. $36^{3/2}$

72. $27^{2/3}$

73. $81^{3/4}$

74. $100^{3/2}$

75. $144^{3/2}$

76. $1{,}000^{2/3}$

77. $\left(\frac{1}{8}\right)^{2/3}$

78. $\left(\frac{4}{9}\right)^{3/2}$

79. $(25x^4)^{3/2}$

80. $(27a^3b^3)^{2/3}$

81. $\left(\frac{8x^3}{27}\right)^{2/3}$

82. $\left(\frac{27}{64y^6}\right)^{2/3}$

Write each expression without using negative exponents. Assume that all variables represent positive numbers.

83. $4^{-1/2}$

84. $8^{-1/3}$

85. $(4)^{-3/2}$

86. $25^{-5/2}$

87. $(16x^2)^{-3/2}$

88. $(81c^4)^{-3/2}$

89. $(-27y^3)^{-2/3}$

90. $(-8z^9)^{-2/3}$

91. $(-32p^5)^{-2/5}$

92. $(16q^6)^{-5/2}$

93. $\left(\frac{1}{4}\right)^{-3/2}$

94. $\left(\frac{4}{25}\right)^{-3/2}$

95. $\left(\frac{27}{8}\right)^{-4/3}$

96. $\left(\frac{25}{49}\right)^{-3/2}$

97. $\left(-\frac{8x^3}{27}\right)^{-1/3}$

98. $\left(\frac{16}{81y^4}\right)^{-3/4}$

Perform the operations. Write answers without negative exponents. Assume that all variables represent positive numbers.

99. $5^{4/9}5^{4/9}$

100. $4^{2/5}4^{2/5}$

101. $(4^{1/5})^3$

102. $(3^{1/3})^5$

103. $\dfrac{9^{4/5}}{9^{3/5}}$

104. $\dfrac{7^{2/3}}{7^{1/2}}$

105. $\dfrac{7^{1/2}}{7^{0}}$

106. $5^{1/3}5^{-5/3}$

107. $6^{-2/3}6^{-4/3}$

108. $\dfrac{3^{4/3}3^{1/3}}{3^{2/3}}$

109. $\dfrac{2^{5/6}2^{1/3}}{2^{1/2}}$

110. $\dfrac{5^{1/3}5^{1/2}}{5^{1/3}}$

111. $a^{2/3}a^{1/3}$

112. $b^{3/5}b^{1/5}$

113. $(a^{2/3})^{1/3}$

114. $(t^{4/5})^{10}$

115. $(a^{1/2}b^{1/3})^{3/2}$

116. $(a^{3/5}b^{3/2})^{2/3}$

117. $(mn^{-2/3})^{-3/5}$

118. $(r^{-2}s^{3})^{1/3}$

119. $\dfrac{(4x^3y)^{1/2}}{(9xy)^{1/2}}$

120. $\dfrac{(27x^3y)^{1/3}}{(8xy^2)^{2/3}}$

121. $(27x^{-3})^{-1/3}$

122. $(16a^{-2})^{-1/2}$

123. $y^{1/3}(y^{2/3} + y^{5/3})$

124. $y^{2/5}(y^{-2/5} + y^{3/5})$

125. $x^{3/5}(x^{7/5} - x^{2/5} + 1)$

126. $x^{4/3}(x^{2/3} + 3x^{5/3} - 4)$

127. $(x^{1/2} + 2)(x^{1/2} - 2)$

128. $(x^{1/2} + y^{1/2})(x^{1/2} - y^{1/2})$

129. $(x^{2/3} - x)(x^{2/3} + x)$

130. $(x^{1/3} + x^2)(x^{1/3} - x^2)$

131. $(x^{2/3} + y^{2/3})^2$

132. $(a^{1/2} - b^{2/3})^2$

133. $(a^{3/2} - b^{3/2})^2$

134. $(x^{-1/2} - x^{1/2})^2$

Use rational exponents to simplify each radical. Assume that all variables represent positive numbers.

135. $\sqrt[6]{p^3}$

136. $\sqrt[8]{q^2}$

137. $\sqrt[4]{25b^2}$

138. $\sqrt[9]{-8x^6}$

WRITING

139. Explain how you would decide whether $a^{1/n}$ is a real number.

140. The expression $(a^{1/2} + b^{1/2})^2$ is not equal to $a + b$. Explain.

SOMETHING TO THINK ABOUT

141. The fraction $\frac{2}{4}$ is equal to $\frac{1}{2}$. Is $16^{2/4}$ equal to $16^{1/2}$? Explain.

142. How would you evaluate an expression with a mixed-number exponent? For example, what is $8^{1\frac{1}{3}}$? What is $25^{2\frac{1}{2}}$? Discuss.

9.4 Simplifying and Combining Radical Expressions

- Properties of Radicals
- Simplifying Radical Expressions
- Adding and Subtracting Radical Expressions
- Some Special Triangles

Getting Ready *Simplify each radical. Assume that all variables represent positive numbers.*

1. $\sqrt{225}$
2. $\sqrt{576}$
3. $\sqrt[3]{125}$
4. $\sqrt[3]{343}$
5. $\sqrt{16x^4}$
6. $\sqrt{\dfrac{64}{121}x^6}$
7. $\sqrt[3]{27a^3b^9}$
8. $\sqrt[3]{-8a^{12}}$

In this section, we will introduce several properties of radicals and use them to simplify radical expressions. Then we will add and subtract radical expressions.

Properties of Radicals

Many properties of exponents have counterparts in radical notation. For example, because $a^{1/n}b^{1/n} = (ab)^{1/n}$, we have

(1) $$\sqrt[n]{a}\sqrt[n]{b} = \sqrt[n]{ab}$$

For example,

$$\sqrt{5}\sqrt{5} = \sqrt{5 \cdot 5} = \sqrt{5^2} = 5$$

$$\sqrt[3]{7x}\sqrt[3]{49x^2} = \sqrt[3]{7x \cdot 7^2x^2} = \sqrt[3]{7^3 \cdot x^3} = 7x$$

$$\sqrt[4]{2x^3}\sqrt[4]{8x} = \sqrt[4]{2x^3 \cdot 2^3x} = \sqrt[4]{2^4 \cdot x^4} = 2x \quad (x > 0)$$

If we rewrite Equation 1, we have the following rule.

Multiplication Property of Radicals

If $\sqrt[n]{a}$ and $\sqrt[n]{b}$ are real numbers, then

$$\sqrt[n]{ab} = \sqrt[n]{a}\sqrt[n]{b}$$

As long as all radicals represent real numbers, *the nth root of the product of two numbers is equal to the product of their nth roots.*

Comment The multiplication property of radicals applies to the *n*th root of the product of two numbers. There is no such property for sums or differences. For example,

$$\sqrt{9+4} \neq \sqrt{9} + \sqrt{4} \qquad \sqrt{9-4} \neq \sqrt{9} - \sqrt{4}$$

$$\sqrt{13} \neq 3 + 2 \qquad \sqrt{5} \neq 3 - 2$$

$$\sqrt{13} \neq 5 \qquad \sqrt{5} \neq 1$$

Thus, $\sqrt{a+b} \neq \sqrt{a} + \sqrt{b}$ and $\sqrt{a-b} \neq \sqrt{a} - \sqrt{b}$.

A second property of radicals involves quotients. Because

$$\frac{a^{1/n}}{b^{1/n}} = \left(\frac{a}{b}\right)^{1/n}$$

it follows that

(2) $$\frac{\sqrt[n]{a}}{\sqrt[n]{b}} = \sqrt[n]{\frac{a}{b}} \quad (b \neq 0)$$

For example,

$$\frac{\sqrt{8x^3}}{\sqrt{2x}} = \sqrt{\frac{8x^3}{2x}} = \sqrt{4x^2} = 2x \quad (x > 0)$$

$$\frac{\sqrt[3]{54x^5}}{\sqrt[3]{2x^2}} = \sqrt[3]{\frac{54x^5}{2x^2}} = \sqrt[3]{27x^3} = 3x$$

If we rewrite Equation 2, we have the following rule.

Division Property of Radicals

If $\sqrt[n]{a}$ and $\sqrt[n]{b}$ are real numbers, then

$$\sqrt[n]{\frac{a}{b}} = \frac{\sqrt[n]{a}}{\sqrt[n]{b}} \quad (b \neq 0)$$

As long as all radicals represent real numbers, *the nth root of the quotient of two numbers is equal to the quotient of their nth roots.*

Simplifying Radical Expressions

A radical expression is said to be in simplest form when each of the following statements is true.

Simplified Form of a Radical Expression

A radical expression is in simplest form when

1. Each prime and variable factor in the radicand appears to a power that is less than the index of the radical.
2. The radicand contains no fractions or negative numbers.
3. No radicals appear in the denominator of a fraction.

EXAMPLE 1 Simplify: **a.** $\sqrt{12}$, **b.** $\sqrt{98}$, and **c.** $\sqrt[3]{54}$.

Solution **a.** Recall that squares of integers, such as 1, 4, 9, 16, 25, and 36, are *perfect squares.* To simplify $\sqrt{12}$, we factor 12 so that one factor is the largest perfect square that divides 12. Since 4 is the largest perfect-square factor of 12, we write 12 as $4 \cdot 3$, use the multiplication property of radicals, and simplify.

$$\begin{aligned} \sqrt{12} &= \sqrt{4 \cdot 3} && \text{Write 12 as } 4 \cdot 3. \\ &= \sqrt{4}\sqrt{3} && \sqrt{4 \cdot 3} = \sqrt{4}\sqrt{3}. \\ &= 2\sqrt{3} && \sqrt{4} = 2. \end{aligned}$$

b. Since the largest perfect-square factor of 98 is 49, we have

$$\begin{aligned} \sqrt{98} &= \sqrt{49 \cdot 2} && \text{Write 98 as } 49 \cdot 2. \\ &= \sqrt{49}\sqrt{2} && \sqrt{49 \cdot 2} = \sqrt{49}\sqrt{2}. \\ &= 7\sqrt{2} && \sqrt{49} = 7. \end{aligned}$$

c. Numbers that are cubes of integers, such as 1, 8, 27, 64, 125, and 216, are called *perfect cubes.* Since the largest perfect-cube factor of 54 is 27, we have

$$\begin{aligned} \sqrt[3]{54} &= \sqrt[3]{27 \cdot 2} && \text{Write 54 as } 27 \cdot 2. \\ &= \sqrt[3]{27}\sqrt[3]{2} && \sqrt[3]{27 \cdot 2} = \sqrt[3]{27}\sqrt[3]{2}. \\ &= 3\sqrt[3]{2} && \sqrt[3]{27} = 3. \end{aligned}$$

Self Check Simplify: **a.** $\sqrt{20}$ and **b.** $\sqrt[3]{24}$.

EXAMPLE 2 Simplify: **a.** $\sqrt{\dfrac{15}{49x^2}}$ $(x > 0)$ and **b.** $\sqrt[3]{\dfrac{10x^2}{27y^6}}$ $(y \neq 0)$.

Solution **a.** We can write the square root of the quotient as the quotient of the square roots and simplify the denominator. Since $x > 0$, we have

$$\sqrt{\frac{15}{49x^2}} = \frac{\sqrt{15}}{\sqrt{49x^2}} = \frac{\sqrt{15}}{7x}$$

b. We can write the cube root of the quotient as the quotient of two cube roots. Since $y \neq 0$, we have

$$\sqrt[3]{\frac{10x^2}{27y^6}} = \frac{\sqrt[3]{10x^2}}{\sqrt[3]{27y^6}} = \frac{\sqrt[3]{10x^2}}{3y^2}$$

Self Check Simplify: **a.** $\sqrt{\dfrac{11}{36a^2}}$ $(a > 0)$ and **b.** $\sqrt[3]{\dfrac{8a^2}{125y^3}}$ $(y \neq 0)$.

EXAMPLE 3 Simplify each expression. Assume that all variables represent positive numbers.

a. $\sqrt{128a^5}$, **b.** $\sqrt[3]{24x^5}$, **c.** $\dfrac{\sqrt{45xy^2}}{\sqrt{5x}}$, and **d.** $\dfrac{\sqrt[3]{-432x^5}}{\sqrt[3]{8x}}$.

Solution **a.** We write $128a^5$ as $64a^4 \cdot 2a$ and use the multiplication property of radicals.

$$\sqrt{128a^5} = \sqrt{64a^4 \cdot 2a} \quad \text{$64a^4$ is the largest perfect square that divides $128a^5$.}$$
$$= \sqrt{64a^4}\sqrt{2a} \quad \text{Use the multiplication property of radicals.}$$
$$= 8a^2\sqrt{2a} \quad \sqrt{64a^4} = 8a^2.$$

b. We write $24x^5$ as $8x^3 \cdot 3x^2$ and use the multiplication property of radicals.

$$\sqrt[3]{24x^5} = \sqrt[3]{8x^3 \cdot 3x^2} \quad \text{$8x^3$ is the largest perfect cube that divides $24x^5$.}$$
$$= \sqrt[3]{8x^3}\sqrt[3]{3x^2} \quad \text{Use the multiplication property of radicals.}$$
$$= 2x\sqrt[3]{3x^2} \quad \sqrt[3]{8x^3} = 2x.$$

c. We can write the quotient of the square roots as the square root of a quotient.

$$\frac{\sqrt{45xy^2}}{\sqrt{5x}} = \sqrt{\frac{45xy^2}{5x}} \quad \text{Use the quotient property of radicals.}$$
$$= \sqrt{9y^2} \quad \text{Simplify the fraction.}$$
$$= 3y$$

d. We can write the quotient of the cube roots as the cube root of a quotient.

$$\frac{\sqrt[3]{-432x^5}}{\sqrt[3]{8x}} = \sqrt[3]{\frac{-432x^5}{8x}}$$ Use the quotient property of radicals.

$$= \sqrt[3]{-54x^4}$$ Simplify the fraction.

$$= \sqrt[3]{-27x^3 \cdot 2x}$$ $-27x^3$ is the largest perfect cube that divides $-54x^4$.

$$= \sqrt[3]{-27x^3}\sqrt[3]{2x}$$ Use the multiplication property of radicals.

$$= -3x\sqrt[3]{2x}$$

Self Check Simplify: **a.** $\sqrt{98b^3}$, **b.** $\sqrt[3]{54y^5}$, and **c.** $\frac{\sqrt{50ab^2}}{\sqrt{2a}}$. Assume that all variables represent positive numbers.

To simplify more complicated radicals, we can use the prime factorization of the radicand to find its perfect-square factors. For example, to simplify $\sqrt{3{,}168x^5y^7}$, we first find the prime factorization of $3{,}168x^5y^7$.

$$3{,}168x^5y^7 = 2^5 \cdot 3^2 \cdot 11 \cdot x^5 \cdot y^7$$

Then we have

$$\sqrt{3{,}168x^5y^7} = \sqrt{2^4 \cdot 3^2 \cdot x^4 \cdot y^6 \cdot 2 \cdot 11 \cdot x \cdot y}$$

$$= \sqrt{2^4 \cdot 3^2 \cdot x^4 \cdot y^6}\sqrt{2 \cdot 11 \cdot x \cdot y}$$ Write each perfect square under the left radical and each nonperfect square under the right radical.

$$= 2^2 \cdot 3x^2y^3\sqrt{22xy}$$

$$= 12x^2y^3\sqrt{22xy}$$

Adding and Subtracting Radical Expressions

Radical expressions with the same index and the same radicand are called **like** or **similar radicals.** For example, $3\sqrt{2}$ and $2\sqrt{2}$ are like radicals. However,

$3\sqrt{5}$ and $4\sqrt{2}$ are not like radicals, because the radicands are different.

$3\sqrt{5}$ and $2\sqrt[3]{5}$ are not like radicals, because the indexes are different.

We can often combine like terms. For example, to simplify the expression $3\sqrt{2} + 2\sqrt{2}$, we use the distributive property to factor out $\sqrt{2}$ and simplify.

$$3\sqrt{2} + 2\sqrt{2} = (3 + 2)\sqrt{2}$$

$$= 5\sqrt{2}$$

Radicals with the same index but different radicands can often be written as like radicals. For example, to simplify the expression $\sqrt{27} - \sqrt{12}$, we simplify both radicals and combine the like radicals.

$$\sqrt{27} - \sqrt{12} = \sqrt{9 \cdot 3} - \sqrt{4 \cdot 3}$$

$$= \sqrt{9}\sqrt{3} - \sqrt{4}\sqrt{3}$$ $\sqrt{ab} = \sqrt{a}\sqrt{b}$.

$$= 3\sqrt{3} - 2\sqrt{3} \qquad \sqrt{9} = 3 \text{ and } \sqrt{4} = 2.$$
$$= (3 - 2)\sqrt{3} \qquad \text{Factor out } \sqrt{3}.$$
$$= \sqrt{3}$$

As the previous examples suggest, we can use the following rule to add or subtract radicals.

Adding and Subtracting Radicals

To add or subtract radicals, simplify each radical and combine all like radicals. To combine like radicals, add the coefficients and keep the common radical.

EXAMPLE 4 Simplify: $2\sqrt{12} - 3\sqrt{48} + 3\sqrt{3}$.

Solution We simplify each radical separately and combine like radicals.

$$\begin{aligned} 2\sqrt{12} - 3\sqrt{48} + 3\sqrt{3} &= 2\sqrt{4 \cdot 3} - 3\sqrt{16 \cdot 3} + 3\sqrt{3} \\ &= 2\sqrt{4}\sqrt{3} - 3\sqrt{16}\sqrt{3} + 3\sqrt{3} \\ &= 2(2)\sqrt{3} - 3(4)\sqrt{3} + 3\sqrt{3} \\ &= 4\sqrt{3} - 12\sqrt{3} + 3\sqrt{3} \\ &= (4 - 12 + 3)\sqrt{3} \\ &= -5\sqrt{3} \end{aligned}$$

Self Check Simplify: $3\sqrt{75} - 2\sqrt{12} + 2\sqrt{48}$.

EXAMPLE 5 Simplify: $\sqrt[3]{16} - \sqrt[3]{54} + \sqrt[3]{24}$.

Solution We simplify each radical separately and combine like radicals:

$$\begin{aligned} \sqrt[3]{16} - \sqrt[3]{54} + \sqrt[3]{24} &= \sqrt[3]{8 \cdot 2} - \sqrt[3]{27 \cdot 2} + \sqrt[3]{8 \cdot 3} \\ &= \sqrt[3]{8}\sqrt[3]{2} - \sqrt[3]{27}\sqrt[3]{2} + \sqrt[3]{8}\sqrt[3]{3} \\ &= 2\sqrt[3]{2} - 3\sqrt[3]{2} + 2\sqrt[3]{3} \\ &= -\sqrt[3]{2} + 2\sqrt[3]{3} \end{aligned}$$

Comment We cannot combine $-\sqrt[3]{2}$ and $2\sqrt[3]{3}$, because the radicals have different radicands.

Self Check Simplify: $\sqrt[3]{24} - \sqrt[3]{16} + \sqrt[3]{54}$.

EXAMPLE 6 Simplify: $\sqrt[3]{16x^4} + \sqrt[3]{54x^4} - \sqrt[3]{-128x^4}$.

Solution We simplify each radical separately, factor out $\sqrt[3]{2x}$, and simplify.

$$\begin{aligned} &\sqrt[3]{16x^4} + \sqrt[3]{54x^4} - \sqrt[3]{-128x^4} \\ &= \sqrt[3]{8x^3 \cdot 2x} + \sqrt[3]{27x^3 \cdot 2x} - \sqrt[3]{-64x^3 \cdot 2x} \\ &= \sqrt[3]{8x^3}\sqrt[3]{2x} + \sqrt[3]{27x^3}\sqrt[3]{2x} - \sqrt[3]{-64x^3}\sqrt[3]{2x} \\ &= 2x\sqrt[3]{2x} + 3x\sqrt[3]{2x} + 4x\sqrt[3]{2x} \end{aligned}$$

$$= (2x + 3x + 4x)\sqrt[3]{2x}$$
$$= 9x\sqrt[3]{2x}$$

Self Check Simplify: $\sqrt{32x^3} + \sqrt{50x^3} - \sqrt{18x^3}$ $(x > 0)$.

Some Special Triangles

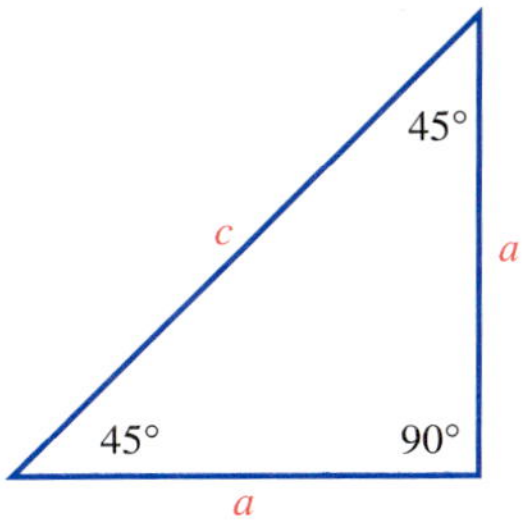

Figure 9-10

An **isosceles right triangle** is a right triangle with two legs of equal length. If we know the length of one leg of an isosceles right triangle, we can use the Pythagorean theorem to find the length of the hypotenuse. Since the triangle shown in Figure 9-10 is a right triangle, we have

$$c^2 = a^2 + a^2 \quad \text{Use the Pythagorean theorem.}$$
$$c^2 = 2a^2 \quad \text{Combine like terms.}$$
$$\sqrt{c^2} = \sqrt{2a^2} \quad \text{Take the positive square root of both sides.}$$
$$c = a\sqrt{2} \quad \sqrt{2a^2} = \sqrt{2}\sqrt{a^2} = \sqrt{2}a = a\sqrt{2}. \text{ No absolute value symbols are needed, because } a \text{ is positive.}$$

Thus, *in an isosceles right triangle, the length of the hypotenuse is the length of one leg times* $\sqrt{2}$.

EXAMPLE 7 If one leg of the isosceles right triangle shown in Figure 9-10 is 10 feet long, find the length of the hypotenuse.

Solution Since the length of the hypotenuse is the length of a leg times $\sqrt{2}$, we have

$$c = 10\sqrt{2}$$

The length of the hypotenuse is $10\sqrt{2}$ feet. To two decimal places, the length is 14.14 feet.

Self Check Find the length of the hypotenuse of an isosceles right triangle if one leg is 12 meters long.

If the length of the hypotenuse of an isosceles right triangle is known, we can use the Pythagorean theorem to find the length of each leg.

EXAMPLE 8 Find the length of each leg of the isosceles right triangle shown in Figure 9-11.

Solution We use the Pythagorean theorem.

$$c^2 = a^2 + a^2$$
$$25^2 = 2a^2 \quad \text{Substitute 25 for } c \text{ and combine like terms.}$$
$$\frac{625}{2} = a^2 \quad \text{Square 25 and divide both sides by 2.}$$
$$\sqrt{\frac{625}{2}} = a \quad \text{Take the positive square root of both sides.}$$
$$a = 17.67766953 \quad \text{Use a calculator.}$$

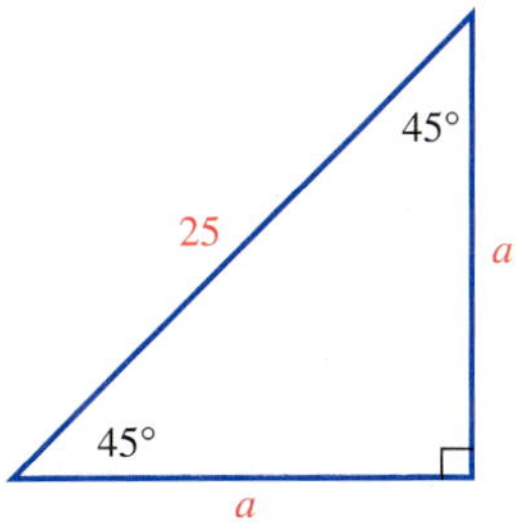

Figure 9-11

To two decimal places, the length is 17.68 units.

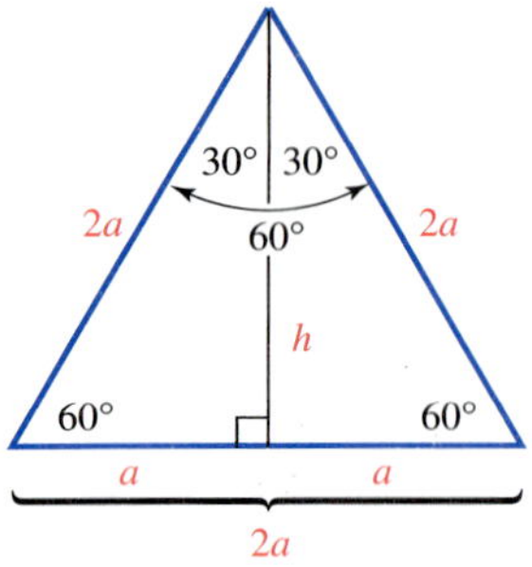

Figure 9-12

From geometry, we know that an **equilateral triangle** is a triangle with three sides of equal length and three 60° angles. If an **altitude** is drawn upon the base of an equilateral triangle, as shown in Figure 9-12, it bisects the base and divides the triangle into two 30°–60°–90° triangles. We can see that the shortest leg of each 30°–60°–90° triangle is a units long. Thus,

The shorter leg of a 30°–60°–90° right triangle is half as long as its hypotenuse.

We can find the length of the altitude, h, by using the Pythagorean theorem.

$$a^2 + h^2 = (2a)^2$$

$$a^2 + h^2 = 4a^2 \qquad (2a)^2 = (2a)(2a) = 4a^2.$$

$$h^2 = 3a^2 \qquad \text{Subtract } a^2 \text{ from both sides.}$$

$$h = \sqrt{3a^2} \qquad \text{Take the positive square root of both sides.}$$

$$h = a\sqrt{3} \qquad \sqrt{3a^2} = \sqrt{3}\sqrt{a^2} = a\sqrt{3}.$$ No absolute value symbols are needed, because a is positive.

Thus,

The length of the longer leg is the length of the shorter side times $\sqrt{3}$.

EXAMPLE 9 Find the length of the hypotenuse and the longer leg of the right triangle shown in Figure 9-13.

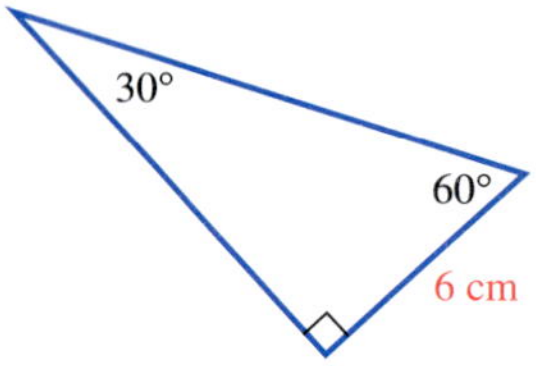

Figure 9-13

Solution Since the shorter leg of a 30°–60°–90° right triangle is half as long as its hypotenuse, the hypotenuse is 12 centimeters long.

Since the length of the longer leg is the length of the shorter leg times $\sqrt{3}$, the longer leg is $6\sqrt{3}$ (about 10.39) centimeters long.

Self Check Find the length of the hypotenuse and the longer leg of a 30°–60°–90° right triangle if the shorter leg is 8 centimeters long.

EXAMPLE 10 Find the length of each leg of the triangle shown in Figure 9-14.

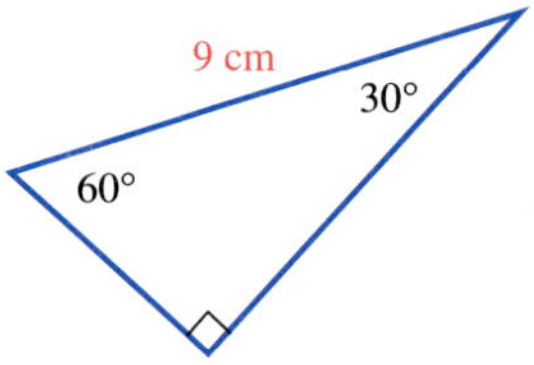

Figure 9-14

Solution Since the shorter leg of a 30°–60°–90° right triangle is half as long as its hypotenuse, the shorter leg is $\frac{9}{2}$ centimeters long.

Since the length of the longer leg is the length of the shorter leg times $\sqrt{3}$, the longer leg is $\frac{9}{2}\sqrt{3}$ (or about 7.79) centimeters long.

Self Check Answers

1. a. $2\sqrt{5}$, **b.** $2\sqrt[3]{3}$ **2. a.** $\frac{\sqrt{11}}{6a}$, **b.** $\frac{2\sqrt[3]{a^2}}{5y}$ **3. a.** $7b\sqrt{2b}$, **b.** $3y\sqrt[3]{2y^2}$, **c.** $5b$ **4.** $19\sqrt{3}$ **5.** $2\sqrt[3]{3} + \sqrt[3]{2}$ **6.** $6x\sqrt{2x}$ **7.** $12\sqrt{2}$ m **9.** 16 cm, $8\sqrt{3}$ cm

Orals *Simplify:*

1. $\sqrt{7}\sqrt{7}$
2. $\sqrt[3]{4^2}\sqrt[3]{4}$
3. $\dfrac{\sqrt[3]{54}}{\sqrt[3]{2}}$

Simplify each expression. Assume that $b \neq 0$.

4. $\sqrt{18}$
5. $\sqrt[3]{16}$
6. $\sqrt[3]{\dfrac{3x^2}{64b^6}}$

Combine like terms:

7. $3\sqrt{3} + 4\sqrt{3}$
8. $5\sqrt{7} - 2\sqrt{7}$
9. $2\sqrt[3]{9} + 3\sqrt[3]{9}$
10. $10\sqrt[5]{4} - 2\sqrt[5]{4}$

9.4 EXERCISES

REVIEW *Perform each operation.*

1. $3x^2y^3(-5x^3y^{-4})$
2. $-2a^2b^{-2}(4a^{-2}b^4 - 2a^2b + 3a^3b^2)$
3. $(3t + 2)^2$
4. $(5r - 3s)(5r + 2s)$
5. $2p - 5\overline{)6p^2 - 7p - 25}$
6. $3m + n\overline{)6m^3 - m^2n + 2mn^2 + n^3}$

VOCABULARY AND CONCEPTS *Fill in the blanks.*

7. $\sqrt[n]{ab} =$ ______
8. $\sqrt[n]{\dfrac{a}{b}} =$ ______

PRACTICE *Simplify each expression. Assume that all variables represent positive numbers.*

9. $\sqrt{6}\sqrt{6}$
10. $\sqrt{11}\sqrt{11}$
11. $\sqrt{t}\sqrt{t}$
12. $-\sqrt{z}\sqrt{z}$
13. $\sqrt[3]{5x^2}\sqrt[3]{25x}$
14. $\sqrt[4]{25a}\sqrt[4]{25a^3}$
15. $\dfrac{\sqrt{500}}{\sqrt{5}}$
16. $\dfrac{\sqrt{128}}{\sqrt{2}}$
17. $\dfrac{\sqrt{98x^3}}{\sqrt{2x}}$
18. $\dfrac{\sqrt{75y^5}}{\sqrt{3y}}$
19. $\dfrac{\sqrt{180ab^4}}{\sqrt{5ab^2}}$
20. $\dfrac{\sqrt{112ab^3}}{\sqrt{7ab}}$
21. $\dfrac{\sqrt[3]{48}}{\sqrt[3]{6}}$
22. $\dfrac{\sqrt[3]{64}}{\sqrt[3]{8}}$
23. $\dfrac{\sqrt[3]{189a^4}}{\sqrt[3]{7a}}$
24. $\dfrac{\sqrt[3]{243x^7}}{\sqrt[3]{9x}}$

Simplify each radical. Assume that all variables represent positive numbers.

25. $\sqrt{20}$
26. $\sqrt{8}$
27. $-\sqrt{200}$
28. $-\sqrt{250}$
29. $\sqrt[3]{80}$
30. $\sqrt[3]{270}$
31. $\sqrt[3]{-81}$
32. $\sqrt[3]{-72}$
33. $\sqrt[4]{32}$
34. $\sqrt[4]{48}$
35. $\sqrt[5]{96}$
36. $\sqrt[7]{256}$
37. $\sqrt{\dfrac{7}{9}}$
38. $\sqrt{\dfrac{3}{4}}$
39. $\sqrt[3]{\dfrac{7}{64}}$
40. $\sqrt[3]{\dfrac{4}{125}}$
41. $\sqrt[4]{\dfrac{3}{10{,}000}}$
42. $\sqrt[5]{\dfrac{4}{243}}$
43. $\sqrt[5]{\dfrac{3}{32}}$
44. $\sqrt[6]{\dfrac{5}{64}}$
45. $\sqrt{50x^2}$
46. $\sqrt{75a^2}$
47. $\sqrt{32b}$
48. $\sqrt{80c}$
49. $-\sqrt{112a^3}$
50. $\sqrt{147a^5}$

51. $\sqrt{175a^2b^3}$

52. $\sqrt{128a^3b^5}$

53. $-\sqrt{300xy}$

54. $\sqrt{200x^2y}$

55. $\sqrt[3]{-54x^6}$

56. $-\sqrt[3]{-81a^3}$

57. $\sqrt[3]{16x^{12}y^3}$

58. $\sqrt[3]{40a^3b^6}$

59. $\sqrt[4]{32x^{12}y^4}$

60. $\sqrt[5]{64x^{10}y^5}$

61. $\sqrt{\dfrac{z^2}{16x^2}}$

62. $\sqrt{\dfrac{b^4}{64a^8}}$

63. $\sqrt[4]{\dfrac{5x}{16z^4}}$

64. $\sqrt[3]{\dfrac{11a^2}{125b^6}}$

Simplify and combine like radicals. All variables represent positive numbers.

65. $4\sqrt{2x} + 6\sqrt{2x}$

66. $6\sqrt[3]{5y} + 3\sqrt[3]{5y}$

67. $8\sqrt[5]{7a^2} - 7\sqrt[5]{7a^2}$

68. $10\sqrt[6]{12xyz} - \sqrt[6]{12xyz}$

69. $\sqrt{3} + \sqrt{27}$

70. $\sqrt{8} + \sqrt{32}$

71. $\sqrt{2} - \sqrt{8}$

72. $\sqrt{20} - \sqrt{125}$

73. $\sqrt{98} - \sqrt{50}$

74. $\sqrt{72} - \sqrt{200}$

75. $3\sqrt{24} + \sqrt{54}$

76. $\sqrt{18} + 2\sqrt{50}$

77. $\sqrt[3]{24} + \sqrt[3]{3}$

78. $\sqrt[3]{16} + \sqrt[3]{128}$

79. $\sqrt[3]{32} - \sqrt[3]{108}$

80. $\sqrt[3]{80} - \sqrt[3]{10{,}000}$

81. $2\sqrt[3]{125} - 5\sqrt[3]{64}$

82. $3\sqrt[3]{27} + 12\sqrt[3]{216}$

83. $14\sqrt[4]{32} - 15\sqrt[4]{162}$

84. $23\sqrt[4]{768} + \sqrt[4]{48}$

85. $3\sqrt[4]{512} + 2\sqrt[4]{32}$

86. $4\sqrt[4]{243} - \sqrt[4]{48}$

87. $\sqrt{98} - \sqrt{50} - \sqrt{72}$

88. $\sqrt{20} + \sqrt{125} - \sqrt{80}$

89. $\sqrt{18} + \sqrt{300} - \sqrt{243}$

90. $\sqrt{80} - \sqrt{128} + \sqrt{288}$

91. $2\sqrt[3]{16} - \sqrt[3]{54} - 3\sqrt[3]{128}$

92. $\sqrt[4]{48} - \sqrt[4]{243} - \sqrt[4]{768}$

93. $\sqrt{25y^2z} - \sqrt{16y^2z}$

94. $\sqrt{25yz^2} + \sqrt{9yz^2}$

95. $\sqrt{36xy^2} + \sqrt{49xy^2}$

96. $3\sqrt{2x} - \sqrt{8x}$

97. $2\sqrt[3]{64a} + 2\sqrt[3]{8a}$

98. $3\sqrt[4]{x^4y} - 2\sqrt[4]{x^4y}$

99. $\sqrt{y^5} - \sqrt{9y^5} - \sqrt{25y^5}$

100. $\sqrt{8y^7} + \sqrt{32y^7} - \sqrt{2y^7}$

101. $\sqrt[5]{x^6y^2} + \sqrt[5]{32x^6y^2} + \sqrt[5]{x^6y^2}$

102. $\sqrt[3]{xy^4} + \sqrt[3]{8xy^4} - \sqrt[3]{27xy^4}$

103. $\sqrt{x^2 + 2x + 1} + \sqrt{x^2 + 2x + 1}$

104. $\sqrt{4x^2 + 12x + 9} + \sqrt{9x^2 + 6x + 1}$

Find the missing lengths in each triangle. Give each answer to two decimal places.

105.

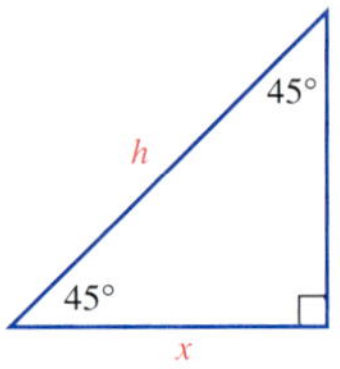

106.

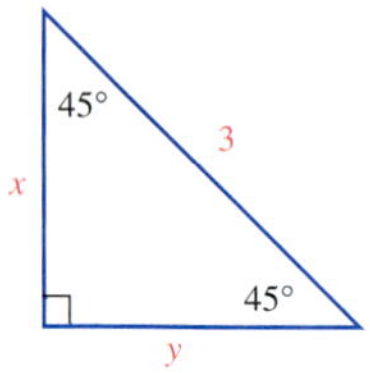

107.

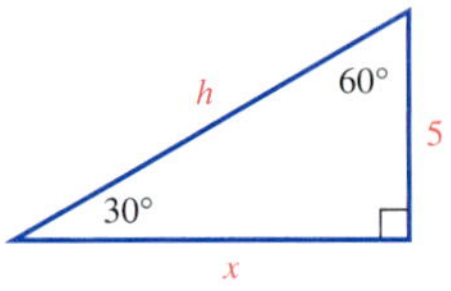

108.

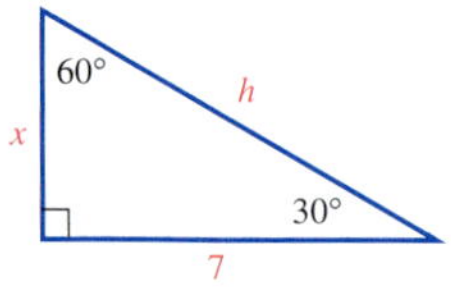

109.

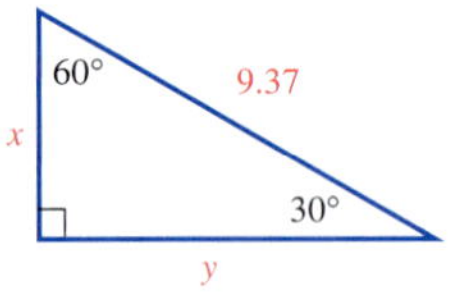

110.

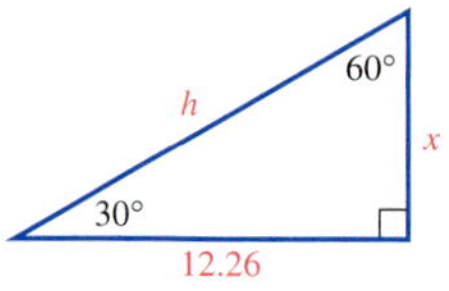

111.

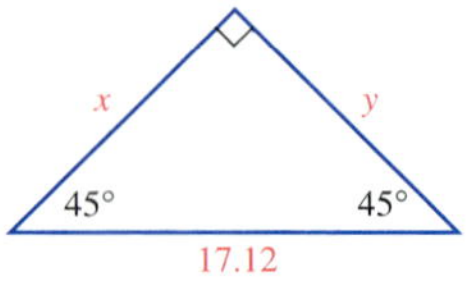

112.

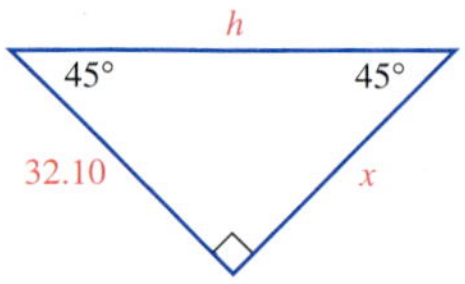

WRITING

113. Explain how to recognize like radicals.

114. Explain how to combine like radicals.

SOMETHING TO THINK ABOUT

115. Can you find any numbers a and b such that $\sqrt{a+b} = \sqrt{a} + \sqrt{b}$?

116. Find the sum: $\sqrt{3} + \sqrt{3^2} + \sqrt{3^3} + \sqrt{3^4} + \sqrt{3^5}$.

9.5 Multiplying and Dividing Radical Expressions

- Multiplying a Monomial by a Monomial
- Multiplying a Polynomial by a Monomial
- Multiplying a Polynomial by a Polynomial
- Problem Solving
- Rationalizing Denominators
- Rationalizing Numerators

Getting Ready *Perform each operation and simplify, if possible.*

1. a^3a^4 **2.** $\frac{b^5}{b^2}$ **3.** $a(a-2)$ **4.** $3b^2(2b+3)$

5. $(a+2)(a-5)$ **6.** $(2a+3b)(2a-3b)$

We now learn how to multiply and divide radical expressions. Then, we will use these skills to solve problems.

Multiplying a Monomial by a Monomial

Radical expressions with the same index can be multiplied and divided.

EXAMPLE 1 Multiply $3\sqrt{6}$ by $2\sqrt{3}$.

Solution We use the commutative and associative properties of multiplication to multiply the coefficients and the radicals separately. Then we simplify any radicals in the product, if possible.

$$
\begin{aligned}
3\sqrt{6}\cdot 2\sqrt{3} &= 3(2)\sqrt{6}\sqrt{3} && \text{Multiply the coefficients and multiply the radicals.}\\
&= 6\sqrt{18} && 3(2) = 6 \text{ and } \sqrt{6}\sqrt{3} = \sqrt{18}.\\
&= 6\sqrt{9}\sqrt{2} && \sqrt{18} = \sqrt{9\cdot 2} = \sqrt{9}\sqrt{2}.\\
&= 6(3)\sqrt{2} && \sqrt{9} = 3.\\
&= 18\sqrt{2}
\end{aligned}
$$

Self Check Multiply $-2\sqrt{7}$ by $5\sqrt{2}$.

Multiplying a Polynomial by a Monomial

To multiply a polynomial by a monomial, we use the distributive property to remove parentheses and then simplify each resulting term, if possible.

EXAMPLE 2 Multiply: $3\sqrt{3}(4\sqrt{8} - 5\sqrt{10})$.

Solution

$$3\sqrt{3}(4\sqrt{8} - 5\sqrt{10})$$

$= 3\sqrt{3} \cdot 4\sqrt{8} - 3\sqrt{3} \cdot 5\sqrt{10}$ Use the distributive property.

$= 12\sqrt{24} - 15\sqrt{30}$ Multiply the coefficients and multiply the radicals.

$= 12\sqrt{4}\sqrt{6} - 15\sqrt{30}$

$= 12(2)\sqrt{6} - 15\sqrt{30}$

$= 24\sqrt{6} - 15\sqrt{30}$

Self Check Multiply: $4\sqrt{2}(3\sqrt{5} - 2\sqrt{8})$.

Multiplying a Polynomial by a Polynomial

To multiply a binomial by a binomial, we use the FOIL method.

EXAMPLE 3 Multiply: $(\sqrt{7} + \sqrt{2})(\sqrt{7} - 3\sqrt{2})$.

Solution

$$(\sqrt{7} + \sqrt{2})(\sqrt{7} - 3\sqrt{2})$$

$= (\sqrt{7})^2 - 3\sqrt{7}\sqrt{2} + \sqrt{2}\sqrt{7} - 3\sqrt{2}\sqrt{2}$

$= 7 - 3\sqrt{14} + \sqrt{14} - 3(2)$

$= 7 - 2\sqrt{14} - 6$

$= 1 - 2\sqrt{14}$

Self Check Multiply: $(\sqrt{5} + 2\sqrt{3})(\sqrt{5} - \sqrt{3})$.

Technically, the expression $\sqrt{3x} - \sqrt{5}$ is not a polynomial, because the variable does not have a whole-number exponent $(\sqrt{3x} = 3^{1/2}x^{1/2})$. However, we will multiply such expressions as if they were polynomials.

EXAMPLE 4 Multiply: $(\sqrt{3x} - \sqrt{5})(\sqrt{2x} + \sqrt{10})$.

Solution

$$(\sqrt{3x} - \sqrt{5})(\sqrt{2x} + \sqrt{10})$$

$= \sqrt{3x}\sqrt{2x} + \sqrt{3x}\sqrt{10} - \sqrt{5}\sqrt{2x} - \sqrt{5}\sqrt{10}$

$= \sqrt{6x^2} + \sqrt{30x} - \sqrt{10x} - \sqrt{50}$

$= \sqrt{6}\sqrt{x^2} + \sqrt{30x} - \sqrt{10x} - \sqrt{25}\sqrt{2}$

$= \sqrt{6}x + \sqrt{30x} - \sqrt{10x} - 5\sqrt{2}$

Self Check Multiply: $(\sqrt{x} + 1)(\sqrt{x} - 3)$.

Comment It is important to draw radical signs carefully so that they completely cover the radicand, but no more than the radicand. To avoid confusion, we often write an expression such as $\sqrt{6}x$ in the form $x\sqrt{6}$.

Problem Solving

EXAMPLE 5 **Photography** Many camera lenses (see Figure 9-15) have an adjustable opening called the *aperture,* which controls the amount of light passing through the lens. The *f-number* of a lens is its *focal length* divided by the diameter of its circular aperture:

Figure 9-15

$$f\text{-number} = \frac{f}{d} \qquad f \text{ is the focal length, and } d \text{ is the diameter of the aperture.}$$

A lens with a focal length of 12 centimeters and an aperture with a diameter of 6 centimeters has an f-number of $\frac{12}{6}$ and is an $f/2$ lens. If the area of the aperture is reduced to admit half as much light, the f-number of the lens will change. Find the new f-number.

Solution We first find the area of the aperture when its diameter is 6 centimeters.

$$A = \pi r^2 \qquad \text{The formula for the area of a circle.}$$
$$A = \pi(3)^2 \qquad \text{Since a radius is half the diameter, substitute 3 for } r.$$
$$A = 9\pi$$

When the size of the aperture is reduced to admit half as much light, the area of the aperture will be $\frac{9\pi}{2}$ square centimeters. To find the diameter of a circle with this area, we proceed as follows:

$$A = \pi r^2 \qquad \text{The formula for the area of a circle.}$$
$$\frac{9\pi}{2} = \pi\left(\frac{d}{2}\right)^2 \qquad \text{Substitute } \frac{9\pi}{2} \text{ for } A \text{ and } \frac{d}{2} \text{ for } r.$$
$$\frac{9\pi}{2} = \frac{\pi d^2}{4} \qquad \left(\frac{d}{2}\right)^2 = \frac{d^2}{4}.$$
$$18 = d^2 \qquad \text{Multiply both sides by 4, and divide both sides by } \pi.$$
$$d = 3\sqrt{2} \qquad \sqrt{18} = \sqrt{9}\sqrt{2} = 3\sqrt{2}.$$

Since the focal length of the lens is still 12 centimeters and the diameter is now $3\sqrt{2}$ centimeters, the new f-number of the lens is

$$f\text{-number} = \frac{f}{d} = \frac{12}{3\sqrt{2}} \qquad \text{Substitute 12 for } f \text{ and } 3\sqrt{2} \text{ for } d.$$
$$\approx 2.828427125 \qquad \text{Use a calculator.}$$

The lens is now an $f/2.8$ lens.

Rationalizing Denominators

To divide radical expressions, we **rationalize the denominator** of a fraction to replace the denominator with a rational number. For example, to divide $\sqrt{70}$ by $\sqrt{3}$, we write the division as the fraction

$$\frac{\sqrt{70}}{\sqrt{3}}$$

To eliminate the radical in the denominator, we multiply the numerator and the denominator by a number that will give a perfect square under the radical in the denominator. Because $3 \cdot 3 = 9$ and 9 is a perfect square, $\sqrt{3}$ is such a number.

$$\frac{\sqrt{70}}{\sqrt{3}} = \frac{\sqrt{70} \cdot \sqrt{3}}{\sqrt{3} \cdot \sqrt{3}}$$ Multiply numerator and denominator by $\sqrt{3}$.

$$= \frac{\sqrt{210}}{3}$$ Multiply the radicals.

Since there is no radical in the denominator and $\sqrt{210}$ cannot be simplified, the expression $\frac{\sqrt{210}}{3}$ is in simplest form, and the division is complete.

EXAMPLE 6 Rationalize the denominator: **a.** $\sqrt{\frac{20}{7}}$ and **b.** $\frac{4}{\sqrt[3]{2}}$.

Solution **a.** We write the square root of the quotient as the quotient of two square roots

$$\sqrt{\frac{20}{7}} = \frac{\sqrt{20}}{\sqrt{7}}$$

and proceed as follows:

$$\frac{\sqrt{20}}{\sqrt{7}} = \frac{\sqrt{20} \cdot \sqrt{7}}{\sqrt{7} \cdot \sqrt{7}}$$ Multiply numerator and denominator by $\sqrt{7}$.

$$= \frac{\sqrt{140}}{7}$$ Multiply the radicals.

$$= \frac{2\sqrt{35}}{7}$$ Simplify $\sqrt{140}$: $\sqrt{140} = \sqrt{4 \cdot 35} = \sqrt{4}\sqrt{35} = 2\sqrt{35}$.

b. Since the denominator is a cube root, we multiply the numerator and the denominator by a number that will give a perfect cube under the radical sign. Since $2 \cdot 4 = 8$ is a perfect cube, $\sqrt[3]{4}$ is such a number.

$$\frac{4}{\sqrt[3]{2}} = \frac{4 \cdot \sqrt[3]{4}}{\sqrt[3]{2} \cdot \sqrt[3]{4}}$$ Multiply numerator and denominator by $\sqrt[3]{4}$.

$$= \frac{4\sqrt[3]{4}}{\sqrt[3]{8}}$$ Multiply the radicals in the denominator.

$$= \frac{4\sqrt[3]{4}}{2}$$ $\sqrt[3]{8} = 2$.

$$= 2\sqrt[3]{4}$$ Simplify.

Self Check Rationalize the denominator: $\frac{5}{\sqrt[4]{3}}$.

EXAMPLE 7 Rationalize the denominator: $\dfrac{\sqrt[3]{5}}{\sqrt[3]{18}}$.

Solution We multiply the numerator and the denominator by a number that will result in a perfect cube under the radical sign in the denominator.

Since 216 is the smallest perfect cube that is divisible by 18 ($216 \div 18 = 12$), multiplying the numerator and the denominator by $\sqrt[3]{12}$ will give the smallest possible perfect cube under the radical in the denominator.

$$\frac{\sqrt[3]{5}}{\sqrt[3]{18}} = \frac{\sqrt[3]{5} \cdot \sqrt[3]{12}}{\sqrt[3]{18} \cdot \sqrt[3]{12}} \quad \text{Multiply numerator and denominator by } \sqrt[3]{12}.$$

$$= \frac{\sqrt[3]{60}}{\sqrt[3]{216}} \quad \text{Multiply the radicals.}$$

$$= \frac{\sqrt[3]{60}}{6} \quad \sqrt[3]{216} = 6.$$

Self Check Rationalize the denominator: $\dfrac{\sqrt{5}}{\sqrt{18}}$.

EXAMPLE 8 Rationalize the denominator of $\dfrac{\sqrt{5xy^2}}{\sqrt{xy^3}}$ (x and y are positive numbers).

Solution

Method 1

$$\frac{\sqrt{5xy^2}}{\sqrt{xy^3}} = \sqrt{\frac{5xy^2}{xy^3}}$$

$$= \sqrt{\frac{5}{y}}$$

$$= \frac{\sqrt{5}}{\sqrt{y}}$$

$$= \frac{\sqrt{5}\sqrt{y}}{\sqrt{y}\sqrt{y}}$$

$$= \frac{\sqrt{5y}}{y}$$

Method 2

$$\frac{\sqrt{5xy^2}}{\sqrt{xy^3}} = \sqrt{\frac{5xy^2}{xy^3}}$$

$$= \sqrt{\frac{5}{y}}$$

$$= \sqrt{\frac{5 \cdot y}{y \cdot y}}$$

$$= \frac{\sqrt{5y}}{\sqrt{y^2}}$$

$$= \frac{\sqrt{5y}}{y}$$

Self Check Rationalize the denominator: $\dfrac{\sqrt{4ab^3}}{\sqrt{2a^2b^2}}$.

To rationalize the denominator of a fraction with square roots in a binomial denominator, we multiply its numerator and denominator by the *conjugate* of its denominator. Conjugate binomials are binomials with the same terms but with opposite signs between their terms.

Conjugate Binomials The conjugate of $a + b$ is $a - b$, and the conjugate of $a - b$ is $a + b$.

EXAMPLE 9 Rationalize the denominator: $\dfrac{1}{\sqrt{2}+1}$.

Solution We multiply the numerator and denominator of the fraction by $\sqrt{2}-1$, which is the conjugate of the denominator.

$$\frac{1}{\sqrt{2}+1} = \frac{1(\sqrt{2}-1)}{(\sqrt{2}+1)(\sqrt{2}-1)} \qquad \frac{\sqrt{2}-1}{\sqrt{2}-1}=1.$$

$$= \frac{\sqrt{2}-1}{(\sqrt{2})^2-1} \qquad (\sqrt{2}+1)(\sqrt{2}-1)=(\sqrt{2})^2-1.$$

$$= \frac{\sqrt{2}-1}{2-1} \qquad (\sqrt{2})^2=2.$$

$$= \sqrt{2}-1 \qquad \frac{\sqrt{2}-1}{2-1}=\frac{\sqrt{2}-1}{1}=\sqrt{2}-1.$$

Self Check Rationalize the denominator: $\dfrac{2}{\sqrt{3}+1}$.

EXAMPLE 10 Rationalize the denominator: $\dfrac{\sqrt{x}+\sqrt{2}}{\sqrt{x}-\sqrt{2}}$ $(x > 0)$.

Solution We multiply the numerator and denominator by $\sqrt{x}+\sqrt{2}$, which is the conjugate of the denominator, and simplify.

$$\frac{\sqrt{x}+\sqrt{2}}{\sqrt{x}-\sqrt{2}} = \frac{(\sqrt{x}+\sqrt{2})(\sqrt{x}+\sqrt{2})}{(\sqrt{x}-\sqrt{2})(\sqrt{x}+\sqrt{2})}$$

$$= \frac{x+\sqrt{2x}+\sqrt{2x}+2}{x-2} \qquad \text{Use the FOIL method.}$$

$$= \frac{x+2\sqrt{2x}+2}{x-2}$$

Self Check Rationalize the denominator: $\dfrac{\sqrt{x}-\sqrt{2}}{\sqrt{x}+\sqrt{2}}$.

Rationalizing Numerators

In calculus, we sometimes have to rationalize a numerator by multiplying the numerator and denominator of the fraction by the conjugate of the numerator.

EXAMPLE 11 Rationalize the numerator: $\dfrac{\sqrt{x}-3}{\sqrt{x}}$ $(x > 0)$.

Solution We multiply the numerator and denominator by $\sqrt{x}+3$, which is the conjugate of the numerator.

$$\frac{\sqrt{x}-3}{\sqrt{x}} = \frac{(\sqrt{x}-3)(\sqrt{x}+3)}{\sqrt{x}(\sqrt{x}+3)}$$
$$= \frac{x+3\sqrt{x}-3\sqrt{x}-9}{x+3\sqrt{x}}$$
$$= \frac{x-9}{x+3\sqrt{x}}$$

The final expression is not in simplified form. However, this nonsimplified form is sometimes desirable in calculus.

Self Check Rationalize the numerator: $\frac{\sqrt{x}+3}{\sqrt{x}}$.

Self Check Answers

1. $-10\sqrt{14}$ **2.** $12\sqrt{10}-32$ **3.** $-1+\sqrt{15}$ **4.** $x-2\sqrt{x}-3$ **6.** $\frac{5\sqrt[4]{27}}{3}$ **7.** $\frac{\sqrt{10}}{6}$ **8.** $\frac{\sqrt{2ab}}{a}$ **9.** $\sqrt{3}-1$ **10.** $\frac{x-2\sqrt{2x}+2}{x-2}$ **11.** $\frac{x-9}{x-3\sqrt{x}}$

Orals *Simplify:*

1. $\sqrt{3}\sqrt{3}$
2. $\sqrt[3]{2}\sqrt[3]{2}\sqrt[3]{2}$
3. $\sqrt{3}\sqrt{9}$
4. $\sqrt{a^3b}\sqrt{ab}\ (b>0)$
5. $3\sqrt{2}(\sqrt{2}+1)$
6. $(\sqrt{2}+1)(\sqrt{2}-1)$
7. $\frac{1}{\sqrt{2}}$
8. $\frac{1}{\sqrt{3}-1}$

9.5 EXERCISES

REVIEW *Solve each equation.*

1. $\frac{2}{3-a}=1$

2. $5(s-4)=-5(s-4)$

3. $\frac{8}{b-2}+\frac{3}{2-b}=-\frac{1}{b}$

4. $\frac{2}{x-2}+\frac{1}{x+1}=\frac{1}{(x+1)(x-2)}$

VOCABULARY AND CONCEPTS *Fill in the blanks.*

5. To multiply $2\sqrt{7}$ by $3\sqrt{5}$, we multiply __ by 3 and then multiply ____ by ____.

6. To multiply $2\sqrt{5}(3\sqrt{8}+\sqrt{3})$, we use the __________ property to remove parentheses and simplify each resulting term.

7. To multiply $(\sqrt{3}+\sqrt{2})(\sqrt{3}-2\sqrt{2})$, we can use the ______ method.

8. The conjugate of $\sqrt{x}+1$ is _______.

9. To rationalize the denominator of $\frac{1}{\sqrt{3}-1}$, multiply both the numerator and denominator by the ________ of the denominator.

10. To rationalize the numerator of $\frac{\sqrt{5}+2}{\sqrt{5}-2}$, multiply both the numerator and denominator by _______.

PRACTICE *Perform each multiplication and simplify, if possible. All variables represent positive numbers.*

11. $\sqrt{2}\sqrt{8}$
12. $\sqrt{3}\sqrt{27}$
13. $\sqrt{5}\sqrt{10}$
14. $\sqrt{7}\sqrt{35}$

15. $2\sqrt{3}\sqrt{6}$

16. $3\sqrt{11}\sqrt{33}$

17. $\sqrt[3]{5}\sqrt[3]{25}$

18. $\sqrt[3]{7}\sqrt[3]{49}$

19. $(3\sqrt[3]{9})(2\sqrt[3]{3})$

20. $(2\sqrt[3]{16})(-\sqrt[3]{4})$

21. $\sqrt[3]{2}\sqrt[3]{12}$

22. $\sqrt[3]{3}\sqrt[3]{18}$

23. $\sqrt{ab^3}\sqrt{ab}$

24. $\sqrt{8x}\sqrt{2x^3y}$

25. $\sqrt{5ab}\sqrt{5a}$

26. $\sqrt{15rs^2}\sqrt{10r}$

27. $\sqrt[3]{5r^2s}\sqrt[3]{2r}$

28. $\sqrt[3]{3xy^2}\sqrt[3]{9x^3}$

29. $\sqrt[3]{a^5b}\sqrt[3]{16ab^5}$

30. $\sqrt[3]{3x^4y}\sqrt[3]{18x}$

31. $\sqrt{x(x+3)}\sqrt{x^3(x+3)}$

32. $\sqrt{y^2(x+y)}\sqrt{(x+y)^3}$

33. $\sqrt[3]{6x^2(y+z)^2}\sqrt[3]{18x(y+z)}$

34. $\sqrt[3]{9x^2y(z+1)^2}\sqrt[3]{6xy^2(z+1)}$

35. $3\sqrt{5}(4-\sqrt{5})$

36. $2\sqrt{7}(3\sqrt{7}-1)$

37. $3\sqrt{2}(4\sqrt{3}+2\sqrt{7})$

38. $-\sqrt{3}(\sqrt{7}-\sqrt{5})$

39. $-2\sqrt{5x}(4\sqrt{2x}-3\sqrt{3})$

40. $3\sqrt{7t}(2\sqrt{7t}+3\sqrt{3t^2})$

41. $(\sqrt{2}+1)(\sqrt{2}-3)$

42. $(2\sqrt{3}+1)(\sqrt{3}-1)$

43. $(4\sqrt{x}+3)(2\sqrt{x}-5)$

44. $(7\sqrt{y}+2)(3\sqrt{y}-5)$

45. $(\sqrt{5z}+\sqrt{3})(\sqrt{5z}+\sqrt{3})$

46. $(\sqrt{3p}-\sqrt{2})(\sqrt{3p}+\sqrt{2})$

47. $(\sqrt{3x}-\sqrt{2y})(\sqrt{3x}+\sqrt{2y})$

48. $(\sqrt{3m}+\sqrt{2n})(\sqrt{3m}+\sqrt{2n})$

49. $(2\sqrt{3a}-\sqrt{b})(\sqrt{3a}+3\sqrt{b})$

50. $(5\sqrt{p}-\sqrt{3q})(\sqrt{p}+2\sqrt{3q})$

51. $(3\sqrt{2r}-2)^2$

52. $(2\sqrt{3t}+5)^2$

53. $-2(\sqrt{3x}+\sqrt{3})^2$

54. $3(\sqrt{5x}-\sqrt{3})^2$

Rationalize each denominator. All variables represent positive numbers.

55. $\sqrt{\frac{1}{7}}$

56. $\sqrt{\frac{5}{3}}$

57. $\sqrt{\frac{2}{3}}$

58. $\sqrt{\frac{3}{2}}$

59. $\frac{\sqrt{5}}{\sqrt{8}}$

60. $\frac{\sqrt{3}}{\sqrt{50}}$

61. $\frac{\sqrt{8}}{\sqrt{2}}$

62. $\frac{\sqrt{27}}{\sqrt{3}}$

63. $\frac{1}{\sqrt[3]{2}}$

64. $\frac{2}{\sqrt[3]{6}}$

65. $\frac{3}{\sqrt[3]{9}}$

66. $\frac{2}{\sqrt[3]{a}}$

67. $\frac{\sqrt[3]{2}}{\sqrt[3]{9}}$

68. $\frac{\sqrt[3]{9}}{\sqrt[3]{54}}$

69. $\frac{\sqrt{8x^2y}}{\sqrt{xy}}$

70. $\frac{\sqrt{9xy}}{\sqrt{3x^2y}}$

71. $\frac{\sqrt{10xy^2}}{\sqrt{2xy^3}}$

72. $\frac{\sqrt{5ab^2c}}{\sqrt{10abc}}$

73. $\frac{\sqrt[3]{4a^2}}{\sqrt[3]{2ab}}$

74. $\frac{\sqrt[3]{9x}}{\sqrt[3]{3xy}}$

75. $\frac{1}{\sqrt[4]{4}}$

76. $\frac{1}{\sqrt[5]{2}}$

77. $\frac{1}{\sqrt[5]{16}}$

78. $\frac{4}{\sqrt[4]{32}}$

79. $\frac{1}{\sqrt{2}-1}$

80. $\frac{3}{\sqrt{3}-1}$

81. $\frac{\sqrt{2}}{\sqrt{5}+3}$

82. $\frac{\sqrt{3}}{\sqrt{3}-2}$

83. $\frac{\sqrt{3}+1}{\sqrt{3}-1}$

84. $\frac{\sqrt{2}-1}{\sqrt{2}+1}$

85. $\dfrac{\sqrt{7}-\sqrt{2}}{\sqrt{2}+\sqrt{7}}$

86. $\dfrac{\sqrt{3}+\sqrt{2}}{\sqrt{3}-\sqrt{2}}$

87. $\dfrac{2}{\sqrt{x}+1}$

88. $\dfrac{3}{\sqrt{x}-2}$

89. $\dfrac{x}{\sqrt{x}-4}$

90. $\dfrac{2x}{\sqrt{x}+1}$

91. $\dfrac{2z-1}{\sqrt{2z}-1}$

92. $\dfrac{3t-1}{\sqrt{3t}+1}$

93. $\dfrac{\sqrt{x}-\sqrt{y}}{\sqrt{x}+\sqrt{y}}$

94. $\dfrac{\sqrt{x}+\sqrt{y}}{\sqrt{x}-\sqrt{y}}$

Rationalize each numerator. All variables represent positive numbers.

95. $\dfrac{\sqrt{3}+1}{2}$

96. $\dfrac{\sqrt{5}-1}{2}$

97. $\dfrac{\sqrt{x}+3}{x}$

98. $\dfrac{2+\sqrt{x}}{5x}$

99. $\dfrac{\sqrt{x}+\sqrt{y}}{\sqrt{x}}$

100. $\dfrac{\sqrt{x}-\sqrt{y}}{\sqrt{x}+\sqrt{y}}$

APPLICATIONS

101. Photography We have seen that a lens with a focal length of 12 centimeters and an aperture $3\sqrt{2}$ centimeters in diameter is an $f/2.8$ lens. Find the f-number if the area of the aperture is again cut in half.

102. Photography A lens with a focal length of 12 centimeters and an aperture 3 centimeters in diameter is an $f/4$ lens. Find the f-number if the area of the aperture is cut in half.

WRITING

103. Explain how to simplify a fraction with the monomial denominator $\sqrt[3]{3}$.

104. Explain how to simplify a fraction with the monomial denominator $\sqrt[3]{9}$.

SOMETHING TO THINK ABOUT

Assume that x is a rational number.

105. Change the numerator of $\dfrac{\sqrt{x}-3}{4}$ to a rational number.

106. Rationalize the numerator: $\dfrac{2\sqrt{3x}+4}{\sqrt{3x}-1}$.

9.6 Radical Equations

- The Power Rule
- Equations Containing One Radical
- Equations Containing Two Radicals
- Equations Containing Three Radicals

Getting Ready *Find each power.*

1. $(\sqrt{a})^2$
2. $(\sqrt{5x})^2$
3. $(\sqrt{x+4})^2$
4. $(\sqrt[4]{y-3})^4$

In this section, we will solve equations that contain radicals. To do so, we will use the **power rule.**

The Power Rule

The Power Rule

If x, y, and n are real numbers and $x = y$, then

$$x^n = y^n$$

If we raise both sides of an equation to the same power, the resulting equation might not be equivalent to the original equation. For example, if we square both sides of the equation

(1) $x = 3$ With a solution set of $\{3\}$.

we obtain the equation

(2) $x^2 = 9$ With a solution set of $\{3, -3\}$.

Equations 1 and 2 are not equivalent, because they have different solution sets, and the solution -3 of Equation 2 does not satisfy Equation 1. Since raising both sides of an equation to the same power can produce an equation with roots that don't satisfy the original equation, we must check each suspected solution in the original equation.

Equations Containing One Radical

EXAMPLE 1 Solve: $\sqrt{x+3} = 4$.

Solution To eliminate the radical, we apply the power rule by squaring both sides of the equation, and proceed as follows:

$$\begin{aligned} \sqrt{x+3} &= 4 \\ \left(\sqrt{x+3}\right)^2 &= (4)^2 && \text{Square both sides.} \\ x + 3 &= 16 \\ x &= 13 && \text{Subtract 3 from both sides.} \end{aligned}$$

We must check the apparent solution of 13 to see whether it satisfies the original equation.

Check:

$$\begin{aligned} \sqrt{x+3} &= 4 \\ \sqrt{13+3} &\stackrel{?}{=} 4 && \text{Substitute 13 for } x. \\ \sqrt{16} &\stackrel{?}{=} 4 \\ 4 &= 4 \end{aligned}$$

Since 13 satisfies the original equation, it is a solution.

Self Check Solve: $\sqrt{a-2} = 3$.

To solve an equation with radicals, we follow these steps.

Solving an Equation Containing Radicals

1. Isolate one radical expression on one side of the equation.
2. Raise both sides of the equation to the power that is the same as the index of the radical.
3. Solve the resulting equation. If it still contains a radical, go back to Step 1.
4. Check the possible solutions to eliminate the ones that do not satisfy the original equation.

EXAMPLE 2 **Height of a bridge** The distance d (in feet) that an object will fall in t seconds is given by the formula

$$t = \sqrt{\frac{d}{16}}$$

To find the height of a bridge, a man drops a stone into the water (see Figure 9-16). If it takes the stone 3 seconds to hit the water, how far above the river is the bridge?

Figure 9-16

Solution We substitute 3 for t in the formula and solve for d.

$$t = \sqrt{\frac{d}{16}}$$

$$3 = \sqrt{\frac{d}{16}}$$

$$9 = \frac{d}{16} \quad \text{Square both sides.}$$

$$144 = d \quad \text{Multiply both sides by 16.}$$

The bridge is 144 feet above the river.

Self Check How high is the bridge if it takes 4 seconds for the stone to hit the water?

EXAMPLE 3 Solve: $\sqrt{3x + 1} + 1 = x$.

Solution We first subtract 1 from both sides to isolate the radical. Then, to eliminate the radical, we square both sides of the equation and proceed as follows:

$$\sqrt{3x + 1} + 1 = x$$

$$\sqrt{3x + 1} = x - 1 \qquad \text{Subtract 1 from both sides.}$$

$$\left(\sqrt{3x + 1}\right)^2 = (x - 1)^2 \qquad \text{Square both sides to eliminate the square root.}$$

$$3x + 1 = x^2 - 2x + 1 \qquad (x - 1)^2 \neq x^2 - 1. \text{ Instead, } (x - 1)^2 = (x - 1)(x - 1) = x^2 - x - x + 1 = x^2 - 2x + 1.$$

$$0 = x^2 - 5x \qquad \text{Subtract } 3x \text{ and 1 from both sides.}$$

$$0 = x(x - 5) \qquad \text{Factor } x^2 - 5x.$$

$$x = 0 \quad \text{or} \quad x - 5 = 0 \qquad \text{Set each factor equal to 0.}$$

$$x = 0 \quad | \quad x = 5$$

We must check each apparent solution to see whether it satisfies the original equation.

Check:

$$\sqrt{3x + 1} + 1 = x \qquad\qquad \sqrt{3x + 1} = x$$

$$\sqrt{3(0) + 1} + 1 \stackrel{?}{=} 0 \qquad\qquad \sqrt{3(5) + 1} + 1 \stackrel{?}{=} 5$$

$$\sqrt{1} + 1 \stackrel{?}{=} 0 \qquad\qquad \sqrt{16} + 1 \stackrel{?}{=} 5$$

$$2 \neq 0 \qquad\qquad 5 = 5$$

Since 0 does not check, it must be discarded. The only solution of the original equation is 5.

Self Check Solve: $\sqrt{4x + 1} + 1 = x$.

Accent on Technology

SOLVING EQUATIONS CONTAINING RADICALS

To find approximate solutions for $\sqrt{3x + 1} + 1 = x$ with a graphing calculator, we can use window settings of $[-5, 10]$ for x and $[-2, 8]$ for y and graph the functions $f(x) = \sqrt{3x + 1} + 1$ and $g(x) = x$, as in Figure 9-17(a). We then trace to find the approximate x-coordinate of their intersection point, as in Figure 9-17(b). After repeated zooms, we will see that $x = 5$.

We can also find the x-coordinate of the intersection point by using the INTERSECT command found in the CALC menu.

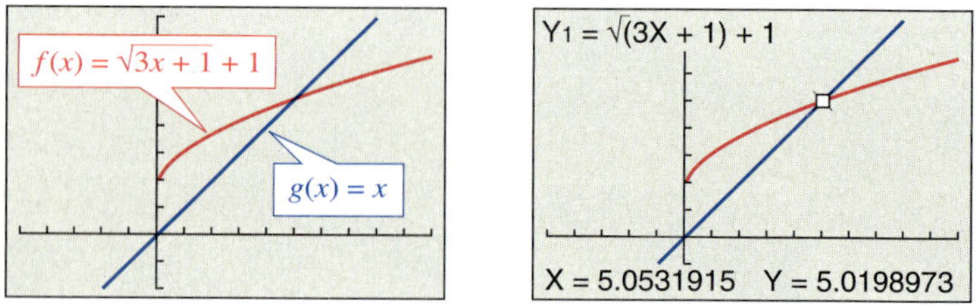

(a) (b)

Figure 9-17

EXAMPLE 4 Solve: $\sqrt[3]{x^3 + 7} = x + 1$.

Solution To eliminate the radical, we cube both sides of the equation and proceed as follows:

$$\sqrt[3]{x^3 + 7} = x + 1$$

$$\left(\sqrt[3]{x^3 + 7}\right)^3 = (x + 1)^3 \qquad \text{Cube both sides to eliminate the cube root.}$$

$$x^3 + 7 = x^3 + 3x^2 + 3x + 1$$

$$0 = 3x^2 + 3x - 6 \qquad \text{Subtract } x^3 \text{ and 7 from both sides.}$$

$$0 = x^2 + x - 2 \qquad \text{Divide both sides by 3.}$$

$$0 = (x + 2)(x - 1)$$

$$x + 2 = 0 \quad \text{or} \quad x - 1 = 0$$

$$x = -2 \quad | \quad x = 1$$

We check each apparent solution to see whether it satisfies the original equation.

Check:

$$\sqrt[3]{x^3 + 7} = x + 1 \qquad\qquad \sqrt[3]{x^3 + 7} = x + 1$$

$$\sqrt[3]{(-2)^3 + 7} \stackrel{?}{=} -2 + 1 \qquad\qquad \sqrt[3]{1 + 7} \stackrel{?}{=} 1 + 1$$

$$\sqrt[3]{-8 + 7} \stackrel{?}{=} -1 \qquad\qquad \sqrt[3]{8} \stackrel{?}{=} 2$$

$$\sqrt[3]{-1} \stackrel{?}{=} -1 \qquad\qquad 2 = 2$$

$$-1 = -1$$

Both solutions satisfy the original equation.

Self Check Solve: $\sqrt[3]{x^3 + 8} = x + 2$.

Equations Containing Two Radicals

When more than one radical appears in an equation, it is often necessary to apply the power rule more than once.

EXAMPLE 5 Solve: $\sqrt{x} + \sqrt{x + 2} = 2$.

Solution To remove the radicals, we square both sides of the equation. Since this is easier to do if one radical is on each side of the equation, we subtract $\sqrt{x}$ from both sides to isolate one radical on one side of the equation.

$$\sqrt{x} + \sqrt{x + 2} = 2$$

$$\sqrt{x + 2} = 2 - \sqrt{x} \qquad \text{Subtract } \sqrt{x} \text{ from both sides.}$$

$$\left(\sqrt{x + 2}\right)^2 = \left(2 - \sqrt{x}\right)^2 \qquad \text{Square both sides to eliminate the square root.}$$

$$x + 2 = 4 - 4\sqrt{x} + x \qquad \left(2 - \sqrt{x}\right)\left(2 - \sqrt{x}\right) = 4 - 2\sqrt{x} - 2\sqrt{x} + x = 4 - 4\sqrt{x} + x.$$

$$2 = 4 - 4\sqrt{x} \qquad \text{Subtract } x \text{ from both sides.}$$

$$-2 = -4\sqrt{x} \qquad \text{Subtract 4 from both sides.}$$

$$\frac{1}{2} = \sqrt{x} \qquad \text{Divide both sides by } -4.$$

$$\frac{1}{4} = x \qquad \text{Square both sides.}$$

Check:

$$\sqrt{x} + \sqrt{x + 2} = 2$$

$$\sqrt{\frac{1}{4}} + \sqrt{\frac{1}{4} + 2} \stackrel{?}{=} 2$$

$$\frac{1}{2} + \sqrt{\frac{9}{4}} \stackrel{?}{=} 2$$

$$\frac{1}{2} + \frac{3}{2} \stackrel{?}{=} 2$$

$$2 = 2$$

The solution checks.

Self Check Solve: $\sqrt{a} + \sqrt{a + 3} = 3$.

Accent on Technology — SOLVING EQUATIONS CONTAINING RADICALS

To find approximate solutions for $\sqrt{x} + \sqrt{x + 2} = 5$ with a graphing calculator, we use window settings of $[-2, 10]$ for x and $[-2, 8]$ for y and graph the functions $f(x) = \sqrt{x} + \sqrt{x + 2}$ and $g(x) = 5$, as in Figure 9-18(a). We then trace to find an approximation of the x-coordinate of their intersection point, as in Figure 9-18(b). From the figure, we can see that $x \approx 5.15$. We can zoom to get better results.

We can also find the x-coordinate of the intersection point by using the INTERSECT command found in the CALC menu.

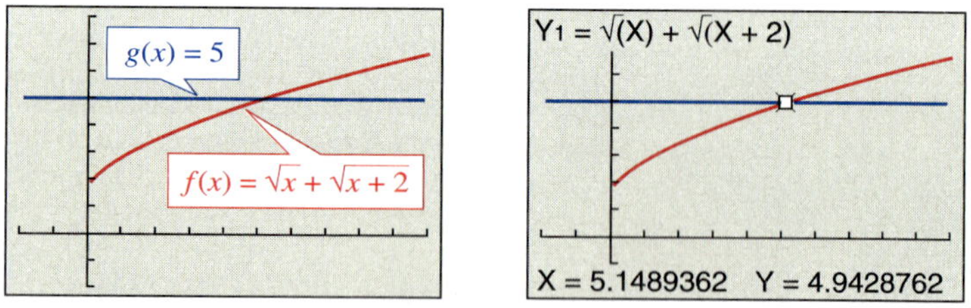

(a) (b)

Figure 9-18

Equations Containing Three Radicals

EXAMPLE 6 Solve: $\sqrt{x + 2} + \sqrt{2x} = \sqrt{18 - x}$.

Solution In this case, it is impossible to isolate one radical on each side of the equation, so we begin by squaring both sides. Then we proceed as follows.

$$\sqrt{x + 2} + \sqrt{2x} = \sqrt{18 - x}$$

$$\left(\sqrt{x + 2} + \sqrt{2x}\right)^2 = \left(\sqrt{18 - x}\right)^2$$ Square both sides to eliminate one square root.

$$x + 2 + 2\sqrt{x + 2}\sqrt{2x} + 2x = 18 - x$$

$$2\sqrt{x+2}\sqrt{2x} = 16 - 4x$$ Subtract $3x$ and 2 from both sides.

$$\sqrt{x+2}\sqrt{2x} = 8 - 2x$$ Divide both sides by 2.

$$\left(\sqrt{x+2}\sqrt{2x}\right)^2 = (8 - 2x)^2$$ Square both sides to eliminate the other square roots.

$$(x+2)2x = 64 - 32x + 4x^2$$
$$2x^2 + 4x = 64 - 32x + 4x^2$$
$$0 = 2x^2 - 36x + 64$$ Write the equation in quadratic form.

$$0 = x^2 - 18x + 32$$ Divide both sides by 2.

$$0 = (x - 16)(x - 2)$$ Factor the trinomial.

$$x - 16 = 0 \quad \text{or} \quad x - 2 = 0$$ Set each factor equal to 0.

$$x = 16 \quad | \quad x = 2$$

Verify that 2 satisfies the equation, but 16 does not. Thus, the only solution is 2.

Self Check Solve: $\sqrt{3x+4} + \sqrt{x+9} = \sqrt{x+25}$.

Self Check Answers

1. 11 **2.** 256 ft **3.** 6; 0 is extraneous **4.** 0, −2 **5.** 1 **6.** 0

Orals *Solve each equation.*

1. $\sqrt{x+2} = 3$

2. $\sqrt{x-2} = 1$

3. $\sqrt[3]{x+1} = 1$

4. $\sqrt[3]{x-1} = 2$

5. $\sqrt[4]{x-1} = 2$

6. $\sqrt[5]{x+1} = 2$

9.6 EXERCISES

REVIEW *If $f(x) = 3x^2 - 4x + 2$, find each quantity.*

1. $f(0)$

2. $f(-3)$

3. $f(2)$

4. $f\left(\frac{1}{2}\right)$

VOCABULARY AND CONCEPTS *Fill in the blanks.*

5. If x, y, and n are real numbers and $x = y$, then ______.

6. When solving equations containing radicals, try to ______ one radical expression on one side of the equation.

7. To solve the equation $\sqrt{x+4} = 5$, we first ______ both sides.

8. To solve the equation $\sqrt[3]{x+4} = 2$, we first ______ both sides.

9. Squaring both sides of an equation can introduce ______ solutions.

10. Always remember to ______ the solutions of an equation containing radicals to eliminate any ______ solutions.

PRACTICE *Solve each equation. Write all solutions and cross out those that are extraneous.*

11. $\sqrt{5x-6} = 2$

12. $\sqrt{7x-10} = 12$

13. $\sqrt{6x+1} + 2 = 7$

14. $\sqrt{6x+13} - 2 = 5$

15. $2\sqrt{4x+1} = \sqrt{x+4}$

16. $\sqrt{3(x+4)} = \sqrt{5x-12}$

17. $\sqrt[3]{7n-1} = 3$

18. $\sqrt[3]{12m+4} = 4$

19. $\sqrt[4]{10p + 1} = \sqrt[4]{11p - 7}$

20. $\sqrt[4]{10y + 2} = 2\sqrt[4]{2}$

21. $x = \dfrac{\sqrt{12x - 5}}{2}$

22. $x = \dfrac{\sqrt{16x - 12}}{2}$

23. $\sqrt{x + 2} = \sqrt{4 - x}$

24. $\sqrt{6 - x} = \sqrt{2x + 3}$

25. $2\sqrt{x} = \sqrt{5x - 16}$

26. $3\sqrt{x} = \sqrt{3x + 12}$

27. $r - 9 = \sqrt{2r - 3}$

28. $-s - 3 = 2\sqrt{5 - s}$

29. $\sqrt{-5x + 24} = 6 - x$

30. $\sqrt{-x + 2} = x - 2$

31. $\sqrt{y + 2} = 4 - y$

32. $\sqrt{22y + 86} = y + 9$

33. $\sqrt{x}\sqrt{x + 16} = 15$

34. $\sqrt{x}\sqrt{x + 6} = 4$

35. $\sqrt[3]{x^3 - 7} = x - 1$

36. $\sqrt[3]{x^3 + 56} - 2 = x$

37. $\sqrt[4]{x^4 + 4x^2 - 4} = -x$

38. $\sqrt[4]{8x - 8} + 2 = 0$

39. $\sqrt[4]{12t + 4} + 2 = 0$

40. $u = \sqrt[4]{u^4 - 6u^2 + 24}$

41. $\sqrt{2y + 1} = 1 - 2\sqrt{y}$

42. $\sqrt{u} + 3 = \sqrt{u - 3}$

43. $\sqrt{y + 7} + 3 = \sqrt{y + 4}$

44. $1 + \sqrt{z} = \sqrt{z + 3}$

45. $\sqrt{v} + \sqrt{3} = \sqrt{v + 3}$

46. $\sqrt{x} + 2 = \sqrt{x + 4}$

47. $2 + \sqrt{u} = \sqrt{2u + 7}$

48. $5r + 4 = \sqrt{5r + 20} + 4r$

49. $\sqrt{6t + 1} - 3\sqrt{t} = -1$

50. $\sqrt{4s + 1} - \sqrt{6s} = -1$

51. $\sqrt{2x + 5} + \sqrt{x + 2} = 5$

52. $\sqrt{2x + 5} + \sqrt{2x + 1} + 4 = 0$

53. $\sqrt{z - 1} + \sqrt{z + 2} = 3$

54. $\sqrt{16v + 1} + \sqrt{8v + 1} = 12$

55. $\sqrt{x - 5} - \sqrt{x + 3} = 4$

56. $\sqrt{x + 8} - \sqrt{x - 4} = -2$

57. $\sqrt{x + 1} + \sqrt{3x} = \sqrt{5x + 1}$

58. $\sqrt{3x} - \sqrt{x + 1} = \sqrt{x - 2}$

59. $\sqrt{\sqrt{a} + \sqrt{a + 8}} = 2$

60. $\sqrt{\sqrt{2y} - \sqrt{y - 1}} = 1$

61. $\dfrac{6}{\sqrt{x + 5}} = \sqrt{x}$

62. $\dfrac{\sqrt{2x}}{\sqrt{x + 2}} = \sqrt{x - 1}$

63. $\sqrt{x + 2} + \sqrt{2x - 3} = \sqrt{11 - x}$

64. $\sqrt{8 - x} - \sqrt{3x - 8} = \sqrt{x - 4}$

APPLICATIONS

65. **Highway design** A curve banked at 8° will accommodate traffic traveling s mph if the radius of the curve is r feet, according to the formula $s = 1.45\sqrt{r}$. If engineers expect 65-mph traffic, what radius should they specify?

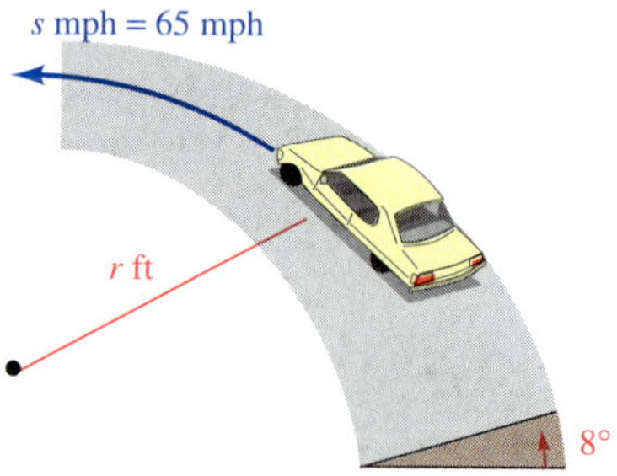

66. **Horizon distance** The higher a lookout tower is built, the farther an observer can see. That distance d (called the *horizon distance,* measured in miles) is related to the height h of the observer (measured in feet) by the formula $d = 1.4\sqrt{h}$. How tall must a lookout tower be to see the edge of the forest, 25 miles away?

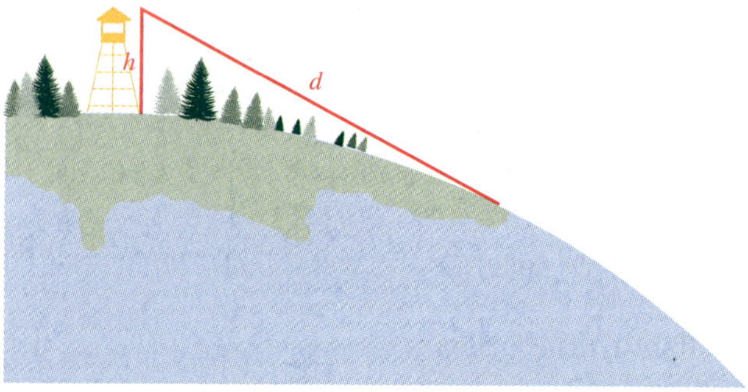

67. **Generating power** The power generated by a windmill is related to the velocity of the wind by the formula

$$v = \sqrt[3]{\frac{P}{0.02}}$$

where P is the power (in watts) and v is the velocity of the wind (in mph). Find the speed of the wind when the windmill is generating 500 watts of power.

68. **Carpentry** During construction, carpenters often brace walls as shown in the illustration, where the length of the brace is given by the formula

$$l = \sqrt{f^2 + h^2}$$

If a carpenter nails a 10-ft brace to the wall 6 feet above the floor, how far from the base of the wall should he nail the brace to the floor?

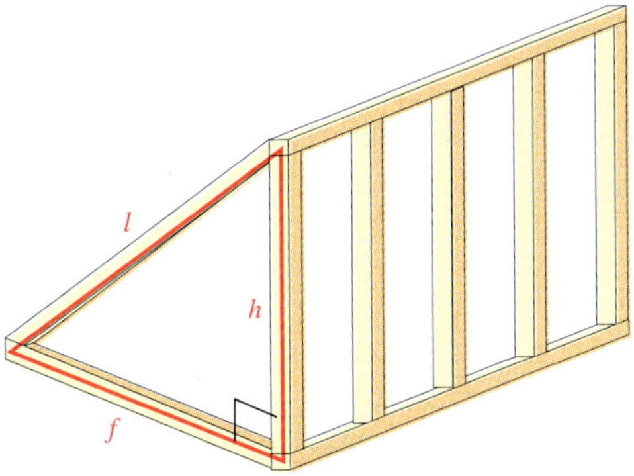

Use a graphing calculator.

69. **Depreciation** The formula

$$r = 1 - \sqrt[n]{\frac{T}{C}}$$

gives the annual depreciation rate r of a car that had an original cost of C dollars, a useful life of n years, and a trade-in value of T dollars. Find the annual depreciation rate of a car that cost \$22,000 and was sold 5 years later for \$9,000. Give the result to the nearest percent.

70. **Savings accounts** The interest rate r earned by a savings account after n compoundings is given by the formula

$$\sqrt[n]{\frac{V}{P}} - 1 = r$$

where V is the current value and P is the original principal. What interest rate r was paid on an account in which a deposit of \$1,000 grew to \$1,338.23 after 5 compoundings?

71. **Marketing** The number of wrenches that will be produced at a given price can be predicted by the formula $s = \sqrt{5x}$, where s is the supply (in thousands) and x is the price (in dollars). If the demand, d, for wrenches can be predicted by the formula $d = \sqrt{100 - 3x^2}$, find the equilibrium price.

72. **Marketing** The number of footballs that will be produced at a given price can be predicted by the formula $s = \sqrt{23x}$, where s is the supply (in thousands) and x is the price (in dollars). If the demand, d, for footballs can be predicted by the formula $d = \sqrt{312 - 2x^2}$, find the equilibrium price.

73. **Medicine** The resistance R to blood flow through an artery can be found using the formula

$$r = \sqrt[4]{\frac{8kl}{\pi R}}$$

where r is the radius of the artery, k is the viscosity of blood, and l is the length of the artery. Solve the formula for R.

74. **Generating power** The power P generated by a windmill is given by the formula

$$s = \sqrt[3]{\frac{P}{0.02}}$$

where s is the speed of the wind. Solve the formula for P.

WRITING

75. If both sides of an equation are raised to the same power, the resulting equation might not be equivalent to the original equation. Explain.

76. Explain why you must check each apparent solution of a radical equation.

SOMETHING TO THINK ABOUT

77. Solve: $\sqrt[3]{2x} = \sqrt{x}$. (*Hint:* Square and then cube both sides.)

78. Solve: $\sqrt[4]{x} = \sqrt{\frac{x}{4}}$.

9.7 Complex Numbers

- Imaginary Numbers
- Simplifying Imaginary Numbers
- Complex Numbers
- Arithmetic of Complex Numbers
- Rationalizing the Denominator
- Powers of *i*
- Absolute Value of a Complex Number

Getting Ready *Perform the following operations.*

1. $(3x + 5) + (4x - 5)$

2. $(3x + 5) - (4x - 5)$

3. $(3x + 5)(4x - 5)$

4. $(3x + 5)(3x - 5)$

We have seen that square roots of negative numbers are not real numbers. However, there is a broader set of numbers, called the **complex numbers,** in which negative numbers do have square roots. In this section, we will discuss this broader set of numbers.

Imaginary Numbers

Consider the number $\sqrt{-3}$. Since no real number squared is -3, $\sqrt{-3}$ is not a real number. For years, people believed that numbers like

$$\sqrt{-1}, \qquad \sqrt{-3}, \qquad \sqrt{-4}, \qquad \text{and} \qquad \sqrt{-9}$$

were nonsense. In the 17th century, René Descartes (1596–1650) called them **imaginary numbers.** Today, imaginary numbers have many important uses, such as describing the behavior of alternating current in electronics.

The imaginary number $\sqrt{-1}$ is often denoted by the letter i:

$$i = \sqrt{-1}$$

Because i represents the square root of -1, it follows that

$$i^2 = -1$$

PERSPECTIVE

The Pythagoreans (ca. 500 B.C.) understood the universe as a harmony of whole numbers. They did not classify fractions as numbers, and were upset that $\sqrt{2}$ was not the ratio of whole numbers. For 2,000 years, little progress was made in the understanding of the various kinds of numbers.

The father of algebra, François Vieta (1540–1603), understood the whole numbers, fractions, and certain irrational numbers. But he was unable to accept negative numbers, and certainly not imaginary numbers.

René Descartes (1596–1650) thought these numbers to be nothing more than figments of his imagination, so he called them *imaginary numbers.* Leonhard Euler (1707–1783) used the letter *i* for $\sqrt{-1}$; Augustin Cauchy (1789–1857) used the term *conjugate;* and Carl Gauss (1777–1855) first used the word *complex.*

Today, we accept complex numbers without question, but it took many centuries and the work of many mathematicians to make them respectable.

Simplifying Imaginary Numbers

If we assume that multiplication of imaginary numbers is commutative and associative, then

$$\begin{aligned}(2i)^2 &= 2^2 i^2 \\ &= 4(-1) \quad i^2 = -1. \\ &= -4\end{aligned}$$

Since $(2i)^2 = -4$, $2i$ is a square root of -4, and we can write

$$\sqrt{-4} = 2i$$

This result can also be obtained by using the multiplication property of radicals:

$$\sqrt{-4} = \sqrt{4(-1)} = \sqrt{4}\sqrt{-1} = 2i$$

We can use the multiplication property of radicals to simplify any imaginary number. For example,

$$\sqrt{-25} = \sqrt{25(-1)} = \sqrt{25}\sqrt{-1} = 5i$$

$$\sqrt{\frac{-100}{49}} = \sqrt{\frac{100}{49}(-1)} = \frac{\sqrt{100}}{\sqrt{49}}\sqrt{-1} = \frac{10}{7}i$$

These examples illustrate the following rule.

Properties of Radicals

If at least one of a and b is a nonnegative real number, then

$$\sqrt{ab} = \sqrt{a}\sqrt{b} \quad \text{and} \quad \sqrt{\frac{a}{b}} = \frac{\sqrt{a}}{\sqrt{b}} \quad (b \neq 0)$$

Comment If a and b are negative, then $\sqrt{ab} \neq \sqrt{a}\sqrt{b}$. For example, if $a = -16$ and $b = -4$,

$$\sqrt{(-16)(-4)} = \sqrt{64} = 8 \quad \text{but} \quad \sqrt{(-16)}\sqrt{(-4)} = (4i)(2i) = 8i^2 = 8(-1) = -8$$

The correct solution is -8.

Complex Numbers

The imaginary numbers are a subset of a set of numbers called the **complex numbers.**

Complex Numbers

A **complex number** is any number that can be written in the form $a + bi$, where a and b are real numbers and $i = \sqrt{-1}$.

In the complex number $a + bi$, a is called the **real part,** and b is called the **imaginary part.**

If $b = 0$, the complex number $a + bi$ is a real number. If $b \neq 0$ and $a = 0$, the complex number $0 + bi$ (or just bi) is an imaginary number.

Any imaginary number can be expressed in *bi* form. For example,

$$\sqrt{-1} = i$$
$$\sqrt{-9} = \sqrt{9(-1)} = \sqrt{9}\sqrt{-1} = 3i$$
$$\sqrt{-3} = \sqrt{3(-1)} = \sqrt{3}\sqrt{-1} = \sqrt{3}i$$

Comment The expression $\sqrt{3}i$ is often written as $i\sqrt{3}$ to make it clear that i is not part of the radicand. Don't confuse $\sqrt{3}i$ with $\sqrt{3i}$.

The relationship between the real numbers, the imaginary numbers, and the complex numbers is shown in Figure 9-19.

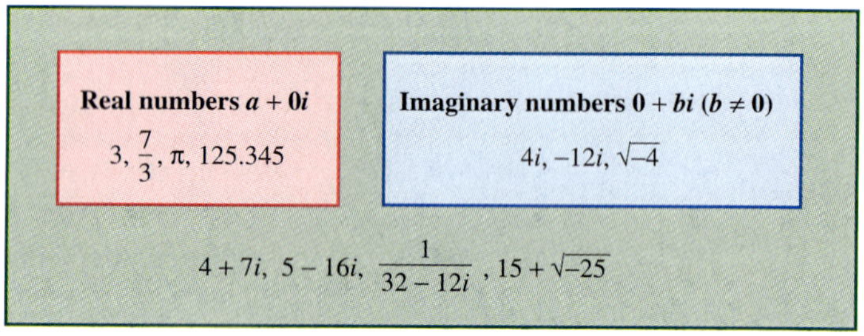

Figure 9-19

Equality of Complex Numbers

The complex numbers $a + bi$ and $c + di$ are equal if and only if

$a = c$ and $b = d$

Because of the previous definition, complex numbers are equal when their real parts are equal and their imaginary parts are equal.

EXAMPLE 1

a. $2 + 3i = \sqrt{4} + \frac{6}{2}i$ because $2 = \sqrt{4}$ and $3 = \frac{6}{2}$.

b. $4 - 5i = \frac{12}{3} - \sqrt{25}i$ because $4 = \frac{12}{3}$ and $-5 = -\sqrt{25}$.

c. $x + yi = 4 + 7i$ if and only if $x = 4$ and $y = 7$.

Arithmetic of Complex Numbers

Addition and Subtraction of Complex Numbers

Complex numbers are added and subtracted as if they were binomials:

$$(a + bi) + (c + di) = (a + c) + (b + d)i$$
$$(a + bi) - (c + di) = (a + bi) + (-c - di) = (a - c) + (b - d)i$$

EXAMPLE 2 Perform the operations:

a. $$\begin{aligned}(8 + 4i) + (12 + 8i) &= 8 + 4i + 12 + 8i\\ &= 20 + 12i\end{aligned}$$

b. $$\begin{aligned}(7 - 4i) + (9 + 2i) &= 7 - 4i + 9 + 2i\\ &= 16 - 2i\end{aligned}$$

c. $$\begin{aligned}(-6 + i) - (3 - 4i) &= -6 + i - 3 + 4i\\ &= -9 + 5i\end{aligned}$$

d. $$\begin{aligned}(2 - 4i) - (-4 + 3i) &= 2 - 4i + 4 - 3i\\ &= 6 - 7i\end{aligned}$$

Self Check Perform the operations: **a.** $(3 - 5i) + (-2 + 7i)$ and **b.** $(3 - 5i) - (-2 + 7i)$.

To multiply a complex number by an imaginary number, we use the distributive property to remove parentheses and simplify. For example,

$$\begin{aligned}-5i(4 - 8i) &= -5i(4) - (-5i)8i && \text{Use the distributive property.}\\ &= -20i + 40i^2 && \text{Simplify.}\\ &= -20i + 40(-1) && \text{Remember that } i^2 = -1.\\ &= -40 - 20i\end{aligned}$$

To multiply two complex numbers, we use the following definition.

Multiplying Complex Numbers

Complex numbers are multiplied as if they were binomials, with $i^2 = -1$:

$$\begin{aligned}(a + bi)(c + di) &= ac + adi + bci + bdi^2\\ &= ac + adi + bci + bd(-1)\\ &= (ac - bd) + (ad + bc)i\end{aligned}$$

EXAMPLE 3 Multiply the complex numbers:

a. $$\begin{aligned}(2 + 3i)(3 - 2i) &= 6 - 4i + 9i - 6i^2 && \text{Use the FOIL method.}\\ &= 6 + 5i + 6 && i^2 = -1\text{, and combine } -4i \text{ and } 9i.\\ &= 12 + 5i\end{aligned}$$

b. $$\begin{aligned}(3 + i)(1 + 2i) &= 3 + 6i + i + 2i^2 && \text{Use the FOIL method.}\\ &= 3 + 7i - 2 && i^2 = -1\text{, and combine } 6i \text{ and } i.\\ &= 1 + 7i\end{aligned}$$

c. $$\begin{aligned}(-4 + 2i)(2 + i) &= -8 - 4i + 4i + 2i^2 && \text{Use the FOIL method.}\\ &= -8 - 2 && i^2 = -1\text{, and combine } -4i \text{ and } 4i.\\ &= -10\end{aligned}$$

Self Check Multiply: $(-2 + 3i)(3 - 2i)$.

The next two examples show how to write complex numbers in $a + bi$ form. It is common to use $a - bi$ as a substitute for $a + (-b)i$.

EXAMPLE 4 Write each number in $a + bi$ form:

a. $7 = 7 + 0i$

b. $3i = 0 + 3i$

c. $4 - \sqrt{-16} = 4 - \sqrt{-1(16)}$
$= 4 - \sqrt{16}\sqrt{-1}$
$= 4 - 4i$

d. $5 + \sqrt{-11} = 5 + \sqrt{-1(11)}$
$= 5 + \sqrt{11}\sqrt{-1}$
$= 5 + \sqrt{11}i$

Self Check Write $3 - \sqrt{-25}$ in $a + bi$ form.

Complex Conjugates The complex numbers $a + bi$ and $a - bi$ are called **complex conjugates.**

For example,

$3 + 4i$ and $3 - 4i$ are complex conjugates.

$5 - 7i$ and $5 + 7i$ are complex conjugates.

EXAMPLE 5 Find the product of $3 + i$ and its complex conjugate.

Solution The complex conjugate of $3 + i$ is $3 - i$. We can find the product as follows:

$$\begin{aligned}(3 + i)(3 - i) &= 9 - 3i + 3i - i^2 && \text{Use the FOIL method.}\\ &= 9 - i^2 && \text{Combine like terms.}\\ &= 9 - (-1) && i^2 = -1.\\ &= 10\end{aligned}$$

Self Check Multiply: $(2 + 3i)(2 - 3i)$.

The product of the complex number $a + bi$ and its complex conjugate $a - bi$ is the real number $a^2 + b^2$, as the following work shows:

$$\begin{aligned}(a + bi)(a - bi) &= a^2 - abi + abi - b^2i^2 && \text{Use the FOIL method.}\\ &= a^2 - b^2(-1) && i^2 = -1.\\ &= a^2 + b^2\end{aligned}$$

Rationalizing the Denominator

To divide complex numbers, we often have to rationalize a denominator.

EXAMPLE 6 Divide and write the result in $a + bi$ form: $\dfrac{1}{3 + i}$.

Solution We can rationalize the denominator by multiplying the numerator and the denominator by the complex conjugate of the denominator.

$$\frac{1}{3+i} = \frac{1}{3+i} \cdot \frac{3-i}{3-i} \qquad \frac{3-i}{3-i} = 1.$$

$$= \frac{3-i}{9-3i+3i-i^2} \qquad \text{Multiply the numerators and multiply the denominators.}$$

$$= \frac{3-i}{9-(-1)} \qquad i^2 = -1.$$

$$= \frac{3-i}{10}$$

$$= \frac{3}{10} - \frac{1}{10}i$$

Self Check Rationalize the denominator: $\frac{1}{5-i}$.

EXAMPLE 7 Write $\frac{3-i}{2+i}$ in $a + bi$ form.

Solution We multiply the numerator and the denominator of the fraction by the complex conjugate of the denominator.

$$\frac{3-i}{2+i} = \frac{3-i}{2+i} \cdot \frac{2-i}{2-i} \qquad \frac{2-i}{2-i} = 1.$$

$$= \frac{6-3i-2i+i^2}{4-2i+2i-i^2} \qquad \text{Multiply the numerators and multiply the denominators.}$$

$$= \frac{5-5i}{4-(-1)} \qquad 6+i^2 = 6-1 = 5.$$

$$= \frac{5(1-i)}{5} \qquad \text{Factor out 5 in the numerator.}$$

$$= 1 - i \qquad \text{Simplify.}$$

Self Check Rationalize the denominator: $\frac{2+i}{5-i}$.

EXAMPLE 8 Write $\frac{4+\sqrt{-16}}{2+\sqrt{-4}}$ in $a + bi$ form.

Solution

$$\frac{4+\sqrt{-16}}{2+\sqrt{-4}} = \frac{4+4i}{2+2i} \qquad \text{Write each number in } a + bi \text{ form.}$$

$$= \frac{2\overset{1}{\cancel{(2+2i)}}}{\underset{1}{\cancel{2+2i}}} \qquad \text{Factor out 2 in the numerator and simplify.}$$

$$= 2 + 0i$$

Self Check Divide: $\frac{3+\sqrt{-25}}{2+\sqrt{-9}}$.

Comment To avoid mistakes, always put complex numbers in $a + bi$ form before doing any complex number arithmetic.

Powers of i

The powers of i produce an interesting pattern:

$i = \sqrt{-1} = i$	$i^5 = i^4 i = 1i = i$
$i^2 = \left(\sqrt{-1}\right)^2 = -1$	$i^6 = i^4 i^2 = 1(-1) = -1$
$i^3 = i^2 i = -1i = -i$	$i^7 = i^4 i^3 = 1(-i) = -i$
$i^4 = i^2 i^2 = (-1)(-1) = 1$	$i^8 = i^4 i^4 = (1)(1) = 1$

The pattern continues: $i, -1, -i, 1, \ldots.$.

EXAMPLE 9 Simplify: i^{29}.

Solution We note that 29 divided by 4 gives a quotient of 7 and a remainder of 1. Thus, $29 = 4 \cdot 7 + 1$, and

$$
\begin{aligned}
i^{29} &= i^{4\cdot 7+1} && 4 \cdot 7 + 1 = 29. \\
&= (i^4)^7 \cdot i && i^{4\cdot 7+1} = i^{4\cdot 7} \cdot i^1 = (i^4)^7 \cdot i. \\
&= 1^7 \cdot i && i^4 = 1. \\
&= i
\end{aligned}
$$

Self Check Simplify: i^{31}.

The results of Example 9 illustrate the following fact.

Powers of i

If n is a natural number that has a remainder of r when divided by 4, then

$$i^n = i^r$$

When n is divisible by 4, the remainder r is 0 and $i^0 = 1$.

EXAMPLE 10 Simplify: i^{55}.

Solution We divide 55 by 4 and get a remainder of 3. Therefore,

$$i^{55} = i^3 = -i$$

Self Check Simplify: i^{62}.

EXAMPLE 11 Simplify each expression:

a. $2i^2 + 4i^3 = 2(-1) + 4(-i)$

$= -2 - 4i$

b. $\frac{3}{2i} = \frac{3}{2i} \cdot \frac{i}{i}$ $\quad \frac{i}{i} = 1.$

$= \frac{3i}{2i^2}$

$= \frac{3i}{2(-1)}$

$= \frac{3i}{-2}$

$= 0 - \frac{3}{2}i$

c. $-\frac{5}{i} = -\frac{5}{i} \cdot \frac{i^3}{i^3}$ $\quad \frac{i^3}{i^3} = 1.$

$= -\frac{5(-i)}{1}$

$= 5i$

$= 0 + 5i$

d. $\frac{6}{i^3} = \frac{6i}{i^3 i}$ $\quad \frac{i}{i} = 1.$

$= \frac{6i}{i^4}$

$= \frac{6i}{1}$

$= 6i$

$= 0 + 6i$

Self Check Simplify: **a.** $3i^3 - 2i^2$ and **b.** $\frac{2}{3i}$.

Absolute Value of a Complex Number

Absolute Value of a Complex Number

The **absolute value** of the complex number $a + bi$ is $\sqrt{a^2 + b^2}$. In symbols,

$$|a + bi| = \sqrt{a^2 + b^2}$$

EXAMPLE 12 Find each absolute value:

a. $|3 + 4i| = \sqrt{3^2 + 4^2}$

$= \sqrt{9 + 16}$

$= \sqrt{25}$

$= 5$

b. $|3 - 4i| = \sqrt{3^2 + (-4)^2}$

$= \sqrt{9 + 16}$

$= \sqrt{25}$

$= 5$

c. $|-5 - 12i| = \sqrt{(-5)^2 + (-12)^2}$

$= \sqrt{25 + 144}$

$= \sqrt{169}$

$= 13$

d. $|a + 0i| = \sqrt{a^2 + 0^2}$

$= \sqrt{a^2}$

$= |a|$

Self Check Evaluate: $|5 + 12i|$.

Comment Note that $|a + bi| = \sqrt{a^2 + b^2}$, not

$$|a + bi| = \sqrt{a^2 + (bi)^2}$$

Self Check Answers

2. a. $1 + 2i$, **b.** $5 - 12i$ **3.** $13i$ **4.** $3 - 5i$ **5.** 13 **6.** $\frac{5}{26} + \frac{1}{26}i$ **7.** $\frac{9}{26} + \frac{7}{26}i$ **8.** $\frac{21}{13} + \frac{1}{13}i$

9. $-i$ **10.** -1 **11. a.** $2 - 3i$, **b.** $0 - \frac{2}{3}i$ **12.** 13

Orals *Write each imaginary number in bi form:*

1. $\sqrt{-49}$ **2.** $\sqrt{-64}$ **3.** $\sqrt{-100}$ **4.** $\sqrt{-81}$

Simplify each power of i:

5. i^3 **6.** i^2 **7.** i^4 **8.** i^5

Find each absolute value:

9. $|3 + 4i|$ **10.** $|5 - 12i|$

9.7 EXERCISES

REVIEW *Perform each operation.*

1. $\dfrac{x^2 - x - 6}{9 - x^2} \cdot \dfrac{x^2 + x - 6}{x^2 - 4}$

2. $\dfrac{3x + 4}{x - 2} + \dfrac{x - 4}{x + 2}$

3. Wind speed A plane that can fly 200 mph in still air makes a 330-mile flight with a tail wind and returns, flying into the same wind. Find the speed of the wind if the total flying time is $3\frac{1}{3}$ hours.

4. Finding rates A student drove a distance of 135 miles at an average speed of 50 mph. How much faster would he have to drive on the return trip to save 30 minutes of driving time?

VOCABULARY AND CONCEPTS *Fill in the blanks.*

5. $\sqrt{-1}$, $\sqrt{-3}$, and $\sqrt{-4}$ are examples of ________ numbers.

6. $\sqrt{-1} =$ __ **7.** $i^2 =$ ___

8. $i^3 =$ ___ **9.** $i^4 =$ __

10. $\sqrt{ab} =$ _______, provided a and b are not both negative.

11. $\sqrt{\dfrac{a}{b}} = \dfrac{}{\rule{2em}{0.4pt}}$ $(b \neq 0)$, provided a and b are not both negative.

12. $3 + 5i$, $2 - 7i$, and $5 - \frac{1}{2}i$ are examples of _______ numbers.

13. The real part of $5 + 7i$ is __. The imaginary part is __.

14. $a + bi = c + di$ if and only if $a =$ __ and $b =$ __.

15. $a + bi$ and $a - bi$ are called complex ________.

16. $|a + bi| =$ ________

PRACTICE *Write each imaginary number in bi form.*

17. $\sqrt{-9}$ **18.** $\sqrt{-16}$

19. $\sqrt{-36}$ **20.** $\sqrt{-81}$

21. $\sqrt{-7}$ **22.** $\sqrt{-11}$

Tell whether the complex numbers are equal.

23. $3 + 7i$, $\sqrt{9} + (5 + 2)i$

24. $\sqrt{4} + \sqrt{25}i$, $2 - (-5)i$

25. $8 + 5i$, $2^3 + \sqrt{25}i^3$

26. $4 - 7i$, $-4i^2 + 7i^3$

27. $\sqrt{4} + \sqrt{-4}$, $2 - 2i$

28. $\sqrt{-9} - i$, $4i$

Perform the operations. Write all answers in a + bi form.

29. $(3 + 4i) + (5 - 6i)$

30. $(5 + 3i) - (6 - 9i)$

31. $(7 - 3i) - (4 + 2i)$

32. $(8 + 3i) + (-7 - 2i)$

33. $(8 + 5i) + (7 + 2i)$

34. $(-7 + 9i) - (-2 - 8i)$

35. $(1 + i) - 2i + (5 - 7i)$

36. $(-9 + i) - 5i + (2 + 7i)$

37. $(5 + 3i) - (3 - 5i) + \sqrt{-1}$

38. $(8 + 7i) - \left(-7 - \sqrt{-64}\right) + (3 - i)$

39. $\left(-8 - \sqrt{3}i\right) - \left(7 - 3\sqrt{3}i\right)$

40. $\left(2 + 2\sqrt{2}i\right) + \left(-3 - \sqrt{2}i\right)$

41. $3i(2 - i)$

42. $-4i(3 + 4i)$

43. $-5i(5 - 5i)$

44. $2i(7 + 2i)$

45. $(2 + i)(3 - i)$

46. $(4 - i)(2 + i)$

47. $(2 - 4i)(3 + 2i)$

48. $(3 - 2i)(4 - 3i)$

49. $\left(2 + \sqrt{2}i\right)\left(3 - \sqrt{2}i\right)$

50. $\left(5 + \sqrt{3}i\right)\left(2 - \sqrt{3}i\right)$

51. $\left(8 - \sqrt{-1}\right)\left(-2 - \sqrt{-16}\right)$

52. $\left(-1 + \sqrt{-4}\right)\left(2 + \sqrt{-9}\right)$

53. $(2 + i)^2$

54. $(3 - 2i)^2$

55. $(2 + 3i)^2$

56. $(1 - 3i)^2$

57. $i(5 + i)(3 - 2i)$

58. $i(-3 - 2i)(1 - 2i)$

59. $(2 + i)(2 - i)(1 + i)$

60. $(3 + 2i)(3 - 2i)(i + 1)$

61. $(3 + i)[(3 - 2i) + (2 + i)]$

62. $(2 - 3i)[(5 - 2i) - (2i + 1)]$

Write each expression in a + bi form.

63. $\dfrac{1}{i}$

64. $\dfrac{1}{i^3}$

65. $\dfrac{4}{5i^3}$

66. $\dfrac{3}{2i}$

67. $\dfrac{3i}{8\sqrt{-9}}$

68. $\dfrac{5i^3}{2\sqrt{-4}}$

69. $\dfrac{-3}{5i^5}$

70. $\dfrac{-4}{6i^7}$

71. $\dfrac{5}{2 - i}$

72. $\dfrac{26}{3 - 2i}$

73. $\dfrac{13i}{5 + i}$

74. $\dfrac{2i}{5 + 3i}$

75. $\dfrac{-12}{7 - \sqrt{-1}}$

76. $\dfrac{4}{3 + \sqrt{-1}}$

77. $\dfrac{5i}{6 + 2i}$

78. $\dfrac{-4i}{2 - 6i}$

79. $\dfrac{3 - 2i}{3 + 2i}$

80. $\dfrac{2 + 3i}{2 - 3i}$

81. $\dfrac{3 + 2i}{3 + i}$

82. $\dfrac{2 - 5i}{2 + 5i}$

83. $\dfrac{\sqrt{5} - \sqrt{3}i}{\sqrt{5} + \sqrt{3}i}$

84. $\dfrac{\sqrt{3} + \sqrt{2}i}{\sqrt{3} - \sqrt{2}i}$

85. $\left(\dfrac{i}{3 + 2i}\right)^2$

86. $\left(\dfrac{5 + i}{2 + i}\right)^2$

87. $\dfrac{i(3 - i)}{3 + i}$

88. $\dfrac{5 + 3i}{i(3 - 5i)}$

89. $\dfrac{(2 - 5i) - (5 - 2i)}{5 - i}$

90. $\dfrac{5i}{(5 + 2i) + (2 + i)}$

Simplify each expression.

91. i^{21}

92. i^{19}

93. i^{27}

94. i^{22}

95. i^{100}

96. i^{42}

97. i^{97}

98. i^{200}

Find each value.

99. $|6 + 8i|$

100. $|12 + 5i|$

101. $|12 - 5i|$

102. $|3 - 4i|$

103. $|5 + 7i|$

104. $|6 - 5i|$

105. $\left|\dfrac{3}{5} - \dfrac{4}{5}i\right|$

106. $\left|\dfrac{5}{13} + \dfrac{12}{13}i\right|$

107. Show that $1 - 5i$ is a solution of $x^2 - 2x + 26 = 0$.

108. Show that $3 - 2i$ is a solution of $x^2 - 6x + 13 = 0$.

109. Show that i is a solution of $x^4 - 3x^2 - 4 = 0$.

110. Show that $2 + i$ is *not* a solution of $x^2 + x + 1 = 0$.

APPLICATIONS *In electronics, the formula $V = IR$ is called **Ohm's law.** It gives the relationship in a circuit between the voltage V (in volts), the current I (in amperes), and the resistance R (in ohms).*

111. Electronics Find V when $I = 2 - 3i$ amperes and $R = 2 + i$ ohms.

112. Electronics Find R when $I = 3 - 2i$ amperes and $V = 18 + i$ volts.

*In electronics, the formula $Z = \frac{V}{I}$ is used to find the **impedance** Z of a circuit, where V is the voltage and I is the current.*

113. Electronics Find the impedance of a circuit when the voltage is $1.7 + 0.5i$ and the current is $0.5i$.

114. Electronics Find the impedance of a circuit when the voltage is $1.6 - 0.4i$ and the current is $-0.2i$.

WRITING

115. Tell how to decide whether two complex numbers are equal.

116. Define the complex conjugate of a complex number.

SOMETHING TO THINK ABOUT

117. Rationalize the numerator: $\frac{3 - i}{2}$.

118. Rationalize the numerator: $\frac{2 + 3i}{2 - 3i}$

Projects

Project 1

The size of a television screen is measured along the diagonal of its screen, as shown in the illustrations. The screen of a traditional TV has an aspect ratio of 4:3. This means that the ratio of the width of the screen to its height is $\frac{4}{3}$. The screen of a wide-screen set has an aspect ratio of 16:9. This means that the ratio of the width of the screen to its is height is $\frac{16}{9}$.

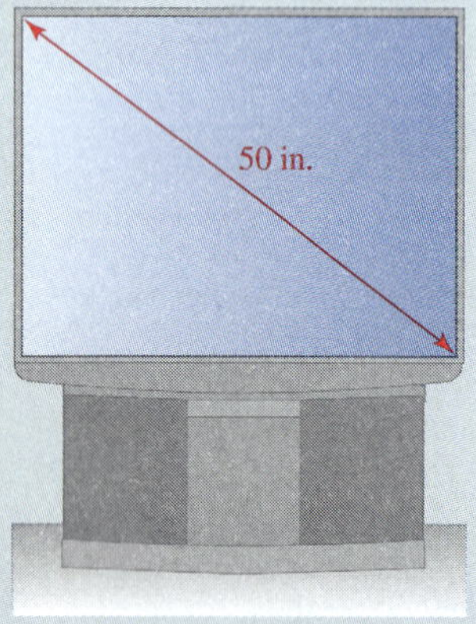

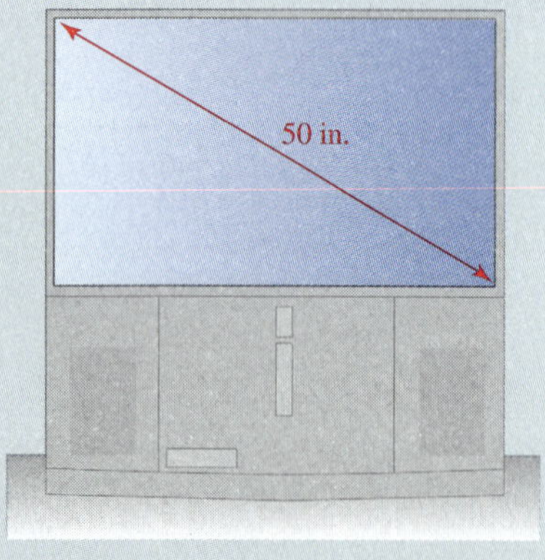

a. Find the width and height of the traditional-screen set shown in the illustration. $\left(\textit{Hint}: \frac{4}{3} = \frac{4x}{3x}.\right)$
b. Find the viewing area of the traditional-screen set in square inches.
c. Find the width and height of the wide-screen set shown in the illustration.
d. Find the viewing area of the wide-screen set in square inches.
e. Which set has the larger viewing area? Give the answer as a percent.

Project 2

Tom and Brian arrange to have a bicycle race. Each leaves his own house at the same time and rides to the other's house, whereupon the winner of the race calls his own house and leaves a message for the loser. A map of the race is shown in the illustration. Brian stays on the highway, averaging 21 mph. Tom knows that he and Brian are evenly matched when biking on the highway, so he cuts across country for the first part of his trip, averaging 15 mph. When Tom reaches the highway at point A, he turns right and follows the highway, averaging 21 mph.

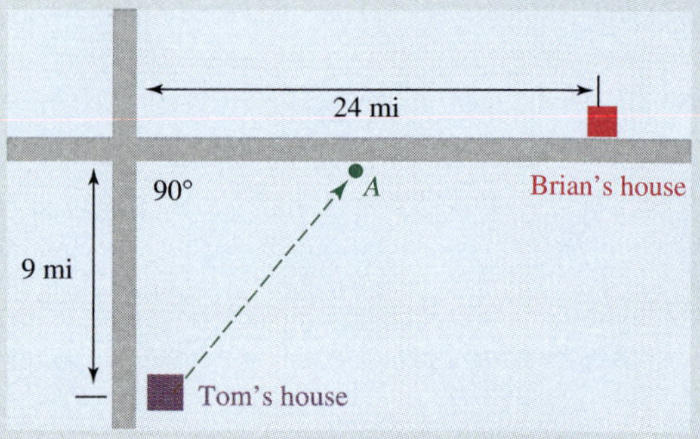

Tom and Brian never meet during the race, and amazingly, the race is a tie. Each of them calls the other at exactly the same moment!

a. How long (to the nearest second) did it take each person to complete the race?

b. How far from the intersection of the two highways is point A? (*Hint:* Set the travel times for Brian and Tom equal to each other. You may find two answers, but only one of them matches all of the information.)

c. Show that if Tom had started straight across country for Brian's house (in order to minimize the distance he had to travel), he would have lost the race. By how much time (to the nearest second) would he have lost? Then show that if Tom had biked across country to a point 9 miles from the intersection of the two highways, he would have won the race. By how much time (to the nearest second) would he have won?

Chapter Summary

CONCEPTS

9.1

If n is an even natural number,

$$\sqrt[n]{a^n} = |a|$$

If n is an odd natural number, greater than 1,

$$\sqrt[n]{a^n} = a$$

If n is a natural number greater than 1 and x is a real number, then

If $x > 0$, then $\sqrt[n]{x}$ is the positive number such that $(\sqrt[n]{x})^n = x$.

If $x = 0$, then $\sqrt[n]{x} = 0$.

If $x < 0$, and n is odd, $\sqrt[n]{x}$ is the real number such that $(\sqrt[n]{x})^n = x$.

If $x < 0$, and n is even, $\sqrt[n]{x}$ is not a real number.

REVIEW EXERCISES

Radical Expressions

Simplify each radical. Assume that x can be any number.

1. $\sqrt{49}$

2. $-\sqrt{121}$

3. $-\sqrt{36}$

4. $\sqrt{225}$

5. $\sqrt[3]{-27}$

6. $-\sqrt[3]{216}$

7. $\sqrt[4]{625}$

8. $\sqrt[5]{-32}$

9. $\sqrt{25x^2}$

10. $\sqrt{x^2 + 4x + 4}$

11. $\sqrt[3]{27a^6b^3}$

12. $\sqrt[4]{256x^8y^4}$

Graph each function.

13. $f(x) = \sqrt{x + 2}$

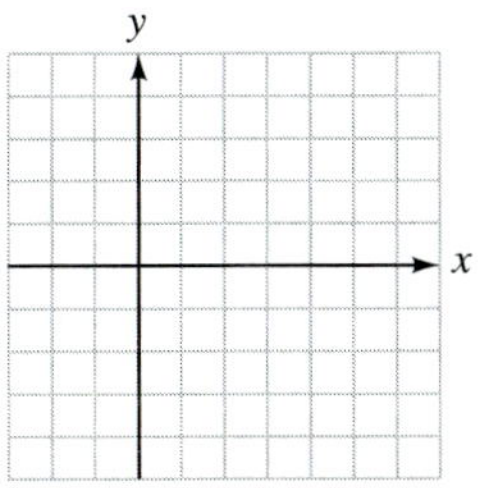

14. $f(x) = -\sqrt{x - 1}$

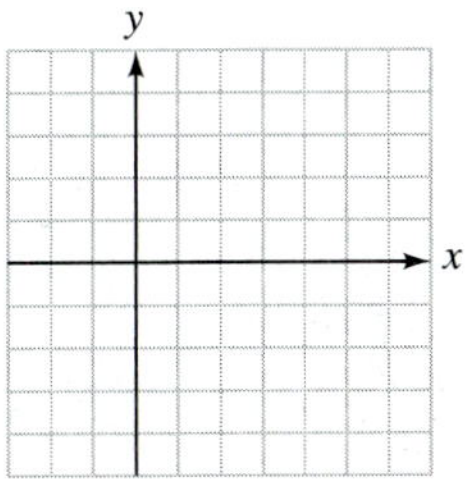

15. $f(x) = -\sqrt{x} + 2$

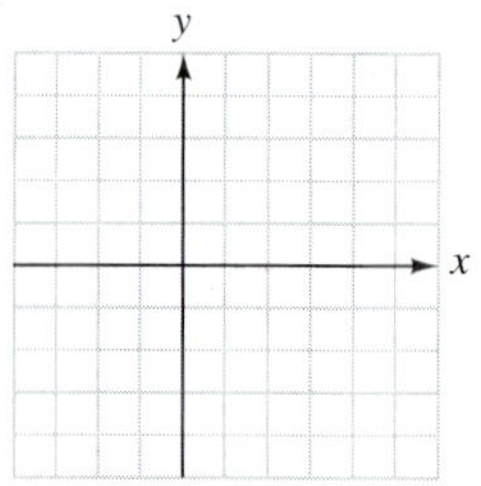

16. $f(x) = -\sqrt[3]{x} + 3$

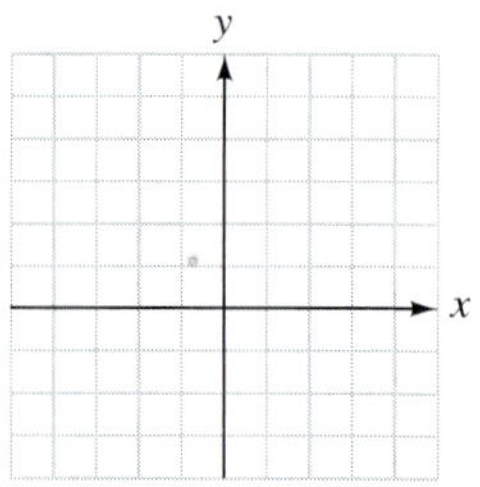

Consider the distribution 4, 8, 12, 16, 20.

17. Find the mean of the distribution.

18. Find the standard deviation.

9.2 Applications of Radicals

The Pythagorean theorem: If a and b are the lengths of the legs of a right triangle and c is the length of the hypotenuse, then

$$a^2 + b^2 = c^2$$

In Exercises 19–20, the horizon distance d (measured in miles) is related to the height h (measured in feet) of the observer by the formula $d = 1.4\sqrt{h}$.

19. **View from a submarine** A submarine's periscope extends 4.7 feet above the surface. How far away is the horizon?

20. **View from a submarine** How far out of the water must a submarine periscope extend to provide a 4-mile horizon?

21. **Sailing** A technique called *tacking* allows a sailboat to make progress into the wind. A sailboat follows the course in the illustration. Find d, the distance the boat advances into the wind.

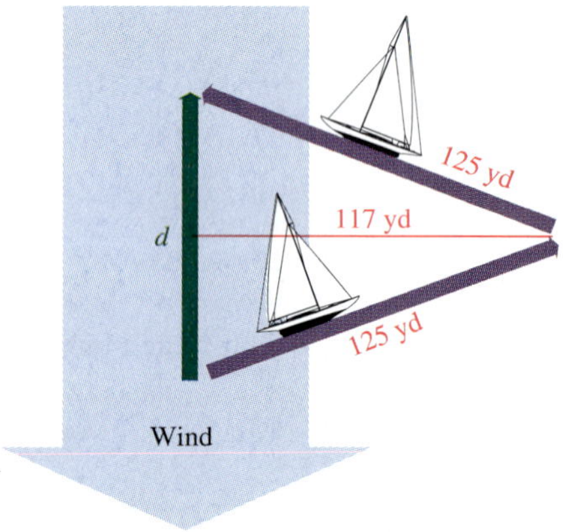

22. Communications Some campers 3,900 yards from a highway are talking to truckers on a citizen's band radio with an 8,900-yard range. Over what length of highway can these conversations take place?

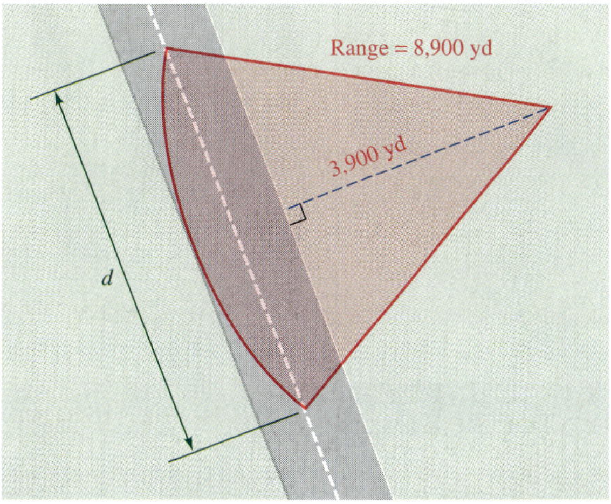

The distance formula:

$$d(PQ) = \sqrt{(x_2 - x_1)^2 + (y_2 - y_1)^2}$$

23. Find the distance between points (0, 0) and (5, −12).

24. Find the distance between points (−4, 6) and (−2, 8). Give the result to the nearest hundredth.

9.3 Rational Exponents

If n ($n > 1$) is a natural number and $\sqrt[n]{x}$ is a real number, then $x^{1/n} = \sqrt[n]{x}$.

If n is even, $(x^n)^{1/n} = |x|$.

If n is a natural number greater than 1 and x is a real number, then

- If $x > 0$, then $x^{1/n}$ is the positive number such that $(x^{1/n})^n = x$.
- If $x = 0$, then $x^{1/n} = 0$.
- If $x < 0$, and n is odd, then $x^{1/n}$ is the real number such that $(x^{1/n})^n = x$.
- If $x < 0$ and n is even, then $x^{1/n}$ is not a real number.

Simplify each expression, if possible. Assume that all variables represent positive numbers.

25. $25^{1/2}$

26. $-36^{1/2}$

27. $9^{3/2}$

28. $16^{3/2}$

29. $(-8)^{1/3}$

30. $-8^{2/3}$

31. $8^{-2/3}$

32. $8^{-1/3}$

33. $-49^{5/2}$

34. $\frac{1}{25^{5/2}}$

35. $\left(\frac{1}{4}\right)^{-3/2}$

36. $\left(\frac{4}{9}\right)^{-3/2}$

37. $(27x^3y)^{1/3}$

38. $(81x^4y^2)^{1/4}$

39. $(25x^3y^4)^{3/2}$

40. $(8u^2v^3)^{-2/3}$

If m and n are positive integers and $x > 0$,

$$x^{m/n} = \sqrt[n]{x^m} = (\sqrt[n]{x})^m$$

$$x^{-m/n} = \frac{1}{x^{m/n}}$$

$$\frac{1}{x^{-m/n}} = x^{m/n} \quad (x \neq 0)$$

Perform the multiplications. Assume that all variables represent positive numbers and write all answers without negative exponents.

41. $5^{1/4}5^{1/2}$
42. $a^{3/7}a^{2/7}$
43. $u^{1/2}(u^{1/2} - u^{-1/2})$
44. $v^{2/3}(v^{1/3} + v^{4/3})$
45. $(x^{1/2} + y^{1/2})^2$
46. $(a^{2/3} + b^{2/3})(a^{2/3} - b^{2/3})$

Simplify each expression. Assume that all variables are positive.

47. $\sqrt[6]{5^2}$
48. $\sqrt[8]{x^4}$
49. $\sqrt[9]{27a^3b^6}$
50. $\sqrt[4]{25a^2b^2}$

9.4 Simplifying and Combining Radical Expressions

Properties of radicals:

$$\sqrt[n]{ab} = \sqrt[n]{a}\sqrt[n]{b}$$

$$\sqrt[n]{\frac{a}{b}} = \frac{\sqrt[n]{a}}{\sqrt[n]{b}} \quad (b \neq 0)$$

Simplify each expression. Assume all variables represent positive numbers.

51. $\sqrt{240}$
52. $\sqrt[3]{54}$
53. $\sqrt[4]{32}$
54. $\sqrt[5]{96}$
55. $\sqrt{8x^3}$
56. $\sqrt{18x^4y^3}$
57. $\sqrt[3]{16x^5y^4}$
58. $\sqrt[3]{54x^7y^3}$
59. $\dfrac{\sqrt{32x^3}}{\sqrt{2x}}$
60. $\dfrac{\sqrt[3]{16x^5}}{\sqrt[3]{2x^2}}$
61. $\sqrt[3]{\dfrac{2a^2b}{27x^3}}$
62. $\sqrt{\dfrac{17xy}{64a^4}}$

Like radicals can be combined by addition and subtraction:

$$3\sqrt{2} + 5\sqrt{2} = 8\sqrt{2}$$

Radicals that are not similar can often be simplified to radicals that are similar and then combined:

$$\sqrt{2} + \sqrt{8} = \sqrt{2} + \sqrt{4}\sqrt{2}$$
$$= \sqrt{2} + 2\sqrt{2}$$
$$= 3\sqrt{2}$$

In an isosceles right triangle, the length of the hypotenuse is the length of one leg times $\sqrt{2}$.

Simplify and combine like radicals. Assume that all variables represent positive numbers.

63. $\sqrt{2} + \sqrt{8}$
64. $\sqrt{20} - \sqrt{5}$
65. $2\sqrt[3]{3} - \sqrt[3]{24}$
66. $\sqrt[4]{32} + 2\sqrt[4]{162}$
67. $2x\sqrt{8} + 2\sqrt{200x^2} + \sqrt{50x^2}$
68. $3\sqrt{27a^3} - 2a\sqrt{3a} + 5\sqrt{75a^3}$
69. $\sqrt[3]{54} - 3\sqrt[3]{16} + 4\sqrt[3]{128}$
70. $2\sqrt[4]{32x^5} + 4\sqrt[4]{162x^5} - 5x\sqrt[4]{512x}$
71. Find the length of the hypotenuse of an isosceles right triangle whose legs measure 7 meters.
72. The hypotenuse of a 30°–60°–90° triangle measures $12\sqrt{3}$ centimeters. Find the length of each leg.

The shorter leg of a 30°–60°–90° triangle is half as long as the hypotenuse. The longer leg is the length of the shorter leg times $\sqrt{3}$.

Find x to two decimal places.

73.

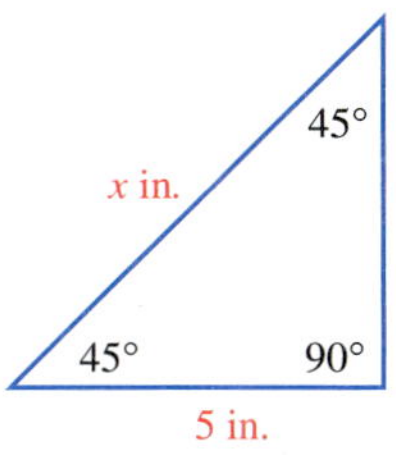

74.

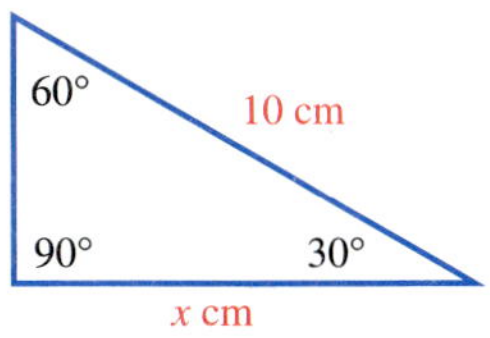

9.5 Multiplying and Dividing Radical Expressions

If two radicals have the same index, they can be multiplied:

$$\sqrt{3x}\sqrt{6x} = \sqrt{18x^2} \ (x > 0)$$
$$= 3x\sqrt{2}$$

Simplify each expression. Assume that all variables represent positive numbers.

75. $(2\sqrt{5})(3\sqrt{2})$

76. $2\sqrt{6}\sqrt{216}$

77. $\sqrt{9x}\sqrt{x}$

78. $\sqrt[3]{3}\sqrt[3]{9}$

79. $-\sqrt[3]{2x^2}\sqrt[3]{4x}$

80. $-\sqrt[4]{256x^5y^{11}}\sqrt[4]{625x^9y^3}$

81. $\sqrt{2}(\sqrt{8} - 3)$

82. $\sqrt{2}(\sqrt{2} + 3)$

83. $\sqrt{5}(\sqrt{2} - 1)$

84. $\sqrt{3}(\sqrt{3} + \sqrt{2})$

85. $(\sqrt{2} + 1)(\sqrt{2} - 1)$

86. $(\sqrt{3} + \sqrt{2})(\sqrt{3} + \sqrt{2})$

87. $(\sqrt{x} + \sqrt{y})(\sqrt{x} - \sqrt{y})$

88. $(2\sqrt{u} + 3)(3\sqrt{u} - 4)$

To rationalize the binomial denominator of a fraction, multiply the numerator and the denominator by the conjugate of the binomial in the denominator.

Rationalize each denominator

89. $\dfrac{1}{\sqrt{3}}$

90. $\dfrac{\sqrt{3}}{\sqrt{5}}$

91. $\dfrac{x}{\sqrt{xy}}$

92. $\dfrac{\sqrt[3]{uv}}{\sqrt[3]{u^5v^7}}$

93. $\dfrac{2}{\sqrt{2} - 1}$

94. $\dfrac{\sqrt{2}}{\sqrt{3} - 1}$

95. $\dfrac{2x - 32}{\sqrt{x} + 4}$

96. $\dfrac{\sqrt{a} + 1}{\sqrt{a} - 1}$

Rationalize each numerator.

97. $\dfrac{\sqrt{3}}{5}$

98. $\dfrac{\sqrt[3]{9}}{3}$

99. $\dfrac{3 - \sqrt{x}}{2}$

100. $\dfrac{\sqrt{a} - \sqrt{b}}{\sqrt{a}}$

9.6 Radical Equations

The power rule:

If $x = y$, then $x^n = y^n$.

Raising both sides of an equation to the same power can lead to extraneous solutions. Be sure to check all suspected solutions.

Solve each equation.

101. $\sqrt{y + 3} = \sqrt{2y - 19}$

102. $u = \sqrt{25u - 144}$

103. $r = \sqrt{12r - 27}$

104. $\sqrt{z + 1} + \sqrt{z} = 2$

105. $\sqrt{2x + 5} - \sqrt{2x} = 1$

106. $\sqrt[3]{x^3 + 8} = x + 2$

9.7 Complex Numbers

Complex numbers: If a, b, c, and d are real numbers and $i^2 = -1$,

$a + bi = c + di$ if and only if $a = c$ and $b = d$

$(a + bi) + (c + di) = (a + c) + (b + d)i$

$(a + bi)(c + di) = (ac - bd) + (ad + bc)i$

$|a + bi| = \sqrt{a^2 + b^2}$

Perform the operations and give all answers in $a + bi$ form.

107. $(5 + 4i) + (7 - 12i)$

108. $(-6 - 40i) - (-8 + 28i)$

109. $\left(-32 + \sqrt{-144}\right) - \left(64 + \sqrt{-81}\right)$

110. $\left(-8 + \sqrt{-8}\right) + \left(6 - \sqrt{-32}\right)$

111. $(2 - 7i)(-3 + 4i)$

112. $(-5 + 6i)(2 + i)$

113. $\left(5 - \sqrt{-27}\right)\left(-6 + \sqrt{-12}\right)$

114. $\left(2 + \sqrt{-128}\right)\left(3 - \sqrt{-98}\right)$

115. $\dfrac{3}{4i}$

116. $\dfrac{-2}{5i^3}$

117. $\dfrac{6}{2 + i}$

118. $\dfrac{7}{3 - i}$

119. $\dfrac{4 + i}{4 - i}$

120. $\dfrac{3 - i}{3 + i}$

121. $\dfrac{3}{5 + \sqrt{-4}}$

122. $\dfrac{2}{3 - \sqrt{-9}}$

123. $|9 + 12i|$

124. $|24 - 10i|$

Simplify.

125. i^{12}

126. i^{583}

Chapter Test

Find each root.

1. $\sqrt{49}$

2. $\sqrt[3]{64}$

3. $\sqrt{4x^2}$

4. $\sqrt[3]{8x^3}$

Graph each function and find its domain and range.

5. $f(x) = \sqrt{x - 2}$

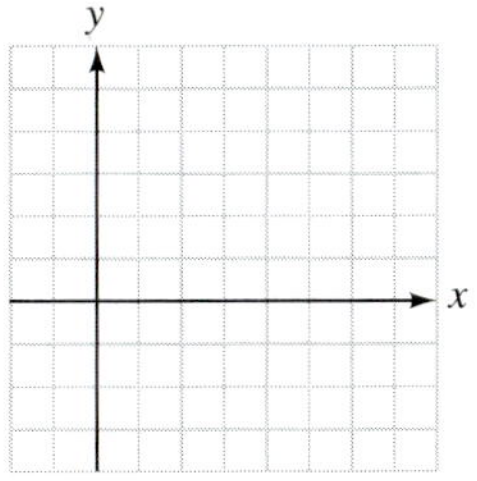

6. $f(x) = \sqrt[3]{x} + 3$

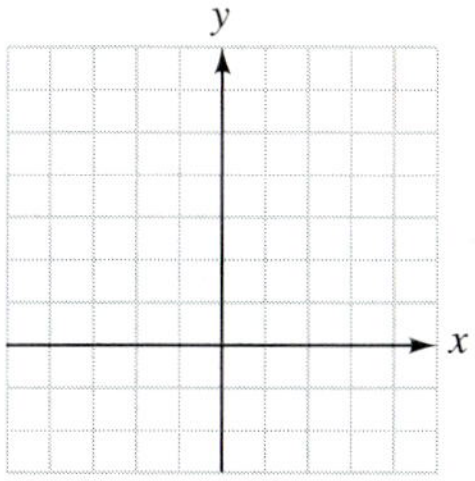

Use a calculator.

7. Shipping crates The diagonal brace on the shipping crate shown in the illustration is 53 inches. Find the height, h, of the crate.

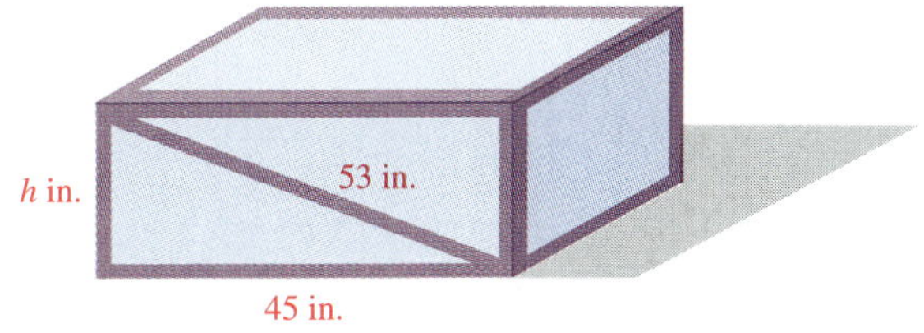

8. Pendulums The 2-meter pendulum rises 0.1 meter at the extremes of its swing. Find the width w of the swing.

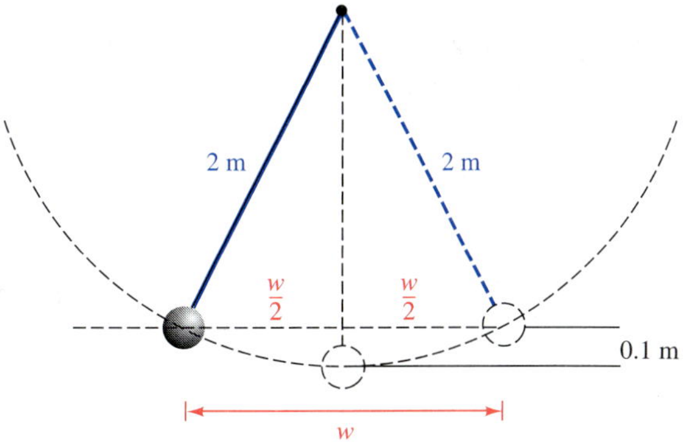

Find the distance between the points.

9. (6, 8), (0, 0)

10. (−2, 5), (22, 12)

Simplify each expression. Assume that all variables represent positive numbers, and write answers without using negative exponents.

11. $16^{1/4}$

12. $27^{2/3}$

13. $36^{-3/2}$

14. $\left(-\dfrac{8}{27}\right)^{-2/3}$

15. $\dfrac{2^{5/3}2^{1/6}}{2^{1/2}}$

16. $\dfrac{(8x^3y)^{1/2}(8xy^5)^{1/2}}{(x^3y^6)^{1/3}}$

Simplify each expression. Assume that all variables represent positive numbers.

17. $\sqrt{48}$
18. $\sqrt{250x^3y^5}$
19. $\dfrac{\sqrt[3]{24x^{15}y^4}}{\sqrt[3]{y}}$
20. $\sqrt{\dfrac{3a^5}{48a^7}}$

Simplify each expression. Assume that the variables are unrestricted.

21. $\sqrt{12x^2}$
22. $\sqrt{8x^6}$
23. $\sqrt[3]{81x^3}$
24. $\sqrt{18x^4y^9}$

Simplify and combine like radicals. Assume that all variables represent positive numbers.

25. $\sqrt{12} - \sqrt{27}$
26. $2\sqrt[3]{40} - \sqrt[3]{5{,}000} + 4\sqrt[3]{625}$
27. $2\sqrt{48y^5} - 3y\sqrt{12y^3}$
28. $\sqrt[4]{768z^5} + z\sqrt[4]{48z}$

Perform each operation and simplify, if possible. All variables represent positive numbers.

29. $-2\sqrt{xy}\left(3\sqrt{x} + \sqrt{xy^3}\right)$
30. $\left(3\sqrt{2} + \sqrt{3}\right)\left(2\sqrt{2} - 3\sqrt{3}\right)$

Rationalize each denominator.

31. $\dfrac{1}{\sqrt{5}}$
32. $\dfrac{3t - 1}{\sqrt{3t} - 1}$

Rationalize each numerator.

33. $\dfrac{\sqrt{3}}{\sqrt{7}}$
34. $\dfrac{\sqrt{a} + \sqrt{b}}{\sqrt{a} - \sqrt{b}}$

Solve and check each equation.

35. $\sqrt[3]{6n + 4} - 4 = 0$
36. $1 - \sqrt{u} = \sqrt{u - 3}$

Perform the operations. Give all answers in a + bi form.

37. $(2 + 4i) + (-3 + 7i)$
38. $\left(3 - \sqrt{-9}\right) - \left(-1 + \sqrt{-16}\right)$
39. $2i(3 - 4i)$
40. $(3 + 2i)(-4 - i)$
41. $\dfrac{1}{i\sqrt{2}}$
42. $\dfrac{2 + i}{3 - i}$

Symmetries of Graphs

There are several ways that a graph can exhibit symmetry about the coordinate axes and the origin. It is often easier to draw graphs of equations if we first find the x- and y-intercepts and find any of the following symmetries of the graph:

1. **y-axis symmetry:** If the point $(-x, y)$ lies on a graph whenever the point (x, y) does, as in Figure I-1(a), we say that the graph is **symmetric about the y-axis.**
2. **Symmetry about the origin:** If the point $(-x, -y)$ lies on the graph whenever the point (x, y) does, as in Figure I-1(b), we say that the graph is **symmetric about the origin.**
3. **x-axis symmetry:** If the point $(x, -y)$ lies on the graph whenever the point (x, y) does, as in Figure I-1(c), we say that the graph is **symmetric about the x-axis.**

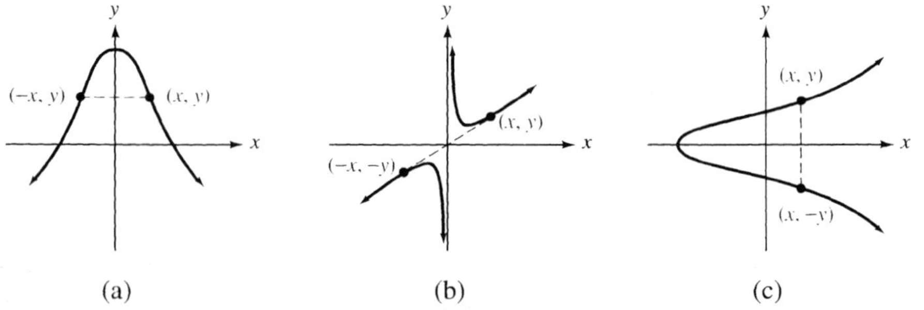

Figure I-1

Tests for Symmetry for Graphs in x and y

- To test a graph for y-axis symmetry, replace x with $-x$. If the new equation is equivalent to the original equation, the graph is symmetric about the y-axis. Symmetry about the y-axis will occur whenever x appears with only even exponents.
- To test a graph for symmetry about the origin, replace x with $-x$ and y with $-y$. If the resulting equation is equivalent to the original equation, the graph is symmetric about the origin.
- To test a graph for x-axis symmetry, replace y with $-y$. If the resulting equation is equivalent to the original equation, the graph is symmetric about the x-axis. The only function that is symmetric about the x-axis is $f(x) = 0$.

EXAMPLE 1 Find the intercepts and the symmetries of the graph of $y = f(x) = x^3 - 9x$. Then graph the function.

Solution **x-intercepts:** To find the x-intercepts, we let $y = 0$ and solve for x:

$$y = x^3 - 9x$$

$$0 = x^3 - 9x \quad \text{Substitute 0 for } y.$$

$$0 = x(x^2 - 9) \quad \text{Factor out } x.$$

$$0 = x(x + 3)(x - 3) \quad \text{Factor } x^2 - 9.$$

$$x = 0 \quad \text{or} \quad x + 3 = 0 \quad \text{or} \quad x - 3 = 0 \quad \text{Set each factor equal to 0.}$$

$$| \quad x = -3 \quad | \quad x = 3$$

Since the x-coordinates of the x-intercepts are 0, -3, and 3, the graph intersects the x-axis at (0, 0), $(-3, 0)$, and (3, 0).

y-intercepts: To find the y-intercepts, we let $x = 0$ and solve for y.

$$y = x^3 - 9x$$

$$y = 0^3 - 9(0) \quad \text{Substitute 0 for } x.$$

$$y = 0$$

Since the y-coordinate of the y-intercept is 0, the graph intersects the y-axis at (0, 0).

Symmetry: We test for symmetry about the y-axis by replacing x with $-x$, simplifying, and comparing the result to the original equation.

(1) $y = x^3 - 9x$ The original equation.

$y = (-x)^3 - 9(-x)$ Replace x with $-x$.

(2) $y = -x^3 + 9x$ Simplify.

Because Equation 2 is not equivalent to Equation 1, the graph is not symmetric about the y-axis.

We test for symmetry about the origin by replacing x and y with $-x$ and $-y$, respectively, and comparing the result to the original equation.

(1) $y = x^3 - 9x$ The original equation.

$-y = (-x)^3 - 9(-x)$ Replace x with $-x$, and y with $-y$.

$-y = -x^3 + 9x$ Simplify.

(3) $y = x^3 - 9x$ Multiply both sides by -1 to solve for y.

Because Equation 3 is equivalent to Equation 1, the graph is symmetric about the origin. Because the equation is the equation of a nonzero function, there is no symmetry about the x-axis.

To graph the equation, we plot the x-intercepts of $(-3, 0)$, (0, 0), and (3, 0) and the y-intercept of (0, 0). We also plot other points for positive values of x and use the symmetry about the origin to draw the rest of the graph, as in Figure I-2(a). (Note that the scale on the x-axis is different from the scale on the y-axis.)

If we graph the equation with a graphing calculator, with window settings of $[-10, 10]$ for x and $[-10, 10]$ for y, we will obtain the graph shown in Figure I-2(b).

From the graph, we can see that the domain is the interval $(-\infty, \infty)$, and the range is the interval $(-\infty, \infty)$.

$y = x^3 - 9x$

x	y
0	0
1	-8
2	-10
3	0
4	28

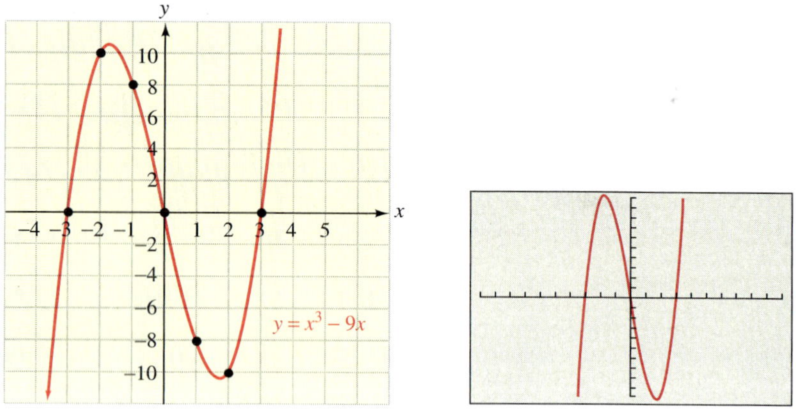

(a) (b)

Figure I-2

EXAMPLE 2 Graph the function $y = f(x) = |x| - 2$.

Solution **x-intercepts:** To find the x-intercepts, we let $y = 0$ and solve for x:

$$y = |x| - 2$$
$$0 = |x| - 2$$
$$2 = |x|$$
$$x = -2 \quad \text{or} \quad x = 2$$

Since -2 and 2 are solutions, the points $(-2, 0)$ and $(2, 0)$ are the x-intercepts, and the graph passes through $(-2, 0)$ and $(2, 0)$.

y-intercepts: To find the y-intercepts, we let $x = 0$ and solve for y:

$$y = |x| - 2$$
$$y = |0| - 2$$
$$y = -2$$

Since $y = -2$, $(0, -2)$ is the y-intercept, and the graph passes through the point $(0, -2)$.

Symmetry: To test for y-axis symmetry, we replace x with $-x$.

(4) $y = |x| - 2$ The original equation.

$y = |-x| - 2$ Replace x with $-x$.

(5) $y = |x| - 2$ $|-x| = |x|$.

Since Equation 5 is equivalent to Equation 4, the graph is symmetric about the y-axis. The graph has no other symmetries.

We plot the x- and y-intercepts and several other points (x, y), and use the y-axis symmetry to obtain the graph shown in Figure I-3(a).

If we graph the equation with a graphing calculator, with window settings of $[-10, 10]$ for x and $[-10, 10]$ for y, we will obtain the graph shown in Figure I-3(b).

From the graph, we see that the domain is the interval $(-\infty, \infty)$, and the range is the interval $[-2, \infty)$.

$y = |x| - 2$

x	y
0	−2
1	−1
2	0
3	1
4	2

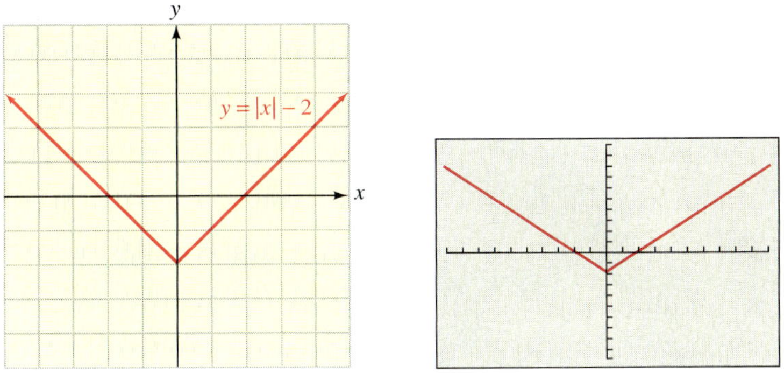

(a) (b)

Figure I-3

I.1 EXERCISES

Find the symmetries of the graph of each equation. ***Do not draw the graph.***

1. $y = x^2 - 1$

2. $y = x^3$

3. $y = x^5$

4. $y = x^4$

5. $y = -x^2 + 2$

6. $y = x^3 + 1$

7. $y = x^2 - x$

8. $y^2 = x + 7$

9. $y = -|x + 2|$

10. $y = |x| - 3$

11. $|y| = x$

12. $y = 2\sqrt{x}$

Graph each function and give its domain and range. Check each graph with a graphing calculator.

13. $f(x) = x^4 - 4$

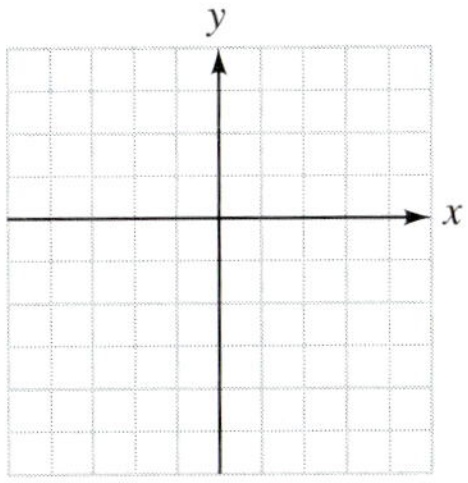

14. $f(x) = \frac{1}{2}x^4 - 1$

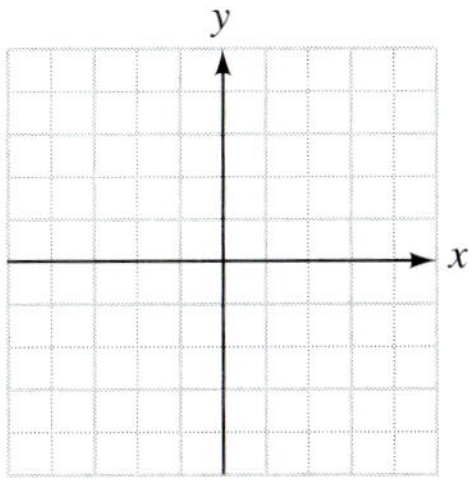

15. $f(x) = -x^3$

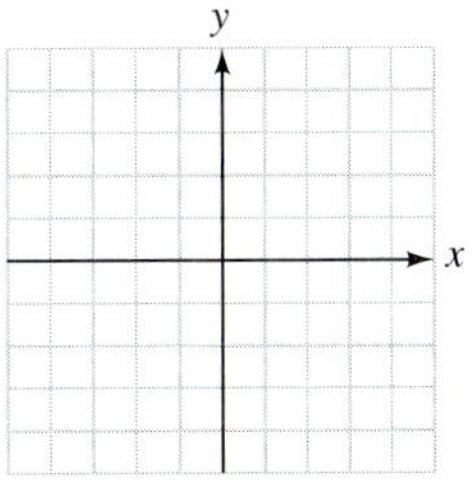

16. $f(x) = x^3 + 2$

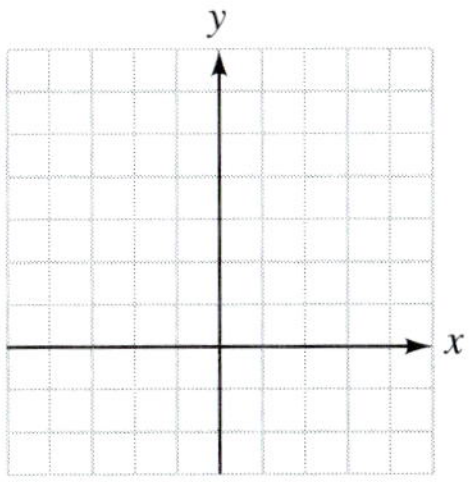

17. $f(x) = x^4 + x^2$

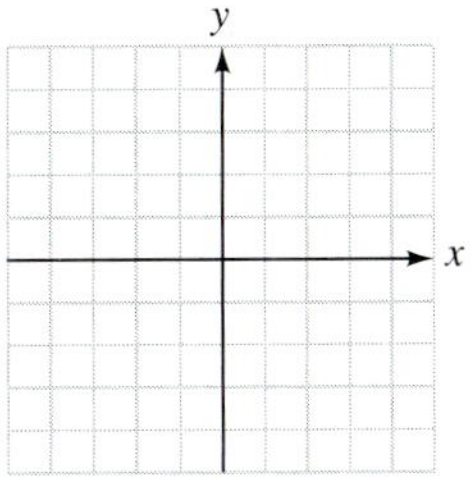

18. $f(x) = 3 - x^4$

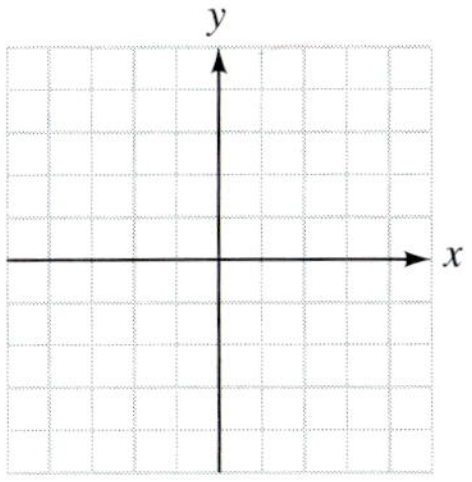

19. $f(x) = x^3 - x$

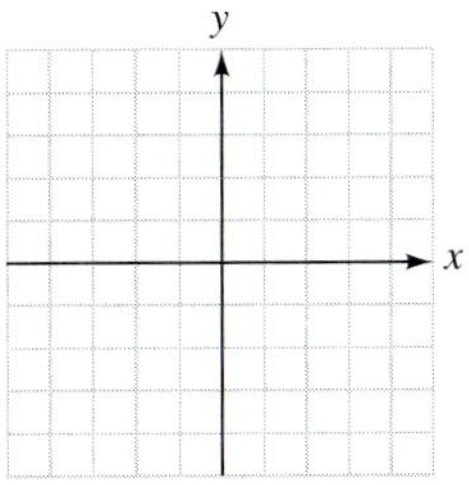

20. $f(x) = x^3 + x$

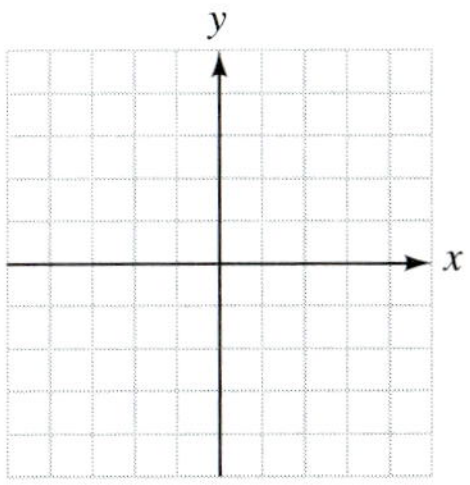

21. $f(x) = \frac{1}{2}|x| - 1$

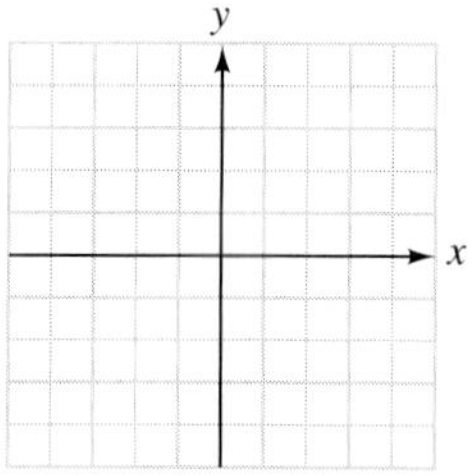

22. $f(x) = -|x| + 1$

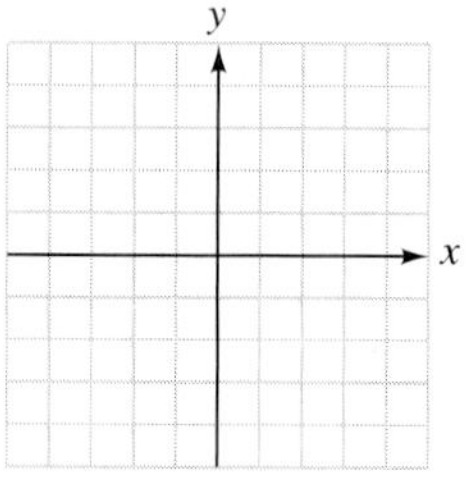

23. $f(x) = -|x + 2|$

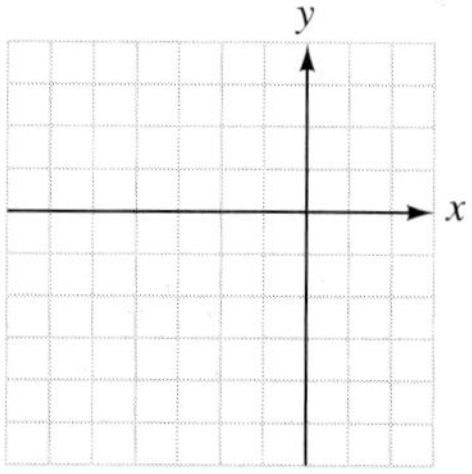

24. $f(x) = |x - 2|$

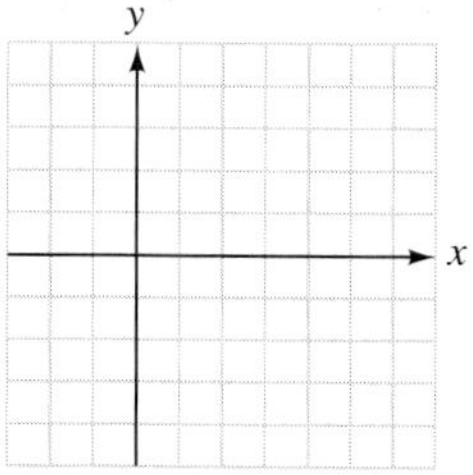

APPENDIX II

Tables

Table A Powers and Roots

n	n^2	$\sqrt{n}$	n^3	$\sqrt[3]{n}$	n	n^2	$\sqrt{n}$	n^3	$\sqrt[3]{n}$
1	1	1.000	1	1.000	51	2,601	7.141	132,651	3.708
2	4	1.414	8	1.260	52	2,704	7.211	140,608	3.733
3	9	1.732	27	1.442	53	2,809	7.280	148,877	3.756
4	16	2.000	64	1.587	54	2,916	7.348	157,464	3.780
5	25	2.236	125	1.710	55	3,025	7.416	166,375	3.803
6	36	2.449	216	1.817	56	3,136	7.483	175,616	3.826
7	49	2.646	343	1.913	57	3,249	7.550	185,193	3.849
8	64	2.828	512	2.000	58	3,364	7.616	195,112	3.871
9	81	3.000	729	2.080	59	3,481	7.681	205,379	3.893
10	100	3.162	1,000	2.154	60	3,600	7.746	216,000	3.915
11	121	3.317	1,331	2.224	61	3,721	7.810	226,981	3.936
12	144	3.464	1,728	2.289	62	3,844	7.874	238,328	3.958
13	169	3.606	2,197	2.351	63	3,969	7.937	250,047	3.979
14	196	3.742	2,744	2.410	64	4,096	8.000	262,144	4.000
15	225	3.873	3,375	2.466	65	4,225	8.062	274,625	4.021
16	256	4.000	4,096	2.520	66	4,356	8.124	287,496	4.041
17	289	4.123	4,913	2.571	67	4,489	8.185	300,763	4.062
18	324	4.243	5,832	2.621	68	4,624	8.246	314,432	4.082
19	361	4.359	6,859	2.668	69	4,761	8.307	328,509	4.102
20	400	4.472	8,000	2.714	70	4,900	8.367	343,000	4.121
21	441	4.583	9,261	2.759	71	5,041	8.426	357,911	4.141
22	484	4.690	10,648	2.802	72	5,184	8.485	373,248	4.160
23	529	4.796	12,167	2.844	73	5,329	8.544	389,017	4.179
24	576	4.899	13,824	2.884	74	5,476	8.602	405,224	4.198
25	625	5.000	15,625	2.924	75	5,625	8.660	421,875	4.217
26	676	5.099	17,576	2.962	76	5,776	8.718	438,976	4.236
27	729	5.196	19,683	3.000	77	5,929	8.775	456,533	4.254
28	784	5.292	21,952	3.037	78	6,084	8.832	474,552	4.273
29	841	5.385	24,389	3.072	79	6,241	8.888	493,039	4.291
30	900	5.477	27,000	3.107	80	6,400	8.944	512,000	4.309
31	961	5.568	29,791	3.141	81	6,561	9.000	531,441	4.327
32	1,024	5.657	32,768	3.175	82	6,724	9.055	551,368	4.344
33	1,089	5.745	35,937	3.208	83	6,889	9.110	571,787	4.362
34	1,156	5.831	39,304	3.240	84	7,056	9.165	592,704	4.380
35	1,225	5.916	42,875	3.271	85	7,225	9.220	614,125	4.397
36	1,296	6.000	46,656	3.302	86	7,396	9.274	636,056	4.414
37	1,369	6.083	50,653	3.332	87	7,569	9.327	658,503	4.431
38	1,444	6.164	54,872	3.362	88	7,744	9.381	681,472	4.448
39	1,521	6.245	59,319	3.391	89	7,921	9.434	704,969	4.465
40	1,600	6.325	64,000	3.420	90	8,100	9.487	729,000	4.481
41	1,681	6.403	68,921	3.448	91	8,281	9.539	753,571	4.498
42	1,764	6.481	74,088	3.476	92	8,464	9.592	778,688	4.514
43	1,849	6.557	79,507	3.503	93	8,649	9.644	804,357	4.531
44	1,936	6.633	85,184	3.530	94	8,836	9.695	830,584	4.547
45	2,025	6.708	91,125	3.557	95	9,025	9.747	857,375	4.563
46	2,116	6.782	97,336	3.583	96	9,216	9.798	884,736	4.579
47	2,209	6.856	103,823	3.609	97	9,409	9.849	912,673	4.595
48	2,304	6.928	110,592	3.634	98	9,604	9.899	941,192	4.610
49	2,401	7.000	117,649	3.659	99	9,801	9.950	970,299	4.626
50	2,500	7.071	125,000	3.684	100	10,000	10.000	1,000,000	4.642

Table B Base-10 Logarithms

N	0	1	2	3	4	5	6	7	8	9
1.0	.0000	.0043	.0086	.0128	.0170	.0212	.0253	.0294	.0334	.0374
1.1	.0414	.0453	.0492	.0531	.0569	.0607	.0645	.0682	.0719	.0755
1.2	.0792	.0828	.0864	.0899	.0934	.0969	.1004	.1038	.1072	.1106
1.3	.1139	.1173	.1206	.1239	.1271	.1303	.1335	.1367	.1399	.1430
1.4	.1461	.1492	.1523	.1553	.1584	.1614	.1644	.1673	.1703	.1732
1.5	.1761	.1790	.1818	.1847	.1875	.1903	.1931	.1959	.1987	.2014
1.6	.2041	.2068	.2095	.2122	.2148	.2175	.2201	.2227	.2253	.2279
1.7	.2304	.2330	.2355	.2380	.2405	.2430	.2455	.2480	.2504	.2529
1.8	.2553	.2577	.2601	.2625	.2648	.2672	.2695	.2718	.2742	.2765
1.9	.2788	.2810	.2833	.2856	.2878	.2900	.2923	.2945	.2967	.2989
2.0	.3010	.3032	.3054	.3075	.3096	.3118	.3139	.3160	.3181	.3201
2.1	.3222	.3243	.3263	.3284	.3304	.3324	.3345	.3365	.3385	.3404
2.2	.3424	.3444	.3464	.3483	.3502	.3522	.3541	.3560	.3579	.3598
2.3	.3617	.3636	.3655	.3674	.3692	.3711	.3729	.3747	.3766	.3784
2.4	.3802	.3820	.3838	.3856	.3874	.3892	.3909	.3927	.3945	.3962
2.5	.3979	.3997	.4014	.4031	.4048	.4065	.4082	.4099	.4116	.4133
2.6	.4150	.4166	.4183	.4200	.4216	.4232	.4249	.4265	.4281	.4298
2.7	.4314	.4330	.4346	.4362	.4378	.4393	.4409	.4425	.4440	.4456
2.8	.4472	.4487	.4502	.4518	.4533	.4548	.4564	.4579	.4594	.4609
2.9	.4624	.4639	.4654	.4669	.4683	.4698	.4713	.4728	.4742	.4757
3.0	.4771	.4786	.4800	.4814	.4829	.4843	.4857	.4871	.4886	.4900
3.1	.4914	.4928	.4942	.4955	.4969	.4983	.4997	.5011	.5024	.5038
3.2	.5051	.5065	.5079	.5092	.5105	.5119	.5132	.5145	.5159	.5172
3.3	.5185	.5198	.5211	.5224	.5237	.5250	.5263	.5276	.5289	.5302
3.4	.5315	.5328	.5340	.5353	.5366	.5378	.5391	.5403	.5416	.5428
3.5	.5441	.5453	.5465	.5478	.5490	.5502	.5514	.5527	.5539	.5551
3.6	.5563	.5575	.5587	.5599	.5611	.5623	.5635	.5647	.5658	.5670
3.7	.5682	.5694	.5705	.5717	.5729	.5740	.5752	.5763	.5775	.5786
3.8	.5798	.5809	.5821	.5832	.5843	.5855	.5866	.5877	.5888	.5899
3.9	.5911	.5922	.5933	.5944	.5955	.5966	.5977	.5988	.5999	.6010
4.0	.6021	.6031	.6042	.6053	.6064	.6075	.6085	.6096	.6107	.6117
4.1	.6128	.6138	.6149	.6160	.6170	.6180	.6191	.6201	.6212	.6222
4.2	.6232	.6243	.6253	.6263	.6274	.6284	.6294	.6304	.6314	.6325
4.3	.6335	.6345	.6355	.6365	.6375	.6385	.6395	.6405	.6415	.6425
4.4	.6435	.6444	.6454	.6464	.6474	.6484	.6493	.6503	.6513	.6522
4.5	.6532	.6542	.6551	.6561	.6571	.6580	.6590	.6599	.6609	.6618
4.6	.6628	.6637	.6646	.6656	.6665	.6675	.6684	.6693	.6702	.6712
4.7	.6721	.6730	.6739	.6749	.6758	.6767	.6776	.6785	.6794	.6803
4.8	.6812	.6821	.6830	.6839	.6848	.6857	.6866	.6875	.6884	.6893
4.9	.6902	.6911	.6920	.6928	.6937	.6946	.6955	.6964	.6972	.6981
5.0	.6990	.6998	.7007	.7016	.7024	.7033	.7042	.7050	.7059	.7067
5.1	.7076	.7084	.7093	.7101	.7110	.7118	.7126	.7135	.7143	.7152
5.2	.7160	.7168	.7177	.7185	.7193	.7202	.7210	.7218	.7226	.7235
5.3	.7243	.7251	.7259	.7267	.7275	.7284	.7292	.7300	.7308	.7316
5.4	.7324	.7332	.7340	.7348	.7356	.7364	.7372	.7380	.7388	.7396

Table B *(continued)*

N	0	1	2	3	4	5	6	7	8	9
5.5	.7404	.7412	.7419	.7427	.7435	.7443	.7451	.7459	.7466	.7474
5.6	.7482	.7490	.7497	.7505	.7513	.7520	.7528	.7536	.7543	.7551
5.7	.7559	.7566	.7574	.7582	.7589	.7597	.7604	.7612	.7619	.7627
5.8	.7634	.7642	.7649	.7657	.7664	.7672	.7679	.7686	.7694	.7701
5.9	.7709	.7716	.7723	.7731	.7738	.7745	.7752	.7760	.7767	.7774
6.0	.7782	.7789	.7796	.7803	.7810	.7818	.7825	.7832	.7839	.7846
6.1	.7853	.7860	.7868	.7875	.7882	.7889	.7896	.7903	.7910	.7917
6.2	.7924	.7931	.7938	.7945	.7952	.7959	.7966	.7973	.7980	.7987
6.3	.7993	.8000	.8007	.8014	.8021	.8028	.8035	.8041	.8048	.8055
6.4	.8062	.8069	.8075	.8082	.8089	.8096	.8102	.8109	.8116	.8122
6.5	.8129	.8136	.8142	.8149	.8156	.8162	.8169	.8176	.8182	.8189
6.6	.8195	.8202	.8209	.8215	.8222	.8228	.8235	.8241	.8248	.8254
6.7	.8261	.8267	.8274	.8280	.8287	.8293	.8299	.8306	.8312	.8319
6.8	.8325	.8331	.8338	.8344	.8351	.8357	.8363	.8370	.8376	.8382
6.9	.8388	.8395	.8401	.8407	.8414	.8420	.8426	.8432	.8439	.8445
7.0	.8451	.8457	.8463	.8470	.8476	.8482	.8488	.8494	.8500	.8506
7.1	.8513	.8519	.8525	.8531	.8537	.8543	.8549	.8555	.8561	.8567
7.2	.8573	.8579	.8585	.8591	.8597	.8603	.8609	.8615	.8621	.8627
7.3	.8633	.8639	.8645	.8651	.8657	.8663	.8669	.8675	.8681	.8686
7.4	.8692	.8698	.8704	.8710	.8716	.8722	.8727	.8733	.8739	.8745
7.5	.8751	.8756	.8762	.8768	.8774	.8779	.8785	.8791	.8797	.8802
7.6	.8808	.8814	.8820	.8825	.8831	.8837	.8842	.8848	.8854	.8859
7.7	.8865	.8871	.8876	.8882	.8887	.8893	.8899	.8904	.8910	.8915
7.8	.8921	.8927	.8932	.8938	.8943	.8949	.8954	.8960	.8965	.8971
7.9	.8976	.8982	.8987	.8993	.8998	.9004	.9009	.9015	.9020	.9025
8.0	.9031	.9036	.9042	.9047	.9053	.9058	.9063	.9069	.9074	.9079
8.1	.9085	.9090	.9096	.9101	.9106	.9112	.9117	.9122	.9128	.9133
8.2	.9138	.9143	.9149	.9154	.9159	.9165	.9170	.9175	.9180	.9186
8.3	.9191	.9196	.9201	.9206	.9212	.9217	.9222	.9227	.9232	.9238
8.4	.9243	.9248	.9253	.9258	.9263	.9269	.9274	.9279	.9284	.9289
8.5	.9294	.9299	.9304	.9309	.9315	.9320	.9325	.9330	.9335	.9340
8.6	.9345	.9350	.9355	.9360	.9365	.9370	.9375	.9380	.9385	.9390
8.7	.9395	.9400	.9405	.9410	.9415	.9420	.9425	.9430	.9435	.9440
8.8	.9445	.9450	.9455	.9460	.9465	.9469	.9474	.9479	.9484	.9489
8.9	.9494	.9499	.9504	.9509	.9513	.9518	.9523	.9528	.9533	.9538
9.0	.9542	.9547	.9552	.9557	.9562	.9566	.9571	.9576	.9581	.9586
9.1	.9590	.9595	.9600	.9605	.9609	.9614	.9619	.9624	.9628	.9633
9.2	.9638	.9643	.9647	.9652	.9657	.9661	.9666	.9671	.9675	.9680
9.3	.9685	.9689	.9694	.9699	.9703	.9708	.9713	.9717	.9722	.9727
9.4	.9731	.9736	.9741	.9745	.9750	.9754	.9759	.9763	.9768	.9773
9.5	.9777	.9782	.9786	.9791	.9795	.9800	.9805	.9809	.9814	.9818
9.6	.9823	.9827	.9832	.9836	.9841	.9845	.9850	.9854	.9859	.9863
9.7	.9868	.9872	.9877	.9881	.9886	.9890	.9894	.9899	.9903	.9908
9.8	.9912	.9917	.9921	.9926	.9930	.9934	.9939	.9943	.9948	.9952
9.9	.9956	.9961	.9965	.9969	.9974	.9978	.9983	.9987	.9991	.9996

Table C Base-*e* Logarithms

N	0	1	2	3	4	5	6	7	8	9
1.0	.0000	.0100	.0198	.0296	.0392	.0488	.0583	.0677	.0770	.0862
1.1	.0953	.1044	.1133	.1222	.1310	.1398	.1484	.1570	.1655	.1740
1.2	.1823	.1906	.1989	.2070	.2151	.2231	.2311	.2390	.2469	.2546
1.3	.2624	.2700	.2776	.2852	.2927	.3001	.3075	.3148	.3221	.3293
1.4	.3365	.3436	.3507	.3577	.3646	.3716	.3784	.3853	.3920	.3988
1.5	.4055	.4121	.4187	.4253	.4318	.4383	.4447	.4511	.4574	.4637
1.6	.4700	.4762	.4824	.4886	.4947	.5008	.5068	.5128	.5188	.5247
1.7	.5306	.5365	.5423	.5481	.5539	.5596	.5653	.5710	.5766	.5822
1.8	.5878	.5933	.5988	.6043	.6098	.6152	.6206	.6259	.6313	.6366
1.9	.6419	.6471	.6523	.6575	.6627	.6678	.6729	.6780	.6831	.6881
2.0	.6931	.6981	.7031	.7080	.7129	.7178	.7227	.7275	.7324	.7372
2.1	.7419	.7467	.7514	.7561	.7608	.7655	.7701	.7747	.7793	.7839
2.2	.7885	.7930	.7975	.8020	.8065	.8109	.8154	.8198	.8242	.8286
2.3	.8329	.8372	.8416	.8459	.8502	.8544	.8587	.8629	.8671	.8713
2.4	.8755	.8796	.8838	.8879	.8920	.8961	.9002	.9042	.9083	.9123
2.5	.9163	.9203	.9243	.9282	.9322	.9361	.9400	.9439	.9478	.9517
2.6	.9555	.9594	.9632	.9670	.9708	.9746	.9783	.9821	.9858	.9895
2.7	.9933	.9969	1.0006	.0043	.0080	.0116	.0152	.0188	.0225	.0260
2.8	1.0296	.0332	.0367	.0403	.0438	.0473	.0508	.0543	.0578	.0613
2.9	.0647	.0682	.0716	.0750	.0784	.0818	.0852	.0886	.0919	.0953
3.0	1.0986	.1019	.1053	.1086	.1119	.1151	.1184	.1217	.1249	.1282
3.1	.1314	.1346	.1378	.1410	.1442	.1474	.1506	.1537	.1569	.1600
3.2	.1632	.1663	.1694	.1725	.1756	.1787	.1817	.1848	.1878	.1909
3.3	.1939	.1969	.2000	.2030	.2060	.2090	.2119	.2149	.2179	.2208
3.4	.2238	.2267	.2296	.2326	.2355	.2384	.2413	.2442	.2470	.2499
3.5	1.2528	.2556	.2585	.2613	.2641	.2669	.2698	.2726	.2754	.2782
3.6	.2809	.2837	.2865	.2892	.2920	.2947	.2975	.3002	.3029	.3056
3.7	.3083	.3110	.3137	.3164	.3191	.3218	.3244	.3271	.3297	.3324
3.8	.3350	.3376	.3403	.3429	.3455	.3481	.3507	.3533	.3558	.3584
3.9	.3610	.3635	.3661	.3686	.3712	.3737	.3762	.3788	.3813	.3838
4.0	1.3863	.3888	.3913	.3938	.3962	.3987	.4012	.4036	.4061	.4085
4.1	.4110	.4134	.4159	.4183	.4207	.4231	.4255	.4279	.4303	.4327
4.2	.4351	.4375	.4398	.4422	.4446	.4469	.4493	.4516	.4540	.4563
4.3	.4586	.4609	.4633	.4656	.4679	.4702	.4725	.4748	.4770	.4793
4.4	.4816	.4839	.4861	.4884	.4907	.4929	.4951	.4974	.4996	.5019
4.5	1.5041	.5063	.5085	.5107	.5129	.5151	.5173	.5195	.5217	.5239
4.6	.5261	.5282	.5304	.5326	.5347	.5369	.5390	.5412	.5433	.5454
4.7	.5476	.5497	.5518	.5539	.5560	.5581	.5602	.5623	.5644	.5665
4.8	.5686	.5707	.5728	.5748	.5769	.5790	.5810	.5831	.5851	.5872
4.9	.5892	.5913	.5933	.5953	.5974	.5994	.6014	.6034	.6054	.6074
5.0	1.6094	.6114	.6134	.6154	.6174	.6194	.6214	.6233	.6253	.6273
5.1	.6292	.6312	.6332	.6351	.6371	.6390	.6409	.6429	.6448	.6467
5.2	.6487	.6506	.6525	.6544	.6563	.6582	.6601	.6620	.6639	.6658
5.3	.6677	.6696	.6715	.6734	.6752	.6771	.6790	.6808	.6827	.6845
5.4	.6864	.6882	.6901	.6919	.6938	.6956	.6974	.6993	.7011	.7029

Use the properties of logarithms and ln 10 ≈ 2.3026 to find logarithms of numbers less than 1 or greater than 10.

Table C *(continued)*

N	0	1	2	3	4	5	6	7	8	9
5.5	1.7047	.7066	.7084	.7102	.7120	.7138	.7156	.7174	.7192	.7210
5.6	.7228	.7246	.7263	.7281	.7299	.7317	.7334	.7352	.7370	.7387
5.7	.7405	.7422	.7440	.7457	.7475	.7492	.7509	.7527	.7544	.7561
5.8	.7579	.7596	.7613	.7630	.7647	.7664	.7681	.7699	.7716	.7733
5.9	.7750	.7766	.7783	.7800	.7817	.7834	.7851	.7867	.7884	.7901
6.0	1.7918	.7934	.7951	.7967	.7984	.8001	.8017	.8034	.8050	.8066
6.1	.8083	.8099	.8116	.8132	.8148	.8165	.8181	.8197	.8213	.8229
6.2	.8245	.8262	.8278	.8294	.8310	.8326	.8342	.8358	.8374	.8390
6.3	.8405	.8421	.8437	.8453	.8469	.8485	.8500	.8516	.8532	.8547
6.4	.8563	.8579	.8594	.8610	.8625	.8641	.8656	.8672	.8687	.8703
6.5	1.8718	.8733	.8749	.8764	.8779	.8795	.8810	.8825	.8840	.8856
6.6	.8871	.8886	.8901	.8916	.8931	.8946	.8961	.8976	.8991	.9006
6.7	.9021	.9036	.9051	.9066	.9081	.9095	.9110	.9125	.9140	.9155
6.8	.9169	.9184	.9199	.9213	.9228	.9242	.9257	.9272	.9286	.9301
6.9	.9315	.9330	.9344	.9359	.9373	.9387	.9402	.9416	.9430	.9445
7.0	1.9459	.9473	.9488	.9502	.9516	.9530	.9544	.9559	.9573	.9587
7.1	.9601	.9615	.9629	.9643	.9657	.9671	.9685	.9699	.9713	.9727
7.2	.9741	.9755	.9769	.9782	.9796	.9810	.9824	.9838	.9851	.9865
7.3	.9879	.9892	.9906	.9920	.9933	.9947	.9961	.9974	.9988	2.0001
7.4	2.0015	.0028	.0042	.0055	.0069	.0082	.0096	.0109	.0122	.0136
7.5	2.0149	.0162	.0176	.0189	.0202	.0215	.0229	.0242	.0255	.0268
7.6	.0281	.0295	.0308	.0321	.0334	.0347	.0360	.0373	.0386	.0399
7.7	.0412	.0425	.0438	.0451	.0464	.0477	.0490	.0503	.0516	.0528
7.8	.0541	.0554	.0567	.0580	.0592	.0605	.0618	.0631	.0643	.0656
7.9	.0669	.0681	.0694	.0707	.0719	.0732	.0744	.0757	.0769	.0782
8.0	2.0794	.0807	.0819	.0832	.0844	.0857	.0869	.0882	.0894	.0906
8.1	.0919	.0931	.0943	.0956	.0968	.0980	.0992	.1005	.1017	.1029
8.2	.1041	.1054	.1066	.1078	.1090	.1102	.1114	.1126	.1138	.1150
8.3	.1163	.1175	.1187	.1199	.1211	.1223	.1235	.1247	.1258	.1270
8.4	.1282	.1294	.1306	.1318	.1330	.1342	.1353	.1365	.1377	.1389
8.5	2.1401	.1412	.1424	.1436	.1448	.1459	.1471	.1483	.1494	.1506
8.6	.1518	.1529	.1541	.1552	.1564	.1576	.1587	.1599	.1610	.1622
8.7	.1633	.1645	.1656	.1668	.1679	.1691	.1702	.1713	.1725	.1736
8.8	.1748	.1759	.1770	.1782	.1793	.1804	.1815	.1827	.1838	.1849
8.9	.1861	.1872	.1883	.1894	.1905	.1917	.1928	.1939	.1950	.1961
9.0	2.1972	.1983	.1994	.2006	.2017	.2028	.2039	.2050	.2061	.2072
9.1	.2083	.2094	.2105	.2116	.2127	.2138	.2148	.2159	.2170	.2181
9.2	.2192	.2203	.2214	.2225	.2235	.2246	.2257	.2268	.2279	.2289
9.3	.2300	.2311	.2322	.2332	.2343	.2354	.2364	.2375	.2386	.2396
9.4	.2407	.2418	.2428	.2439	.2450	.2460	.2471	.2481	.2492	.2502
9.5	2.2513	.2523	.2534	.2544	.2555	.2565	.2576	.2586	.2597	.2607
9.6	.2618	.2628	.2638	.2649	.2659	.2670	.2680	.2690	.2701	.2711
9.7	.2721	.2732	.2742	.2752	.2762	.2773	.2783	.2793	.2803	.2814
9.8	.2824	.2834	.2844	.2854	.2865	.2875	.2885	.2895	.2905	.2915
9.9	.2925	.2935	.2946	.2956	.2966	.2976	.2986	.2996	.3006	.3016

APPENDIX III

Answers to Selected Exercises

Getting Ready (page 2)

1. 1, 2, 3, etc. **2.** $\frac{1}{2}, \frac{2}{3}$, etc. **3.** $-3, -21$, etc.

Orals (page 10)

11. -15 **12.** 25

Exercise 1.1 (page 10)

1. set **3.** whole **5.** subset **7.** prime
9. is not equal to **11.** is greater than or equal to
13. number **15.** 1, 2, 6, 9 **17.** 1, 2, 6, 9
19. $-3, -1, 0, 1, 2, 6, 9$
21. $-3, -\frac{1}{2}, -1, 0, 1, 2, \frac{5}{3}, \sqrt{7}, 3.25, 6, 9$ **23.** $-3, -1, 1, 9$
25. 6, 9 **27.** 9; natural, odd, composite, and whole number
29. 0; even integer, whole number
31. 24; natural, even, composite, and whole number
33. 3; natural, odd, prime, and whole number **35.** $=$
37. $<$ **39.** $>$ **41.** $=$ **43.** $=$ **45.** $<$ **47.** $=$
49. $7 > 3$ **51.** $8 \leq 8$ **53.** $3 + 4 = 7$
55. $\sqrt{2} \approx 1.41$ **57.** $7 \geq 3$ **59.** $0 < 6$ **61.** $8 < 3 + 8$
63. $10 - 4 > 6 - 2$ **65.** $3 \cdot 4 > 2 \cdot 3$ **67.** $\frac{24}{6} > \frac{12}{4}$

69. 

; 6, 6

71. 

; 11, 11

73. 

; 2, 2

75. 

; 8, 8

77. 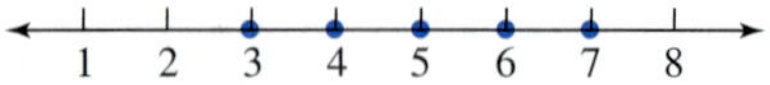

79. 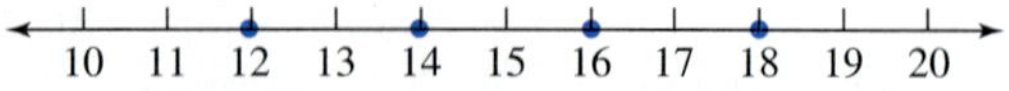

81. 

83. 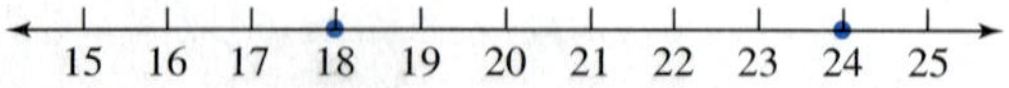

85. 

87. 

89. 36 **91.** 0 **93.** 230 **95.** 8

Getting Ready (page 12)

1. 250 **2.** 148 **3.** 16,606 **4.** 105

Orals (page 24)

1. $\frac{1}{2}$ **2.** $\frac{1}{2}$ **3.** $\frac{1}{2}$ **4.** $\frac{1}{3}$ **5.** $\frac{5}{12}$ **6.** $\frac{9}{20}$ **7.** $\frac{4}{9}$ **8.** $\frac{6}{25}$
9. $\frac{11}{9}$ **10.** $\frac{3}{7}$ **11.** $\frac{1}{6}$ **12.** $\frac{5}{4}$ **13.** 2.86 **14.** 1.24
15. 0.5 **16** 3.9 **17.** 3.24 **18.** 3.25

Exercise 1.2 (page 24)

1. true **3.** false **5.** true **7.** true **9.** $=$ **11.** $=$
13. numerator **15.** simplify **17.** proper **19.** 1
21. multiply **23.** numerators, denominator **25.** plus
27. repeating **29.** $\frac{1}{2}$ **31.** $\frac{3}{4}$ **33.** $\frac{4}{3}$ **35.** $\frac{9}{8}$ **37.** $\frac{3}{10}$
39. $\frac{8}{5}$ **41.** $\frac{3}{2}$ **43.** $\frac{1}{4}$ **45.** 10 **47.** $\frac{20}{3}$ **49.** $\frac{9}{10}$ **51.** $\frac{5}{8}$
53. $\frac{1}{4}$ **55.** $\frac{14}{5}$ **57.** 28 **59.** $\frac{1}{5}$ **61.** $\frac{6}{5}$ **63.** $\frac{1}{13}$ **65.** $\frac{5}{24}$
67. $\frac{19}{15}$ **69.** $\frac{17}{12}$ **71.** $\frac{22}{35}$ **73.** $\frac{9}{4}$ **75.** $\frac{29}{3}$ **77.** $5\frac{1}{5}$
79. $1\frac{2}{3}$ **81.** $1\frac{1}{4}$ **83.** $\frac{5}{9}$ **85.** 158.65 **87.** 44.785
89. 44.88 **91.** 4.55 **93.** 350.49 **95.** 55.21
97. 3,337.52 **99.** 10.02 **101.** $31\frac{1}{6}$ acres **103.** $45\frac{1}{2}$ yd
105. \$10,200 **107.** 13,475 **109.** \$20,944,000
111. \$1,026.57 **113.** \$1,170
115. the high-capacity boards **117.** 205,200 lb
119. the high-efficiency furnace

Getting Ready (page 27)

1. 4 **2.** 9 **3.** 27 **4.** 8 **5.** $\frac{1}{4}$ **6.** $\frac{1}{27}$ **7.** $\frac{8}{125}$
8. $\frac{27}{1,000}$

59. 9 **61.** −1 **63.** 8 **65.** 5 **67.** 4 **69.** −3
71. 1 **73.** identity **75.** contradiction **77.** 16
79. contradiction **81.** identity **83.** identity **89.** 0

Getting Ready (page 106)

1. $(2x + 2)$ ft **2.** $4x$ ft **3.** $P = 2l + 2w$

Orals (page 112)

1. 70° **2.** 110° **3.** 20 ft **4.** $18,000$r$

Exercise 2.4 (page 112)

1. 200 cm^3 **3.** $7x - 6$ **5.** $-\frac{3}{2}$ **7.** $1,488 **9.** $2l + 2w$
11. vertex **13.** complementary **15.** 180°
17. 4 ft and 8 ft **19.** 15 ft, 10 ft, 5 ft **21.** 6 ft, 8 ft, 10 ft
23. 10° **25.** 159° **27.** 47° **29.** 27° **31.** 53°
33. 130° **35.** 19 ft **37.** 29 m by 18 m
39. 17 in. by 39 in. **41.** 60° **43.** $6,000
45. $4,500 at 9% and $19,500 at 14%
47. $3,750 in each account **49.** $5,000 **51.** 6% and 7%

Getting Ready (page 115)

1. 60 mi **2.** 385 mi **3.** 5.6 gal **4.** 9 lbs

Orals (page 121)

1. $50h$ mi **2.** $(65h - 55h)$ mi or $10h$ mi **3.** 4.8 oz
4. 7d$

Exercise 2.5 (page 121)

1. −17 **3.** −1 **5.** −3 **7.** 6 **9.** $d = rt$ **11.** $v = pn$
13. 3 hr **15.** 3.5 days **17.** 6.5 hr **19.** $\frac{2}{7}$ hr **21.** 10 hr
23. 7.5 hr **25.** 65 mph and 45 mph **27.** 4 hr
29. 2.5 liters **31.** 20 gal **33.** 50 gal **35.** 7.5 oz
37. 40 lb lemon drops and 60 lb jelly beans **39.** $1.20
41. 80 lb **43.** $5.60

Getting Ready (page 124)

1. 3 **2.** −5 **3.** r **4.** $-a$ **5.** 7 **6.** 12 **7.** d
8. s

Orals (page 128)

1. $a = \frac{d-c}{b}$ **2.** $b = \frac{d-c}{a}$ **3.** $c = d - ab$
4. $d = ab + c$ **5.** $a = \frac{c}{d} - b$ **6.** $b = \frac{c}{d} - a$
7. $c = d(a + b)$ **8.** $d = \frac{c}{a+b}$

Exercise 2.6 (page 128)

1. $5x - 5y$ **3.** $-x - 13$ **5.** literal **7.** isolate
9. subtract **11.** $I = \frac{E}{R}$ **13.** $w = \frac{V}{lh}$ **15.** $b = P - a - c$
17. $w = \frac{P - 2l}{2}$ **19.** $t = \frac{A - P}{Pr}$ **21.** $r = \frac{C}{2\pi}$
23. $w = \frac{2gK}{v^2}$ **25.** $R = \frac{P}{I^2}$ **27.** $g = \frac{wv^2}{2K}$ **29.** $M = \frac{Fd^2}{Gm}$
31. $d^2 = \frac{GMm}{F}$ **33.** $r = \frac{G}{2b} + 1$ or $r = \frac{G + 2b}{2b}$
35. $t = \frac{d}{r}$; $t = 3$ **37.** $t = \frac{i}{pr}$; $t = 2$
39. $c = P - a - b$; $c = 3$ **41.** $h = \frac{2K}{a+b}$; $h = 8$
43. $I = \frac{E}{R}$; $I = 4$ amp **45.** $r = \frac{C}{2\pi}$; $r = 2.28$ ft
47. $R = \frac{P}{I^2}$; $R = 13.78$ ohms **49.** $m = \frac{Fd^2}{GM}$
51. $D = \frac{L - 3.25r - 3.25R}{2}$; $D = 6$ ft
53. $C \approx 0.1304T$, about 13% of taxable income
57. 90,000,000,000 joules

Getting Ready (page 131)

1. 

2. 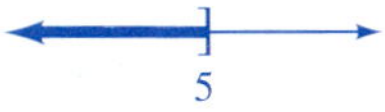

3. 

4. 

Orals (page 136)

1. $x < 2$ **2.** $x \geq 1$ **3.** $x \geq 2$ **4.** $x < -2$ **5.** $x < 6$
6. $x > -1$

Exercise 2.7 (page 136)

1. $5x^2 - 2y^2$ **3.** $-x + 14$ **5.** is less than **7.** ≥
9. inequality **11.** $x > 3$; 

13. $x \geq -10$; 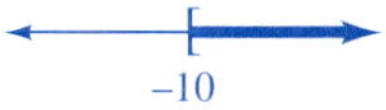

15. $x < -1$; 

17. $x \leq 4$; 

19. $x < -2$; 

21. $x < -4$; 

23. $x \leq -1$; 

25. $x \geq -13$; 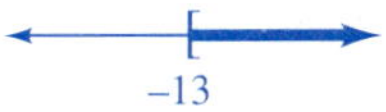

27. $x > -3$; 

29. $x < -2$; 

31. $x \geq 2$; 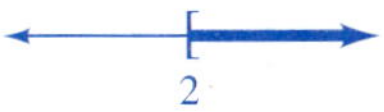

33. $x > 3$; 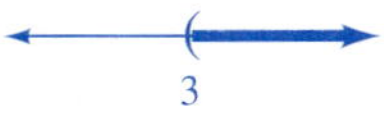

35. $x > -15$; 

37. $x \leq 20$; 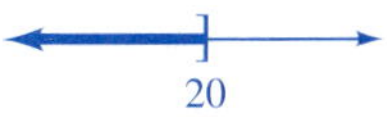

39. $x \geq 3$; 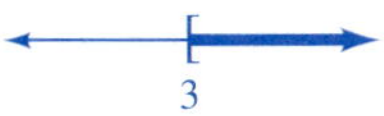

41. $x > -7$;

43. $x \geq 4$; 

45. $7 < x < 10$; 

47. $-9 < x \leq 3$; 

49. $-10 \leq x \leq 0$; 

51. $-5 < x < -2$; 

53. $-6 \leq x \leq 10$; 

55. $2 \leq x < 3$; 

57. $-1 \leq x < 2$; 

59. $-4 < x < 1$; 

61. $-6 < x < 14$; 

63. $s \geq 98$ **65.** $r \geq 27$ mpg **67.** 0 ft $< s \leq$ 19 ft
69. 0.1 mi $\leq x \leq$ 2.5 mi **71.** 3.3 mi $< x <$ 4.1 mi
73. $66.2° < F < 71.6°$ **75.** 37.052 in. $< C <$ 38.308 in.
77. 68.18 kg $< w <$ 86.36 kg **79.** 5 ft $< w <$ 9 ft

Chapter Summary (page 140)

1. yes **2.** no **3.** no **4.** yes **5.** yes **6.** no **7.** 1
8. 9 **9.** 16 **10.** 0 **11.** 4 **12.** −2 **13.** $105.40
14. $97.70 **15.** 5 **16.** −2 **17.** $\frac{1}{2}$ **18.** $\frac{3}{2}$ **19.** 18
20. −35 **21.** $-\frac{1}{2}$ **22.** 6 **23.** 245 **24.** 1,300
25. 37% **26.** 12.5% **27.** 3 **28.** 2 **29.** 1 **30.** 1
31. 1 **32.** −2 **33.** 2 **34.** 7 **35.** −2 **36.** −1
37. 5 **38.** 3 **39.** 13 **40.** −12 **41.** 5 **42.** 7
43. 8 **44.** 30 **45.** $\frac{15}{2}$ **46.** 44 **47.** $320 **48.** 6.5%
49. 96.4% **50.** 53.8% **51.** $14x$ **52.** $19a$ **53.** $5b$
54. $-2x$ **55.** $-2y$ **56.** not like terms **57.** $9x$
58. $6 - 7x$ **59.** $4y^2 - 6$ **60.** 4 **61.** 7 **62.** 13
63. −3 **64.** −41 **65.** 9 **66.** −7 **67.** 7 **68.** 4
69. −8 **70.** −18 **71.** identity **72.** contradiction
73. identity **74.** 5 ft from one end **75.** 15° **76.** 45°
77. 21° **78.** 111° **79.** 13 in.
80. $16,000 at 7%, $11,000 at 9% **81.** 20 min
82. 24 liters **83.** 10 lb of each **84.** 147 kwh **85.** 85 ft
86. $R = \frac{E}{I}$ **87.** $t = \frac{i}{pr}$ **88.** $R = \frac{P}{I^2}$ **89.** $r = \frac{d}{t}$
90. $h = \frac{V}{lw}$ **91.** $m = \frac{y - b}{x}$ **92.** $h = \frac{V}{\pi r^2}$ **93.** $r = \frac{a}{2\pi h}$
94. $G = \frac{Fd^2}{Mm}$ **95.** $m = \frac{RT}{PV}$ **96.** 

97. 

98. 

99. 

100. 

101. 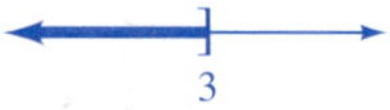

102. 

103. 

104. $l \leq 20$ ft

Chapter 2 Test (page 144)

1. solution **2.** solution **3.** not a solution **4.** solution
5. −36 **6.** 47 **7.** −12 **8.** −7 **9.** −2 **10.** 1
11. 7 **12.** −2 **13.** −3 **14.** 0 **15.** $6x - 15$
16. $8x - 10$ **17.** $-18x$ **18.** $-36x^2 + 13x$ **19.** 75°
20. 75° **21.** $6,000 at 6%, $4,000 at 5% **22.** $\frac{3}{5}$ hr
23. $7\frac{1}{2}$ liters **24.** $t = \frac{d}{r}$ **25.** $l = \frac{P - 2w}{2}$ **26.** $h = \frac{A}{2\pi r}$
27. $r = \frac{A - P}{Pt}$
28. 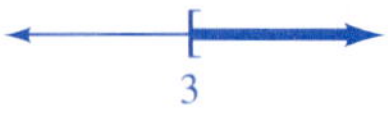

29. 

30. 

31. 

Cumulative Review Exercises (page 145)

1. integer, rational, real, positive **2.** rational, real, negative
3. 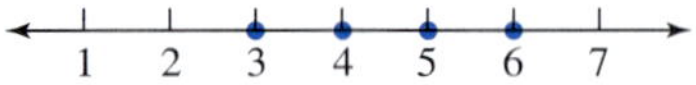

4. 

5. 0 **6.** $\frac{10}{3}$ **7.** $8\frac{1}{10}$ **8.** 35.65 **9.** 0 **10.** −2
11. 16 **12.** 0 **13.** 24.75 **14.** 5,275 **15.** 5
16. 37, y **17.** $-2x + 2y$ **18.** $x - 5$ **19.** x^2y^3
20. $4x^2$ **21.** 13 **22.** 41 **23.** $\frac{7}{4}$ **24.** −11
25. $22,814.56 **26.** $900 **27.** $12,650 **28.** 125 lb
29. no **30.** 7.3 and 10.7 ft **31.** $h = \frac{2A}{b + B}$ **32.** $x = \frac{y - b}{m}$
33. −9 **34.** 1 **35.** 280 **36.** −564
37. 

38. 

Getting Ready (page 148)

1. 

2. 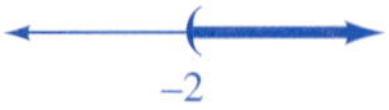

3. 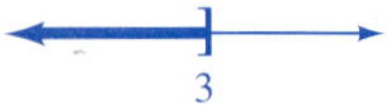

4. 

Orals (page 156)

2. the origin **3.** IV **4.** y-axis

Exercise 3.1 (page 156)

1. 12 **3.** 8 **5.** 7 **7.** −49 **9.** ordered pair
11. origin **13.** rectangular coordinate **15.** no
17. origin, left, up **19.** II **21.** 3 or −3, 5 or −5, 4 or −4, 5 or −5, 3 or −3, 5 or −5, 4 or −4 **23.** 10 minutes before the workout, her heart rate was 60 beats per min.

25. 150 beats per min **27.** approximately 5 min and 50 min after starting **29.** 10 beats per min faster after cooldown

31.

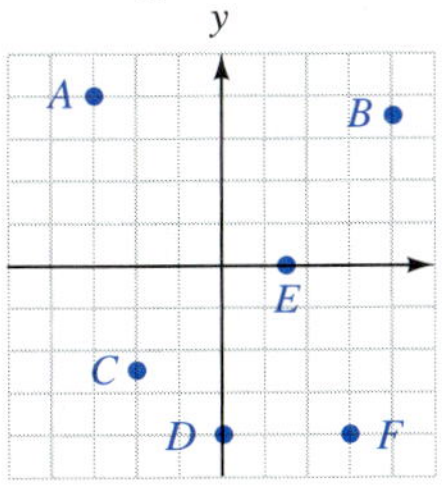

33. Carbondale (3, J), Champaign $(4\frac{1}{2}, \text{D})$, Chicago $(5\frac{1}{2}, \text{B})$, Peoria $(3\frac{1}{2}, \text{C})$, Rockford $(3\frac{1}{2}, \text{A})$, Springfield $(2\frac{1}{4}, \text{E})$, St. Louis (2, H)
35. a. 60°; 4 ft **b.** 30°; 4 ft
37. a. \$2 **b.** \$4 **c.** \$7 **d.** \$9

39.

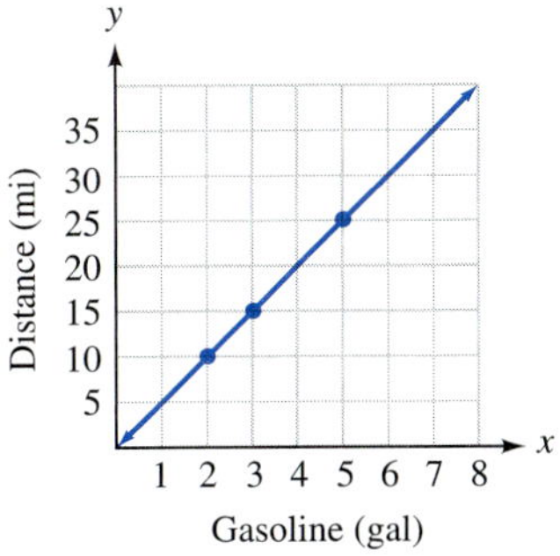

a. 35 mi **b.** 4 gal **c.** 32.5 mi

41.

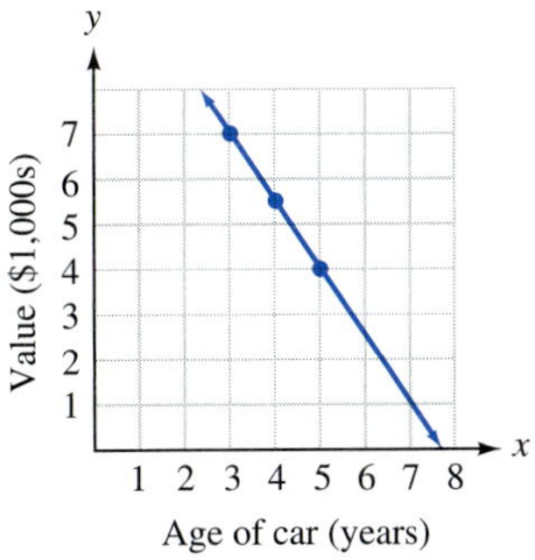

a. A 3-yr-old car is worth \$7,000. **b.** \$1,000 **c.** 6 yr

Getting Ready (page 160)

1. 1 **2.** 5 **3.** −3 **4.** 2

Orals (page 170)

1. 3

Exercise 3.2 (page 170)

1. −96 **3.** an expression **5.** 1.25 **7.** 0.1 **9.** two **11.** independent, dependent **13.** linear **15.** *y*-intercept **17.** yes **19.** no **21.** −3, −2, −5, −7 **23.** 0, −2, −6, 4

25.

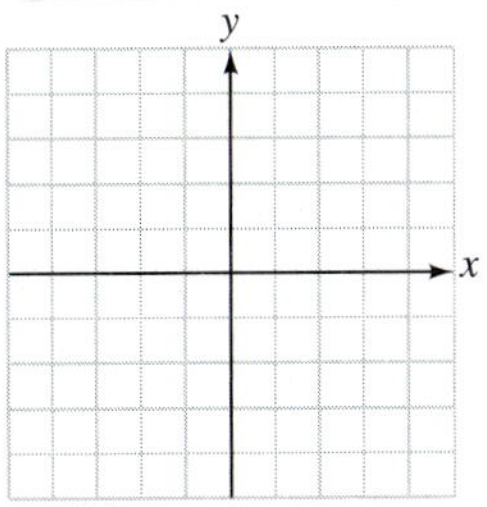

27.

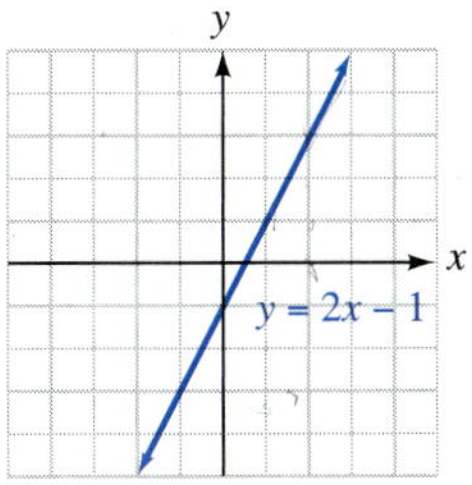

29.

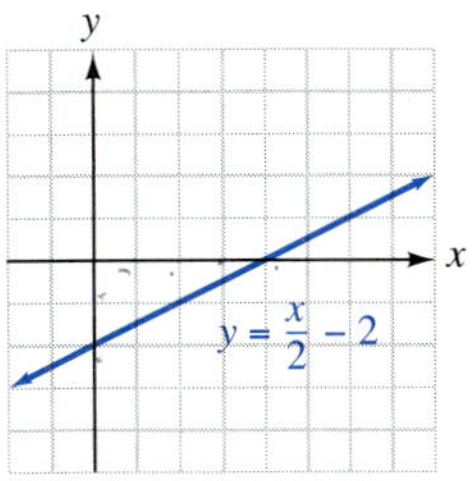

31.

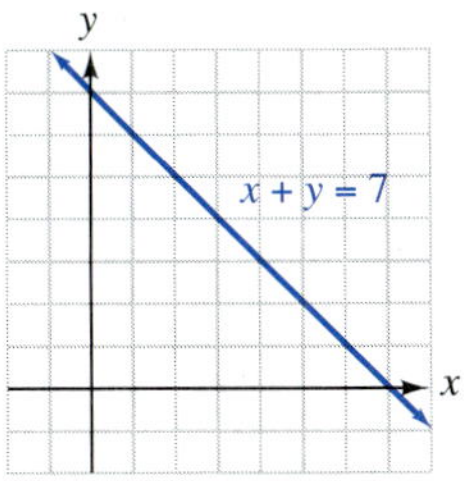

33.

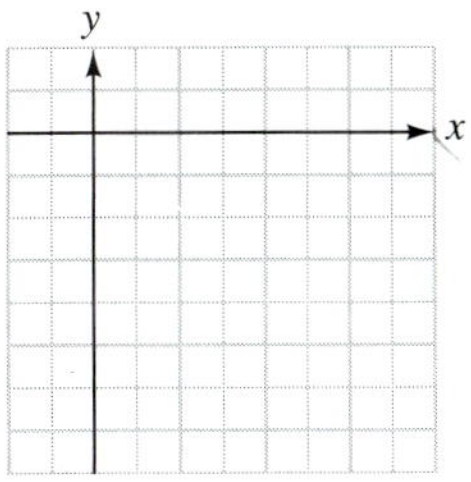

35.

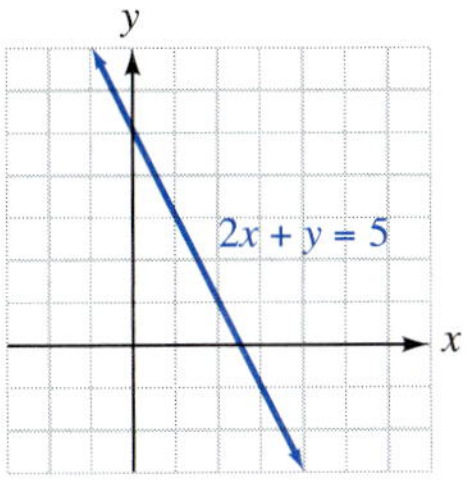

37.

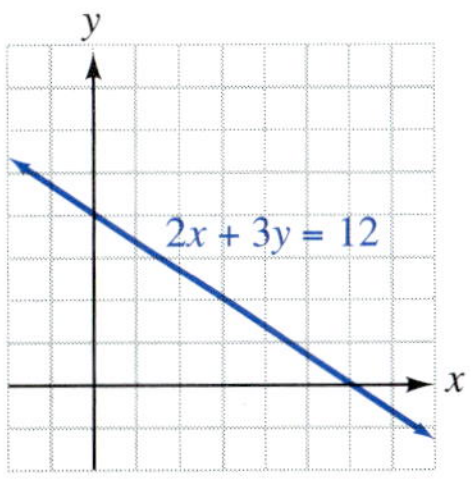

39.

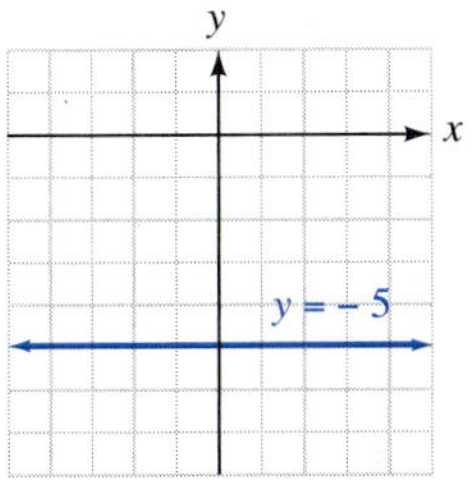

41.

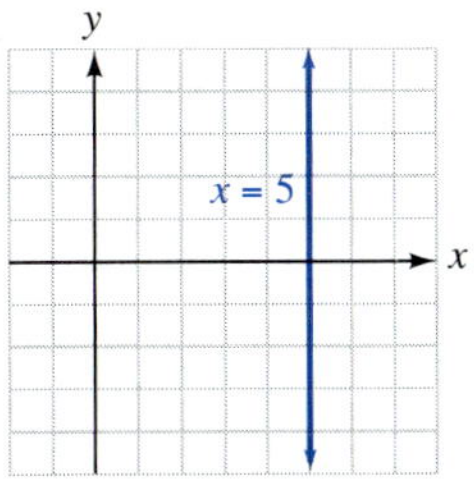

43.

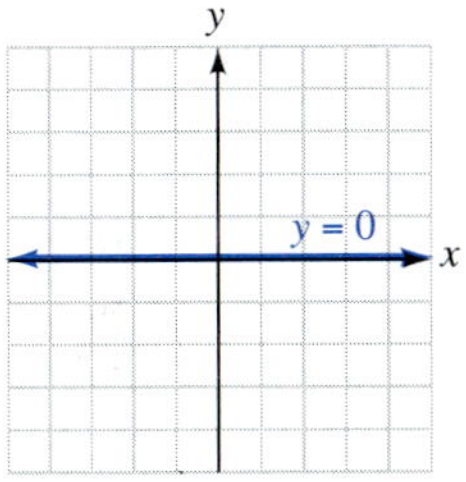

45.

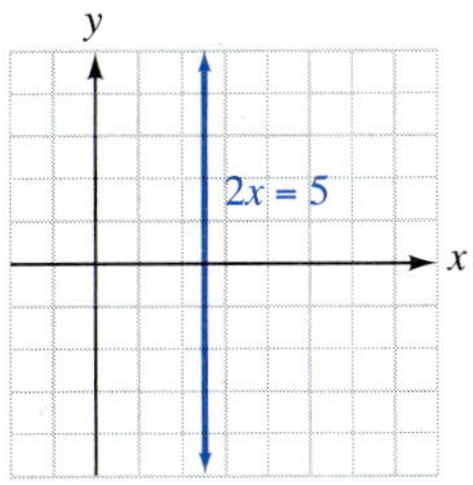

47.

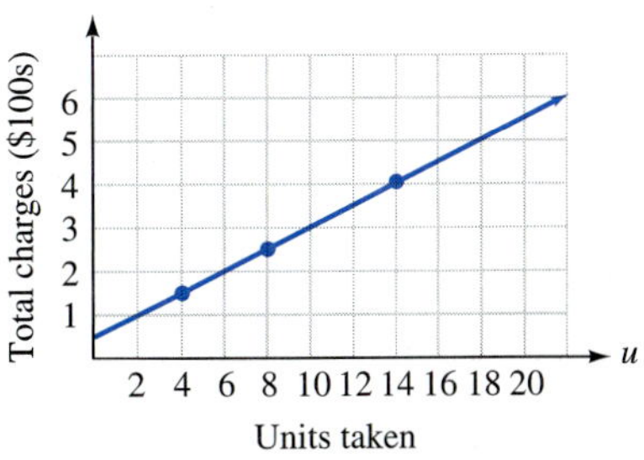

a. $c = 50 + 25u$ **b.** 150, 250, 400 **c.** The service fee is \$50. **d.** \$850

49.

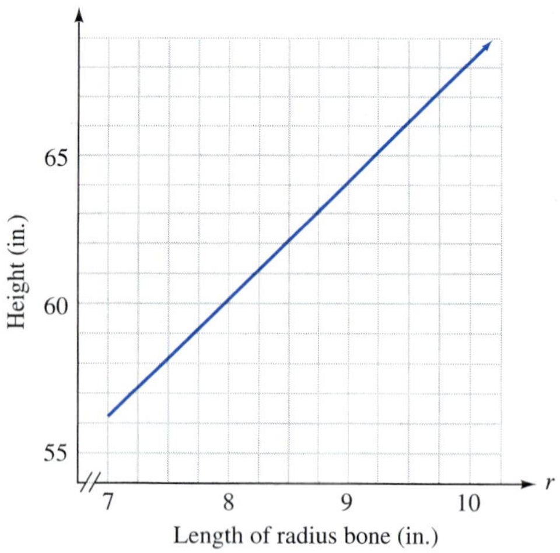

a. 56.2, 62.1, 64.0 **b.** taller the woman is **c.** 58 in.

57. (6, 6) **59.** $\left(-\frac{1}{2}, \frac{5}{2}\right)$ **61.** (7, 6)

Getting Ready (page 174)

1. −3 **2.** −2 **3.** 1 **4.** 6

Orals (page 182)

1. yes **2.** yes **3.** no **4.** no

Exercise 3.3 (page 182)

1. 16 **3.** −18 **5.** system **7.** independent **9.** inconsistent **11.** yes **13.** yes **15.** no **17.** yes **19.** no **21.** no

23.

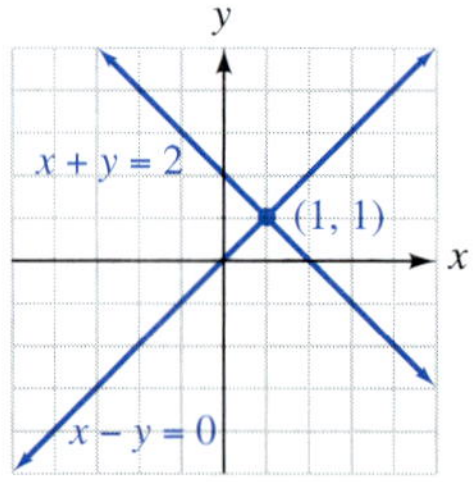

25.

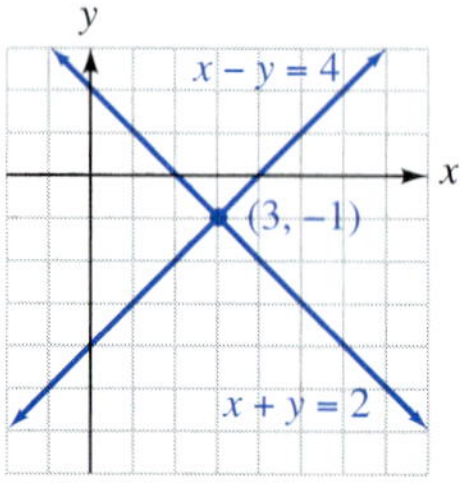

27.

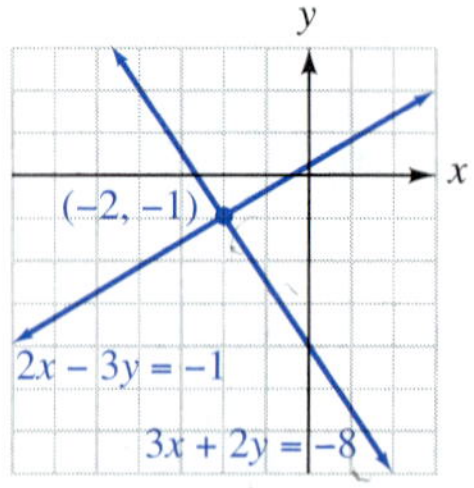

29.

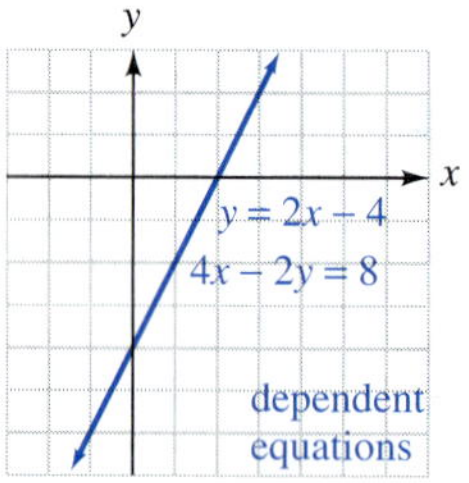

31.

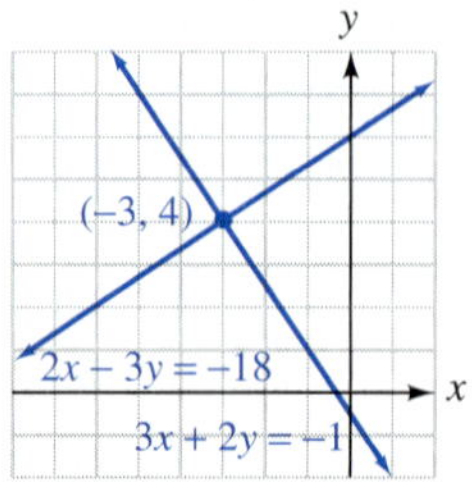

33.

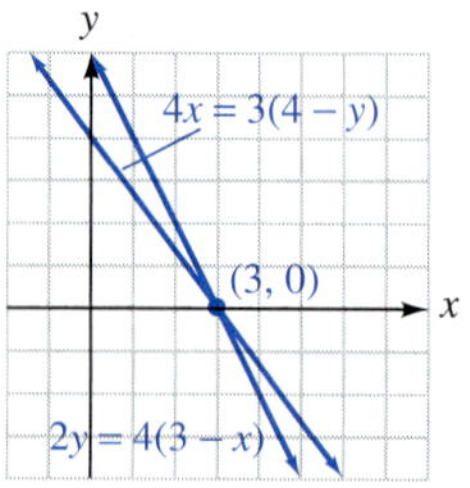

35.

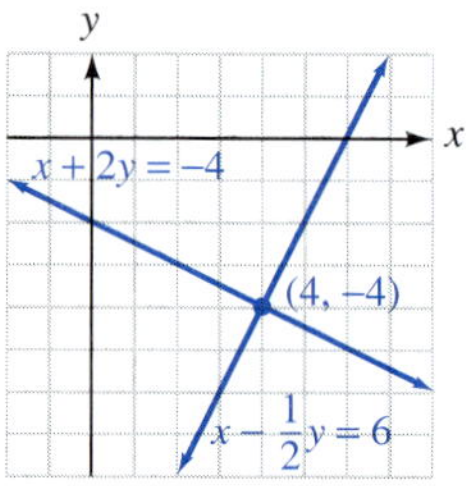

37.

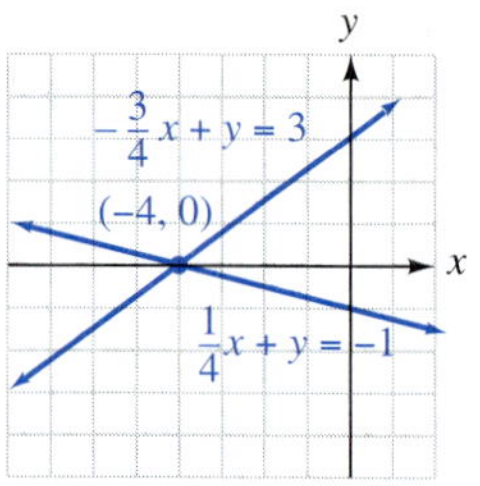

39.

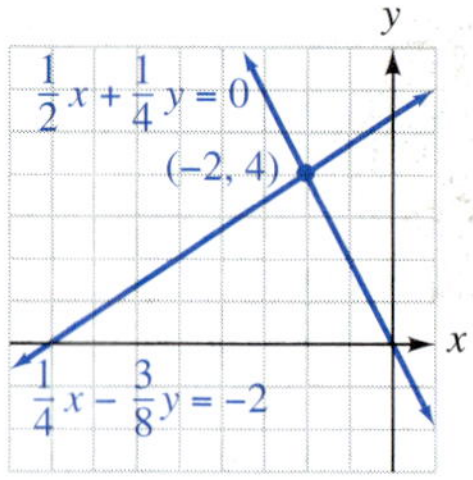

41.

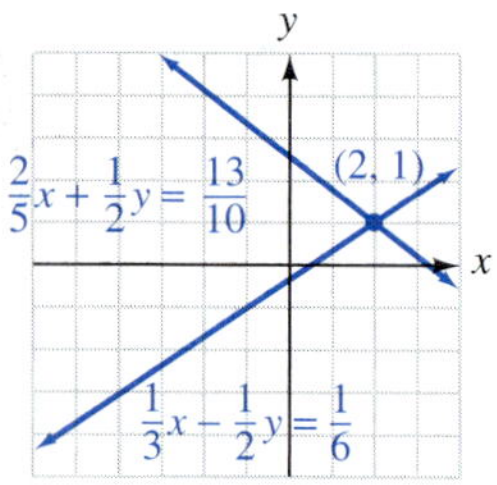

43. (1, 3) **45.** (0.67, −0.33) **47. a.** Donors outnumbered patients. **b.** 1999; 4,100 **c.** The patients outnumber the donors. **49. a.** Houston, New Orleans, St. Augustine **b.** St. Louis, Memphis, New Orleans **c.** New Orleans

Getting Ready (page 185)

1. $6x + 4$ **2.** $-25 - 10x$ **3.** $2x - 4$ **4.** $3x - 12$

Orals (page 190)

1. $2z + 1$ **2.** $z + 2$ **3.** $3t + 3$ **4.** $\frac{t}{3} + 4$

Exercise 3.4 (page 190)

1. 5 **3.** 12 **5.** 10 **7.** *y*, terms **9.** remove **11.** infinitely many **13.** (2, 4) **15.** (3, 0) **17.** (−3, −1) **19.** inconsistent system **21.** (−2, 3) **23.** $\left(\frac{1}{2}, 3\right)$ **25.** (3, 2) **27.** (3, −2) **29.** (−1, 2) **31.** (−1, −1) **33.** dependent equations **35.** (1, 1) **37.** inconsistent system **39.** (4, −2) **41.** (−3, −1) **43.** (−1, −3) **45.** $\left(\frac{1}{2}, \frac{1}{3}\right)$ **47.** (1, 4) **49.** (4, 2) **51.** (−5, −5) **53.** (−6, 4) **55.** $\left(\frac{1}{5}, 4\right)$ **57.** (5, 5)

Getting Ready (page 191)

1. $5x = 10$ **2.** $y = 6$ **3.** $2x = 33$ **4.** $18y = 28$

Orals (page 197)

1. 1 **2.** 2 **3.** 3 **4.** 5

Exercise 3.5 (page 197)

1. 4 **3.** 4 **5.**

7. coefficient **9.** general **11.** 15 **13.** (1, 4) **15.** (−2, 3) **17.** (−1, 1) **19.** (2, 5) **21.** (−3, 4) **23.** (0, 8) **25.** (2, 3) **27.** (3, −2) **29.** (2, 7) **31.** inconsistent system **33.** dependent equations **35.** (4, 0) **37.** $\left(\frac{10}{3}, \frac{10}{3}\right)$ **39.** (5, −6) **41.** $\left(\frac{1}{2}, \frac{1}{3}\right)$ **43.** (−5, 0) **45.** (−1, 2) **47.** $\left(1, -\frac{5}{2}\right)$ **49.** dependent equations **51.** (−1, 2) **53.** (0, 1) **55.** (−2, 3) **57.** (2, 2) **61.** (1, 4)

Getting Ready (page 199)

1. $x + y$ **2.** $x - y$ **3.** xy **4.** $\frac{x}{y}$ **5.** $A = lw$
6. $P = 2l + 2w$

Orals (page 206)

1. $2x$ **2.** $y + 1$ **3.** $2x + 3y$ **4.** $\$(3x + 2y)$
5. $\$(4x + 5y)$

Exercise 3.6 (page 206)

1. 

3. 

5. 8^3c

7. a^2b^2 **9.** variable **11.** system **13.** 32, 64 **15.** 8, 5
17. 140 **19.** \$15, \$5 **21.** \$5.40, \$6.20 **23.** 10 ft, 15 ft
25. \$100,000 **27.** causes: 24 min; outcome: 6 min
29. 90,000 accidents; 540,000 cancer **31.** 25 ft by 30 ft
33. 60 ft^2 **35.** 9.9 yr **37.** 80+ **39.** \$2,000 **41.** 250
43. 10 mph **45.** 50 mph **47.** 5 L of 40% solution, 10 L of 55% solution **49.** 32 lb peanuts, 16 lb cashews **51.** 15
53. 9% **55.** \$800

Getting Ready (page 210)

1. below **2.** above **3.** below **4.** on **5.** on
6. above **7.** below **8.** above

Orals (page 218)

1. no **2.** no **3.** yes **4.** yes **5.** no **6.** yes **7.** no
8. yes

Exercise 3.7 (page 218)

1. 3 **3.** $t = \frac{A - P}{Pr}$ **5.** $7a - 15$ **7.** $-2a + 7b$
9. inequality **11.** boundary **13.** inequalities
15. doubly shaded **17. a.** yes **b.** no **c.** yes **d.** no
19. a. no **b.** yes

21.
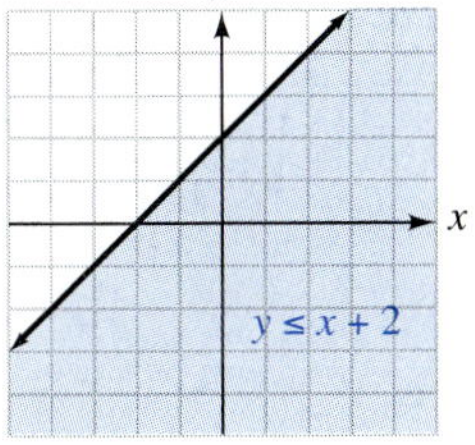

23.
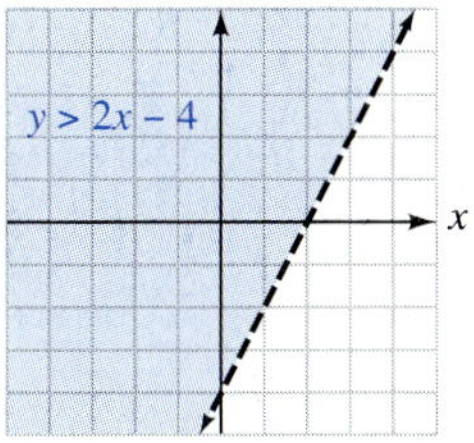

25.
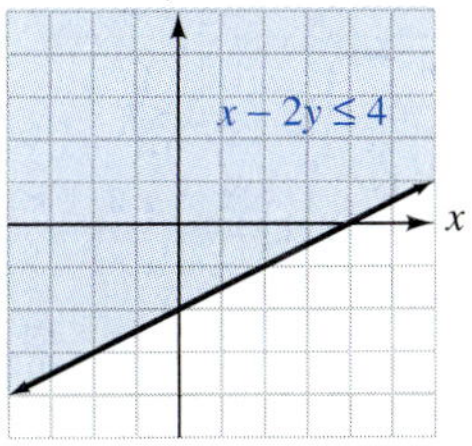

27.
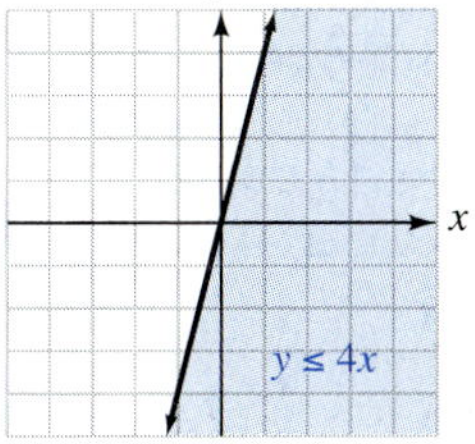

29.
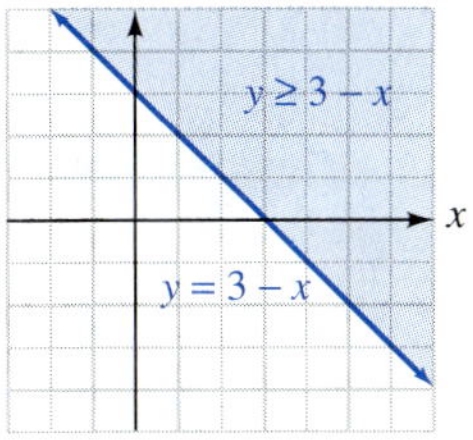

31.
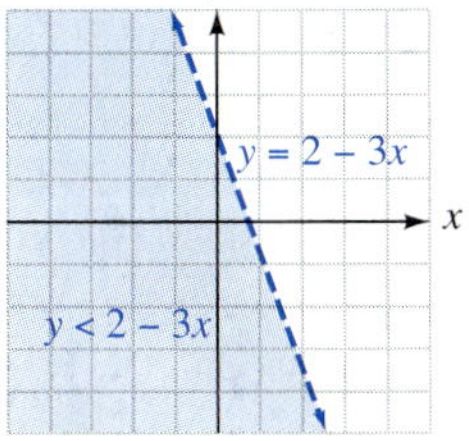

33.
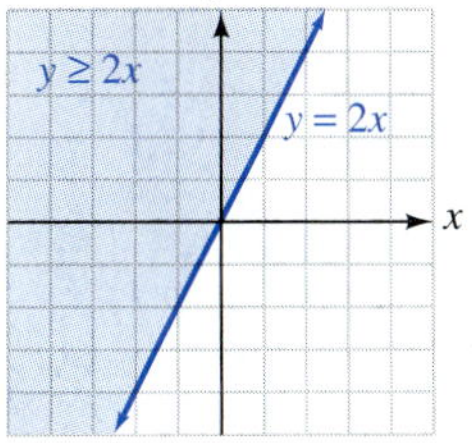

35.
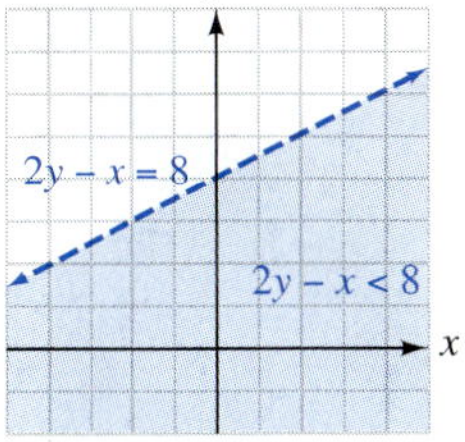

37.
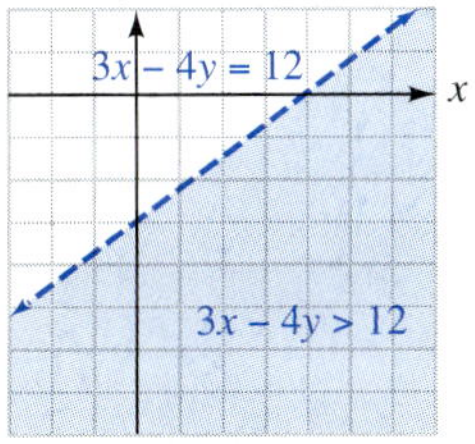

39.
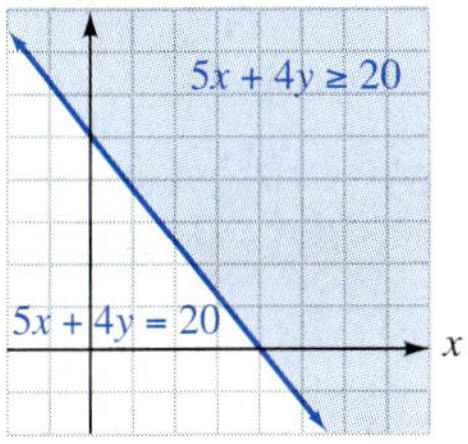

41.
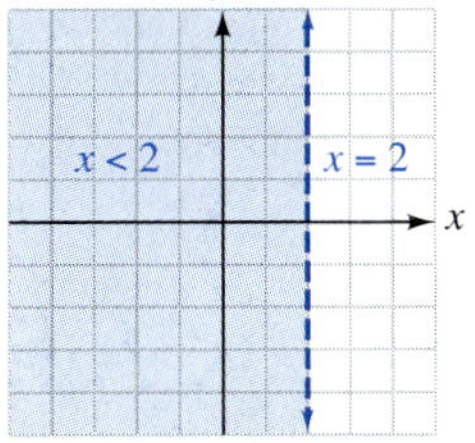

43.
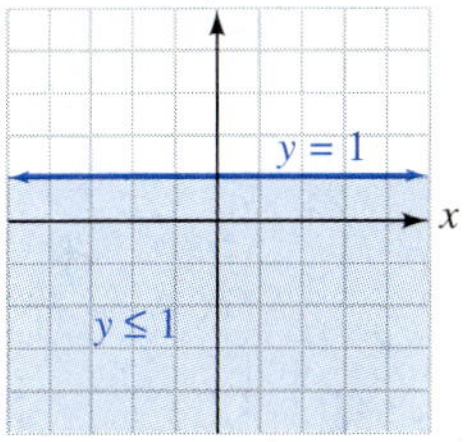

45.
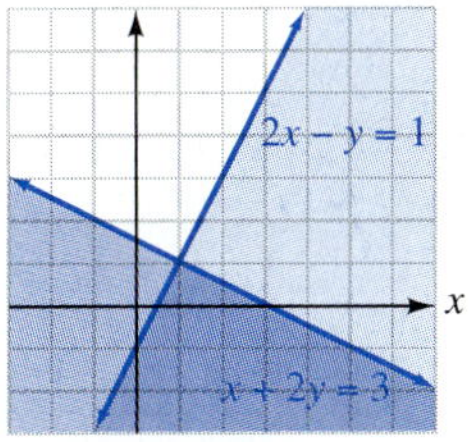

47.
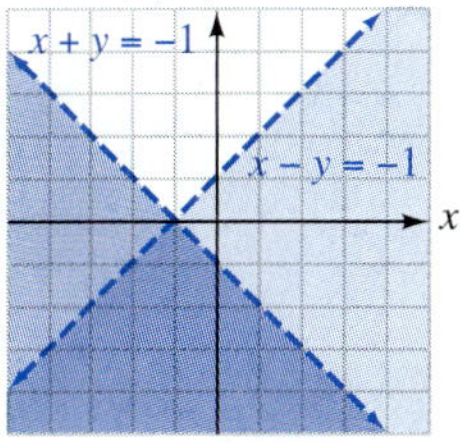

49.
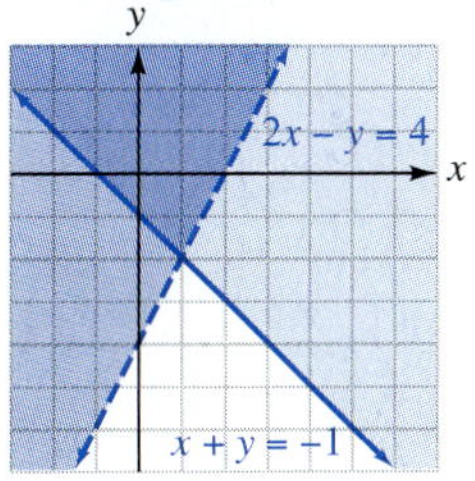

51.
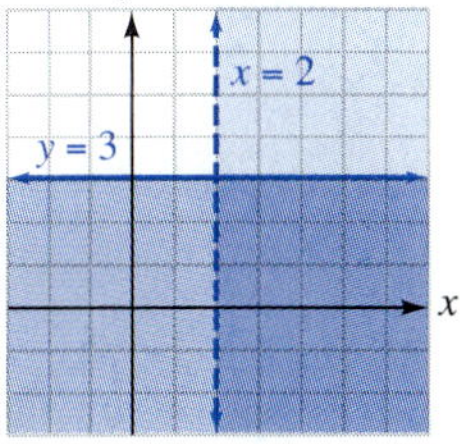

53.

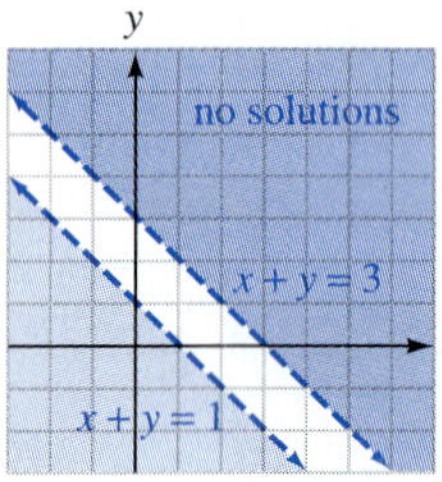

55.

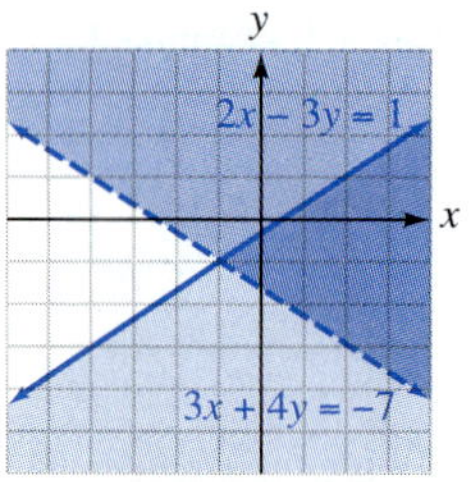

57.

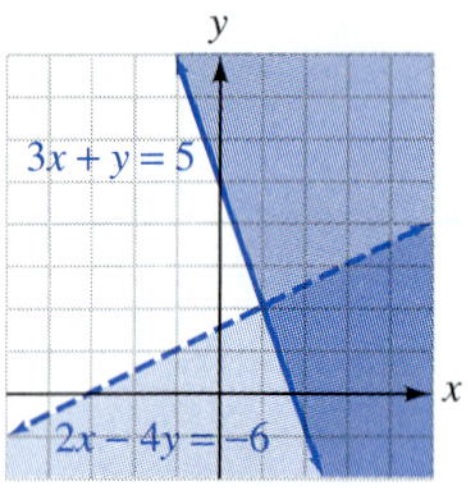

59.

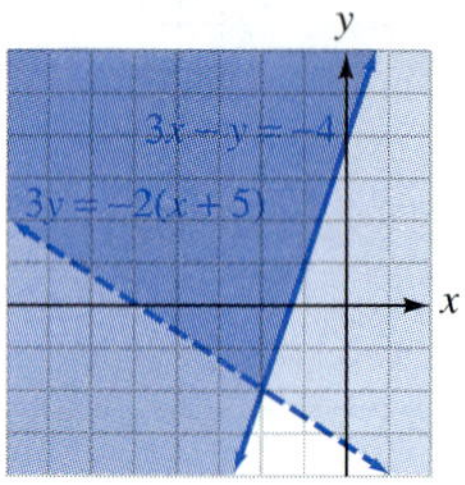

61.

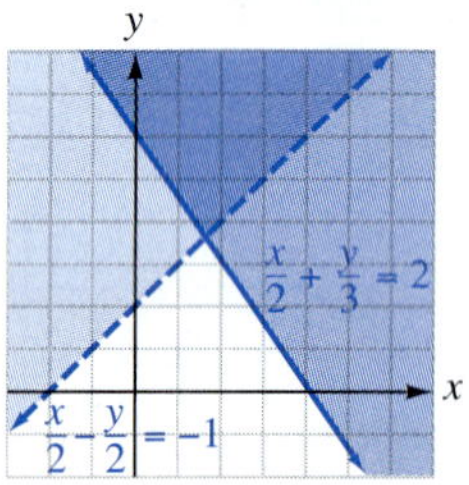

63. (10, 10), (20, 10), (10, 20)

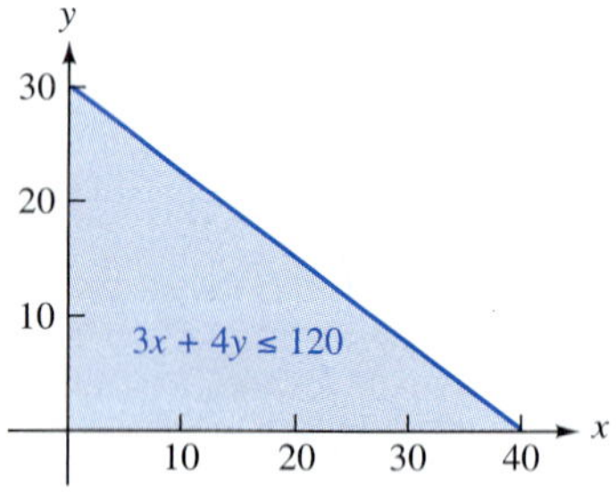

65. (50, 50), (30, 40), (40, 40)

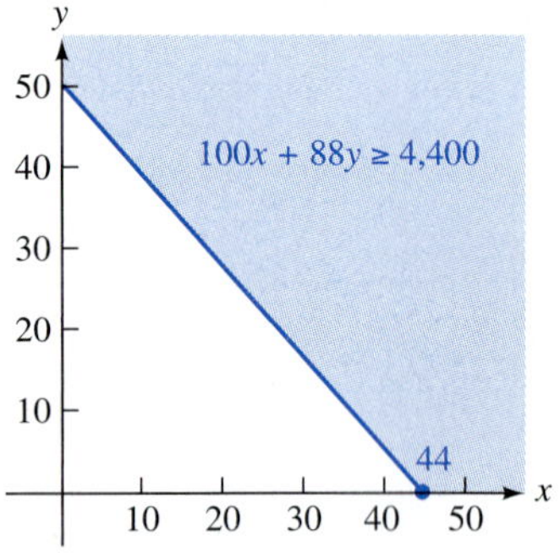

67. (80, 40), (80, 80), (120, 40)

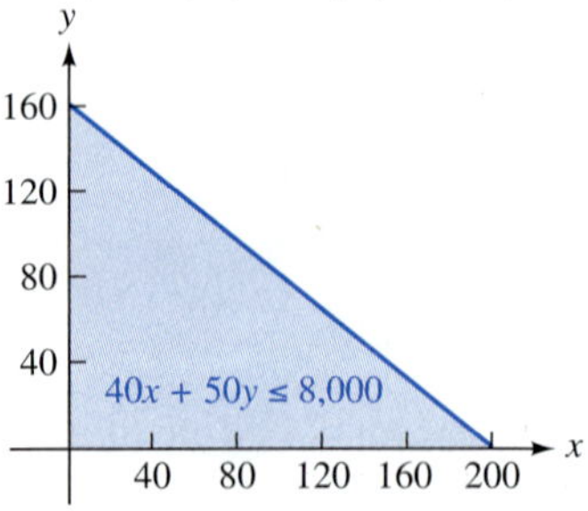

69.

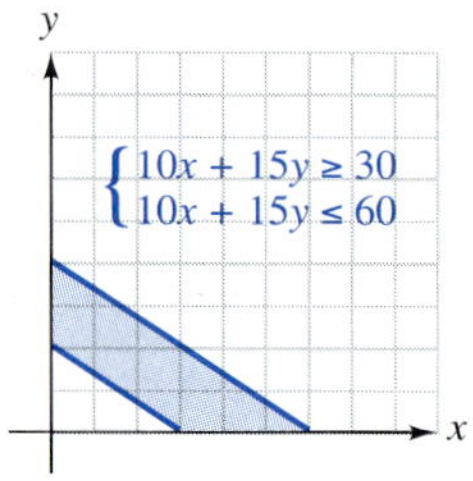

1 \$10 CD and 2 \$15 CDs; 4 \$10 CDs and 1 \$15 CD

71.

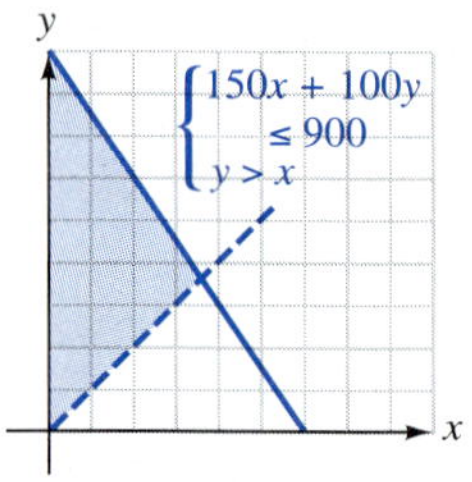

2 desk chairs and 4 side chairs; 1 desk chair and 5 side chairs

Chapter Summary (page 224)

1–6.

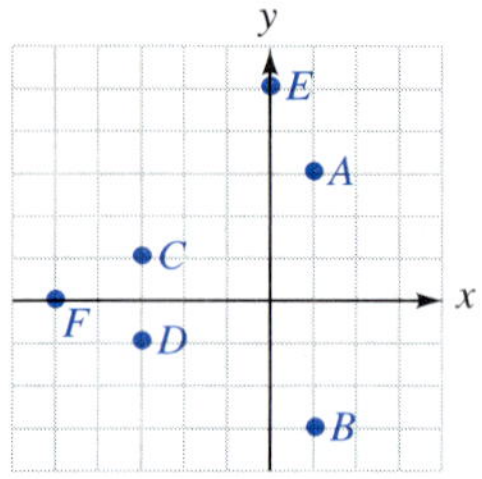

7. (3, 1) **8.** (−4, 5) **9.** (−3, −4) **10.** (2, −3) **11.** (0, 0) **12.** (0, 4) **13.** (−5, 0) **14.** (0, −3) **15.** no **16.** yes

17.

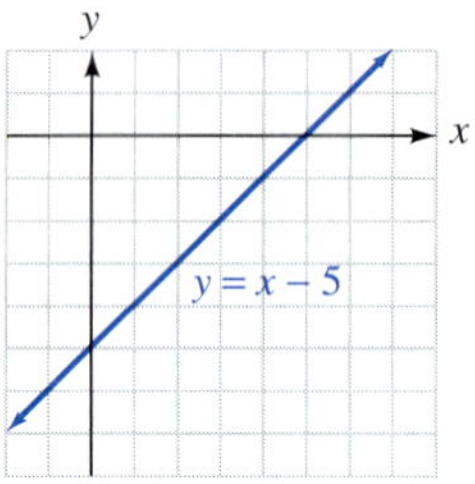

18.

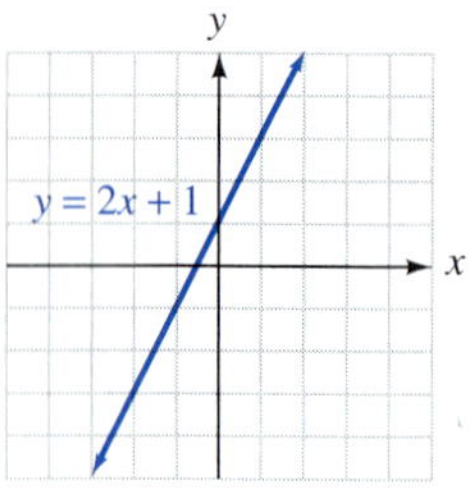

19.

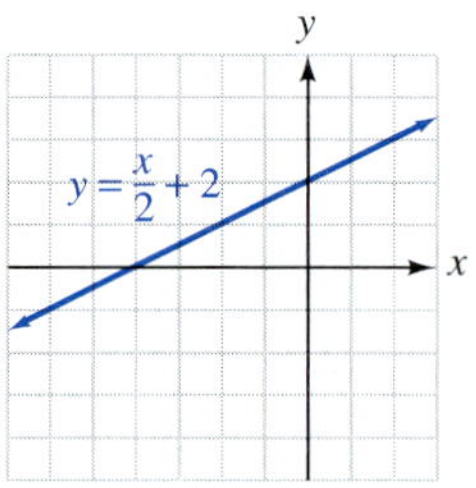

20.

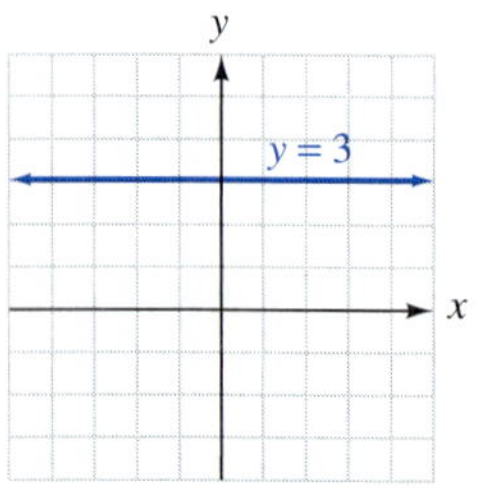

21.

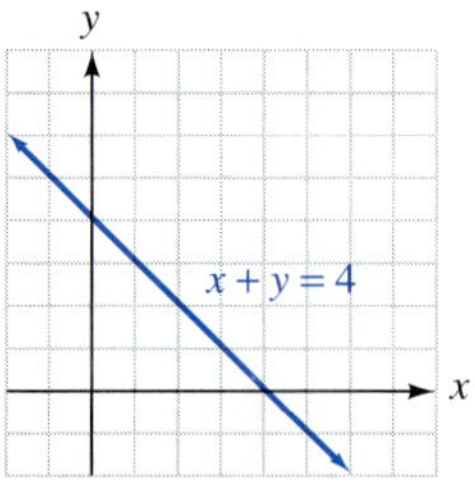

22.

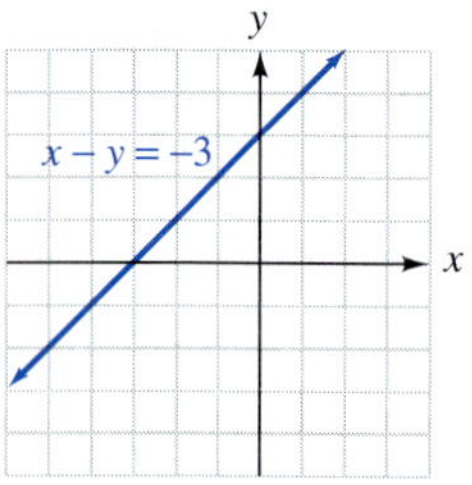

23.

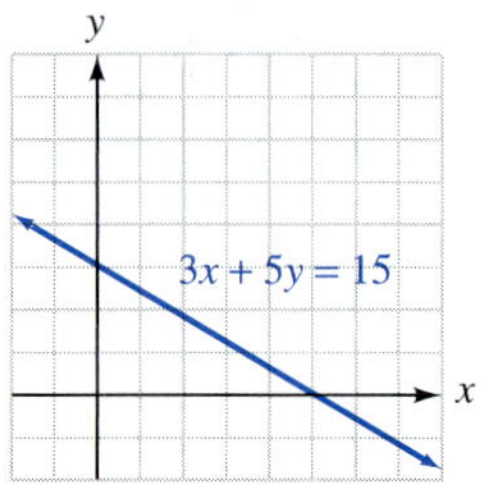

24.

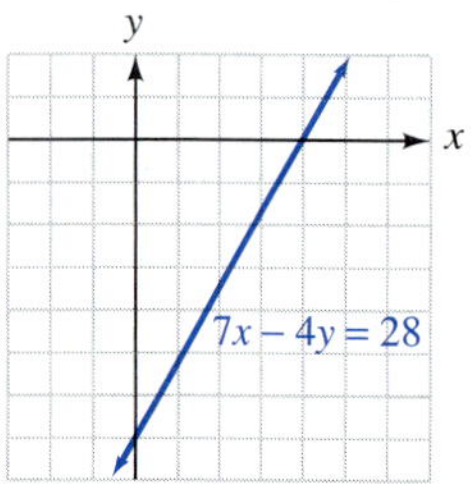

25. yes **26.** no **27.** yes **28.** yes

29.
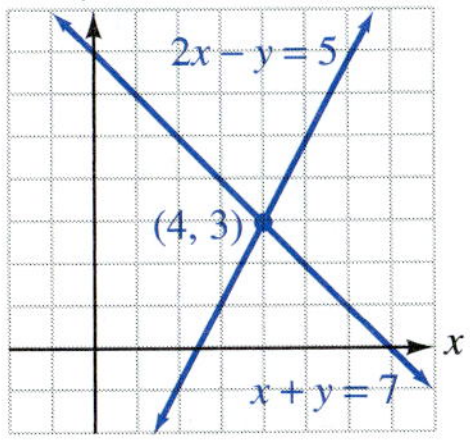

30.
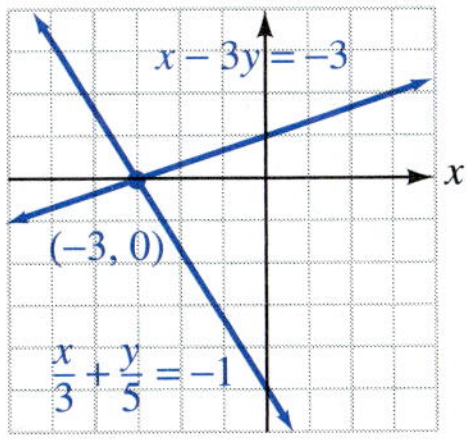

31.
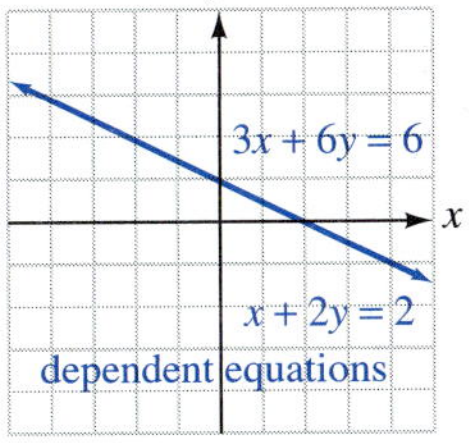

32.
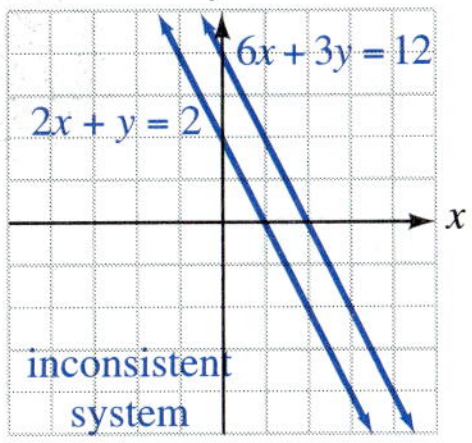

33. $(-1, -2)$ **34.** $(-2, 5)$ **35.** $(1, -1)$ **36.** $(-2, 1)$
37. $(3, -5)$ **38.** $\left(3, \frac{1}{2}\right)$ **39.** $(-1, 7)$ **40.** $\left(-\frac{1}{2}, \frac{7}{2}\right)$
41. $(0, 9)$ **42.** inconsistent system
43. dependent equations **44.** $(0, 0)$ **45.** 3, 15
46. 3 ft by 9 ft **47.** 50¢ **48.** \$66 **49.** \$1.69
50. \$750

51.
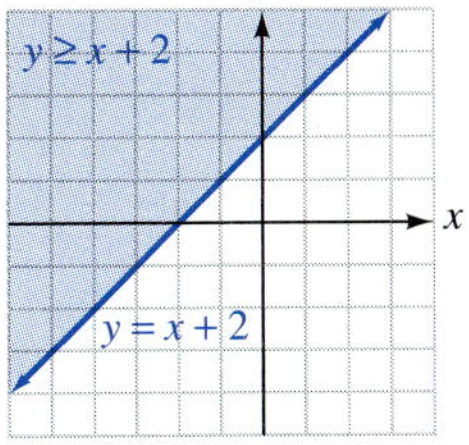

52.
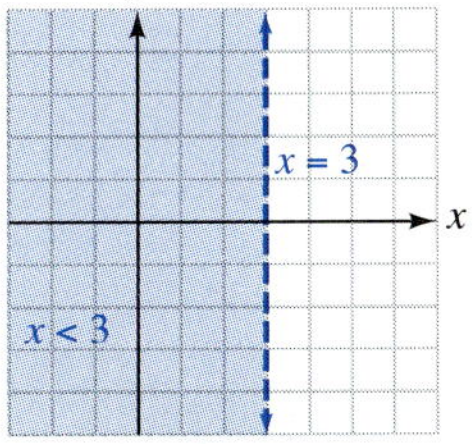

53.
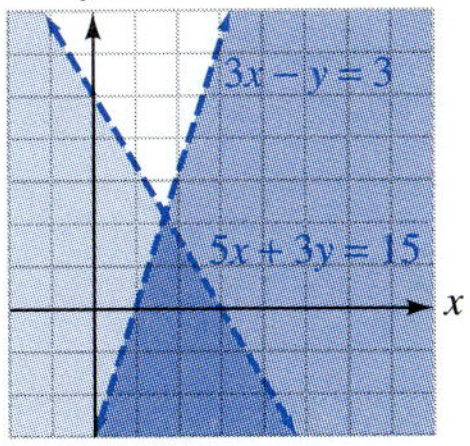

54.
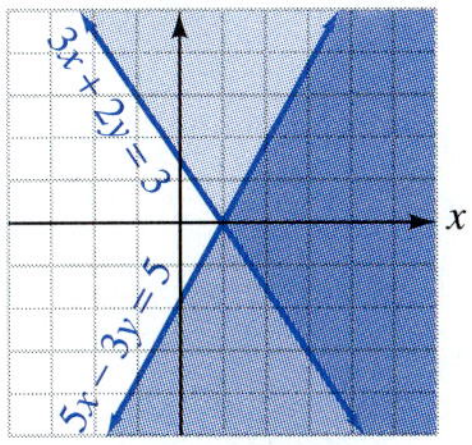

55.
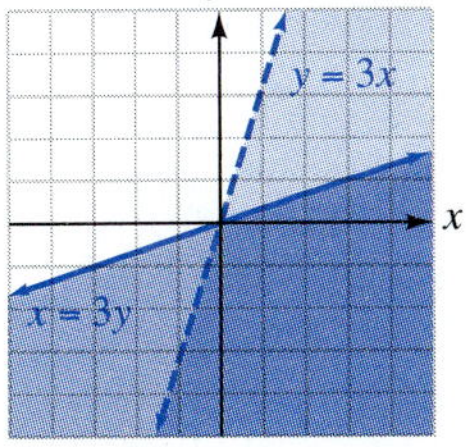

56.
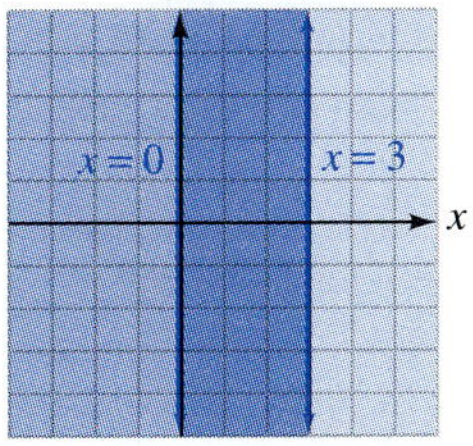

Chapter 3 Test (page 228)

1.
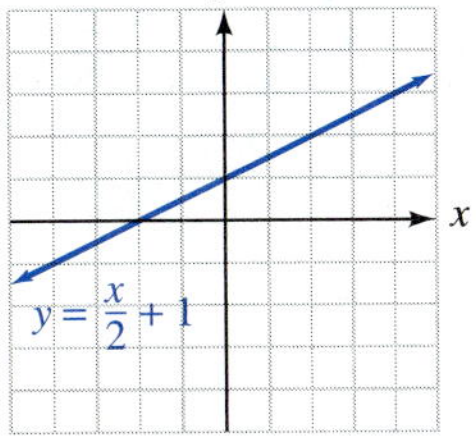

2.
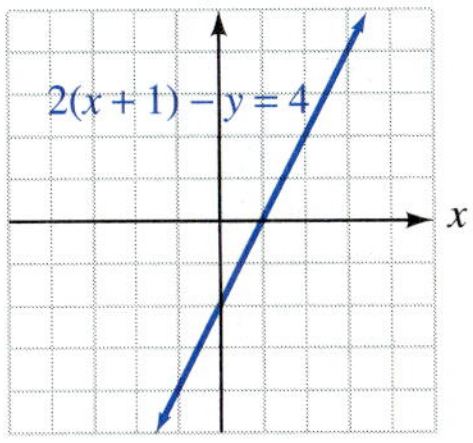

3.
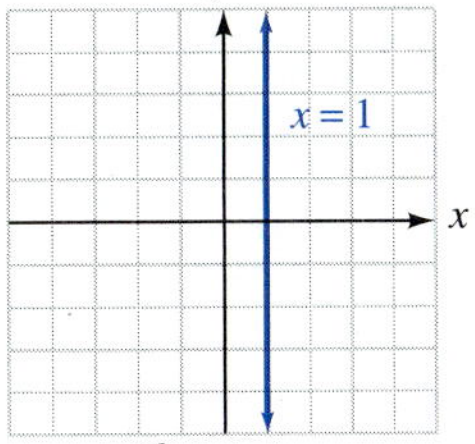

4.
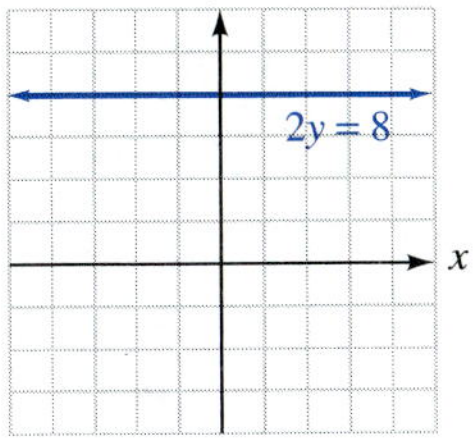

5. yes **6.** no

7.
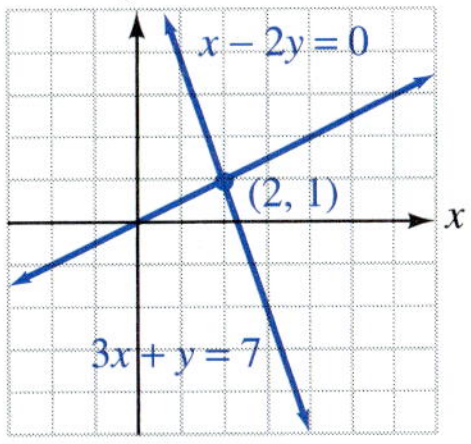

8.
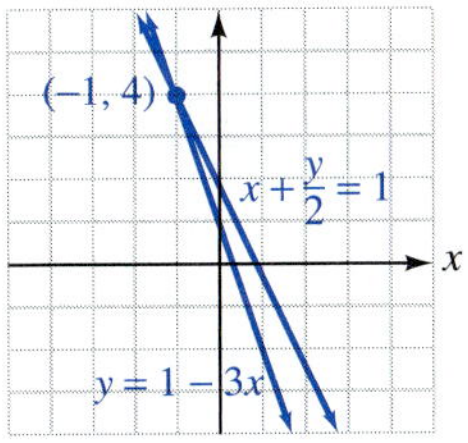

9. $(-3, -4)$ **10.** $(12, 10)$ **11.** $(2, 4)$ **12.** $(-3, 3)$
13. inconsistent **14.** consistent **15.** 65 **16.** \$4,000

17.
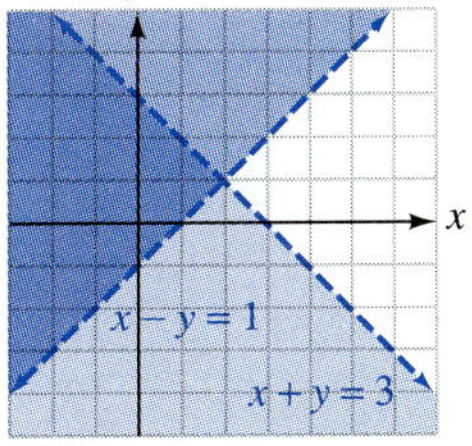

18.
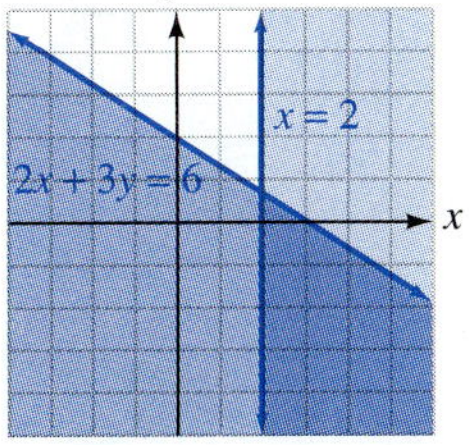

Getting Ready (page 231)

1. 8 **2.** 9 **3.** 6 **4.** 6 **5.** 12 **6.** 32 **7.** 18 **8.** 3

Orals (page 236)

1. base x, exponent 3 **2.** base 3, exponent x
3. base b, exponent c **4.** base ab, exponent c **5.** 36
6. 36 **7.** 9 **8.** 27

Exercise 4.1 (page 237)

1.
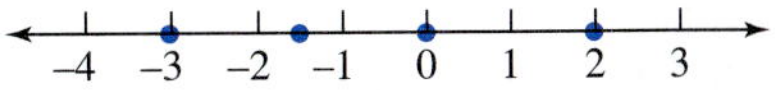

3. the product of 3 and the sum of x and y **5.** $|2x| + 3$
7. $-5, 3$ **9.** $(3x)(3x)(3x)(3x)$ **11.** $y \cdot y \cdot y \cdot y \cdot y$
13. $x^n y^n$ **15.** $a^{b \cdot c}$ **19.** base 4, exponent 3

21. base x, exponent 5 **23.** base $2y$, exponent 3 **25.** base x, exponent 4 **27.** base x, exponent 1 **29.** base x, exponent 3 **31.** $5 \cdot 5 \cdot 5$ **33.** $x \cdot x \cdot x \cdot x \cdot x \cdot x \cdot x$ **35.** $-4 \cdot x \cdot x \cdot x \cdot x \cdot x$ **37.** $(3t)(3t)(3t)(3t)(3t)$ **39.** 2^3 **41.** x^4 **43.** $(2x)^3$ **45.** $-4t^4$ **47.** 625 **49.** 13 **51.** 561 **53.** -725 **55.** x^7 **57.** x^{10} **59.** t^3 **61.** a^{12} **63.** y^9 **65.** $12x^7$ **67.** $-4y^5$ **69.** $6x^9$ **71.** 3^8 **73.** y^{15} **75.** a^{21} **77.** x^{25} **79.** $243z^{30}$ **81.** x^{31} **83.** r^{36} **85.** s^{33} **87.** x^3y^3 **89.** r^6s^4 **91.** $16a^2b^4$ **93.** $-8r^6s^9t^3$ **95.** $\frac{a^3}{b^3}$ **97.** $\frac{x^{10}}{y^{15}}$ **99.** $\frac{-32a^5}{b^5}$ **101.** $\frac{b^6}{27a^3}$ **103.** x^2 **105.** y^4 **107.** $3a$ **109.** ab^4 **111.** $\frac{10r^{13}s^3}{3}$ **113.** $\frac{x^{12}y^{16}}{2}$ **115.** $\frac{y^3}{8}$ **117.** $-\frac{8r^3}{27}$ **119.** 2 ft **121.** \$16,000 **123.** \$45,947.93

Getting Ready (page 239)

1. $\frac{1}{3}$ **2.** $\frac{1}{y}$ **3.** 1 **4.** $\frac{1}{xy}$

Orals (page 242)

1. $\frac{1}{2}$ **2.** $\frac{1}{4}$ **3.** 2 **4.** 1 **5.** x **6.** $\frac{1}{y^7}$ **7.** 1 **8.** $\frac{y}{x}$

Exercise 4.2 (page 242)

1. 2 **3.** $\frac{6}{5}$ **5.** $s = \frac{f(P-L)}{i}$ or $s = \frac{fP - fL}{i}$ **7.** 1 **9.** 1 **11.** 8 **13.** 1 **15.** 1 **17.** 512 **19.** 2 **21.** 1 **23.** 1 **25.** -2 **27.** $\frac{1}{x^2}$ **29.** $\frac{1}{b^5}$ **31.** $\frac{1}{16y^4}$ **33.** $\frac{1}{a^3b^6}$ **35.** $\frac{1}{y}$ **37.** $\frac{1}{r^6}$ **39.** y^5 **41.** 1 **43.** $\frac{1}{a^2b^4}$ **45.** $\frac{1}{x^6y^3}$ **47.** $\frac{1}{x^3}$ **49.** $\frac{1}{y^2}$ **51.** a^8b^{12} **53.** $-\frac{y^{10}}{32x^{15}}$ **55.** a^{14} **57.** $\frac{1}{b^{14}}$ **59.** $\frac{256x^{28}}{81}$ **61.** $\frac{16y^{14}}{z^{10}}$ **63.** $\frac{x^{14}}{128y^{28}}$ **65.** $\frac{16u^4v^8}{81}$ **67.** $\frac{1}{9a^2b^2}$ **69.** $\frac{c^{15}}{216a^9b^3}$ **71.** $\frac{1}{512}$ **73.** $\frac{17y^{27}z^5}{x^{35}}$ **75.** x^{3m} **77.** u^{5m} **79.** y^{2m+2} **81.** y^m **83.** $\frac{1}{x^{3n}}$ **85.** x^{2m+2} **87.** x^{8n-12} **89.** y^{4n-8} **91.** \$6,678.04 **93.** \$3,183.76 **95.** \$9,875.85

Getting Ready (page 244)

1. 100 **2.** 1,000 **3.** 10 **4.** $\frac{1}{100}$ **5.** 500 **6.** 8,000 **7.** 30 **8.** $\frac{7}{100}$

Orals (page 248)

1. 3.72×10^2 **2.** 37.2 **3.** 4.72×10^3 **4.** 3.72×10^3 **5.** 3.72×10^{-1} **6.** 2.72×10^{-2}

Exercise 4.3 (page 249)

1. 5 **3.** comm. prop. of add. **5.** 6 **7.** scientific notation **9.** 2.3×10^4 **11.** 1.7×10^6 **13.** 6.2×10^{-2} **15.** 5.1×10^{-6} **17.** 4.25×10^3 **19.** 2.5×10^{-3} **21.** 230 **23.** 812,000 **25.** 0.00115 **27.** 0.000976 **29.** 25,000,000 **31.** 0.00051 **33.** 714,000 **35.** 30,000 **37.** 200,000 **39.** 2.57×10^{13} mi **41.** 114,000,000 mi **43.** 6.22×10^{-3} mi **45.** 1.9008×10^{11} ft **47.** 3.3×10^{-1} km/sec **49.** 1.5×10^{-4}, 2.5×10^{13}

Getting Ready (page 250)

1. $2x^2y^3$ **2.** $3xy^3$ **3.** $2x^2 + 3y^2$ **4.** $x^3 + y^3$ **5.** $6x^3y^3$ **6.** $5x^2y^2z^4$ **7.** $5x^2y^2$ **8.** $x^3y^3z^3$

Exercise 4.4 (page 258)

1. 8 **3.** 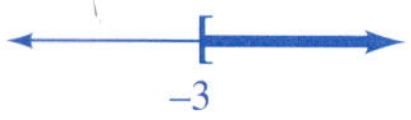

5. x^{18} **7.** y^9 **9.** algebraic **11.** polynomial **13.** trinomial **15.** degree **17.** function **19.** domain **21.** yes **23.** yes **25.** binomial **27.** trinomial **29.** monomial **31.** binomial **33.** trinomial **35.** none of these **37.** 4th **39.** 3rd **41.** 8th **43.** 6th **45.** 12th **47.** 0th **49.** 7 **51.** -8 **53.** -4 **55.** -5 **57.** 1 **59.** -9 **61.** 3 **63.** 11 **65.** 1, -2, -3, -2, 1 **67.** -6, 1, 2, 3, 10

69.

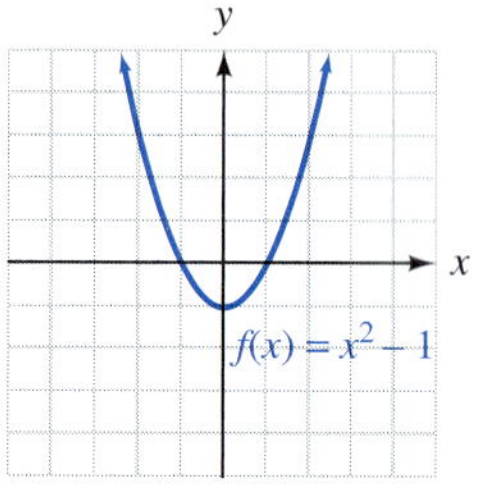

71.

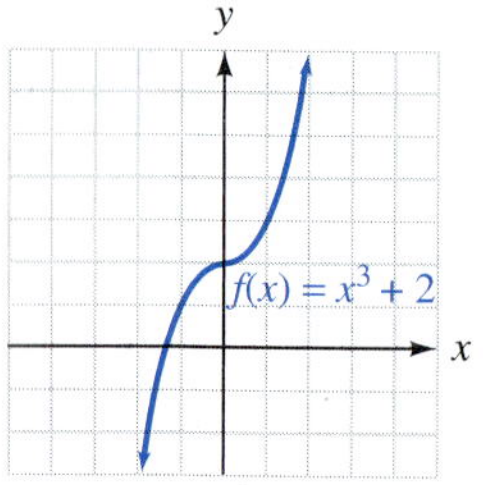

73. 64 ft **75.** \$28,362 **77.** 63 ft

Getting Ready (page 260)

1. $5x$ **2.** $2y$ **3.** $25x$ **4.** $5z$ **5.** $12r$ **6.** not possible **7.** 0 **8.** not possible

Orals (page 264)

1. $4x^3$ **2.** $4xy$ **3.** $2y$ **4.** $2 - 2x$ **5.** $-2y^2$ **6.** $6x^2 + 4y$ **7.** $4x^2 + y$ **8.** $2y$

Exercise 4.5 (page 264)

1. -8 **3.** -9 **5.** 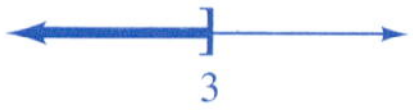

7. monomial **9.** coefficients, variables **11.** like terms **13.** like terms, $7y$ **15.** unlike terms **17.** like terms, $13x^3$ **19.** like terms, $8x^3y^2$ **21.** like terms, $65t^6$ **23.** unlike terms **25.** $9y$ **27.** $-12t^2$ **29.** $16u^3$ **31.** $7x^5y^2$ **33.** $14rst$ **35.** $-6a^2bc$ **37.** $15x^2$ **39.** $4x^2y^2$ **41.** $95x^8y^4$ **43.** $7x + 4$ **45.** $2a + 7$ **47.** $7x - 7y$ **49.** $-19x - 4y$ **51.** $6x^2 + x - 5$ **53.** $7b + 4$ **55.** $3x + 1$ **57.** $5x + 15$ **59.** $3x - 3y$ **61.** $5x^2 - 25x - 20$ **63.** $5x^2 + x + 11$

65. $-7x^3 - 7x^2 - x - 1$ **67.** $2x^2y + xy + 13y^2$
69. $5x^2 + 6x - 8$ **71.** $-x^3 + 6x^2 + x + 14$
73. $-12x^2y^2 - 13xy + 36y^2$ **75.** $6x^2 - 2x - 1$
77. $t^3 + 3t^2 + 6t - 5$ **79.** $-3x^2 + 5x - 7$ **81.** $6x - 2$
83. $-5x^2 - 8x - 19$ **85.** $4y^3 - 12y^2 + 8y + 8$
87. $3a^2b^2 - 6ab + b^2 - 6ab^2$
89. $-6x^2y^2 + 4xy^2z - 20xy^3 + 2y$ **91.** \$114,000
93. \$132,000 **95. a.** \$263,000 **b.** \$263,000
97. $y = -1{,}100x + 6{,}600$ **99.** $y = -2{,}800x + 15{,}800$
103. $6x + 3h - 10$ **105.** 49

Getting Ready (page 267)

1. $6x$ **2.** $3x^4$ **3.** $5x^3$ **4.** $8x^5$ **5.** $3x + 15$
6. $-2x - 10$ **7.** $4y - 12$ **8.** $-2y^2 + 6$

Orals (page 274)

1. $6x^3 - 2x^2$ **2.** $10y^3 - 15y$ **3.** $7x^2y + 7xy^2$
4. $-4xy + 6y^2$ **5.** $x^2 + 5x + 6$ **6.** $x^2 - x - 6$
7. $2x^2 + 7x + 6$ **8.** $9x^2 - 1$ **9.** $x^2 + 6x + 9$
10. $x^2 - 10x + 25$

Exercise 4.6 (page 274)

1. distrib. prop. **3.** comm. prop. of mult. **5.** 0
7. monomial **9.** trinomial **11.** $6x^2$ **13.** $15x$
15. $12x^5$ **17.** $-24b^6$ **19.** $6x^5y^5$ **21.** $-3x^4y^7z^8$
23. $x^{10}y^{15}$ **25.** $a^5b^4c^7$ **27.** $3x + 12$ **29.** $-4t - 28$
31. $3x^2 - 6x$ **33.** $-6x^4 + 2x^3$ **35.** $3x^2y + 3xy^2$
37. $6x^4 + 8x^3 - 14x^2$ **39.** $2x^7 - x^2$ **41.** $-6r^3t^2 + 2r^2t^3$
43. $-6x^4y^4 - 6x^3y^5$ **45.** $a^2 + 9a + 20$
47. $3x^2 + 10x - 8$ **49.** $6a^2 + 2a - 20$
51. $6x^2 - 7x - 5$ **53.** $2x^2 + 3x - 9$
55. $6s^2 + 7st - 3t^2$ **57.** $x^2 + xz + xy + yz$
59. $u^2 + 2tu + uv + 2tv$ **61.** $-4r^2 - 20rs - 21s^2$
63. $4x^2 + 11x + 6$ **65.** $12x^2 + 14xy - 10y^2$ **67.** $x^3 - 1$
69. $2x^3 + 7x^2 + x - 1$ **71.** $x^2 + 8x + 16$
73. $t^2 - 6t + 9$ **75.** $r^2 - 16$ **77.** $x^2 + 10x + 25$
79. $4s^2 + 4s + 1$ **81.** $16x^2 - 25$ **83.** $x^2 - 4xy + 4y^2$
85. $4a^2 - 12ab + 9b^2$ **87.** $16x^2 - 25y^2$
89. $2x^2 - 6x - 8$ **91.** $3a^3 - 3ab^2$
93. $4t^3 + 11t^2 + 18t + 9$
95. $-3x^3 + 25x^2y - 56xy^2 + 16y^3$ **97.** $x^3 - 8y^3$
99. $5t^2 - 11t$ **101.** $x^2y + 3xy^2 + 2x^2$
103. $2x^2 + xy - y^2$ **105.** $8x$ **107.** $5s^2 - 7s - 9$
109. -3 **111.** -8 **113.** -1 **115.** 0 **117.** 1
119. 4 m **121.** 90 ft

Getting Ready (page 276)

1. $2xy^2$ **2.** y **3.** $\frac{3xy}{2}$ **4.** $\frac{x}{y}$ **5.** xy **6.** 3

Orals (page 280)

1. $2x^2$ **2.** $2y$ **3.** $5bc^2$ **4.** $-2pq$ **5.** 1 **6.** $3x$

Exercise 4.7 (page 281)

1. binomial **3.** none of these **5.** 2 **7.** polynomial
9. two **11.** $\frac{a}{b}$ **13.** $\frac{1}{3}$ **15.** $-\frac{5}{3}$ **17.** $\frac{3}{4}$ **19.** 1
21. $-\frac{1}{4}$ **23.** $\frac{42}{19}$ **25.** $\frac{x}{z}$ **27.** $\frac{r^2}{s}$ **29.** $\frac{2x^2}{y}$ **31.** $-\frac{3u^3}{v^2}$
33. $\frac{4r}{y^2}$ **35.** $-\frac{13}{3rs}$ **37.** $\frac{x^4}{y^6}$ **39.** a^8b^8 **41.** $-\frac{3r}{s^9}$
43. $-\frac{x^3}{4y^3}$ **45.** $\frac{125}{8b^3}$ **47.** $\frac{xy^2}{3}$ **49.** a^8 **51.** z^3
53. $\frac{2}{y} + \frac{3}{x}$ **55.** $\frac{1}{5y} - \frac{2}{5x}$ **57.** $\frac{1}{y^2} + \frac{2y}{x^2}$
59. $3a - 2b$ **61.** $\frac{1}{y} - \frac{1}{2x} + \frac{2z}{xy}$ **63.** $3x^2y - 2x - \frac{1}{y}$
65. $5x - 6y + 1$ **67.** $\frac{10x^2}{y} - 5x$ **69.** $-\frac{4x}{3} + \frac{3x^2}{2}$
71. $xy - 1$ **73.** $\frac{x}{y} - \frac{11}{6} + \frac{y}{2x}$ **75.** 2 **77.** yes
79. yes

Getting Ready (page 282)

1. 13 **2.** 21 **3.** 19 **4.** 13

Orals (page 287)

1. $2 + \frac{3}{x}$ **2.** $3 - \frac{5}{x}$ **3.** $2 + \frac{1}{x+1}$ **4.** $3 + \frac{2}{x+1}$
5. x **6.** x

Exercise 4.8 (page 287)

1. 21, 22, 24, 25, 26, 27, 28 **3.** 5 **5.** -5
7. $8x^2 - 6x + 1$ **9.** divisor, dividend **11.** remainder
13. $4x^3 - 2x^2 + 7x + 6$ **15.** $6x^4 - x^3 + 2x^2 + 9x$
17. $0x^3$ and $0x$ **19.** $x + 2$ **21.** $y + 12$ **23.** $a + b$
25. $3a - 2$ **27.** $b + 3$ **29.** $x - 3y$ **31.** $2x + 1$
33. $x - 7$ **35.** $3x + 2y$ **37.** $2x - y$ **39.** $x + 5y$
41. $x - 5y$ **43.** $x^2 + 2x - 1$ **45.** $2x^2 + 2x + 1$
47. $x^2 + xy + y^2$ **49.** $x + 1 + \frac{-1}{2x+3}$
51. $2x + 2 + \frac{-3}{2x+1}$ **53.** $x^2 + 2x + 1$
55. $x^2 + 2x - 1 + \frac{6}{2x+3}$ **57.** $2x^2 + 8x + 14 + \frac{31}{x-2}$
59. $x + 1$ **61.** $2x - 3$ **63.** $x^2 - x + 1$
65. $a^2 - 3a + 10 + \frac{-30}{a+3}$ **67.** $5x^2 - x + 4 + \frac{16}{3x-4}$

Chapter Summary (page 289)

1. $(-3x)(-3x)(-3x)(-3x)$ **2.** $\left(\frac{1}{2}pq\right)\left(\frac{1}{2}pq\right)\left(\frac{1}{2}pq\right)$ **3.** 125
4. 243 **5.** 64 **6.** -64 **7.** 13 **8.** 25 **9.** x^5
10. x^9 **11.** y^{21} **12.** x^{42} **13.** a^3b^3 **14.** $81x^4$
15. b^{12} **16.** $-y^2z^5$ **17.** $256s^3$ **18.** $-3y^6$ **19.** x^{15}
20. $4x^4y^2$ **21.** x^4 **22.** $\frac{x^2}{y^2}$ **23.** $\frac{2y^2}{x^2}$ **24.** $5yz^4$ **25.** 1
26. 1 **27.** 9 **28.** $9x^4$ **29.** $\frac{1}{x^3}$ **30.** x **31.** y
32. x^{10} **33.** $\frac{1}{x^2}$ **34.** $\frac{a^6}{b^3}$ **35.** $\frac{1}{x^5}$ **36.** $\frac{1}{9z^2}$
37. 7.28×10^2 **38.** 9.37×10^3 **39.** 1.36×10^{-2}
40. 9.42×10^{-3} **41.** 7.73×10^0 **42.** 7.53×10^5
43. 1.8×10^{-4} **44.** 6×10^4 **45.** 726,000
46. 0.000391 **47.** 2.68 **48.** 57.6 **49.** 7.39
50. 0.000437 **51.** 0.03 **52.** 160 **53.** 7th, monomial
54. 2nd, binomial **55.** 5th, trinomial **56.** 5th, binomial

57. 11 **58.** 2 **59.** -4 **60.** 4 **61.** 402 **62.** 0
63. 82 **64.** 0.3405 **65.** -4 **66.** 21 **67.** 0
68. $-\frac{15}{4}$
69.

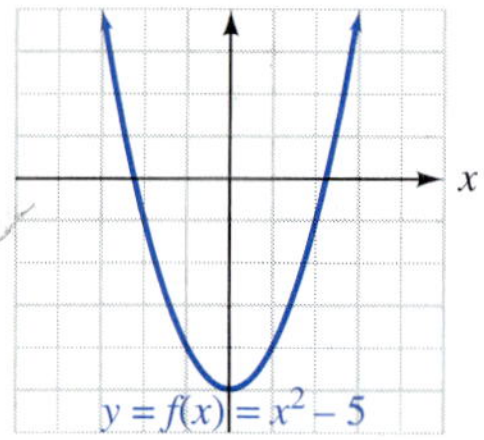

70.

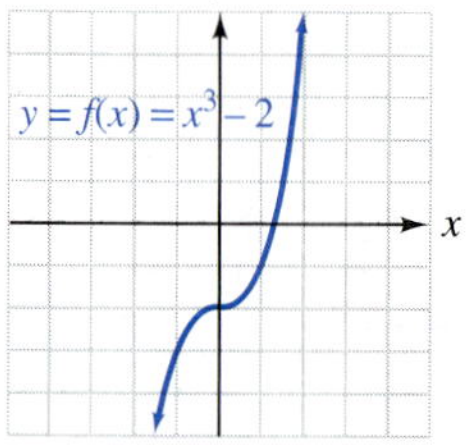

71. $7x$ **72.** in simplest terms **73.** $4x^2y^2$ **74.** x^2yz
75. $8x^2 - 6x$ **76.** $4a^2 + 4a - 6$ **77.** $5x^2 + 19x + 3$
78. $6x^3 + 8x^2 + 3x - 72$ **79.** $10x^3y^5$ **80.** x^7yz^5
81. $5x + 15$ **82.** $6x + 12$ **83.** $3x^4 - 5x^2$
84. $2y^4 + 10y^3$ **85.** $-x^2y^3 + x^3y^2$ **86.** $-3x^2y^2 + 3x^2y$
87. $x^2 + 5x + 6$ **88.** $2x^2 - x - 1$ **89.** $6a^2 - 6$
90. $6a^2 - 6$ **91.** $2a^2 - ab - b^2$ **92.** $6x^2 + xy - y^2$
93. $x^2 + 6x + 9$ **94.** $x^2 - 25$ **95.** $y^2 - 4$
96. $x^2 + 8x + 16$ **97.** $x^2 - 6x + 9$ **98.** $y^2 - 2y + 1$
99. $4y^2 + 4y + 1$ **100.** $y^4 - 1$
101. $3x^3 + 7x^2 + 5x + 1$ **102.** $8a^3 - 27$ **103.** 1
104. -1 **105.** 7 **106.** 5 **107.** 1 **108.** 0
109. $\frac{3}{2y} + \frac{3}{x}$ **110.** $2 - \frac{3}{y}$ **111.** $-3a - 4b + 5c$
112. $-\frac{x}{y} - \frac{y}{x}$ **113.** $x + 1 + \frac{3}{x + 2}$ **114.** $x - 5$
115. $2x + 1$ **116.** $x + 5 + \frac{3}{3x - 1}$
117. $3x^2 + 2x + 1 + \frac{2}{2x - 1}$ **118.** $3x^2 - x - 4$

Chapter 4 Test (page 293)

1. $2x^3y^4$ **2.** 134 **3.** y^6 **4.** $6b^7$ **5.** $32x^{21}$ **6.** $8r^{18}$
7. 3 **8.** $\dfrac{2}{y^3}$ **9.** y^3 **10.** $\dfrac{64a^3}{b^3}$ **11.** 2.8×10^4
12. 2.5×10^{-3} **13.** 7,400 **14.** 0.000093 **15.** binomial
16. 10th degree **17.** 0 **18.**

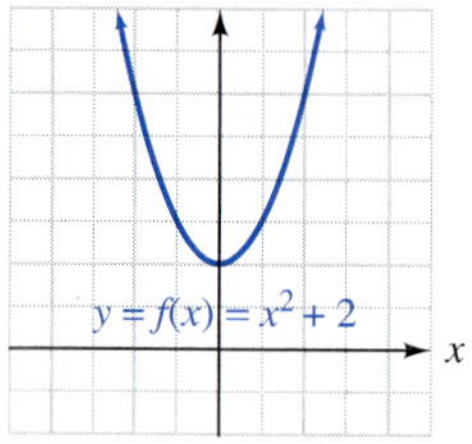

19. $-7x + 2y$ **20.** $-3x + 6$ **21.** $5x^3 + 2x^2 + 2x - 5$
22. $-x^2 - 5x + 4$ **23.** $-4x^5y$ **24.** $3y^4 - 6y^3 + 9y^2$
25. $6x^2 - 7x - 20$ **26.** $2x^3 - 7x^2 + 14x - 12$
27. $\frac{1}{2}$ **28.** $\frac{y}{2x}$ **29.** $\frac{a}{4b} - \frac{b}{2a}$ **30.** $x - 2$

Cumulative Review Exercises (page 294)

1. 11 **2.** 71 **3.** $-\frac{11}{10}$ **4.** 7 **5.** 15 **6.** 4 **7.** -10
8. -6 **9.**

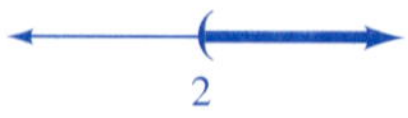

10.

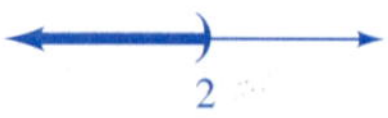

11.

12.

13. $r = \frac{A - p}{pt}$ **14.** $h = \frac{2A}{b}$

15.

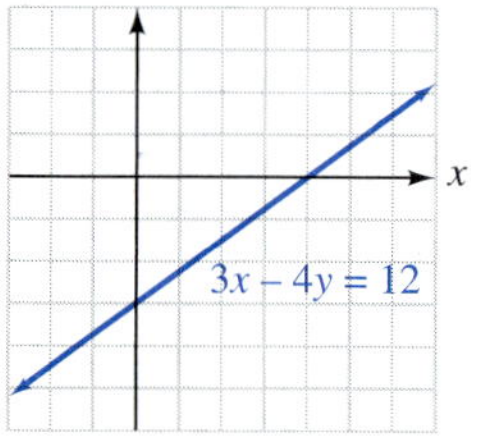

16.

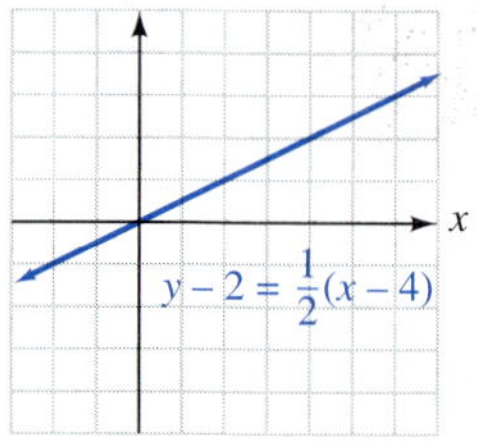

17. -2 **18.** 13 **19.** -12 **20.** -1 **21.** y^{14} **22.** xy
23. $\dfrac{a^7}{b^6}$ **24.** x^2y^2 **25.** $x^2 + 4x - 14$
26. $6x^2 + 10x - 56$ **27.** $x^3 - 8$ **28.** $2x + 1$
29. 4.8×10^{18} m **30.** 4 in. **31.** 879.6 in.^{2} **32.** \$512

Getting Ready (page 297)

1. $5x + 15$ **2.** $7y - 56$ **3.** $3x^2 - 2x$ **4.** $5y^2 + 9y$
5. $3x + 3y + ax + ay$ **6.** $xy + x + 5y + 5$
7. $5x + 5 - yx - y$ **8.** $x^2 + 2x - yx - 2y$

Orals (page 303)

1. $2^2 \cdot 3^2$ **2.** 3^3 **3.** 3^4 **4.** $3^2 \cdot 5$ **5.** 3 **6.** $3ab$
7. $x + 3$ **8.** $a - 1$ **9.** $5(3xy + 2)$ **10.** $5xy(3 + 2y)$
11. $(x + 3)(a - 3)$ **12.** $(x - 2)(b + 1)$

Exercise 5.1 (page 303)

1. 7 **3.** 11 **5.** prime **7.** largest **9.** $a + b$
11. $2^2 \cdot 3$ **13.** $3 \cdot 5$ **15.** $2^3 \cdot 5$ **17.** $2 \cdot 7^2$ **19.** $3^2 \cdot 5^2$
21. $2^5 \cdot 3^2$ **23.** 4 **25.** r^2 **27.** $4, x$ **29.** $a + b$
31. $p - q$ **33.** $3(x + 2)$ **35.** $x(y - z)$ **37.** $t^2(t + 2)$
39. $r^2(r^2 - 1)$ **41.** $a^2b^3z^2(az - 1)$ **43.** $8xy^2z^3(3xyz + 1)$
45. $6uvw^2(2w - 3v)$ **47.** $3(x + y - 2z)$
49. $a(b + c - d)$ **51.** $2y(2y + 4 - x)$
53. $3r(4r - s + 3rs^2)$ **55.** $abx(1 - b + x)$
57. $2xyz^2(2xy - 3y + 6)$ **59.** $7a^2b^2c^2(10a + 7bc - 3)$
61. $-(a + b)$ **63.** $-(2x - 5y)$ **65.** $-(2a - 3b)$
67. $-(3m + 4n - 1)$ **69.** $-(3xy - 2z - 5w)$
71. $-(3ab + 5ac - 9bc)$ **73.** $-3xy(x + 2y)$
75. $-4a^2b^2(b - 3a)$ **77.** $-2ab^2c(2ac - 7a + 5c)$
79. $-7ab(2a^5b^5 - 7ab^2 + 3)$
81. $-5a^2b^3c(1 - 3abc + 5a^2)$ **83.** $(x + y)(2 + b)$
85. $(x + y)(3 - a)$ **87.** $(r - 2s)(3 - x)$
89. $(x - 3)(x - 2)$ **91.** $2(a^2 + b)(x + y)$
93. $3(r + 3s)(x^2 - 2y^2)$ **95.** $(a + b + c)(3x - 2y)$
97. $7xy(r + 2s - t)(2x - 3)$ **99.** $(x + 1)(x + 3 - y)$
101. $(x^2 - 2)(3x - y + 1)$ **103.** $(x + y)(2 + a)$
105. $(r + s)(7 - k)$ **107.** $(r + s)(x + y)$
109. $(2x + 3)(a + b)$ **111.** $(b + c)(2a + 3)$
113. $(x + y)(2x - 3)$ **115.** $(v - 3w)(3t + u)$
117. $(3p + q)(3m - n)$ **119.** $(m - n)(p - 1)$
121. $(a - b)(x - y)$ **123.** $x^2(a + b)(x + 2y)$
125. $4a(b + 3)(a - 2)$ **127.** $(x^2 + 1)(x + 2)$
129. $y(x^2 - y)(x - 1)$ **131.** $(r - s)(2 + b)$
133. $(x + y)(a + b)$ **135.** $(a - b)(c - d)$

137. $r(r + s)(a - b)$ **139.** $(b + 1)(a + 3)$
141. $(r - s)(p - q)$

Getting Ready (page 306)

1. $a^2 - b^2$ **2.** $4r^2 - s^2$ **3.** $9x^2 - 4y^2$ **4.** $16x^4 - 9$

Orals (page 309)

1. $(x + 3)(x - 3)$ **2.** $(y + 6)(y - 6)$ **3.** $(z + 2)(z - 2)$ **4.** $(p + q)(p - q)$ **5.** $(5 + t)(5 - t)$ **6.** $(6 + r)(6 - r)$ **7.** $(10 + y)(10 - y)$ **8.** $(10 + y^2)(10 - y^2)$

Exercise 5.2 (page 309)

1. $p = w\left(k - h - \dfrac{v^2}{2g}\right)$ **3.** difference of two squares
5. $(p - q)$ **7.** $(x - 3)$ **9.** $(2m - 3n)$
11. $(x + 4)(x - 4)$ **13.** $(y + 7)(y - 7)$
15. $(2y + 7)(2y - 7)$ **17.** $(3x + y)(3x - y)$
19. $(5t + 6u)(5t - 6u)$ **21.** $(4a + 5b)(4a - 5b)$
23. prime **25.** $(a^2 + 2b)(a^2 - 2b)$
27. $(7y + 15z^2)(7y - 15z^2)$ **29.** $(14x^2 + 13y)(14x^2 - 13y)$
31. $8(x + 2y)(x - 2y)$ **33.** $2(a + 2y)(a - 2y)$
35. $3(r + 2s)(r - 2s)$ **37.** $x(x + y)(x - y)$
39. $x(2a + 3b)(2a - 3b)$ **41.** $3m(m + n)(m - n)$
43. $x^2(2x + y)(2x - y)$ **45.** $2ab(a + 11b)(a - 11b)$
47. $(x^2 + 9)(x + 3)(x - 3)$ **49.** $(a^2 + 4)(a + 2)(a - 2)$
51. $(a^2 + b^2)(a + b)(a - b)$
53. $(9r^2 + 16s^2)(3r + 4s)(3r - 4s)$
55. $(a^2 + b^4)(a + b^2)(a - b^2)$
57. $(x^4 + y^4)(x^2 + y^2)(x + y)(x - y)$
59. $2(x^2 + y^2)(x + y)(x - y)$
61. $b(a^2 + b^2)(a + b)(a - b)$
63. $3n(4m^2 + 9n^2)(2m + 3n)(2m - 3n)$
65. $3ay(a^4 + 2y^4)$ **67.** $3a^2(a^4 + b^2)(a^2 + b)(a^2 - b)$
69. $2y^2(x^4 + 4y^2)(x^2 + 2y)(x^2 - 2y)$
71. $a^2b^2(a^2 + b^2c^2)(a + bc)(a - bc)$
73. $a^2b^3(b^2 + 25)(b + 5)(b - 5)$
75. $3rs(9r^2 + 4s^2)(3r + 2s)(3r - 2s)$
77. $(4x - 4y + 3)(4x - 4y - 3)$ **79.** $(a + 3)^2(a - 3)$
81. $(y + 4)(y - 4)(y - 3)$ **83.** $3(x + 2)(x - 2)(x + 1)$
85. $3(m + n)(m - n)(m + a)$
87. $2(m + 4)(m - 4)(mn^2 + 4)$

Getting Ready (page 310)

1. $x^2 + 12x + 36$ **2.** $y^2 - 14y + 49$ **3.** $a^2 - 6a + 9$
4. $x^2 + 9x + 20$ **5.** $r^2 - 7r + 10$ **6.** $m^2 - 4m - 21$
7. $a^2 + ab - 12b^2$ **8.** $u^2 - 8uv + 15v^2$
9. $x^2 - 2xy - 24y^2$

Orals (page 318)

1. 4 **2.** −, − **3.** −, 3 **4.** −, 2 **5.** 6, 1 **6.** 6, 1

Exercise 5.3 (page 318)

1. 

3. 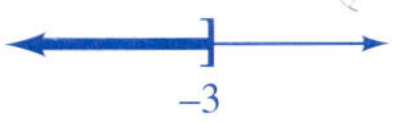

5. 

7. 

9. $(x + y)^2$ **11.** 4, 2 **13.** y, $2y$ **15.** $(x + 2)(x + 1)$
17. $(z + 11)(z + 1)$ **19.** $(a - 5)(a + 1)$
21. $(t - 7)(t - 2)$ **23.** prime **25.** $(y - 6)(y + 5)$
27. $(a + 8)(a - 2)$ **29.** $(t - 10)(t + 5)$ **31.** prime
33. $(y + z)(y + z)$ **35.** $(x + 2y)(x + 2y)$
37. $(m + 5n)(m - 2n)$ **39.** $(a - 6b)(a + 2b)$
41. $(u + 5v)(u - 3v)$ **43.** $-(x + 5)(x + 2)$
45. $-(y + 5)(y - 3)$ **47.** $-(t + 17)(t - 2)$
49. $-(r - 10)(r - 4)$ **51.** $-(a + 3b)(a + b)$
53. $-(x - 7y)(x + y)$ **55.** $(x - 4)(x - 1)$
57. $(y + 9)(y + 1)$ **59.** $(c + 5)(c - 1)$
61. $-(r - 2s)(r + s)$ **63.** $(r + 3x)(r + x)$
65. $(a - 2b)(a - b)$ **67.** $2(x + 3)(x + 2)$
69. $3y(y + 1)(y + 1)$ **71.** $-5(a - 3)(a - 2)$
73. $3(z - 4t)(z - t)$ **75.** $4y(x + 6)(x - 3)$
77. $-4x(x + 3y)(x - 2y)$ **79.** $(x + 2 + y)(x + 2 - y)$
81. $(b - 3 + c)(b - 3 - c)$ **83.** $(a + b + 2)(a + b - 2)$
85. $(b + y + 2)(b - y - 2)$ **87.** $(x + 3)(x + 3)$
89. $(y - 4)(y - 4)$ **91.** $(t + 10)(t + 10)$
93. $(u - 9)(u - 9)$ **95.** $(x + 2y)(x + 2y)$
97. $(r - 5s)(r - 5s)$

Getting Ready (page 320)

1. $6x^2 + 7x + 2$ **2.** $6y^2 - 19y + 10$ **3.** $8t^2 + 6t - 9$
4. $4r^2 + 4r - 15$ **5.** $6m^2 - 13m + 6$
6. $16a^2 + 16a + 3$

Orals (page 326)

1. 2, 3 **2.** 2 **3.** +, − **4.** +, − **5.** 3, 1 **6.** 3, 1

Exercise 5.4 (page 326)

1. $n = \frac{l - f + d}{d}$ **3.** descending **5.** opposites **7.** 2, 1
9. y, y **11.** $(2x - 1)(x - 1)$ **13.** $(3a + 1)(a + 4)$
15. $(z + 3)(4z + 1)$ **17.** $(3y + 2)(2y + 1)$
19. $(3x - 2)(2x - 1)$ **21.** $(3a + 2)(a - 2)$
23. $(2x + 1)(x - 2)$ **25.** $(2m - 3)(m + 4)$
27. $(5y + 1)(2y - 1)$ **29.** $(3y - 2)(4y + 1)$
31. $(5t + 3)(t + 2)$ **33.** $(8m - 3)(2m - 1)$
35. $(3x - y)(x - y)$ **37.** $(2u + 3v)(u - v)$
39. $(2a - b)(2a - b)$ **41.** $(3r + 2s)(2r - s)$
43. $(2x + 3y)(2x + y)$ **45.** $(4a - 3b)(a - 3b)$
47. $(3x + 2)(x - 5)$ **49.** $(2a - 5)(4a - 3)$
51. $(4y - 3)(3y - 4)$ **53.** prime **55.** $(2a + 3b)(a + b)$
57. $(3p - q)(2p + q)$ **59.** prime **61.** $(4x - 5y)(3x - 2y)$
63. $2(2x - 1)(x + 3)$ **65.** $y(y + 12)(y + 1)$
67. $3x(2x + 1)(x - 3)$ **69.** $3r^3(5r - 2)(2r + 5)$
71. $4(a - 2b)(a + b)$ **73.** $4(2x + y)(x - 2y)$
75. $-2mn(4m + 3n)(2m + n)$ **77.** $-2uv^3(7u - 3v)(2u - v)$
79. $(2x + 3)^2$ **81.** $(3x + 2)^2$ **83.** $(4x - y)^2$
85. $(2x + y + 4)(2x + y - 4)$
87. $(3 + a + 2b)(3 - a - 2b)$
89. $(3p + 1 + q)(3p + 1 - q)$

Getting Ready (page 328)

1. $x^3 - 27$ **2.** $x^3 + 8$ **3.** $y^3 + 64$ **4.** $r^3 - 125$
5. $a^3 - b^3$ **6.** $a^3 + b^3$

Orals (page 331)

1. $(x - y)(x^2 + xy + y^2)$ **2.** $(x + y)(x^2 - xy + y^2)$
3. $(a + 2)(a^2 - 2a + 4)$ **4.** $(b - 3)(b^2 + 3b + 9)$
5. $(1 + 2x)(1 - 2x + 4x^2)$ **6.** $(2 - r)(4 + 2r + r^2)$
7. $(xy + 1)(x^2y^2 - xy + 1)$ **8.** $(5 - 2t)(25 + 10t + 4t^2)$

Exercise 5.5 (page 331)

1. 0.0000000000001 cm **3.** $(x^2 - xy + y^2)$
5. $(y + 1)(y^2 - y + 1)$ **7.** $(a - 3)(a^2 + 3a + 9)$
9. $(2 + x)(4 - 2x + x^2)$ **11.** $(s - t)(s^2 + st + t^2)$
13. $(3x + y)(9x^2 - 3xy + y^2)$
15. $(a + 2b)(a^2 - 2ab + 4b^2)$
17. $(4x - 3)(16x^2 + 12x + 9)$
19. $(3x - 5y)(9x^2 + 15xy + 25y^2)$
21. $(a^2 - b)(a^4 + a^2b + b^2)$ **23.** $(x^3 + y^2)(x^6 - x^3y^2 + y^4)$
25. $2(x + 3)(x^2 - 3x + 9)$ **27.** $-(x - 6)(x^2 + 6x + 36)$
29. $8x(2m - n)(4m^2 + 2mn + n^2)$
31. $xy(x + 6y)(x^2 - 6xy + 36y^2)$
33. $3rs^2(3r - 2s)(9r^2 + 6rs + 4s^2)$
35. $a^3b^2(5a + 4b)(25a^2 - 20ab + 16b^2)$
37. $yz(y^2 - z)(y^4 + y^2z + z^2)$
39. $2mp(p + 2q)(p^2 - 2pq + 4q^2)$
41. $(x + 1)(x^2 - x + 1)(x - 1)(x^2 + x + 1)$
43. $(x^2 + y)(x^4 - x^2y + y^2)(x^2 - y)(x^4 + x^2y + y^2)$
45. $(x + y)(x^2 - xy + y^2)(3 - z)$
47. $(m + 2n)(m^2 - 2mn + 4n^2)(1 + x)$
49. $(a + 3)(a^2 - 3a + 9)(a - b)$
51. $(y + 1)(y - 1)(y - 3)(y^2 + 3y + 9)$

Getting Ready (page 332)

1. $3ax(x + a)$ **2.** $(x + 3y)(x - 3y)$
3. $(x - 2)(x^2 + 2x + 4)$ **4.** $2(x + 2)(x - 2)$
5. $(x - 5)(x + 2)$ **6.** $(2x - 3)(3x - 2)$
7. $2(3x - 1)(x - 2)$ **8.** $(a + b)(x + y)(x - y)$

Orals (page 334)

1. common factor **2.** difference of two squares
3. sum of two cubes **4.** grouping **5.** none, prime
6. common factor **7.** difference of two squares
8. difference of two cubes

Exercise 5.6 (page 335)

1. $\frac{8}{3}$ **3.** 0 **5.** factors **7.** binomials **9.** $3(2x + 1)$
11. $(x - 7)(x + 1)$ **13.** $(3t - 1)(2t + 3)$
15. $(2x + 5)(2x - 5)$ **17.** $(t - 1)(t - 1)$
19. $(a - 2)(a^2 + 2a + 4)$ **21.** $(y^2 - 2)(x + 1)(x - 1)$
23. $7p^4q^2(10q - 5 + 7p)$ **25.** $2a(b + 6)(b - 2)$
27. $-4p^2q^3(2pq^4 + 1)$ **29.** $(2a - b + 3)(2a - b - 3)$
31. prime **33.** $-2x^2(x - 4)(x^2 + 4x + 16)$
35. $2t^2(3t - 5)(t + 4)$ **37.** $(x - a)(a + b)(a - b)$
39. $(2p^2 - 3q^2)(4p^4 + 6p^2q^2 + 9q^4)$
41. $(5p - 4y)(25p^2 + 20py + 16y^2)$
43. $-x^2y^2z(16x^2 - 24x^3yz^3 + 15yz^6)$
45. $(9p^2 + 4q^2)(3p + 2q)(3p - 2q)$ **47.** prime
49. $2(3x + 5y^2)(9x^2 - 15xy^2 + 25y^4)$ **51.** prime
53. $t(7t - 1)(3t - 1)$
55. $(x + y)(x - y)(x + y)(x^2 - xy + y^2)$
57. $2(a + b)(a - b)(c + 2d)$

Getting Ready (page 336)

1. 1 **2.** 13 **3.** 3 **4.** 2

Orals (page 340)

1. 7, 8 **2.** 2, −9 **3.** 0, −7 **4.** 0, 12 **5.** 1, 1
6. 4, −5

Exercise 5.7 (page 340)

1. u^9 **3.** $\frac{a}{b}$ **5.** quadratic **7.** second **9.** 2, −3
11. 4, −1 **13.** $\frac{5}{2}, -2$ **15.** 1, −2, 3 **17.** 0, 3 **19.** $0, \frac{5}{2}$
21. 0, 7 **23.** $0, -\frac{8}{3}$ **25.** 0, 2 **27.** $0, -\frac{1}{5}$ **29.** 5, −5
31. 7, −7 **33.** $\frac{1}{2}, -\frac{1}{2}$ **35.** $\frac{2}{3}, -\frac{2}{3}$ **37.** 7, −7 **39.** $\frac{9}{2}, -\frac{9}{2}$
41. 12, 1 **43.** 5, −3 **45.** −3, 7 **47.** 8, 1
49. −3, −5 **51.** −4, 2 **53.** $\frac{1}{2}, 2$ **55.** $\frac{1}{2}, -\frac{2}{3}$ **57.** $\frac{1}{5}, 1$
59. $-\frac{1}{7}, -\frac{3}{2}$ **61.** 1, −2, −3 **63.** 3, −3, −5
65. $1, 5, -\frac{3}{2}$ **67.** 9, −9, −2 **69.** 0, −1, −2
71. 0, 9, −3 **73.** $-\frac{1}{3}, 3$ **75.** $\frac{2}{3}, -\frac{1}{5}$ **77.** $-\frac{3}{2}, \frac{2}{3}$ **79.** $\frac{1}{8}, 1$
81. $0, -3, -\frac{1}{3}$ **83.** 0, −3, −3

Getting Ready (page 342)

1. s^2 **2.** $2w + 4$ **3.** $x(x + 1)$ **4.** $w(w + 3)$

Orals (page 345)

1. $A = lw$ **2.** $A = \frac{1}{2}bh$ **3.** $A = s^2$ **4.** $V = lwh$
5. $P = 2l + 2w$ **6.** $P = 4s$

Exercise 5.8 (page 346)

1. −10 **3.** 675 cm^2 **5.** analyze **7.** 5, 7 **9.** 9
11. 9 sec **13.** $\frac{15}{4}$ sec and 10 sec **15.** 2 sec **17.** 21 ft
19. 4 m by 9 m **21.** 48 ft **23.** $b = 4$ in., $h = 18$ in.
25. 18 sq units **27.** 1 m **29.** 3 cm **31.** 4 cm by 7 cm
33. 12 **35.** 841π sq m

Chapter Summary (page 349)

1. $5 \cdot 7$ **2.** $3^2 \cdot 5$ **3.** $2^5 \cdot 3$ **4.** $2 \cdot 3 \cdot 17$ **5.** $3 \cdot 29$
6. $3^2 \cdot 11$ **7.** $2 \cdot 5^2 \cdot 41$ **8.** 2^{12} **9.** $3(x + 3y)$
10. $5a(x^2 + 3)$ **11.** $7x(x + 2)$ **12.** $3x(x - 1)$
13. $2x(x^2 + 2x - 4)$ **14.** $a(x + y - z)$ **15.** $a(x + y - 1)$
16. $xyz(x + y)$ **17.** $(x + y)(a + b)$
18. $(x + y)(x + y + 1)$ **19.** $2x(x + 2)(x + 3)$
20. $3x(y + z)(1 - 3y - 3z)$ **21.** $(p + 3q)(3 + a)$
22. $(r - 2s)(a + 7)$ **23.** $(x + a)(x + b)$
24. $(y + 2)(x - 2)$ **25.** $(x + y)(a + b)$
26. $(x + 3)(x - 3)$ **27.** $(xy + 4)(xy - 4)$
28. $(x + 2 + y)(x + 2 - y)$ **29.** $(z + x + y)(z - x - y)$
30. $6y(x + 2y)(x - 2y)$ **31.** $(x + y + z)(x + y - z)$
32. $(x + 3)(x + 7)$ **33.** $(x - 3)(x + 7)$
34. $(x + 6)(x - 4)$ **35.** $(x - 6)(x + 2)$
36. $(2x + 1)(x - 3)$ **37.** $(3x + 1)(x - 5)$
38. $(2x + 3)(3x - 1)$ **39.** $3(2x - 1)(x + 1)$
40. $x(x + 3)(6x - 1)$ **41.** $x(4x + 3)(x - 2)$

42. $(c - 3)(c^2 + 3c + 9)$ **43.** $(d + 2)(d^2 - 2d + 4)$
44. $2(x + 3)(x^2 - 3x + 9)$ **45.** $2ab(b - 1)(b^2 + b + 1)$
46. $y(3x - y)(x - 2)$ **47.** $5(x + 2)(x - 3y)$
48. $a(a + b)(2x + a)$ **49.** $(x + a + y)(x + a - y)$
50. $(x + 2)(x - 2 + b)$
51. $a(x + y)(x^2 - xy + y^2)(x - y)(x^2 + xy + y^2)$
52. $0, -2$ **53.** $0, 3$ **54.** $0, \frac{2}{3}$ **55.** $0, -5$ **56.** $3, -3$
57. $5, -5$ **58.** $3, 4$ **59.** $5, -3$ **60.** $-4, 6$ **61.** $2, 8$
62. $3, -\frac{1}{2}$ **63.** $1, -\frac{3}{2}$ **64.** $\frac{1}{2}, -\frac{1}{2}$ **65.** $\frac{2}{3}, -\frac{2}{3}$ **66.** $0, 3, 4$
67. $0, -2, -3$ **68.** $0, \frac{1}{2}, -3$ **69.** $0, -\frac{2}{3}, 1$ **70.** 5 and 7
71. $\frac{1}{3}$ **72.** 6 ft by 8 ft **73.** 3 ft by 9 ft **74.** 3 ft by 6 ft

Chapter 5 Test (page 352)

1. $2^2 \cdot 7^2$ **2.** $3 \cdot 37$ **3.** $5a(12b^2c^3 + 6a^2b^2c - 5)$
4. $3x(a + b)(x - 2y)$ **5.** $(x + y)(a + b)$
6. $(x + 5)(x - 5)$ **7.** $3(a + 3b)(a - 3b)$
8. $(4x^2 + 9y^2)(2x + 3y)(2x - 3y)$ **9.** $(x + 3)(x + 1)$
10. $(x - 11)(x + 2)$ **11.** $(x + 9y)(x + y)$
12. $6(x - 4y)(x - y)$ **13.** $(3x + 1)(x + 4)$
14. $(2a - 3)(a + 4)$ **15.** $(2x - y)(x + 2y)$
16. $(4x - 3)(3x - 4)$ **17.** $6(2a - 3b)(a + 2b)$
18. $(x - 4)(x^2 + 4x + 16)$ **19.** $8(3 + a)(9 - 3a + a^2)$
20. $z^3(x^3 - yz)(x^6 + x^3yz + y^2z^2)$ **21.** $0, -3$
22. $-1, -\frac{3}{2}$ **23.** $3, -3$ **24.** $3, -6$ **25.** $\frac{9}{5}, -\frac{1}{2}$
26. $-\frac{9}{10}, 1$ **27.** $\frac{1}{5}, -\frac{9}{2}$ **28.** $-\frac{1}{10}, 9$ **29.** 12 sec **30.** 10 m

Getting Ready (page 355)

1. $\frac{1}{2}$ **2.** $\frac{2}{3}$ **3.** $-\frac{4}{5}$ **4.** $-\frac{5}{9}$

Orals (page 359)

1. $\frac{5}{7}$ **2.** $\frac{50}{1}$ **3.** $\frac{1}{3}$ **4.** $\frac{7}{10}$

Exercise 6.1 (page 359)

1. 17 **3.** 6 **5.** $2(x + 3)$ **7.** $(2x + 3)(x - 2)$
9. quotient **11.** equal **15.** $\frac{5}{7}$ **17.** $\frac{1}{2}$ **19.** $\frac{2}{3}$ **21.** $\frac{2}{7}$
23. $\frac{1}{3}$ **25.** $\frac{1}{5}$ **27.** $\frac{3}{7}$ **29.** $\frac{3}{4}$ **31.** \$1,825 **33.** $\frac{22}{365}$
35. \$8,725 **37.** $\frac{336}{1,745}$ **39.** $\frac{1}{16}$ **41.** $\frac{\$30.09}{17 \text{ gal}}$; \$1.77/gal
43. 7¢/oz **45.** the 6-oz can **47.** the first student
49. $\frac{11,880 \text{ gal}}{27 \text{ min}}$; 440 gal/min **51.** 5% **53.** 65 mph
55. the second car

Getting Ready (page 361)

1. 10 **2.** $\frac{7}{3}$ **3.** $\frac{20}{7}$ **4.** 5 **5.** $\frac{14}{3}$ **6.** 5 **7.** 21 **8.** $\frac{3}{2}$

Orals (page 369)

1. proportion **2.** not a proportion **3.** not a proportion
4. proportion

Exercise 6.2 (page 369)

1. 90% **3.** $\frac{1}{3}$ **5.** 480 **7.** \$73.50 **9.** proportion, ratios
11. means **13.** shape **15.** ad, bc **17.** triangle
19. no **21.** yes **23.** no **25.** yes **27.** 4 **29.** 6
31. -3 **33.** 9 **35.** 0 **37.** -17 **39.** $-\frac{3}{2}$ **41.** $\frac{83}{2}$
43. -6 **45.** \$17 **47.** \$6.50 **49.** 24 **51.** about $4\frac{1}{4}$

53. 47 **55.** $7\frac{1}{2}$ gal **57.** \$309 **59.** 49 ft $3\frac{1}{2}$ in. **61.** 162
63. not exactly, but close **65.** 39 ft **67.** $46\frac{7}{8}$ ft
69. 6,750 ft **71.** 15,840 ft

Getting Ready (page 372)

1. $\frac{3}{4}$ **2.** 2 **3.** $\frac{5}{11}$ **4.** $\frac{1}{2}$

Orals (page 378)

1. $\frac{2}{3}$ **2.** 2 **3.** $\frac{z}{w}$ **4.** $2x$ **5.** $\frac{x}{y}$ **6.** $\frac{1}{y}$ **7.** 1 **8.** -1

Exercise 6.3 (page 378)

1. $(a + b) + c = a + (b + c)$ **3.** 0 **5.** $\frac{5}{3}$
7. numerator **9.** 0 **11.** negatives **13.** $\frac{a}{b}$
15. factor, common **17.** 2 **19.** $2, -1$ **21.** $\frac{3}{2}, -1$
23. $\frac{4}{5}$ **25.** $\frac{4}{5}$ **27.** $\frac{2}{13}$ **29.** $\frac{2}{9}$ **31.** $-\frac{1}{3}$ **33.** $2x$
35. $-\frac{x}{3}$ **37.** $\frac{5}{a}$ **39.** $\frac{2}{z}$ **41.** $\frac{a}{3}$ **43.** $\frac{2}{3}$ **45.** $\frac{3}{2}$
47. in lowest terms **49.** $\frac{3x}{y}$ **51.** $\frac{7x}{8y}$ **53.** $\frac{1}{3}$ **55.** 5
57. $\frac{x}{2}$ **59.** $\frac{3x}{5y}$ **61.** $\frac{2}{3}$ **63.** -1 **65.** -1 **67.** -1
69. $\frac{x + 1}{x - 1}$ **71.** $\frac{x - 5}{x + 2}$ **73.** $\frac{2x}{x - 2}$ **75.** $\frac{x}{y}$ **77.** $\frac{x + 2}{x^2}$
79. $\frac{x - 4}{x + 4}$ **81.** $\frac{2(x + 2)}{x - 1}$ **83.** in lowest terms
85. $\frac{3 - x}{3 + x}$ or $-\frac{x - 3}{x + 3}$ **87.** $\frac{4}{3}$ **89.** $x + 3$ **91.** $x + 1$
93. $a - 2$ **95.** $\frac{b + 2}{b + 1}$ **97.** $\frac{y + 3}{x - 3}$

Getting Ready (page 380)

1. $\frac{2}{3}$ **2.** $\frac{14}{3}$ **3.** 3 **4.** 6 **5.** $\frac{5}{2}$ **6.** 1 **7.** $\frac{3}{4}$ **8.** 2

Orals (page 386)

1. $\frac{3}{2}$ **2.** $\frac{7}{5}$ **3.** 5 **4.** 1 **5.** $\frac{1}{4}$ **6.** x

Exercise 6.4 (page 386)

1. $-6x^5y^6z$ **3.** $\frac{1}{81y^4}$ **5.** $\frac{1}{x^m}$ **7.** $4y^3 + 4y^2 - 8y + 32$
9. numerator **11.** numerators, denominators **13.** 1
15. divisor, multiply **17.** $\frac{45}{91}$ **19.** $-\frac{3}{11}$ **21.** $\frac{5}{7}$ **23.** $\frac{3x}{2}$
25. $\frac{yx}{z}$ **27.** $\frac{14}{9}$ **29.** x^2y^2 **31.** $2xy^2$ **33.** $-3y^2$ **35.** $\frac{b^3c}{a^4}$
37. $\frac{r^3t^4}{s}$ **39.** $\frac{(z + 7)(z + 2)}{7z}$ **41.** x **43.** $\frac{x}{5}$ **45.** $x + 2$
47. $\frac{3}{2x}$ **49.** $x - 2$ **51.** x **53.** $\frac{(x - 2)^2}{x}$ **55.** $\frac{(m - 2)(m - 3)}{2(m + 2)}$
57. 1 **59.** $\frac{x + y}{x(x - y)}$ **61.** $\frac{-(x - y)(x^2 + xy + y^2)}{a + b}$ **63.** $\frac{c^2}{ab}$
65. $\frac{x + 1}{2(x - 2)}$ **67.** $\frac{2}{3}$ **69.** $\frac{3}{5}$ **71.** $\frac{3}{2y}$ **73.** 3 **75.** $\frac{6}{y}$
77. 6 **79.** $\frac{2x}{3}$ **81.** $\frac{2}{y}$ **83.** $\frac{2}{3x}$ **85.** $\frac{2(z - 2)}{z}$ **87.** $\frac{5z(z - 7)}{z + 2}$
89. $\frac{x + 2}{3}$ **91.** 1 **93.** $\frac{x - 2}{x - 3}$ **95.** $x + 5$ **97.** $a + 2$
99. $-\frac{p}{m + n}$ **101.** $\frac{9}{2x}$ **103.** $\frac{x}{36}$ **105.** $\frac{(x + 1)(x - 1)}{5(x - 3)}$
107. 2 **109.** $\frac{2x(1 - x)}{5(x - 2)}$ **111.** $\frac{y^2}{3}$ **113.** $\frac{x + 2}{x - 2}$

Getting Ready (page 389)

1. $\frac{4}{5}$ **2.** 1 **3.** $\frac{7}{8}$ **4.** 2 **5.** $\frac{1}{9}$ **6.** $\frac{1}{2}$ **7.** $-\frac{2}{13}$ **8.** $\frac{13}{10}$

Orals (page 398)

1. equal **2.** equal **3.** not equal **4.** equal **5.** equal
6. not equal **7.** equal **8.** equal

Exercise 6.5 (page 398)

1. 7^2 **3.** $2^3 \cdot 17$ **5.** $2 \cdot 3 \cdot 17$ **7.** $2^4 \cdot 3^2$ **9.** LCD **11.** numerators, common denominator **13.** $\frac{2}{3}$ **15.** $\frac{1}{3}$ **17.** $\frac{4x}{y}$ **19.** $\frac{2}{y}$ **21.** $\frac{2y + 6}{5z}$ **23.** 9 **25.** $\frac{1}{7}$ **27.** $-\frac{1}{8}$ **29.** $\frac{x}{y}$ **31.** $\frac{y}{x}$ **33.** $\frac{1}{y}$ **35.** 1 **37.** $\frac{4x}{3}$ **39.** $\frac{2x}{3y}$ **41.** $\frac{2(2x - y)}{y + 2}$ **43.** $\frac{2(x + 5)}{x - 2}$ **45.** $\frac{125}{20}$ **47.** $\frac{8xy}{x^2y}$ **49.** $\frac{3x(x + 1)}{(x + 1)^2}$ **51.** $\frac{2y(x + 1)}{x^2 + x}$ **53.** $\frac{z(z + 1)}{z^2 - 1}$ **55.** $\frac{2(x + 2)}{x^2 + 3x + 2}$ **57.** $6x$ **59.** $18xy$ **61.** $x^2 - 1$ **63.** $x^2 + 6x$ **65.** $(x + 1)(x + 5)(x - 5)$ **67.** $\frac{7}{6}$ **69.** $\frac{5y}{9}$ **71.** $\frac{53x}{42}$ **73.** $\frac{4xy + 6x}{3y}$ **75.** $\frac{2 - 3x^2}{x}$ **77.** $\frac{4y + 10}{15y}$ **79.** $\frac{x^2 + 4x + 1}{x^2y}$ **81.** $\frac{2x^2 - 1}{x(x + 1)}$ **83.** $\frac{2xy + x - y}{xy}$ **85.** $\frac{x + 2}{x - 2}$ **87.** $\frac{2x^2 + 2}{(x - 1)(x + 1)}$ **89.** $\frac{2(2x + 1)}{x - 2}$ **91.** $\frac{x}{x - 2}$ **93.** $\frac{5x + 3}{x + 1}$ **95.** $-\frac{1}{2(x - 2)}$

Getting Ready (page 400)

1. 4 **2.** −18 **3.** 7 **4.** −8 **5.** $3 + 3x$ **6.** $2 - y$ **7.** $12x - 2$ **8.** $3y + 2x$

Orals (page 405)

1. $\frac{4}{3}$ **2.** 4 **3.** $\frac{1}{4}$ **4.** 3

Exercise 6.6 (page 405)

1. t^9 **3.** $-2r^7$ **5.** $\frac{81}{256r^8}$ **7.** $\frac{r^{10}}{9}$ **9.** complex fraction **11.** single, divide **13.** $\frac{8}{9}$ **15.** $\frac{3}{8}$ **17.** $\frac{5}{4}$ **19.** $\frac{5}{7}$ **21.** $\frac{x^2}{y}$ **23.** $\frac{5t^2}{27}$ **25.** $\frac{1 - 3x}{5 + 2x}$ **27.** $\frac{1 + x}{2 + x}$ **29.** $\frac{3 - x}{x - 1}$ **31.** $\frac{1}{x + 2}$ **33.** $\frac{1}{x + 3}$ **35.** $\frac{xy}{y + x}$ **37.** $\frac{y}{x - 2y}$ **39.** $\frac{x^2}{(x - 1)^2}$ **41.** $\frac{7x + 3}{-x - 3}$ **43.** $\frac{x - 2}{x + 3}$ **45.** −1 **47.** $\frac{y}{x^2}$ **49.** $\frac{x + 1}{1 - x}$ **51.** $\frac{a^2 - a + 1}{a^2}$ **53.** 2 **55.** $\frac{y - 5}{y + 5}$ **59.** $\frac{1}{2}, \frac{2}{3}, \frac{3}{5}, \frac{5}{8}$

Getting Ready (page 407)

1. $3x + 1$ **2.** $8x - 1$ **3.** $3 + 2x$ **4.** $y - 6$ **5.** 19 **6.** $7x + 6$ **7.** y **8.** $3x + 5$

Orals (page 412)

1. Multiply by 10. **2.** Multiply by $x(x - 1)$. **3.** Multiply by 9. **4.** Multiply by 15.

Exercise 6.7 (page 412)

1. $x(x + 4)$ **3.** $(2x + 3)(x - 1)$ **5.** $(x^2 + 4)(x + 2)(x - 2)$ **7.** extraneous **9.** LCD **11.** xy **13.** 4 **15.** −20 **17.** 6 **19.** 60 **21.** −12 **23.** 0 **25.** −7 **27.** −1 **29.** 12 **31.** 0 **33.** −3 **35.** 3 **37.** 0 is extraneous **39.** 1 **41.** 5 **43.** −2 is extraneous **45.** 5 is extraneous **47.** −1 **49.** 6 **51.** 2 **53.** −3 **55.** 1 **57.** −2 is extraneous **59.** 1, 2 **61.** 3; −3 is extraneous **63.** 3, −4 **65.** 1 **67.** 0 **69.** −2, 1 **71.** $a = \frac{b}{b - 1}$ **73.** $f = \frac{d_1d_2}{d_1 + d_2}$ **77.** 1

Getting Ready (page 414)

1. $\frac{1}{5}$ **2.** $\$(0.05x)$ **3.** $\$\left(\frac{y}{0.05}\right)$ **4.** $\frac{y}{52}$ hr

Orals (page 417)

1. $i = pr$ **2.** $d = rt$ **3.** $C = qd$

Exercise 6.8 (page 417)

1. −1, 6 **3.** −2, −3, −4 **5.** 0, 0, 1 **7.** 1, −1, 2, −2 **11.** 2 **13.** 5 **15.** $\frac{2}{3}, \frac{3}{2}$ **17.** $2\frac{2}{9}$ hr **19.** $2\frac{6}{11}$ days **21.** $7\frac{1}{2}$ hr **23.** 4 mph **25.** 7% and 8% **27.** 5 **29.** 30 **31.** 25 mph

Chapter Summary (page 420)

1. $\frac{1}{2}$ **2.** $\frac{4}{5}$ **3.** $\frac{2}{3}$ **4.** $\frac{5}{6}$ **5.** \$2.93 **6.** 568.75 kwh per week **7.** no **8.** yes **9.** $\frac{9}{2}$ **10.** 0 **11.** 7 **12.** 1 **13.** 20 ft **14.** 3, −3 **15.** $\frac{2}{5}$ **16.** $-\frac{2}{3}$ **17.** $-\frac{1}{3}$ **18.** $\frac{7}{3}$ **19.** $\frac{1}{2x}$ **20.** $\frac{5}{2x}$ **21.** $\frac{x}{x + 1}$ **22.** $\frac{1}{x}$ **23.** 2 **24.** 1 **25.** −1 **26.** $\frac{x + 7}{x + 3}$ **27.** $\frac{x}{x - 1}$ **28.** $\frac{a + 2}{a + b}$ **29.** $\frac{3x}{y}$ **30.** $\frac{6}{x^2}$ **31.** 1 **32.** $\frac{2x}{x + 1}$ **33.** $\frac{3y}{2}$ **34.** $\frac{1}{x}$ **35.** $x + 2$ **36.** 1 **37.** $x + 2$ **38.** 1 **39.** $\frac{2(x + 1)}{x - 7}$ **40.** $\frac{x^2 + x - 1}{x(x - 1)}$ **41.** $\frac{x - 7}{7x}$ **42.** $\frac{x - 2}{x(x + 1)}$ **43.** $\frac{x^2 + 4x - 4}{2x^2}$ **44.** $\frac{x + 1}{x}$ **45.** 0 **46.** $\frac{9}{4}$ **47.** $\frac{3}{2}$ **48.** $\frac{1 + x}{1 - x}$ **49.** $\frac{x(x + 3)}{2x^2 - 1}$ **50.** $x^2 + 3$ **51.** $\frac{a(a + bc)}{b(b + ac)}$ **52.** 3 **53.** 1 **54.** 3 **55.** 4, $-\frac{3}{2}$ **56.** −2 **57.** 0 **58.** $r_1 = \frac{rr_2}{r_2 - r}$ **59.** $T_1 = \frac{T_2}{1 - E}$ or $T_1 = \frac{-T_2}{E - 1}$ **60.** $R = \frac{HB}{B - H}$ or $R = \frac{-HB}{H - B}$ **61.** $9\frac{9}{19}$ hr **62.** $5\frac{5}{6}$ days **63.** 5 mph **64.** 40 mph

Chapter 6 Test (page 424)

1. $\frac{2}{3}$ **2.** yes **3.** $\frac{2}{3}$ **4.** 45 ft **5.** $\frac{8x}{9y}$ **6.** $\frac{x + 1}{2x + 3}$ **7.** 3 **8.** $\frac{5y^2}{4t}$ **9.** $\frac{x + 1}{3(x - 2)}$ **10.** $\frac{3t^2}{5y}$ **11.** $\frac{x^2}{3}$ **12.** $x + 2$ **13.** $\frac{10x - 1}{x - 1}$ **14.** $\frac{13}{2y + 3}$ **15.** $\frac{2x^2 + x + 1}{x(x + 1)}$ **16.** $\frac{2x + 6}{x - 2}$ **17.** $\frac{2x^3}{y^3}$ **18.** $\frac{x + y}{y - x}$ **19.** −5 **20.** 6 **21.** 4 **22.** $B = \frac{RH}{R - H}$ **23.** $3\frac{15}{16}$ hr **24.** 5 mph **25.** 8,050 ft

Cumulative Review Exercises (page 425)

1. x^7 **2.** x^{10} **3.** x^3 **4.** 1 **5.** $6x^3 - 2x - 1$ **6.** $2x^3 + 2x^2 + x - 1$ **7.** $13x^2 - 8x + 1$ **8.** $16x^2 - 24x + 2$ **9.** $-12x^5y^5$ **10.** $-35x^5 + 10x^4 + 10x^2$ **11.** $6x^2 + 14x + 4$

12. $15x^2 - 2xy - 8y^2$ **13.** $x + 4$ **14.** $x^2 + x + 1$
15. $3xy(x - 2y)$ **16.** $(a + b)(3 + x)$ **17.** $(a + b)(2 + b)$
18. $(5p^2 + 4q)(5p^2 - 4q)$ **19.** $(x - 12)(x + 1)$
20. $(x - 3y)(x + 2y)$ **21.** $(3a + 4)(2a - 5)$
22. $(4m + n)(2m - 3n)$ **23.** $(p - 3q)(p^2 + 3pq + 9q^2)$
24. $8(r + 2s)(r^2 - 2rs + 4s^2)$ **25.** 15 **26.** 4
27. $\frac{2}{3}, -\frac{1}{2}$ **28.** 0, 2 **29.** −1, −2 **30.** $\frac{3}{2}, -4$

31. 

32. 

33. 

34. 

35. 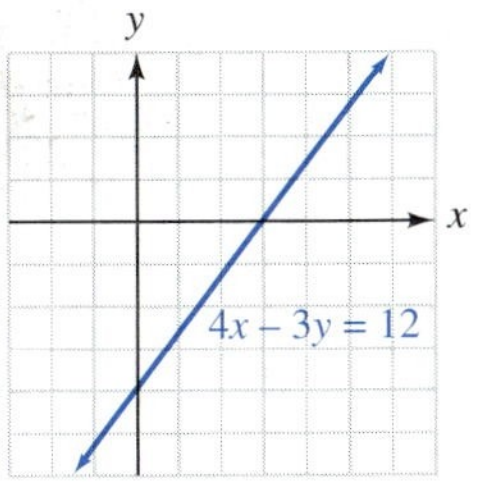

36. 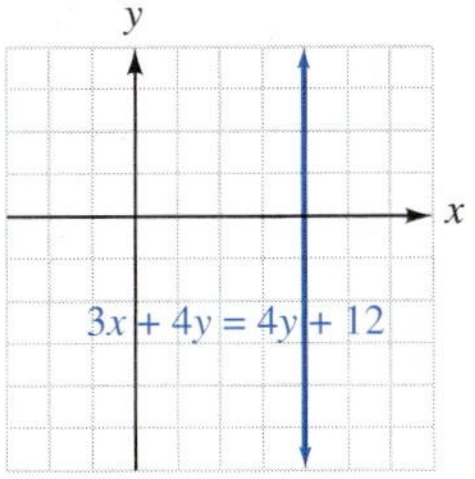

37. −3 **38.** 15 **39.** 5 **40.** $8x^2 - 3$ **41.** $\frac{x + 1}{x - 1}$
42. $\frac{x - 3}{x - 2}$ **43.** $\frac{(x - 2)^2}{x - 1}$ **44.** $\frac{(p + 2)(p - 3)}{3(p + 3)}$
45. 1 **46.** $\frac{2(x^2 + 1)}{(x + 1)(x - 1)}$ **47.** $\frac{-1}{2(a - 2)}$ **48.** $\frac{y + x}{y - x}$

Getting Ready (page 428)

1. 2 **2.** 8 **3.** 10 **4.** −7

Orals (page 437)

1. 1 **2.** 4 **3.** $x < 2$ **4.** $x \geq 3$ **5.** $x < -4$
6. $x \geq -8$

Exercise 7.1 (page 438)

1. $\frac{1}{t^{12}}$ **3.** 471 or more **5.** equation
7. multiplied, divided **9.** contradiction **11.** is greater than
13. half-open **15.** positive **17.** 6 **19.** 4 **21.** 2
23. 6 **25.** −4 **27.** −6 **29.** −11 **31.** 24 **33.** 6
35. 0 **37.** identity **39.** −6 **41.** contradiction
43. identity **45.** $B = \frac{3V}{h}$ **47.** $w = \frac{P - 2l}{2}$
49. $x = z\sigma + \mu$ **51.** $x = \frac{y - b}{m}$ **53.** $s = \frac{f(P - L)}{i}$
55. $(2, \infty)$ 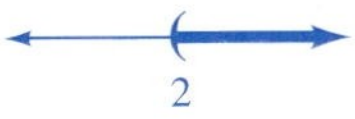

57. $[-2, \infty)$ 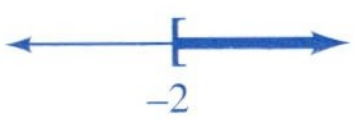

59. $(-\infty, 1]$ 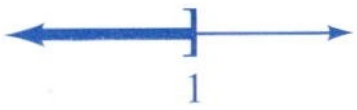

61. $\left(-\infty, -\frac{8}{5}\right)$

63. $[-36, \infty)$ 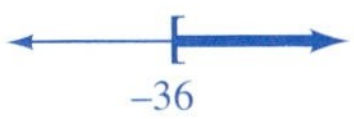

65. $(-2, 5)$ 

67. $(8, 11)$ 

69. $[-4, 6)$ 

71. no solution **73.** $[-2, 4]$ 

75. $(-\infty, 2) \cup (7, \infty)$ 

77. $(-\infty, 1)$ 

79. no solution
81. 7 ft, 15 ft
83. 12 m by 24 m
85. 20 ft by 45 ft **87.** $p < \$15$ **89.** 18

Getting Ready (page 441)

1. 6 **2.** 5 **3.** $2 - x$ **4.** $\pi - 2$

Orals (page 445)

1. 5 **2.** −5 **3.** −6 **4.** −4 **5.** 8 or −8
6. no solution **7.** 5 **8.** 0, −2

Exercise 7.2 (page 445)

1. $\frac{3}{4}$ **3.** 6 **5.** x **7.** 0 **9.** $a = -b$ **11.** 8 **13.** 12
15. −2 **17.** −30 **19.** 50 **21.** $4 - \pi$ **23.** $|2|$
25. $|5|$ **27.** $|-2|$ **29.** $-|-4|$ **31.** $-|-7|$ **33.** $-x$
35. 8, −8 **37.** 9, −3 **39.** 4, −1 **41.** $\frac{14}{3}, -6$
43. no solution **45.** 8, −4 **47.** $2, -\frac{1}{2}$ **49.** −8
51. −4, −28 **53.** 0, −6 **55.** $\frac{20}{3}$ **57.** $-2, -\frac{4}{5}$
59. 3, −1 **61.** 0, −2 **63.** 0 **65.** $\frac{4}{3}$ **67.** no solution

Getting Ready (page 446)

1. $x > 1$ **2.** $x > -2$ **3.** $x \leq 7$

Orals (page 449)

1. $-8 < x < 8$ **2.** $x < -8$ or $x > 8$ **3.** $x \leq -4$ or $x \geq 4$
4. $-7 \leq x \leq 7$ **5.** $-3 < x < 1$ **6.** $x < -3$ or $x > 1$

Exercise 7.3 (page 450)

1. $t = \frac{A - p}{pr}$ **3.** $l = \frac{P - 2w}{2}$ **5.** $-k < x < k$
7. $x < -k$ or $x > k$ **9.** $(-4, 4)$ 

11. $[-21, 3]$ 

13. no solution
15. $\left[-\frac{3}{2}, 2\right]$ 

17. $(-2, 5)$ 

19. $(-\infty, -1) \cup (1, \infty)$ 

21. $(-\infty, -12) \cup (36, \infty)$ 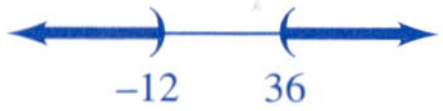

23. $\left(-\infty, -\frac{16}{3}\right) \cup (4, \infty)$ 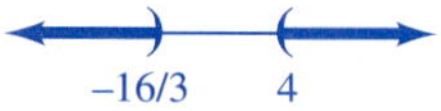

25. $(-\infty, \infty)$ 

27. $(-\infty, -2] \cup \left[\frac{10}{3}, \infty\right)$ 

29. $(-\infty, -2) \cup (5, \infty)$ 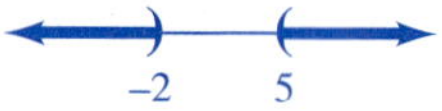

31. $\left(-\infty, \frac{3}{8}\right) \cup \left(\frac{3}{8}, \infty\right)$ 

33. $[-10, 14]$ 

35. $\left(-\frac{5}{3}, 1\right)$ 

37. $(-\infty, -4] \cup [-1, \infty)$ 

39. no solution

41. $(-\infty, -24) \cup (-18, \infty)$ 

43. $(-\infty, 25) \cup (25, \infty)$ 

45. $\left(-\frac{65}{9}, -\frac{5}{9}\right)$ 

47. $(-\infty, -4) \cup (-4, \infty)$ 

49. $[-7, -7]$  **51.** $[5, 5]$ 

Getting Ready (page 451)

1. $6x^3y - 3x^2y^2$ **2.** $x^2 - 4$ **3.** $x^2 - x - 6$ **4.** $6x^2 + 7x - 3$ **5.** $x^3 - 27$ **6.** $x^3 + 8$

Orals (page 461)

1. $3xy(y - 2x)$ **2.** $(x + 1)(x - 1)$ **3.** $(x + 6)(x - 1)$ **4.** $(2x + 1)(x - 1)$ **5.** $(a + 2)(a^2 - 2a + 4)$ **6.** $(b - 3)(b^2 + 3b + 9)$

Exercise 7.4 (page 461)

1. 1,100 ft/sec **3.** 12 **5.** $ab + ac$ **7.** perfect **9.** $(x^2 - xy + y^2)$ **11.** $2(x + 4)$ **13.** $2x(x - 3)$ **15.** $5x^2y(3 - 2y)$ **17.** $13ab^2c(c^2 - 2a^2)$ **19.** $3z(9z^2 + 4z + 1)$ **21.** $6s(4s^2 - 2st + t^2)$ **23.** $-3(a + 2)$ **25.** $-3x(2x + y)$ **27.** $-7u^2v^3z^2(9uv^3z^7 - 4v^4 + 3uz^2)$ **29.** $x^2(x^n + x^{n+1})$

$y^n(2y^2 - 3y^3)$ **33.** $(x + y)(a + b)$ **35.** $(x + 2)(x + y)$

$- c)(c + d)$ **39.** $2(2 + x)(y + 2)$ **41.** $r_1 = \dfrac{rr_2}{r_2 - r}$

43. $r = \dfrac{S - a}{S - l}$ **45.** $(x + 2)(x - 2)$ **47.** $(3y + 8)(3y - 8)$ **49.** $(9a^2 + 7b)(9a^2 - 7b)$ **51.** $(x + y + z)(x + y - z)$ **53.** $(x^2 + y^2)(x + y)(x - y)$ **55.** $2(x + 12)(x - 12)$ **57.** $2x(x + 4)(x - 4)$ **59.** $(x^m + y^{2n})(x^m - y^{2n})$ **61.** $(a + b)(a - b + 1)$ **63.** $(2x + y)(1 + 2x - y)$ **65.** $(x + 3)(x + 2)$ **67.** prime **69.** $(2x - 1)(x - 5)$ **71.** $(3y + 2)(2y + 1)$ **73.** $(4a - 3)(2a + 3)$ **75.** $(a + b)(a - 4b)$ **77.** $(2y - 3t)(y + 2t)$ **79.** prime **81.** $(4x - 3)(2x - 1)$ **83.** $3(x + 7)(x - 3)$ **85.** $x(3x - 1)(x - 3)$ **87.** $b^2(a - 11)(a - 2)$ **89.** $-(a - 8)(a + 4)$ **91.** $-3(x - 3)(x - 2)$ **93.** $-(3a + 2b)(a - b)$ **95.** $-x(2x - 3)^2$ **97.** $(x^2 + 5)(x^2 + 3)$ **99.** $(y^2 - 10)(y^2 - 3)$ **101.** $(a + 3)(a - 3)(a + 2)(a - 2)$ **103.** $(x^n + 1)^2$ **105.** $(2a^{3n} + 1)(a^{3n} - 2)$ **107.** $(x^{2n} + y^{2n})^2$ **109.** $(3x^n - 1)(2x^n + 3)$ **111.** $(x + 2 + y)(x + 2 - y)$ **113.** $(x + 1 + 3z)(x + 1 - 3z)$ **115.** $(c + 2a - b)(c - 2a + b)$ **117.** $(y + 1)(y^2 - y + 1)$ **119.** $(a - 3)(a^2 + 3a + 9)$ **121.** $(2 + x)(4 - 2x + x^2)$ **123.** $(s - t)(s^2 + st + t^2)$ **125.** $(3x + y)(9x^2 - 3xy + y^2)$ **127.** $(a + 2b)(a^2 - 2ab + 4b^2)$ **129.** $(3x - 5y)(9x^2 + 15xy + 25y^2)$ **131.** $(a^2 - b)(a^4 + a^2b + b^2)$ **133.** $(x^3 + y^2)(x^6 - x^3y^2 + y^4)$ **135.** $2(x + 3)(x^2 - 3x + 9)$ **137.** $-(x - 6)(x^2 + 6x + 36)$ **139.** $8x(2m - n)(4m^2 + 2mn + n^2)$ **141.** $xy(x + 6y)(x^2 - 6xy + 36y^2)$ **143.** $3rs^2(3r - 2s)(9r^2 + 6rs + 4s^2)$ **145.** $a^3b^2(5a + 4b)(25a^2 - 20ab + 16b^2)$ **147.** $2mp(p + 2q)(p^2 - 2pq + 4q^2)$ **149.** $(x + y)(x^2 - xy + y^2)(3 - z)$ **151.** $(m + 2n)(m^2 - 2mn + 4n^2)(1 + x)$ **153.** $(a + 3)(a^2 - 3a + 9)(a - b)$ **155.** $(y + 1)(y - 1)(y - 3)(y^2 + 3y + 9)$ **157.** $(x + 1)(x^2 - x + 1)(x - 1)(x^2 + x + 1)$ **159.** $(x^2 + y)(x^4 - x^2y + y^2)(x^2 - y)(x^4 + x^2y + y^2)$

Getting Ready (page 464)

1. $\frac{5}{3}$ **2.** $\frac{4}{15}$ **3.** $\frac{19}{6}$ **4.** $-\frac{11}{6}$

Orals (page 471)

1. $\frac{2}{3}$ **2.** $\frac{2}{3}$ **3.** $-\frac{5}{6}$ **4.** $-\frac{2}{5}$ **5.** $\frac{x}{y}$ **6.** 2 **7.** -1 **8.** $x - 1$

Exercise 7.5 (page 475)

1.  **3.** $w = \dfrac{P - 2l}{2}$ **5.** $2, -2, 3, -3$ **7.** $\dfrac{a}{b}$ **9.** $\dfrac{ad}{bc}$ **11.** $4x^2$ **13.** $-\dfrac{4y}{3x}$ **15.** $\dfrac{3y}{7(y - z)}$ **17.** 1 **19.** $\dfrac{1}{x - y}$ **21.** $\dfrac{-3(x + 2)}{x + 1}$ **23.** $x + 2$ **25.** $\dfrac{x + 1}{x + 3}$ **27.** $\dfrac{m - 2n}{n - 2m}$ **29.** $\dfrac{x + 4}{2(2x - 3)}$ **31.** $\dfrac{3(x - y)}{x + 2}$ **33.** $\dfrac{1}{x^2 + xy + y^2 - 1}$ **35.** $\dfrac{xy^2d}{c^3}$ **37.** $-\dfrac{x^{10}}{y^2}$ **39.** $x + 1$

41. 1 **43.** $\frac{x-4}{x+5}$ **45.** $\frac{(a+7)^2(a-5)}{12x^2}$ **47.** $\frac{t-1}{t+1}$ **49.** $\frac{n+2}{n+1}$ **51.** $x-5$ **53.** $\frac{x+y}{x-y}$ **55.** $-\frac{x+3}{x+2}$ **57.** $\frac{a+b}{(x-3)(c+d)}$ **59.** $-\frac{x+1}{x+3}$ **61.** $-\frac{2x+y}{3x+y}$ **63.** $-\frac{x^7}{18y^4}$ **65.** $x^2(x+3)$ **67.** $\frac{3x}{2}$ **69.** $\frac{x-7}{x+7}$ **71.** 1 **73.** $\frac{3-a}{a+b}$ **75.** 2 **77.** 3 **79.** $\frac{6x}{(x-3)(x-2)}$ **81.** $\frac{9a}{10}$ **83.** $\frac{17}{12x}$ **85.** $\frac{10a+4b}{21}$ **87.** $\frac{8x-2}{(x+2)(x-4)}$ **89.** $\frac{7x+29}{(x+5)(x+7)}$ **91.** $\frac{x^2+1}{x}$ **93.** 2 **95.** $\frac{2x^2+x}{(x+3)(x+2)(x-2)}$ **97.** $\frac{-4x^2+14x+54}{x(x+3)(x-3)}$ **99.** $\frac{x^2-5x-5}{x-5}$ **101.** $\frac{2}{x+1}$ **103.** $\frac{3x+1}{x(x+3)}$ **105.** $\frac{2y}{3z}$ **107.** $-\frac{1}{y}$ **109.** $\frac{b+a}{b}$ **111.** $y-x$ **113.** $\frac{-1}{a+b}$ **115.** $\frac{x+2}{x-3}$ **117.** $\frac{a-1}{a+1}$ **119.** $\frac{y+x}{y-x}$ **121.** xy **123.** $\frac{x^2(xy^2-1)}{y^2(x^2y-1)}$ **125.** $\frac{(b+a)(b-a)}{b(b-a-ab)}$ **127.** $\frac{3a^2+2a}{2a+1}$ **133.** yes **135.** a, d

Getting Ready (page 478)

1. $x+1$ with a remainder of 1, 1
2. $x+3$ with a remainder of 9, 9

Orals (page 483)

1. 9 **2.** -3 **3.** yes **4.** no

Exercise 7.6 (page 483)

1. 4 **3.** $12a^2+4a-1$ **5.** $8x^2+2x+4$ **7.** $P(r)$ **9.** $x+2$ **11.** $x-3$ **13.** $x+2$ **15.** $x-7+\frac{28}{x+2}$ **17.** $3x^2-x+2$ **19.** $2x^2+4x+3$ **21.** $6x^2-x+1+\frac{3}{x+1}$ **23.** $7.2x-0.66+\frac{0.368}{x-0.2}$ **25.** $2.7x-3.59+\frac{0.903}{x+1.7}$ **27.** $9x^2-513x+29{,}241+\frac{-1{,}666{,}762}{x+57}$ **29.** -1 **31.** -37 **33.** 23 **35.** -1 **37.** 2 **39.** -1 **41.** 18 **43.** 174 **45.** -8 **47.** 59 **49.** 44 **51.** $\frac{29}{32}$ **53.** yes **55.** no **57.** 64 **61.** 1

Chapter Summary (page 486)

1. 8 **2.** 7 **3.** 19 **4.** 8 **5.** identity **6.** contradiction **7.** $h=\frac{3V}{\pi r^2}$ **8.** $x=\frac{6V}{ab}-y$ **9.** 5 ft from one end **10.** 45 m^2 **11.** $(-\infty, -24]$

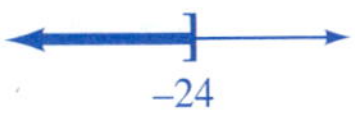

12. $\left(-\infty, -\frac{51}{11}\right)$

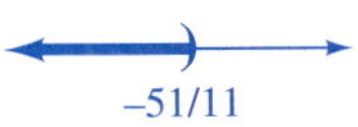

13. $\left(-\frac{1}{3}, 2\right)$

14. $(2, \infty)$

15. $3, -\frac{11}{3}$ **16.** $\frac{26}{3}, -\frac{10}{3}$ **17.** $\frac{1}{5}, -5$ **18.** $-1, 1$ **19.** $(-5, -2)$

20. $\left(-\infty, \frac{4}{3}\right] \cup [4, \infty)$

21. $(-\infty, \infty)$

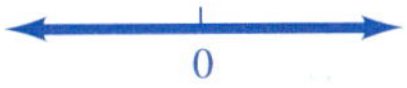

22. no solutions **23.** $4(x+2)$ **24.** $5xy^2(xy-2)$ **25.** $-4x^2y^3z^2(2z^2+3x^2)$ **26.** $3a^2b^4c^2(4a^4+5c^4)$ **27.** $(x+2)(y+4)$ **28.** $(a+b)(c+3)$ **29.** $x^n(x^n+1)$ **30.** $y^{2n}(1-y^n)$ **31.** $(x^2+4)(x^2+y)$ **32.** $(a^3+c)(a^2+b^2)$ **33.** $(z+4)(z-4)$ **34.** $(y+11)(y-11)$ **35.** $2(x^2+7)(x^2-7)$ **36.** $3x^2(x^2+10)(x^2-10)$ **37.** $(y+20)(y+1)$ **38.** $(z-5)(z-6)$ **39.** $-(x+7)(x-4)$ **40.** $-(y-8)(y+3)$ **41.** $y(y+2)(y-1)$ **42.** $2a^2(a+3)(a-1)$ **43.** $3(5x+y)(x-4y)$ **44.** $5(6x+y)(x+2y)$ **45.** $(x+2+2p^2)(x+2-2p^2)$ **46.** $(y+2)(y+1+x)$ **47.** $(x+7)(x^2-7x+49)$ **48.** $(a-5)(a^2+5a+25)$ **49.** $8(y-4)(y^2+4y+16)$ **50.** $4y(x+3z)(x^2-3xz+9z^2)$ **51.** $\frac{31x}{72y}$ **52.** $\frac{x-7}{x+7}$ **53.** 1 **54.** 1 **55.** $\frac{5y-3}{x-y}$ **56.** $\frac{6x-7}{x^2+2}$ **57.** $\frac{5x+13}{(x+2)(x+3)}$ **58.** $\frac{4x^2+9x+12}{(x-4)(x+3)}$ **59.** $\frac{3x(x-1)}{(x-3)(x+1)}$ **60.** 1 **61.** $\frac{5x^2+11x}{(x+1)(x+2)}$ **62.** $\frac{2(3x+1)}{x-3}$ **63.** $\frac{5x^2+23x+4}{(x+1)(x-1)(x-1)}$ **64.** $\frac{-x^4-4x^3+3x^2+18x+16}{(x-2)(x+2)^2}$ **65.** $\frac{3y-2x}{x^2y^2}$ **66.** $\frac{y+2x}{2y-x}$ **67.** $\frac{2x+1}{x+1}$ **68.** $\frac{y-x}{y+x}$ **69.** 20 **70.** -1 **71.** yes **72.** no

Chapter 7 Test (page 490)

1. -12 **2.** 6 **3.** $i=\frac{f(P-L)}{s}$ **4.** $a=\frac{180n-360}{n}$ **5.** $13\frac{1}{3}$ ft **6.** 36 cm^2 **7.** $(-\infty, -5]$

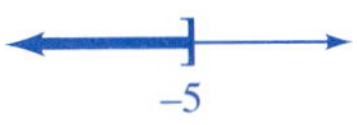

8. $(-2, 16)$

9. $4, -7$ **10.** $4, -4$ **11.** $[-7, 1]$

12. $(-\infty, -9) \cup (13, \infty)$

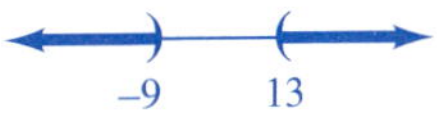

13. $3xy(y + 2x)$ **14.** $3abc(4a^2b - abc + 2c^2)$
15. $(a - y)(x + y)$ **16.** $(x + y)(a + b - c)$
17. $(x + 7)(x - 7)$ **18.** $2(x + 4)(x - 4)$
19. $4(y^2 + 4)(y + 2)(y - 2)$ **20.** $(b + 5)(b^2 - 5b + 25)$
21. $(b - 3)(b^2 + 3b + 9)$ **22.** $3(u - 2)(u^2 + 2u + 4)$
23. $(x + 5)(x + 3)$ **24.** $(3b + 2)(2b - 1)$
25. $3(u + 2)(2u - 1)$ **26.** $(x + 3 + y)(x + 3 - y)$
27. $\frac{-2}{3xy}$ **28.** $\frac{2x + 1}{4}$ **29.** $\frac{xz}{y^4}$ **30.** 1 **31.** $\frac{(x + y)^2}{2}$
32. $\frac{2x + 3}{(x + 1)(x + 2)}$ **33.** $\frac{u^2}{2vw}$ **34.** $\frac{2x + y}{xy - 2}$ **35.** -7
36. 47

Getting Ready (page 493)

1. 1 **2.** 9 **3.** 5 **4.** 2

Orals (page 499)

1. (3, 0), (0, 3) **2.** (2, 0), (0, 6) **3.** (8, 0), (0, 2)
4. (4, 0), (0, −3) **5.** (4, 6) **6.** (0, −1)

Exercise 8.1 (page 499)

1.

3. $x(x - 1)$

5. $(x - 1)(x^2 + x + 1)$ **7.** origin **9.** y-coordinate
11. x-axis **13.** horizontal

15–21.

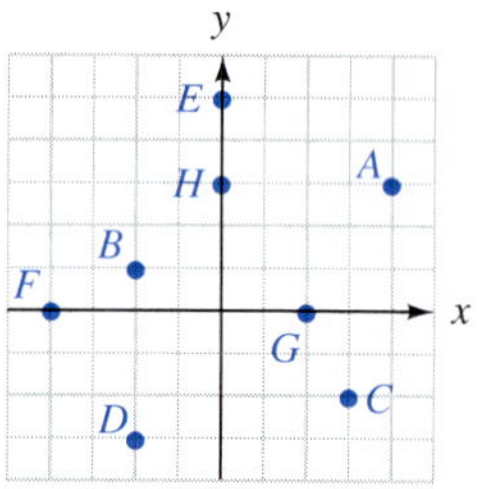

23. (2, 4)
25. (−2, −1)
27. (4, 0)
29. (0, 0)

31.

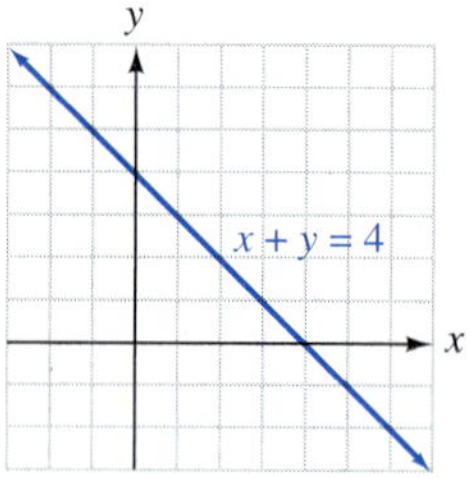

33.

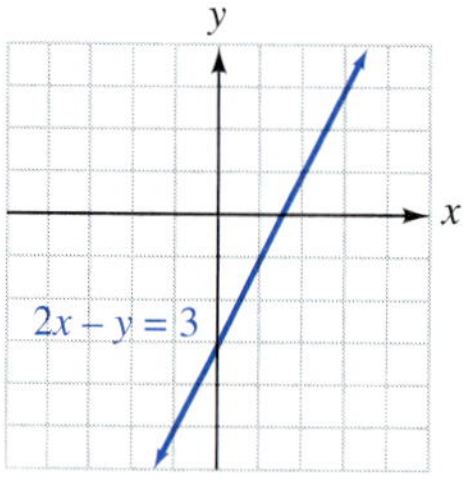

35.

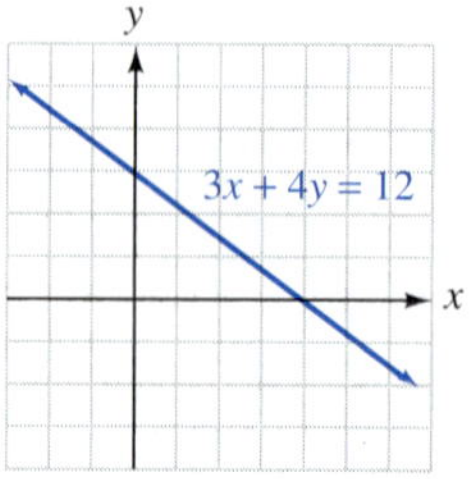

37.

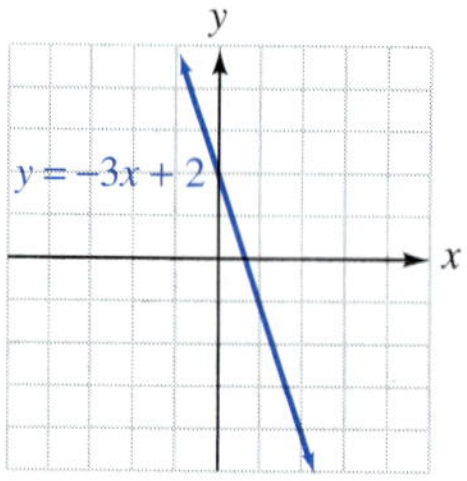

39.

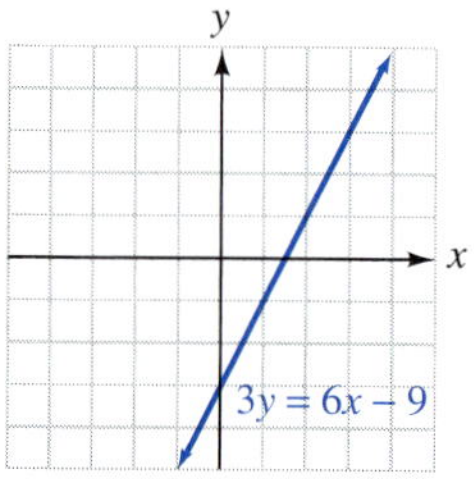

41.

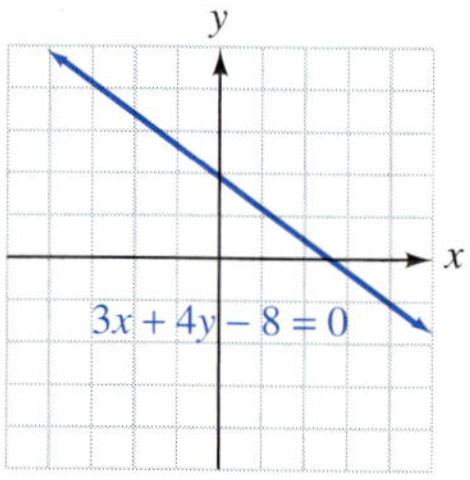

43.

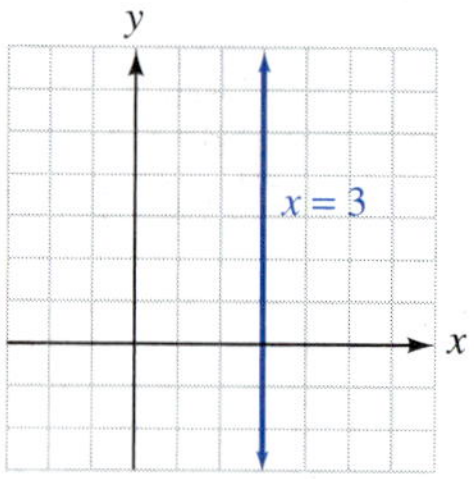

45.

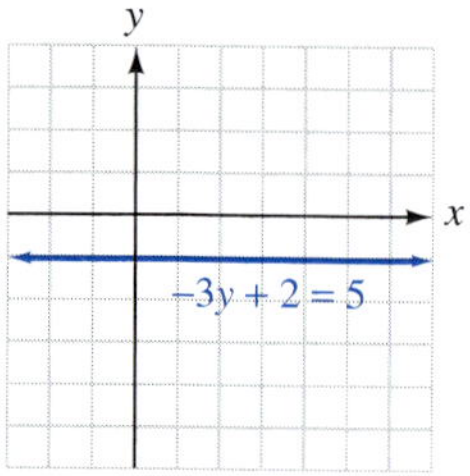

47. (3, 4) **49.** (9, 12) **51.** $\left(\frac{7}{2}, 6\right)$ **53.** $\left(\frac{1}{2}, -2\right)$
55. (−4, 0) **57.** $\left(\frac{5a}{2}, 2b\right)$ **59.** $(a, 2b)$ **61.** (4, 1)
63. \$162,500, \$200,000 **65.** 200 **67.** 100 rpm
71. $a = 0, b > 0$

Getting Ready (page 502)

1. 1 **2.** −1 **3.** $\frac{13}{14}$ **4.** $\frac{3}{14}$

Orals (page 509)

1. 3 **2.** 2 **3.** yes **4.** 5 **5.** yes

Exercise 8.2 (page 509)

1. x^9y^6 **3.** $\frac{x^{12}}{y^8}$ **5.** 1 **7.** y, x **9.** $\frac{y_2 - y_1}{x_2 - x_1}$ **11.** run
13. vertical **15.** parallel **17.** 2 **19.** 3 **21.** −1
23. $-\frac{1}{3}$ **25.** 0 **27.** undefined **29.** −1 **31.** $-\frac{3}{2}$
33. $\frac{3}{4}$ **35.** $\frac{1}{2}$ **37.** 0 **39.** negative **41.** positive
43. undefined **45.** perpendicular **47.** neither
49. parallel **51.** parallel **53.** perpendicular **55.** neither
57. not the same line **59.** not the same line **61.** same line
63. $y = 0, m = 0$ **71.** $\frac{1}{165}$ **73.** 0.04, 0.1, 0.16 **75.** $\frac{18}{5}$
77. \$20,000 per yr **83.** 4

Getting Ready (page 512)

1. 14 **2.** $-\frac{5}{3}$ **3.** $y = 3x - 4$ **4.** $x = \frac{-By - 3}{A}$

Orals (page 521)

1. $y - 3 = 2(x - 2)$ **2.** $y - 8 = 2(x + 3)$
3. $y = -3x + 5$ **4.** $y = -3x - 7$ **5.** parallel
6. perpendicular

Exercise 8.3 (page 522)

1. 6 **3.** −1 **5.** 20 oz **7.** $y - y_1 = m(x - x_1)$
9. $Ax + By = C$ **11.** perpendicular **13.** $5x - y = -7$
15. $3x + y = 6$ **17.** $2x - 3y = -11$ **19.** $y = x$
21. $y = \frac{7}{3}x - 3$ **23.** $y = -\frac{7}{4}x + \frac{1}{2}$ **25.** $y = 3x + 17$

27. $y = -7x + 54$ **29.** $y = -4$ **31.** $y = -\frac{1}{2}x + 11$
33. 1, (0, −1) **35.** $\frac{2}{3}$, (0, 2)

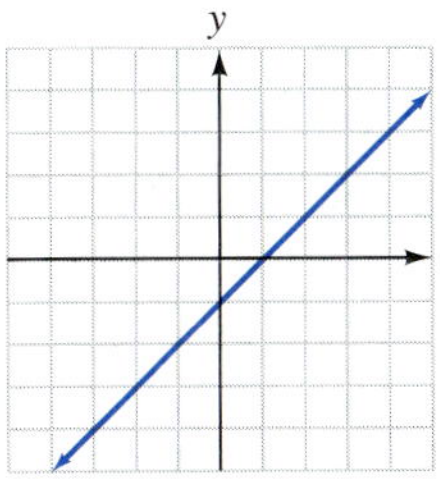

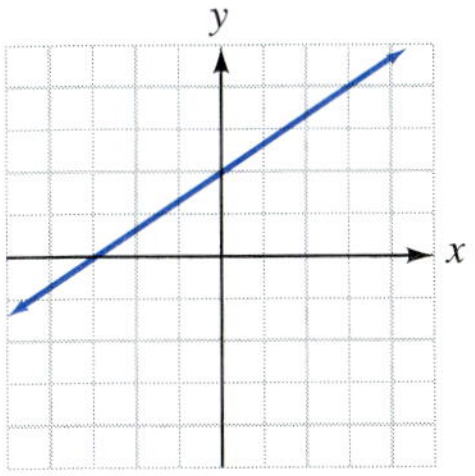

37. $-\frac{2}{3}$, (0, 6)

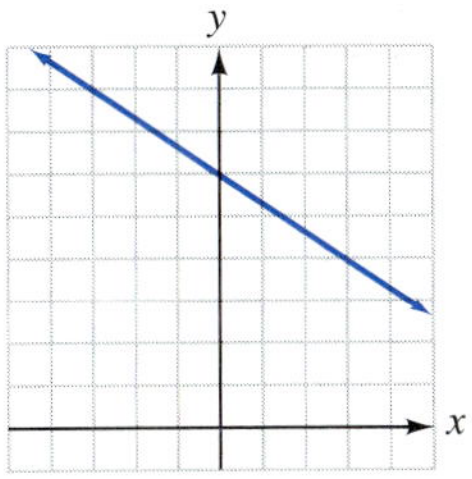

39. $\frac{3}{2}$, (0, −4) **41.** $-\frac{1}{3}$, $\left(0, -\frac{5}{6}\right)$
43. $\frac{7}{2}$, (0, 2) **45.** parallel
47. perpendicular **49.** parallel
51. perpendicular
53. perpendicular
55. perpendicular **57.** $y = 4x$
59. $y = 4x - 3$
61. $y = \frac{4}{5}x - \frac{26}{5}$
63. $y = -\frac{1}{4}x$ **65.** $y = -\frac{1}{4}x + \frac{11}{2}$ **67.** $y = -\frac{5}{4}x + 3$
69. perpendicular **71.** parallel **73.** $x = -2$ **75.** $x = 5$
77. $y = -\frac{A}{B}x + \frac{C}{B}$ **79.** $y = -2{,}298x + 19{,}984$
81. $y = 50{,}000x + 250{,}000$ **83.** $y = -\frac{950}{3}x + 1{,}750$
85. \$490 **87.** \$154,000 **89.** \$180 **99.** $a < 0, b > 0$

Getting Ready (page 525)

1. 1 **2.** 7 **3.** −20 **4.** $-\frac{11}{4}$

Orals (page 532)

1. yes **2.** no **3.** no **4.** 1 **5.** 3 **6.** −3

Exercise 8.4 (page 532)

1. −2 **3.** 2 **5.** input **7.** x **9.** function, input, output
11. range **13.** 0 **15.** $y = mx + b$ **17.** yes **19.** yes
21. yes **23.** no **25.** 9, −3 **27.** 3, −5 **29.** 22, 2
31. 3, 11 **33.** 4, 9 **35.** 7, 26 **37.** 9, 16 **39.** 6, 15
41. 4, 4 **43.** 2, 2 **45.** $\frac{1}{5}$, 1 **47.** −2, $\frac{2}{5}$
49. $2w$, $2w + 2$ **51.** $3w - 5$, $3w - 2$ **53.** 12
55. $2b - 2a$ **57.** $2b$ **59.** 1
61. D = {−2, 4, 6}; R = {3, 5, 7}
63. D = $(-\infty, 4) \cup (4, \infty)$; R = $(-\infty, 0) \cup (0, \infty)$
65. not a function
67. a function; D = $(-\infty, \infty)$; R = $(-\infty, \infty)$
69. D = $(-\infty, \infty)$; R = $(-\infty, \infty)$ **71.** D = $(-\infty, \infty)$; R = $(-\infty, \infty)$

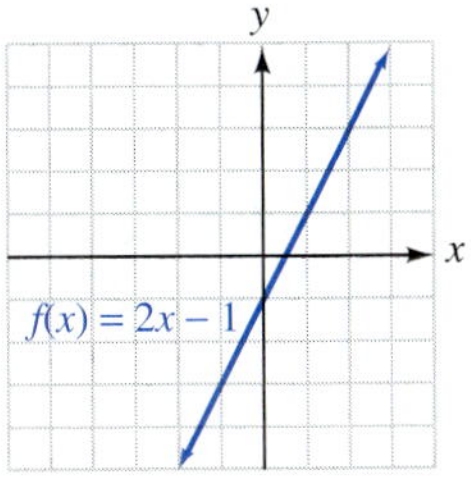

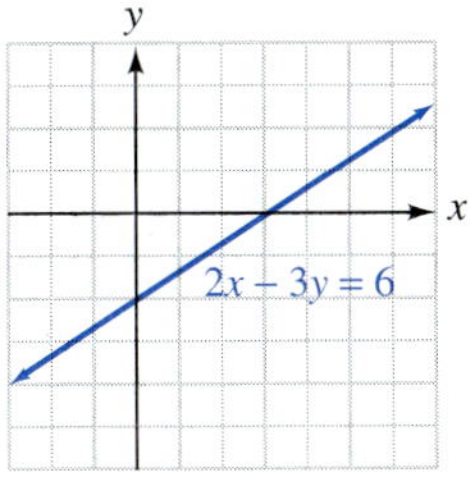

73. no **75.** yes **77.** 624 ft **79.** 1 sec **81.** 77° F
83. 192 **87.** yes

Getting Ready (page 535)

1. 2, (0, −3) **2.** −3, (0, 4) **3.** 6, −9 **4.** 4, $\frac{5}{2}$

Exercise 8.5 (page 546)

1. 41, 43, 47 **3.** $a \cdot b = b \cdot a$ **5.** 1 **7.** squaring
9. absolute value **11.** horizontal **13.** 2, down
15. 4, to the left **17.** rational

19.

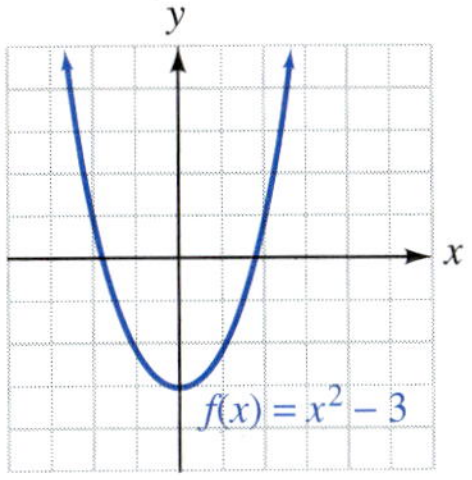

21.

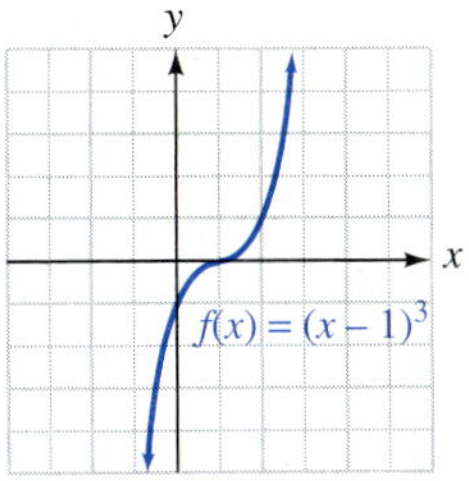

23.

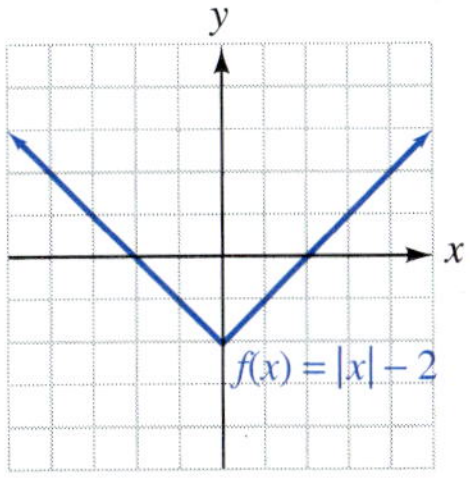

25.

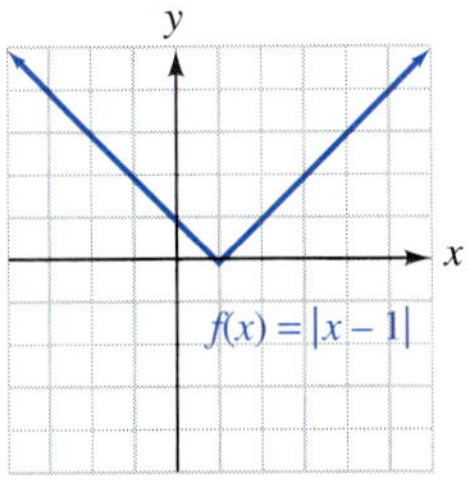

27.

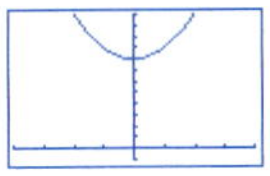

29.

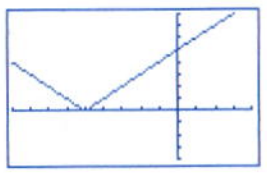

31.

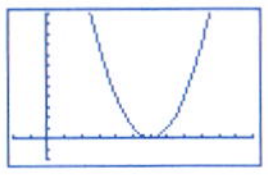

33.

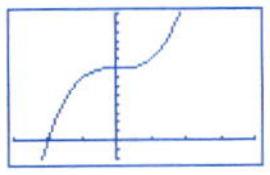

35.

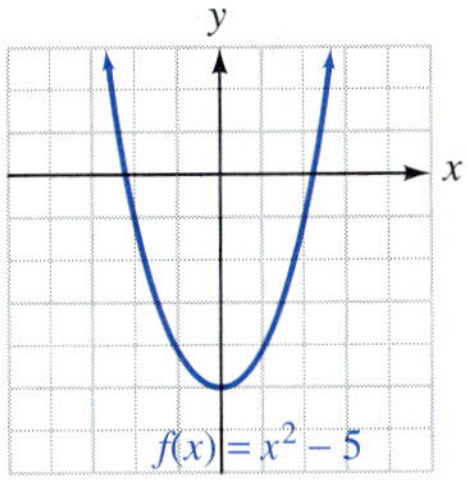

37.

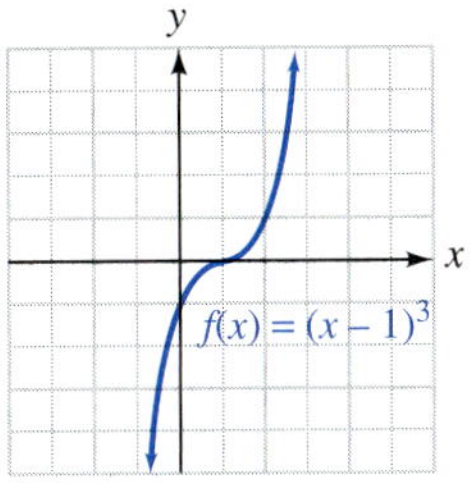

39.

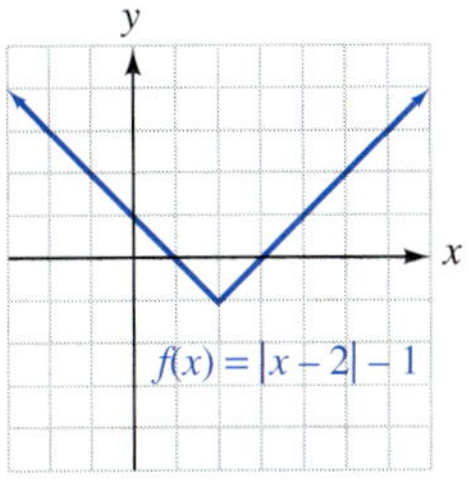

41.

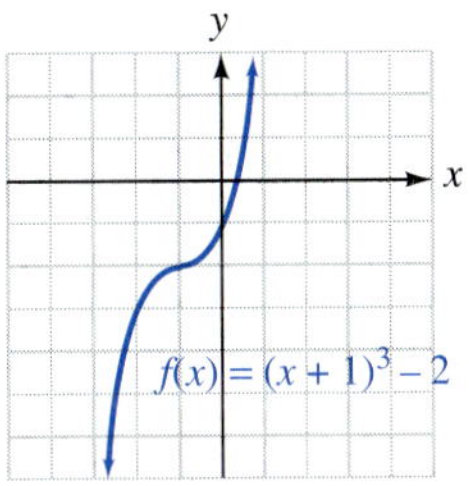

43.

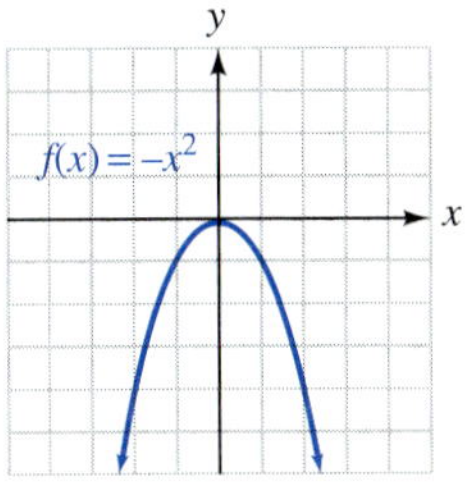

45.

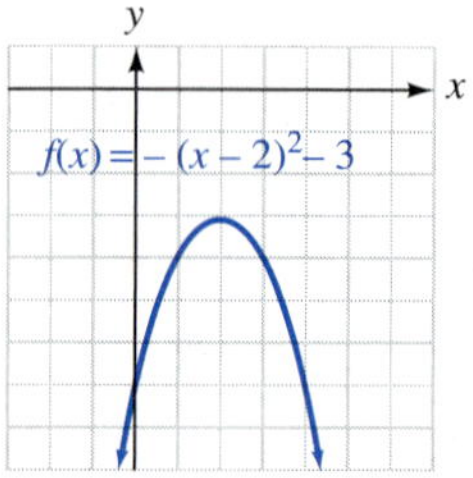

47. 20 hr **49.** 12 hr **51.** \$5,555.56 **53.** \$50,000 **55.** all reals but 2; all reals but 1 **57.** all reals but ± 2; all reals **59.** $c = f(x) = 1.25x + 700$ **61.** \$1,325 **63.** \$1.95 **65.** $c = f(n) = 0.09n + 7.50$ **67.** \$77.25 **69.** 9.75¢ **71.** $c = f(x) = 350x + 5{,}000$ **73.** \$47,000

Getting Ready (page 548)

1. $\frac{3}{2}$ **2.** $\frac{10}{7}$ **3.** 4 **4.** 36

Orals (page 554)

1. 1 **2.** 9 **3.** $\frac{14}{5}$ **4.** $a = kb$ **5.** $a = \frac{k}{b}$ **6.** $a = kbc$ **7.** $a = \frac{kb}{c}$

Exercise 8.6 (page 554)

1. x^{10} **3.** -1 **5.** 3.5×10^4 **7.** 0.0025 **9.** proportion **11.** direct **13.** rational **15.** joint **17.** direct **19.** neither **21.** 3 **23.** 5 **25.** -3 **27.** 5 **29.** 4, -1 **31.** 2, -2 **33.** 39 **35.** $-\frac{5}{2}, -1$ **37.** $A = kp^2$ **39.** $v = \dfrac{k}{r^3}$ **41.** $B = kmn$ **43.** $P = \dfrac{ka^2}{j^3}$ **45.** L varies jointly with m and n. **47.** E varies jointly with a and the square of b. **49.** X varies directly with x^2 and inversely with y^2. **51.** R varies directly with L and inversely with d^2. **53.** 36π in.^{2} **55.** 432 mi **57.** 25 days **59.** 12 in.^{3} **61.** 85.3 **63.** 12 **65.** 26,437.5 gal **67.** 3 ohms **69.** 0.275 in. **71.** 546 Kelvin

Chapter Summary (page 559)

1.

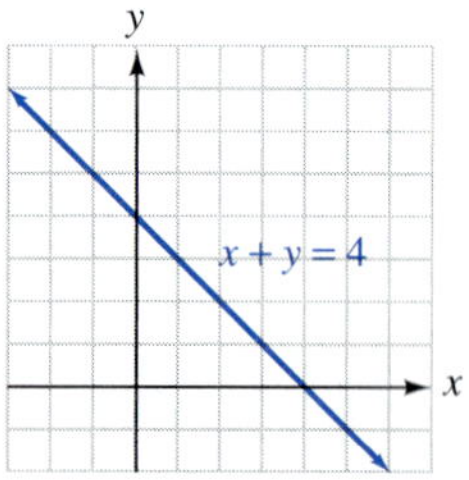

2.

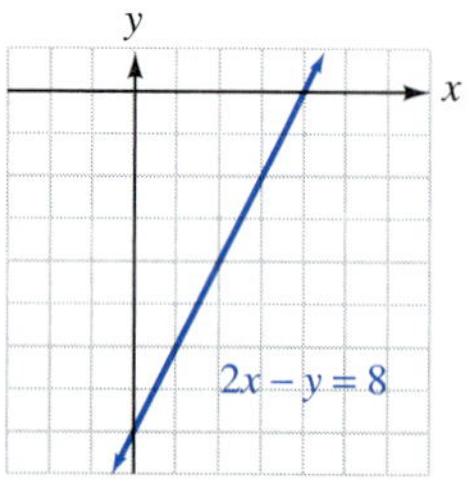

3.

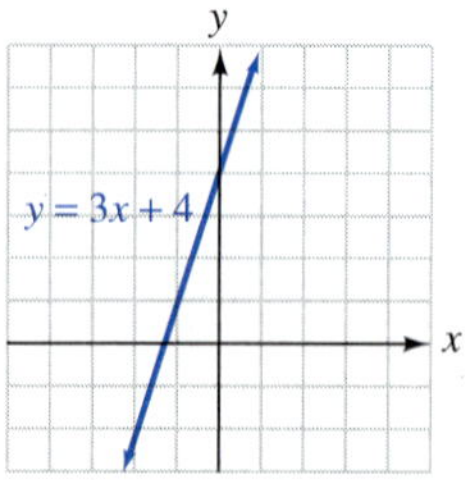

4.

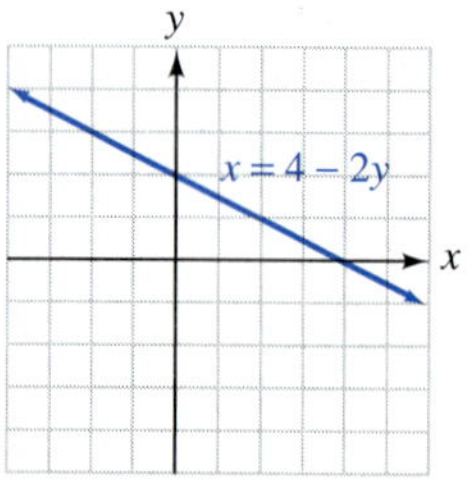

5.

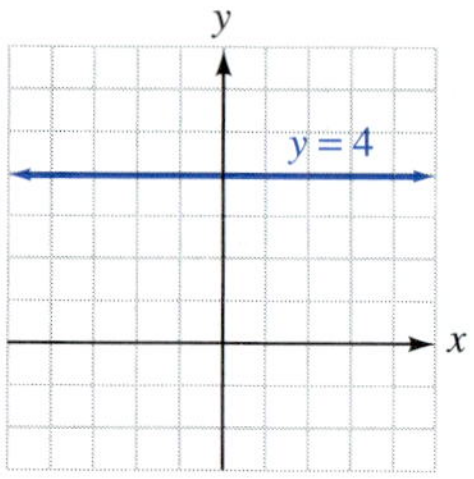

6.

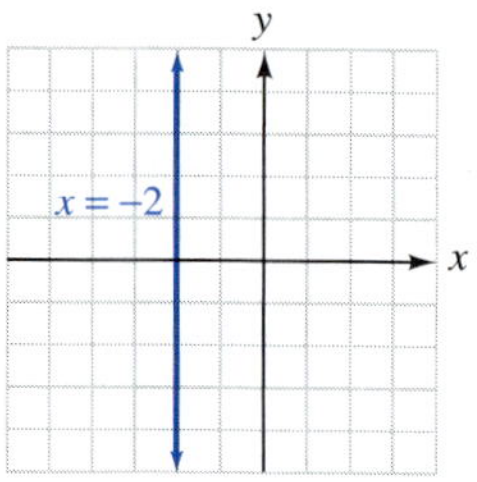

7.

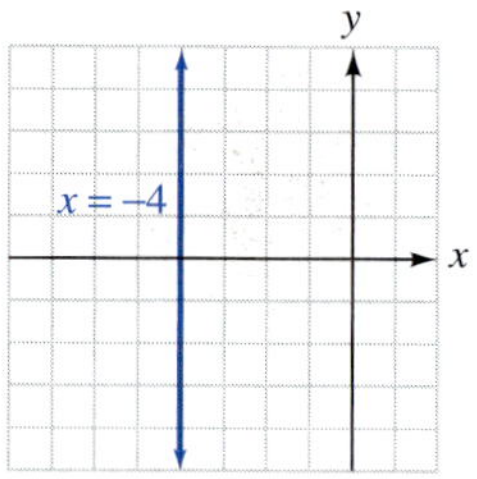

8.

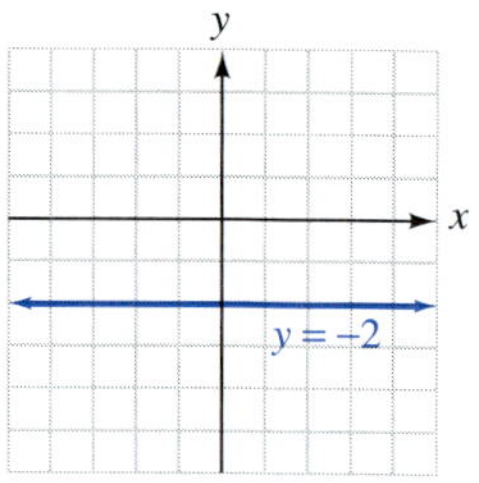

9. $\left(\frac{3}{2}, 8\right)$ **10.** 1 **11.** $\frac{14}{9}$ **12.** 5 **13.** $\frac{5}{11}$ **14.** 0 **15.** no defined slope **16.** $\frac{2}{3}$ **17.** -2 **18.** no defined slope **19.** 0 **20.** perpendicular **21.** parallel **22.** neither **23.** perpendicular **24.** \$21,666.67 **25.** $3x - y = -29$ **26.** $13x + 8y = 6$ **27.** $3x - 2y = 1$ **28.** $2x + 3y = -21$

29.

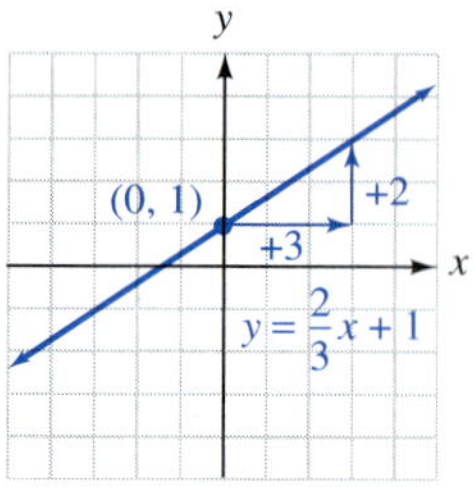

30. perpendicular **31.** $y = -1{,}720x + 8{,}700$ **32.** yes **33.** yes **34.** no **35.** no **36.** -7 **37.** 60 **38.** 0 **39.** 17 **40.** $D = (-\infty, \infty)$; $R = (-\infty, \infty)$ **41.** $D = (-\infty, \infty)$; $R = (-\infty, \infty)$ **42.** $D = (-\infty, \infty)$; $R = [1, \infty)$ **43.** $D = (-\infty, 2) \cup (2, \infty)$; $R = (-\infty, 0) \cup (0, \infty)$ **44.** $D = (-\infty, 3) \cup (3, \infty)$; $R = (-\infty, 0) \cup (0, \infty)$ **45.** $D = (-\infty, \infty)$; $R = \{7\}$ **46.** a function **47.** not a function **48.** not a function **49.** a function

50.

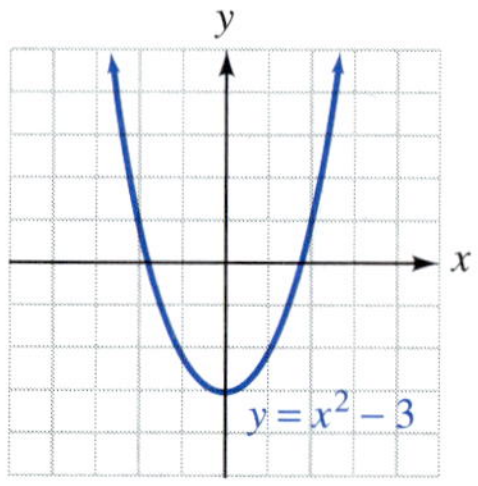

51.

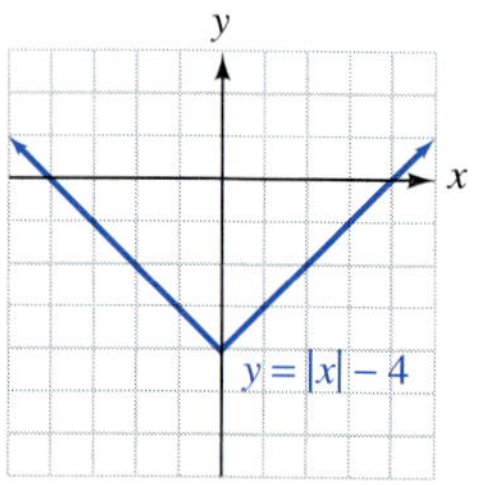

52.

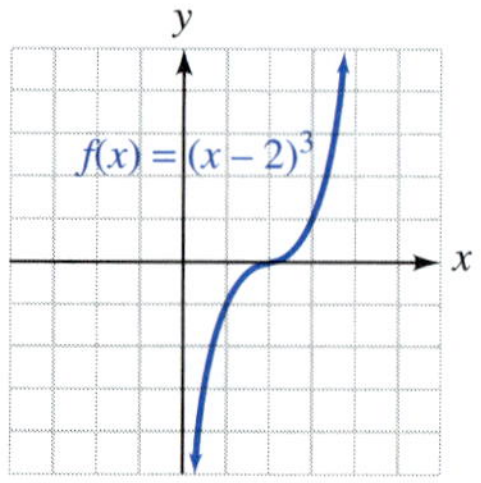

53.

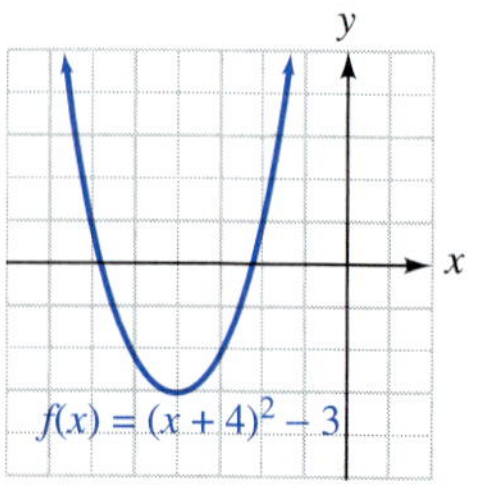

54.

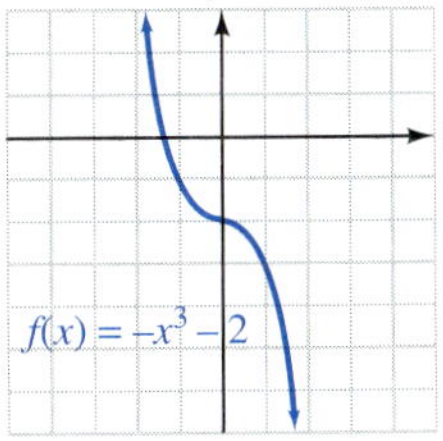

55.

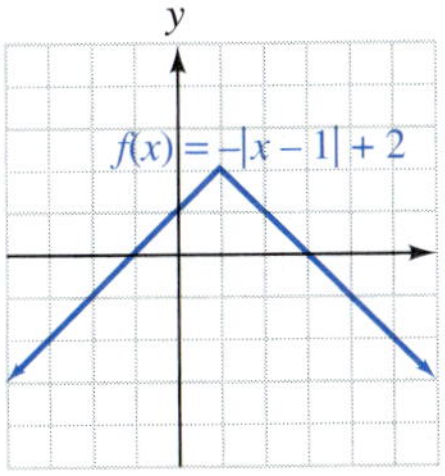

56.

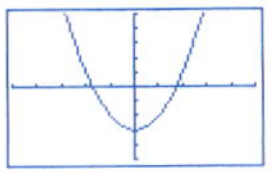

57.

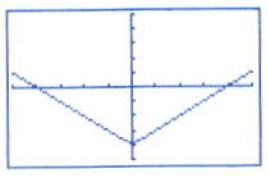

58.

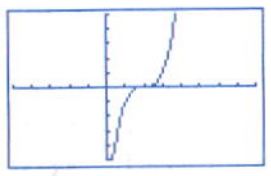

59.

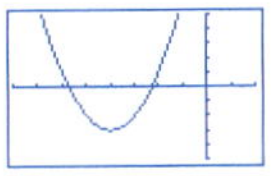

60.

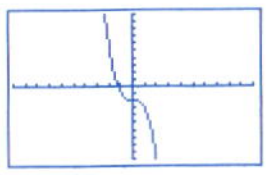

61.

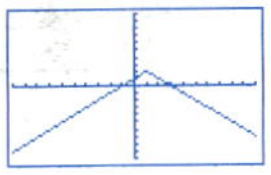

62. D: $(-\infty, 2) \cup (2, \infty)$;
R: $(-\infty, 0) \cup (0, \infty)$

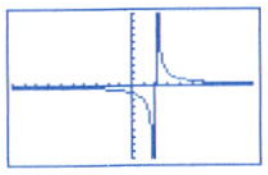

63. D: $(-\infty, -3) \cup (-3, \infty)$;
R: $(-\infty, 1) \cup (1, \infty)$

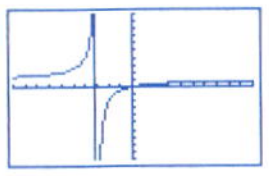

64. $\frac{13}{3}$
65. $-4, -12$
66. 2 **67.** 6
68. $\frac{1}{32}$ **69.** 16

Chapter 8 Test (page 564)

1.

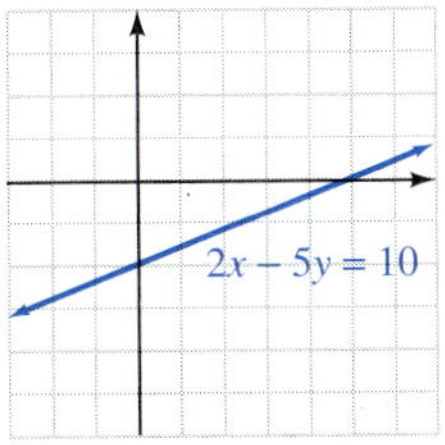

2. $\left(\frac{1}{2}, \frac{1}{2}\right)$
3. x-intercept $(3, 0)$, y-intercept $\left(0, -\frac{3}{5}\right)$ **4.** vertical **5.** $\frac{1}{2}$
6. $\frac{2}{3}$ **7.** undefined **8.** 0
9. $y = \frac{2}{3}x - \frac{23}{3}$
10. $8x - y = -22$
11. $m = -\frac{1}{3}, \left(0, -\frac{3}{2}\right)$
12. neither **13.** perpendicular
14. $y = \frac{3}{2}x$ **15.** $y = \frac{3}{2}x + \frac{21}{2}$
16. no **17.** D = $(-\infty, \infty)$; R = $[0, \infty)$
18. D = $(-\infty, \infty)$; R = $(-\infty, \infty)$ **19.** 10 **20.** -2
21. $3a + 1$ **22.** $x^2 - 2$ **23.** yes **24.** no

25.

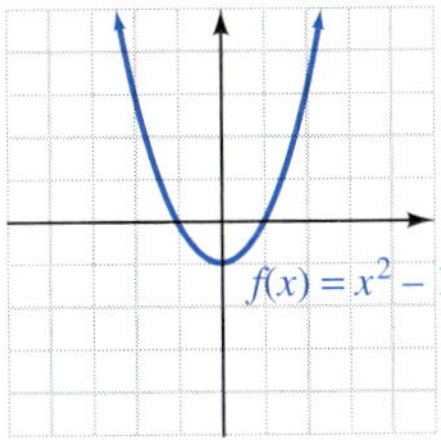

26.

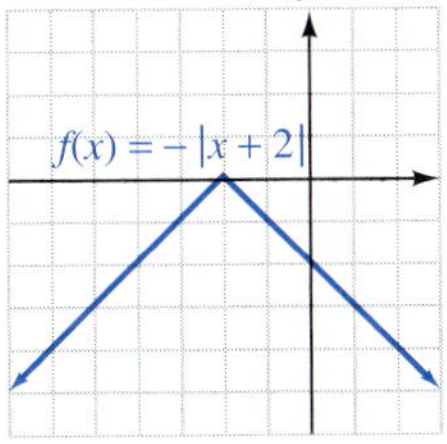

27. $6, -1$ **28.** $\frac{6}{5}$ **29.** $\frac{44}{3}$ **30.** yes

Cumulative Review Exercises (page 566)

1. 1, 2, 6, 7 **2.** 0, 1, 2, 6, 7 **3.** $-2, 0, 1, 2, \frac{13}{12}, 6, 7$
4. $\sqrt{5}, \pi$ **5.** -2 **6.** $-2, 0, 1, 2, \frac{13}{12}, 6, 7, \sqrt{5}, \pi$ **7.** 2, 7
8. 6 **9.** $-2, 0, 2, 6$ **10.** 1, 7

11.

12.

13. -2 **14.** -2
15. 22 **16.** -2 **17.** -4 **18.** -3 **19.** 4 **20.** -5
21. assoc. prop. of add. **22.** distrib. prop.
23. comm. prop. of add. **24.** assoc. prop. of mult.
25. x^8y^{12} **26.** c^2 **27.** $-\dfrac{b^3}{a^2}$ **28.** 1 **29.** 4.97×10^{-6}
30. 932,000,000

Getting Ready (page 568)

1. 0 **2.** 16 **3.** 16 **4.** -16 **5.** $\frac{8}{125}$ **6.** $\frac{81}{256}$
7. $49x^2y^2$ **8.** $343x^3y^3$

Orals (page 577)

1. 3 **2.** -4 **3.** -2 **4.** 2 **5.** $8|x|$ **6.** $-3x$
7. not a real number **8.** $(x + 1)^2$

Exercise 9.1 (page 577)

1. $\frac{x + 3}{x - 4}$ **3.** 1 **5.** $\frac{3(m^2 + 2m - 1)}{(m + 1)(m - 1)}$ **7.** $(5x^2)^2$ **9.** positive
11. 3, up **13.** x **15.** odd **17.** 0 **19.** $3x^2$
21. $a^2 + b^3$ **23.** 11 **25.** -8 **27.** $\frac{1}{3}$ **29.** $-\frac{5}{7}$
31. not real **33.** 0.4 **35.** 4 **37.** not real **39.** 3.4641
41. 26.0624 **43.** $2|x|$ **45.** $3a^2$ **47.** $|t + 5|$ **49.** $5|b|$
51. $|a + 3|$ **53.** $|t + 12|$ **55.** 1 **57.** -5 **59.** $-\frac{2}{3}$
61. 0.4 **63.** $2a$ **65.** $-10pq$ **67.** $-\frac{1}{2}m^2n$ **69.** $0.2z^3$
71. 3 **73.** -3 **75.** -2 **77.** $\frac{2}{5}$ **79.** $\frac{1}{2}$ **81.** not real
83. $2|x|$ **85.** $2a$ **87.** $\frac{1}{2}|x|$ **89.** $|x^3|$ **91.** $-x$
93. $-3a^2$ **95.** $x + 2$ **97.** $0.1x^2|y|$ **99.** 0 **101.** 4
103. 1 **105.** 0.5 **107.** 4.1231 **109.** 2.5539
111. D: $[-4, \infty)$, R: $[0, \infty)$ **113.** D: $[0, \infty)$, R: $(-\infty, -3]$,

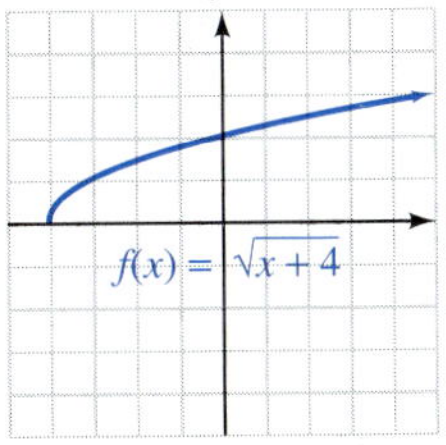

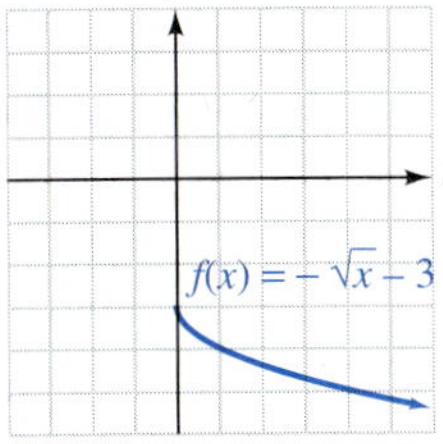

115. 1.67 **117.** 11.8673 **119.** 3 units **121.** 4 sec
123. about 7.4 amperes

Getting Ready (page 580)

1. 25 **2.** 169 **3.** 18 **4.** 11,236

Orals (page 584)

1. 5 **2.** 10 **3.** 13 **4.** 5 **5.** 10 **6.** 13 **7.** 4
8. 3 **9.** 5

Exercise 9.2 (page 584)

1. $12x^2 - 14x - 10$ **3.** $15t^2 + 2ts - 8s^2$ **5.** hypotenuse
7. $a^2 + b^2 = c^2$ **9.** distance **11.** 10 ft **13.** 80 m
15. 48 in. **17.** 9.9 cm **19.** 5 **21.** 5 **23.** 13 **25.** 10

27. 10.2 **33.** 13 ft **35.** about 127 ft **37.** about 135 ft **39.** not quite **41.** yes **43.** 173 yd **45.** 0.05 ft **47.** 24 cm^2 **51.** about 25

Getting Ready (page 587)

1. x^7 **2.** a^{12} **3.** a^4 **4.** 1 **5.** $\frac{1}{x^4}$ **6.** a^3b^6 **7.** $\frac{b^6}{c^9}$ **8.** a^{10}

Orals (page 593)

1. 2 **2.** 3 **3.** 3 **4.** 1 **5.** 8 **6.** 4 **7.** $\frac{1}{2}$ **8.** 2 **9.** $2x$ **10.** $2x^2$

Exercise 9.3 (page 593)

1. $x < 3$ **3.** $r > 28$ **5.** $1\frac{2}{3}$ pints **7.** $a \cdot a \cdot a \cdot a$ **9.** a^{mn} **11.** $\frac{a^n}{b^n}$ **13.** $\frac{1}{a^n}, 0$ **15.** $\left(\frac{b}{a}\right)^n$ **17.** $|x|$ **19.** $\sqrt[3]{7}$ **21.** $\sqrt[5]{8}$ **23.** $\sqrt[4]{3x}$ **25.** $\sqrt[4]{\frac{1}{2}x^3y}$ **27.** $\sqrt[5]{4a^2b^3}$ **29.** $\sqrt{x^2+y^2}$ **31.** $11^{1/2}$ **33.** $(3a)^{1/4}$ **35.** $3a^{1/5}$ **37.** $\left(\frac{1}{7}abc\right)^{1/6}$ **39.** $\left(\frac{1}{2}mn\right)^{1/5}$ **41.** $(a^2 - b^2)^{1/3}$ **43.** 2 **45.** 3 **47.** 2 **49.** 2 **51.** $\frac{1}{2}$ **53.** $\frac{1}{2}$ **55.** -2 **57.** -3 **59.** not real **61.** 0 **63.** $5|y|$ **65.** $2|x|$ **67.** $3x$ **69.** not real **71.** 216 **73.** 27 **75.** 1,728 **77.** $\frac{1}{4}$ **79.** $125x^6$ **81.** $\frac{4x^2}{9}$ **83.** $\frac{1}{2}$ **85.** $\frac{1}{8}$ **87.** $\frac{1}{64x^3}$ **89.** $\frac{1}{9y^2}$ **91.** $\frac{1}{4p^2}$ **93.** 8 **95.** $\frac{16}{81}$ **97.** $-\frac{3}{2x}$ **99.** $5^{8/9}$ **101.** $4^{3/5}$ **103.** $9^{1/5}$ **105.** $7^{1/2}$ **107.** $\frac{1}{36}$ **109.** $2^{2/3}$ **111.** a **113.** $a^{2/9}$ **115.** $a^{3/4}b^{1/2}$ **117.** $\frac{n^{2/5}}{m^{3/5}}$ **119.** $\frac{2x}{3}$ **121.** $\frac{1}{3}x$ **123.** $y + y^2$ **125.** $x^2 - x + x^{3/5}$ **127.** $x - 4$ **129.** $x^{4/3} - x^2$ **131.** $x^{4/3} + 2x^{2/3}y^{2/3} + y^{4/3}$ **133.** $a^3 - 2a^{3/2}b^{3/2} + b^3$ **135.** $\sqrt{p}$ **137.** $\sqrt{5b}$ **141.** yes

Getting Ready (page 595)

1. 15 **2.** 24 **3.** 5 **4.** 7 **5.** $4x^2$ **6.** $\frac{8}{11}x^3$ **7.** $3ab^3$ **8.** $-2a^4$

Orals (page 603)

1. 7 **2.** 4 **3.** 3 **4.** $3\sqrt{2}$ **5.** $2\sqrt[3]{2}$ **6.** $\frac{\sqrt[3]{3x^2}}{4b^2}$ **7.** $7\sqrt{3}$ **8.** $3\sqrt{7}$ **9.** $5\sqrt[3]{9}$ **10.** $8\sqrt[5]{4}$

Exercise 9.4 (page 603)

1. $\frac{-15x^5}{y}$ **3.** $9t^2 + 12t + 4$ **5.** $3p + 4 + \frac{-5}{2p - 5}$ **7.** $\sqrt[n]{a}\sqrt[n]{b}$ **9.** 6 **11.** t **13.** $5x$ **15.** 10 **17.** $7x$ **19.** $6b$ **21.** 2 **23.** $3a$ **25.** $2\sqrt{5}$ **27.** $-10\sqrt{2}$ **29.** $2\sqrt[3]{10}$ **31.** $-3\sqrt[3]{3}$ **33.** $2\sqrt[4]{2}$ **35.** $2\sqrt[5]{3}$ **37.** $\frac{\sqrt{7}}{3}$ **39.** $\frac{\sqrt[3]{7}}{4}$ **41.** $\frac{\sqrt[3]{3}}{10}$ **43.** $\frac{\sqrt[5]{3}}{2}$ **45.** $5x\sqrt{2}$ **47.** $4\sqrt{2b}$ **49.** $-4a\sqrt{7a}$ **51.** $5ab\sqrt{7b}$ **53.** $-10\sqrt{3xy}$ **55.** $-3x^2\sqrt[3]{2}$ **57.** $2x^4y\sqrt[3]{2}$ **59.** $2x^3y\sqrt[4]{2}$ **61.** $\frac{z}{4x}$ **63.** $\frac{\sqrt[4]{5x}}{2z}$ **65.** $10\sqrt{2x}$ **67.** $\sqrt[5]{7a^2}$ **69.** $4\sqrt{3}$ **71.** $-\sqrt{2}$ **73.** $2\sqrt{2}$ **75.** $9\sqrt{6}$ **77.** $3\sqrt[3]{3}$ **79.** $-\sqrt[3]{4}$ **81.** -10 **83.** $-17\sqrt[4]{2}$ **85.** $16\sqrt[4]{2}$ **87.** $-4\sqrt{2}$ **89.** $3\sqrt{2} + \sqrt{3}$ **91.** $-11\sqrt[3]{2}$ **93.** $y\sqrt{z}$ **95.** $13y\sqrt{x}$ **97.** $12\sqrt[3]{a}$ **99.** $-7y^2\sqrt{y}$ **101.** $4x\sqrt[5]{xy^2}$ **103.** $2x + 2$ **105.** $h = 2.83, x = 2.00$ **107.** $x = 8.66, h = 10.00$ **109.** $x = 4.69, y = 8.11$ **111.** $x = 12.11, y = 12.11$ **115.** If $a = 0$, then b can be any nonnegative real number. If $b = 0$, then a can be any nonnegative real number.

Getting Ready (page 605)

1. a^7 **2.** b^3 **3.** $a^2 - 2a$ **4.** $6b^3 + 9b^2$ **5.** $a^2 - 3a - 10$ **6.** $4a^2 - 9b^2$

Orals (page 611)

1. 3 **2.** 2 **3.** $3\sqrt{3}$ **4.** a^2b **5.** $6 + 3\sqrt{2}$ **6.** 1 **7.** $\frac{\sqrt{2}}{2}$ **8.** $\frac{\sqrt{3}+1}{2}$

Exercise 9.5 (page 611)

1. 1 **3.** $\frac{1}{3}$ **5.** $2, \sqrt{7}, \sqrt{5}$ **7.** FOIL **9.** conjugate **11.** 4 **13.** $5\sqrt{2}$ **15.** $6\sqrt{2}$ **17.** 5 **19.** 18 **21.** $2\sqrt[3]{3}$ **23.** ab^2 **25.** $5a\sqrt{b}$ **27.** $r\sqrt[3]{10s}$ **29.** $2a^2b^2\sqrt[3]{2}$ **31.** $x^2(x + 3)$ **33.** $3x(y + z)\sqrt[3]{4}$ **35.** $12\sqrt{5} - 15$ **37.** $12\sqrt{6} + 6\sqrt{14}$ **39.** $-8x\sqrt{10} + 6\sqrt{15x}$ **41.** $-1 - 2\sqrt{2}$ **43.** $8x - 14\sqrt{x} - 15$ **45.** $5z + 2\sqrt{15z} + 3$ **47.** $3x - 2y$ **49.** $6a + 5\sqrt{3ab} - 3b$ **51.** $18r - 12\sqrt{2r} + 4$ **53.** $-6x - 12\sqrt{x} - 6$ **55.** $\frac{\sqrt{7}}{7}$ **57.** $\frac{\sqrt{6}}{3}$ **59.** $\frac{\sqrt{10}}{4}$ **61.** 2 **63.** $\frac{\sqrt[3]{4}}{2}$ **65.** $\sqrt[3]{3}$ **67.** $\frac{\sqrt[3]{6}}{3}$ **69.** $2\sqrt{2x}$ **71.** $\frac{\sqrt{5y}}{y}$ **73.** $\frac{\sqrt[3]{2ab^2}}{b}$ **75.** $\frac{\sqrt[4]{4}}{2}$ **77.** $\frac{\sqrt[5]{2}}{2}$ **79.** $\sqrt{2} + 1$ **81.** $\frac{3\sqrt{2} - \sqrt{10}}{4}$ **83.** $2 + \sqrt{3}$ **85.** $\frac{9 - 2\sqrt{14}}{5}$ **87.** $\frac{2(\sqrt{x} - 1)}{x - 1}$ **89.** $\frac{x(\sqrt{x} + 4)}{x - 16}$ **91.** $\sqrt{2z} + 1$ **93.** $\frac{x - 2\sqrt{xy} + y}{x - y}$ **95.** $\frac{1}{\sqrt{3} - 1}$ **97.** $\frac{x - 9}{x(\sqrt{x} - 3)}$ **99.** $\frac{x - y}{\sqrt{x}(\sqrt{x} - \sqrt{y})}$ **101.** $f/4$ **105.** $\frac{x - 9}{4(\sqrt{x} + 3)}$

Getting Ready (page 613)

1. a **2.** $5x$ **3.** $x + 4$ **4.** $y - 3$

Orals (page 619)

1. 7 **2.** 3 **3.** 0 **4.** 9 **5.** 17 **6.** 31

Exercise 9.6 (page 619)

1. 2 **3.** 6 **5.** $x^n = y^n$ **7.** square **9.** extraneous **11.** 2 **13.** 4 **15.** 0 **17.** 4 **19.** 8 **21.** $\frac{5}{2}, \frac{1}{2}$ **23.** 1 **25.** 16 **27.** 14, ~~6~~ **29.** 4, 3 **31.** 2, ~~7~~ **33.** 9, ~~−25~~ **35.** 2, −1 **37.** −1, ~~1~~ **39.** ~~1~~, no solutions **41.** 0, ~~4~~ **43.** ~~−3~~, no solutions **45.** 0 **47.** 1, 9 **49.** 4, ∅ **51.** 2, ~~142~~ **53.** 2 **55.** ~~6~~, no solutions **57.** 0, ~~$-\frac{12}{11}$~~ **59.** 1 **61.** 4, ~~−9~~ **63.** 2, ~~$\frac{2}{2}$~~ **65.** 2,010 ft **67.** about 29 mph **69.** 16% **71.** $5 **73.** $R = \frac{8kl}{\pi r^4}$ **77.** 0, 4

Getting Ready (page 622)

1. $7x$ **2.** $-x + 10$ **3.** $12x^2 + 5x - 25$ **4.** $9x^2 - 25$

Orals (page 630)

1. $7i$ **2.** $8i$ **3.** $10i$ **4.** $9i$ **5.** $-i$ **6.** −1 **7.** 1 **8.** i **9.** 5 **10.** 13

Exercise 9.7 (page 630)

1. −1 **3.** 20 mph **5.** imaginary **7.** −1 **9.** 1 **11.** $\frac{\sqrt{a}}{\sqrt{b}}$ **13.** 5, 7 **15.** conjugates **17.** $3i$ **19.** $6i$ **21.** $i\sqrt{7}$ **23.** yes **25.** no **27.** no **29.** $8 - 2i$ **31.** $3 - 5i$ **33.** $15 + 7i$ **35.** $6 - 8i$ **37.** $2 + 9i$ **39.** $-15 + 2\sqrt{3}i$ **41.** $3 + 6i$ **43.** $-25 - 25i$ **45.** $7 + i$ **47.** $14 - 8i$ **49.** $8 + \sqrt{2}i$ **51.** $-20 - 30i$ **53.** $3 + 4i$ **55.** $-5 + 12i$ **57.** $7 + 17i$ **59.** $5 + 5i$ **61.** $16 + 2i$ **63.** $0 - i$ **65.** $0 + \frac{4}{5}i$ **67.** $\frac{1}{8} - 0i$ **69.** $0 + \frac{3}{5}i$ **71.** $2 + i$ **73.** $\frac{1}{2} + \frac{5}{2}i$ **75.** $-\frac{42}{25} - \frac{6}{25}i$ **77.** $\frac{1}{4} + \frac{3}{4}i$ **79.** $\frac{5}{13} - \frac{12}{13}i$ **81.** $\frac{11}{10} + \frac{3}{10}i$ **83.** $\frac{1}{4} - \frac{\sqrt{15}}{4}i$ **85.** $-\frac{5}{169} + \frac{12}{169}i$ **87.** $\frac{3}{5} + \frac{4}{5}i$ **89.** $-\frac{6}{13} - \frac{9}{13}i$ **91.** i **93.** $-i$ **95.** 1 **97.** i **99.** 10 **101.** 13 **103.** $\sqrt{74}$ **105.** 1 **111.** $7 - 4i$ **113.** $1 - 3.4i$ **117.** $\frac{5}{3 + i}$

Chapter Summary (page 633)

1. 7 **2.** −11 **3.** −6 **4.** 15 **5.** −3 **6.** −6 **7.** 5 **8.** −2 **9.** $5|x|$ **10.** $|x + 2|$ **11.** $3a^2b$ **12.** $4x^2|y|$

13.

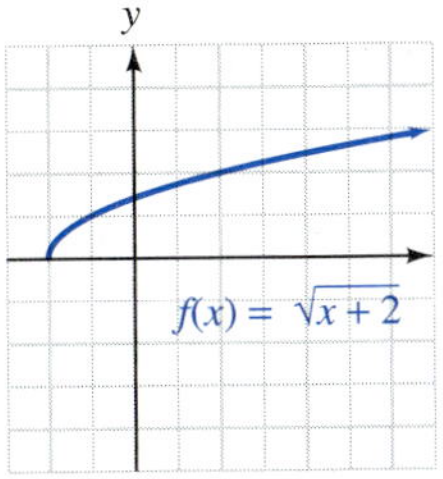

14.

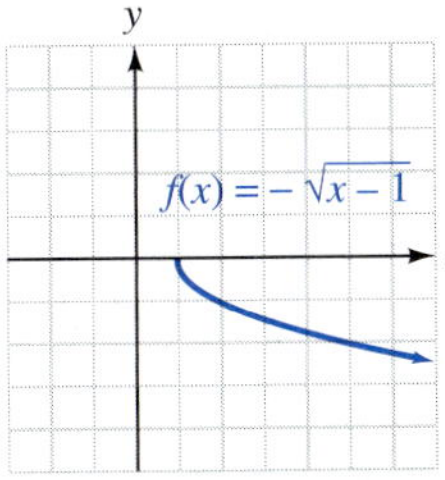

15.

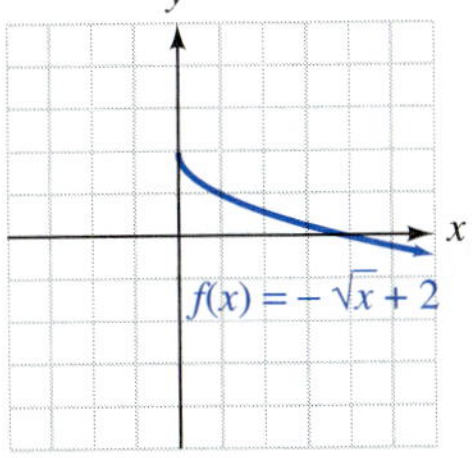

16.

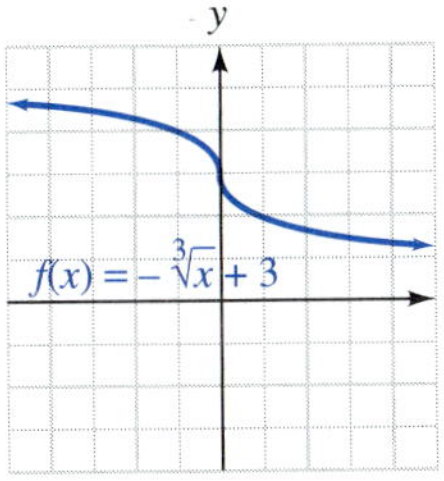

17. 12 **18.** about 5.7 **19.** 3 mi **20.** 8.2 ft **21.** 88 yd **22.** 16,000 yd, or about 9 mi **23.** 13 **24.** 2.83 units **25.** 5 **26.** −6 **27.** 27 **28.** 64 **29.** −2 **30.** −4 **31.** $\frac{1}{4}$ **32.** $\frac{1}{2}$ **33.** −16,807 **34.** $\frac{1}{3{,}125}$ **35.** 8 **36.** $\frac{27}{8}$ **37.** $3xy^{1/3}$ **38.** $3xy^{1/2}$ **39.** $125x^{9/2}y^6$ **40.** $\frac{1}{4u^{4/3}v^2}$ **41.** $5^{3/4}$ **42.** $a^{5/7}$ **43.** $u - 1$ **44.** $v + v^2$ **45.** $x + 2x^{1/2}y^{1/2} + y$ **46.** $a^{4/3} - b^{4/3}$ **47.** $\sqrt[3]{5}$ **48.** $\sqrt{x}$ **49.** $\sqrt[3]{3ab^2}$ **50.** $\sqrt{5ab}$ **51.** $4\sqrt{15}$ **52.** $3\sqrt[3]{2}$ **53.** $2\sqrt[4]{2}$ **54.** $2\sqrt[5]{3}$ **55.** $2x\sqrt{2x}$ **56.** $3x^2y\sqrt{2y}$ **57.** $2xy\sqrt[3]{2x^2y}$ **58.** $3x^2y\sqrt[3]{2x}$ **59.** $4x$ **60.** $2x$ **61.** $\frac{\sqrt[3]{2a^2b}}{3x}$ **62.** $\frac{\sqrt{17xy}}{8a^2}$ **63.** $3\sqrt{2}$ **64.** $\sqrt{5}$ **65.** 0 **66.** $8\sqrt[4]{2}$ **67.** $29x\sqrt{2}$ **68.** $32a\sqrt{3a}$ **69.** $13\sqrt[3]{2}$ **70.** $-4x\sqrt[4]{2x}$ **71.** $7\sqrt{2}$ m **72.** $6\sqrt{3}$ cm, 18 cm **73.** 7.07 in. **74.** 8.66 cm **75.** $6\sqrt{10}$ **76.** 72 **77.** $3x$ **78.** 3 **79.** $-2x$ **80.** $-20x^3y^3\sqrt{xy}$ **81.** $4 - 3\sqrt{2}$ **82.** $2 + 3\sqrt{2}$ **83.** $\sqrt{10} - \sqrt{5}$ **84.** $3 + \sqrt{6}$ **85.** 1 **86.** $5 + 2\sqrt{6}$ **87.** $x - y$ **88.** $6u + \sqrt{u} - 12$ **89.** $\frac{\sqrt{3}}{3}$ **90.** $\frac{\sqrt{15}}{5}$ **91.** $\frac{\sqrt{xy}}{y}$ **92.** $\frac{\sqrt[3]{u^2}}{u^2v^2}$ **93.** $2(\sqrt{2} + 1)$ **94.** $\frac{\sqrt{6} + \sqrt{2}}{2}$ **95.** $2(\sqrt{x} - 4)$ **96.** $\frac{a + 2\sqrt{a} + 1}{a - 1}$ **97.** $\frac{3}{5\sqrt{3}}$ **98.** $\frac{1}{\sqrt[3]{3}}$ **99.** $\frac{9 - x}{2(3 + \sqrt{x})}$ **100.** $\frac{a - b}{a + \sqrt{ab}}$ **101.** 22 **102.** 16, 9 **103.** 3, 9 **104.** $\frac{9}{16}$ **105.** 2 **106.** 0, −2 **107.** $12 - 8i$ **108.** $2 - 68i$ **109.** $-96 + 3i$ **110.** $-2 - 2\sqrt{2}i$ **111.** $22 + 29i$ **112.** $-16 + 7i$ **113.** $-12 + 28\sqrt{3}i$ **114.** $118 + 10\sqrt{2}i$ **115.** $0 - \frac{3}{4}i$ **116.** $0 - \frac{2}{5}i$ **117.** $\frac{12}{5} - \frac{6}{5}i$ **118.** $\frac{21}{10} + \frac{7}{10}i$ **119.** $\frac{15}{17} + \frac{8}{17}i$ **120.** $\frac{4}{5} - \frac{3}{5}i$ **121.** $\frac{15}{29} - \frac{6}{29}i$ **122.** $\frac{1}{3} + \frac{1}{3}i$ **123.** $15 + 0i$ **124.** $26 + 0i$ **125.** 1 **126.** $-i$

Chapter 9 Test (page 639)

1. 7 **2.** 4 **3.** $2|x|$ **4.** $2x$ **5.** $D = [2, \infty), R = [0, \infty)$ **6.** $D = (-\infty, \infty), R = (-\infty, \infty)$

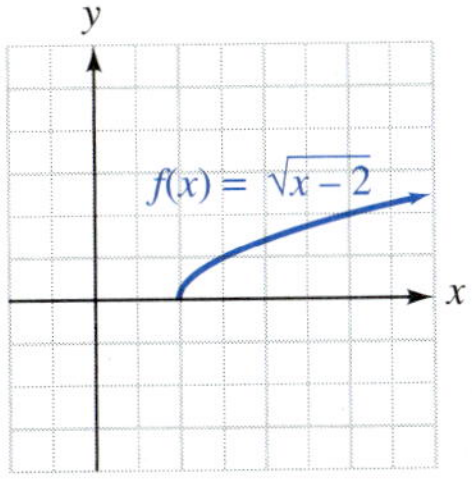

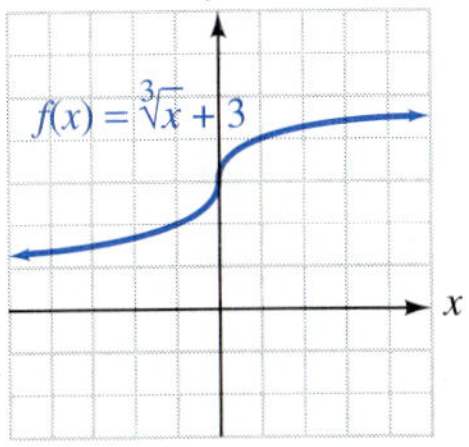

7. 28 in. **8.** 1.25 m **9.** 10 **10.** 25 **11.** 2 **12.** 9 **13.** $\frac{1}{216}$ **14.** $\frac{9}{4}$ **15.** $2^{4/3}$ **16.** $8xy$ **17.** $4\sqrt{3}$ **18.** $5xy^2\sqrt{10xy}$ **19.** $2x^5y\sqrt[3]{3}$ **20.** $\frac{1}{4a}$ **21.** $2|x|\sqrt{3}$ **22.** $2|x^3|\sqrt{2}$ **23.** $3x\sqrt[3]{3}$ **24.** $3x^2y^4\sqrt{2y}$ **25.** $-\sqrt{3}$

26. $14\sqrt[3]{5}$ **27.** $2y^2\sqrt[4]{3y}$ **28.** $6z\sqrt[4]{3z}$
29. $-6x\sqrt{y} - 2xy^2$ **30.** $3 - 7\sqrt{6}$ **31.** $\frac{\sqrt{5}}{5}$
32. $\sqrt{3t} + 1$ **33.** $\dfrac{3}{\sqrt{21}}$ **34.** $\dfrac{a - b}{a - 2\sqrt{ab} + b}$ **35.** 10
36. $\not 4$, no solutions **37.** $-1 + 11i$ **38.** $4 - 7i$
39. $8 + 6i$ **40.** $-10 - 11i$ **41.** $0 - \frac{\sqrt{2}}{2}i$ **42.** $\frac{1}{2} + \frac{1}{2}i$

Getting Ready (page 642)

1. $(x + 5)(x - 5)$ **2.** $(b + 9)(b - 9)$ **3.** $(3x + 2)(2x - 1)$
4. $(2x - 3)(2x + 1)$

Orals (page 650)

1. ± 7 **2.** $\pm\sqrt{10}$ **3.** 4 **4.** 9 **5.** $\frac{9}{4}$ **6.** $\frac{25}{4}$

Exercise 10.1 (page 650)

1. 1 **3.** $t \le 4$ **5.** $x = \sqrt{c}, x = -\sqrt{c}$
7. positive or negative **9.** $0, -2$ **11.** $5, -5$
13. $-2, -4$ **15.** $6, 1$ **17.** $2, \frac{1}{2}$ **19.** $\frac{2}{3}, -\frac{5}{2}$ **21.** ± 6
23. $\pm\sqrt{5}$ **25.** $\pm\dfrac{4\sqrt{3}}{3}$ **27.** $0, -2$ **29.** $4, 10$
31. $-5 \pm \sqrt{3}$ **33.** $2 \pm \sqrt{5}$ **35.** $\pm 4i$ **37.** $\pm\frac{9}{2}i$
39. $2, -4$ **41.** $2, 4$ **43.** $-1, -4$ **45.** $1, -\frac{1}{2}$
47. $-\frac{1}{3}, -\frac{3}{2}$ **49.** $\frac{3}{4}, -\frac{3}{2}$ **51.** $-\dfrac{7}{10} \pm \dfrac{\sqrt{29}}{10}$ **53.** $-1 \pm i$
55. $-4 \pm i\sqrt{2}$ **57.** $\dfrac{1}{3} \pm \dfrac{2\sqrt{2}}{3}i$ **59.** $-\dfrac{1}{4} \pm \dfrac{\sqrt{41}}{4}$
61. $-\dfrac{1}{2} \pm \dfrac{\sqrt{13}}{2}$ **63.** 4 sec **65.** 72.5 mph
67. 4% **71.** $\frac{3}{4}$

Getting Ready (page 652)

1. $x^2 + 12x + 36, (x + 6)^2$ **2.** $x^2 - 7x + \frac{49}{4}, \left(x - \frac{7}{2}\right)^2$
3. 7 **4.** 8

Orals (page 656)

1. $3, -4, 7$ **2.** $-2, 1, -5$

Exercise 10.2 (page 656)

1. $y = \dfrac{-Ax + C}{B}$ **3.** $2\sqrt{6}$ **5.** $\sqrt{3}$ **7.** $3, -2, 6$
9. $-1, -2$ **11.** $-3, 5$ **13.** $-6, -6$ **15.** $-1, \frac{3}{2}$
17. $-\dfrac{1}{2} \pm \dfrac{\sqrt{5}}{10}$ **19.** $-\frac{3}{2}, -\frac{1}{2}$ **21.** $\frac{1}{4}, -\frac{3}{4}$ **23.** $-\dfrac{5}{2} \pm \dfrac{\sqrt{17}}{2}$
25. $-1 \pm i$ **27.** $-\dfrac{1}{4} \pm \dfrac{\sqrt{7}}{4}i$ **29.** $\dfrac{2}{3} \pm \dfrac{\sqrt{2}}{3}i$
31. $\dfrac{1}{3} \pm \dfrac{2\sqrt{2}}{3}i$ **33.** $-\frac{1}{2} \pm \sqrt{5}$ **35.** $-\dfrac{1}{3} \pm \dfrac{\sqrt{5}}{3}i$
37. $8.98, -3.98$ **39.** $N = \dfrac{1 \pm \sqrt{1 + 8C}}{2}$ **41.** 16, 18
43. 6, 7 **45.** $x^2 - 8x + 15 = 0$
47. $x^3 - x^2 - 14x + 24 = 0$ **49.** 8 ft by 12 ft **51.** 4 units
53. $\frac{4}{3}$ cm **55.** 30 mph **57.** \$4.80 or \$5.20 **59.** 4,000

61. 2.26 in. **63.** about 6.13×10^{-3} M **67.** $\sqrt{2}, -3\sqrt{2}$
69. $i, 2i$

Getting Ready (page 659)

1. 17 **2.** -8

Orals (page 663)

1. -3 **2.** -7 **3.** irrational and unequal
4. complex conjugates **5.** no **6.** yes

Exercise 10.3 (page 664)

1. -2 **3.** $\frac{9}{5}$ **5.** $b^2 - 4ac$ **7.** rational, unequal
9. rational, equal **11.** complex conjugates **13.** irrational, unequal **15.** rational, unequal **17.** $6, -6$ **19.** $12, -12$
21. 5 **23.** $12, -3$ **25.** yes **27.** $k < -\frac{4}{3}$
29. $1, -1, 4, -4$ **31.** $1, -1, \sqrt{2}, -\sqrt{2}$
33. $1, -1, \sqrt{5}, -\sqrt{5}$ **35.** $1, -1, 2, -2$
37. $1, -1, 2\sqrt{3}, -2\sqrt{3}$ **39.** 1 **41.** no solution
43. $-8, -27$ **45.** $-1, 27$ **47.** $-1, -4$ **49.** $4, -5$
51. $0, 2$ **53.** $-1, -\frac{27}{13}$ **55.** $1, 1, -1, -1$ **57.** $1 \pm i$
59. $x = \pm\sqrt{r^2 - y^2}$ **61.** $d = \pm\sqrt{\dfrac{k}{I}} = \pm\dfrac{\sqrt{kI}}{I}$
63. $y = \dfrac{-3x \pm \sqrt{9x^2 - 28x}}{2x}$ **65.** $\mu^2 = \dfrac{\Sigma x^2}{N} - \sigma^2$
67. $\frac{2}{3}, -\frac{1}{4}$ **69.** $-\dfrac{5}{4} \pm \dfrac{\sqrt{17}}{4}$ **71.** $\dfrac{1}{3} \pm \dfrac{\sqrt{11}}{3}i$ **73.** $-1 \pm 2i$
77. no

Getting Ready (page 665)

1. -2 **2.** 2 **3.** 0 **4.** 8 **5.** 1 **6.** 4

Orals (page 674)

1. down **2.** up **3.** up **4.** down **5.** $(3, -1)$
6. $(-2, 2)$

Exercise 10.4 (page 674)

1. 10 **3.** $3\frac{3}{5}$ hr **5.** $f(x) = ax^2 + bx + c, a \neq 0$
7. vertex **9.** upward **11.** to the right **13.** upward

15.
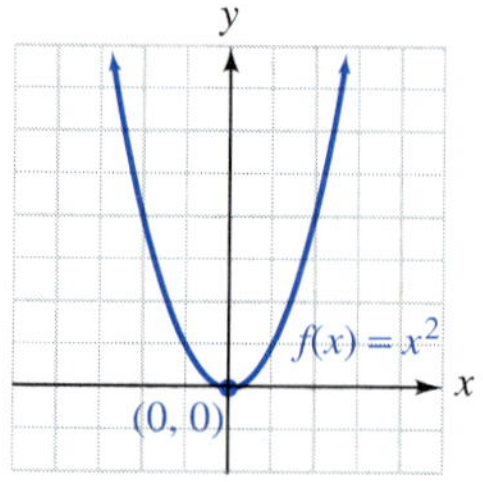

17.
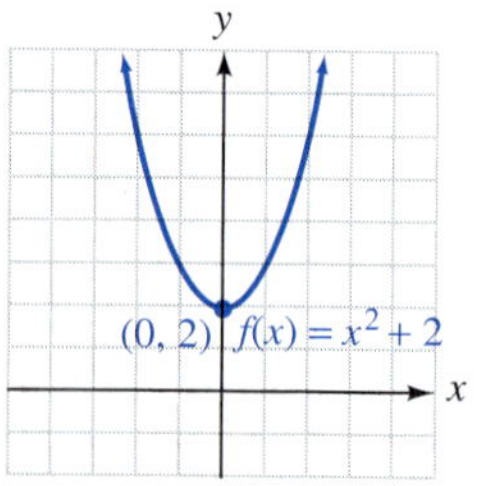

19.
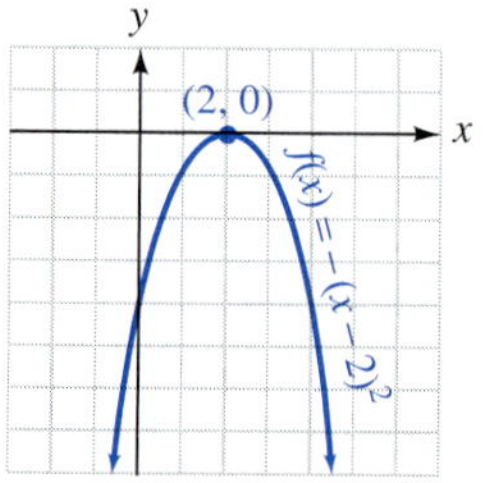

21.
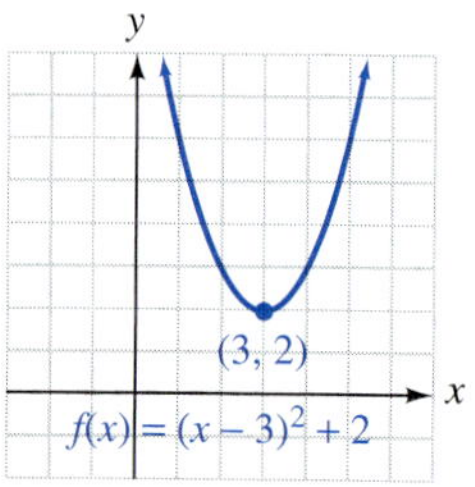

23.

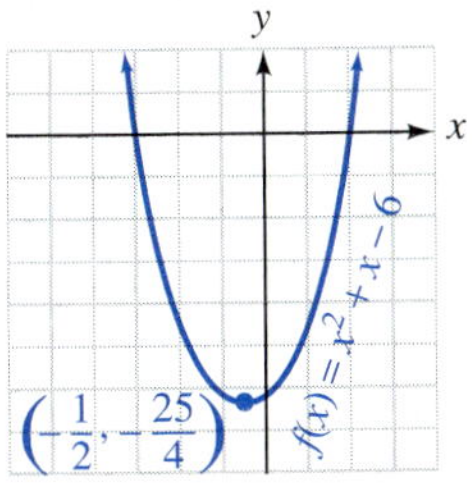

25.

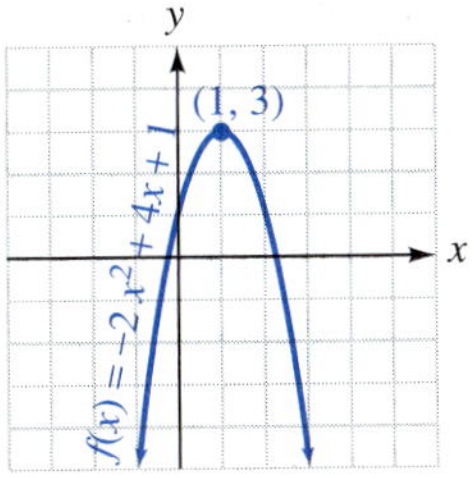

27. $(1, 2), x = 1$ **29.** $(-3, -4), x = -3$ **31.** $(0, 0), x = 0$
33. $(1, -2), x = 1$ **35.** $(2, 21), x = 2$ **37.** $\left(\frac{5}{12}, \frac{143}{24}\right), x = \frac{5}{12}$
39. $(5, 2)$ **41.** $(0.25, 0.88)$ **43.** $(0.5, 7.25)$ **45.** $2, -3$
47. $-1.85, 3.25$ **49.** 36 ft, 3 sec
51. 50 ft by 50 ft, 2,500 ft^2 **53.** 75 ft by 75 ft, 5,625 ft^2
55. 5,000 **57.** 3,276, \$14,742 **59.** \$35
61. 0.25 and 0.75

Getting Ready (page 677)

1. $(x + 5)(x - 3)$ **2.** $(x - 2)(x - 1)$

Orals (page 683)

1. $x = 2$ **2.** $x > 2$ **3.** $x < 2$ **4.** $x = -3$ **5.** $x > -3$
6. $x < -3$ **7.** $1 < 2x$ **8.** $1 > 2x$

Exercise 10.5 (page 684)

1. $y = kx$ **3.** $t = kxy$ **5.** 3 **7.** greater **9.** undefined

11. (1, 4)

13. (−∞, 3) ∪ (5, ∞)

15. [−4, 3]

17. (−∞, −5] ∪ [3, ∞)

19. no solutions

21. (−∞, −3] ∪ [3, ∞)

23. (−5, 5)

25. (−∞, 0) ∪ (1/2, ∞)

27. (0, 2]

29. (−∞, −5/3) ∪ (0, ∞)

31. (−∞, −3) ∪ (1, 4)

33. [−5, −2) ∪ [4, ∞)

35. (−∞, −4)

37. (−1/2, 1/3) ∪ (1/2, ∞)

39. (0, 2) ∪ (8, ∞)

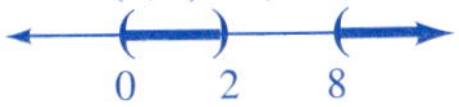

41. (−∞, −2) ∪ (2, 18]

43. [−34/5, −4) ∪ (3, ∞)

45. (−4, −2] ∪ (−1, 2]

47. (−∞, −16) ∪ (−4, −1) ∪ (4, ∞)

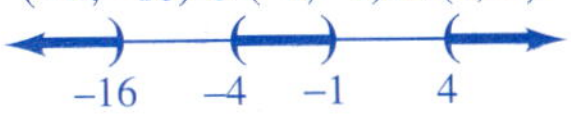

49. (−∞, −2) ∪ (−2, ∞)

51. $(-1, 3)$ **53.** $(-\infty, -3) \cup (2, \infty)$

55.

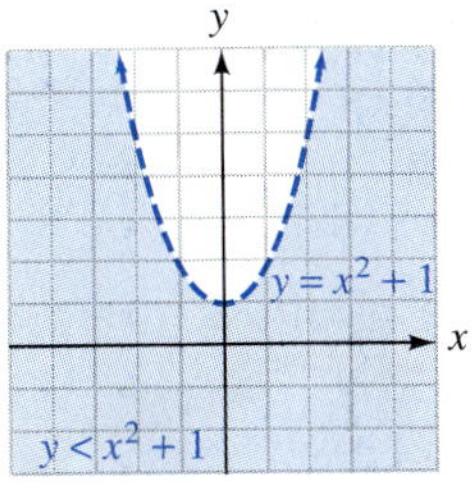

57.

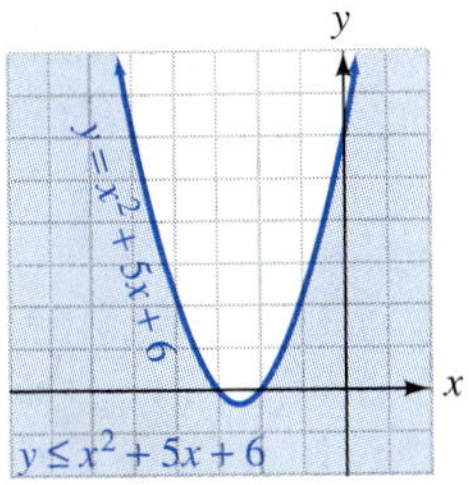

59.

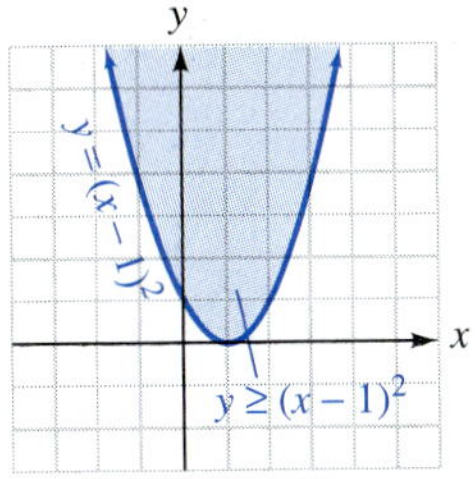

61.

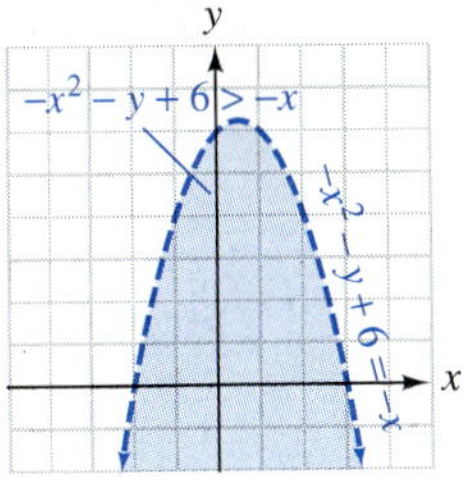

63.

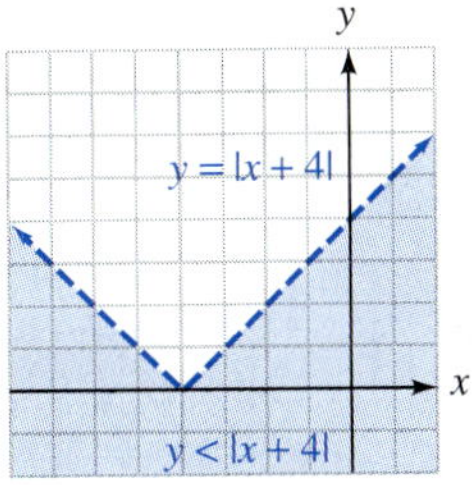

65.

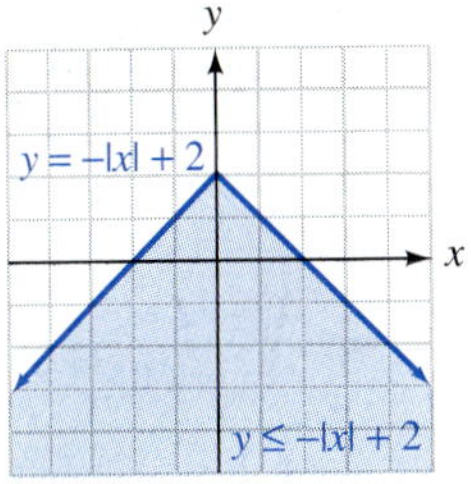

69. when 4 factors are negative, 2 factors are negative, or no factors are negative

Getting Ready (page 686)

1. $3x - 1$ **2.** $x + 3$ **3.** $2x^2 - 3x - 2$ **4.** $\frac{2x + 1}{x - 2}$

Orals (page 692)

1. $5x$ **2.** x **3.** $8x^2$ **4.** $\frac{3}{2}$ **5.** 2 **6.** $12x^2$ **7.** $8x$
8. $6x$

Exercise 10.6 (page 692)

1. $-\frac{3x + 7}{x + 2}$ **3.** $\frac{x - 4}{3x^2 - x - 12}$ **5.** $f(x) + g(x)$ **7.** $f(x)g(x)$
9. domain **11.** $f(x)$ **13.** $7x, (-\infty, \infty)$
15. $12x^2, (-\infty, \infty)$ **17.** $x, (-\infty, \infty)$
19. $\frac{4}{3}, (-\infty, 0) \cup (0, \infty)$ **21.** $3x - 2, (-\infty, \infty)$
23. $2x^2 - 5x - 3, (-\infty, \infty)$ **25.** $-x - 4, (-\infty, \infty)$
27. $\frac{x - 3}{2x + 1}, \left(-\infty, -\frac{1}{2}\right) \cup \left(-\frac{1}{2}, \infty\right)$
29. $-2x^2 + 3x - 3, (-\infty, \infty)$
31. $(3x - 2)/(2x^2 + 1), (-\infty, \infty)$ **33.** $3, (-\infty, \infty)$
35. $(x^2 - 4)/(x^2 - 1), (-\infty, -1) \cup (-1, 1) \cup (1, \infty)$ **37.** 7
39. 24 **41.** -1 **43.** $-\frac{1}{2}$ **45.** $2x^2 - 1$ **47.** $16x^2 + 8x$
49. 58 **51.** 110 **53.** 2 **55.** $9x^2 - 9x + 2$ **57.** 2
59. $2x + h$ **61.** $4x + 2h$ **63.** $2x + h + 1$
65. $2x + h + 3$ **67.** $4x + 2h + 3$ **69.** 2 **71.** $x + a$

73. $2x + 2a$ **75.** $x + a + 1$ **77.** $x + a + 3$
79. $2x + 2a + 3$ **85.** $3x^2 + 3xh + h^2$
87. $C(t) = \frac{5}{9}(2{,}668 - 200t)$

Getting Ready (page 694)

1. $y = \dfrac{x - 2}{3}$ **2.** $y = \dfrac{2x - 10}{3}$

Orals (page 701)

1. $\{(2, 1), (3, 2), (10, 5)\}$ **2.** $\{(1, 1), (8, 2), (64, 4)\}$
3. $f^{-1}(x) = 2x$ **4.** $f^{-1}(x) = \frac{1}{2}x$ **5.** no **6.** yes

Exercise 10.7 (page 701)

1. $3 - 8i$ **3.** $18 - i$ **5.** 10 **7.** one-to-one **9.** 2
11. x **13.** yes **15.** no

17.

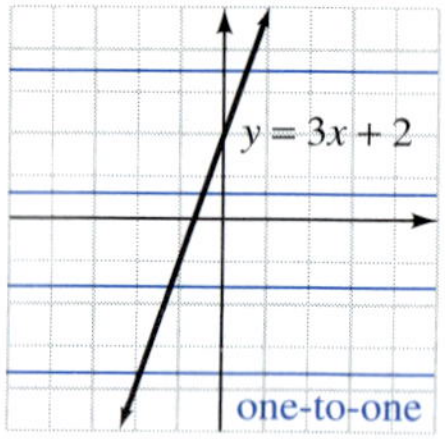

19.

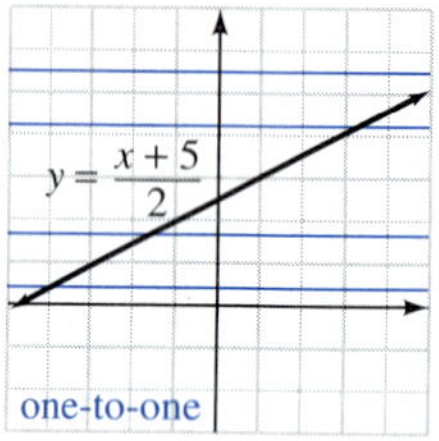

21.

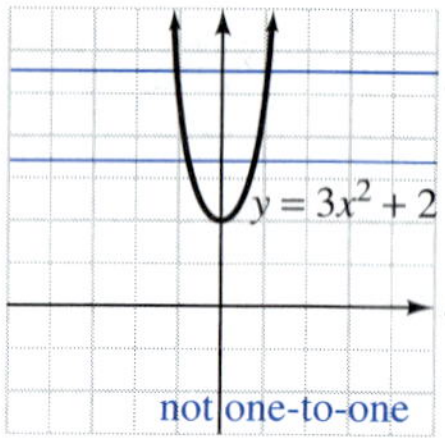

23.

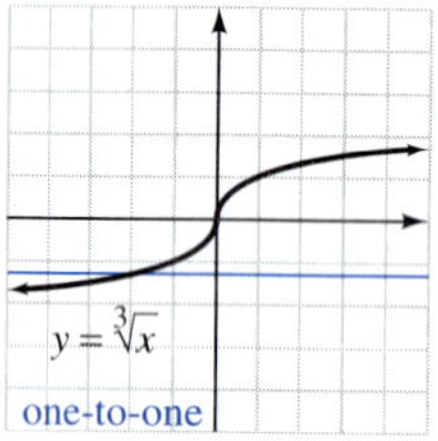

25. $\{(2, 3), (1, 2), (0, 1)\}$; yes
27. $\{(2, 1), (3, 2), (3, 1), (5, 1)\}$; no
29. $\{(1, 1), (4, 2), (9, 3), (16, 4)\}$; yes **31.** $f^{-1}(x) = \frac{1}{3}x - \frac{1}{3}$
33. $f^{-1}(x) = 5x - 4$ **35.** $f^{-1}(x) = 5x + 4$
37. $f^{-1}(x) = \frac{5}{4}x + 5$

39.

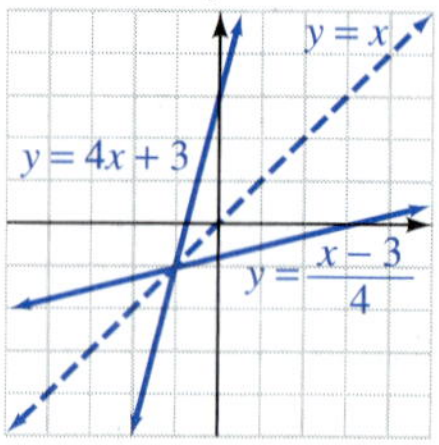

41.

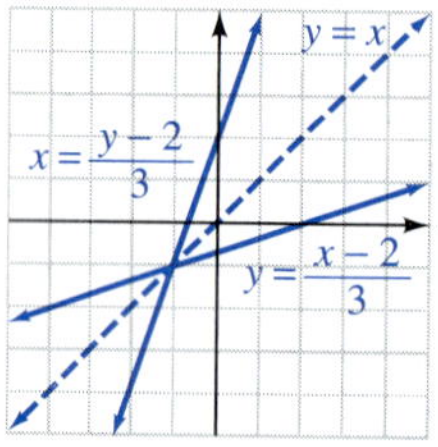

43.

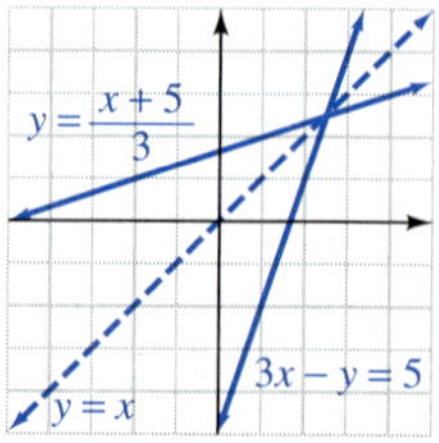

45.

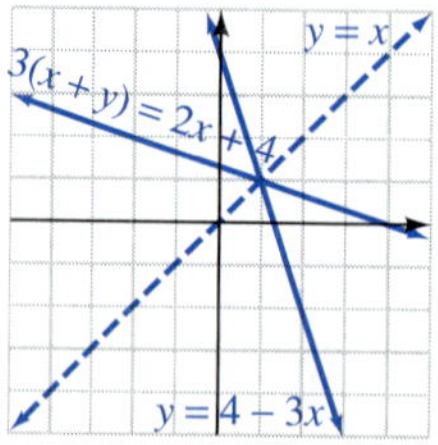

47. $y = \pm\sqrt{x - 4}$, no **49.** $y = \sqrt[3]{x}$, yes
51. $x = |y|$, no **53.** $f^{-1}(x) = \sqrt[3]{\frac{x + 3}{2}}$

55.

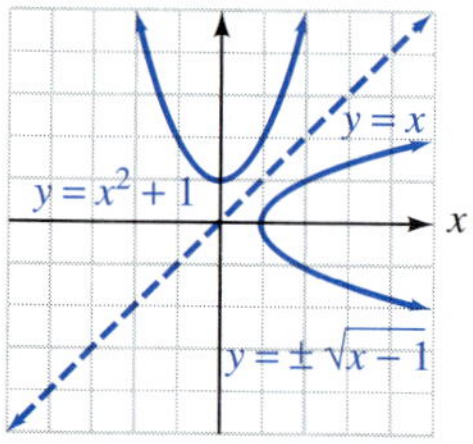

57.

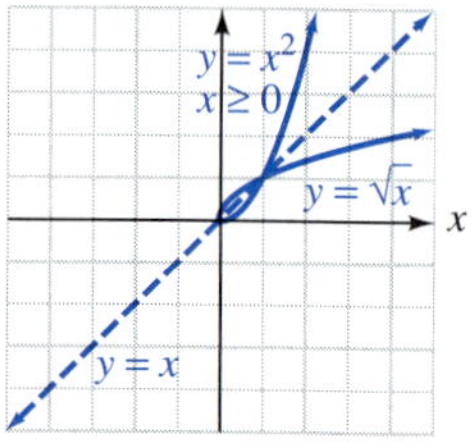

61. $f^{-1}(x) = \frac{x + 1}{x - 1}$

Chapter Summary (page 705)

1. $\frac{2}{3}, -\frac{3}{4}$ **2.** $-\frac{1}{3}, -\frac{5}{2}$ **3.** $\frac{2}{3}, -\frac{4}{5}$ **4.** $4, -8$ **5.** $-4, -2$
6. $\frac{7}{2}, 1$ **7.** $\dfrac{1}{4} \pm \dfrac{\sqrt{41}}{4}$ **8.** $9, -1$ **9.** $0, 10$ **10.** $\frac{1}{2}, -7$
11. $-7, \frac{1}{3}$ **12.** $\dfrac{1}{4} \pm \dfrac{\sqrt{17}}{4}$ **13.** $-\dfrac{1}{2} \pm \dfrac{\sqrt{7}}{2}i$
14. 4 cm by 6 cm **15.** 2 ft by 3 ft **16.** 7 sec **17.** 196 ft
18. irrational, unequal **19.** complex conjugates
20. 12, 152 **21.** $k \geq -\frac{7}{3}$ **22.** 1, 144 **23.** $8, -27$
24. 1 **25.** $1, -\frac{8}{5}$ **26.** $\frac{14}{3}$ **27.** 1

28.

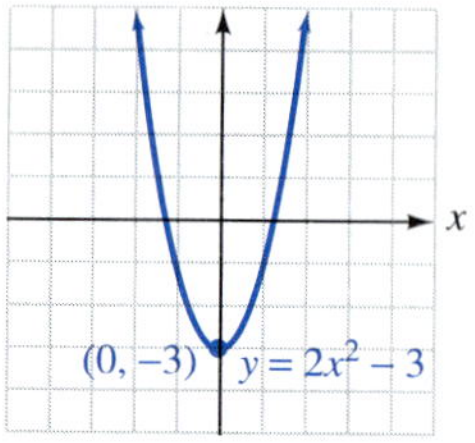

29.

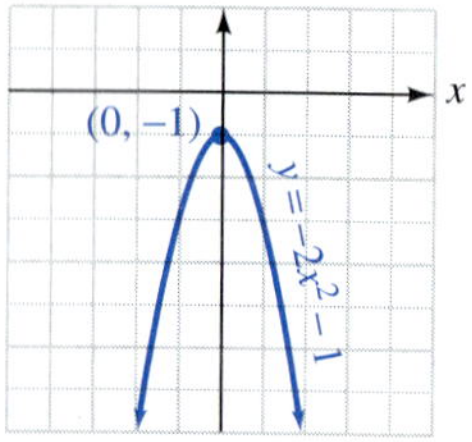

30.

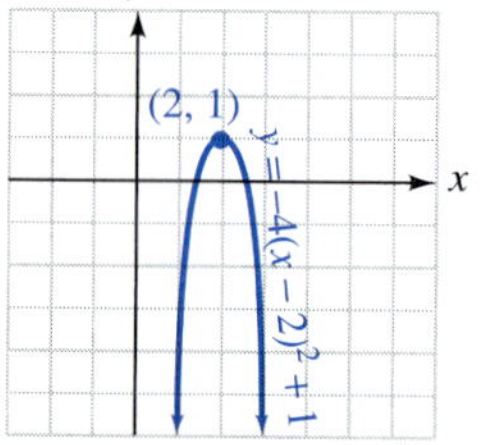

31.

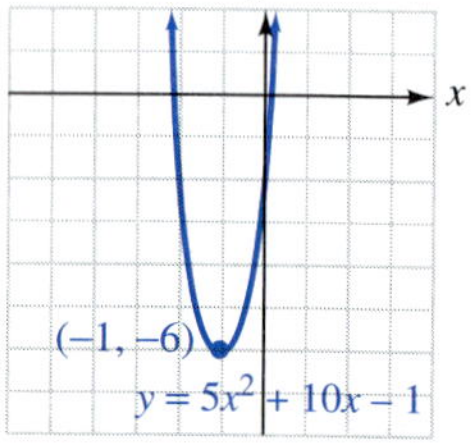

32.

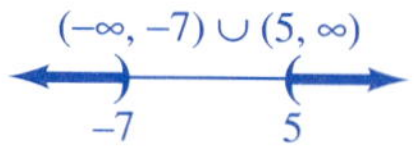

33.

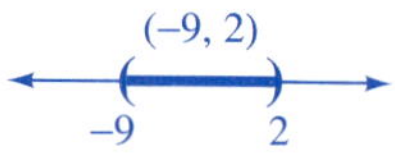

34.

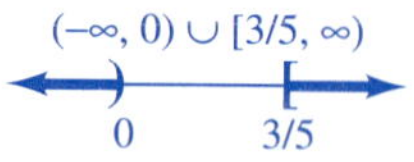

35.

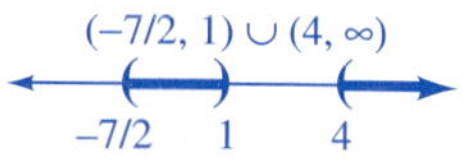

36. $x < -7$ or $x > 5$ **37.** $-9 < x < 2$

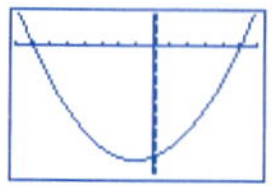

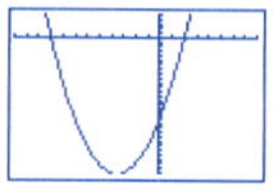

38. $x < 0$ or $x \geq \frac{3}{5}$ **39.** $-\frac{7}{2} < x < 1$ or $x > 4$

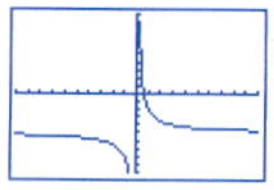

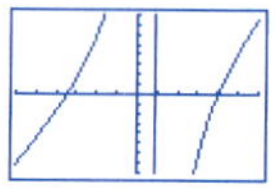

40.

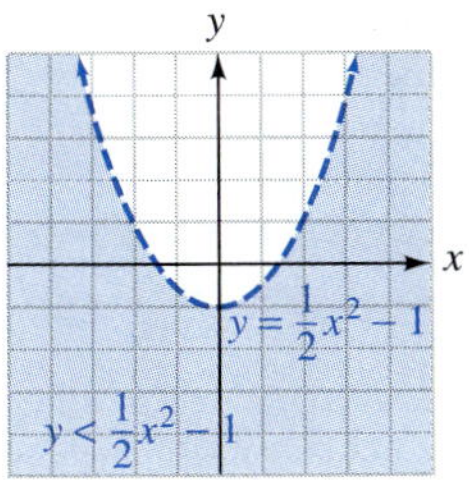

41.

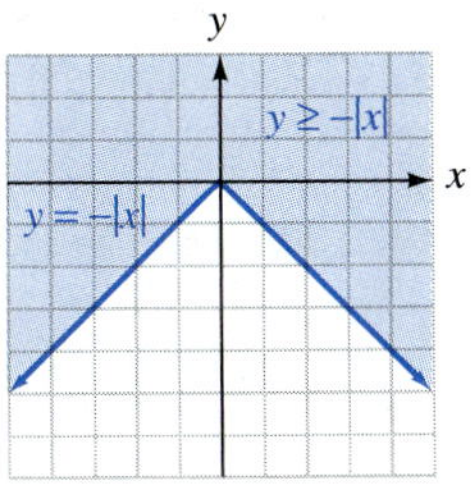

42. $(f + g)(x) = 3x + 1$ **43.** $(f - g)(x) = x - 1$

44. $(f \cdot g)(x) = 2x^2 + 2x$ **45.** $(f/g)(x) = \dfrac{2x}{x + 1}$ **46.** 6

47. -1 **48.** $2(x + 1)$ **49.** $2x + 1$

50. yes **51.** no

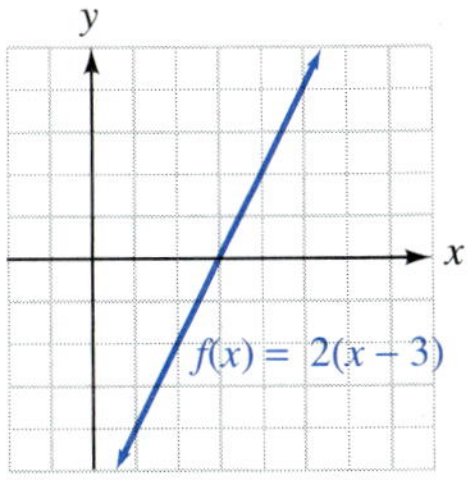

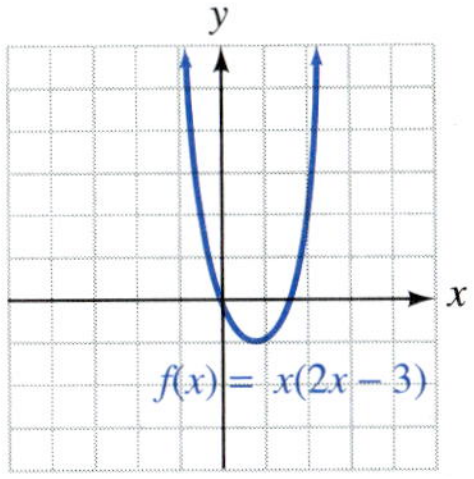

52. no **53.** no

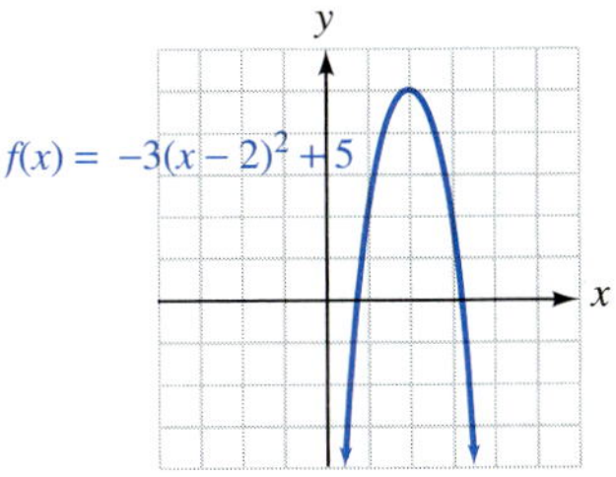

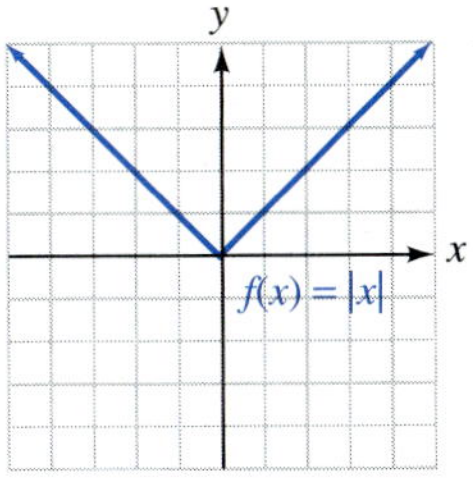

54. $f^{-1}(x) = \dfrac{x + 3}{6}$ **55.** $f^{-1}(x) = \dfrac{x - 5}{4}$

56. $y = \sqrt{\dfrac{x + 1}{2}}$ **57.** $x = |y|$

Chapter 10 Test (page 709)

1. $3, -6$ **2.** $-\frac{3}{2}, -\frac{5}{3}$ **3.** 144 **4.** 625 **5.** $-2 \pm \sqrt{3}$

6. $\dfrac{5}{2} \pm \dfrac{\sqrt{37}}{2}$ **7.** $-\dfrac{5}{4} \pm \dfrac{\sqrt{17}}{4}$ **8.** $\dfrac{1}{2} \pm \dfrac{\sqrt{11}}{2}i$

9. nonreal **10.** 2 **11.** 10 in. **12.** $1, \frac{1}{4}$

13.

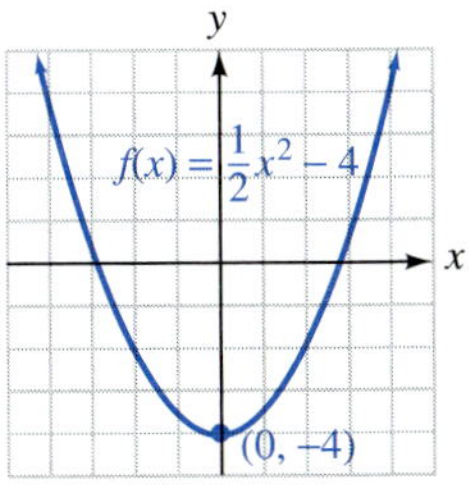

14.

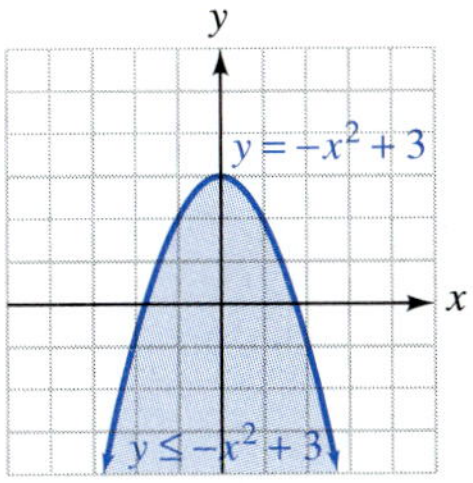

15.

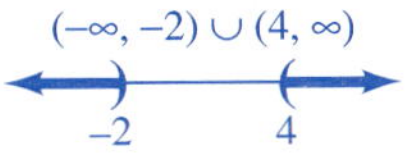

16.

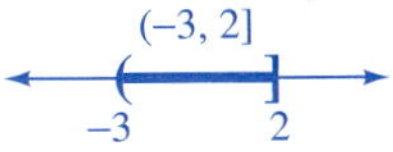

17. $(g + f)(x) = 5x - 1$ **18.** $(f - g)(x) = 3x + 1$

19. $(g \cdot f)(x) = 4x^2 - 4x$ **20.** $(g/f)(x) = \dfrac{x - 1}{4x}$ **21.** 3

22. -4 **23.** -8 **24.** -9 **25.** $4(x - 1)$ **26.** $4x - 1$

27. $y = \frac{12 - 2x}{3}$ **28.** $y = -\sqrt{\dfrac{x - 4}{3}}$

Cumulative Review Exercises (page 710)

1. D = $(-\infty, \infty)$; R = $[-3, \infty)$

2. D = $(-\infty, \infty)$; R = $(-\infty, 0]$ **3.** $y = 3x + 2$

4. $y = -\frac{2}{3}x - 2$ **5.** $-4a^2 + 12a - 7$ **6.** $6x^2 - 5x - 6$

7. $(x^2 + 4y^2)(x + 2y)(x - 2y)$ **8.** $(3x + 2)(5x - 4)$

9. $6, -1$ **10.** $0, \frac{2}{3}, -\frac{1}{2}$ **11.** $5x^2$ **12.** $4t\sqrt{3t}$ **13.** $-3x$

14. $4x$ **15.** $\frac{1}{2}$ **16.** 16 **17.** y^2 **18.** $x^{17/12}$

19.

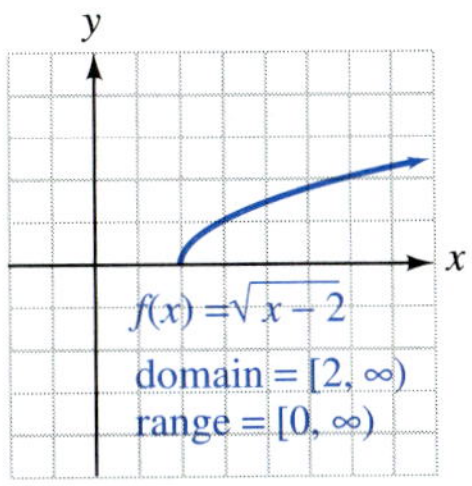

20.

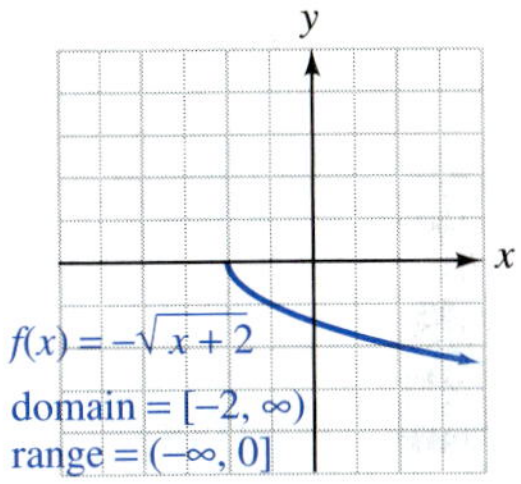

21. $x^{4/3} - x^{2/3}$ **22.** $\frac{1}{x} + 2 + x$ **23.** $7\sqrt{2}$

24. $-12\sqrt[4]{2} + 10\sqrt[4]{3}$ **25.** $-18\sqrt{6}$ **26.** $\dfrac{5\sqrt[3]{x^2}}{x}$

27. $\dfrac{x + 3\sqrt{x} + 2}{x - 1}$ **28.** $\sqrt{xy}$ **29.** 2, 7 **30.** $\frac{1}{4}$

31. $3\sqrt{2}$ in. **32.** $2\sqrt{3}$ in. **33.** 10 **34.** 9

35. $1, -\frac{3}{2}$ **36.** $-\frac{2}{3} \pm \frac{\sqrt{7}}{3}$

37.

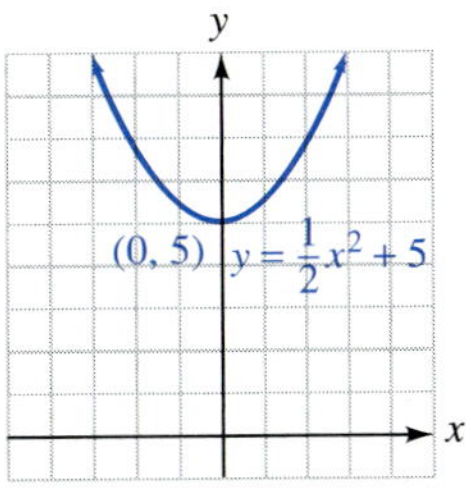

38.

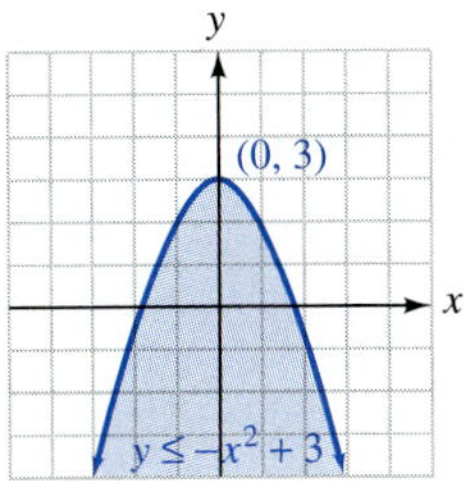

39. $7 + 2i$ **40.** $-5 - 7i$ **41.** 13 **42.** $12 - 6i$

43. $-12 - 10i$ **44.** $\frac{3}{2} + \frac{1}{2}i$ **45.** $\sqrt{13}$ **46.** $\sqrt{61}$

47. -2 **48.** 9, 16

49.

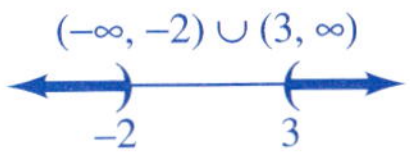

50.

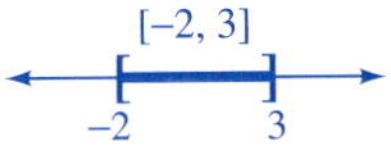

51. 5 **52.** 27 **53.** $12x^2 - 12x + 5$ **54.** $6x^2 + 3$

55. $f^{-1}(x) = \frac{x - 2}{3}$ **56.** $f^{-1}(x) = \sqrt[3]{x - 4}$

Getting Ready (page 713)

1. 8 **2.** 5 **3.** $\frac{1}{25}$ **4.** $\frac{8}{27}$

Orals (page 722)

1. 4 **2.** 25 **3.** 18 **4.** 3 **5.** $\frac{1}{4}$ **6.** $\frac{1}{25}$ **7.** $\frac{2}{9}$ **8.** $\frac{1}{27}$

Exercise 11.1 (page 722)

1. 40 **3.** 120° **5.** exponential **7.** $(0, \infty)$
9. increasing **11.** $P\left(1 + \frac{r}{k}\right)^{kt}$ **13.** 2.6651 **15.** 36.5548
17. 8 **19.** $7^{3\sqrt{3}}$

21.

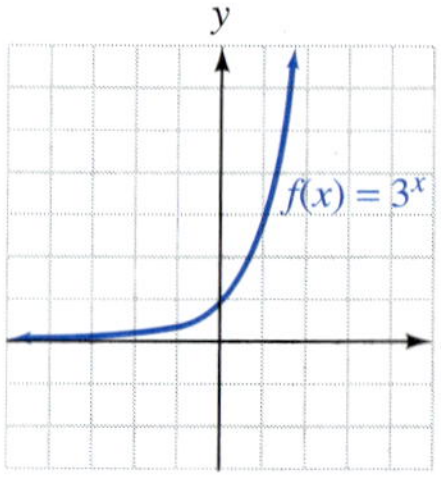

23.

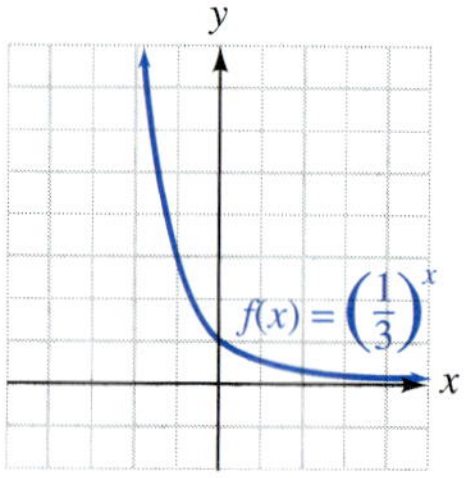

25.

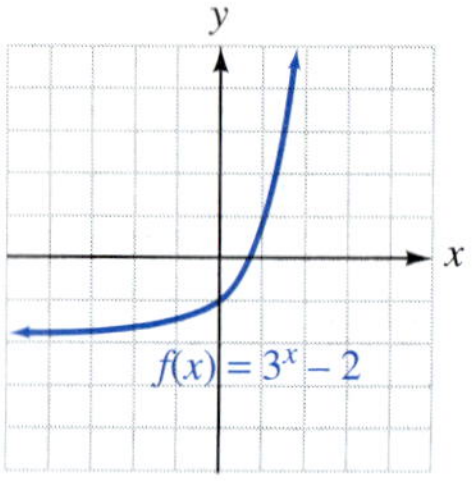

27.

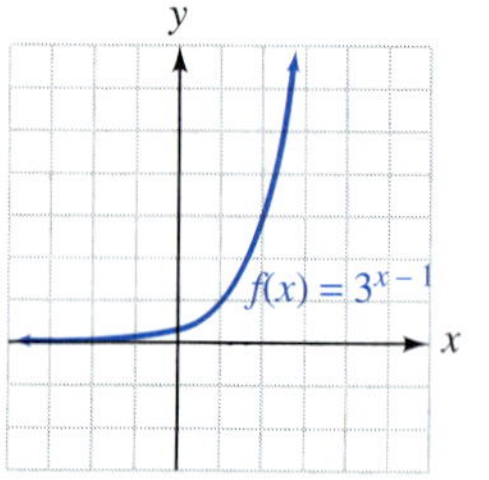

29. $b = \frac{1}{2}$ **31.** no value **33.** $b = 2$ **35.** $b = 3$
37. increasing function **39.** decreasing function

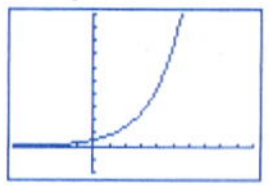

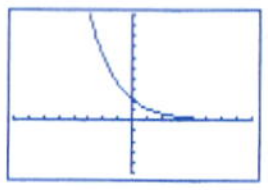

41. \$22,080.40 **43.** \$32.03 **45.** \$2,273,996.13
47. $\frac{32}{243}A_0$ **49.** 5.0421×10^{-5} coulombs **51.** \$1,115.33

Getting Ready (page 724)

1. 2 **2.** 2.25 **3.** 2.44 **4.** 2.59

Orals (page 729)

1. 1 **2.** 2.72 **3.** 7.39 **4.** 20.09 **5.** 2 **6.** 2

Exercise 11.2 (page 729)

1. $4x^2\sqrt{15x}$ **3.** $10y\sqrt{3y}$ **5.** 2.72 **7.** increasing
9. $A = Pe^{rt}$

11.

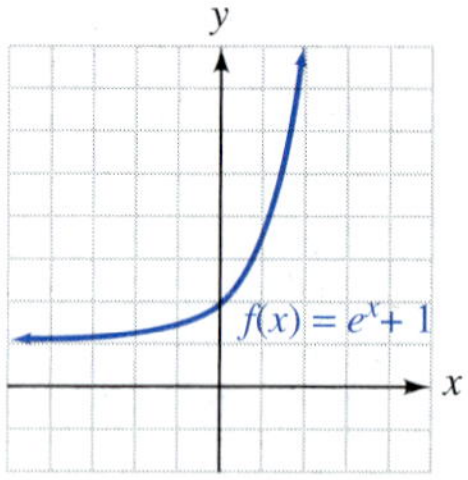

13.

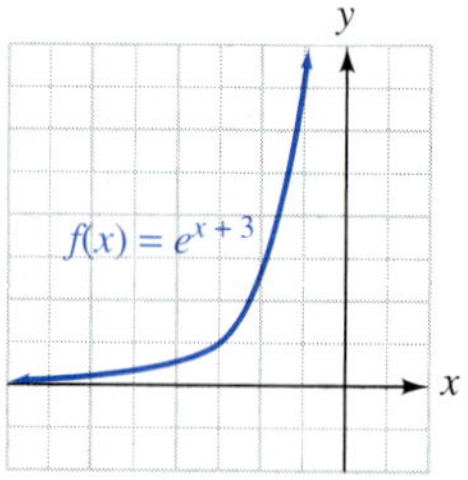

15.

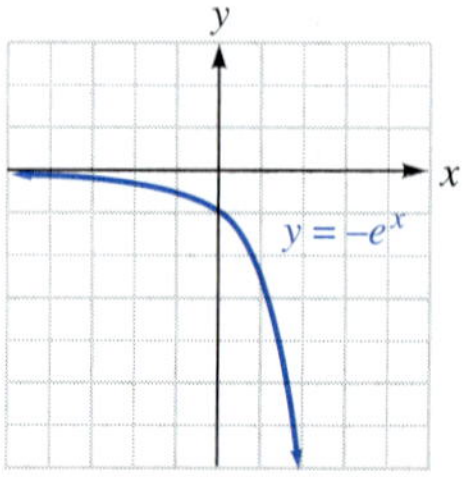

17.

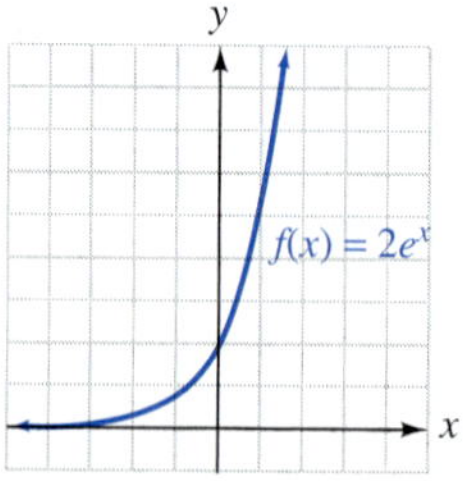

19. no **21.** no **23.** \$10,272.17 **25.** \$6,391.10
27. \$7,518.28 from annual compounding; \$7,647.95 from continuous compounding **29.** 10.6 billion **31.** 2.6
33. 6,817 **35.** 0.16 **37.** 0 **39.** 49 mps
41. \$3,094.15 **43.** 72 yr **49.** $k = e^5$

Getting Ready (page 732)

1. 1 **2.** 25 **3.** $\frac{1}{25}$ **4.** 4

Orals (page 739)

1. 3 **2.** 2 **3.** 5 **4.** 2 **5.** 2 **6.** 2 **7.** $\frac{1}{4}$ **8.** $\frac{1}{2}$
9. 2

Exercise 11.3 (page 739)

1. 10 **3.** 0; $-\frac{5}{9}$ does not check **5.** $x = b^y$ **7.** range
9. inverse **11.** exponent **13.** $(b, 1), (1, 0)$ **15.** $20 \log \frac{E_O}{E_I}$
17. $3^3 = 27$ **19.** $\left(\frac{1}{2}\right)^2 = \frac{1}{4}$ **21.** $4^{-3} = \frac{1}{64}$ **23.** $\left(\frac{1}{2}\right)^3 = \frac{1}{8}$
25. $\log_6 36 = 2$ **27.** $\log_5 \frac{1}{25} = -2$ **29.** $\log_{1/2} 32 = -5$
31. $\log_x z = y$ **33.** 4 **35.** 2 **37.** 3 **39.** $\frac{1}{2}$ **41.** -3
43. 49 **45.** 6 **47.** 5 **49.** $\frac{1}{25}$ **51.** $\frac{1}{6}$ **53.** $-\frac{3}{2}$ **55.** $\frac{2}{3}$
57. 5 **59.** $\frac{3}{2}$ **61.** 4 **63.** 8 **65.** 4 **67.** 4 **69.** 3
71. 100 **73.** 0.9165 **75.** -2.0620 **77.** 25.25
79. 17,378.01 **81.** 0.00 **83.** 8
85. increasing **87.** decreasing

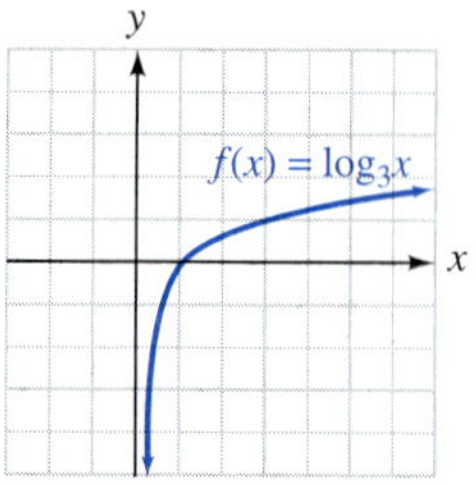

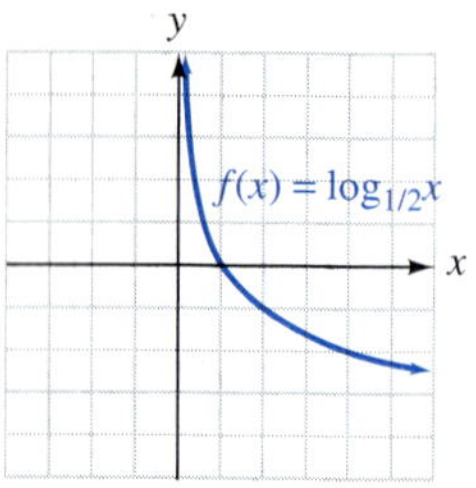

89.

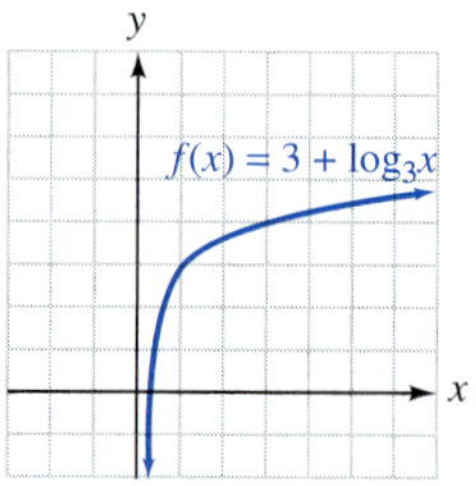

91.

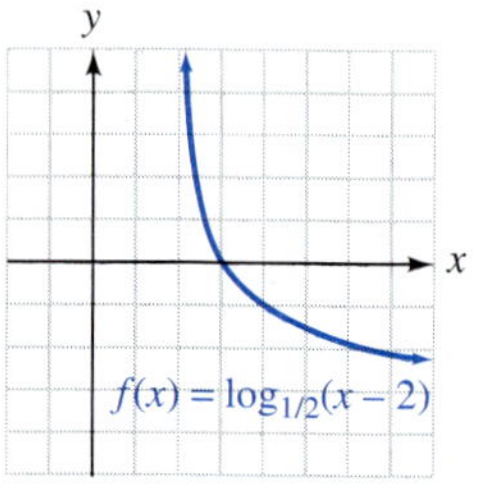

93.

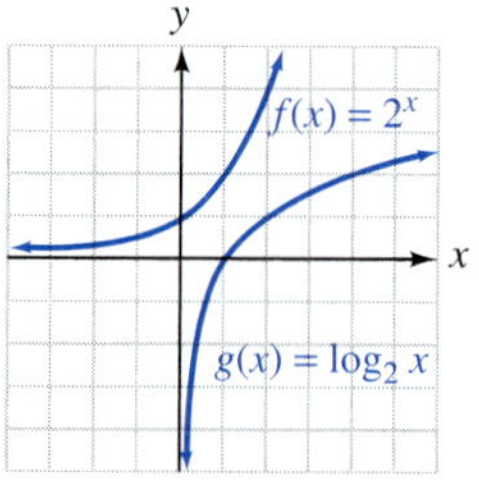

95.

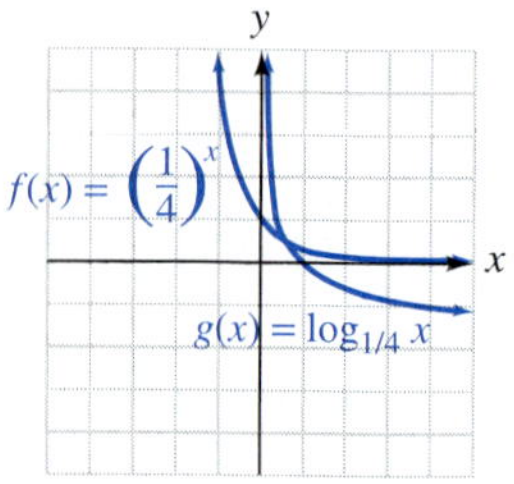

97. no value of b **99.** 3 **101.** 29.0 dB **103.** 49.5 dB
105. 4.4 **107.** 4 **109.** 4.2 yr **111.** 10.8 yr

Getting Ready (page 742)

1. 2 **2.** -3 **3.** 1 **4.** 0

Orals (page 746)

1. $e^y = x$ **2.** $\ln b = a$ **3.** $t = \frac{\ln 2}{r}$

Exercise 11.4 (page 746)

1. $y = 5x$ **3.** $y = -\frac{3}{2}x + \frac{13}{2}$ **5.** $x = 5$ **7.** $\dfrac{1}{2x - 3}$

9. $\dfrac{x + 1}{3(x - 2)}$ **11.** $\log_e x$ **13.** $(-\infty, \infty)$ **15.** 10

17. $\dfrac{\ln 2}{r}$ **19.** 3.2288 **21.** 2.2915 **23.** −0.1592

25. none **27.** 9.9892 **29.** 23.8075 **31.** 0.0089

33. 61.9098 **35.** no **37.** no

39.

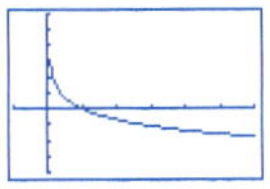

41.

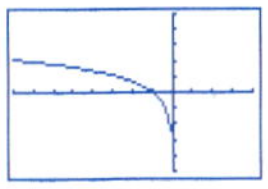

43. 5.8 yr
45. 9.2 yr
47. about 2.9 hr

Getting Ready (page 748)

1. x^{m+n} **2.** 1 **3.** x^{mn} **4.** x^{m-n}

Orals (page 755)

1. 2 **2.** 5 **3.** 343 **4.** $\frac{1}{4}$ **5.** 2 **6.** 2 **7.** $\frac{1}{4}$ **8.** $\frac{1}{2}$
9. 2

Exercise 11.5 (page 755)

1. $-\frac{7}{6}$ **3.** $\left(1, -\frac{1}{2}\right)$ **5.** 0 **7.** M, N **9.** x, y **11.** x

13. $\neq$ **15.** 0 **17.** 7 **19.** 10 **21.** 1 **23.** 0 **25.** 7

27. 10 **29.** 1 **37.** $\log_b x + \log_b y + \log_b z$

39. $\log_b 2 + \log_b x - \log_b y$ **41.** $3 \log_b x + 2 \log_b y$

43. $\frac{1}{2}(\log_b x + \log_b y)$ **45.** $\log_b x + \frac{1}{2} \log_b z$

47. $\frac{1}{3} \log_b x - \frac{1}{4} \log_b y - \frac{1}{4} \log_b z$ **49.** $\log_b \dfrac{x + 1}{x}$

51. $\log_b x^2 y^{1/2}$ **53.** $\log_b \dfrac{z^{1/2}}{x^3 y^2}$ **55.** $\log_b \dfrac{\frac{x}{z} + x}{\frac{y}{z} + y} = \log_b \dfrac{x}{y}$

57. false **59.** false **61.** true **63.** false **65.** true

67. true **69.** 1.4472 **71.** 0.3521 **73.** 1.1972

75. 2.4014 **77.** 2.0493 **79.** 0.4682 **81.** 1.7712

83. −1.0000 **85.** 1.8928 **87.** 2.3219 **89.** 4.77

91. from 2.5119×10^{-8} to 1.585×10^{-7}

93. It will increase by $k \ln 2$.

95. The intensity must be cubed.

Getting Ready (page 757)

1. $2 \log x$ **2.** $\frac{1}{2} \log x$ **3.** 0 **4.** $2b \log a$

Orals (page 764)

1. $x = \frac{\log 5}{\log 3}$ **2.** $x = \frac{\log 3}{\log 5}$ **3.** $x = -\frac{\log 7}{\log 2}$ **4.** $x = -\frac{\log 1}{\log 6}$

5. $x = 2$ **6.** $x = \frac{1}{2}$ **7.** $x = 10$ **8.** $x = 10$

Exercise 11.6 (page 764)

1. 0, 5 **3.** $\frac{2}{3}, -4$ **5.** exponential **7.** $A_0 2^{-t/h}$

9. 1.1610 **11.** 3.9120 **13.** 22.1184 **15.** 1.2702

17. 1.7095 **19.** 0 **21.** ± 1.0878 **23.** 0, 1.0566

25. 3, −1 **27.** −2, −2 **29.** 0 **31.** 0.2789 **33.** 1.8

35. 3, −1 **37.** 2 **39.** 3 **41.** −7 **43.** 4

45. 10, −10 **47.** 50 **49.** 20 **51.** 10 **53.** 10

55. 1, 100 **57.** no solution **59.** 6 **61.** 9 **63.** 4

65. 1, 7 **67.** 20 **69.** 8 **71.** 5.1 yr **73.** 42.7 days

75. about 4,200 yr **77.** 5.6 yr **79.** 5.4 yr

81. because $\ln 2 \approx 0.7$ **83.** 25.3 yr **85.** 2.828 times larger

87. 13.3 **91.** $x \leq 3$

Chapter Summary (page 768)

1. $5^{2\sqrt{2}}$ **2.** $2^{\sqrt{10}}$

3.

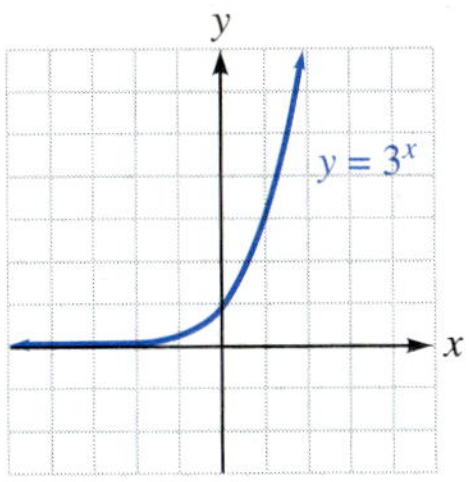

4.

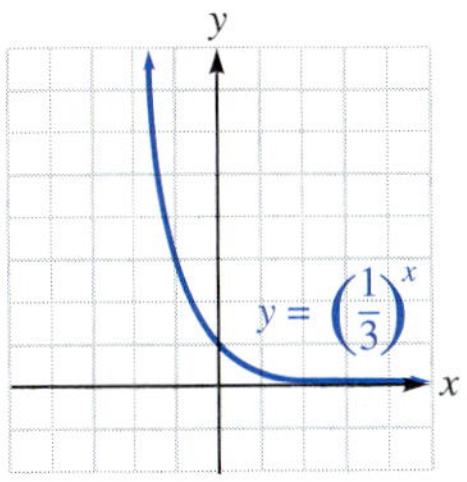

5. $x = 1, y = 6$ **6.** D: $(-\infty, \infty)$, R: $(0, \infty)$

7.

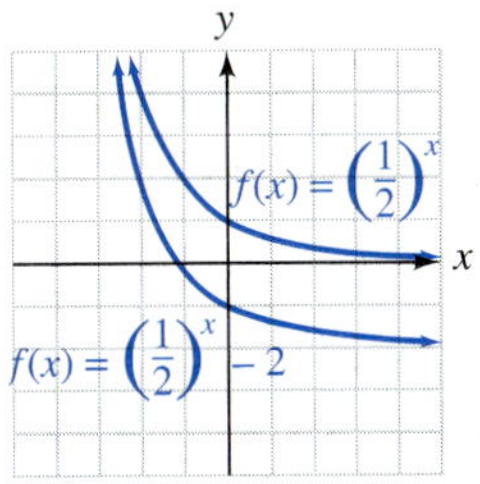

8.

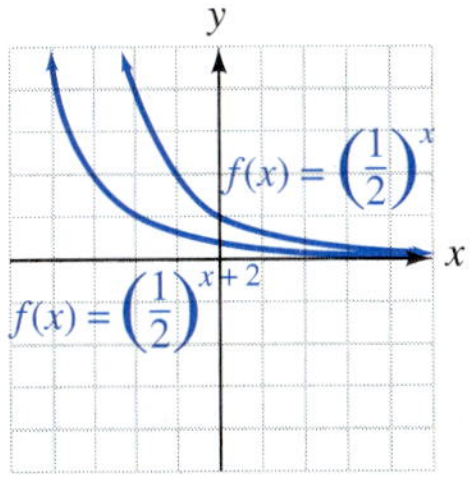

9. \$2,189,703.45 **10.** \$2,324,767.37

11.

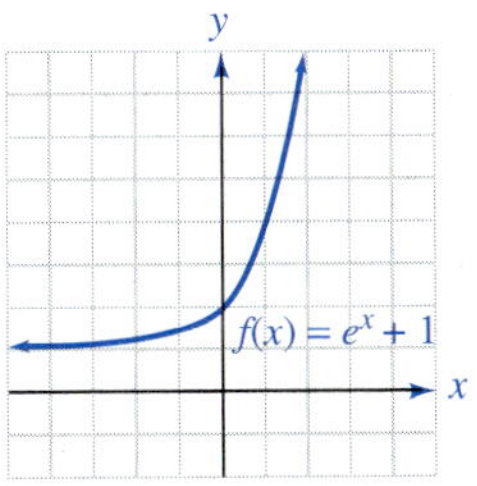

12.

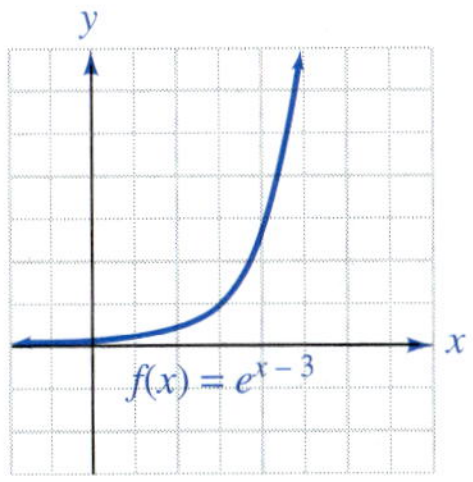

13. about 582,000,000 **14.** D: $(0, \infty)$, R: $(-\infty, \infty)$ **15.** 2

16. $-\frac{1}{2}$ **17.** 0 **18.** −2 **19.** $\frac{1}{2}$ **20.** $\frac{1}{3}$ **21.** 32

22. 9 **23.** 27 **24.** −1 **25.** $\frac{1}{8}$ **26.** 2 **27.** 4 **28.** 2

29. 10 **30.** $\frac{1}{25}$ **31.** 5 **32.** 3

33.

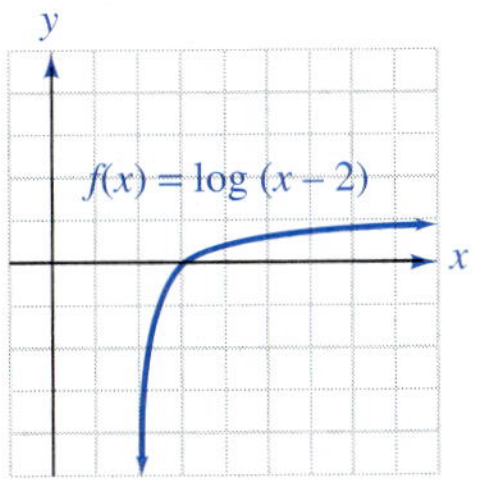

34.

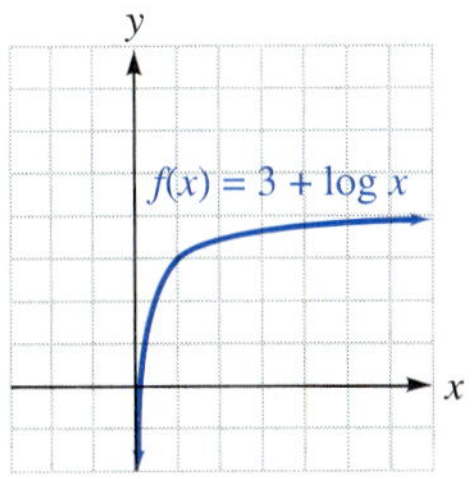

35.

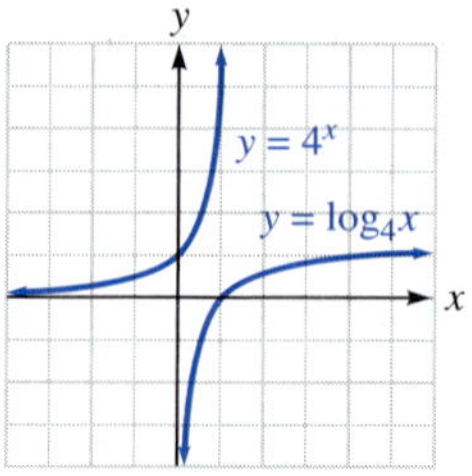

36.

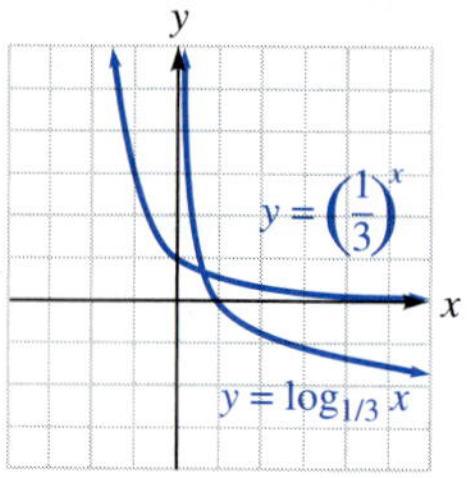

37. 53 dB **38.** 4.4 **39.** 6.1137 **40.** −0.1111
41. 10.3398 **42.** 2.5715
43.

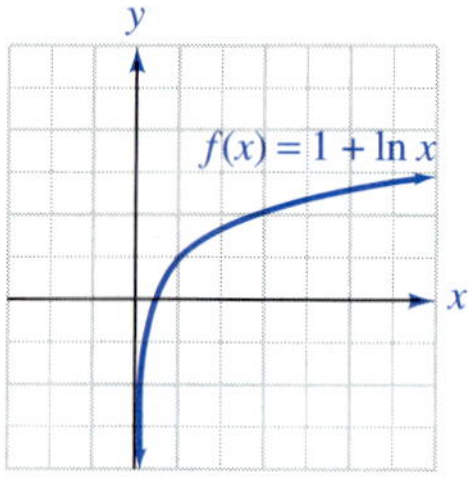

44.

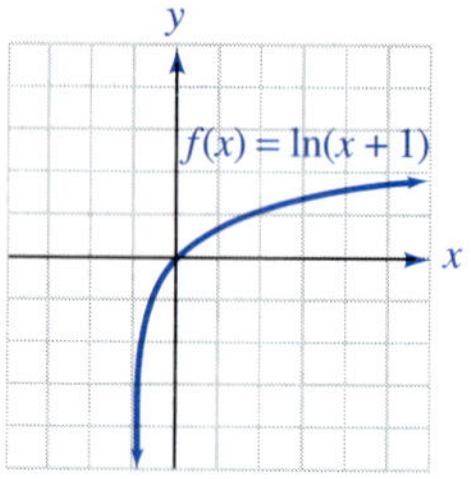

45. 23 yr **46.** 0 **47.** 1 **48.** 3 **49.** 4
50. 4 **51.** 0 **52.** 7 **53.** 3 **54.** 4
55. 9 **56.** $2 \log_b x + 3 \log_b y - 4 \log_b z$
57. $\frac{1}{2}(\log_b x - \log_b y - 2 \log_b z)$ **58.** $\log_b \frac{x^3 z^7}{y^5}$
59. $\log_b \frac{y^3\sqrt{x}}{z^7}$ **60.** 3.36 **61.** 1.56
62. 2.64 **63.** −6.72 **64.** 1.7604
65. about 7.94×10^{-4} gram-ions per liter **66.** $k \ln 2$ less
67. $\frac{\log 7}{\log 3} \approx 1.7712$ **68.** 2 **69.** 31.0335 **70.** 2
71. $\frac{\log 3}{\log 3 - \log 2} \approx 2.7095$ **72.** −1, −3 **73.** 25, 4 **74.** 4
75. 2 **76.** 4, 3 **77.** 6 **78.** 31 **79.** $\frac{\ln 9}{\ln 2} \approx 3.1699$
80. no solution **81.** $\frac{e}{e - 1} \approx 1.5820$ **82.** 1
83. about 3,300 yr

Chapter 11 Test (page 773)

1.

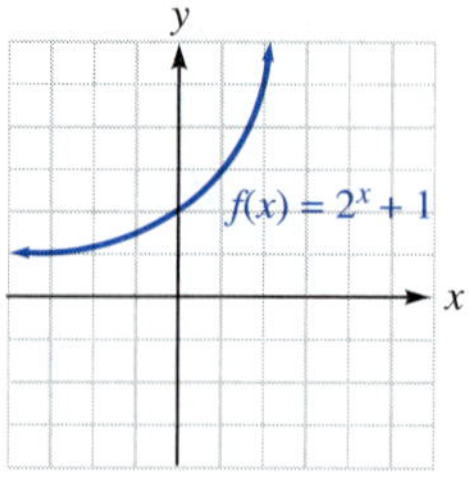

2.

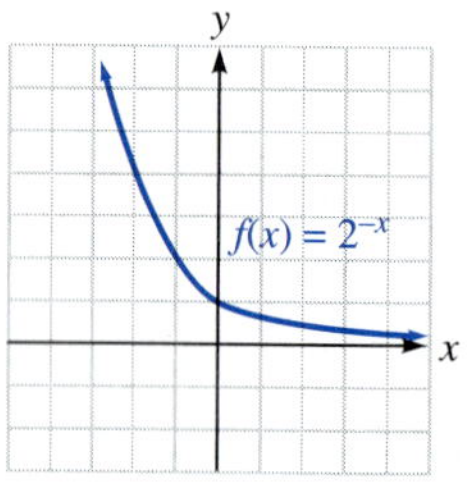

3. $\frac{3}{64}$ g **4.** \$1,060.90 **5.**

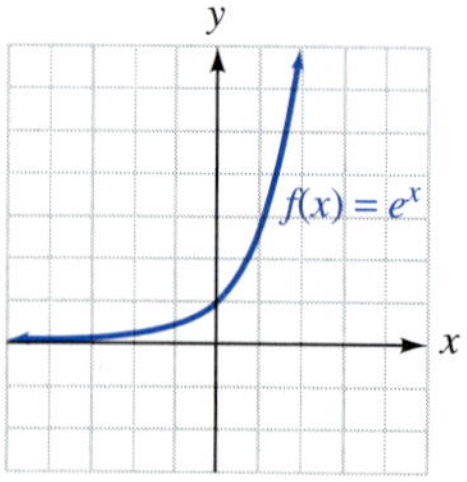

6. \$4,451.08 **7.** 2
8. 3 **9.** $\frac{1}{27}$ **10.** 10
11. 2 **12.** $\frac{27}{8}$

13.

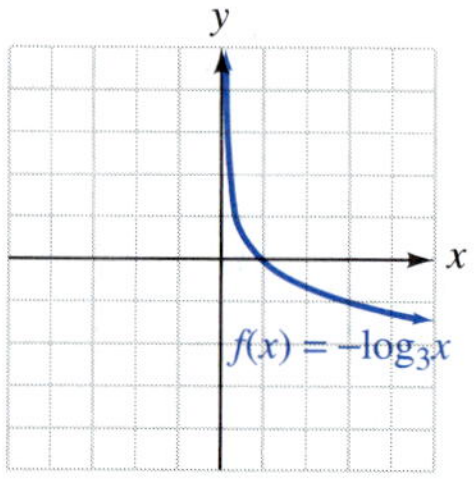

14.

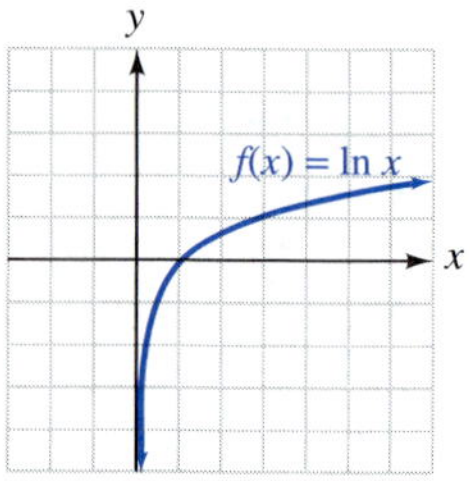

15. $2 \log a + \log b + 3 \log c$ **16.** $\frac{1}{2}(\ln a - 2 \ln b - \ln c)$
17. $\log \frac{b\sqrt{a + 2}}{c^3}$ **18.** $\log \frac{\sqrt[3]{a}}{c\sqrt[3]{b^2}}$ **19.** 1.3801
20. 0.4259 **21.** $\frac{\log 3}{\log 7}$ or $\frac{\ln 3}{\ln 7}$ **22.** $\frac{\log e}{\log \pi}$ or $\frac{\ln e}{\ln \pi}$ **23.** true
24. false **25.** false **26.** false **27.** 6.4 **28.** 46
29. $\frac{\log 3}{\log 5}$ **30.** $\frac{\log 3}{(\log 3) - 2}$ **31.** 1 **32.** 10

Getting Ready (page 776)

1. $x^2 - 4x + 4$ **2.** $x^2 + 8x + 16$ **3.** $\frac{81}{4}$ **4.** 36

Orals (page 784)

1. (0, 0), 12 **2.** (0, 0), 11 **3.** (2, 0), 4 **4.** (0, −1), 3
5. down **6.** up **7.** left **8.** right

Exercise 12.1 (page 785)

1. 5, $-\frac{7}{3}$ **3.** 3, $-\frac{1}{4}$ **5.** circle, plane **7.** $r^2 < 0$
9. parabola, (3, 2), right
11.

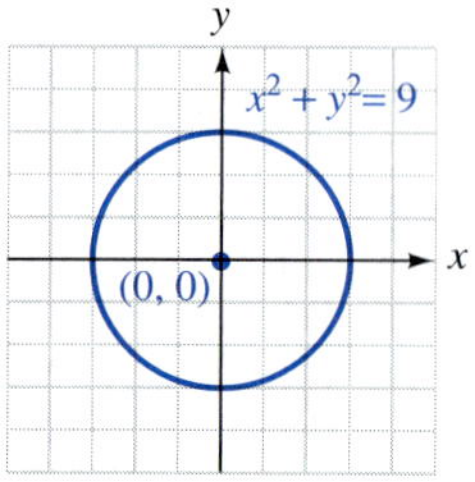

13.

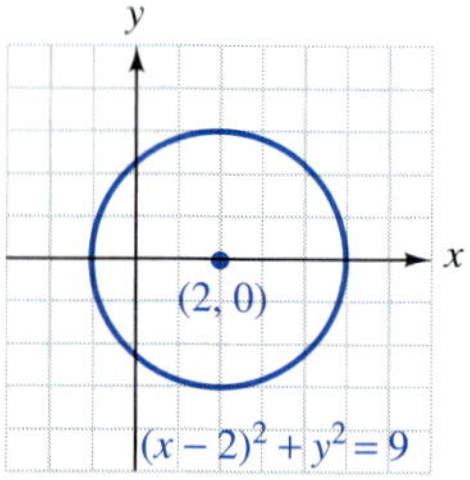

15.

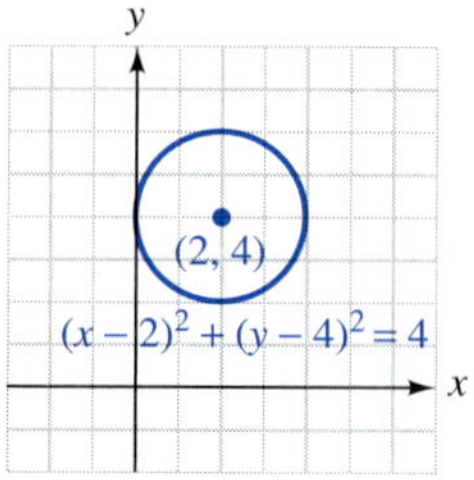

17.

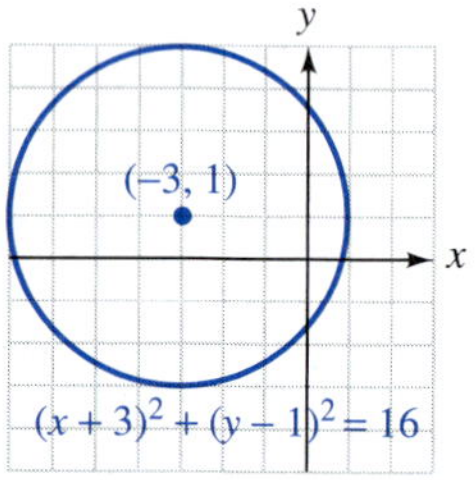

19.

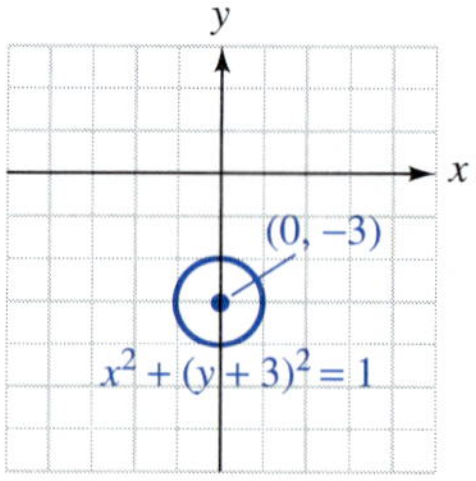

21.

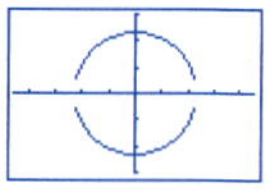

23.

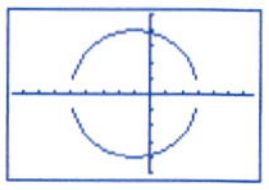

25. $x^2 + y^2 = 1$ **27.** $(x - 6)^2 + (y - 8)^2 = 25$
29. $(x + 2)^2 + (y - 6)^2 = 144$ **31.** $x^2 + y^2 = 2$

33.
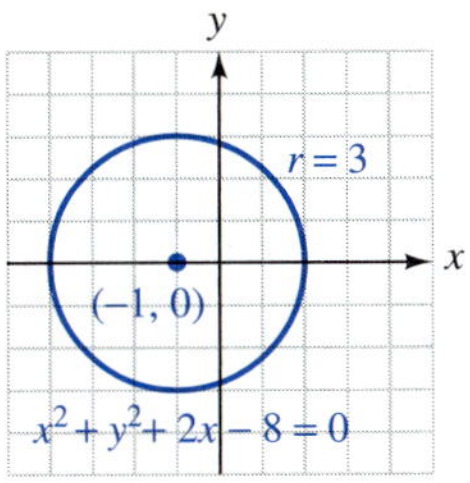

35.
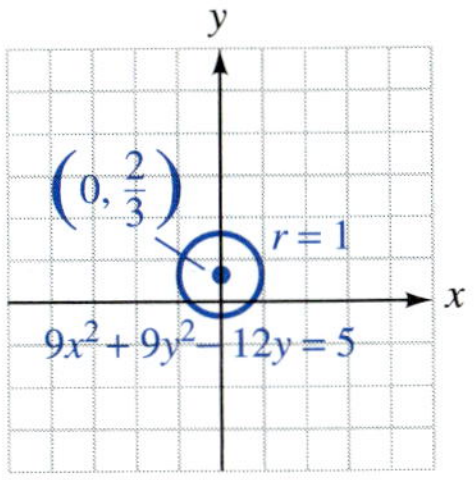

37.
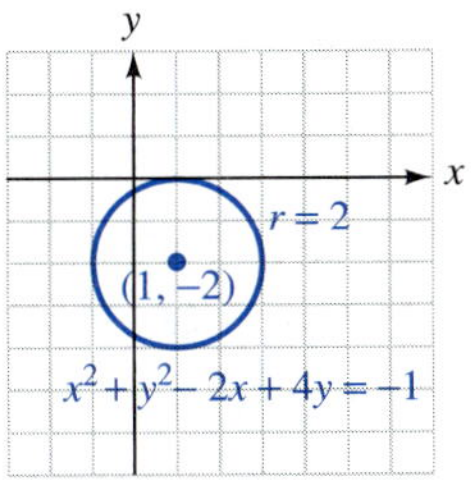

39.
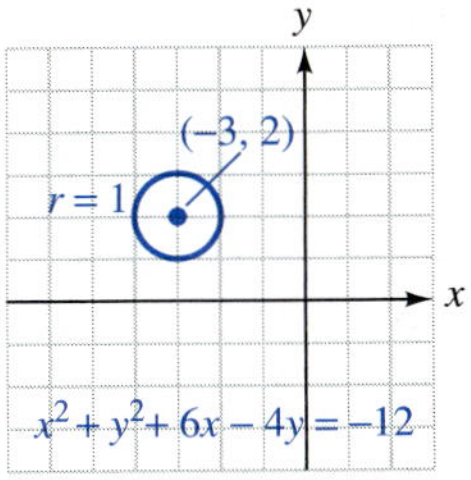

41.
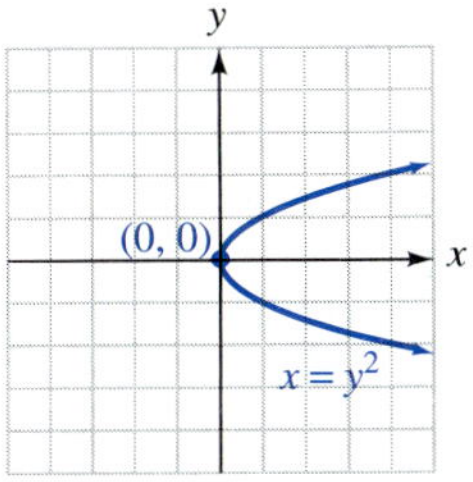

43.
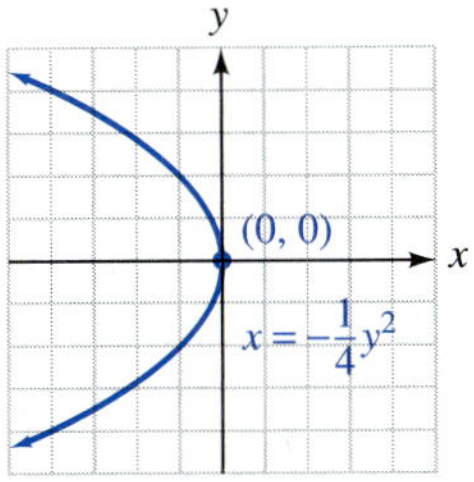

45.
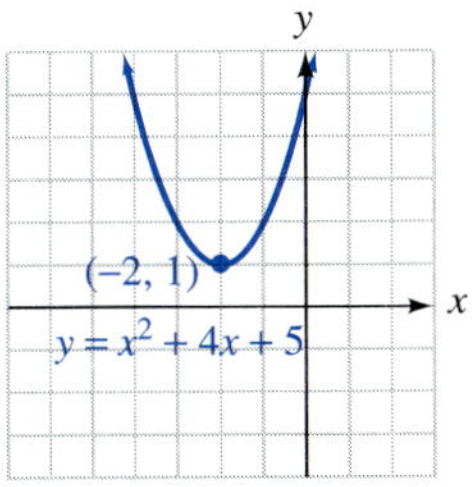

47.
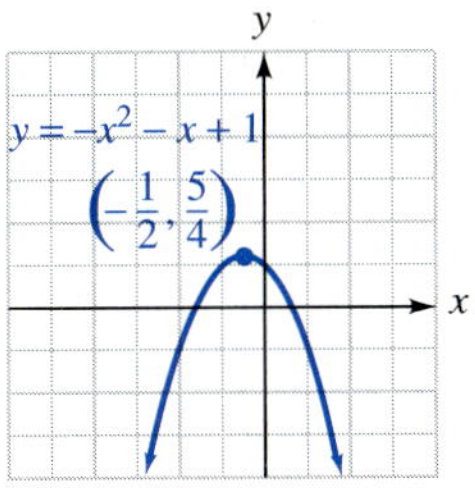

49.
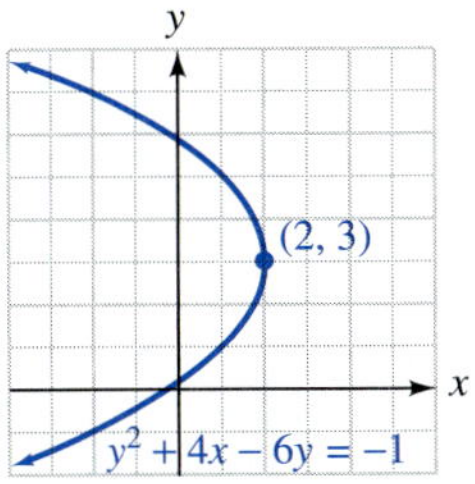

51.
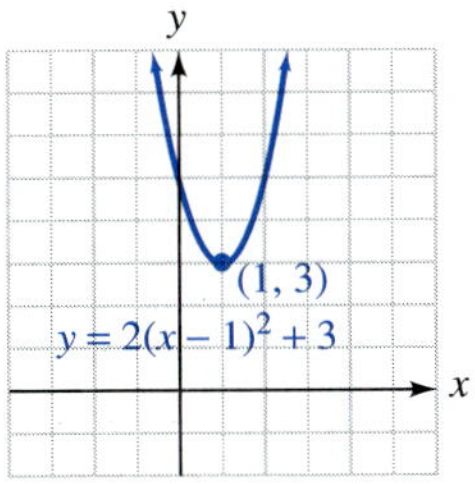

53.
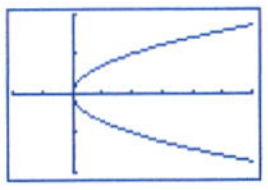

55.
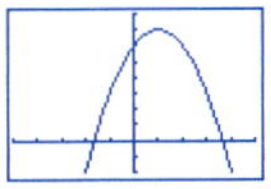

57. $(x - 7)^2 + y^2 = 9$ **59.** no **61.** 30 ft away
63. 2 AU

Getting Ready (page 789)

1. $y = \pm b$ **2.** $x = \pm a$

Orals (page 795)

1. $(\pm 3, 0), (0, \pm 4)$ **2.** $(\pm 5, 0), (0, \pm 6)$ **3.** $(2, 0)$
4. $(0, -1)$

Exercise 12.2 (page 796)

1. $12y^2 + \frac{9}{x^2}$ **3.** $\frac{y^2 + x^2}{y^2 - x^2}$ **5.** ellipse, sum **7.** center
9. $(0, 0)$

11.
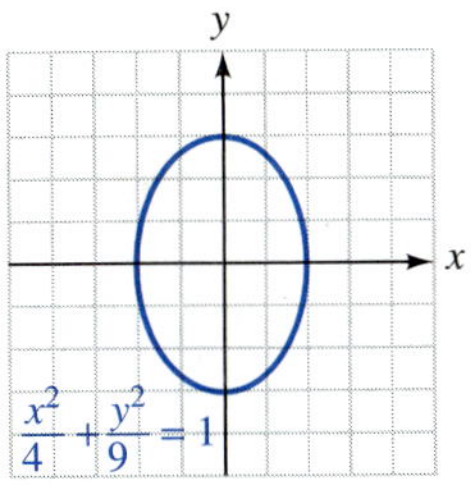

13.
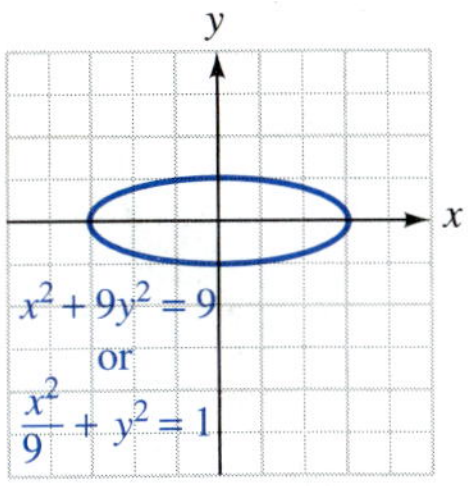

15.
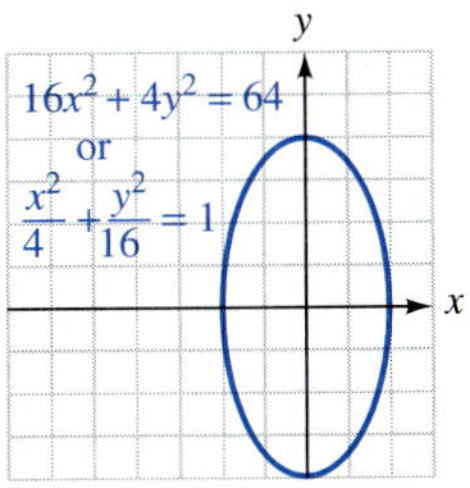

17.
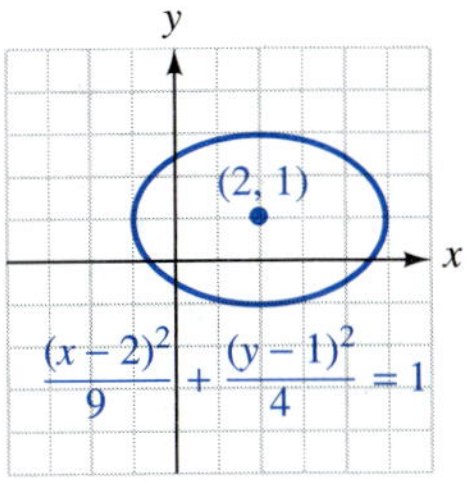

19.
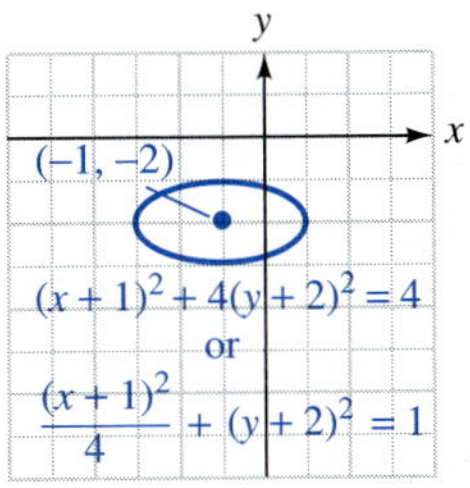

21.
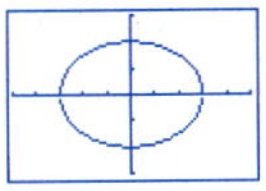

23.
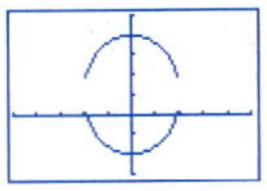

25.
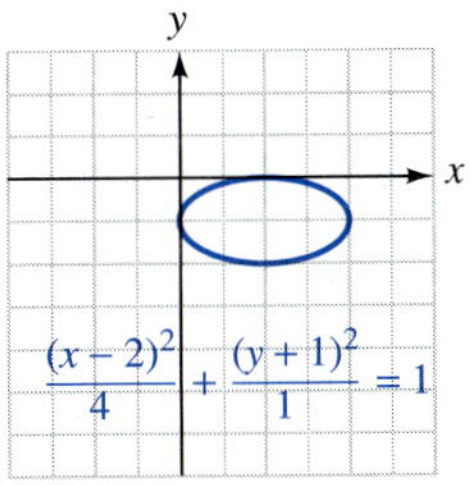

27.
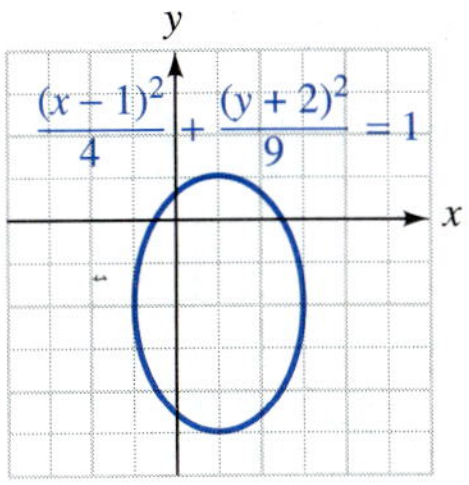

29. $y = \frac{1}{2}\sqrt{400 - x^2}$ **31.** $12\,\pi$ sq. units

Getting Ready (page 798)

1. $y = \pm 2.0$ **2.** $y = \pm 2.9$

Orals (page 804)

1. $(\pm 3, 0)$ **2.** $(0, \pm 5)$

Exercise 12.3 (page 805)

1. $-3x^2(2x^2 - 3x + 2)$ **3.** $(5a + 2b)(3a - 2b)$
5. hyperbola, difference **7.** center **9.** $(0, 0)$

11.

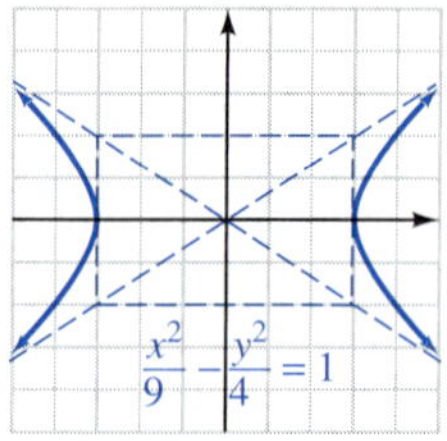

13.

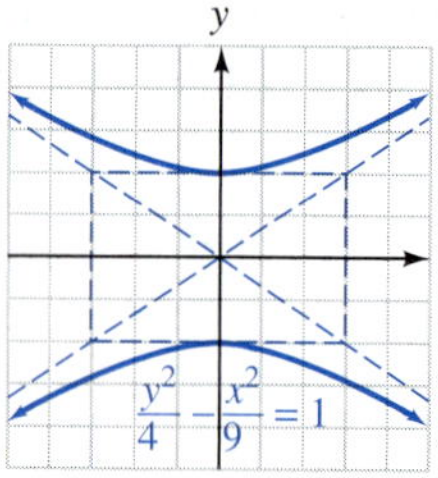

15.

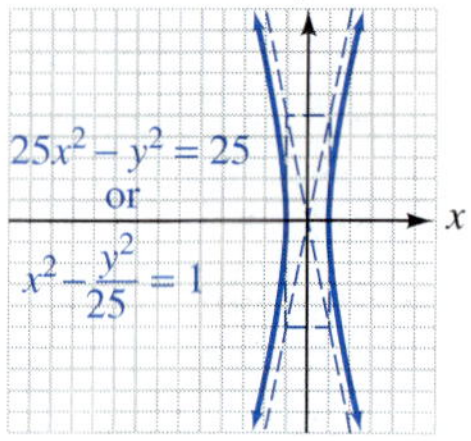

17.

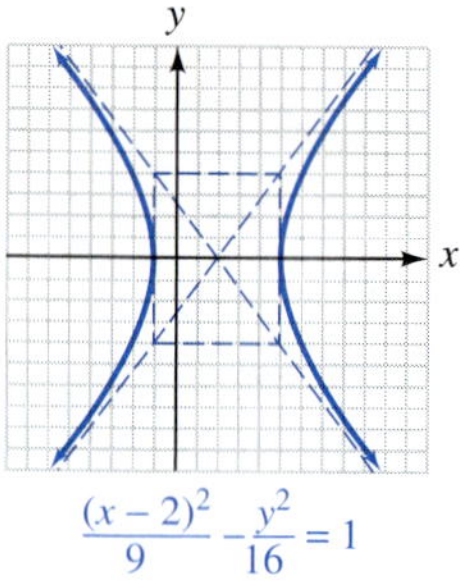

19.

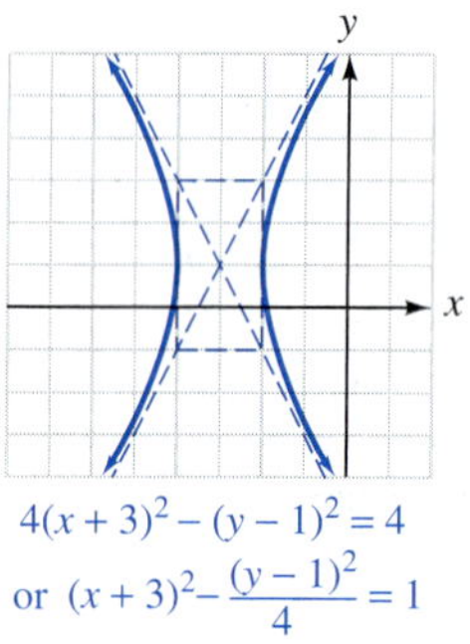

21.

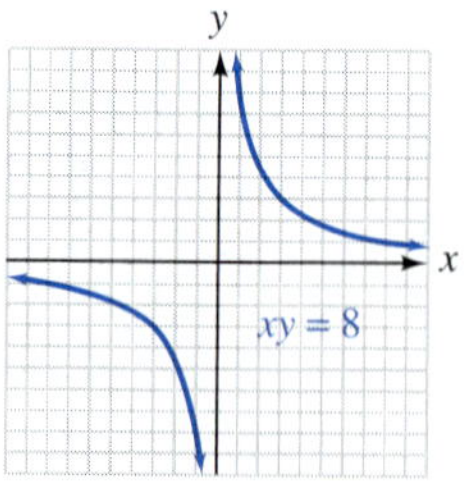

23.

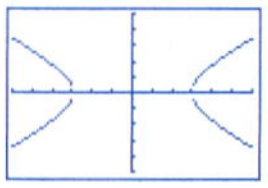

25.

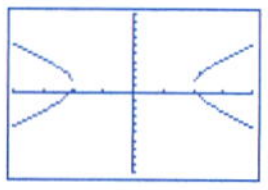

27.

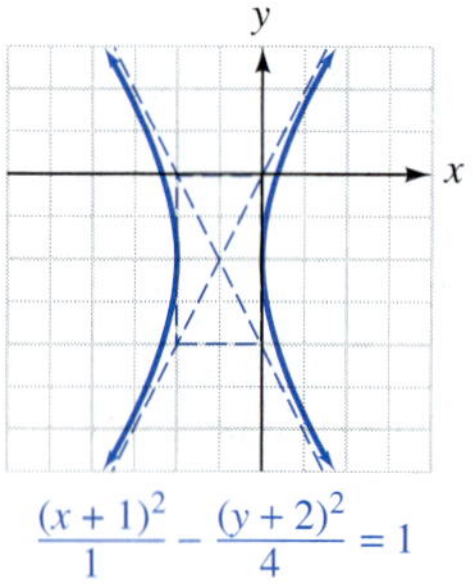

29.

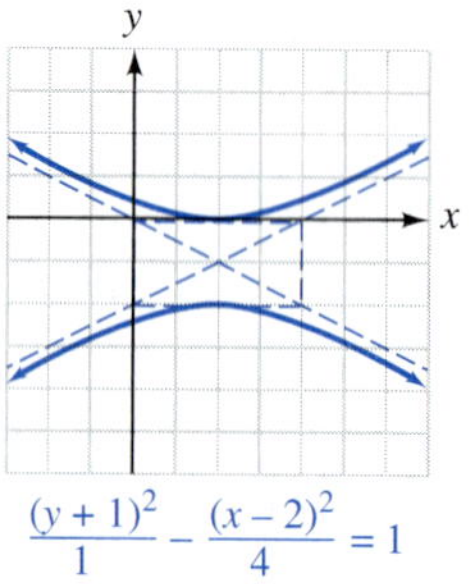

31. 3 units **33.** $10\sqrt{3}$ units

Getting Ready (page 808)

1. positive **2.** negative **3.** 98 **4.** −3

Orals (page 811)

1. increasing **2.** decreasing **3.** constant **4.** increasing

Exercise 12.4 (page 811)

1. 20 **3.** domains **5.** constant, $f(x)$ **7.** step
9. increasing on $(-\infty, 0)$, decreasing on $(0, \infty)$
11. decreasing on $(-\infty, 0)$, constant on $(0, 2)$, increasing on $(2, \infty)$

13. constant on $(-\infty, 0)$, increasing on $(0, \infty)$

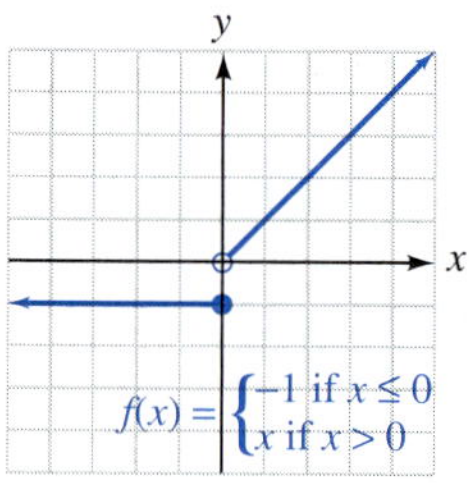

15. decreasing on $(-\infty, 0)$, increasing on $(0, 2)$, decreasing on $(2, \infty)$

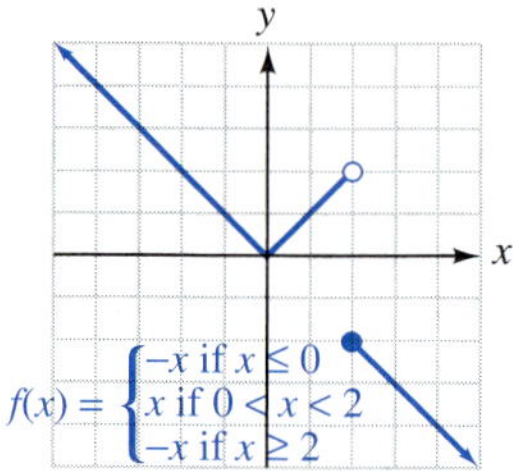

17.

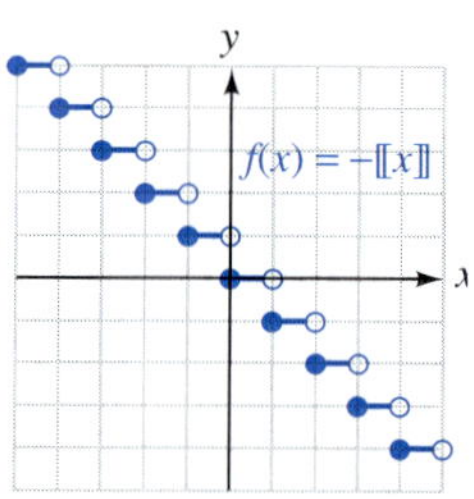

19.

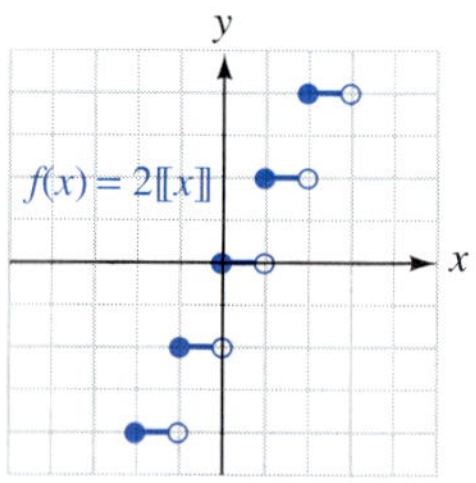

21.

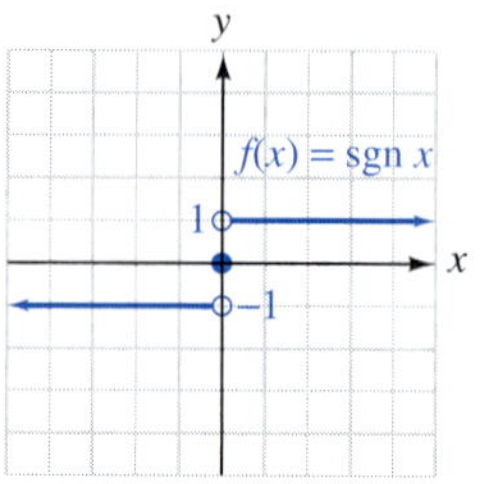

23. $30

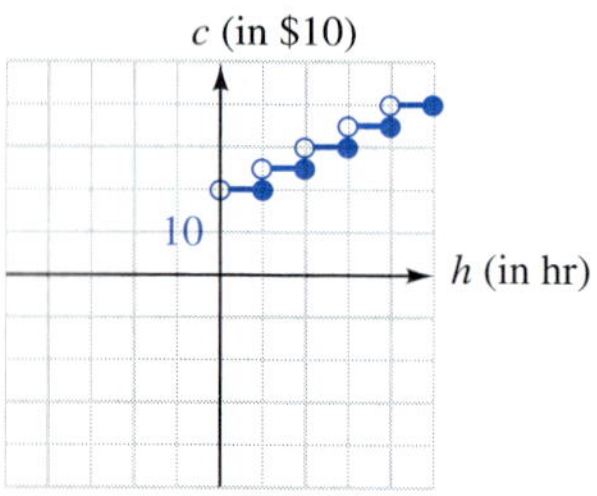

25. After 2 hours, network B is cheaper.

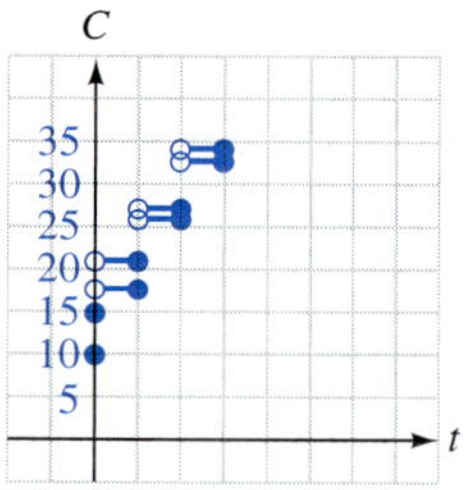

Chapter Summary (page 814)

1.

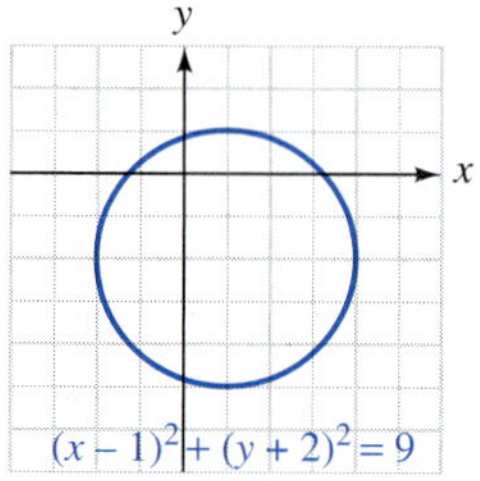

2.

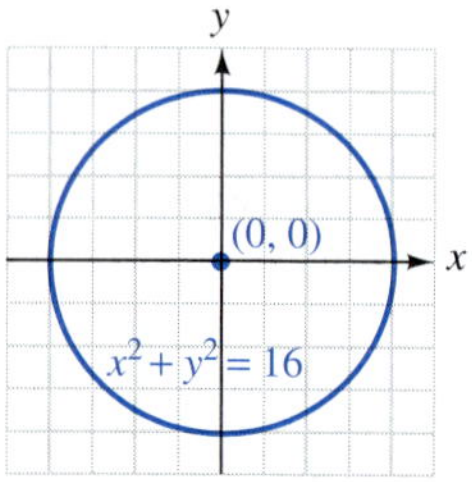

3.

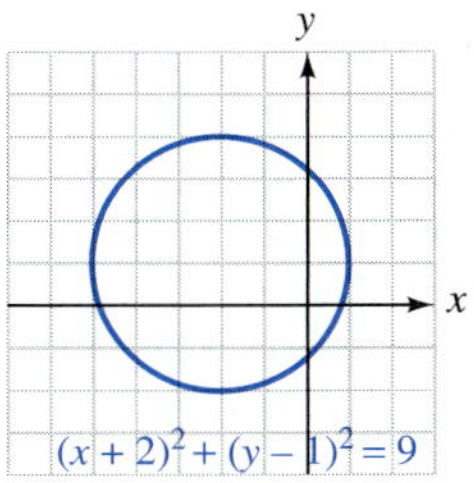

4.

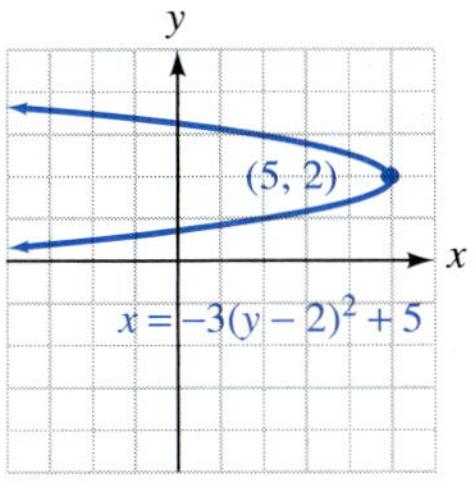

5.

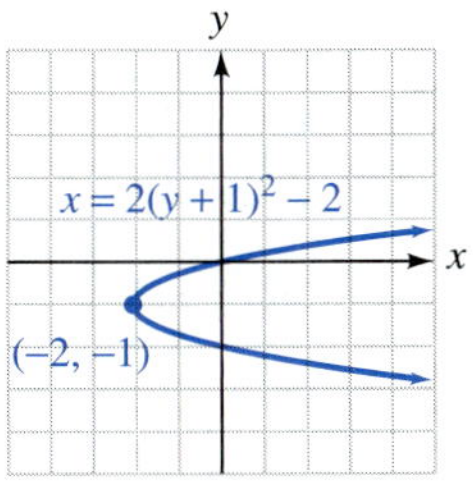

6.

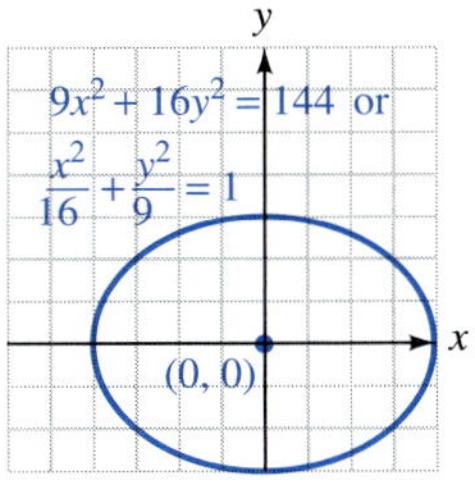

7.

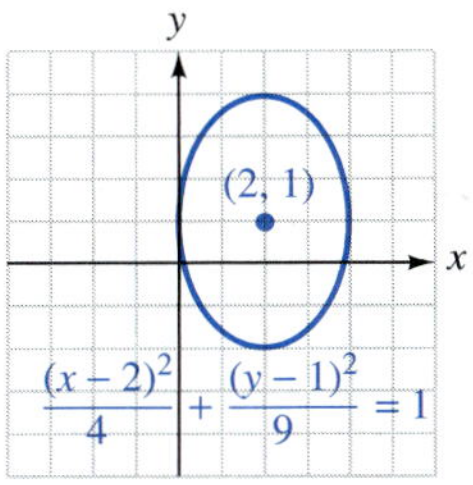

8.

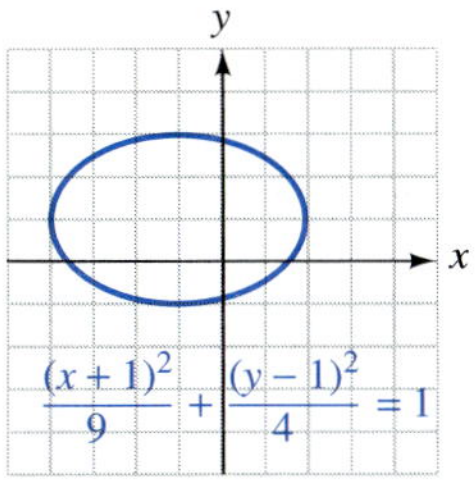

9.

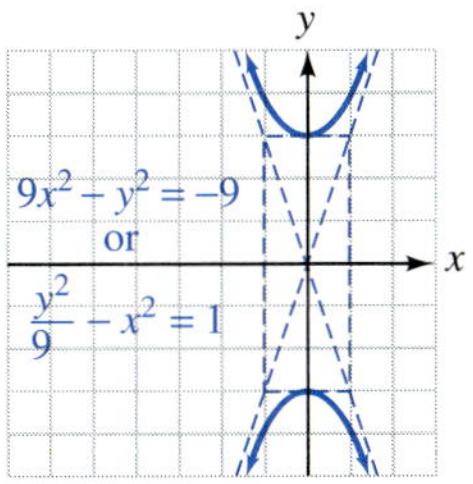

10.

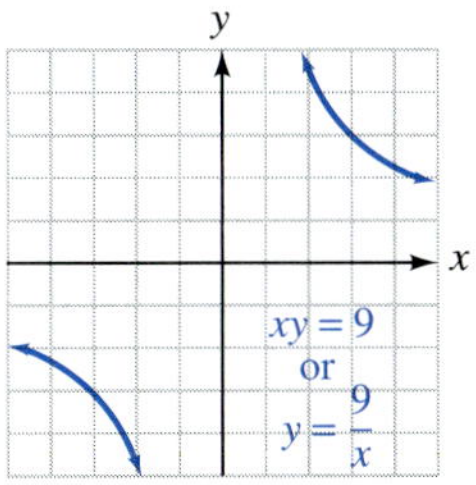

11. hyperbola **12.**

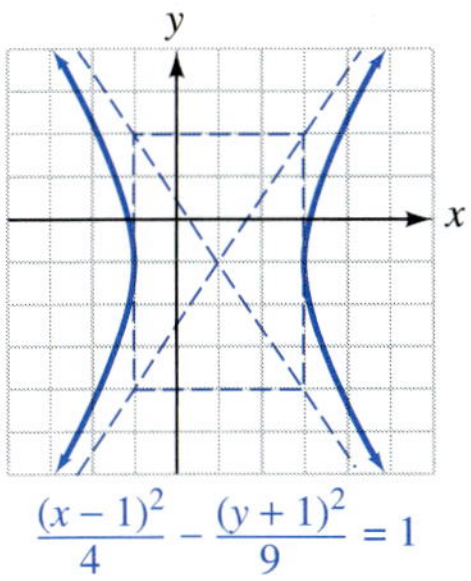

13. increasing on $(-\infty, -2)$, constant on $(-2, 1)$, decreasing on $(1, \infty)$

14.

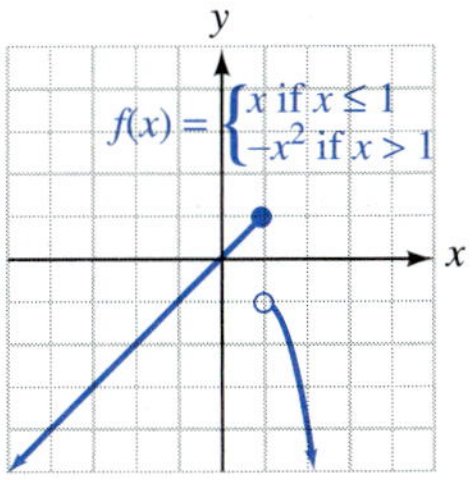

15.

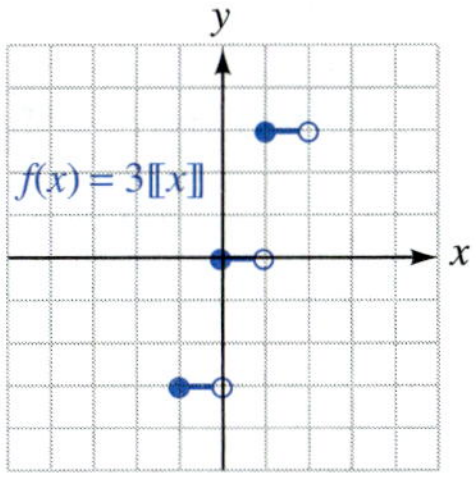

Chapter 12 Test (page 817)

1. $(2, -3)$, 2 **2.** $(-2, 3)$, 4 **3.**

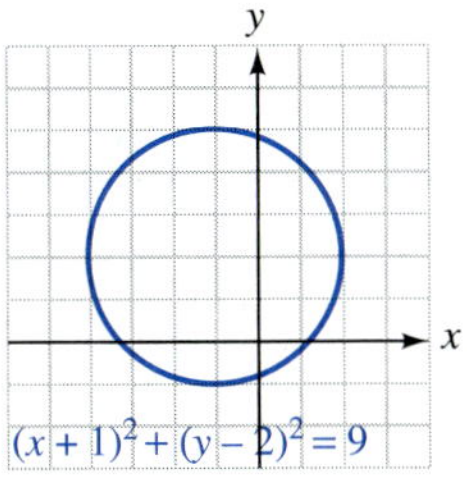

4.

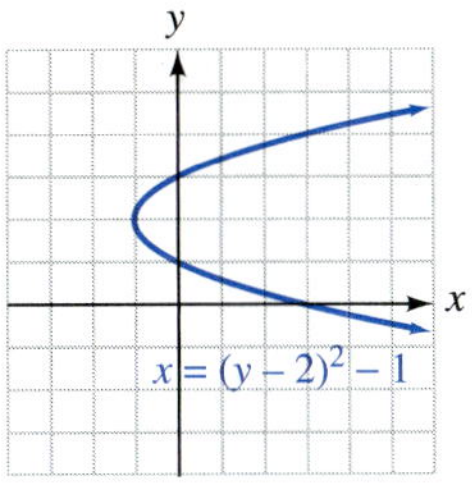

5.

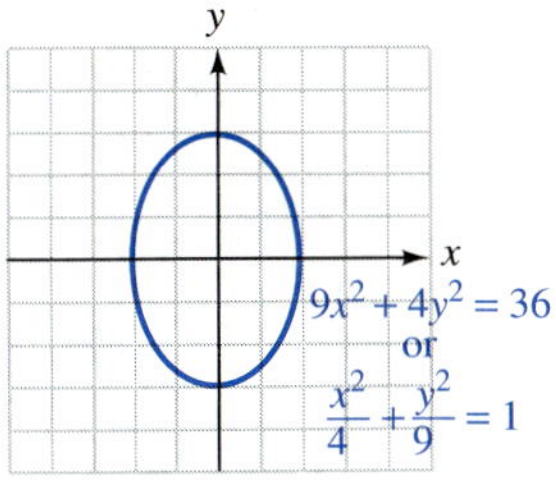

6.

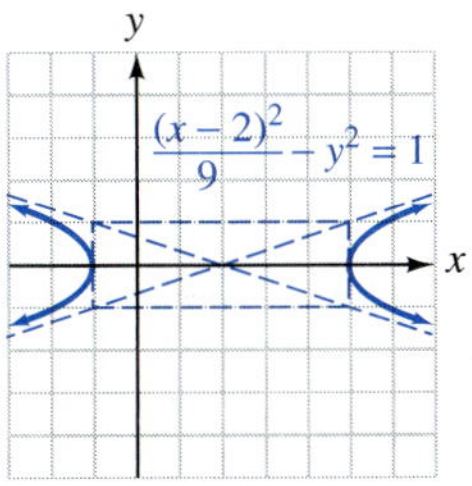

7.

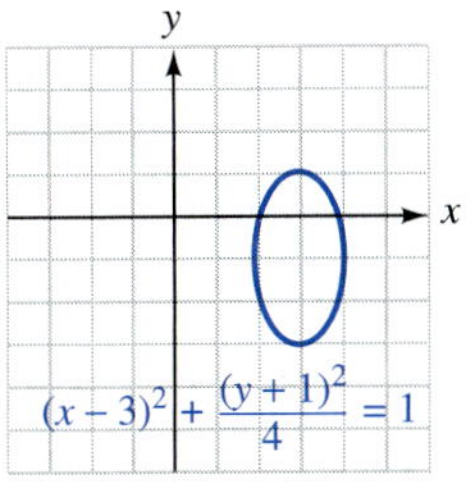

8.

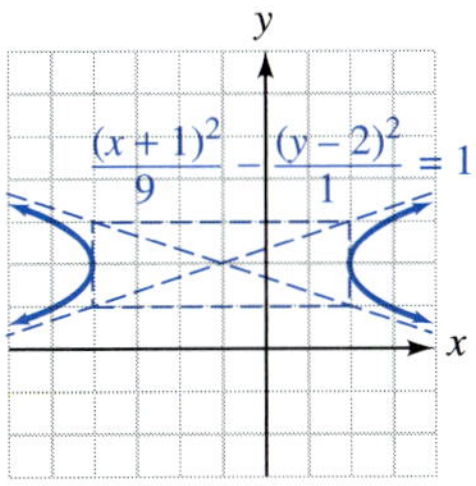

9. increasing on $(-3, 0)$, decreasing on $(0, 3)$

10.

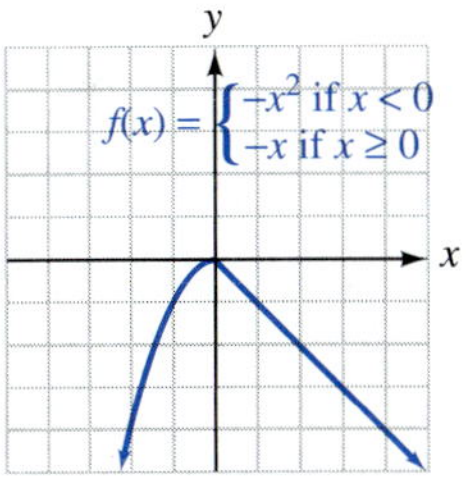

Cumulative Review Exercises (page 818)

1. $12x^2 - 5xy - 3y^2$ **2.** $a^{2n} - 2a^n - 3$ **3.** $\frac{5}{a - 2}$ **4.** $a^2 - 3a + 2$ **5.** 1 **6.** $\frac{4a - 1}{(a + 2)(a - 2)}$ **7.** parallel **8.** perpendicular **9.** $y = -2x + 5$ **10.** $y = -\frac{9}{13}x + \frac{7}{13}$

11.

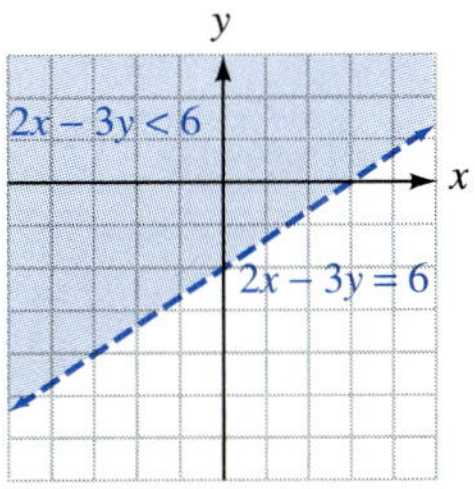

12.

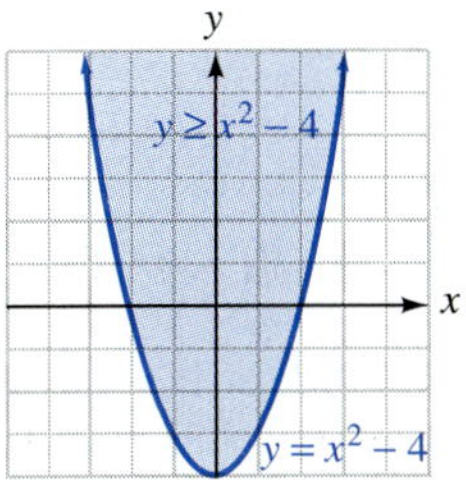

13. $5\sqrt{2}$ **14.** $81x\sqrt[3]{3x}$ **15.** 0, 5 **16.** 0 **17.** $\frac{2}{3}, -\frac{3}{2}$

18. $-\frac{4}{3} \pm \frac{\sqrt{19}}{3}$ **19.** $4x^2 + 4x - 1$

20. $f^{-1}(x) = \sqrt[3]{\frac{x+1}{2}}$

21.

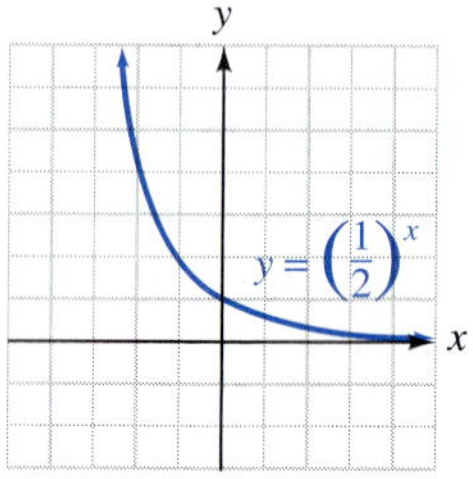

22. $2^y = x$

23. $\frac{2 \log 2}{\log 3 - \log 2}$

24. 16

25.

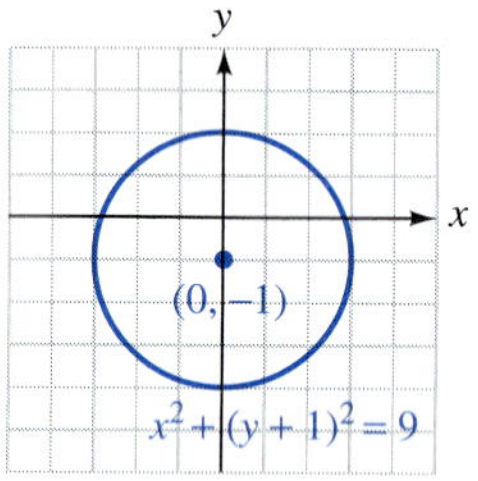

26.

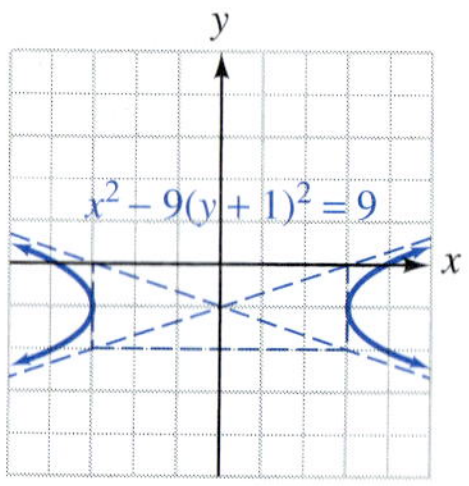

Getting Ready (page 821)

1. 2 **2.** 5 **3.** −4 **4.** −3

Orals (page 830)

1. no solution **2.** infinitely many solutions **3.** one solution **4.** no solution **5.** 2 **6.** 4 **7.** 4 **8.** 3

Exercise 13.1 (page 831)

1. a^{22} **3.** $\frac{1}{81x^{32}y^4}$ **5.** $r = \frac{A - p}{pt}$ **7.** $r = \frac{r_1 r_2}{r_2 + r_1}$

9. consistent **11.** independent

13.

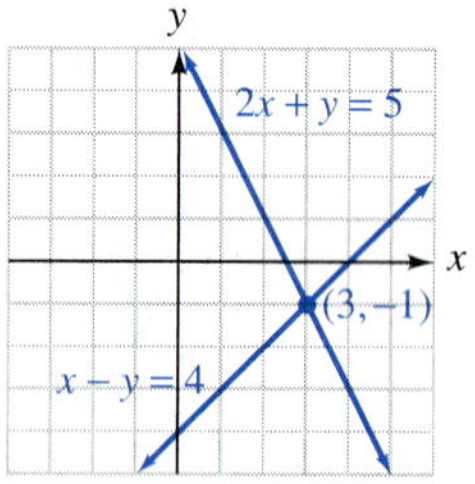

15.

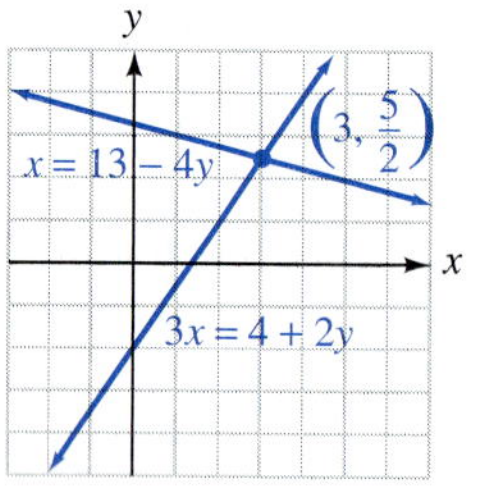

17.

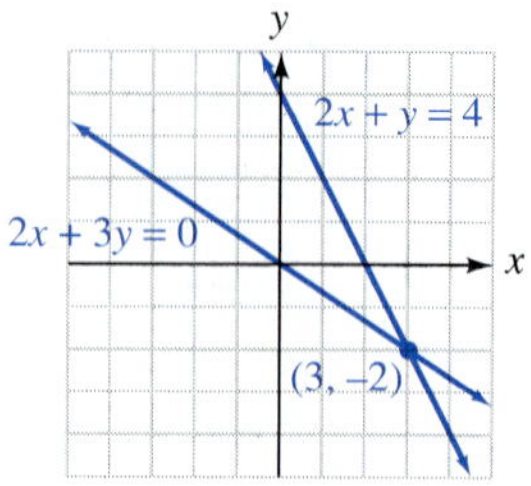

19.

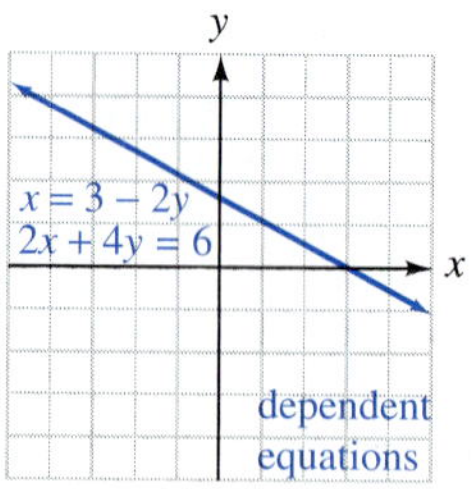

21.

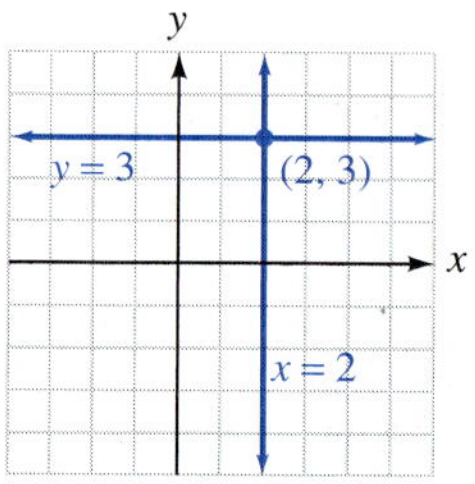

23.

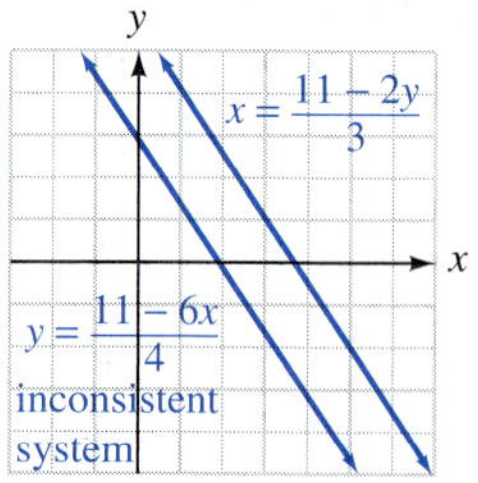

25.

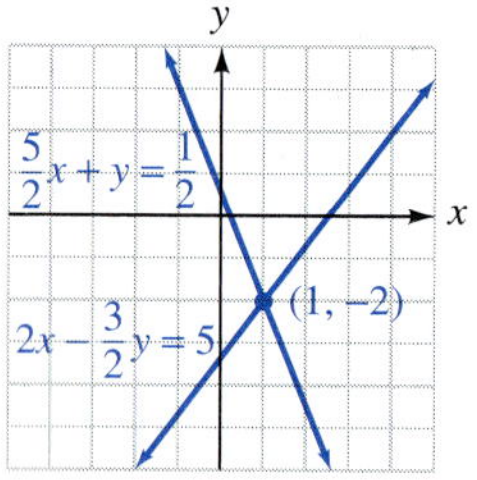

27. (−0.37, −2.69) **29.** (−7.64, 7.04) **31.** (2, 2) **33.** (5, 3) **35.** (−2, 4) **37.** no solution **39.** $\left(5, \frac{3}{2}\right)$ **41.** $\left(-2, \frac{3}{2}\right)$ **43.** (5, 2) **45.** (−4, −2) **47.** (1, 2) **49.** $\left(\frac{1}{2}, \frac{2}{3}\right)$ **51.** dependent equations

53. no solution **55.** (4, 8) **57.** (20, −12)

59.

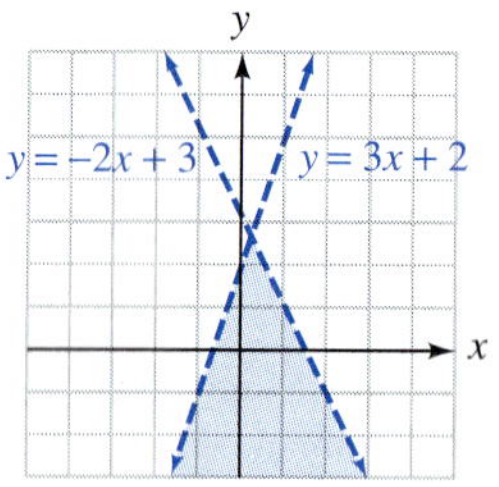

61.

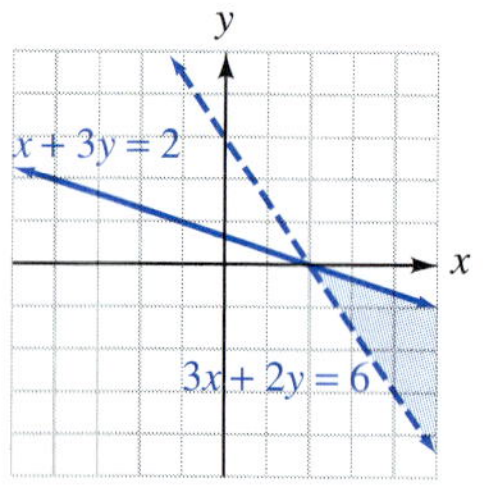

63. a. \$2 million **b.** \$3 million **c.** 10,000 cameras **65.** \$57 **67.** 625 ohms and 750 ohms **69.** 16 m by 20 m **77.** no

Getting Ready (page 834)

1. yes **2.** yes **3.** no **4.** yes

Orals (page 841)

1. yes **2.** no

Exercise 13.2 (page 841)

1. $\frac{9}{5}$ **3.** 1 **5.** $2s^2 + 1$ **7.** plane **9.** infinitely **11.** yes **13.** (1, 1, 2) **15.** (0, 2, 2) **17.** (3, 2, 1) **19.** inconsistent system, no solution **21.** $\left(\frac{3}{4}, \frac{1}{2}, \frac{1}{3}\right)$ **23.** dependent equations, infinitely many solutions **25.** (2, 6, 9) **27.** −2, 4, 16 **29.** $A = 40°, B = 60°, C = 80°$ **31.** 1, 2, 3 **33.** 30 expensive, 50 middle-priced, 100 inexpensive **35.** 250 \$5 tickets, 375 \$3 tickets, 125 \$2 tickets **37.** 3 poles, 2 bears, 4 deer **39.** $y = x^2 - 4x$ **41.** $x^2 + y^2 - 2x - 2y - 2 = 0$ **45.** (1, 1, 0, 1)

Getting Ready (page 844)

1. 5 8 13 **2.** 0 3 7 **3.** −1 −1 −2 **4.** 3 3 −5

Orals (page 850)

1. $\begin{bmatrix} 3 & 2 \\ 4 & -3 \end{bmatrix}$ **2.** $\begin{bmatrix} 3 & 2 & 8 \\ 4 & -3 & 6 \end{bmatrix}$ **3.** yes **4.** no

Exercise 13.3 (page 851)

1. 9.3×10^7 **3.** 6.3×10^4 **5.** matrix **7.** 3, columns
9. augmented **11.** type 1 **13.** nonzero **15.** 0
17. 8 **19.** (1, 1) **21.** (2, −3) **23.** (0, −3) **25.** (8, 8)
27. (1, 2, 3) **29.** (−1, −1, 2) **31.** (2, 1, 0) **33.** (1, 2)
35. (2, 0) **37.** no solution **39.** (1, 2)
41. $(-6 - z, 2 - z, z)$ **43.** $(2 - z, 1 - z, z)$ **45.** 22°, 68°
47. 40°, 65°, 75° **49.** $y = 2x^2 - x + 1$ **53.** $k \neq 0$

Getting Ready (page 853)

1. −22 **2.** 22 **3.** −13 **4.** −13

Orals (page 860)

1. 1 **2.** −2 **3.** 0 **4.** $\begin{vmatrix} 1 & 2 \\ 2 & -1 \end{vmatrix}$ **5.** $\begin{vmatrix} 5 & 2 \\ 4 & -1 \end{vmatrix}$
6. $\begin{vmatrix} 1 & 5 \\ 2 & 4 \end{vmatrix}$

Exercise 13.4 (page 860)

1. −3 **3.** 0 **5.** number **7.** $\begin{vmatrix} a_2 & c_2 \\ a_3 & c_3 \end{vmatrix}$ **9.** $\begin{vmatrix} 3 & 4 \\ 2 & -3 \end{vmatrix}$
11. 8 **13.** −2 **15.** $x^2 - y^2$ **17.** 0 **19.** −13
21. 26 **23.** 0 **25.** $10a$ **27.** 0 **29.** (4, 2)
31. (−1, 3) **33.** $\left(-\frac{1}{2}, \frac{1}{3}\right)$ **35.** (2, −1) **37.** no solution
39. $\left(5, \frac{14}{5}\right)$ **41.** (1, 1, 2) **43.** (3, 2, 1) **45.** no solution
47. (3, −2, 1) **49.** dependent equations **51.** (−2, 3, 1)
53. no solution **55.** 2 **57.** 2
59. \$5,000 in HiTech, \$8,000 in SaveTel, \$7,000 in HiGas
61. −23 **63.** 26 **69.** −4

Getting Ready (page 863)

1. $7x^2 - y^2 = 44$ **2.** $5x^2 - 3y^2 = 8$

Orals (page 867)

1. 0, 1, 2 **2.** 0, 1, 2 **3.** 0, 1, 2, 3, 4 **4.** 0, 1, 2, 3, 4

Exercise 13.5 (page 867)

1. $-11x\sqrt{2}$ **3.** $\frac{t}{2}$ **5.** graphing, substitution

7.
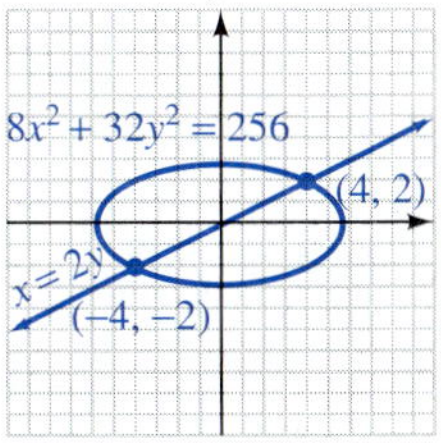

9.
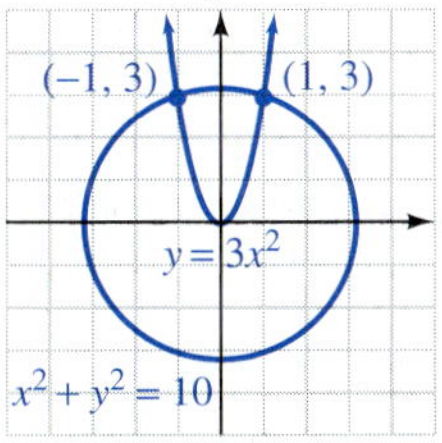

11.
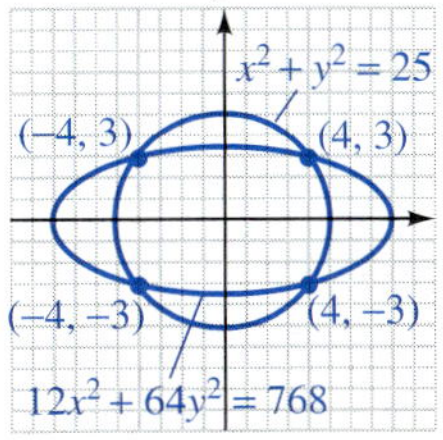

13.
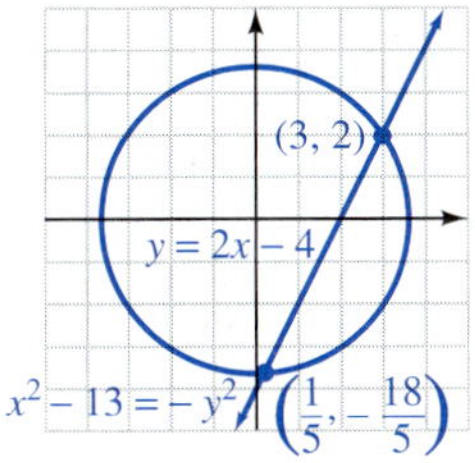

15. (1, 0), (5, 0) **17.** (3, 0), (0, 5) **19.** (1, 1)

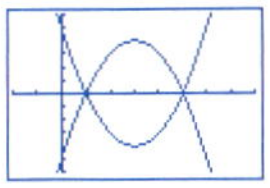

21. (1, 2), (2, 1) **23.** (−2, 3), (2, 3)
25. $(\sqrt{5}, 5), (-\sqrt{5}, 5)$
27. (3, 2), (3, −2), (−3, 2), (−3, −2)
29. (2, 4), (2, −4), (−2, 4), (−2, −4)
31. $(-\sqrt{15}, 5), (\sqrt{15}, 5), (-2, -6), (2, -6)$
33. (0, −4), (−3, 5), (3, 5)
35. (−2, 3), (2, 3), (−2, −3), (2, −3) **37.** (3, 3)
39. $(6, 2), (-6, -2), (\sqrt{42}, 0), (-\sqrt{42}, 0)$
41. $\left(\frac{1}{2}, \frac{1}{3}\right), \left(\frac{1}{3}, \frac{1}{2}\right)$ **43.** 4 and 8

45.
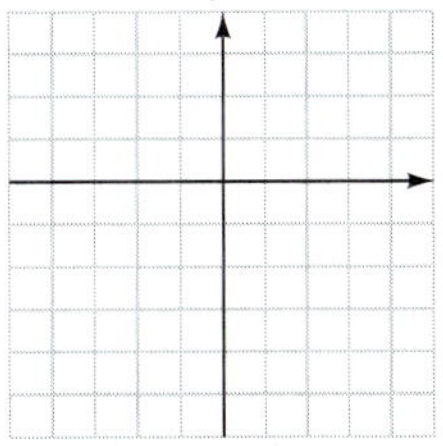

47.
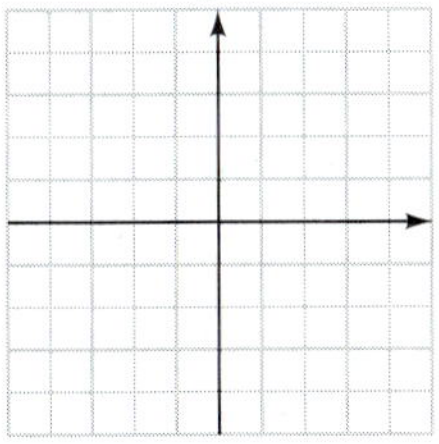

49. 7 cm by 9 cm **51.** either \$750 at 9% or \$900 at 7.5%
53. 68 mph, 4.5 hr **57.** 0, 1, 2, 3, 4

Chapter Summary (page 871)

1.
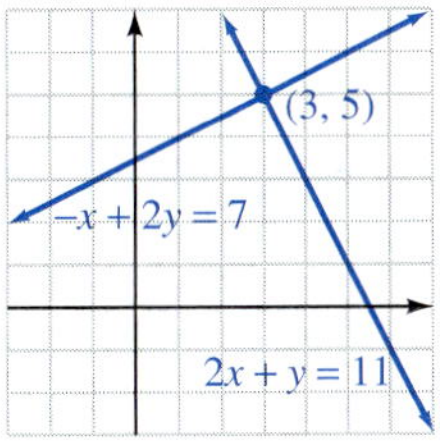

2.
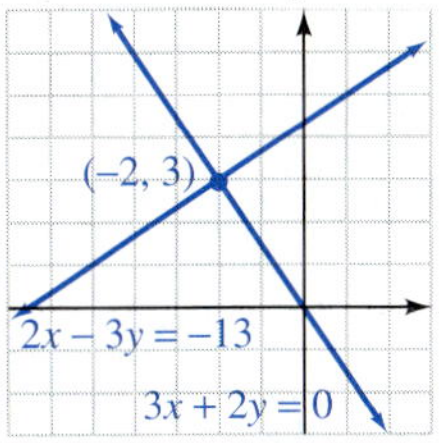

3.
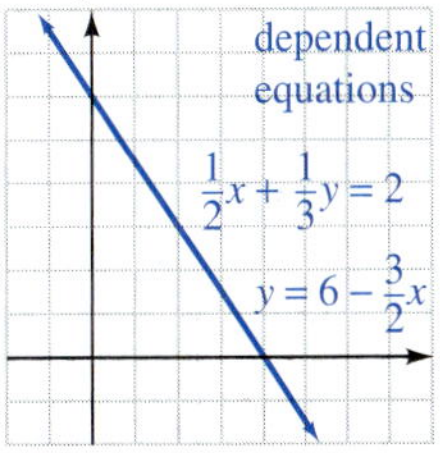

4.
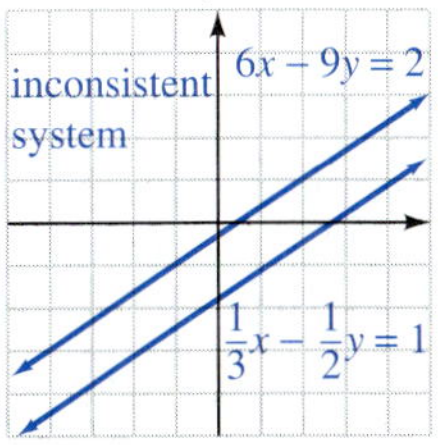

5. (−1, 3) **6.** (−3, −1) **7.** (3, 4) **8.** (−4, 2)
9. (−3, 1) **10.** (1, −1) **11.** (9, −4) **12.** $\left(4, \frac{1}{2}\right)$

13. (1, 2, 3) **14.** inconsistent system **15.** (2, 1)
16. (1, 3, 2) **17.** (1, 2) **18.** $(3z, 1 - 2z, z)$ **19.** 18
20. 38 **21.** −3 **22.** 28 **23.** (2, 1) **24.** (−1, 3)
25. (1, −2, 3) **26.** (−3, 2, 2)
27. (4, 2), (4, −2), (−4, 2), (−4, −2)
28. (2, 3), (2, −3), (−2, 3), (−2, −3)
29.

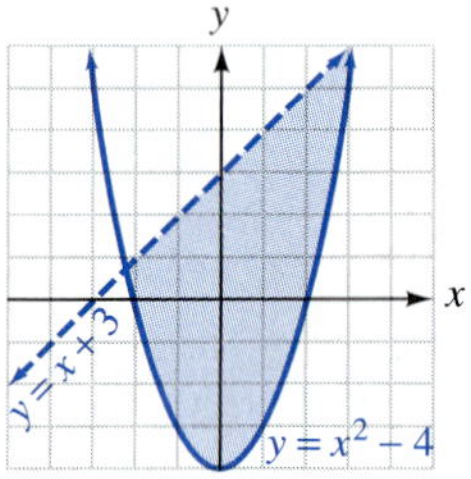

Chapter 13 Test (page 874)

1.

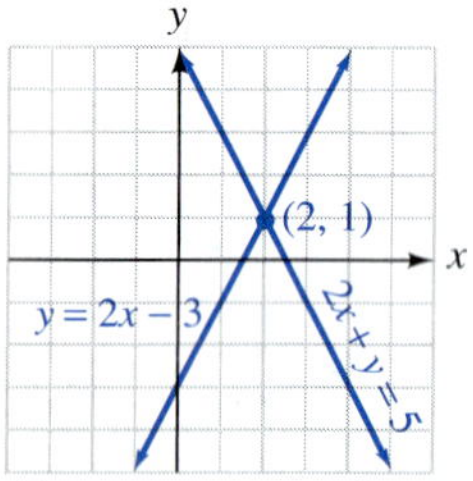

2. (7, 0) **3.** (2, −3)
4. (−6, 4) **5.** dependent
6. consistent **7.** 6 **8.** −8

9. $\begin{bmatrix} 1 & 1 & 1 & 4 \\ 1 & 1 & -1 & 6 \\ 2 & -3 & 1 & -1 \end{bmatrix}$

10. $\begin{bmatrix} 1 & 1 & 1 \\ 1 & 1 & -1 \\ 2 & -3 & 1 \end{bmatrix}$ **11.** (2, 2) **12.** (−1, 3) **13.** 22

14. −17 **15.** 4 **16.** 13 **17.** $\begin{vmatrix} -6 & -1 \\ -6 & 1 \end{vmatrix}$

18. $\begin{vmatrix} 1 & -1 \\ 3 & 1 \end{vmatrix}$ **19.** −3 **20.** 3 **21.** 3 **22.** −1

23. (2, 1), (2, −1), (−2, 1), (−2, −1) **24.** (3, 4), (−3, 4)
25.

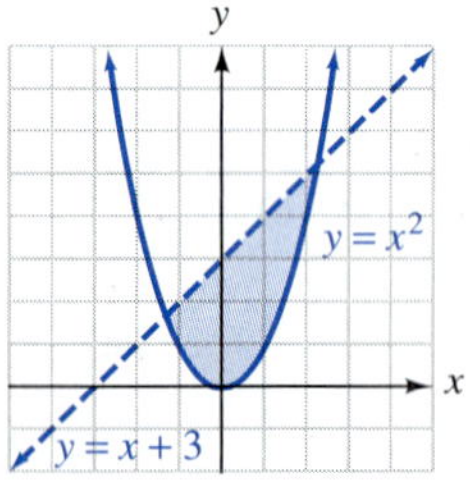

Getting Ready (page 877)

1. $x^2 + 4x + 4$ **2.** $x^2 - 6x + 9$ **3.** $x^3 + 3x^2 + 3x + 1$
4. $x^3 - 6x^2 + 12x - 8$

Orals (page 881)

1. 1 **2.** 24 **3.** 1 **4.** 120 **5.** $m^2 + 2mn + n^2$
6. $m^2 - 2mn + n^2$ **7.** $p^2 + 4pq + 4q^2$
8. $4p^2 - 4pq + q^2$

Exercise 14.1 (page 881)

1. 2 **3.** 5 **5.** one **7.** Pascal's **9.** 6! **11.** 1
13. 6 **15.** −120 **17.** 30 **19.** 144 **21.** 40,320
23. $\frac{1}{110}$ **25.** 2,352 **27.** 10 **29.** 21 **31.** $\frac{1}{168}$
33. $x^3 + 3x^2y + 3xy^2 + y^3$
35. $x^4 - 4x^3y + 6x^2y^2 - 4xy^3 + y^4$
37. $8x^3 + 12x^2y + 6xy^2 + y^3$
39. $x^3 - 6x^2y + 12xy^2 - 8y^3$
41. $8x^3 + 36x^2y + 54xy^2 + 27y^3$
43. $\frac{x^3}{8} - \frac{x^2y}{4} + \frac{xy^2}{6} - \frac{y^3}{27}$
45. $81 + 216y + 216y^2 + 96y^3 + 16y^4$
47. $\frac{x^4}{81} - \frac{2x^3y}{27} + \frac{x^2y^2}{6} - \frac{xy^3}{6} + \frac{y^4}{16}$
51. 1, 1, 2, 3, 5, 8, 13, . . . ; beginning with 2, each number is the sum of the previous two numbers.

Getting Ready (page 883)

1. 6 **2.** 10 **3.** −20 **4.** −4

Orals (page 884)

1. 2 **2.** 3 **3.** 6 **4.** 6 **5.** 5 **6.** 2 **7.** 1 **8.** 8

Exercise 14.2 (page 884)

1. (2, 3) **3.** 8 **5.** 3 **7.** 7 **9.** $3a^2b$ **11.** $-4xy^3$
13. $15x^2y^4$ **15.** $28x^6y^2$ **17.** $90x^3$ **19.** $640x^3y^2$
21. $-12x^3y$ **23.** $-70{,}000x^4$ **25.** $810xy^4$ **27.** $180x^4y^2$
29. $-\frac{1}{6}x^3y$ **31.** $\frac{n!}{3!(n-3)!}a^{n-3}b^3$ **33.** $\frac{n!}{4!(n-4)!}a^{n-4}b^4$
35. $\frac{n!}{(r-1)!(n-r+1)!}a^{n-r+1}b^{r-1}$ **39.** 252

Getting Ready (page 885)

1. 3, 5, 7, 9 **2.** 4, 7, 10, 13

Orals (page 890)

1. 14 **2.** 1 **3.** 5 **4.** −6 **5.** 3 **6.** 5

Exercise 14.3 (page 890)

1. $18x^2 + 8x - 3$ **3.** $\frac{6a^2 + 16}{(a+2)(a-2)}$ **5.** sequence
7. arithmetic, difference **9.** arithmetic mean
11. 1 + 2 + 3 + 4 + 5 **13.** 3, 5, 7, 9, 11
15. −5, −8, −11, −14, −17 **17.** 5, 11, 17, 23, 29
19. −4, −11, −18, −25, −32
21. −118, −111, −104, −97, −90 **23.** 34, 31, 28, 25, 22
25. 5, 12, 19, 26, 33 **27.** 355 **29.** −179 **31.** −23
33. 12 **35.** $\frac{17}{4}, \frac{13}{2}, \frac{35}{4}$ **37.** 12, 14, 16, 18 **39.** $\frac{29}{2}$ **41.** $\frac{5}{4}$
43. 1,335 **45.** 459 **47.** 354 **49.** 255 **51.** 1,275
53. 2,500 **55.** 60 **57.** 31 **59.** 12
61. 60, 110, 160, 210, 260, 310; $6,060 **63.** 11,325
65. 368 ft **69.** $\frac{3}{2}, 2, \frac{5}{2}, 3, \frac{7}{2}, 4$

Getting Ready (page 893)

1. 10, 20, 40 **2.** 18, 54, 162

Orals (page 898)

1. 27 **2.** $\frac{1}{27}$ **3.** 2.5 **4.** $\sqrt{3}$ **5.** 6 **6.** 1

Exercise 14.4 (page 898)

1. $[-1, 6]$ **3.** $(-\infty, -3) \cup (4, \infty)$ **5.** geometric
7. common ratio **9.** $s_n = \dfrac{a - ar^n}{1 - r}$ **11.** 3, 6, 12, 24, 48
13. $-5, -1, -\frac{1}{5}, -\frac{1}{25}, -\frac{1}{125}$ **15.** 2, 8, 32, 128, 512
17. $-3, -12, -48, -192, -768$ **19.** $-64, 32, -16, 8, -4$
21. $-64, -32, -16, -8, -4$ **23.** 2, 10, 50, 250, 1,250
25. 3,584 **27.** $\frac{1}{27}$ **29.** 3 **31.** 6, 18, 54
33. $-20, -100, -500, -2{,}500$ **35.** -16 **37.** $10\sqrt{2}$
39. No geometric mean exists. **41.** 728 **43.** 122
45. -255 **47.** 381 **49.** $\frac{156}{25}$ **51.** $-\frac{21}{4}$
53. about 669 people **55.** \$1,469.74 **57.** \$140,853.75
59. $\left(\frac{1}{2}\right)^{11} \approx 0.0005$ **61.** \$4,309.14 **67.** arithmetic mean

Getting Ready (page 900)

1. 4 **2.** 4.5 **3.** 3 **4.** 2.5

Orals (page 903)

1. 8 **2.** $\frac{1}{8}$ **3.** $\frac{1}{2}$ **4.** $\frac{1}{8}$ **5.** 27 **6.** 16

Exercise 14.5 (page 903)

1. yes **3.** no **5.** infinite **7.** $S = \frac{a}{1 - r}$ **9.** 16
11. 81 **13.** 8 **15.** $-\frac{135}{4}$ **17.** no sum **19.** $-\frac{81}{2}$
21. $\frac{1}{9}$ **23.** $-\frac{1}{3}$ **25.** $\frac{4}{33}$ **27.** $\frac{25}{33}$ **29.** 30 m **31.** 5,000
35. $\frac{4}{5}$ **39.** no; $0.999999 = \frac{999{,}999}{1{,}000{,}000} < 1$

Getting Ready (page 904)

1. 24 **2.** 120 **3.** 30 **4.** 168

Orals (page 912)

1. 15 **2.** 120 **3.** 3 **4.** 6 **5.** 1 **6.** 1

Exercise 14.6 (page 912)

1. 6, -3 **3.** 8 **5.** $p \cdot q$ **7.** $P(n, r)$ **9.** $n!$
11. $\binom{n}{r}$, combinations **13.** 1 **15.** 6 **17.** 60 **19.** 12
21. 5 **23.** 1,260 **25.** 10 **27.** 20 **29.** 50
31. 2 **33.** 1 **35.** $\frac{n!}{2!(n - 2)!}$
37. $x^4 + 4x^3y + 6x^2y^2 + 4xy^3 + y^4$
39. $8x^3 + 12x^2y + 6xy^2 + y^3$
41. $81x^4 - 216x^3 + 216x^2 - 96x + 16$ **43.** $-1{,}250x^2y^3$
45. $-4x^6y^3$ **47.** 35 **49.** 1,000,000 **51.** 136,080
53. 8,000,000 **55.** 720 **57.** 2,880 **59.** 13,800
61. 720 **63.** 900 **65.** 364 **67.** 5 **69.** 1,192,052,400
71. 18 **73.** 7,920 **77.** 48

Chapter Summary (page 916)

1. 144 **2.** 20 **3.** 15 **4.** 220 **5.** 1 **6.** 8
7. $x^5 + 5x^4y + 10x^3y^2 + 10x^2y^3 + 5xy^4 + y^5$
8. $x^4 - 4x^3y + 6x^2y^2 - 4xy^3 + y^4$
9. $64x^3 - 48x^2y + 12xy^2 - y^3$
10. $x^3 + 12x^2y + 48xy^2 + 64y^3$ **11.** $6x^2y^2$ **12.** $-10x^2y^3$
13. $-108x^2y$ **14.** $864x^2y^2$ **15.** 42
16. 122, 137, 152, 167, 182 **17.** $\frac{41}{3}, \frac{58}{3}$ **18.** 1,550
19. $-\frac{45}{2}$ **20.** $\frac{15}{2}$ **21.** 378 **22.** 14 **23.** 360
24. 24, 12, 6, 3, $\frac{3}{2}$ **25.** 4 **26.** 24, -96 **27.** $\frac{2{,}186}{9}$
28. $-\frac{85}{8}$ **29.** 125 **30.** $\frac{5}{99}$ **31.** 136 **32.** 5,040 **33.** 1
34. 20,160 **35.** $\frac{1}{10}$ **36.** 1 **37.** 1 **38.** 28 **39.** 84
40. 700 **41.** $\frac{7}{4}$ **42.** \$1,638.40 **43.** \$134,509.57
44. 12 yr **45.** 1,600 ft **46.** 120 **47.** 720 **48.** 120
49. 150

Chapter 14 Test (page 919)

1. 210 **2.** 1 **3.** $-5x^4y$ **4.** $24x^2y^2$ **5.** 66 **6.** 306
7. 34, 66 **8.** 3 **9.** -81 **10.** $\frac{364}{27}$ **11.** 18, 108
12. $\frac{27}{2}$ **13.** 120 **14.** 40,320 **15.** 15 **16.** 56
17. 720 **18.** 322,560 **19.** 24 **20.** $\frac{7}{30}$ **21.** 35
22. 30

Cumulative Review Exercises (page 920)

1. (2, 1) **2.** (1, 1) **3.** (2, -2) **4.** (3, 1) **5.** -1
6. -1 **7.** $(-1, -1, 3)$ **8.** 1
9.

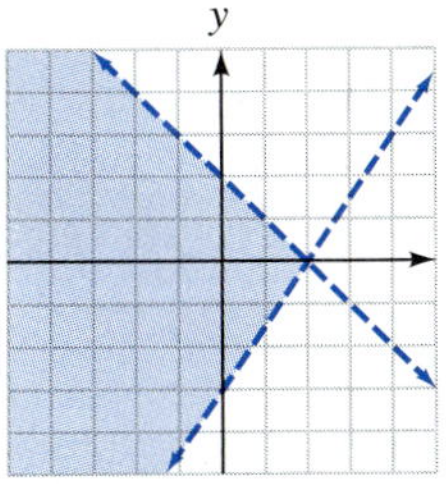

10.

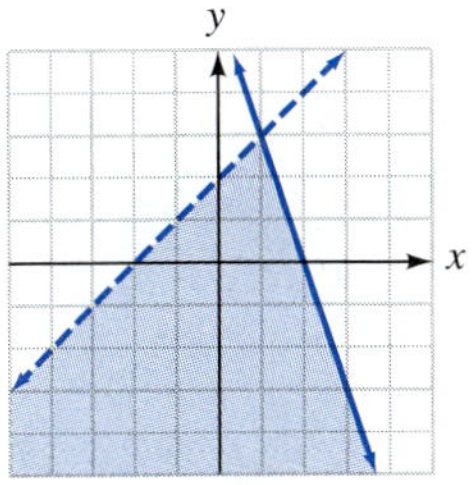

11.

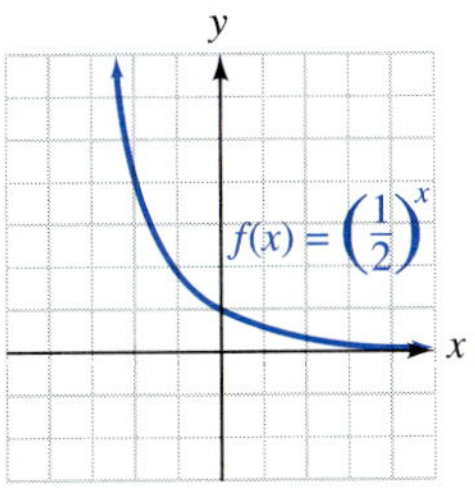

12. $2^y = x$ **13.** 5 **14.** 3
15. $\frac{1}{27}$ **16.** 1 **17.** $y = 2^x$
18. x **19.** 1.9912
20. 0.3010 **21.** 1.6902
22. 0.1461 **23.** $\frac{5 \log 2}{\log 3 - \log 2}$
24. 8 **25.** \$2,848.31
26. 1.16056 **27.** 30,240
28. $81a^4 - 108a^3b + 54a^2b^2 - 12ab^3 + b^4$ **29.** $112x^2y^6$
30. 103 **31.** 690 **32.** 8 and 19 **33.** 42 **34.** 27
35. 27 **36.** $\frac{1{,}023}{64}$ **37.** 12, -48 **38.** $\frac{27}{2}$ **39.** 504
40. 35 **41.** $\frac{7}{3}$ **42.** $C(n, n)$ **43.** 5,040 **44.** 84

Exercise I.1 (page A-3)

1. y-axis **3.** origin **5.** y-axis **7.** none **9.** none
11. x-axis
13.

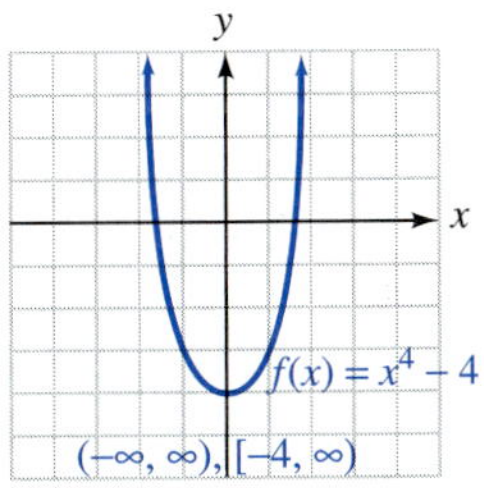

15.

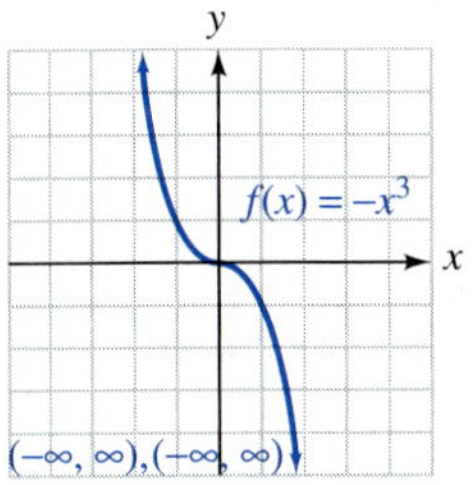

17.

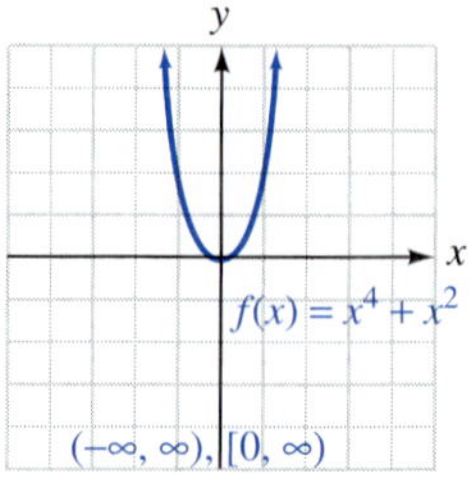

19.

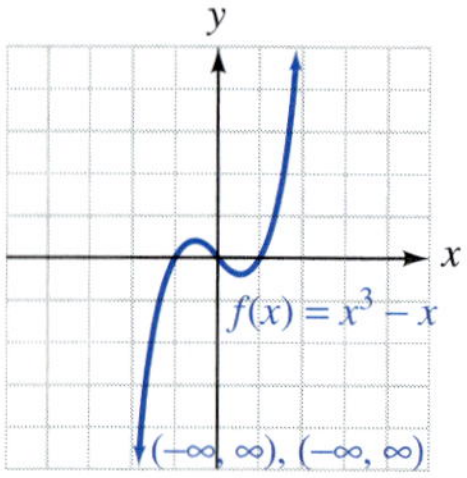

21.

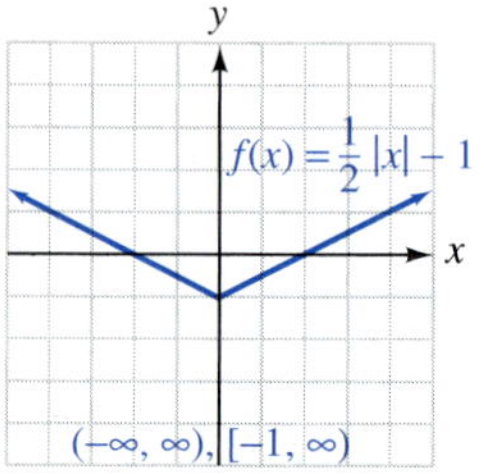

23.

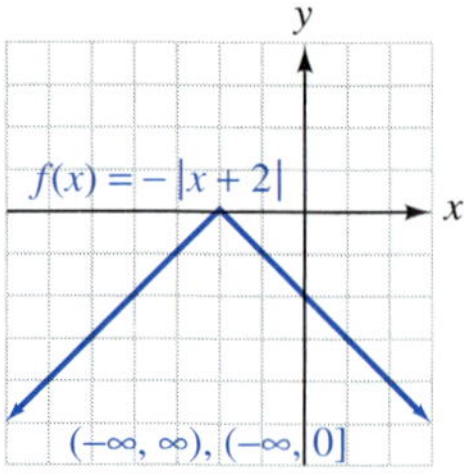

Index

CHAPTER 1 REAL NUMBERS AND THEIR BASIC PROPERTIES

Natural numbers: $\{1, 2, 3, 4, 5, \ldots\}$

Whole numbers: $\{0, 1, 2, 3, 4, 5, \ldots\}$

Integers: $\{\ldots, -3, -2, -1, 0, 1, 2, 3, \ldots\}$

Rational numbers: {All numbers that can be written as a fraction with an integer numerator and a nonzero integer denominator}

Real numbers: {All numbers that are either a rational number or an irrational number}

Prime numbers: $\{2, 3, 5, 7, 11, 13, 17, \ldots\}$

Composite numbers: $\{4, 6, 8, 9, 10, 12, 14, 15, \ldots\}$

Even integers: $\{\ldots, -6, -4, -2, 0, 2, 4, 6, \ldots\}$

Odd integers: $\{\ldots, -5, -3, -1, 1, 3, 5, \ldots\}$

Fractions:

If there are no divisions by 0, then

$$\frac{ax}{bx} = \frac{a}{b} \qquad \frac{a}{b} \cdot \frac{c}{d} = \frac{ac}{bd}$$

$$\frac{a}{b} \div \frac{c}{d} = \frac{ad}{bc} \qquad \frac{a}{d} + \frac{b}{d} = \frac{a+b}{d}$$

$$\frac{a}{d} - \frac{b}{d} = \frac{a-b}{d}$$

Exponents and order of operations:

If n is a natural number, then

$$x^n = \overbrace{x \cdot x \cdot x \cdot \cdots \cdot x}^{n \text{ factors of } x}$$

To simplify expressions, do all calculations within each pair of grouping symbols, working from the innermost pair to the outermost pair.

1. Find the values of any exponential expressions.
2. Do all multiplications and divisions from left to right.
3. Do all additions and subtractions from left to right.

In a fraction, simplify the numerator and denominator separately and then simplify the fraction, if possible.

Figure	Perimeter	Area
Square	$P = 4s$	$A = s^2$
Rectangle	$P = 2l + 2w$	$A = lw$
Triangle	$P = a + b + c$	$A = \frac{1}{2}bh$
Trapezoid	$P = a + b + c + d$	$A = \frac{1}{2}h(b + d)$
Circle	$C = \pi D = 2\pi r$	$A = \pi r^2$

Figure	Volume
Rectangular solid	$V = lwh$
Cylinder	$V = Bh$*
Pyramid	$V = \frac{1}{3}Bh$*
Cone	$V = \frac{1}{3}Bh$*
Sphere	$V = \frac{4}{3}\pi r^3$

*B is the area of the base.

If a, b, and c are real numbers, then

Closure properties:

$a + b$ is a real number.

$a - b$ is a real number.

ab is a real number.

$\frac{a}{b}$ is a real number $(b \neq 0)$.

Commutative properties:

$a + b = b + a$

$ab = ba$

Associative properties:

$(a + b) + c = a + (b + c)$

$(ab)c = a(bc)$

Distributive property:

$a(b + c) = ab + ac$

CHAPTER 2 EQUATIONS AND INEQUALITIES

Let a, b, and c be real numbers.

If $a = b$, then $a + c = b + c$.

If $a = b$, then $a - c = b - c$.

If $a = b$, then $\frac{a}{c} = \frac{b}{c}$ $(c \neq 0)$.

If $a = b$, then $ca = cb$.

Sale price = regular price − markdown

Retail price = wholesale cost + markup

Percentage = rate · base

Solving inequalities:

Let a, b, and c be real numbers.

If $a < b$, then $a + c < b + c$.

If $a < b$, then $a - c < b - c$.

If $a < b$ and $c > 0$, then $ac < bc$.

If $a < b$ and $c < 0$, then $ac > bc$.

If $a < b$ and $c > 0$, then $\frac{a}{c} < \frac{b}{c}$.

If $a < b$ and $c < 0$, then $\frac{a}{c} > \frac{b}{c}$.

CHAPTER 3 GRAPHING AND SOLVING SYSTEMS OF EQUATIONS AND INEQUALITIES

General form of the equation of a line:

$Ax + By = C$

Equation of a vertical line: $x = a$

Equation of a horizontal line: $y = b$

CHAPTER 4 POLYNOMIALS

Properties of exponents: If x is a number, m and n are integers, and there are no divisions by 0, then

$$x^n = \overbrace{x \cdot x \cdot x \cdot \cdots \cdot x}^{n \text{ factors of } x}$$

$x^m x^n = x^{m+n}$ $\quad$ $(x^m)^n = x^{mn}$

$(xy)^n = x^n y^n$ $\quad$ $\left(\frac{x}{y}\right)^n = \frac{x^n}{y^n}$

$\frac{x^m}{x^n} = x^{m-n}$ $\quad$ $x^0 = 1$

$x^{-n} = \frac{1}{x^n}$ $\quad$ $\frac{x^m}{x^n} = x^{m-n}$

Special products:

$(x + y)^2 = x^2 + 2xy + y^2$

$(x - y)^2 = x^2 - 2xy + y^2$

$(x + y)(x - y) = x^2 - y^2$

CHAPTER 5 FACTORING POLYNOMIALS

Factoring the difference of two squares:

$a^2 - b^2 = (a + b)(a - b)$

Factoring perfect-square trinomials:

$a^2 + 2ab + b^2 = (a + b)^2$

$a^2 - 2ab + b^2 = (a - b)^2$

Factoring the sum and difference of two cubes:

$x^3 + y^3 = (x + y)(x^2 - xy + y^2)$

$x^3 - y^3 = (x - y)(x^2 + xy + y^2)$

Zero-factor property: Let a and b be real numbers.

If $ab = 0$, then $a = 0$ or $b = 0$.

CHAPTER 6 PROPORTIONS AND RATIONAL EXPRESSIONS

If $\frac{a}{b} = \frac{c}{d}$, then $ad = bc$.

If there are no divisions by 0, then

$\frac{ac}{bc} = \frac{a}{b}$ $\quad$ $\frac{a}{1} = a$ $\quad$ $\frac{a}{0}$ is undefined $\quad$ $\frac{a}{b} \cdot \frac{c}{d} = \frac{ac}{bd}$

$\frac{a}{b} \div \frac{c}{d} = \frac{ad}{bc}$ $\quad$ $\frac{a}{d} + \frac{b}{d} = \frac{a + b}{d}$

$\frac{a}{d} - \frac{b}{d} = \frac{a - b}{d}$

CHAPTER 7 MORE EQUATIONS, INEQUALITIES, AND FACTORING

If $x \geq 0$, then $|x| = x$.

If $x < 0$, then $|x| = -x$.

If $k > 0$, then

$|x| = k$ is equivalent to $x = k$ or $x = -k$.

$|x| < k$ is equivalent to $-k < x < k$.

$|x| > k$ is equivalent to $x < -k$ or $x > k$.

CHAPTER 8 WRITING EQUATIONS OF LINES, FUNCTIONS, AND VARIATION

Slope of a nonvertical line: If $P(x_1, y_1)$ and $Q(x_2, y_2)$ are two points on a line, the **slope** of the line is

$$m = \frac{y_2 - y_1}{x_2 - x_1} \quad (x_2 \neq x_1)$$

Equations of a line:

$y - y_1 = m(x - x_1)$ point-slope form

$y = mx + b$ slope-intercept form

$Ax + By = C$ general form

$y = b$ a horizontal line (slope is 0)

$x = a$ a vertical line (slope is undefined)

Two lines with the same slope are parallel.

If the product of the slopes of two lines is -1, the lines are perpendicular.

Direct variation: $y = kx$

Inverse variation: $y = \frac{k}{x}$

Joint variation: $y = kxz$

Combined variation: $y = \frac{kx}{z}$

CHAPTER 9 RADICALS AND RATIONAL EXPONENTS

Pythagorean theorem: If a and b are the lengths of two legs of a right triangle and c is the length of the hypotenuse, then

$$a^2 + b^2 = c^2$$

Distance formula: The distance d between the points (x_1, y_1) and (x_2, y_2) is given by the formula

$$d = \sqrt{(x_2 - x_1)^2 + (y_2 - y_1)^2}$$

Rational exponents:

If n is a positive integer ($n > 1$) and all radicals represent real numbers, then

$$(a^{1/n})^n = a \qquad a^{m/n} = (a^{1/n})^m = (a^m)^{1/n}$$

$$a^{-m/n} = \frac{1}{a^{m/n}} \qquad \frac{1}{a^{-m/n}} = a^{m/n}$$

$$a^{1/n} = \sqrt[n]{a}$$

Properties of radicals: If a and b are not both negative, then

$$\sqrt[n]{ab} = \sqrt[n]{a}\sqrt[n]{b} \quad \text{and} \quad \sqrt[n]{\frac{a}{b}} = \frac{\sqrt[n]{a}}{\sqrt[n]{b}} \quad (b \neq 0)$$

Radical equations:

Let a and b be real numbers.

If $a = b$, then $a^2 = b^2$.

Complex numbers:

$i^2 = -1$

$a + bi = c + di$ when $a = c$ and $b = d$

$(a + bi) + (c + di) = (a + c) + (b + d)i$

$(a + bi)(c + di) = (ac - bd) + (ad + bc)i$

$a + bi$ and $a - bi$ are complex conjugates.

$(a + bi)(a - bi) = a^2 + b^2$

$|a + bi| = \sqrt{a^2 + b^2}$

CHAPTER 10 QUADRATIC FUNCTIONS, INEQUALITIES, AND ALGEBRA OF FUNCTIONS

The quadratic formula:

$$x = \frac{-b \pm \sqrt{b^2 - 4ac}}{2a} \quad (a \neq 0)$$

The discriminant:

If	then the roots of $ax^2 + bx + c = 0$ are
$b^2 - 4ac > 0$	real and unequal
$b^2 - 4ac = 0$	rational and equal
$b^2 - 4ac < 0$	nonreal

If r_1 and r_2 are the roots of $ax^2 + bx + c = 0$,

$$r_1 + r_2 = -\frac{b}{a} \quad \text{and} \quad r_1 r_2 = \frac{c}{a}$$

Graphs of quadratic functions:

The graph of $y = ax^2 + bx + c \quad (a \neq 0)$ is a parabola. It opens

$$\begin{Bmatrix} \text{upward} \\ \text{downward} \end{Bmatrix} \text{ when } \begin{Bmatrix} a > 0 \\ a < 0 \end{Bmatrix}$$

The graph of $y = a(x - h)^2 + k \quad (a \neq 0)$ is a parabola with vertex at (h, k). It opens

$$\begin{Bmatrix} \text{upward} \\ \text{downward} \end{Bmatrix} \text{ when } \begin{Bmatrix} a > 0 \\ a < 0 \end{Bmatrix}$$

Algebra of functions:

$(f + g)(x) = f(x) + g(x)$

$(f - g)(x) = f(x) - g(x)$

$(f \cdot g)(x) = f(x)g(x)$

$(f/g)(x) = \dfrac{f(x)}{g(x)} \quad (g(x) \neq 0)$

$(f \circ g)(x) = f(g(x))$

CHAPTER 11 EXPONENTIAL AND LOGARITHMIC FUNCTIONS

$y = \log_b x$ if and only if $x = b^y \quad (b > 0, b \neq 1)$

$y = b^x$ and $y = \log_b x$ are inverse functions.

Compound interest: $A = P\left(1 + \dfrac{r}{k}\right)^{kt}$

Continuous compound interest: $A = Pe^{rt}$

Malthusian population growth: $A = Pe^{rt}$

Decibel voltage gain: db gain $= 20 \log \dfrac{E_O}{E_I}$

Richter scale: $R = \log \dfrac{A}{P}$

Doubling time: $t = \dfrac{\ln 2}{r}$

Properties of logarithms: If M, N, and b are positive numbers and $b \neq 1$, then

$\log_b 1 = 0 \qquad \log_b b = 1$

$\log_b b^x = x \qquad b^{\log_b x} = x$

$\log_b MN = \log_b M + \log_b N$

$\log_b \dfrac{M}{N} = \log_b M - \log_b N$

$\log_b M^p = p \log_b M \qquad$ If $\log_b x = \log_b y$, then $x = y$.

Change-of-base formula: $\log_b y = \dfrac{\log_a y}{\log_a b}$

pH scale: $\text{pH} = -\log[\text{H}^+]$

Weber–Fechner law: $L = k \ln I$

Carbon dating: $A = A_0 2^{-t/h}$

CHAPTER 12 CONIC SECTIONS AND MORE GRAPHING

Equations of a circle with radius r:

$(x - h)^2 + (y - k)^2 = r^2$ center at (h, k)

$x^2 + y^2 = r^2$ center at $(0, 0)$

Equations of a parabola:

Parabola opening	Vertex at origin	Vertex at (h, k)
Up	$y = ax^2 \ (a > 0)$	$y = a(x - h)^2 + k \ (a > 0)$
Down	$y = ax^2 \ (a < 0)$	$y = a(x - h)^2 + k \ (a < 0)$
Right	$x = ay^2 \ (a > 0)$	$x = a(y - k)^2 + h \ (a > 0)$
Left	$x = ay^2 \ (a < 0)$	$x = a(y - k)^2 + h \ (a < 0)$

Equations of an ellipse:

Center at $(0, 0)$ If $a > b > 0$,

$$\frac{x^2}{a^2} + \frac{y^2}{b^2} = 1 \qquad \frac{y^2}{a^2} + \frac{x^2}{b^2} = 1$$

Center at (h, k)

$$\frac{(x - h)^2}{a^2} + \frac{(y - k)^2}{b^2} = 1 \quad (a > b > 0)$$

$$\frac{(y - k)^2}{a^2} + \frac{(x - h)^2}{b^2} = 1 \quad (a > b > 0)$$

Equations of a hyperbola:

Center at $(0, 0)$

$$\frac{x^2}{a^2} - \frac{y^2}{b^2} = 1 \qquad \frac{y^2}{a^2} - \frac{x^2}{b^2} = 1$$

Center at (h, k)

$$\frac{(x - h)^2}{a^2} - \frac{(y - k)^2}{b^2} = 1$$

$$\frac{(y - k)^2}{a^2} - \frac{(x - h)^2}{b^2} = 1$$

CHAPTER 13 MORE ON SYSTEMS OF EQUATIONS AND INEQUALITIES

$$\begin{vmatrix} a & b \\ c & d \end{vmatrix} = ad - bc$$

CHAPTER 14 MISCELLANEOUS TOPICS

Factorial notation:

$n! = n(n - 1)(n - 2) \cdot \cdots \cdot 3 \cdot 2 \cdot 1$

$0! = 1$

The binomial theorem:

$$(a + b)^n = a^n + \frac{n!}{1!(n - 1)!}a^{n-1}b + \frac{n!}{2!(n - 2)!}a^{n-2}b^2 + \cdots + b^n$$

Arithmetic sequence:

$$S_n = \frac{n(a + l)}{2} \quad \text{(sum of the first } n \text{ terms)}$$

Summation notation:

$$\sum_{k=1}^{4} k = 1 + 2 + 3 + 4$$

Geometric sequence:

$$S_n = \frac{a - ar^n}{1 - r} \quad (r \neq 1) \quad \text{(sum of the first } n \text{ terms)}$$

The sum of all the terms of a geometric sequence is

$$S = \frac{a}{1 - r} \quad (|r| < 1)$$

Formulas for permutations and combinations:

$$P(n, r) = \frac{n!}{(n - r)!}$$

$P(n, n) = n!$

$P(n, 0) = 1$

$$C(n, r) = \binom{n}{r} = \frac{n!}{r!(n - r)!}$$

$$C(n, n) = \binom{n}{n} = 1$$

$$C(n, 0) = \binom{n}{0} = 1$$